# *Quantifying Sustainable Development*

# *Quantifying Sustainable Development*

## THE FUTURE OF TROPICAL ECONOMIES

Edited by
**Charles A. S. Hall**
*State University of New York*
*Syracuse, New York*

Associate Editors

**Carlos Leon Perez**
*Ministry of Agriculture*
*San Jose, Costa Rica*

**Gregoire Leclerc**
*Laboratory of Geographical Analysis*
*CATIE, Turrialba 7170, Costa Rica*

*San Diego San Francisco New York Boston London Sydney Tokyo*

*Front cover photographs*: Computer model output of forested areas at several points in time projected on top of it. (Images by Charles Hall and Joseph Cornell.) See CD for details. *Background photograph*: Braulio Carillio National Park. (Photograph by Charles Hall.)

For technical support on the CD enclosed, contact the Harcourt Technical Support Center at the numbers indicated below. Service is available in English only between the hours of 7am and 6pm US Central Time (15:00 to 02:00 GMT), Mondays through Fridays.
Toll free in the US and Canada (877)809-6433
Direct dial (817)820-3710
Toll free fax in the US only (800)354-1774
Direct fax (817)820-5100
E-mail: tscap@hbtechsupport.com

This book is printed on acid-free paper. ∞



Academic Press
*A Harcourt Science and Technology Company*
525 B Street, Suite 1900, San Diego, California 92101-4495, USA
http://www.academicpress.com

Academic Press
Harcourt Place, 32 Jamestown Road, London NW1 7BY, UK
http://www.hbuk.co.uk/ap/

Library of Congress Catalog Card Number: 98-85623

International Standard Book Number: 0-12-318860-1

PRINTED IN THE UNITED STATES OF AMERICA
00 01 02 03 04 05 EB 9 8 7 6 5 4 3 2 1

# CONTENTS

## SECTION II
## *Development and Sustainability*

## SECTION III
## *Adding a Spatial Dimension: Tools for Dynamic Geographical Analysis*

# SECTION IV
# *Building a Geographical Database for Costa Rica*

# CONTRIBUTORS

*Numbers in parentheses indicate the pages on which the authors' contributions begin.*

**SERGIO ABARCA** (423), Ministry of Agriculture and Livestock, San José, Costa Rica

**BERNARDO AGUILAR** (19, 595), Present address: Department of Ecological Economics and Environmental Law, Prescott College, Prescott, Arizona 86301

**ALFREDO ALVARADO** (265), Agronomy Investigation Center, University of Costa Rica, San José, Costa Rica

**FLORIA BERTSCH** (265), Agronomy Investigation Center, University of Costa Rica, San José, Costa Rica

**MARK BROWN** (695), Department of Environmental Engineering Sciences, University of Florida, Gainesville, Florida 32611

**SANDRA BROWN** (503), U.S. Environmental Protection Agency, National Health and Environmental Effects Research Laboratory, Corvallis, Oregon 97333; Present address: Winrocle International, Arlington, Virginia 22209

**MEEGHAN CARROLL** (527), Department of Forestry, State University of New York, Syracuse, New York 13210

**JOSEPH D. CORNELL** (543), Department of Environmental and Forest Biology, College of Environmental Science and Forestry, State University of New York, Syracuse, New York 13210

**OSCAR FLORES** (423), University of San Carlos, Guatemala

**CYNTHIA FRIDGEN** (563), Department of Resource Development, College of Agriculture and Natural Resources, Michigan State University, East Lansing, Michigan 48824

**CHARLES A. S. HALL** (3, 19, 45, 91, 121, 159, 177, 223, 295, 349, 563, 647, 695, 715), Department of Environmental and Forest Biology and Program of Environmental Science, College of Environmental Science and Forestry, State University of New York, Syracuse, New York 13078

**MYRNA HALL** (19, 177), College of Environmental Science and Forestry, State University of New York, Syracuse, New York 13078

**E. H. HELMER** (503), Department of Forest Science, Oregon State University, Corvallis, Oregon 97331; Present address: Institute of Tropical Forestry, Southern Forest Experiment Station, U. S. Forest Service, Rio Piedras, Puerto Rico

**CARLOS HENRIQUEZ** (265), Agronomy Investigation Center, University of Costa Rica, San José, Costa Rica

**CARLOS HERNANDEZ** (563), Central Administration, Earth College, Guapiles, Costa Rica

**GERRIT HOOGENBOOM** (403), Department of Biological and Agricultural Engineering, University of Georgia, Griffin, Georgia 30223

**GLENN HYMAN** (205, 449), Centro de Computo, CATIE, Turrialba, Costa Rica; Present address: CIAT, Cali, Colombia

**MUHAMMED IBRAHIM** (423), Department of Agroforestry, CATIE, Turrialba, Costa Rica

**JULIE KLOCKER** (595), Comparative Extension, University of Minnesota, Minneapolis, Minnesota 55455

**JAE-YOUNG KO** (91), College of Environmental Science and Forestry, State University of New York, Syracuse, New York 13078; Present address: Center for Wetlands, Louisiana State University, Baton Rouge, Louisiana 70803

**TIMM KROEGER** (629, 665), Graduate Program in Environmental Science, College of Environmental Science and Forestry, State University of New York, Syracuse, New York 13210

**LOIS LEVITAN** (121), Environmental Risk Analysis Program, Center for the Environment; Cornell University, Ithaca, New York 14850

**GREGOIRE LECLERC** (159, 223, 295), Laboratory of Geographical Analysis, CATIE, Turrialba 7170, Costa Rica; Present address: CIAT, Cali, Colombia

**CARLOS LEÓN PÉREZ** (349, 403), Ministry of Agriculture, San José, Costa Rica

**PABLO MARTINEZ** (527), Departmento Ambiental, Avila Catholic University, Los Cantero sln 05005, Avila, Spain

**RAFAEL MATA** (265), Agronomy Investigation Center, University of Costa Rica, San José, Costa Rica

**DAWN R. MONTANYE** (647, 665), Graduate Program in Environmental Science, College of Environmental Science and Forestry, State University of New York, Syracuse, New York 13210; Present address: Friends of the Earth, Washington, DC 20005

**PATRICK G. MOTEL** (527), Schenectady, New York 12303

**WILL RAVENSCROFT** (91, 349), Department of Environmental and Forest Biology, College of Environmental Science and Forestry, State University of New York, Syracuse, New York 13210; Present address: Cooperative Fish and Wildlife Unit, College of Agriculture and Forestry, West Virginia University, Morgantown, West Virginia 26506

**TENG REYES** (223), CATIE, Turrialba, Costa Rica

**DAVID ROSSITER** (403), Department of Agronomy, Cornell University, Ithaca, New York 14850; Present address: International Institute for Aerospace Survey and Earth Sciences, Enshede, The Netherlands

**BENJAMIN RUBIN** (449), Department of Environmental and Forest Biology, College of Environmental Science and Forestry, State University of New York, Syracuse, New York 13210

**OLEGARIO SAENZ** (91), Department of Economics, University of Costa Rica, San José, Costa Rica

**G. ARTURO SANCHEZ-AZOFEIFA** (473), Department of Civil Engineering, University of Costa Rica, San José, Costa Rica; Present address: Department of Earth and Atmospheric Sciences, University of Albera, Edmonton, Canada T6G 2E3

**TOMÁS SCHLICHTER** (121), Program on Sustainable Development, CATIE, Turrialba, Costa Rica; Present address: Instituto Nacional de Techologia Agropecuaria (INTA), EEA Bariloche, CC 277, San Carlos de Bariloche, Argentina

**MARSHALL TAYLOR** (177), Resources Planning Associates, Ithaca, New York 14850

**PATRICK VAN LAAKE** (159, 205, 403), Forest Resources Division, Food and Agriculture Organization of the United Nations, San José, Costa Rica; Present address: Department of Earth and Atmospheric Sciences, University of Albera, Edmonton, Canada T6G 2E3

**JUAN-RAPHAEL VARGAS** (91, 647), Department of Economics, University of Costa Rica, San José, Costa Rica

**MATHIS WACKERNAGEL** (695), Indicators Program Redefining Progress, San Francisco, California 94108

**HONGQING WANG** (349), Department of Environmental and Forest Biology, College of Environmental Science and Forestry, State University of New York, Syracuse, New York 13210

**SCOTT G. WITTER** (563), Department of Resource Development, College of Agriculture and Natural Resources, Michigan State University, East Lansing, Michigan 48824

# FOREWORD

In the temperate zones north and south of the equator there are, and have been, wealthy countries. South of the equator Australia and New Zealand are wealthy. Argentina and Chile were two of the twenty wealthiest countries in the world in the late 19th century. Since the industrial revolution, no country within the tropics—with the exception of one very small city state (Singapore)—has been anything but poor. Even within the same country this iron law of experience seems to hold. Those parts of Brazil south of the Tropic of Capricorn are much richer than those parts of Brazil within the tropics.

Understanding why this should be so is a bit like a murder mystery. There are lots of suspects (the tropics have more problems with diseases because cold weather controls many diseases; leached soils from too much rainfall may present problems more difficult to solve than those of the deserts with inadequate rainfall found outside the equatorial regions), but no obviously guilty killer. If anything, the tropics have more than their share of natural resources. Big oil producers include Indonesia, Nigeria, and Venezuela. Paying for oil is a big problem for Costa Rica (the country most analyzed in this book), but lots of tropical countries with lots of oil are much poorer than Costa Rica.

Tropical population growth rates are certainly a problem. No country has ever become rich without at least a century or more of population growth rates of 1% or less. Rates much above this level cause a poor country to devote too many of its resources to bringing new people up to existing economic levels, and essentially no resources are left after this is done to devote to raising the income levels of those people who already exist. But almost universally, countries within the temperate zones have managed to limit their population growth rates, while almost none within the tropics have been able to do so. Why?

It is easy to attack existing models of economic development (this book does some of that) since they haven't worked in the tropics. Developmental economists would probably say that the people who tried to make them work in the tropics have not been disciplined enough to make them work. Using standard economic models, China after all has leapt ahead while Africa, not using those models, has fallen behind—poorer per capita now than it was in 1965. When it comes to the social discipline to get village schools organized, levels of illiteracy are orders of magnitude higher in the tropics than they are in the most illiterate countries in the temperate zones. Why? So too within countries—education levels in northern Brazil are far below those in southern Brazil.

There is a fundamental problem for all those, including the authors of this book, who would produce alternatives. Since the industrial revolution, no form of economic organization other than competitive markets has managed to generate high standards of living anywhere in the world. Many alternatives have been tried—socialism, communism, feudalism, cooperatives—all have failed. Even with all of the positive ideology that they once had behind them, the kibbutzes are dying out in Israel. Finding an alternative to the existing conventional wisdom is not easy—and maybe not even possible. In any case, it has yet to be discovered, much less tested. The burden of proof is on those who call for other solutions.

As this book points out, one of the alternatives is to call for sustainable development. But the definition of sustainability is a slippery concept. If sustainability means a much lower standard of living than that which now exists, "sustainability" isn't sustainable. If sustainability means the ability to sustain economic development, the ability differs radically from country to country. From space, Argentina and the United States look very similar. Both were given similar natural endowments, and Argentina if anything inherited a more favorable natural position than the United States. Indeed in the 19th century, it was one point wealthier per capita than the United States. Yet Argentina's economic development was not sustainable while America's was.

If one visits agricultural villages in a poor country (I recently did so in Laos), one finds villages with very different standards of living situated right next to each other with the same natural endowments. Some are well organized and some are completely disorganized. Their ability to get organized determines their ability to generate high or low standards of living for their inhabitants. Sociology in the end may be more important than biology.

If a detective has an unsolved murder case, he or she interviews all of the potential witnesses (reliable or unreliable) to see if something somewhere will yield a clue to who the murderer was. This is the right frame of reference with which to approach any book on tropical development. Within it are clues to why tropical economic development is almost an oxymoron.

Like you, the reader, I do not claim to know the exact reasons that tropical countries are universally economic failures. But some of the clues that may lead you and me to find the murderer are certainly found within the new perspectives and extensive data analysis contained within this book.

*Lester C. Thurow*
*Sloan School of Management*
*Massachusetts Institute of Technology*

# PREFACE

This book is about the myths and realities of the developing tropics. It is also about the inadequacy of our dominant intellectual paradigms for understanding and guiding that development. It considers the continuing failure, despite large efforts, of programs based on these dominant paradigms to resolve major economic and environmental problems there. Our book also uses new tools to develop a more comprehensive and realistic framework for analysis and decision making. These include biophysical and systems perspectives that integrate environment and economics, and some cutting-edge technology that combines GIS, modeling, and extensive geographical and time series databases.

We have two underlying objectives: first, to examine the possibilities and limitations of "sustainable development" for guiding the future of national economies, and second, to understand more precisely the relation of the biological and physical properties of a region to important components of its economy, including its development and conservation possibilities. We offer no silver bullets to magically resolve difficult problems, but offer instead a suite of conceptual and computer tools that enable a systems approach and consequently, we believe, a more rigorous and detailed analysis than has been possible in the past.

We focus on these issues for two reasons: One is that I, and most of the contributors to this book, believe that the conceptual basis for current economic and development theory is in many respects fatally flawed. If we are to understand the principles of operation and the potential for development—sustainable or otherwise—of tropical economies, we must use a different kind of economics. We can do this by expanding the range of what we do under its aegis by again focusing on explaining, in accounting terms, what happens in society and then using the scientific method, rather than ideology, to ask why. Another way of saying this is that we need to question and test what economics as a discipline has assumed and promoted. We believe this to be essential if any sort of sustainable approach to economic health and well-being is ever to be generated in developing countries. We also believe that most of the "environmental" or "green" alternatives offered to date, including many under the rubric of "sustainability," are not particularly useful for solving the deficiencies in the traditional economic approach because they tend to focus on issues that are economically trivial. Thus, our second reason for writing this book is to examine the potential of an alternative to the dominant economic paradigm, namely, a biophysical perspective. To paraphrase Von Clauswitz, who said the "war is too important to leave to the generals," we believe that "economics is too important to leave to the economists."

Our intended audience is anyone interested in a fresh and rigorous approach to understanding how developing nations operate, or might operate, as systems. The book is interdisciplinary, analytical, and academic and is appropriate for graduate courses in the environmental sciences, public administration, international relations, and economics. We especially hope that it will find an interested audience in graduate schools in the developing world and in development and governmental agencies. If we are truly successful it will serve as a blueprint for how economics "should" be done in developing countries, and the chapter-by-chapter procedures given here can serve as a basis for many countries that want to generate a full resource inventory and incorporate it into economic analysis. But the level of writing and the importance of the subject material make the book appropriate for any intelligent person interested in the intertwining of environments and economies in developing countries. Finally, because the approach is in many respects very different from what has been written about previously, we hope that anyone who wants a different perspective on the developing tropics will find the book interesting.

I write as a professional ecologist. This introduces several potential misconceptions because of the use of the term in the popular media. "Ecologist" means to me a professionally trained person who studies some aspect of ecosystems, in much the same way as we might use "chemist" or "mathematician." As such the word "ecology" includes no implied advocacy or political position. I have no problem with advocates of certain environmental policies calling themselves ecologists, but for that usage I prefer the term "environmentalist." A second issue relates to the subject matter pursued by professional ecologists. My own interest has always been in asking how ecosystems work, especially those ecosystems dominated by human activities. Curiously, most professional ecologists do not study human-dominated ecosystems, such as cities or agricultural regions, except as they impinge upon "natural" ecosystems. Instead they seek out remnant and often remote "natural ecosystems." Thus, the structure and function of human-dominated systems, the ecosystems that increasingly and overwhelmingly dominate the surface of this planet, are rarely examined as *ecosystems* using the tools of natural science, but instead are studied mostly using the approaches of the social sciences. To me this is a serious omission, as they are just as much "ecosystems" as "social systems," and as such are characterized and constrained by energetics and material fluxes just as are the remaining "natural" ecosystems.

On the other hand, the one great disappointment that I have had in writing this book is that despite my large efforts to include a chapter on the important issue of natural biodiversity, a subject in which I have little particular strength, no such chapter was forthcoming. The issue of biodiversity is treated, but not as completely as I would like, in other chapters.

Our reason for undertaking this analysis is scientific and not to suggest any particular policy. It is our strong belief that the policy by which any nation should be run is the business of that country alone. Thus, with one exception (in the last chapter), we make no explicit policy recommendations in this book, although we were strongly encouraged to write a shorter, policy-oriented book by one of our reviewers. Ours is a difficult tightwire to tread, for our analysis includes examination and critiques of past policy and in some cases possible future policies. But we do so only from the perspective of analyzing consequences and of trying to understand how the system has and might work. However, if leaders in the developing world find our analyses helpful in formulating their own policies, we are of course happy about that. We have undertaken similar analyses previously for the United States (Cleveland *et al.*, 1984; Hall *et al.*, 1986; Ko *et al.*, 1998) and Argentina (Paruelo *et al.*, 1987) without explicitly recommending any policy.

The first section of the book gives the reader a nonquantitative introduction to the economic and environmental situation in the developing tropics in general and in Costa Rica in particular. The second section reviews conditions of sustainability, considers whether the discipline of economics is the most appropriate set of tools for examining the issues that surround sustainability, and puts forth the alternative biophysical approach. The third and fourth sections are mostly technical and give in-depth treatments of particular techniques and resource bases. The final sections use the technical chapters and historical information to examine the potential for sustainability. As a synthesis we include an empirically based computer model that includes its own decision-making framework, based on real data and a synthetic perspective. This model and others are contained on the enclosed CD, so that if the reader is unconvinced by our arguments or assumptions he or she can devise some other, we hope realistic, scenario and run a new set of assumptions. The CD also contains all of the databases that we use in the book as well as unique software to display them. We believe that this tool and the in-depth analyses that we have done for Costa Rica should be broadly applicable to most other developing nations.

The reader may be surprised at the lack of models of a *theoretical* or densely mathematical nature of this book, even though it is focused on several areas (economics and ecology) that often are characterized by intense theoretical work. This is no accident, as the editor and most of the contributors are of the opinion that purely theoretical work in the past has tended to mislead more than lead in both disciplines (e.g., Hall, 1988, 1991, 1992). We offer instead a comprehensive data-based approach linked with straightforward *empirical* models (i.e., data-based simulations). We do believe in empirical models, which we consider to be the formalization of our assumptions about a system, because they allow the generation of explicit and testable predictions from those assumptions. This is an essential part of applying the scientific method to complex situations such as national environments and economies. In addition, as part of the scientific process, each chapter was peer reviewed by two or more appropriate reviewers.

Although much can be learned about the possibilities and limitations of various economic plans using a biophysical approach, it is important to approach this endeavor with more than a little humility. Many pronouncements about economic possibilities and limitations have been made in the past, and many, perhaps most, have been wrong. Will our "more scientific" approach work any better? Does the biophysical snapshot we take today have any relevance in a society changing as rapidly as most economies of the developing tropics are today? We believe so, but ultimately only time will tell.

The book focuses on Costa Rica for reasons given in Chapter 2. Costa Ricans have many reasons for being especially proud of their country. It has a democratic process that is widely considered to be as good as any in the world. Its health, education, and human rights records are more than exemplary. The nation is also renowned for its natural beauty, and its people enjoy the highest standard of living in Central America (Panama, with a slightly higher standard of living, is considered part of South America or, sometimes, the Caribbean). Its biotic and cultural richness, borrowing in each case from systems to the north and to the south, helps to make it a bridge between North and South America. But there have been difficulties in writing this book for, in many respects, it can be construed as critical of Costa Rica. Our target of criticism, however, is only the way in which economic analyses and policies are undertaken, often by people outside of Costa Rica. The problem is exacerbated in that the book is edited and written to a large degree by "northerners," although we tried very hard to include as many Costa Ricans as possible. Fortunately, when we asked our Costa Rican collaborators about this issue, they responded uniformly: "Tell it as you see it. Hiding from unpleasant reality does us no good." I thank all my good Costa Rican friends and colleagues, not all of whom agree with all of our analyses, and the many other wonderful ticos we met while living and working in Costa Rica. I also acknowledge that there are a number of professionals in Costa Rica who have analyzed many of these same problems with great skill.

This book builds upon and continues many of the most important research interests of my professional life. These include the use of empirically based models as tools to analyze the interaction of humans and ecosystems (Hall and Day, 1977), the assessment of economies from a material and energy (rather than only economic) basis (Cleveland *et al.*, 1984; Hall *et al.*, 1986; Hall and Hall, 1992), and, in general, a systems approach to all of these analyses (Hall and Day, 1977; Hall, 1995). My goal has been to show that (1.) what is commonly called economic analysis is by itself quite insufficient to make large-scale economic decisions, and (2.) to install a systems and biophysical approach as a legitimate equal partner for analyzing these decisions. What is new here is the focus on an entire nation using geographical analysis combined with intensive data analysis and modeling, and the application of these tools to an entire economy. In this endeavor I have learned, and have been helped greatly, by the enormous GIS and computational skills and competence of my loving wife, Myrna. My visits to Costa Rica have been financed by several programs administered through CATIE (Turrialba, Costa Rica) and through a sabbatical grant from the State University of New York. The Randolph Pack Foundation provided some important funds, and the Flathead Lake Biological Station of the University of Montana provided excellent facilities for thinking and writing. At Academic Press, I acknowledge both the tremendous insight and support of David Packer, who is a true believer in systems science, and the editorial expertise of Monique Larson. Finally, the ideas of Howard Odum, my great teacher, underlie most of this book.

*Charles A. S. Hall*
*Syracuse, NY*

*Note.* We use the metric system throughout. Common abbreviations are ha for hectare, m for meter, cm for centimeter, kg for kilogram, T for metric ton, and MJ for megajoule (239.01 kcal).

# SECTION I

# *Introduction*

## THE TROPICS—PARADISE IMPERILED?

Section I gives a qualitative analysis of the major changes occurring in the tropics today, both in general and for the specific case of Costa Rica. It focuses on the importance of the biophysical (that is, biological and physical, sometimes used as an opposite to social) changes that underlie the economic changes, especially population growth and rapid increases in energy use and energy-based trade to and from these regions. It also begins the process of examining what the biophysical constraints on development might be. Chapter 2 examines Costa Rica and includes a brief history, which accentuates how large and how rapid the changes have been. This chapter is complemented with a series of illustrative photographs on the accompanying CD. These can be accessed by clicking on "Ecoview," then "Demonstrations."

CHAPTER 1

# The Changing Tropics

CHARLES A. S. HALL



## I. INTRODUCTION

Tropical nations and their economies are changing extremely rapidly compared to most "developed" nations to the north. Population growth, *in situ* and de facto industrialization, international trade, tourism, oil-based transportation, and many other aspects of "global change" are completely transforming these countries. It is hardly possible to tell what changes are good or what are bad because we do not have adequately tested and agreed-upon criteria to judge, nor can we even agree on which of the many operating and often conflicting criteria to apply. More importantly, we have not addressed the long-term dynamic nature of what these changes imply, but have dealt, at best, with only the immediate consequences. Meanwhile the changes accelerate.

## II. THE INDUSTRIALIZATION OF AGRICULTURE

Tropical nations are often poor and sometimes hungry. The normal approach to alleviating these problems has been "development," that is, the investment

*Quantifying Sustainable Development*

of money into the economy to build, to modernize, to supply inputs to agriculture, and to build capital equipment and governmental infrastructure (roads, bridges, schools, etc.). These would increase the ability of a nation to feed itself, accelerate general economic activity, and/or generate foreign exchange. This process has been undertaken by individual farmers or communities, the nations themselves, and national or international development agencies and banks, including the World Bank, the International Monetary Fund (IMF), the U.S. Agency for International Development (USAID), and many international assistance programs funded by, e.g., the European industrial nations. Some of these agencies have provided outright aid, and others have acted as lending institutions. Generally the most important goals of development for tropical countries have been to increase agricultural output and, often, to increase the role of international trade.

I have argued elsewhere that the term "development" is rather a euphemism, because it usually means in practice almost exactly "industrialization" (Hall, 1992). *Industrialization,* that is, the increase in the use of fossil and other fuels and the machines that allow their use, is the basic driver that makes most of the other changes of development possible. There has been surprisingly little discussion about the occurrence, meaning, or desirability of this industrialization. Indeed it just happens, propelled by mostly unplanned population growth, technological development in other parts of the world, the personal economic interests of corporations as well as of millions of individual players, the growing acceptance of free enterprise, and the spreading impact of media advertisement for more material-intensive lifestyles. While much has been written about this economic development as if it were a process that changes only human actions or policy, in reality it is better to call a spade a spade. Development means, almost always, increasing the rate of exploitation of resources through the increasing use of energy, materials, and land area. We view the terms "developing," "less developed," and so on not as good or bad, nor complementary or pejorative, but simply as descriptive of the degree of industrialization and its rate of change.

It is important to examine these questions about development and industrialization within a historical context. Until relatively recently, the human population of the tropics has fed itself more or less on local agriculture, while sometimes exporting "breakfast" (coffee, tea, cocoa, sugar, oranges, lemons, and so on) to the temperate areas of the world. Traditional tropical agriculture tended to be extensive, low yielding, and independent of other regions of the world. *Shifting cultivation,* the most widespread and common type, was the traditional way that people in the tropics fed themselves (Land Use on CD). This approach appears to have evolved independently on all three major tropical continents: a forested area of roughly 1 ha is cleared, allowed to dry, and then burned. This leaves behind an open field fertilized with the ashes of the trees that once grew there. The farmer typically would plant from 3 to as many as 70 different crop varieties, and get good yields for a year or two. Then, as fertility declined and agricultural pests increased, the area was abandoned and the forest allowed to recover. The recovering forests would sequester nutrients from rain and deep soils, and agricultural pests would fade away. After 10 to 30 years the cycle was ready to be repeated, and in the meantime the farmer had shifted his efforts to clearing new, neighboring plots.

Traditional shifting cultivation uses roughly 10–20 ha per family (2 to 4 ha per person, including fallow areas) to produce an adequate diet with essentially no inputs from the industrial world beyond a steel machete. At low population densities it is a successful system that feeds human populations indefinitely while still maintaining basic forest cover (Geertz, 1963; Kunstader *et al.*, 1978; Watters, 1971; Lal and Cummins, 1979; Ramakrishna, 1990). Shifting cultivation has been part of most tropical forests for many thousands of years. In a sense it increases biological diversity by changing a monotonous cover of pure forest into a mosaic of different types of ecosystems, each characterized by somewhat different plants and animals, although none may have as many total species as the original forest. Shifting cultivation has worked for about 10 thousand years, and it still feeds from 5 to 10% of the world's people. Clearly it is as close to sustainable agriculture as humans have developed.

Recent population pressures, however, have undermined this system in many, perhaps most, areas where it had been the traditional means of feeding people (Pandey and Singh, 1984; FAO, 1985). There are perhaps 250 million people in the world dependent principally upon shifting cultivation for their food, including shifting cultivators commuting by bus and motor scooter from Bangkok, Thailand, to hillsides hundreds of kilometers away. This many people require roughly 400 million ha, nearly a quarter of the 1.9 billion ha of the earth's remaining tropical wet forests (Hadley and Lanley, 1983). The obvious conclusion is that all over the world population growth is threatening the possibility that shifting cultivation can support the people dependent upon it.

The initial response of farmers to this increased population pressure normally has been to decrease fallow times, so that forests are recut after 8 years instead of the traditional 12 to 20. This reduces the fertility of the new clearing, since the forests have not had time to accumulate their full nutrient inventory, which in turn reduces yields and forces the farmer to clear a larger field to feed his or her family. Perversely this leads to a destructive feedback process that shortens fallow times even more, since even more land is needed to produce a given quality of food. In many countries fallow periods are now only a few years, and yields are poor. Ramakrishna (1990) and his colleagues have done the most thorough analysis of the physical and chemical properties of shifting cultivation systems. They found that yields of shifting cultivation systems in India were 2.6 tons per hectare of cultivated field per year if the traditional 30-year cycle was followed, but were reduced to 1.3 tons when a 10-year cycle was used, and only 0.13 tons when a 5-year cycle was used. As fertility declines, 10 or 20 ha of additional forest land may be exploited for green manure to support 1 ha of land in crops (Pandey and Singh, 1984).

Where possible, that is, where they can afford it, the response of farmers in tropical countries has been to use increasing amounts of inputs such as commercial fertilizers and pesticides in permanent crops. From an energy perspective what is happening is that the solar-powered natural ecosystems previously responsible for the fertility and low pest density of the sites are being replaced by industrially—derived products that perform the same functions. Since these petroleum-derived inputs are rarely produced within the tropical countries themselves, they must be imported from northern industrialized countries and the petroleum fields of Venezuela and the Arabian Peninsula. This industrialization of agriculture has allowed the growing populations to continue to be fed, and in some cases to become better fed and wealthier, since the yields of fossil-fuel-assisted agriculture tend to be considerably higher than those of traditional agriculture (Table 1-1).

Unfortunately, as a rule, food production in the tropics has barely kept up with population growth, so that globally, per capita food production appears to be declining (Figure 1-1). Consequently many tropical populations have become increasingly dependent upon a relatively few industrialized countries for the direct importation of food, mostly in the form of basic grains (corn or maize, rice, wheat, sorghum, and a few more), which constitute the majority of food for most people in poorer countries. In the early 1990s the United States and Canada, for example, exported net calories to 110 different countries, most in the tropics. Most analyses of the food situation in the tropics show that these countries are having increasing difficulty in feeding themselves, or at least are importing an increasing proportion of their food requirements (Brown, 1996). Some people have argued that it is the availability of cheap imports that discourages local farmers from growing as much food as might be possible (Lappe *et al.*, 1998) while others argue that in many areas populations are too high for any possible agriculture to feed existing populations. This is an important debate that needs examination, as we do in Chapters 5 and 12.

Thus the forces that have caused the tropics to become food sinks rather than sources are a combination of human population growth, increased production of export crops, the remarkably successful industrialization of agriculture in the temperate countries—and hence the higher and cheaper yields there—and a general feeling among economists and decision makers that trade is good. But there is another reason. Few people realize that the human population density in the United States and Canada is only about 10% (20% for the United

TABLE 1-1 Yields per Hectare for Different Types of Nonindustrial Tropical Agriculture

| Region and crop | Yield (tons) as harvested | Dry weight |
|---|---|---|
| | Shifting cultivation | |
| Costa Rica | | |
| Rice (hill) | 0.33 | 0.28 |
| Rice (bottom) | 1.3 | 1.11 |
| Maize (hill) | 0.20 | 0.17 |
| Maize (bottom) | 0.75 | 0.64 |
| Bean (hill) | 0.24 | 0.21 |
| Sarawak | | |
| Rice (upland) | 0.80 | 0.68 |
| Columbia | | |
| Rice (upland) | 0.70 | 0.60 |
| Liberia | | |
| Rice (upland) | 1.12 | 0.95 |
| Rice (swamp) | 1.12 | 0.95 |
| Manioc | 5.10 | 1.53 |
| Sugarcane | 37.00 | 7.40 |
| Central African Republic | | |
| Cotton | 0.21 | 0.21 |
| Manioc | 2.70 | 2.30 |
| Sorghum | 0.42 | 0.36 |
| Maize | 0.13 | 0.11 |
| Ground nuts | 0.22 | 0.20 |
| Sesame | 0.02 | 0.20 |
| Beans | 0.06 | 0.05 |
| Eastern Nigeria | | |
| Compound plot | 28.4 | ? |
| Fields | 3.07 | 2.60 |
| | Irrigation holdings | |
| India, Deoria | | |
| Rice | 1.21 | 1.02 |
| Wheat | 1.81 | 1.54 |
| Sugarcane | 42.0 | 8.4 |
| Yemen, Tihama | | |
| Cotton | 1.11 | 1.11 |
| Sorghum | 1.23 | 1.05 |
| Pakistan, Sind | | |
| Cotton | 1.30 | 1.30 |
| Wheat | 1.90 | 1.62 |
| Sugarcane | 57.0 | 11.40 |
| India, Punjab | | |
| Cotton | 0.81 | 0.81 |
| Wheat | 2.45 | 2.08 |
| India, Coimbatore | | |
| Rice | 2.50 | 2.12 |
| Cotton | 1.50 | 1.50 |

(*continues*)

TABLE 1-1 (*Continued*)

| Region and crop | Yield (tons) as harvested | Dry weight |
|---|---|---|
| Sudan, Gezira | | |
| Cotton | 1.84 | 1.84 |
| Wheat | 1.36 | 1.15 |
| Ground nuts | 1.34 | 1.25 |
| Sorghum | 1.98 | 1.68 |

*Source:* Mostly Ruthenberg 1976; for Costa Rica Fernside, Personal Communication.

States alone) compared to that in Central America, the Caribbean, much of Africa, all of Asia, and even Europe (Table 1-2). Fewer would guess that industrially derived fertilizer use per hectare per year is higher in El Salvador and Costa Rica than it is in the United States, or that worldwide, nutrient gain from industrial fertilizer applications about equals the removal of nutrients in crop plant yields (FAO, 1981). Low human population densities (due in part to the wholesale slaughter of the original inhabitants), large initial supplies of

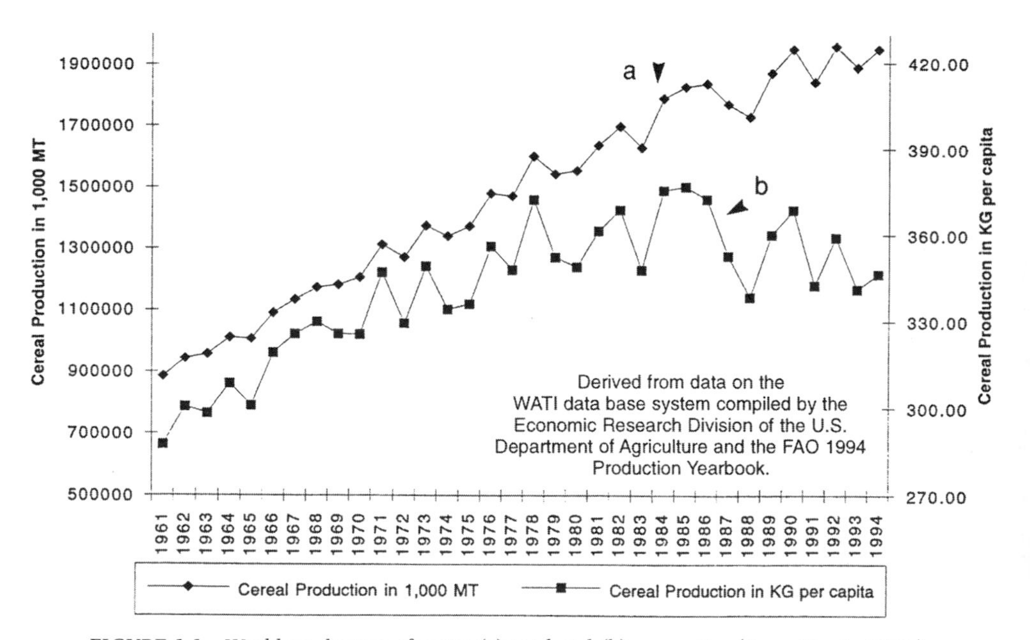


FIGURE 1-1 World production of grain: (a) total and (b) per capita. (From Harris, 1996.)

**TABLE 1-2 Population Density of Selected Parts of the World**

| Region | Number of people (per square mile) | Population growth rate (%) | |
|---|---|---|---|
| | | 1980–1990 | 1990–2000 |
| North America | 33 | | |
| United States | 69 | 1.0 | 0.7 |
| Canada | 7 | 1.0 | 1.0 |
| Middle America | 121 | 2.4 | 2.2 |
| Costa Rica | 155 | 2.7 | 2.4 |
| Carribean | 360 | 1.2 | 1.4 |
| Europe | 266 | 0.3 | 0.3 |
| East Asia | 293 | 1.2 | 1.2 |
| South Asia | 289 | 2.3 | 2.0 |
| World | 101 | 1.7 | 1.6 |

*Source:* Modified from Hall and Hall (1992) and references therein.

fossil fuels, and fertile soils where summer rain is abundant have allowed North America the luxury of food surpluses and hence exports. Only Australia, Argentina, and a few European countries are also important food exporters, although none on the scale of the United States and Canada.

The great hope of tropical agricultural research is that it will be possible to increase tropical yields greatly (as happened in the United States) or at least decrease the costs and environmental impact of the increased production that does take place through research, technology of various kinds, and extension. The net result of the research to date is somewhat ambiguous. Although grain yields have risen in the tropics, generally it has been from the traditional rate of about 0.5 to 1.0 tons (dry) per hectare per year to about 1.5 or 2.0 tons per hectare per year, enough to feed five or six people (Table 1-1). (An exception is the high yields, 5 or 6 tons per hectare per year, in intensely managed and fertilized wet rice cultures, primarily in Asia.) For the tropics as a whole grain production per capita has not increased since 1973, the year of the first petroleum price shock (Harris, 1996). Disturbingly, at the national level most of the increases appear to be no more, and in fact less, than one might expect from simply the quantity of fertilizer added, implying that the only technological improvements came from the industrialization (i.e., use of industrially derived fertilizers) of agriculture, not from some aspect of pure technology such as genetic change alone. One implication is an increasing dependence of all tropical countries on petroleum from other parts of the world.

## III. THE INCREASING NEED FOR FOREIGN EXCHANGE

If food or fertilizer is to be imported it has to be paid for (unless received as a gift). Paying for this food or fertilizer requires what is called *foreign exchange*, that is, a currency such as dollars that will be accepted by international organizations or banks and, eventually, fertilizer manufacturers in Louisiana and farmers in Nebraska and Alberta. In addition, of course, foreign exchange is required to pay for all other inputs from the industrialized world, such as automobiles, trucks, agrochemicals, televisions, computers, and so on. In tropical countries that foreign exchange traditionally has been generated from agricultural products such as coffee or other natural resource products such as shrimp or mahogany. Thus the new reality is that developing countries are increasingly dependent upon exports to pay for their agricultural needs and other imports.

There is another path, that of explicit industrialization. A relatively few tropical countries or regions, including Puerto Rico, Singapore, Thailand, and Indonesia, have been able to develop their own specialized manufacturing economies that have been successful in generating products competitive on the world market. This process has tended to require very cheap labor, stable (and sometimes repressive) political conditions, and, often, a highly educated component of the population. Such economies tend to be very energy intensive (Rios, 1993). Several of these nations have their own petroleum, but most do not. The large volume of exports generated in these successful countries makes this industrialization without domestic fuels feasible. These requirements for industrialization are not often available all at once, and it is difficult for a country that has not already chosen the high-tech path to enter into stiff international competition with those nations that have already developed such industries.

Meanwhile, many tropical countries have tended to continue to have very high birth rates (Figure 1-2). Nearly all have, in addition, a "pyramid-shaped" population age structure, meaning that there are proportionally very many young people who will contribute to population increases in the future even if each family has only two children. A critical question follows: how will these people be fed? Will it be from food grown on the land of their own nation, or instead from food grown elsewhere that is purchased by generating foreign exchange? The answers are not easy to come by, and they become more difficult each year as tropical populations continue to grow.

Within this context it is useful to determine just what the potential agricultural yield might be for a country, how it might change if production and hence erosion is accelerated as technologies and prices of inputs change, and,

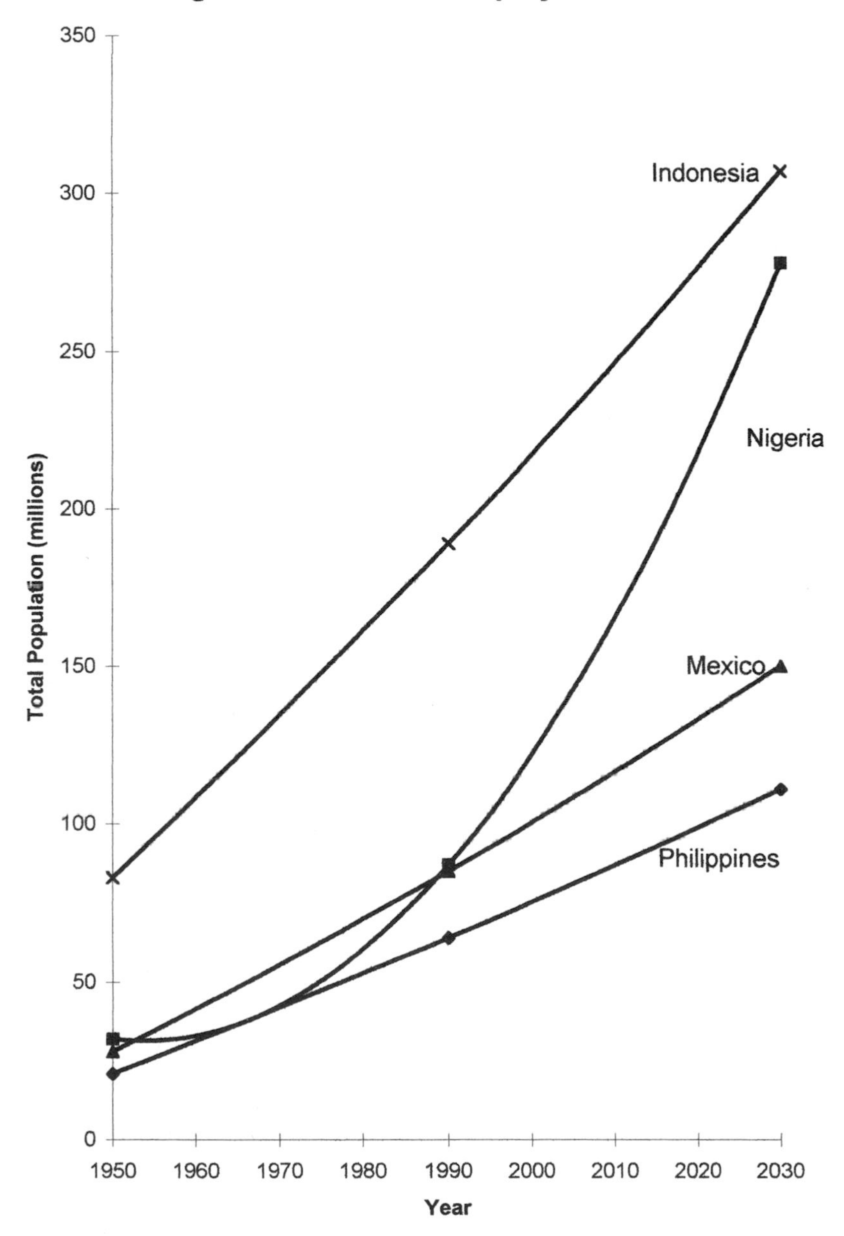


FIGURE 1-2 Population growth of selected tropical countries.

in general, how to optimize a nation's agricultural production in terms of net economic yield. Alternately we may wish to examine the potential of that nation for generating foreign exchange so that food could be purchased on world markets. We examine these issues in Chapters 12 and 23.

## IV. TROPICAL ENVIRONMENTAL ISSUES AND THEIR ECONOMIC CONSEQUENCES

Since about 1980 a much greater emphasis has been placed on the importance of conservation issues in tropical countries, especially by the developed world. This emphasis has arisen because of: (1) the high biodiversity of many tropical countries; (2) the rapid rate of deforestation in many areas; (3) the linkage between the activities of consumers in the developed world and environmental degradation in the tropics (e.g., Myers, 1981; German Bundestag, 1990); (4) various arguments that have been developed about the value that biodiversity (and various other characteristics of natural ecosystems) contribute to human cultures and economies (e.g., Solorzano *et al.,* 1991); and (5) the role of tropical forests in regional and global climate, including especially their role in storing, or releasing through deforestation, carbon and other greenhouse gases (Detwiler and Hall, 1988; Veldkamp, 1994), and the sensitivity of the tropics to possible climate change.

The climatic issues are especially compelling to some people because the tropics are, and always have been, extremely susceptible to climatic extremes, including floods, hurricanes, and especially droughts. Some computer models (Rind *et al.,* 1990) suggest that expected global climate changes will impact the tropics especially hard, particularly with respect to decreasing soil moisture. The models predict that some tropical areas (northern Mexico, the Horn of Africa, southeast Africa, and western Costa Rica) will be especially impacted. Many of these regions in fact have been experiencing more or less unprecedented droughts. While I believe it too early to make any explicit predictions about whether the climate is going to change adversely for the tropics as a function of human activities, I do believe that climates will change, as they have always changed, and that human populations are extremely sensitive to that change. The world is littered with failed civilizations, many of which were undermined by climatic change (Tainter, 1988, 1992). Since many human populations are at the edge of the ability of their region to feed them, and are increasingly sensitive to the disruption of the trade system that brings them either food or fertilizer, human populations in the tropics are likely to continue to be very sensitive to climate changes.

Meanwhile, there is one aspect of these arguments that seems to be unequivocal. While not all elements of natural ecosystems have been demonstrated to have high direct economic utility, some clearly do. Undisturbed natural forests protect water quality, reduce extremes in stream flow, and reduce stream sedimentation, the latter being especially important where there are downstream dams and reservoirs. Thus there are clear economic reasons to maintain undisturbed forests in watersheds that feed reservoirs or that are flood-prone. For another example, trees are valuable as moderators of climatic extremes, reducing the energy required to heat or cool buildings. And, of course, relatively natural ecosystems are valuable recreation and relaxation areas. We call these contributions to human well-being the *public service function* of ecosystems (Hall, 1975). They rarely enter into routine monetary-based economic analysis. An important question concerns the degree to which these public service functions extend to additional components of natural ecosystems. A natural forest protects soil. But economic value to the community may be increased, at least initially, if the forest is cut and crops planted, even if soil erosion is increased enormously. Do we include the soil protection services of the forest in its public service function evaluation? Do we include the economic value of the diversity of plants from which an important new drug might be found? Can a crowded planet maintain its natural capital if it is in the long-term best economic interest while hungry people need food now? Can we possibly live on the interest and not the capital of natural ecosystems? And, finally, who gets the gains and who gets the losses of development?

## V. HAS DEVELOPMENT FAILED?

The degree to which development has succeeded or failed is quite debatable and depends on the criteria one uses and whom you ask. Surely the additional billions of people alive now compared to when I was born in 1943 could not have been fed without the industrialization of agriculture, including tropical agriculture. Nevertheless we have found that there is a tendency, as of the time of this writing, to consider development as a failure overall, although it must be asked, "failed relative to what?" (See Chapters 24 and 26.)

## VI. WAS CONVENTIONAL ECONOMICS THE WRONG TOOL TO GUIDE DEVELOPMENT?

There is a second general problem with "traditional" development, besides its possible failure. Most development agencies are run and staffed by economists, which makes a certain sense because development is usually about lending money. Where scientific or technical people have some power they generally are asked to subjugate their scientific expertise to economists (Goodland and

Ledec, 1987). Most economists these days have been trained in what is called the neoclassical tradition. This approach has been extremely successful in pervading our culture. Various aspects of neoclassical economics are routinely taught and thought of as universal "truths" even though as an intellectual way of thinking about the world the conceptual basis of neoclassical economics is hardly more than a century old, and it is only one of many ways that we have and might think about economics (see Chapter 3). The interested reader can see the degree to which the neoclassic world view has taken the moral high ground (deserved or not) even in the environmental literature (e.g., Cruz *et al.*, 1996). I have no argument with the use of neoclassical economics when it is considered a tool tempered by the use of other approaches. But too often it is applied as the only possible approach or even as a religion (Chapter 24). Chapter 3 examines various types of economic analyses from the perspective of their strengths and weaknesses and their potential to guide the actual operations of tropical economies and ecosystems.

## VII. SUSTAINABLE DEVELOPMENT, THE NEW SILVER BULLET

Given the very real environmental degradation generally associated with development (e.g., Ayres, 1995, 1996) and the real and perceived failure of traditional approaches to solving the economic and social problems of the tropics, a new perspective has arisen in the 1980s and 1990s—that of *sustainable development*. Exactly what sustainable development means seems to vary greatly from user to user (for a list of definitions see Table 1-3; see also Chapter 3). A common theme is that past patterns of development, especially those based heavily on the use of energy-intensive inputs, are destructive of soils, biota, and systems of production, so that new approaches that focus instead on the long-term maintenance of the productive system should be the goal.

But the world is what it is and not what we wish it to be. If there is to be sustainable development, indeed any development, it must be within the context of whatever is biologically and physically possible. Humans can change what is possible through technology. But technology traditionally has been mostly a means of finding ways of making human actions more powerful through the increased exploitation of resources and the increased use of energy, especially fossil energy (Cottrell, 1955; Odum, 1971; Hall *et al.*, 1986). Thus to the degree that any development means the increased exploitation of resources with the use of fossil energy, it is by definition not sustainable, because fossil fuels are finite and are not being made today on any important scale. Other stock resources, including especially soils and many minerals, but also fossil water, many forests, and fishes, are likewise essentially being mined

TABLE III Various Definitions of Sustainability

- ". . . is development that meets the needs of the present without compromising the ability of future generations to meet their own needs." The World Commission on Environment and Development, *Our Common Future* (New York: Oxford University Press, 1987), p. 4.
- ". . . requires meeting the basic needs of all people and extending opportunities for economic and social advancement. Finally, the term also implies the capacity of development projects to endure organizationally and financially. A development initiative is considered sustainable if, in addition to protecting the environment and creating opportunity, it is able to carry out activities and generate its own financial resources after donor contributions have run out." Bread for the World, *Background Paper No. 129,* Washington, DC, March 1993.
- "[improves] . . . the quality of human life while living within the carrying capacity of supporting ecosystems." International Union for the Conservation of Nature and Natural Resources (IUCN), World Conservation Union, United Nations Environment Programme (UNEP), and World Wide Fund for Nature (WWF), *Caring for the Earth* (Gland, Switzerland: IUCN, UNEP, WWF, 1991), p. 10.
- "[uses] . . . natural renewable resources in a manner that does not eliminate or degrade them or otherwise diminish their renewable usefulness for future generations while maintaining effectively constant or nondeclining stocks of natural resources such as soil, groundwater, and biomass." World Resources Institute, "Dimensions of Sustainable Development," *World Resources 1992–93: A Guide to the Global Environment* (New York: Oxford University Press, 1992), p. 2.
- "[maximizes] . . . the net benefits of economic development, subject to maintaining the services and quality of natural resources." R. Goodland and G. Ledec, "Neoclassical Economics and Principles of Sustainable Development," *Ecological Modeling* 38:36, 1987.
- "[is based on the premise that] . . . current decisions should not impair the prospects for maintaining or improving future living standards. . . . This implies that our economic systems should be managed so that we live off the dividend of our resources, maintaining and improving the asset base." R. Repetto, *World Enough and Time* (New Haven, CT: Yale University Press, 1986), pp. 15–16.
- ". . . is taken to mean a positive rate of change in the quality of life of people, based on a system that permits this positive rate of change to be maintained indefinitely." L.M. Eisgruber, "Sustainable Development, Ethics, and the Endangered Species Act," *Choices,* Third Quarter 1993, pp. 4–8.
- ". . . is development without growth—a physically steady-state economy that may continue to develop greater capacity to satisfy human wants by increasing the efficiency of resource use, but not by increasing resource throughput." H.E. Daly, "Steady State Economics: Concepts, Questions, and Policies," *Ecological Economics* 6:333–338, 1992.
- ". . . is the search and the carrying out of rational strategies that allow society to manage, in equilibrium and perpetuity, its interaction with the natural system (biotic/abiotic) such that society, as a whole, benefits and the natural system keeps a level that permits its recuperation." E. Gutierrez-Espeleta, "Indicadores de Sostenibilidad: Instrumentos Para La Evaluacion de las Politicas Nacionales," unpublished paper presented at *50th Anniversary Conference of the Economic Sciences Faculty* sponsored by the University of Costa Rica, San Jose, Costa Rica, Nov. 19, 1993.

Data from OTA (1994).

and used once. By this chain of logic one might conclude that sustainable development is an oxymoron, that is, a phrase that is internally inconsistent, such as "jumbo shrimp." On the other hand perhaps new technologies can decouple development from resource exploitation and the use of nonrenewable resources.

Thus a principal objective of this book is to see if it is possible—based on the information at our disposal—for the concepts of sustainable development to be, in fact, applied to a tropical nation such as Costa Rica, and whether this might enable us to understand what is and what is not possible. If "sustainable development" is not possible, is the concept but one more in a series of exploitations (including "colonialism," "economic development," and "structural adjustment") perpetrated upon the developing world by the developed world? Will it too leave these nations again impoverished, overexploited, and increasingly dependent upon institutions and resources in the developed nations? Or are any or all of the exploitations associated with development necessary to pay for the ever more crucial imports required from the more developed world? Are they necessary simply to take care of the needs of the increasing populations? In other words, are many tropical areas already too densely populated relative to their resource base for truly sustainable development to work on a large scale? The following chapters provide a background for assessing this question, and the final chapters attempt to assess our principal objective: to what degree is sustainable development possible for Costa Rica?

Our first job is to understand in some detail how the economy of a tropical country does in fact work—what are the products and services produced, how has the economy developed historically and for what reasons, what are the goods and services that might be produced, and what are the physical requirements for producing those goods and services? Then we can start asking some other questions about the degree to which these processes are dependent upon, or might be dependent upon, sustainable versus unsustainable resources. We begin this in the next chapter with a close examination of the economic history of one country, Costa Rica.

## REFERENCES

Ayres, R. U. 1995. Economic growth: Politically necessary but not environmentally friendly. *Ecological Economics* 15:97–99.

Ayres, R. U. 1996. Limits to the growth paradigm. *Ecological Economics* 19:117–134.

Brown, L. R. 1996. The acceleration of history. In *State of the World 1996*, pp. 3–20. W. W. Norton, New York.

Cleveland, C., R. Costanza, C. Hall, and R. Kaufmann. 1984. Energy and the United States economy: A biophysical perspective. *Science* 225:890–897.

Cottrell, F. 1955. *Energy and Society*. McGraw–Hill, New York. Reprinted 1970 by Greenwood Press, Westport, CT.

Cruz, W., M. Munasinghe, and J. Warford. 1996. Greening development: Environmental implications of economic policies. *Environment* 38:6–37.

Detwiler, R. P., and C. A. S. Hall. 1988. Tropical forests and the global carbon cycle. *Science* 239:42–47.

FAO. 1981. Crop production levels and fertilizer use. In *FAO Fertilizer and Plant Nutrition Bulletin*. FAO, Rome.

FAO. 1985. *Changes in Shifting Cultivation in Africa. Seven Case Studies*. Forestry Department, Food and Agricultural organization of the United Nations, Rome.

Geertz, C. 1963. *Agricultural Involution*. University of California Press, Berkeley, CA.

German Bundestag. 1990. *Protecting the Tropical Forests. A High Priority International Task*. German Bundestag, Bonn, Germany.

Goodland, R., and G. Ledec. 1987. Neoclassical economics and the principles of sustainable development. *Ecological Modelling* 38:19–46.

Hadley, M., and J. P. Lanley. 1983. Tropical forest ecosystems: Identifying differences, seeking similarities. *IUCN Bulletin* **1983**, 3–19.

Hall, C. A. S. 1975. The biosphere, the industriosphere and their interactions. *Bulletin of Atomic Scientists* 31:11–21.

Hall, C. A. S., and J. Day. 1977. *Ecosystem Modeling in Theory and Practice*. Wiley–Interscience, New York. Reprinted 1990, University Press of Colorado, Boulder, CO.

Hall, C. A. S., C. J. Cleveland, and R. K. Kaufmann. 1986. *Energy and Resource Quality: The Ecology of the Economic Process*. Wiley–Interscience, New York. Reprinted 1992, University Press of Colorado, Boulder, CO.

Hall, C. A. S. 1988. An assessment of several of the historically most influential theoretical models in ecology and of the data provided in their support. *Ecological Modelling* 43:5–31.

Hall, C. A. S. 1991. An idiosyncratic assessment of the role of mathematical models in environmental sciences. *Environment International* 17:507–517.

Hall, C. A. S. 1992. Economic development or developing economics: What are our priorities? In M. Wali (Ed.), *Ecosystem Rehabilitation*, Vol. I. *Policy Issues*, pp. 101–126. SPB Publishing, The Hague, The Netherlands.

Hall, C., and Hall, M. 1993. The efficiency of land and energy use in tropical economies and agriculture. *Ag. Ecosystem Environ.* **46**, 1–30.

Hall, C. A. S. 1995. Maximum Power: The Ideas and Applications of H. T. Odum. University Press of Colorado, Niwot, Colorado.

Harris, J. M. 1996. World agricultural futures: Regional sustainability and ecological limits. *Ecological Economics* 17:95–116.

Ko, J-Y., C. A. S. Hall, and L. G. Lopez Lemus. 1998. Resource use rates and efficiency as indicators of regional sustainability: An examination of five countries. *Environmental Monitoring and Assessment* **51**, 571–593.

Lal, R., and D. J. Cummings. 1979. Clearing a tropical forest. I. Effects on soil and micro-climate. *Field Crops Research* 2:91–107.

Lappe, F. M., Joseph Collins, and Peter Rosset. 1998. *World Hunger: 12 Myths*. Grove Press, New York.

Kunstader, P., E. C. Chapman, and S. Sabhasri. 1978. *Farmers in the Forest: Economic Development and Marginal Agriculture in Northern Thailand*. University Press of Hawaii, Honolulu, HI.

Myers, N. 1981. The hamburger connection: How Central America's forests become North America's hamburgers. *Ambio* 10:2–8.

Odum, H. T. 1971. *Environment, Power and Society*. Wiley–Interscience, New York.

OTA (Office of Technology Assessment) U.S. Congress 1994. Perspectives of the role of science and technology in sustainable development. OTA-ENV-609. Washington, D.C.

Pandey, U., and J. S. Singh. 1984. Energy-flow relationships between agro- and forest ecosystems in Central Himalaya. *Environmental Conservation* 11:45–53.

Paruelo, J. M., P. Aphalo, C. A. S. Hall, and D. Gibson. 1987. Energy use and economic output for Argentina. In G. Pillet (Ed.), *Environmental Economics: The Analysis of a Major Interface,* pp. 169–184. Leimgruber, Geneva, Switzerland.

Ramakrishna, P. S. 1990. Agricultural systems of the northeastern hill region of India. In S. R. Gleisman (Ed.), *Agroecology: Researching the Ecological Basis for Sustainable Agriculture,* pp. 251–274. Springer-Verlag, New York.

Rind, D., R. Goldberg, J. Hansen, C. Rosensweig, and R. Ruedy. 1990. Potential evapotranspiration and the likelihood of future drought. *Journal of Geophysical Research* 95:9983–10004.

Rios, A. 1993. Energy flow analysis for Puerto Rico. Ph.D. dissertation, State University of New York, College of Environmental Science and Forestry, Syracuse, NY.

Ruthenberg, H. 1976. *Farming systems in the tropics.* 2nd ed. Oxford University Press, Oxford, U.K.

Solórzano, R., R. de Camino, R. Woodward, J. Tosi, V. Watson, A. Vásquez, C. Villalobos, J. Jiménez, R. Repetto, and W. Cruz. 1991. *Accounts overdue: Natural resource depletion in Costa Rica.* Tropical Science Center, San Jose C. R. and World Resources Institute, Washington, DC.

Tainter, J. 1988. *The Collapse of Complex Societies.* Cambridge University Press, Cambridge.

Tainter, J. 1992. Evolutionary consequences of war. In G. Ausenda (Ed.), *Effects of War on Society,* pp. 103–130. Center for Interdisciplinary Research on Social Stress, San Marino, CA.

Veldkamp, E., A. M. Weitz, I. G. Staritsky, and E. J. Huising. 1994. Deforestation trends in the Atlantic zone of Costa Rica: A case study. *Land Degradation and Rehabilitation* 3:71–84.

Watters, R. F. 1971. *Shifting cultivation in Latin America.* FAO Forestry Development Paper No. 17. FAO, Rome.

CHAPTER 2

# A Brief Historical and Visual Introduction to Costa Rica

CHARLES A. S. HALL, MYRNA HALL, AND BERNARDO AGUILAR



## I. COSTA RICA, GREEN PARADISE?

Flying into Costa Rica from the northeast, one crosses over sandy beaches fringed with clear blue water on one side and palm trees on the other, and then great expanses of lush verdant forests. Large, beautiful mountains are visible in the distance, and there are many rivers. It certainly looks like the image of tropical paradise promoted on travel posters.

Soon the landscape becomes fragmented with human activities: large banana plantations and coffee plots mixed with small farms. Suddenly they give way to brown, denuded slopes, and then, just as suddenly, green forests appear again. The transition is not often gradual, but instead there is a sharp delineation between the green of forests and plantations and the new areas, which are pastures. In these pastures there is no sign of life—no houses, no cultivated fields—but only grass, very short, clinging to the steep slopes (Land Use, on CD). We are told it is grazed, yet no animal life is visible. The scene is similar throughout much of Central America. Is this land so unproductive that it has been abandoned? Or have the cows been moved elsewhere to feed while the area below recuperates? Or have the cattle been harvested early to avoid the dry season? Perhaps all of these are true, depending on the area visited. But the overall impression is clear, and it is supported by driving in almost any direction in Costa Rica as well as by many scientific studies. Pastures, or at least something resembling pastures, are everywhere. But cows are relatively few. In fact, the undisturbed forested area in Costa Rica has diminished from 3,420,000 ha in 1940 to 870,000 ha in 1983 [from 67 to 17% of the national area, which originally was 99.8% covered by forests (Perez and Protti, 1978; Rodas, 1984; Sader and Joyce, 1985)]. This deforestation, mostly pasture driven, is continuing, although the dry northwest is showing significant signs of natural reforestation and there are some areas of human-imposed reforestation.

Yet Costa Rica remains a very special place from the perspective of both Costa Ricans and North Americans. First and foremost, in many respects it really is a green tropical paradise (on CD; click on Land Use). Nearly 20% of the land area is in verdant national parks, among the highest percentage of any country in the world. Because of its mountains and its close relation to two oceans, Costa Rica may be the wettest country in the world and most of the country receives more than 2.5 m (100 in.) of rainfall annually, making it very green for much of the year (Hartshorn *et al.*, 1982). The changes from hot coasts to cold high mountains and from wet to dry areas provide a tremendous range of ecological conditions, which are reflected by nearly the highest biodiversity in the world. Costa Rica has other attributes of an earthly paradise. Because of its mountains the temperature is generally very comfortable over much of the nation; because the country is a peaceful democracy with no military its citizens live free of armed conflict; because of a strong commitment to health care and other aspects of social welfare (an amazing 18.7% of the GNP is spent on social programs) both citizens and tourists generally feel very safe and secure; because of a reasonably healthy economy and some subsidies from more northerly countries the standard of living is relatively high for the tropics. The life expectancy of 75.3 years, an infant mortality rate of 15.3 per 1000 live births, an unemployment rate of 4.2%, and a literacy rate of 93.1% (Holl *et al.*, 1993; Ministerio de Planificación y Política Económica, 1995) are excellent by nearly any country's standards, and a remarkable achievement for a country that is relatively poor compared to the standards of the industrial world. Corruption, the bane of so many governments in Latin America, is certainly present but not crippling as it often is elsewhere, in part because "Costa Rica is so small that everyone knows what is going on" (Carlos Leon, personal communication). Perhaps more important, when corruption is uncovered those responsible are held accountable and often end up in jail, even if once-high governmental officials.

This positive environmental and social image is a critical aspect of how the country is perceived by many people, in and out of the Costa Rican government, and has been used aggressively to promote development, tourism, and the concept that Costa Rica is a good place to invest. This includes investment in the environment, such as national parks and forests to sequester atmospheric carbon. These attributes of Costa Rica also have been capitalized upon by the tourism and scientific establishments of the industrialized nations, so that Costa Rica is for many "the place to go to experience the tropics." Ecotourism, in particular, has been highly developed in Costa Rica, and ecotour buses are a very common sight everywhere. More than 30,000 U.S. citizens have chosen to retire in Costa Rica, wooed through tax and customs incentives.

For decades Costa Rica has been an important research and training ground for temperate ecologists and biologists (on CD; see Institutions). Five excellent and venerable institutions are the Organization of Tropical Studies (OTS), which offers courses in Costa Rica under the auspices of Duke University; the OTS La Selva Biological Station, a forest research site at Puerto Viejo; the Tropical Science Center (El Centro Científico Tropical or CCT) in San Pedro and the 10,500-ha Monteverde Biological Cloud Forest Preserve which it owns and operates; the Centro de Agricultura Tropical para Investigacion y Ensenanza (CATIE) in Turrialba, an agricultural research station dedicated to helping the small farmers of Central America; and the facilities at Santa Rosa National Park. In addition there are a number of excellent Costa Rican institutions with strong programs focusing on the environment, including the University of Costa Rica at San Pedro and EARTH College at Guapiles, which focuses on lowland tropical agriculture. These institutes collectively provide very sophisticated analyses of Costa Rican natural ecosystems. Much of this information has been synthesized in Daniel Janzen's comprehensive, "A Natural History of Costa Rica" (1983). In short, Costa Rica has been a wonderful place for both local and visiting scientists to work. For all these reasons Costa Ricans are almost universally very proud of their country, as they should be, and nearly all agree that it is a very good place in which to live. For other summaries of Costa Rica see Beletsky, 1998, Coates, 1997; and Rojas *et al.*, 1987.

## A. Is There Trouble in Paradise?

Nevertheless, Costa Rica has, is now, and almost certainly will continue to face a series of very serious problems in addition to deforestation. The population growth rate of 2.4% per year (FAO, 1989; note that other estimates vary from 2 to 3% per year) will, if continued, translate into the 3.5 million people of 1995 totaling 6 million by the year 2015. Costa Ricans are now importing about 30% of the food calories they eat, and this proportion is increasing year by year, implying serious difficulties for feeding the nation in the future. Social conditions for children have been changing dramatically, such that half of the children born in Costa Rica in recent years were born to women who do not have husbands. The national health care and education systems, still in many respects models for the rest of the world, are increasingly being cut back by a government desperate for cash. Despite very large infusions of foreign aid and development funds, Costa Rica had a $3.9 billion foreign economic debt as of 1995, which exceeds its annual gross domestic product and is among the highest per capita debt/earnings load in the world (Annis, 1990; see also Chapter 4). The country, traditionally and still basically middle class, is becoming increasingly a nation of extremes. Huge new houses of a sort rarely seen in the past are now common at the outskirts of many small cities, and for the first time there are large slums. Severe air pollution is a daily occurrence in San Jose, and most rivers below their forested headlands are heavily polluted with sewage, garbage, and agricultural wastes. New fast food restaurants are contributing large quantities of trash to once clean city streets, and life seems increasingly crowded and frantic.

How do we understand these changes? In an overall context are they good or bad, or do we even have a way of judging that? What does the future hold for the economy and the environment? Can the economic progress that has been made be sustained? What decisions can be made now to make the future more agreeable? This book explores many aspects of these questions.

First we examine Costa Rica within a qualitative historical context. Since Costa Rica's economy was traditionally, and in many respects is still, based on agriculture, we focus on this sector. Our review is based on Janzen (1983), other referenced sources, newspapers, and conversations with knowledgeable Costa Ricans. Later chapters address these issues quantitatively and in more depth.

## II. A HISTORICAL PERSPECTIVE ON COSTA RICA AND ITS ECONOMY

In pre-Columbian times the area that is now the country of Costa Rica supported a substantial human population (Seligson, 1980). It was a crossroad of trade routes and a meeting place of cultures, and approximately the boundary between Meso-American and South-American cultures. At that time an important agricultural division corresponded to linguistic and ethnic differences. In the northwest, in what is now the province of Guanacaste, lived the Chorotega, a people whose primary agricultural crop was maize, which they ground and made into tortillas or combined with chocolate to make the drink called "chicha." The rest of Costa Rica was occupied by the Huetares and Talamancas along the Atlantic coast and in the central highlands, and the Diquis of the Terraba Valley. Their basic food crops were yuca and other tubers and the pejibaye palm.

At the time of the Spanish conquest, beginning in 1560, much of the Atlantic coast was cultivated, probably principally by shifting cultivators. But there were no huge empires such as existed in Mexico, Guatemala, and Belize to the north or Peru to the south, so, at least until the time of the Spanish arrival, there were (to our knowledge) no major cultural clashes.

## A. The Colonial Period and the Spanish Legacy

Columbus visited Costa Rica in 1502, and the first Spanish expeditions to explore the coasts started around 1516, under the command of Juan de Castañeda and Hernán Ponce de León (Pinto Soria, 1993). The colonial period began in Costa Rica around 1560, when Juán de Cavallón led the first expeditions from the north (Pinto Soría, 1993). When the first Spanish settlers arrived, the Indian population of Costa Rica was estimated to be from 17,000 to 80,000 [depending upon one's interpretation of the early Spanish records (Seligson, 1980)]. Within a few short years this was reduced by as much as 95% by fighting, disease, and slavery. The Indian cultures were lost, the fields abandoned, and the forest reclaimed most of the once-cultivated areas (Janzen, 1983).

In Costa Rican academic circles there is currently a debate over the settlement pattern and socioeconomic structure of the colonial period. The issue is the degree to which Costa Rica followed the traditional Spanish pattern of settlement characteristic of most of the New World or whether something different happened. The traditional pattern was based on the medieval stratified class structure of Iberia, which was perpetuated in much of Latin America by many institutions, especially "el latifundio," a feudal agrarian system based economically on monoculture and socially on class stratification (Heyck, 1988). El latifundio also implies extensive and inefficient use of labor and land as well as a lack of concern for preserving the quality of the land (Heyck, 1988). It promotes monoculture of crops for export, for therein lie the greatest profits. From the time of the first colonial settlement the system reserved land ownership (and thus power and wealth) for the European colonists and their descendants, called the Creoles. The system remains intact in much of Latin America but less powerfully in Bolivia, Nicaragua, and Mexico, where true "structural" land reform occurred. Land reform in Costa Rica is more correctly classified as "marginal land reform" (Aguilar and Torrealba, 1987, using García, 1982, for the classification). Even today the Gini coefficient of land concentration applied to the latest agrarian census of Costa Rica (0.71) indicates that relatively few people own much of the land.

Some historians believe that before the introduction of coffee, European Costa Rica was quite different and consisted, instead, of a closed, subsistence-based peasantry, internally homogenous and self-sufficient, and was made up of small landholding households (Seligson, 1980; Gudmundson, 1986). According to this view, during colonial times in the area that is today Costa Rica there existed a more or less egalitarian societal structure with a weak elite of office holders who wielded some power over the peasant masses, as well as some larger hacienda owners. This was mostly an egalitarianism of the poor, such that in 1719 the governor reported that he had to sow and reap his own crops or he would starve. There was no national money, and transactions were normally done with cacao beans!

A slightly different view that has been gaining popularity for the last 20 years follows. Research on village birth, death, and tax records suggests a Malthusian development pattern in which one finds not individual households dispersed over the land with small landholdings, as was assumed before, but rather "nucleated settlements" or small communities which tilled the land immediately surrounding the village for subsistence crops, such as wheat and corn (Gudmundson, 1986). There was little exchange among villages and there were no real urban centers. These settlements, according to Gudmundson, quickly reached a point past which land close to the settlement site (within a half day's walk) was fully cultivated. As food supplies became scarcer an increasing number of households either had to accept relative poverty or move beyond the village's land claims. Without the motivation to "seek one's fortune" (there was no external market) and given the difficulty of migrating through dense forest, people preferred to remain together in a more "civilized community." The jungle beyond came to represent an unappealing life of subsistence and isolation.

According to Gudmundson and others this Malthusian-driven development pattern (where societies reached environmental carrying capacity) fostered inequality. Those with extensive landholdings did reasonably well while many others hung on at the barest minimum of existence (i.e., lacking an adequate food supply) and social position. This research also indicates that the common lands and estate structure of colonial lands limited private ownership of small holdings in Costa Rica, just as it did in other Spanish-American colonial societies. Church, state, and the elite that supported these institutions relied on the nucleated settlements to generate taxes, tithes, and laborers.

More recently a view has been developing that accepts that Costa Rica did follow the traditional pattern of Spanish colonization in every major social aspect, including the establishment of "encomiendas"—that is, Indians sold as part of real estate—the use of black slavery, and strong social stratification. But there still remained a difference when compared to many other Latin American nations, due principally to: (1) the lack of precious metals and

(2) the (probably) low numbers of Indians that were in the country when the Spaniards arrived (Quirós, 1993; Pinto Soria, 1993). The latter ignores the area of Nicoya, which was documented to have large amounts of indigenous populations under the "encomienda" system, because it was not part of Costa Rica during the colonial period. According to Stone (1982) power in Costa Rica has been concentrated in several families that are direct descendants of the Spanish conquerors.

The perception of egalitarianism comes more from the specific characteristics of the small farming systems in the Central Valley and the fact that coffee, which can be grown profitably in small plots, adapted very well to this pattern of land tenure. The concentration of wealth and power happened mostly in the processing and exporting operations, within what was probably a more or less benevolent oligarchy. This relative benevolence contributed to the feeling of a small rural democracy and therefore was a social buffer (Winson, 1989). Thus within this system, even though the land reform was quite marginal, it was still sufficient to maintain political stability even while the concentration of land and power remained principally in the hands of the old oligarchs.

The relative poverty of early Costa Rica, attributed to a lack of exploitable gold and silver, had many impacts that continue today. There were few large haciendas or powerful landlords, as was the case further north in Central America, because slave labor, and indeed labor of all kinds, was scarce. This pattern of small holder development led to a belief in a small landholder or "yeoman" heritage, commonly called "the rural democratic model." This was a potent force leading to the 1948 Civil War and the formation of the Partido Liberacion Nacional (The National Liberation Party, initially responsible for many of the democratic principles put into effect in Costa Rica). This heritage, and the strong governmental commitment to social programs begun in 1948, is one of the fundamental reasons that land and class struggles have been avoided in Costa Rica, unlike in its neighboring nations where the discrepancy between rich and poor is much greater (Seligson, 1980). It is also the foundation for the strong respect for government-sponsored social programs and egalitarianism that continue in Costa Rica to this day.

Whatever the precise social system, the control of officials over other people was weaker in Costa Rica compared to elsewhere, due to its isolation and the relative unimportance of world trade. The nation's only export products were cacao, which was not particularly important in the world market, some hides and tallow, and a little tobacco. Therefore those large landholders (latifundistas) employing workers to grow crops for export did not have the degree of direct power over the rest of society that was apparently the case in many other parts of Latin America. It is known that until the time of Costa Rican independence in 1824 the country remained a sparsely inhabited, isolated nation, with settlement occurring primarily in the fertile central valley (meseta central), which is only about 6% of the country's land area. Since slavery was abolished in 1824 there was little incentive for people to own large plots of land. Development, such as it was, meant simply clearing a bit more forest for the slowly increasing population that was about 100,000 people in 1850. This system too was probably close to being sustainable because there was little trade, relatively little land degradation associated with export crops, and so much land that if one region became "farmed out" new fields could be cleared elsewhere.

## 1. The Advent and Development of Coffee and Bananas

Coffee (of the high-quality Arabica type), originally from northeast Africa (Anthony *et al.*, 1993), arrived in Costa Rica in 1808, when governor Tomás de Acosta brought seeds from Jamaica. Two policies facilitated its spread: (a) In 1831 deeds were given to the owners of lands where coffee had been cultivated for at least five years. (b) The first full-scale credit program by the government to coffee growers started in 1832. The first exports were made in that year to Valparaiso, Chile, where they were then reexported to Europe (Rojas *et al.*, 1987).

The degree to which precoffee Costa Rica was egalitarian is debatable, but with the advent of full-scale coffee culture in about 1830 there arrived direct exploitive relations between coffee barons and small holders and laborers, and the birth of a true laborer class (jornaleros) (Seligson, 1980; Gudmundson, 1986). By the middle of the 19th century Costa Rica was exporting 4000 tons and was the world's leading exporter of coffee. Great Britain was the financier of this conversion from subsistence farming to the production of a major export crop. The principal beneficiaries of this new industry were a small elite group tracing their heritage to the conquistadors and deriving their power from land. They grew coffee but more importantly processed it in production units called beneficios and controlled the exports (on CD; see Coffee and Bananas). Social divisions based on who owned the beneficios became increasingly important. Yet the Costa Rican beneficio owners did not tend to buy out small owners as was the case in other parts of Latin America (e.g., Brazil and Colombia). The reasons were several. Beneficios gave credit to small and medium farmers, which was a good vehicle to earn interest and to consolidate a benevolent oligarchic situation over entire regions (Santana, 1975; Winson, 1989). Also, the distance between the farms and the broken topography in many of the best growing areas posed obstacles to managing many cafetales (coffee plantations) under one establishment. Finally the structure of producers/processors/exporters was a good way for the elite to gain value added through processing and trade and to avoid the natural risks of cultivation (Rojas *et al.*, 1987).

The incentive to grow coffee promoted serious forest cutting in central Costa Rica because for the first time large landholdings had value. The next phase in deforestation can be attributed to an American named Minor Keith who, in the late 1800s, built a railroad from the meseta central to Limon on the Atlantic coast to facilitate the export of coffee. In exchange, he received land from the Costa Rican government on which he established banana plantations. This was the beginning of the United Fruit Company. With growing U.S. investment in Central America and with U.S. military interventions to protect those investments, Costa Rica became a classic example of a neocolonial "banana republic" (Janzen, 1983). This period marks the beginning of the transition between two competitive forms of agricultural organization—production for family use (subsistence cropping) and production for exchange or cash cropping. The different market conditions, ecological requirements, ownership, and locations of bananas versus coffee have had a very large effect on Costa Rica. Most important is that for coffee production, much lower levels of energy and other foreign inputs are required and a much higher proportion of money stays in the country (Aguilar, 1994; Aguilar *et al.*, 1996). Another effect is that the large quantities of capital owned by banana companies allowed them to undertake very large plantations, each associated with major deforestation (Figure 2-1) (Aguilar, 1994; Vandermeer and Perfecto, 1995). Since World War II there has been some diversification of export crops, primarily sugar in the 1960s, when the United States stopped buying from Cuba, but more importantly beef. Most recently "nontraditional" crops have been increasingly emphasized.

## B. Cattle Raising in Costa Rica

The history of Costa Rica that is usually presented is that of the main families in the Central Valley producing mostly annual crops, with additional farms in the Atlantic producing mainly cacao. But the actual development pattern was more complex and reflected the complex ecological conditions of Costa Rica as well as the ancestry of the colonialists. At the time of the Spanish conquest Spain was not an agricultural country, but one based on livestock (Smith, 1976). Doctor David Robinson (Syracuse University, personal interview) says there is practically no mention in early Spanish accounts from Latin America of the Spaniards using the agricultural systems of the societies they took over, which is remarkable, considering the many technological advances of those societies that have been revealed through archaeological and historical work. The conquerors engaged in land "inversion," occupying the fertile valleys with cattle and forcing the Indians to make their plantings of corn and beans on the steep slopes of the forested hillsides. The land eroded quickly, rendering

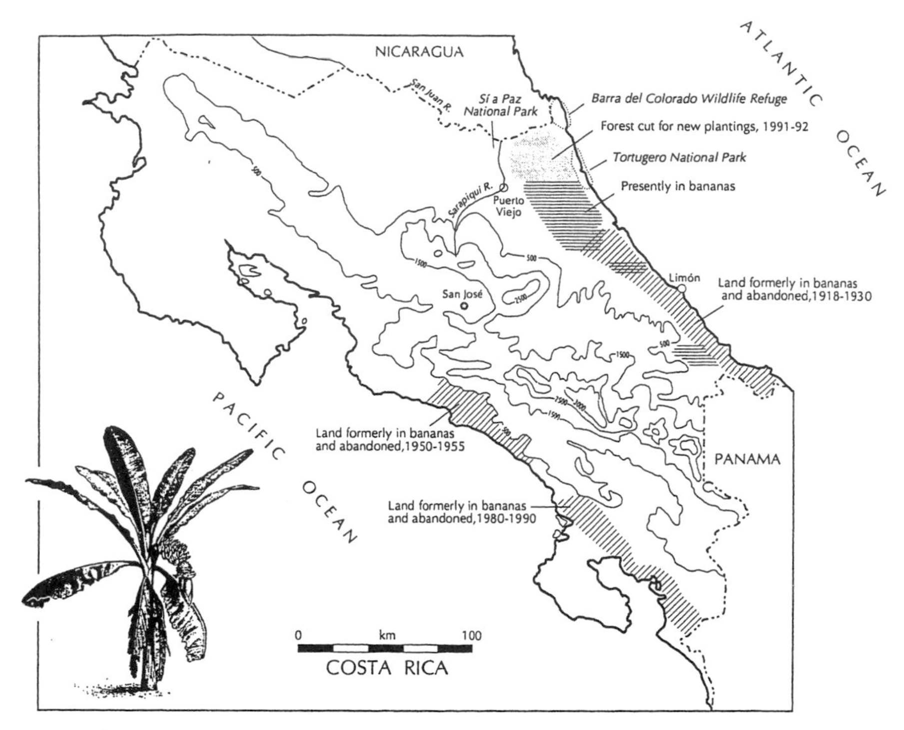


FIGURE 2-1 Land presently in bananas (horizontal lines) and previously in bananas (diagonal lines). Large areas have been abandoned mostly due to banana disease, decreasing soil fertility, and erosion, but also social and political factors such as labor unrest (from Hunter, 1994, reprinted by permission of Blackwell Science, Inc.).

it infertile and driving what few Indians remained even further up the slopes (Smith, 1976). The consequences were devastating throughout Latin America. A thousand years of labor and engineering invested in terracing steep mountain slopes and bringing irrigation water to them were destroyed in decades by free ranging cattle. Horrendous gully washouts often ensued.

The Spaniards introduced cattle ranching in Costa Rica in the Nicoya Peninsula during their first century of occupancy. For the next 400 years people viewed the clearing of forests for cattle a practical means of opening the limitless agricultural frontier (Annis, 1990). By 1759 in Guanacaste there were 20 herds of cattle with 9600 head in the region of Bagaces and 19 herds with 6050 head in Cañas. At the time Bagaces had fewer than 600 inhabitants, indicating how important cattle were to the economy of the region (Morel de Santa Cruz, 1994). By the late 1800s, 47 haciendas covered almost 190,000 ha in Guanacaste, 18.5% of the total area of the province (Sequeira, 1985).

Prior to 1950 Costa Rica had four resources that made it an attractive locale in which to develop a cattle industry. First, it had a relatively small population and an abundant supply of land, although much of it was forested (Tosi, 1975, in Meehan and Whiteford, 1985). Second, it had raised a rugged criollo stock of cattle ever since Spanish colonists arrived in the 16th century (Barlett, 1976). Third, as a result of the Spanish interest in livestock, rural Costa Ricans had acquired a sizable stock of knowledge about cattle raising. Fourth, an aggressive and relatively high-yielding African pasture grass, called jaragua, was introduced into the area in the 1920s and grew well in many places (Parsons, 1976, in Meehan and Whiteford, 1985). "With these resources it is not surprising that the availability of an export market and large amounts of capital would transform the forest into pasture" (Meehan and Whiteford, 1985, pp. 178-195). Other important reasons not mentioned by Meehan are that cattle production did not require much in the way of labor or expensive imported materials, as did most crops, and the fact that cattle could be raised in a variety of conditions, including the dry northwest.

By 1950 pasture occupied about 630,000 ha, or 35% of Costa Rica's total farmland. Beef cattle production more than doubled from 1960 to 1972, and the area in planted pasture increased 62% in 10 years (see Chapter 13). This increase was due primarily to the availability of credit for livestock production, instituted by the World Alliance for Progress initiative (Annis, 1990), and was designed to supply the fast food chains of North America with cheap beef (Myers, 1981). According to FAO (UN Food and Agriculture Organization) statistics the area of permanent pasture has doubled since 1973, mostly at the expense of the forest. As of the mid-1990s, cattle grazed on 2.2 million ha, 54% of all agricultural land and about 44% of Costa Rica's total land mass (FAO annual statistics; Land Use on CD).

## C. Legal Forces Driving the Forest to Pasture Conversion and Their Subsequent Revision

Costa Rican common law, based on Spanish and ultimately Roman law, recognizes explicitly that a person who openly occupies and works land not actively used by another gains "rights due to labor" or "improvements" proven by land clearing and agricultural usage. An unwritten law of rural people is that possession and use are more important than title alone. In the last century the government promoted settlement of public domain lands through a series of laws that guaranteed ownership after 3 to 15 years of occupation. Most settlers never applied for title rights because it was expensive and time consuming. Settlement was largely unorganized and haphazard so that the government lost all control over the public domain. It is doubtful whether the government today can even identify many lands remaining in the public domain (Hartshorn *et al.*, 1982).

Establishing ownership of newly cleared land was formalized in the Civil Code of 1888, which allowed ownership by fencing in a portion of land, and continued under various laws to the present. Legal title to land can be acquired by anyone who possesses the land continually, publicly, and peacefully for 10 years, and who has a "good-faith" title. This means that even though the original squatter cannot claim title, he or she can sell the "improvements" to a person who will qualify under this code for a legal title. The Law of Possession Information of 1942 gave possession to anyone who occupies (uses) private land for a year. After 10 years of continuous possession the possessor can claim legal title (Hartshorn *et al.*, 1982). These laws and the court's interpretation of them have led to serious abuse of intent. Squatting and speculation on both public and private land have become businesses unto themselves. Since squatters cannot claim the land under the good-faith law, they clear only enough land to qualify as "improvements," and then cut a much larger strip around an area of virgin forest. They sell the improvements with the understanding that the "farm" includes all the land within the boundary strip.

The new good-faith owner often continues destroying forest land, often selling stumpage to logging contractors who obtained a concession for their timber from the government. After 5 to 10 years the good-faith owner can obtain legal title. Meanwhile, the original squatter has gone on to repeat the process on another section of virgin forest. In some cases squatters have been financed in their commercial operations by eager purchasers, not infrequently cattlemen desirous of expanding an existing farm or acquiring others (Hartshorn *et al.*, 1982). Although the 1969 Forestry Law prohibited further spontaneous settlements of the public lands, the location of these lands was not registered in government archives and there was no means of enforcement since the forestry department was both understaffed and underorganized. Squatters continued to invade public as well as private land more or less freely, including in some cases the forest reserves established by law, fraudulently claiming that possession originated prior to establishment of the reserve.

The Forestry Law of 1995 maintains the prohibition against squatting, and presently the management of all protected areas is under the jurisdiction of SINAC (National Conservation Area System), associated with the Ministry of the Environment. Squatting does in fact still occur, but good progress has been made in updating the borders of protected areas at the national mapping office. The Agrarian Jurisdiction, created in the early 1990s, has reformed the situation of agrarian property in Costa Rica somewhat. Nevertheless effective acts of possession, such as farming, that are essentially squatting are the most common way to gain possession of agrarian land.

Presently local legal tribunals recognize what is known as "ecological possession," defined as that activity (or lack thereof) which has as its main goal the conservation for future use of areas presently not used. It recognizes the need to give special protection to this type of possession in view of the common good and excludes it from regular agrarian jurisdiction. Yet the problem remains that in view of the lack of formal laws these matters are judged by civil judges that do not necessarily understand the intent of conservation initiatives (Umaña, 1995). Also, the enactment in 1994 of the Reform of Article 50 of the Costa Rican Constitution makes it a constitutional right for Costa Ricans to have a clean and healthy environment. This reform conflicts with the recognition in Article 45 of the Constitution of absolute private property rights which date back to 1871, and has generated conflicting superior court decisions. This has made the enforcement of environmental regulations in private areas even more difficult (Aguilar *et al.*, 1994).

## 1. The Importance of the Building of Transportation Infrastructure to the Development of the Agrarian Sector

The concept of development is hardly a new phenomenon in Costa Rica. Just as the railroad built at the end of the last century facilitated the tremendous expansion of coffee growing for export, the building of roads throughout Costa Rica has opened up the frontier to settlement and influenced land use choices. For example, the building of an all-weather road in Puriscal in 1962 and the construction of a coffee-buying station there brought about access to new markets for tobacco as well as coffee (Barlett, 1976). In the region of Nicoya, cattle ranching increased rapidly in the 1950s as bridges and all-weather roads were completed throughout the peninsula, connecting it with the central valley (Barlett and Harrison, 1985). "Most of the funds available from international institutions over the last few years ($20–30 million) have gone into infrastructure development on ranches" (USDA, 1973, in Meehan and Whiteford, 1985, pp. 178–195).

During the 1980s the largest area of deforestation in Costa Rica can be linked directly to the roads built by the U.S. government leading to the eastern Nicaraguan border. The purpose of the roads was apparently both military and commercial, allowing for a tremendous expansion of banana plantations in northeastern Costa Rica (Jack Ewel, personal communication). Thus road building is the most important component of infrastructure building, and "road building has enormously facilitated this rapid conversion of forest to cropland and pasture" (Parsons, 1976, in Meehan and Whiteford, 1985, p. 123). Other necessary transportation linkages were an international shipping service to the United States (USDA, 1969, in Meehan and Whiteford, 1985) and a long-distance refrigerated trucking service initiated in 1960.

Particularly detailed, although not necessarily characteristic, analyses of the general process of agricultural development have been done for the region of Puriscal by Barlett (1982) and Hueveldop and Espinoza (1983). Barlett, especially, emphasizes the importance of continued population growth, which has shortened or eliminated years of fallow (years when the soil is allowed to rest and be rejuvenated by the growth of vegetation). The reduction in fallow length has led to the degradation of soils and hence greatly reduced agricultural possibilities in the region over time. In 1935 the area was called "el granero del país" (the breadbasket of the nation). But only 44 years later, in 1979, the same region was designated an "emergency area" by the Minister of Agriculture and Cattle Raising due to its eroded land base, which no longer afforded most inhabitants the means to a sustainable livelihood. Ecologist Gideon van Melle, called in to interpret aerial photographs in order to determine the extent of degradation and the possibilities for future use, stated, "Practically the whole canton is not suited for any form of agricultural use, due to its steep slopes in combination with heavy precipitation and unstable soils." (van Melle *in* Hueveldop and Espinoza, p. 15)

## 2. How Farmers Cope with a Difficult Situation

By 1972 land prices and declining soil fertility were making conditions increasingly difficult in Puriscal (and elsewhere), especially for poor tenant farmers. One farmer said, "The land is tired now. We used to get good crops, but now we get little, even with fertilizer." But cattle could still be raised, both on degraded cropland and on new pastures cut from the forest. This process was encouraged through low-cost government loans. These loans were financed in turn by government borrowing from the World Bank and other foreign sources. According to Annis (1990) $1.2 billion of such loans were made between 1969 and 1985. This generated relatively little foreign exchange and very large environmental debt. Even though the production (and hence profits) per hectare was very low for cattle production (about 300 kg live weight, or less than 50 kg edible dry weight per hectare per year compared to roughly 1500 kg dry weight for grain), those with large amounts of land available found it profitable to take land out of fallow or change forest to pastures in order to maximize profits. The investments and risks of cattle raising (weather, pests, etc.) are much less than those for crops, and the necessity to find and maintain a good labor force is small (except for the initial clearing of forest land). This is why, as we have discussed, other creative ways were found for getting this done. These include letting tenant farmers clear a piece and cultivate it for two or three years, and then moving them to a new patch. But total food production did not increase much even though the area converted to food production increased because of the low yield of pastures.

In the Puriscal situation of population pressure and new land shortage, where soils became overworked, the traditional extensive cropping systems did not produce much. With insufficient land available for shifting cultivation, "traditional" corn and bean production using slash–mulch cultivation was abandoned. Many farmers shifted to terrace production where they got three harvests per year, tobacco, beans, and corn, although several times more labor per hectare was required. In many other regions with less population pressure fertility could be maintained simply by cutting new forests and allowing the agricultural land to go to pasture.

### D. The Present Situation for Farmers

Since 1972 tobacco growing in much of Puriscal has given way to coffee, and maize was hardly grown at all once government subsidies were removed in the 1980s. Although farmers have very mixed opinions about credit it is generally conceded that access to credit for fertilizer has acted to maintain the productivity of lands in the community that can no longer lie fallow, thus allowing the community to continue an active, productive life in agriculture. With fertilizer applications of about 100 kg/ha, yields of many food crops are increased from much less than 1 ton to roughly 1.5 tons per hectare, depending on the condition of the site. But "for a nation as a whole, fertilizer use represents not only a drain on foreign exchange but a new kind of vulnerability" (Barlett, 1982). Just as the farmer earlier said, "We are trapped," so too is the nation as a whole trapped by the rising costs of inputs from the industrial world required as substitutes for its exhausted soils. Fortunately much of Costa Rica is less steep and has better soils than Puriscal, but, on the other hand, much of the nation is not so different either.

## III. THE EFFECTS OF FOREIGN AID AND LOANS

Guess (1979) and Annis (1990) conclude that the government, in building roads and offering credit, was wrong with respect to encouraging meat production despite good intentions. The theory was that the export of meat would reduce external debt, dependence on foreign loans, and the level of inflation. In fact, they say, all have increased considerably. There are analysts such as Feder (1975) who would argue that World Bank schemes and U.S. Agency for International Development (USAID) programs were never intended to help the rural poor but rather to enhance world markets for agribusiness industrialists in the first world, and to increase the dependence of the third world. This perspective was inadvertently supported by the words of J. Brian Atwood, administrator of the USAID, when commemorating 50 years of U.S.–Costa Rican cooperation as USAID officially closed its mission in Costa Rica. The U.S. government had provided $1.7 billion to Costa Rica since 1946. But Mr. Atwood also said that the aid had been successful from the perspective of the United States since U.S. exports to Costa Rica had increased to more than $2 billion a year.

The other side of the coin is not so comfortable either: without the industrialization of agriculture brought on in part by foreign aid, Costa Rica almost certainly could not feed itself, let alone generate the foreign exchange necessary to develop the modern society that it has. More and more young people look to the cities for employment, and the agricultural base is increasingly devoted to coffee, bananas, and cattle—in other words, export crops. As of this writing about 30% of the food consumed inside Costa Rica is produced outside its borders, and that percentage is increasing every decade. How is that food to be paid for? Curiously, in large part by agriculture, since agricultural products are the largest component of exports.

Annis (1990) concludes that credit resulting from the World Bank rural lending scheme benefited only a minority of Costa Rican farmers. Cattle raising undertaken by that minority has, in fact, worked against many small and landless farmers in Costa Rica (Barlett, 1982). This heightens the tension between the landed and the landless, and puts ever more pressure on what land and resources are available to a rural population that is still growing. By the early 1980s, when the recession was coming into full swing following the oil price hikes of the 1970s, high interest rates and scarcity of funds were making it difficult for farmers to obtain credit. Indeed the decade of the 1980s was an extremely difficult time for many Costa Ricans.

## A. The Impact of the Oil Price Increases of the 1970s

Although per capita energy consumption in Costa Rica in the 1970s was less than 10% that in the United States, its economy was in many respects more sensitive to oil price changes than more intensively industrialized economies because imports are expensive relative to gross national product. The oil price increases of 1973 and especially 1979 devastated the Costa Rican economy, causing severe inflation, devaluation of the colón, and a tremendous loss of purchasing power. Perhaps the most dramatic illustration of how petroleum imports have affected the Costa Rican economy is that in 1970 only 4% of each working person's wage was spent on petroleum imports. By the late 1970s it was 20%, and by 1981 nearly 50% (Levitan, 1988). Since the price of manufactured goods also reflects fossil energy costs, the relative price of imported manufactured goods also increased considerably. Despite the price increases the economy had become so dependent upon industrial products that their importation continued to increase (Annis, 1990; U.S. Senate, 1983; Oficina de Café, 1985; Commonwealth Secretary, 1984) (Figure 20-9). Five percent of gross national export earnings was spent to import petroleum products in 1970, whereas 21% was spent in 1982, and a large proportion of the remaining on other industrial goods. The increasing quantities of ever more expensive petroleum and petroleum-derived imports were paid for essentially by increasing national debt.

The increase in the price of external products also increased the stress on Costa Rican ecosystems, because more agricultural land was required to generate the foreign exchange necessary to purchase imports and to pay interest on the debt. The farm and grazing land brought into production was of increasingly marginal quality and lower yield potential. In Costa Rica this spiraling pattern of demand for agricultural land led to an annual loss of forested land equal to almost 1% of the total land in the country through the decade of the seventies (Sader and Joyce, 1988; Levitan, 1988; Annis, 1990) (Chapters 15 and 16).

The relaxation of the price of fossil fuels since the mid-1980s has reduced the severity of these problems greatly, but in 1995 the total petroleum import bill for Costa Rica was still about $600 million, or $200 per capita. This can be compared to the average annual salary of about $2500 or the total money earned from the international sale of bananas, also about $600 million. Costa Rica remains very vulnerable to future price increases (Figure 4-5).

Why then, you might ask, did and does Costa Rica continue to increase its use of oil and oil-derived products? The short answer is that it cannot afford not to. The area of good agricultural land, already severely impacted by use, is simply too small to support the food and export-income requirements for three and a half million people without these imports. In addition all sectors of the economy are increasingly industrialized, and it is clear that this is the direction most people prefer. In this way Costa Rica is little different from most other nations which are also industrializing rapidly because that industrialization allows the expanding population to live on the finite quantity of land.

## IV. COSTA RICA IN THE 1990s—RADICAL CHANGES IN THE COUNTRY AND THE ECONOMY

Visitors who were familiar with the "old" Costa Rica and who returned in the 1990s found a very rapidly changing nation. Fifty airplane flights a day now bring in tourists and others to the international airport. Many people are optimistic that the tourist industry, which in 1994 surpassed bananas as the major source of foreign exchange, will save the Costa Rican economy, which continues to be plagued by debt. Government seems increasingly impotent because of the mounting debt and the stringent requirements that the lenders are placing on the government in order to attempt to recover their money. Nevertheless borrowing continues, and in some years the debt increases.

Very expensive hotels are being built, especially along the Pacific Coast, and a new jetport in the middle of nowhere brings tourists in directly from such cities as Minneapolis. Ecotourism, although only a small part of all tourism, is rapidly becoming more important and gives hope that development and foreign exchange generation can be made compatible with preserving Costa Rica's unique and wonderful natural resources. The government is very interested in preserving and expanding the national parks, which constitute about a quarter of the land area of the country. A controversial fee system initiated in 1995 charged foreigners $15 per visit, some 25 times more than Costa Rican nationals to visit national parks. The net result was a great reduction in foreign visitors, and the fees subsequently were dropped to about $6, four times the fee for Costa Ricans. This money is being used to refurbish and maintain the magnificent national parks. Nontraditional export crops, including macadamia nuts, specialty fruits, and cut flowers and ferns, are expanding rapidly due to the ready availability of air shipping. Cattle raising in the northwest is decreasing rapidly as the sons of cattle ranchers seek their fortunes in San Jose, so that at least here the forests are recovering to some degree. Large new computer industries may be important in the future.

But the changes are certainly a two-edged sword. San Jose, traditionally a clean, easygoing and quiet city is alive, many would say clogged, with traffic, commerce, and U.S.-style fast food restaurants and their trash. Air pollution during rush hour can be nearly as bad as anywhere in the world. More than 1 million of the nation's 3.5 million people live in the metropolitan area, and they are not farming. Two enormous, even by U.S. standards, shopping malls were built in 1995 in the suburb of San Pedro. Advertisements for these and other new shopping malls are almost universally focused on products such as luxury foods and automobiles that are not produced in Costa Rica. Consumption here seems much more obvious than production. Security systems are, more or less for the first time, big sellers. Urban slum areas are a new reality. Many traditional farmers can no longer afford rents in rural areas and are forced to move their families to the slums while they take a bus 40 minutes to work on plantations.

Social mores are changing rapidly. Traditionally the culture of most Costa Ricans is deeply Roman Catholic and very family oriented. But television and urbanization are changing that, and many young people derive their values from popular soap operas that promote luxury consumption and romance with no consequences, such as pregnancy. Now the majority of babies are born to mothers who have no husbands, and the outlook for many of these children is bleak indeed. The population growth rate, which had declined by 50% from 1960 to 1975, has remained at a plateau of about 3.2% since then, and babies are born increasingly to poor people (Holl *et al.*, 1993). Governmental and private debt continued to increase through the early and mid-1990s, the latter accelerated by the ready availability of credit on charge cards. The smoothly operating and universally admired systems of health, educational, and other social services are threatened by governmental insolvency, and foreign creditors are demanding reduced government spending while the need for governmental services expands. The government has embarked on an aggressive plan to reduce debt, including an increase in the efficiency of tax collection, an increase in indirect taxes (such as sales taxes), and an "opening" of public utilities to foreign investment.

Increasingly at least some parts of Costa Rican society reflect the highly developed world. Serious crimes such as bank robberies, car thefts, and even murders, until recently essentially unheard of, occur sometimes weekly. The distribution of wealth, once extremely equitable, especially by Central American standards, is now increasingly inequitable, and very fancy houses are no longer rare. Where does the new money come from? Nearly every Tico (the Costa Ricans' name for themselves) has an explanation. A popular view, hard to nail down, is that increasingly, drug money is passing through Costa Rica. As of this writing a number of prominent politicians were found guilty of money laundering and operations with drug cartels from South America. This is not the only case of corruption. Several state banks had to be closed down due to irregular loans that were given to influential people in the country. In recent years problems have arisen about irregularities associated with assigning public works and service concessions. Some people did very well by this corruption, but the average person was badly hurt. Perhaps a better explanation for the increasing gap between the rich and the poor is that the process of industrialization and free marketism, like other past patterns of large-scale economic change, have created winners and losers. In this case, the winners win big, and the losers lose big.

Thus there are a whole new suite of opportunities and problems facing Costa Rica, not completely different from past problems and opportunities. Costa Rica has a modern, well-educated, relatively sophisticated population with needs and desires for industrial products that cannot possibly all be produced in Costa Rica. In the past this problem has been resolved by increasing the export of tropical agricultural products. But now we must question the degree to which Costa Rica is or is not approaching the limits of its ability to grow and export these crops profitably, and whether there will be new sources of foreign exchange.

Perhaps the overriding issue for Costa Rica and many other developing nations is how to pay for the industrial inputs required for modernization. At this time there appear to be three possible ways to continue to do this: (1) the expansion of exports of tropical crops and (via tourism) use of natural ecosystems, (2) further industrialization, perhaps associated with microelectronics, and (3) increasing the debt. Further industrialization makes some sense for the following reasons: (1) there is plenty of skilled and semiskilled labor available, and (2) investment in Costa Rica is attractive to many investors because of its relatively low wages and stable political climate. On the other hand the drawbacks are that Costa Rica does not have (1) indigenous fuel (except for biomass and hydropower), (2) other industrial raw materials, or (3) a very high-tech industrial base. This approach also has its risks. For example, there was an expansion of shirt making recently in Costa Rica. Foreign exchange generated from textiles went from near zero in 1985 to $600 million a year in 1995, and then declined precipitously. It is not clear how much more expansion of this kind is likely, since the textile expansion was based on a one-time decision by the United States to find a market for its surplus cotton. How can a small nation control its destiny when so much is dependent upon the outside world of markets?

Meanwhile much of daily life and political activity in Costa Rica in the 1990s is dominated by "structural adjustment," a package of economic restrictions imposed by loan-granting institutions such as USAID and the International Monetary Fund. The concept of structural adjustment was intended to bring about changes in the Costa Rican economy that would reduce debt. These changes included a privatization of formerly governmental agricultural loan-granting agencies; a reduction in government spending, including subsidies for domestic food production; the encouragement of the importation of comestible grains from the United States; a reduction of import or export tariffs (i.e., an encouragement of all international trade); and the encouragement of export crops at the expense of subsistence crops (Hansen-Kuhn, 1993; Korten, 1993). According to Hansen-Kuhn and Korten the consequences of these harsh new regulations have been devastating to many small Costa Rican farmers, who can no longer compete with cheap wheat from the United States or afford fertilizers for export-oriented agriculture. Who came out ahead from these changes? Again according to these authors it is the already wealthy, including importers, exporters of nontraditional products, and those linked to the financial services sectors. Whatever the case, it is clear that structural adjustment did not meet its principal objective of reducing the trade deficit, which in fact increased from $135 million in 1984 to $569 million in 1990, or the debt, which increased from $2 billion to $4 billion (despite a $1 billion forgiveness) during the same period.

What alternatives are left? In the 1990s Costa Rica billed itself as a "green economy" (Quesada-Mateo and Solis-Rivera, 1990; Ammour, 1993; Figueres-Olsen, 1996), and there has been considerable hope that new industries such as ecotourism could be created that would generate much needed foreign exchange with relatively little destruction of the resource base. Critics argue that Costa Rica will become "green" only to the degree that it is economically useful to do so. Much actual data in fact argues that Costa Rica is currently not very "green." Deforestation continues. Polluting (and nonpolluting) industries are encouraged and most indices of environmental quality are decreasing. Perhaps it is fairest to say that many in Costa Rica would like to build a "green" economy, but that faced with declining agricultural revenues, an increasing population, and increasing debts, the nation must do whatever it can to maintain the health of the economy and the public sector. If that can be green, so much the better. If not, human needs take precedence.

## V. CONCLUSIONS

In conclusion we can see that the "cup," that is, the economic well-being of Costa Rica, is either half empty or half full, depending upon your point of view and which aspects you choose to emphasize. More accurately it is best to say that the future of Costa Rica is very uncertain. We believe that the full examination of this question requires a much fuller biophysical analysis of what resources or industries might be developed and to what extent and at what cost. To what degree is the country already industrially based and hence vulnerable to oil price changes? An important component of our analysis is that the actual resource base is not static but changes over time and over space. It is increased through technology and fossil energy use, and it is depleted as a function of the previous amount of development (e.g., erosion) and as a function of the history of its use. Whatever the total "pie" developed, it will have to be divided by the ever growing population. These are particularly important factors that are not captured by most economic models. These models are often based instead on a static representation of the economic system, and often entirely ignore the basic Ricardian concept that a resource base is of highly differing quality as a function of how much is being developed at any one time.

This leads us to the necessity of using a spatial approach to our analysis of the resource base of Costa Rica. Much of the middle section of this book is about how we do this. The rest is about the development and application of tools to the understanding of Costa Rica's potential for development, sustainable or otherwise.

# REFERENCES


Aguilar, B., and A. Torrealba. 1987. El perfil funcional de la organización subjetiva en la empresa de reforma agraria: Contribuciones a la búsqueda de un modelo teórico para Costa Rica. Thesis, Universidad de Costa Rica, San Jose, Costa Rica. 1181 pp.

Aguilar, B. 1994. Banana production in light of sustainable development in Costa Rica: Eco-friendly banana? In *Proceedings, Third Biennial Meeting of the International Society for Ecological Economics, "Down to Earth, Practical Applications of Ecological Economics," Oct. 24, 1994, San Jose, Costa Rica,* p. 38. Inter-American Institute for Cooperation on Agriculture/International Society for Ecological Economics.

Aguilar, B., L. Umaña, G. Scholl, S. Hannigan, and S. Schantz. 1994. Ajuste dinámico entre el derecho y el desarrollo sostenible en Costa Rica: Dos estudios de caso de conflictos entre la propiedad privada y el ordenamiento territorial-las zonas ribereñas en la Parte Baja de la Cuenca del Río Grande y la Zona Protectora El Rodeo. Acuerdo Bilateral de Desarrollo Sostenible Costa Rica-Holanda/Universidad Nacional de Costa Rica. Conferencia Internacional "Desarrollo Sostenible, Política Regional y Ordenamiento Territorial"; 1994 Nov. 15, San José, Costa Rica. Universidad Nacional de Costa Rica, San José.

Aguilar, B. J., T. Gillespie, C. Waddick, C. Williams, E. Rodman, E. Jones, and D. Fuchman. 1996. A biophysical assessment of tropical crops according to trade levels: Potential implications of multi-cropping nationally consumed and exported crops in developing countries. In *Proceedings, Fourth Biennial Meeting of the International Society for Ecological Economics—Designing Sustainability: Building Partnerships among Society, Business and the Environment, Aug. 4–7, 1996, Boston University, Boston, MA,* pp. 4–7. International Society for Ecological Economics/ Center for Energy and Environmental Studies/Boston University. Boston, Massachusetts.

Ammour, T. 1993. Conservacion y desarrollo sostenible en America Central: Manejo y aprovechamiento de la biodiversidad. *Revista Forestal Centroamericana* 5:20–25.

Annis, S. 1990. Debt and wrong-way resource flows in Costa Rica. *Ethics and International Affairs* 4:107–121.

Anthony, F., M. N. Clifford, and M. Noirot. 1993. Biochemical diversity in the genus Coffea L., chlorogenic acids, caffeine and mazambioside contents. *Genetic Resources and Crop Evolution* 40:61–70.

Barlett, P. F. 1976. Labor efficiency and the mechanism of agricultural evolution. *Journal of Anthropological Research* 32(20): 124–140.

Barlett, P. F. 1982. *Agricultural Choice and Change.* Rutgers University Press, New Brunswick, NJ.

Barlett, P. F., and P. F. Harrison. 1985. Poverty in rural Costa Rica: A conceptual model. In W. Derman and S. Whiteford (Eds.), *Social Impact Analysis and Development Planning in the Third World,* Ser. 12, pp. 141–159. Westview Press, Boulder, CO.

Beletsky, L. 1998. Costa Rica The Ecotraveller's Wildlife Guide. Academic Press, London.

Coates, A. G. 1997. Central America A Natural and Cultural History. Yale University Press, New Haven, CT, and London.

Commonwealth Secretary. 1984. *Fruit and Tropical Products.* Marlborough House, London.

FAO (Food and Agricultural Organization of the United Nations). Each year. *Population Yearbook.* FAO, Rome.

Feder, E. 1975. *The New Penetration of the Agricultures of the Underdeveloped Countries by the Industrial Nations and Their Multinational Concerns.* Institute of Latin American Studies, Occasional Paper 19, University of Glasgow, Scotland.

Figueres-Olsen, J. M. 1996. Sustainable development: A new challenge for Costa Rica. *SAIS Review* 16:187–202.

Garcia, A. 1982. *Modelos operacionales de reforma agraria y desarollo rural en América Latina.* IICA, San José, Costa Rica.

Gudeman, S. 1978. *The Demise of a Rural Economy*. Routledge, London.

Gudmundson, L. 1986. *Costa Rica before Coffee*. Louisiana State University Press, Baton Rouge.

Guess, G. 1979. Pasture expansion, forestry and development contradictions: The case of Costa Rica. *Studies in Comparative International Development* 14:42–45.

Hansen-Kuhn, K. 1993. Sapping the economy. Structural adjustment policies in Costa Rica. *Ecologist* 23:179–184.

Hartshorn, G. S., L. Hartshorn, A. Atmella, L. D. Gomez, A. Mata, R. Morales, R. Ocampo, D. Pool, C. Quesada, C. Solera, R. Solarzano, G. Stiles, J. Tosi, A. Umaña, C. Villalobos, and R. Wells. 1982. *Costa Rica: Country Environmental Profile: A Field Study*. Tropical Science Center, San Jose, Costa Rica.

Heyck, D. L. D. 1988. Tradicion y Cambio, Random House, New York, NY.

Holl, K. D., G. C. Daly, and P. R. Ehrlich. 1993. The fertility plateau in Costa Rica: A review of causes and remedies. *Environmental Conservation* 20:317–323.

Hueveldop, J., and L. Espinoza. 1983. *El componente arboreo an Acosta y Puriscal, Costa Rica*. Departamento de Recursos Naturales Renovables, Turrialba, Costa Rica.

Hunter, J. R. 1994. Is Costa Rica truly conservation-minded? *Conserv. Biol.* 8:592–595.

Janzen, D. H. 1983. *Costa Rican Natural History*. University of Chicago Press, Chicago, IL. 816 pp.

Korten, A. 1993. Cultivating disaster. Structural adjustment and Costa Rican agriculture. *Multinational Monitor*. July/August:20–22.

Levitan, L. C. 1988. Land and energy constraints in the development of Costa Rican agriculture. Unpublished M.S. thesis, Cornell University, Ithaca, NY.

Meehan, P., and M. Whiteford. 1985. Expansion of commercial cattle production and its effects on stratification and migration: A Costa Rican case. In W. Derman and S. Whiteford (Eds.), *Social Impact Analysis and Development in the Third World*, pp. 178–195. Westview Press, Boulder, CO.

MIDEPLAN (Ministerio de Planificacion Nacional de Costa Rica). 1995. *Panorama Economico de Costa Rica 1994*. Gobierno de Costa Rica, San Jose, Costa Rica.

Morel de Santa Cruz, P. 1994. *Costa Rica en 1751. Informe de una Visita*. Convento La Dolorosa, San Jose, Costa Rica. 157 pp.

Myers, N. 1981. The hamburger connection: How Central America's forests become North America's hamburgers. *Ambio* 10:2–8.

Oficina del Cafe. 1985. *Informe sobre la actividad cafetalera de Costa Rica (Information on Coffee-Growing in Costa Rica)*. Preparado en la Oficina del cafe para los delegados al XIV Congreso Nacional Cafetalero, San Jose, Costa Rica. 134 pp.

Perez, S. and F. Protti. 1978. Compaortamiento del sector forestal durante el periodo 1956-1977. Oficina de Planification Sectorial Agropecuaria. Doc - OPSA-15, San Jose, Costa Rica

Pinto Soria, J. 1993. *Historia General de Centroamérica. Tomo II: El Régimen Colonial*. Sociedad Estatal Quinto Centenario, Madrid.

Quesada-Mateo, C., and V. Solis-Rivera. 1990. Costa Rica's national strategy for sustainable development: A summary. *Futures* 22:396–416.

Quirós, C. 1993. *La Era de la Encomienda*. Editorial Universidad de Costa Rica, San Jose, Costa Rica.

Rodas, J. 1984. *Diagnostico del Sector Industrial Forestal*. Editorial Universidad Estatal a Distancia, San Jose, Costa Rica.

Rodas, J. F.G. 1985, Diagnostico del sector industrial forestal. Editorial, Universidad Estatal a Distancia, San Jose, Costa Rica.

Rojas, O. E. 1985. *Estudio agroclimático de Costa Rica*. Instituto Interamericano de Cooperación para la Agricultura, San Jose, Costa Rica.

Rojas, J., *et al.* 1987. *Costa Rica. Su Historia, Tierra y Gentes*, Tomo 2. Editorial Océano, San Jose, Costa Rica. 368 pp.

Sader, S. A., and A. T. Joyce. 1985. Global tropical forest monitoring. In *Proceedings: Advanced Technology for Monitoring and Processing Global Environmental Data Conference, London, England,* pp. 41–50.

Sader, S. A., and A. T. Joyce. 1988. Deforestation rates and trends in Costa Rica 1940 to 1983. *Biotropica* 20:11–14.

Santana, C. 1975. La Formación de la Hacienda Cafetalera Costarricense en el Siglo XIX in Haciendas, Latifundios y Plantaciones en América Latina. Siglo XXI Editores, México, pp. 635–667.

Seligson, M. A. 1980. *Peasants of Costa Rica and the Development of Agrarian Capitalism.* University of Wisconsin Press, Madison, WI.

Sequeira, W. 1985. *La Hacienda Ganadera en Guanacaste: Aspectos Económicos y Sociales 1850–1900.* Editorial UNED, San Jose, Costa Rica. 220 pp.

Smith, T. 1976. *The Race between Population and Food Supply in Latin America.* University of New Mexico Press, Albuquerque.

Stone, S. 1982. La Dinastia de los Conquistadores. La Crisis del Poder en la Costa Rica Contempiranea. San Jose. *Editorial Universitaria Centroamericana.* 623 pp.

Tosi, J. 1975. *Los Recursos Forestales de Costa Rica.* Centro Cientifico Tropical, San Jose, Costa Rica.

Umaña, L. 1995. *La Normativa Conservacionista de las Zonas Ribereñas en Costa Rica.* Tesis para optar por el título de licenciado en Derecho, Universidad de Costa Rica, San Jose, Costa Rica. 109 pp.

United States Department of Agriculture (USDA). 1969. *The Beef Cattle Industries of Central America and Panama.* Foreign Agriculture Service Monograph, Washington, DC.

United States Department of Agriculture (USDA). 1973. *The Beef Cattle Industries of Central America and Panama.* Foreign Agriculture Service Monograph, Washington, DC.

U.S. Senate. 1983. *Enabling Legislation for the 1983 International Coffee Agreement. Hearing before the Subcommittee on International Trade of the Committee on Finance, Washington, DC.* 98th Cong., 1st sess., Sept. 19, 1983. U.S. Government Printing Office, Washington, DC.

Vandermeer, J., and I. Perfecto. 1995. *Breakfast of Biodiversity. The Truth about Rainforest Destruction.* Institute for Food & Development Policy, Oakland, CA. 185 pp.

Winson, A. 1989. *Coffee and Democracy in Modern Costa Rica.* St. Martin's Press, New York.

# SECTION II

# *Development and Sustainability*

The first chapter in Section II reviews some relevant economic terminology and the major approaches that have been used traditionally to guide economic decision making in countries such as Costa Rica, including a series of past programs that were designed to guide and enhance development. Next it reviews and criticizes the dominant neoclassical economic approach and considers alternatives. These include especially a biophysical approach, one that focuses on the land, energy, and material requirements (and only secondarily monetary requirements) for economic production. Chapter 4 analyzes the basic Costa Rican economic data, including the beginning of an analysis of the degree to which it is or is not compatible with sustainability from a biophysical perspective. Chapter 5 analyzes from a biophysical perspective the important factors that determine food production in Costa Rica, both now and in the future. Important questions include whether Costa Rica has the biotic and physical resource base to support continued development, sustainable or otherwise, and the degree to which the answer to that question depends on economic factors outside the control of Costa Ricans.

CHAPTER 3

# The Theories and Myths That Have Guided Development

Charles A. S. Hall



## I. THE ECONOMIC ISSUES

The central issues facing most of the world today are associated with economic development: How do we measure it? How much is good? Good for whom? Is too much bad? Bad for whom? What kind is desirable and what kind is possible for a particular region? What are the costs, environmental, social, and otherwise, of development? What might be the costs that we do not know about? Where might it lead us in the next decade, generation, or century? What kind of economic models should or should not guide us along the development path? Should we emphasize international trade or protect local industries? What works and what does not work? By what criteria? For whom? Are the criteria that governments use the right ones? Is it suicidal to develop? Is it suicidal not to develop? Or should we pursue a very different type of development from what we have done so far? What would be the costs and the gains of that?

Another question that follows logically from those we have just posed concerns the degree to which this development should occur unconstrained by any government regulation, i.e., in response to "market forces" and "free enterprise," or the extent to which those market forces should be regulated by government. Do free markets deliver what their advocates promise? To whom?

Whatever our own personal opinions on these questions it is clear that development is happening and is happening big time in the tropics. In addition, governments, businesses, and international corporations generally are doing everything in their power to encourage economic development, generally as rapidly as possible and often without any particular consideration of the qualitative aspects of that development. This is as true in Costa Rica as anywhere: the most obvious aspect of life in Costa Rica in recent years has been development and its associated changes.

A component of development is that it is encouraged, guided, and often sanctioned by economic analysis, and this economic analysis tends to be based on a particular approach to economics known as neoclassical economics. Neoclassical economics, although but one of a number of possible approaches to understanding and guiding economic systems, is overwhelmingly and increasingly dominant in the world today. It is taught to millions of college undergraduates as the only proper way to understand how economies operate. Its fundamental approach and assumptions are rarely questioned by the general society except (occasionally) from the perspective of those who focus on the sometimes unpleasant social consequences of free markets.

A new twist is the tremendous instability in the once booming and once seemingly bulletproof economies of tropical Asia. Many countries are now looking more cautiously at unconstrained rapid economic growth. The importance of assessing the real resources available or not available, so that estimates of potential economic growth are not based on just inflation or speculation, is becoming increasingly clear.

## II. THE LURE OF SUSTAINABLE DEVELOPMENT

When one hears the word "development" applied today it is very often preceded by the word "sustainable." Whatever "sustainable development" does or does not mean, whatever the possibilities (or lack thereof) for its implementation, just about everyone is for it. There is something for "both sides" in this term: development for those who like development, and sustainable for those who are nervous about it. In this "have your cake and eat it too" world, all sides are able to find at least some of what they want within the connotations of this one term, even though some very thoughtful analyses define it more restrictively. An important point is that there are at least three major groups who use the term for very different and often contradictory goals (meaning economic, social, and environmental) (Goodland and Daly, 1996). So, at least superficially, one main question of this book is simply, "is sustainable development possible in Costa Rica?" I say superficially because I believe that the question is phrased more properly as, "to what degree is development itself sustainable in Costa Rica?"

## A. Some Definitions

The following is borrowed from Ko *et al.* (1998), where we asked some basic questions about development and sustainability for five different countries, including Costa Rica.

The International Union for the Conservation of Nature and Natural Resources (IUCN) introduced the concept of *sustainable development* in 1980 (Lelé, 1991). The most widely used definition of sustainable development was introduced by the World Commission on Environment and Development (WCED, often called the Brundtland Commission): ". . . development that meets the needs of the present without compromising the ability of future generations to meet their own needs" (WCED, 1987, p. 43). The Brundtland Commission considered population control, food security, and energy supply as critical components of sustainability.

Since then there has been a great deal of discussion about the meaning and feasibility of sustainable development within the context of an increasing recognition of both the vulnerability of ecosystems and the ever growing pressures for economic growth. In practice, *sustainable development* is about sustainable *economic* development. The emphasis in many sustainable development studies has been on resource availability, i.e., the biosphere's ability to supply resources, its waste assimilation capacity, its environmental quality, and the possibility of improving the quality of human life. *Conventional economic development* has focused on improving the material quality of life by increasing the *quantitative* outputs of economic activities. Therefore, the concept of sustainable development should be understood in terms of how long economic development can be maintained at some given level without exceeding either the assimilation or the exploitation capacity of a region.

Too frequently, the concepts of "economic development" and "economic growth" are used interchangeably, but there are differences between the two. *Economic growth* is a one-dimensional measurement of quantitative changes in economic capacity, while *economic development* pertains to multiple dimensions such as quality of life, social structure, and industrialization, as well as economic growth (Todaro, 1994; Gillis *et al.*, 1996). Thus, if a country achieves economic growth, it does not necessarily achieve economic development. It is probably extremely difficult, however, to achieve economic development without at least some economic growth. Therefore, sustainable development probably requires at least some economic growth as a precondition.

An "optimistic" group asserts that both goals, growth and sustainability, can be achieved by technologically improving the efficiency of our resource use (e.g., Simon and Kahn, 1984; Goldemberg *et al.*, 1988; Ausubel, 1996) and increasing our market-economy-driven efficiency (Smith, 1994). A second group argues that without population and consumption control, sustainable development is not attainable (e.g., Ehrlich and Ehrlich, 1990; Hall *et al.*, 1994). This view implies that any efficiency gains are insufficient to make up for increasing population and affluence. The extreme "optimists" argue that population growth itself increases technological development, so that the global food base and pollution assimilation technology will be expanded indefinitely to meet the needs of a growing human population (Simon and Kahn, 1984). "Pessimists," ever since Malthus (1798), along with Ehrlich and Ehrlich (1990) and Pimentel and Pimentel (1996), emphasize the adverse consequences of population growth, including famine, due to the finite nature of the earth's carrying and regenerative capacity.

It seems obvious that there is no unequivocal, universal answer to these questions. Instead we need to examine the relative contribution of different factors that influence the functioning of economies, case by case, using specifically defined criteria. Explicitly for the purposes of this book we define *economic development* as an increase in the generation of economic output as imperfectly measured in physical units (e.g., tons) or inflation-corrected dollars of gross domestic product (GDP). This is the general case. For some specific situations we define development in terms of other aspects of quality of life, such as health. We define *sustainability* as not degrading the resource base that makes the economic activity possible. We define *efficiency* as output over input as specified for each particular use. Our approach has a very strong biophysical basis, meaning that it is based in large part on the analysis of the biotic and physical conditions that allow and constrain economic production. In other words, if a proposed economic activity is not biophysically sustainable then there is no good reason to advocate it, no matter how wonderful it sounds. Nevertheless the reader should be aware that there are other uses of the term "sustainable development" that refer to social aspects, including especially the continuity of certain traditional socioeconomic systems. This is not the definition we use here.

## B. Why Costa Rica?

We have chosen the country of Costa Rica to examine the possibilities and limitations of sustainable development from a biophysical as well as conventional economic point of view. We chose this country as a focal point for our analysis for a number of reasons. The first is that it has served as a model for "green" or "sustainable" development. This government-initiated and -sanctioned program includes a series of approaches designed to generate economic benefits from preserving the beautiful natural forests and other aspects of Costa Rica's remarkable natural biotic wealth (Quesada-Mateo and Solis Rivera, 1990). For example, some of Costa Rica's external debt was "purchased" by conservation groups. In these "debt for nature" schemes part of the outstanding international debt was forgiven in exchange for Costa Rica agreeing to protect remaining natural forests. As another example, Costa Rica has been in the forefront of protecting its own natural areas while promoting ecotourism, thus generating revenues from the protected areas. This program is the first by any government, and serves as a model for such programs in other countries.

Perhaps the most conspicuous component of this "greenness" is that Costa Rica has formally endorsed the idea of sustainable development as a focal point for its economic policy. In the words of its president, it is the aim of his administration to turn the country "into a pilot project of sustainable development," with Costa Rica "offering itself to the world as a 'laboratory' for this new development paradigm" (Figures Olsen, 1996, p. 90). This bill is supplemented by Costa Rica's signing of the 1990 Rio conference, in which it pledged that the nation would restrict its annual carbon emissions from industrial fuel burning and deforestation to 1990 levels. All of this, added to the strong efforts of many Costa Rican and other conservation groups to protect natural areas and the Costa Rican tourist industry's attempts to convey the image of an environmental paradise, has given Costa Rica a very "green" image. One might think that if sustainable development could work anywhere it would be here.

It should also be pointed out that there are many who view Costa Rica as perhaps not so green. Hunter (1994) asks, "Is Costa Rica truly conservation-minded?" He states further that "such optimistic and upbeat presentations of how well we are managing the world's environment tend to hide grim reality," (p. 594) and gives many examples of continued deforestation (including in protected areas), illegal exploitation of wildlife, pesticide overuse and poisoning, continued high population growth, and an almost "anticonservation trend" in at least some respects. The paradox existing in Costa Rica is not lost on Costa Rican environmentalists. Quesada-Mateo and Solis-Rivera (1990) begin their paper on Costa Rica's plan for sustainability with, "Costa Rica has one of the highest deforestation rates per unit area in the world; it also has one of the largest percentages of land protected as natural parks and other natural reserves." Of course this series of environmental and development paradoxes make Costa Rica an interesting and fertile region for scientific investigation. But the main point is that of all countries in the world, Costa Rica is the one most explicitly exploring a green path. At the time of this writing new national elections have produced a new president whose commitment to a program of sustainable development is much less obvious.

A second reason that we chose Costa Rica is that it is often held up as an economic and political model for developing nations. It has been much more successful than most of its neighbors, indeed most developing countries, in avoiding conflict and corruption, in providing basic services to all, and in increasing per capita wealth. As one explicit example of this economic success, on July 24, 1996, in San Jose, Costa Rica, J. Brian Atwood, administrator of the U.S. Agency for International Development, commemorated 50 years of U.S.–Costa Rican cooperation as USAID officially closed its mission in Costa Rica. The U.S. government had provided $1.7 billion to Costa Rica since 1946. Mr. Atwood said that the aid program had been successful because the mean per capita income of Costa Ricans had increased from $1000 in 1996 U.S. dollars to nearly $2500 in that 50 years.

A third reason we have chosen to focus on Costa Rica is that from a scientist's point of view its diversity is intriguingly appealing. It has among the most diverse terrain, climate, and biota of any country in the world, and all in an area the size of West Virginia. There is also a very highly developed scientific infrastructure in Costa Rica (staffed by Costa Rican scientists as well as scientists from Europe and the rest of the Americas). Thus many excellent data bases are available; for example, the Instituto de Meteorologia has very comprehensive climate information, and the Ministero de Agricultura has conducted comprehensive agricultural surveys roughly every decade. Highly detailed economic statistics are maintained by the Instituto de Economia at the University of Costa Rica. Finally, both the size of Costa Rica and the data bases available are about the right size to manage comfortably a comprehensive national-level assessment of the biophysical conditions of an entire nation. We have been able to study essentially everything about Costa Rica at a 1-by-1-km resolution, which translates conveniently into a roughly 400 by 400 grid space, a good size for today's computer screens and processors!

## III. THE CENTRAL QUESTION OF THIS BOOK

We do not think it possible to answer, or even to address, the questions of development, including those related to sustainability, with only, or even principally, the tools of conventional economics. We believe this even though these are the methods used most commonly. Our reasons are given below. We offer instead the beginnings of a different perspective, what we call a biophysical perspective, to attempt to answer the questions given at the start of the chapter. We are acutely aware that the tools we will present here also are insufficient to answer fully many of the important questions about development, and that a biophysical analysis sometimes approaches a conventional economic analysis. Nevertheless we believe strongly that these tools can take us much further in the direction of accurate assessments of how economies really do work, including the potential for sustainable development.

## A. The Principal Economic Question Faced by Costa Rica

The principal economic question faced by Costa Rica is whether it has the biophysical resource base to support more than three and a half million people at a modern, moderate Western standard of living. Even given its many advantages of climate, government, education, and social services, does Costa Rica already simply have too many people? Is it possible for the physical resources of the nation, its soil, climate, energy potential, crop production possibilities, and so on, to generate the foreign exchange (i.e., foreign currency such as dollars) required to import the goods and services required for necessities, let alone the relatively high standard of living desired by a majority, and already obtained by a minority, of the population? To what degree are policy options restricted by biophysical constraints? These issues all revolve around "sustainability," however that may be defined (see our previous definition).

## B. Sustainable Development in an Historical Context

As discussed in Chapter 2, the general approach to improving the economic conditions of Costa Ricans in the past has been based in large part on "development." Development in turn has meant the infusion of mostly foreign capital in order to increase the rate of exploitation of forests, soils, and agricultural potential in order to generate increased income, including foreign exchange. In fact we might say that Costa Rica has been a laboratory for the concept of development (see Chapter 2). A summary of previous large-scale development projects undertaken in Costa Rica includes:

(a) The provision of British capital for coffee development in the 1800s
(b) The construction of railroads, docking facilities, and other infrastructure to encourage banana production in the late 1800s and throughout the 1900s, mostly by North American companies
(c) The encouragement of coffee production for U.S. troops (and civilians) during World War II

(d) Investments in banana plantations by United Fruit and others throughout this century
(e) General foreign aid from the developed world, including agricultural research and development (for example, through USAID and CATIE, which is financed principally by foreign aid from European nations)
(f) Loans from the World Bank, IMF, etc., for the development of export cattle production, especially in the 1950s and 1960s
(g) Loans for general development during the economic crises at the beginning of the 1980s (these were often used to maintain government services rather than to develop new production)
(h) Structural adjustment, meaning many things (see Chapter 24), but especially including an emphasis on encouraging "nontraditional exports" such as cut flowers and ferns, macadamia nuts, and fresh fruits during the early 1990s
(i) A government (and private) reemphasis on traditional exports during the late 1990s, although overlapping with a continued encouragement of nontraditional crops
(j) A government (and private) focus on tourism and ecotourism, especially since about the mid-1980s
(k) Government subsidies for foreign-financed high-tech industries in the mid-1990s

A perspective that seems lost on many modern day analysts is that in one way or another each of these development schemes was probably considered as leading to "sustainability" by those who advocated them at the time. Although "sustainability" then certainly had less of an environmental twist than now, the older development literature is full of enthusiasm as to how this or that project will lead to long-term economic well-being, which certainly is an element of sustainable development. As we have become somewhat more sophisticated in our analysis of what "sustainable" might mean we now often denigrate some of these earlier development schemes as "nonsustainable." But almost certainly the perspective of these project's advocates at the time of their inception was that they were contributing to long-term economic well-being.

## C. Has Development Been Successful?

Has this past development been a failure or a success? Was it sustainable? These questions defy any simple answers, and involve the observer's time horizon. One might answer that clearly it has been successful and sustainable to date. The evidence is that large quantities of coffee, bananas, cows, and cut flowers are still grown; an overwhelming majority of Costa Ricans want to get up and go to work in the morning; shops are full of food and basic goods; services are abundant (if not always swift); and incomes are still reasonably high by Central American standards.

Perhaps the most basic way to consider these questions is to think of what life would be like if there were no development since, say, 1940. Life in 1940 may have been considered pretty good, at least if you owned good land, although a lot of hard physical labor was required. If there had been neither population growth nor development since then it might still be considered about the same—pretty good. In reality population did not stay the same; in fact it has increased by at least a factor of seven. It surely would be impossible even to feed the present population, let alone support an increased standard of living, on the technology and infrastructure of the 1940s. Viewed from this perspective there is no question that development has been both sustainable and a success—at least so far. An additional conclusion might be that human population growth necessitates development.

A look at the literature that has assessed the success of development reveals a rather more complex perspective. The answer to whether development has been successful depends both on whom one asks and on the criteria used for judgment. Something judged successful as measured by growth in the GNP or the profits earned by transnational corporations may or may not be considered as contributing to the increased health and well-being of the country. The strongest voices in support of development's success tend to come from the development agencies themselves (Table 3-1).

There is, on the other hand, a large literature attacking development policies and their consequences. The following criticisms are aimed at development worldwide. (Literature specific to Costa Rica is given in Chapter 24.) These critical studies have focused on perceived failures related to society and the environment, including increased poverty, a growing disparity between the rich and the poor, a higher infant mortality rate, devaluation of women, destruction of land-based traditional cultures, dwindling natural resources, and environmental destruction (Bello *et al.*, 1982; Mies and Shiva, 1993; Meier, 1994; Ghai, 1994; Reed, 1996). Mies and Shiva (1993) conclude that economic development through the globalization and privatization of trade has worked to displace small farmers and encourage investment and ownership by transnational corporations at the expense of subsistence producers. Hill (1986) criticizes the loss of localized indigenous knowledge at the expense of global economic development. Reed (1996) believes that liberalized trade regimes such as GATT and NAFTA encourage the externalization of environmental costs. Reed (1996) also believes that expanding nontraditional agricultural exports causes deleterious environmental impacts by encouraging short-term economic growth at the expense of long-term environmental costs.

TABLE 3-1 Email from USAID, Washington, D.C. (Reprinted Here in Full and without Comment)

50 Year Partnership Between USAID and Costa Rica

On July 24, 1996 in San Jose, Costa Rica, J. Brian Atwood, administrator of the U.S. Agency for International Development (USAID), will commemorate 50 years of U.S.–Costa Rica cooperation as USAID officially closes its mission in Costa Rica. The U.S. government has provided $1.7 billion to Costa Rica since 1946. This assistance has funded over 1,000 diverse activities ranging from technical assistance, to constructing water systems to dramatic macroeconomic policy efforts. Costa Rica's per capita income of nearly $2,500 compares to only $1,000 in 1950, and U.S. exports to Costa Rica have increased five times in the last 10 years to more than $2 billion in 1995. USAID's program in Costa Rica encompasses the four areas of sustainable development: promoting economic growth; supporting democracy; protecting the environment; and improving health and population conditions.

Following are some program highlights:

Promoting Economic Growth: Costa Rica's transformation from an economy based largely on traditional agriculture to today's market economy is a proven success. During the turbulent 1980's, USAID provided technical assistance to Costa Rican policy reformers who worked to eliminate price controls and open markets to international competition, as well as privatize state-owned enterprises. Support for private banks financed an export boom, increasing wages and reducing poverty. Overall Costa Rican trade grew sharply in the 1980s, particularly non-traditional exports, which were a central focus of the U.S. economic assistance throughout the 1980's and early 1990s. Costa Rica's exports grew from $844 million in 1986 to $1.5 billion in 1990. Between 1983 and 1992, the value of non-traditional exports soared from $90 million to $781 million, while the proportion of export earnings represent by non-traditional exports quadrupled. Costa Rica's primary resource is its people. USAID contributed to their education in numerous ways, including sending more than 5,000 Costa Ricans to the United States for training. USAID programs provided technical assistance and training to improve the business climate and position Costa Rica's entrance into a more competitive marketplace.

Trade and investment interventions were largely executed by the Costa Rican Coalition of Development Initiatives (CINDE). Investment promotion programs implemented by CINDE were instrumental in attracting $417 million in new investment from firms in Europe, Asia and the United States by 1993. These investments created 56,000 jobs. More than 30,000 people from over 8,000 companies participated in seminars organized by CINDE. Given the remarkable results achieved, multilateral development institutions have recognized Costa Rica as one of Latin America's most successful trade development programs.

Support Democracy: Since the 1950s, USAID has supported the development of Costa Rica's democratic institutions. USAID has supported improvement in public safety, tax reform, budget reform, municipal and community development, privatization, legislative support, administration of justice and legal reforms. More recently, USAID launched a pilot effort in the municipalities of Perez Zelon and Puriscal for local government decentralization and strengthening.

Protecting the Environment: With support from the Kellogg Foundation and the Costa Rican government, USAID financed the Agricultural School for the Humid Tropical Region (EARTH). EARTH is a four-year regional agricultural college with a working farm, and a philosophy of learning by doing, the first such college in the lowland humid tropics. EARTH is educating a corps of creative experts in agriculture in one of the globe's most difficult ecosystems. It is also carrying out basic research, including an experiment recently carried out aboard the Space Shuttle "Discovery" to help reduce the deadly "Chagas" disease. In

(*continues*)

TABLE 3-1 (*Continued*)

partnership with The Nature Conservancy, USAID's Parks in Peril program has actively worked in Costa Rica to recruit, train and equip park rangers; install basic infrastruture; promote community participation in natural resource management; and identify and develop long-term financing for park management.

Improving Health and Population: Costa Rica has placed a great emphasis on public health and family planning, and the payoffs have been considerable. In 1950, the average Costa Rican woman had seven children. Today, the average number of children per couple is three. During the same period, infant mortality fell from 110 per thousand to 13—one of the fastest rates of decrease in the world. These smaller, better cared for, and more productive families are a direct result of declining infant mortality, strong education and the availability of family planning services. USAID support for family planning, begun in the late '60s, included both private organizations and the public sector. Support for innovative private organizations like PROFAMILIA made family planning services available throughout the country. Initial support from USAID helped these organizations get started, and today they are entirely self-supporting.

Braidoitti *et al.* (1994) criticize the "neo-liberal economics of the 1980s" that encouraged free markets and the private sector as the provider of the dynamics for economic growth. This led in turn to a standstill and even regression in social development in many countries of the South where poverty was increasing the pressure on the natural resource base to compensate for the tightening economic situation (Smith *et al.*, 1994). In effect, natural resources were exploited increasingly for debt payment, while governmental expenditures to meet local needs for fuel, water, and income were neglected, resulting in an exacerbation of social inequity.

Case studies in developing countries reveal additional economic failures resulting from international economic development policies. For example, structural adjustment policies implemented in Cameroon led to hardship for cocoa and coffee farmers, the traditional export commodity producers. In 1990 their output prices were reduced and their governmental support services (e.g., extension, input subsidies, credit, and marketing systems) were scaled down drastically. The urban elite and public servants have been better able to defend themselves and have suffered the least. By the end of 1993, however, most sectors of society were experiencing economic hardship (Reed, 1996).

The literature is much thinner in terms of policy makers themselves admitting to misguided policies. It is also quite difficult to find purely economic analyses of development policies that indicate that development has failed. A few examples exist: Taylor (1993) states that development strategies had only modest economic success in Latin America during the 1980s. In the best cases inflation stabilized at about 20% and hyperinflation was brought under control, but even there economic growth lagged, resulting in declines in per capita income. Recent U.S. congressional hearings on IMF and World Bank policies reviewed specifically their structural adjustment policies. One unnamed participant stated that only 7 of the 55 countries which had structural adjustment programs during the 1980s were regarded as being successful (World Bank, 1994). Perhaps it is unfair, however, to attribute all the economic failures to a failure of development policies or structural adjustment, for the development was generally necessitated by preexisting debt.

## D. The Response of Development Agencies to These Criticisms

In the face of these criticisms some development organizations have been changing their approach. The U.S. Agency for International Development revamped many of its development policies to take into account the value of the country's natural resource base in response to public outcry about its policies leading to environmental degradation. The North American Free Trade Agreement treaty of 1993 explicitly incorporated the North American Agreement on Environmental Cooperation, which includes environmental, labor, and broad human rights issues (Johnson and Beaulieu, 1996). The World Bank certainly has been putting a much greener face on the world, even at times encouraging a biophysical framework of analysis at the expense of conventional economics (World Bank, 1996). It seems clear from the literature that there has been an evolution of development policies that now address changes in expectations and objectives (Preston, 1996; Meier, 1994), mainly resulting from the glaring failures most evident in social and environmental factors.

The objective of all development, however, no matter how green and/or sustainable, remains economic growth and (generally) liberalized trade policies. So it is a delicate balance that these new treaties and policies attempt to achieve. One might even consider it a gamble that trade-induced growth, even with a new focus on environmental issues, will not translate into larger environmental degradation brought by industrialization, expansion of cash crops, or a general increase in affluence and consumption (Johnson and Beaulieu, 1996). Indeed much of the economic literature recognizes social and environmental degradation (Meier, 1994; Preston, 1996), but blames the problem on inadequate representation of externalities in the marketplace (Esty, 1994; O'Connor, 1994). Braidotti *et al.* (1994) state that UNICEF, in trying to ameliorate the negative consequences of structural adjustment, proposed "Adjustment with a Human Face." Even here structural adjustment as a disciplinary measure was not fundamentally questioned, and countries were steered further toward export-induced growth, in spite of its inherent vulnerability to world market fluctuations and transnational corporate investment decision making.

Whether development has been good or bad, or simply a necessary evil to deal with increased human population numbers, it is clear that it is guided and sanctioned most frequently by the discipline of economics, which today means the premises of neoclassical economics. We next take a close look at the economic principles that have guided the analysis of tropical economies as well as economic development.

## IV. SOME IMPORTANT HISTORY ABOUT THE DEVELOPMENT OF ECONOMICS

Economic systems exist. People go to work, goods and services are produced and consumed, and people do or do not satisfy their psychological needs through the expenditure of money. From the perspective of an ecologist these economic systems are ecosystems with their own structures and functions including: population dynamics of people, crops, and domestic animals; energy inputs and transfers; material fluxes from the earth to economic products and back again; transfers across boundaries; and so on. There is general consensus that the proper tools for studying these phenomena are found under the aegis (or discipline) of economics. More properly it is probably better to say that the economists have taken over this arena and relatively few investigators from other disciplines have questioned this.

In fact the principal kind of economics that is practiced today, neoclassical economics, is but one of a series of possible approaches that have been used in the past, and might be used in the future, to study and perhaps understand economic systems. Certainly other schools of thought exist, each with their own conceptual frameworks, theories, and assumptions. These include institutional economics, social economics, Marxist economics, ecological economics, biophysical economics, and Keynesian economics, to mention just a few. These contemporary schools of economic thought are based in one way or another on classical economic thought. Interestingly, neoclassical economics also traces its roots to the classical economists. Yet, as will be argued, one of the fundamental differences between classical and neoclassical theory—one that has far-reaching consequences for development strategies—is the underlying theory of value. For a classical economist the value of goods and services was fundamentally based on the *labor* required to produce them (this is especially true for Marxist economists, who developed the labor theory of value). For neoclassical economists the exchange value of goods and services is determined by the supply and demand for a specific good or service that is determined, in turn, by the cost function of the supplier and the utility function of the consumer. This relation then determines the exchange value (i.e., price) assigned to these goods and services in the marketplace.

## A. Classical Economics

Prior to about 1860 there was no consolidated field of neoclassical economics, and economics as a discipline, such as it existed, was based principally on the ideas of classical economists such as Adam Smith, David Ricardo, Karl Marx, and, to a lesser extent, Thomas Malthus. The focus of classical economists such as Smith and Ricardo was principally on economic production, that is, on determining what materials should be used, and how they should be used, in order to be transformed (via the input of human labor and ingenuity) into the goods and services consumers demand to meet their needs or wants. Labor was generally perceived as a critically important source of wealth, and hence an important focus of economics: "Among a nation of hunters, for example, it usually costs twice the labor to kill a beaver which it does to kill a deer, one beaver should exchange for or be worth two deer" (Smith, 1776). Classical economists came to focus on "embodied labor," that is, the labor that it took to generate the economic goods and services being considered. Another important classical economist, David Ricardo, made additional contributions by assessing the increasing costs per unit produced as production expanded. For example, as demand for food in a region expanded, farming had to take place on increasingly poor land, since earlier farmers, being no fools, tended to farm on the most productive land. Thus Ricardo formulated the theory of *decreasing marginal product,* which says that yields per unit input (including land) will decrease as additional inputs are used. This will be an important concept as we attempt to understand the agricultural situation in Costa Rica.

One of the discussions of classical economics revolved around the "diamonds versus water paradox." Water is essential to human welfare and even life itself, yet its value appears very low if one takes the willingness of consumers to pay for water as a reflection of its value. Diamonds, on the other hand, are essentially useless except for jewelry and a few industrial uses, yet they cost a great deal. The solution to this apparent paradox was scarcity, the guiding principle of much of economics. Since water is abundant it is cheap, while diamonds are very rare and thus are valued more highly. Then what about the value of water in different locations? Here a discussion ensued about the difference between *use value* and *exchange value,* that is, the value consumers gain from the use of a good or service and the value that the same good or service can attain in the marketplace.

## B. Neoclassical Economics

The preceding discussion set the stage for neoclassical economics and its focus on market exchange. Leon Walrus, Stanley Jevons, and others had formalized the concept by the 1860s. For example, "decreasing marginal utility" is one explanatory concept for the water versus diamonds phenomenon and the underlying importance of scarcity; it also reconciles the use value vs exchange value problem: people will pay a very high price for the first unit of something (i.e., the first liter of water in a day) but subsequent quantities are worth progressively less and less. Even where water is scarce, it is only the first liter that is really valuable, and liter number six and seven lose value rapidly. These concepts also provided the underpinnings for the formalized concept of supply and demand, which is probably the best known economic principle among ordinary people. Neoclassical economics thus provides an internally consistent theory by which the marginal costs of goods and services determine their supply while consumer desire determines their demand. Both supply and demand, that is, producers and consumers interacting in competitive markets, determine the exchange value assigned to the goods and services through the free interaction of individuals. Who could object to that? Packaged in the mathematical formalisms of differential calculus, neoclassical economics took shape, and it continues to wield an extremely strong influence, particularly in the United States. The collapse of Communism in about 1990 added prestige and power to the neoclassical model since the only other model in use, Marxism, seemed to have failed miserably.

But what about those material inputs that did not have any, or only a very low, cost? And what about those goods and services not traded in markets, such as environmental amenities? Through the lens of neoclassical economics these two consequences are seen as entirely plausible, consistent, and even desirable. Low marginal costs do and should lead to high use in the production process. Low exchange value leads to low price and high use in the production process. Consequently, according to many (but not the purest neoclassical economist) "free" environmental resources are overused and "free" environmental goods, at least those for which no easy exclusion mechanisms can be established that would enforce their scarcity, are neglected in economic analysis and deemed valueless.

Other consequences follow logically from the rationale of neoclassical theory and are put forth by adherents as self-evident: since free markets preserve the efficiency of the market system to allocate resources optimally, there should be no interference with markets to bring about the "right" price as derived from consumer demand and producer supply. Individuals, as decision makers in the market, know best, or so goes the rationale. Any interference by, e.g., governments would bring about suboptimal prices and would reduce overall welfare. There is an almost religious commitment by most neoclassical economists to the concepts of markets as systems for deriving value, free markets as a means of maximizing wealth generation, and market economies as symbolic of a nation's virtue. For many years (for example, the 1980s) the United States intervened militarily in Central America in defense of many things at least nominally related to neoclassical economics, and official U.S. foreign policy continues to emphasize many aspects of neoclassical economics, including open markets, free trade, and the encouragement of trade by all means possible. In the minds of many there is essentially no difference between promoting democracy and promoting free trade and other aspects of neoclassical economics, although they are quite independent issues.

Such policies had a very large impact in Costa Rica due to, for example, the "structural adjustment" programs. Whether or not neoclassical economics is or is not a pipeline to "God's own truth" about how a nation's economy should be run, it seems to me to be very dangerous to have only one dominant perspective by which most decision makers examine economies.

## C. Some Critiques of Neoclassical Economics

Most economists, trained principally in neoclassical economics, seem to be pretty much enamored of the neoclassical school of thought, and often have acted as if all the issues of economics have been essentially resolved within the context of the basic neoclassical paradigm—give or take an externality or two. What is less commonly understood is that there is a large and growing literature, written by both natural and social scientists, that is highly critical of neoclassical economics as the sole, or (sometimes) even a legitimate means of undertaking economic analysis. Many scientists and engineers believe that neoclassical economics simply misses the boat about the necessity of incorporating the underlying biophysical framework into economic analysis (Hall *et al.*, in prep.). Although the arguments often are complex, they are well summarized in the pungent words of Wassily Leontief, himself a Nobel Prize winner in economics. Leontief, in 1982, described many neoclassical economic models as unable "to advance, in any perceptible way, a systematic understanding of the structure and the operations of a real economic system" (p. 107). Instead they are based on "sets of more or less plausible but entirely arbitrary assumptions" leading to "precisely stated but irrelevant theoretical conclusions." Nelson (1997) wrote a scathing blast against neoclassical economics in his article, "In Memorium: On the Death of the 'Market Mechanism'" (or perhaps more accurately, on the death of its supposedly value-neutral status). Additional criticisms come from investigators representing various subdivisions in the field of economics, calling themselves Evolutionary Economists, Social Economists, Institutional Economists, and Feminist Economists. All of these critics believe that there are many reasons why neoclassical economics should not be accepted at face value as a means of making many decisions that effect economies. The interested reader is advised to see Leontief (1982), Mirowski (1988), Makgetla and Seidman (1989), the special issue of *Ecological Modeling*

(Vol. 38), Dung (1992), Hall (1992), Hall and Hall (1993), Daly (1996), O'Hara (1995, 1996), Gowdy and O'Hara (1996), and Hall *et al.* (in preparation).

The next section, derived from Hall (1992a), examines some of the criticisms that have been leveled at neoclassical economics.

## D. What Development Really Is

Although wealth may appear to some to be produced through economic policy, wealth production occurs generally and explicitly only through the increased exploitation of natural resources, normally in an increasingly nonrenewable manner, and almost entirely through the increasing use of fossil fuels. Cheap oil and its derivatives continue to be used to alleviate the principal impacts of depletion and environmental degradation and mismanagement, giving too often the appearance of solutions whereas in reality solutions only are being deferred. In short, economic development, and the models that guide that development (of both the "left" and the "right"), works principally because we extract oil, coal, and other resources out of the ground to make it work (Hall *et al.*, 1986). Therefore the words "development" and "economically successful" do not describe accurately the production processes necessary for economies to grow. In virtually all cases these terms should be replaced by "increased exploitation of resources" and "industrialization," for that is what most development is all about.

## E. A Difference in Philosophy

I develop some specific critiques of contemporary neoclassical economics from the perspective of a natural scientist. My criticisms are both fair and unfair: fair because they apply to most contemporary neoclassical economics as taught and practiced, and unfair because some basically neoclassical economists (e.g., Samuelson, Mishan, and Hoteling) as well as a whole generation of new economists of various stripes (e.g., Daly, O'Hara, and Costanza) have made extremely thoughtful contributions to a consideration of these problems, and because many social scientists use extremely rigorous procedures in difficult terrain. But the influence of such thought on routine economic analysis seems very small (Goodland and Ledec, 1987; XXTIMM), and, as I develop later, is still extremely inadequate.

First, it is important to consider a fundamental question: is the difference in outlook between natural scientists and economists (at least those who ascribe to the "mechanistic" neoclassical model) due to unresolvable philosophical differences in how to approach the problem, or rather is it intrinsic to the different types of problems studied by the two disciplines? Differences between economists and natural scientists arise from their academic outlook and training. A natural scientist tends to view knowledge, especially models of that knowledge, as tentative, even ephemeral—as ideas that are examined, tested, and subjected to rigorous assessment. Natural scientists tend to be suspicious of established knowledge because they have watched some of their most trusted principles crumble. Intensive scrutiny is essential, for the human mind is far better at providing simple, elegant models than is most of nature.

But most economic models are not well tested by the empirical criteria used in science. The economist Milton Freidman has said that "a [economic] theory cannot be tested by comparing its 'assumptions' directly with 'reality'" (Friedman, 1953, p. 41). (Note that he does emphasize appropriately the importance of validating predictions.) In general very few economic textbooks presented economic ideas as hypotheses to be tested. This led Leontief (1982, p. 104) to ask, "How long will researchers working in adjoining fields . . . abstain from expressing serious concern about the splendid isolation in which academic economics now finds itself[?]" This attitude that testing and validation of the basic models are unnecessary often confuses natural scientists. The latter expect theoretical models to be tested before applied or developed further. But major decisions that affect millions of people are often based on economic models that, although elegant and widely accepted, are not validated. Many practitioners of the scientific method find this arrogance of ideology over empiricism, especially where it affects so many lives, to be unconscionable (e.g., Ayres and Nair, 1984; Hall, 1992; Ayres, 1995, 1996).

An additional problem for many natural scientists arises from the fact that neoclassical economics as a discipline pays almost no attention to the physical characteristics of economic systems (e.g., Ayres and Nair, 1984; Cleveland *et al.*, 1984; Peet, 1992; Dung, 1992). In effect there is no such thing in economics as the laws of thermodynamics to constrain economic activity [with the exception of Georgescu-Roegen and his bioeconomist followers (see, e.g., Cleveland and Ruth, 1997)]. Instead the economy is represented as a perpetual motion machine that has no limits.

The fundamental differences between natural scientists and neoclassical economists on this issue are laid out in the first two chapters of Barnett and Morse (1963; see also Smith, 1979; Fisher, 1981), a book that presents the archetypal neoclassic position on resources. In the anthropocentric view of Barnett and Morse, it is incorrect to focus on, or even to consider, the physical characteristics of resources if one is interested in examining future resource availability. In their view, resources are supplied not by nature but by human ingenuity, and the only interesting index of scarcity is the price per unit resource, since that is the only way that resources are important to humans, which is entirely consistent with the concept of exchange value previously explained. Barnett and Morse tested for the effects of scarcity only by examining inflation-corrected prices for resources over the period 1870–1957. They found that, except for forest products, there was no clear pattern of price increases, and hence no increasing scarcity of resources.

Most neoclassical economists have accepted this assessment and therefore have relegated resources to a position of unimportance in their analyses. For them human ingenuity, and its application in technology, engineering, etc., can overcome any possible constraint imposed by nature. It is their view that currently available resources are not all there is; there always exists the possibility of future development to overcome current constraints. At the extreme, Solow (1974) suggested that "the world can, in effect, get along without natural resources." But more recent assessments, some by economists, show that natural resources have become scarcer even by Barnett and Morse's criteria (Slade, 1982; Cleveland *et al.*, 1984; Hall and Hall, 1984).

## F. Fundamental Conceptual Problems with Neoclassical Economics

"The purpose of studying economics is not to acquire a set of ready made answers to economic questions, but to learn how to avoid being deceived by economists" (Joan Robinson, quoted in Galbraith, 1973). The fundamental premise of this section is that economics as a discipline (more specifically, the neoclassical model that dominates economic analysis) has severe and unresolved problems pertaining to its fundamental intellectual basis. As such, it is insufficient and even misleading for guiding development, decision making, or graduate education in either the developing or the developed world. Discussion on this issue is critically needed. Since Marxism faded in importance following the fall of Communism in Eastern Europe in 1989 there has been little discourse on whether the neoclassical model is indeed the appropriate model to follow. What if the neoclassical, Western approach to economics "works" only because it helps nations to run through their resource stocks faster by giving people incentives to do so?

Based in part on Makgetla and Seidman (1989) there are at least five fundamental flaws that underlie the use of contemporary economics as the principal tool for making economic decisions. These are given and discussed next.

### 1. Neoclassical Economics Normally Uses the Gross National Product (GNP) as a Proxy for Human Well-Being

Development projects tend to be evaluated using a theory of value wherein only their projected contributions to gross national product are counted (Goodland and Ledec, 1987). But GNP (or GDP) is often an irrelevant metric for assessing the attenuation of the major problems of the world for the following reasons: First, GNP is only a partial measure of those conditions that contribute to human happiness and well-being, the supposed goal of economic activity. It simply measures the total output of goods and services measured in inflation-corrected dollars—thus it is a measure consistent with the exchange value propagated in neoclassical theory. Any other source or expression of welfare remains unaccounted for. Second, GNP says nothing about whether basic human needs versus, for example, the production of luxury goods for a financial elite are being satisfied. Thus the use of GNP allows leaders to ignore questions about the distribution of wealth (see Schmid, 1987). Third, GNP is an inaccurate measure of production, and this is especially true in areas where development is being introduced. All "informal" nonmarket production remains unaccounted for. Hence it undervalues both environmental services and nonmarket sources of goods, such as subsistence food.

There is no provision made within the use of GNP for including the economic benefits of properly functioning ecosystems [e.g., their "public service functions"; see Hall (1975) and Hall *et al.* (1986)], or their degradation, because such processes normally do not interact with markets. For example, the proper functioning of the hydrologic cycle is the most important input to most of the world's economies, but farmers do not pay rain clouds for their services, despite the fact that they might pay almost all of their income to have the hydrological cycle operate normally. As human activity disrupts the hydrologic cycle, there is no information with which the market can account for these nonmonetized but essential natural processes. Fourth, GNP counts so-called defensive expenditures, that is, expenditures necessary to remedy the effects of pollution, ill health, deterioration of buildings due to air emissions, or the like, as positive (i.e., adding to GNP).

Lastly, since GNP measures only flows of, and not the stocks of, resources available at any point in time, natural resources that are becoming depleted rapidly remain outside the evaluative system of GNP. Although the depreciation of a factory is accounted for in the estimate of GNP, the loss of natural capital such as soil is not. Collectively, these and other failures of GNP to measure real wealth have given us a distorted view of what GNP really is. For example, Pimentel *et al.* (1995) estimate that the annual loss in agricultural and water management capacity caused by soil erosion costs the United States $24 billion each year, but this is not included in the assessment of GNP. A limited analysis of the impacts of some factors not included in GNP, including hours worked, pollution, and congestion, greatly reduced the apparent growth of GNP from 1930 to 1972 (Nordhaus and Tobin, 1972). An updated extension of that analysis would show a further significant decline in wealth production in the United States over the past decade and a half which is not reflected in GNP. The government of France is now attempting to include all capital, including forests, soils, and water quality, in a new annual inventory of national natural resource capital, which is then assessed along with traditional GNP. Likewise Germany is considering a tax on goods and services representing all environmental destruction caused by their production. Finally, the beginnings of such accounting have been done for Costa Rica. When losses of soils, biodiversity, and marine production were factored into estimates of GNP, it declined by almost 30% (Solarzano *et al.*, 1991). These are significant, but still insufficient, steps in the right direction.

Using GNP as an indicator of economic well-being can encourage high turnover and the construction of often inferior products. The GNP rises if people buy replacement goods more frequently because the original goods are poorly made, even though that process produces more pollution today and decreases the resource base available for the future. Development policies whose principal goal is to increase GNP, as opposed to meeting basic human needs, can encourage developing countries and their entrepreneurs to liquidate stocks, such as a forest, as rapidly as possible to gain monetary flow (Repetto, 1988). When rivers flood and droughts prevail as a result of the removal of the buffering impacts of the forest, there is no system to account for the losses to GNP. Paradoxically, GNP actually increases as villages destroyed by floods are rebuilt.

### 2. Economic Models Have Not Been Validated

As briefly described earlier and in more detail in the literature cited, the most fundamental assumptions of neoclassical economics are nearly untested, but practitioners often act as if they were "truth." This is evidenced by the virtual lack of analyses that test economic hypotheses in most economic textbooks, development reports, or political pronouncements on economic policy. In fact, there is often a complete lack of any hypothesis forming and data testing in most economic textbooks, other than the occasional, but useful, time series analysis of one parameter (e.g., Samuelson, 1976; Atkinson, 1982). Where there have been empirical analyses (of, for example, consumer choice), they have shown frequently that the behavior of real people in experimental or laboratory situations was quite different from the assumptions of a given neoclassical model (Schoemaker, 1982; Knetsch and Sinden, 1984; Smith, 1989; Hall and Hall, 1984). Empirical tests to validate economic models have been undertaken even less frequently in the developing countries where they are often and increasingly applied.

Most noneconomists do not appreciate the degree to which contemporary economics is laden with arbitrary assumptions. Nominally objective operations, such as determining the least cost for a project, evaluating costs and benefits, or calculating the total cost of a project, normally use explicit and supposedly objective economic criteria. In theory, all economists might come up with the same conclusions to a given problem. In fact, such "objective" analyses, based on arbitrary and convenient assumptions, produce logically and mathematically tractable, but not necessarily correct, models (Figure 3-1). Of course making such models site-specific would be complex, since they would need to account

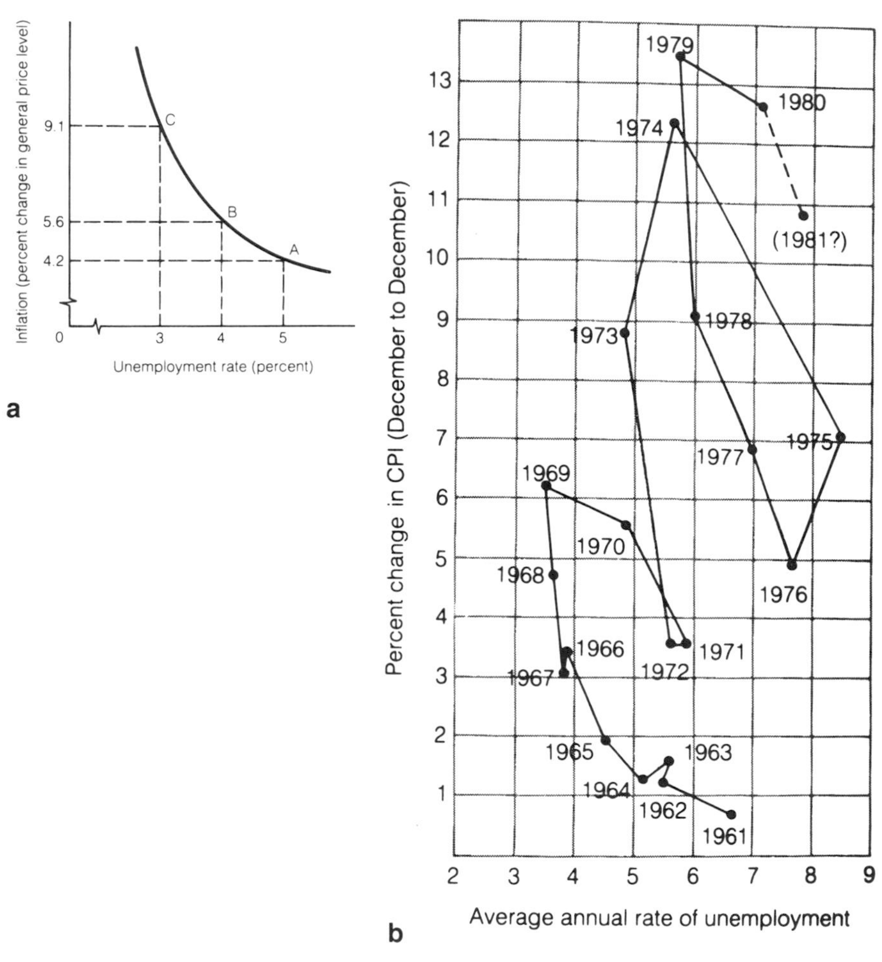


FIGURE 3-1 (a) Phillips curve in theory, representing the supposed inverse relation between inflation rate (CPI) and employment. [From Samuelson (1976). Reprinted by permission of McGraw-Hill Book Company, New York.] (b) Actual data on employment and inflation. Note that after 1969 and especially after the oil price shocks of 1973 and 1979 the theoretical relation breaks down. While more recent economic theory has considered these changes, it is not based on the physical importance of resource availability, which is clearly of paramount importance.

for the social arrangements, cultural differences, and institutional specifics of different markets, particularly different markets in different countries. But these types of complexities are real and have a lot to do with whether the affected people will view the project as good or bad. In a sense it is no more difficult than the complexities faced by ecologists when faced with the complexities of, and differences in, different biotic communities in different parts of the world.

These problems are not distinct only to economics, but are part of a broader tendency in many disciplines to believe in untested hypotheses and to apply a reductionist approach far beyond the limits of its appropriateness. Too often we have science asking us to believe and apply partial truths which then turn into an ethical transformation of society. This approach to science originally asked us, for example, to believe in green revolution technologies, industrialization, energy-intensive development, etc., without a consideration of, or generally even a discussion of, their difficulties, environmental costs, and limitations.

Both classical and neoclassical theories were originally developed using concepts of markets as they existed in agrarian societies. Although there has been much written in economics about how markets have changed, "markets" as originally envisioned (based on the agrarian markets of England 200 years ago) are a very different entity today from what was originally observed. Provisions must be added to the basic theory of industrialization, the consequences of the development of the power of money itself, the development of very large corporations or the development of advertising, each of which characterizes contemporary society, and the "markets" where we buy and sell. Are markets "free" if corporations have to advertise to get people to buy their product?

Neoclassical economics is in the undisputed mainstream of a modern economics education. Economic theory taught and studied in prestigious graduate departments should include natural resources and their relation to money. Instead, it is generally about abstract mathematical models and sophisticated ways of dealing with the formulations of these models

### 3. Economic Analysis Leads to the Destruction of Nature and the Basis of Real Wealth

Neoclassical economics argues implicitly for the destruction of the natural (as opposed to the developed) world because of its noninclusion of nonmarket processes, and as such assists in the destruction of many existing nonmarket economic activities. For example, the services of forested ecosystems (such as controlling hydrologic cycles, building soils, sequestering carbon, maintaining rare species, and moderating climates) are not normally reflected in market prices of timber. Thus neoclassical economics often contributes to the destruction of real economic wealth while not accounting for that destruction, meanwhile encouraging the generation of other, often less important, wealth that happens to enter markets. Development analyses often evaluate the economic advantages of the project in great detail whereas the costs are considered partially, at best, because of a lack of knowledge of the processes and the inadequacy of economics for putting a price on them.

In the developing nations, investment policies based on neoclassical economic analyses encourage borrowing from developed countries and hence growing indebtedness. Pressure by the lending agencies to service the debt encourages the mining of resources to get a quick return on the investment so that the lending banks get their cash return. In the meantime, the long-term productivity of the region may be destroyed. But those assessments are not included in the original analyses, and in the rare cases where they are utilized, their value is heavily discounted.

In one good analysis, Repetto (1988) showed that many tropical countries sell their trees at a price far below their worth to either the buyer or, especially, the seller. The desire for quick profits, and often corruption, results in magnificent forests being sold for 50% to less than 10% of their value by any system of accounting. Many of these sales are sanctioned by nominally objective economic analysis. Canada does much the same (Canadian Broadcasting Corp., 1985), as has the United States in the Tongas and Flathead National Forests. Likewise Hildyard (1989) has documented in the Amazon Basin massive destruction of long-standing renewable economies by various shortsighted development plans. These are just a few recent examples of a continuing saga of poor decision making that compromises the long-term economic welfare of nations by undervaluing natural resources. A previous generation of such studies has been chronicled by Milton (1968). Just as history is written by the winners, economics is written by the monetary winners, meaning generally the ones with power.

The discount rate (a mathematical procedure to weigh the time value of money) used in most calculations is set by the U.S. Federal Government (or by other nations or large banks). The official U.S. federal discount rate reflects primarily the interest rates charged by the nation's largest banks and thus, in general, the cost of borrowing money from commercial banks. It changes as a function of federal monetary policy and other factors that are independent of, and perhaps quite irrelevant to, basic resource decisions.

The use of a discount rate means that a one-time gain of a thousand dollars today could weigh more heavily than tens of thousands of dollars gained slowly over a long time. Since many of the direct benefits of natural ecosystems are gained at low rates (as measured in dollars) but over very long, even indefinite, time scales, their value tends to be heavily discounted. For example, natural areas with limited economic outputs but zero requirements for fossil-fuel-derived inputs (in contrast to virtually all modern developments) are developed based on an expected stream of economic revenues from that development. Environmental losses, if considered, are small for any one year but last indefinitely. Hence the benefit–cost ratio calculated for developing a marsh might be favorable because the long-term benefits are heavily discounted, even though the marsh might serve as a regional flood or hurricane buffer, a nursery for fish or wildlife, or a recreational area. Another aspect is that should the cost of the fuel or other inputs required to make the development work rise dramatically in the future, as it will eventually, it might be found that the monetary value of the costs exceeds the gain. The better approach, even as evaluated according to conventional economic criteria, would have been to leave the area undeveloped.

Thus the use of discount rates also argues for the destruction of natural systems where short-term, high-gain development is the alternative. That is unwise. When the petroleum subsidies are gone or become too expensive, the benefits of development will be gone, and the natural systems may no longer provide their original solar-powered services. This scenario is clear from an example in southern Louisiana where the economic boom of petroleum production has been replaced by economic depression and cancer in an area that has neither much petroleum left nor the original natural environment that once supported people's livelihoods (*Wall Street Journal*, 1984).

The concept of making decisions while discounting the future was taken to its logical extreme by Clark (1976), who cautioned that in many circumstances the "rational" thing to do was to harvest fish stocks to exhaustion. The money so gained should be invested elsewhere (perhaps in some other fishery that was doing the same thing). Eventually this penny-wise, pound-foolish approach would produce lots of money but nothing real left to invest it in.

We have argued in the past that there is good reason to consider a negative discount rate for certain resources, rather than the positive discount rate used routinely (Hall *et al.*, 1979). Due to depletion, a barrel of oil is likely to be more, not less, valuable in the future. The same is true for a ton of soil or a hectare of forest. The use of a power function discount rate makes any of these resources appear essentially worthless in a decade or two as presently evaluated economically, although in fact they are likely to be more valuable in the future.

To what degree can society afford to discount the future? If forests are destroyed, the rainfall, and hence agricultural production, of a region may be diminished. This may be only a small amount for any given year, but the long-term effect over many years could be very large. But if discounting is used in economic analysis, the value of the agricultural loss would appear negligible. Much of the Levant that was in forests and agricultural production in biblical times is now desert, due largely to human activities. The money gained from that original deforestation is almost certainly trivial, even if invested, compared to the integrated loss of biological production for a thousand or more years. Where environmental contributions are long lasting, their importance, even where included, is calculated as trivial although their actual contribution to human welfare may be not only large but essential.

## 4. The Market—The Wrong Yardstick for Large-Scale Decision Making

Neoclassical economics, by giving primacy to "the market," gives the economic and resource decision making of entire nations to the "day to day" commercial tastes of individual consumers. The assumption is that consumers will budget monetary resources in a way that is "best" for them, "best" being defined according to a person's wants. Consumers are assumed to be "rational," rational meaning essentially selfish and entirely materialistic. Since neoclassical economics is based on the assumption that people's wants and needs are best expressed by their behavior (purchases) in the marketplace, then there need be no further discussion of the future investments made for a nation. Those decisions will be determined only by entrepreneurs providing for anticipated routine consumer purchases (even this ignores the fact that consumer tastes may be derived artificially through advertising). Those items that are not explicitly available in the marketplace, such as public health, clean air, or justice before the law, will not be provided (Dohan, 1977). Another problem is that equity among different groups of people, most especially in terms of social justice, cannot be solved by the market. This requires a political process quite independent of market mechanisms (Aguilar, 1997; Prugh, 1995). If distribution is poor, the market will tend to consolidate the unfair situation with the logical consequence of social polarization.

Thus a high reliance on neoclassical economics destroys the necessary discussion of economic means and ends, and replaces it with socially sterile and simplistic objectives based essentially on shortsighted and often manipulated human greed (e.g., Daly, 1979).

## 5. Price Does Not Always Reflect Scarcity

Price, the economist's usual metric of scarcity, reflects many important aspects of scarcity very poorly. Many scientists, especially environmental scientists but some economists as well, have argued vehemently against the perspective of Barnett and Morse that inflation-corrected price changes are the only relevant measure of scarcity. For example, Daly (1977) showed that if all resources become more scarce, then the prices of all goods, including resources, will inflate as a general trend and inflation-corrected values for each material will not increase.

Hall *et al.* (1986) argued that the original analysis of Barnett and Morse (which found little indication of increasing scarcity of raw materials as reflected by their price) was incomplete because the decreasing price of energy, and its increasing use, masked the consequences of resource depletion. They showed for many resources that large increases in energy use have been required to supply society with cheap raw materials as they were depleted and/or mismanaged. Since energy was not scarce in the United States during the period analyzed by Barnett and Morse, and since cheap energy has allowed ever lower grade domestic reserves, as well as foreign resources, increasingly to be exploited, there is no reason for prices to increase even though the highest grades of virtually all major U. S. resources have become exhausted. Should energy become scarce in the future, as it did for a period in the 1970s, then all resources probably would become scarce by Barnett and Morse's criteria, as indeed did occur at that time (Cleveland *et al.*, 1984, Fig. 7; Cleveland, 1991). When international prices of energy came back down, so did the prices of raw materials.

But perhaps technologies will be developed that can compensate for the probable shortages of conventional fuels in the future. That is an article of faith and the essence of the neoclassical economist's lack of concern about resources. The evidence for technology overcoming any scarcity in the past, without increased fuel use, is ambiguous at best (see, e.g., Cleveland *et al.*, 1984; Hall *et al.*, 1986; Cleveland and Ruth, 1997; Ko *et al.*, 1998). Thus, in a sense, the fundamental argument is philosophical, at least until such time when the world will face a major petroleum shortage again. At that point, which we believe to be inevitable, the signals given by prices will have been completely inadequate and indeed misleading relative to the needs of people to prepare for that time.

## 6. The Increasing Globalization of Production

An interesting new twist has been added to the discussion of scarcity. Assume for the moment that prices of raw materials, for example, copper, have remained constant even though their quality (ore grade) in the United States has, in general, declined more rapidly than technological extraction efficiency increased. Also assume that in general the United States and other large consumers of resources have turned increasingly to sources from the developing world—in the case of copper, to Zaire and Chile. Finally assume that the environmental safeguards in the new supply nations are not as strict as they are for the United States, so that more pollution is produced per kilogram of copper extracted. The net effect may be no change in the price of copper in the United States, but an enormous increase in the total cost of providing that copper. That cost increase, however, is paid not by the consumer in the United States but rather by the nationals and ecosystems of the country where the copper is mined. Thus the net effect of increasing scarcity of copper in the United States is increasing social and environmental costs of mining copper elsewhere, although that is not reflected in the price of copper for the U.S. consumer. In fact, this process is occurring routinely.

Despite these and other large problems with the essentials of contemporary economics, the basic concepts of the neoclassical approach have been adopted in principle recently by, for example, the International Monetary Fund and the U.S. Agency for International Development. This approach is also used by other development agencies, as well as by organizations in the developing countries. The general goal of most development projects is to increase GNP.

## G. Many Economists Agree with These Criticisms

Ecologists and other noneconomists are not the only ones who have serious reservations about neoclassical economics as a discipline. The following quote is from an economist who shares these reservations:

> The difference between science and economics is that science aims at an understanding of the behavior of nature, while economics is involved with an understanding of the behavior of models—and many of these models have no relation to any state of nature that has ever existed on this planet, or any that is likely between now and doomsday. The word that comes to my mind when confronted by these fantasies is fraud; but even when not fraudulent they are usually self-deceptive, irrelevant, picayune, and always expensive. The latter is true because the wages of economic researchers, unlike the wages of sin, tend to be paid by the community rather than the perpetrator, and the community still labors under the delusion that the sophistry it sees in the "learned journals" of economics represents a valuable contribution to the general welfare. (Banks, 1988, p. 1)

Banks concludes by suggesting that the most important thing that researchers in economics can do is not to continue in the same old way but to refute that previous work which he says is bogus. Leontief (1982), Bailey (1982), Lekachman (1986), Myrnick (1982), Goodland and Ledec (1987), Makgetla and Seidman (1989), and many others from both within and outside the discipline of economics have chronicled the failure, mutual conflicts, and frustrations of many economic models, and even of the entire approach of neoclassical economics (see also Georgescu-Roegan, 1971; Morgan, 1988; Hausman, 1989; and Smith, 1989). When economists are confronted with these issues they tend to avoid any response, and when they do respond it seems to me the discourse is trivialized (Solow, 1997; Stiglitz, 1997).

Although some economists have attempted to discuss some of the aforementioned inadequacies (e.g., Samuelson, 1976), these problems generally are omitted from theoretical and applied analysis either because they are difficult to quantify in the economist's metric or because they may be mathematically intractable. Yet the mainstream applied economist may assume that the particular problem has been resolved because it has been published, when in fact the concept is operationally relegated to obscurity. And even when the corrections are included into the basic framework of analysis, they may still fail to resolve the issue because the framework itself is based on faulty logic. It is legitimate to ask whether contemporary economics simply reflects the materialistic viewpoints of Western people, or perhaps people in general, and the problem lies not with economics per se but rather with the attitudes of people. This issue may be of particular concern when economic procedures developed in industrialized Western nations are applied in other nations or cultures. Either way the problems remain. But this aspect is beyond the scope of this chapter.

An interesting aside is that neoclassical economics is in a way borrowed from natural sciences, especially the mechanistic, static (i.e., nonevolving), highly mathematized approach of physics. Neoclassical economics, for some reason, chose to borrow the "reductionist" methods of physics (but not its material basis) to expand and formalize the basic ideas about buyers and sellers, and firms and households interacting, rather than a more synthetic approach. Reductionist physics works well only for a certain set of highly specified circumstances. For example, weather systems are essentially entirely physical. But (nontrivial) predictions can no more be made about the state of those systems a month or a year ahead than can be done for economies. Many of the same criticisms we have already raised about economics can be leveled at various natural sciences as well (e.g., Hall, 1988). Therefore, maybe the criticisms here are really not exactly about neoclassical economics but rather concern the faulty basis of using reductionist scientific analysis for problems that do not meet the very restricted criteria required for such analyses. This, of course, in no way excuses neoclassical economics for choosing what might be an inappropriate model. The interested reader might explore the rich literature on this subject, for example, Merchant (1983), Stewart (1994), Gell-Mann (1994), and Lelé and Noorgard (1996).

## H. An Alternate Approach to Examining Economic Systems

Collectively, the arguments thus far presented make a strong case for including the physical and biotic resource base that underlies economic activity explicitly in the study of economics. Today's dominant schools of economics may be in large part irrelevant for solving many of the basic economic problems of the developing world because explicit considerations of resources are absent from virtually all neoclassical analyses and, except for the genuine importance of overwhelming political oppression and inequity in land distribution, economic problems around the world are caused principally by the depletion and mismanagement of the resource base and the changing ratio of people to resources.

## V. BIOPHYSICAL ECONOMICS

An alternative to conventional economics, both classical and neoclassical, exists, and is derived from the physical rather than the social sciences. This tradition of a physical approach to economics is more comprehensive and has a longer history than most economists realize (see, e.g., Ostwald, 1907; Soddy, 1926; Cottrell, 1955; Georgescu-Roegen, 1971; Odum and Odum, 1976; Cleveland, 1987; Martinez-Alier and Schlupmann, 1987; Cleveland and Ruth, 1997).

The source of these ideas is generally attributed to the French Physiocrats of the 18th century, who focused especially on land as the origin of wealth and its proper basis for analysis. While not all of their ideas seem useful or even moral today (such as absolute free markets with an implied lack of environmental controls and a strong support of the landed gentry), they did emphasize the importance of a biophysical basis for economic analysis, that is, the necessity of examining the biological and physical basis for wealth production. Later neo-Physiocrats, also called biophysical economists, which includes this author, have extended the perspective of the early Physiocrats to include the special importance of fossil fuels and other concentrated energy resources, such as hydro- and nuclear power, as a means of generating wealth (see review in Cleveland, 1987). Indeed, as documented in many chapters in this book and in publications such as Cleveland *et al.* (1984), Hall *et al.* (1986), Dung (1992), Cleveland and Ruth (1997), Ko *et al.* (1998), and elsewhere, there is much to be said for the Physiocrat's perspective, the validity of which has been documented scientifically again and again for essentially all nations of the world. A remaining question is the degree to which this approach can or should be integrated with traditional economics.

It is easy for most people to grasp the fundamental approach of biophysical economics by simply thinking of economics this way—each time you spend a dollar nearly a half liter of oil (about a soft drink can's worth) or its equivalent as coal, etc., has been pulled out of the ground somewhere and burned to create the good or service represented by that dollar. No energy use, no economic product. We call the energy required to generate the product its *embodied energy*. Because of the interdependency of different sectors of the economy (i.e., steel companies buy oil, paper, and insurance, as do universities) it is not remarkably important what you buy, with the exception of energy itself—roughly the same amount of energy is required per dollar (see, e.g., Costanza, 1980; Hannon, 1982; and a review in Hall *et al.*, 1986). Thus the Physiocrat's perspective seems even more meaningful now than when it was first derived. Now we understand that before the industrial revolution land was essentially the only way to capture energy, which came into economic systems almost exclusively by being captured by land-requiring biomass. This perspective also explains how, during the industrial revolution, the most wealthy members of society became, for nearly the first time in history, no longer large landowners, but mill owners dependent upon the concentrated energy of dammed water or coal. This led to the redistribution of wealth and political power, not to mention many interesting novels about the ensuing changes in social status.

## A. Models of the Economy

The standard neoclassical perspective of how economies operate is the "circular" model found in the majority of economics textbooks (Figure 3-2a). Although caveats may be added, the ideas implicit in such figures are still embodied in the most sophisticated economic analyses. Essentially this model is a perpetual motion machine, except that it grows. We have restructured this basic diagram to capture the reality of the thermodynamic and material processes associated with real economic systems (Figure 3-2b,c). The following, derived in part from Cleveland *et al.* (1984), develops this new economic model more explicitly.

Because we believe that the implications of resource scarcity and other aspects of the biophysical world are inadequately incorporated into neoclassical analysis, we have approached macroeconomics from a thermodynamic perspective. This analysis emphasizes the production of goods; the neoclassical perspective emphasizes the exchange of goods according to subjective human preferences. Production is the economic process that upgrades the organizational state of matter into lower entropy goods and services. These commodities are allocated according to human wants, needs, and ability to pay. Upgrading matter during the production process involves a unidirectional, one-time throughput of low-entropy fuel that is eventually lost (for economic purposes) as waste heat. Production is explicitly a work process during which materials are concentrated, refined, and otherwise transformed. Like any work process,

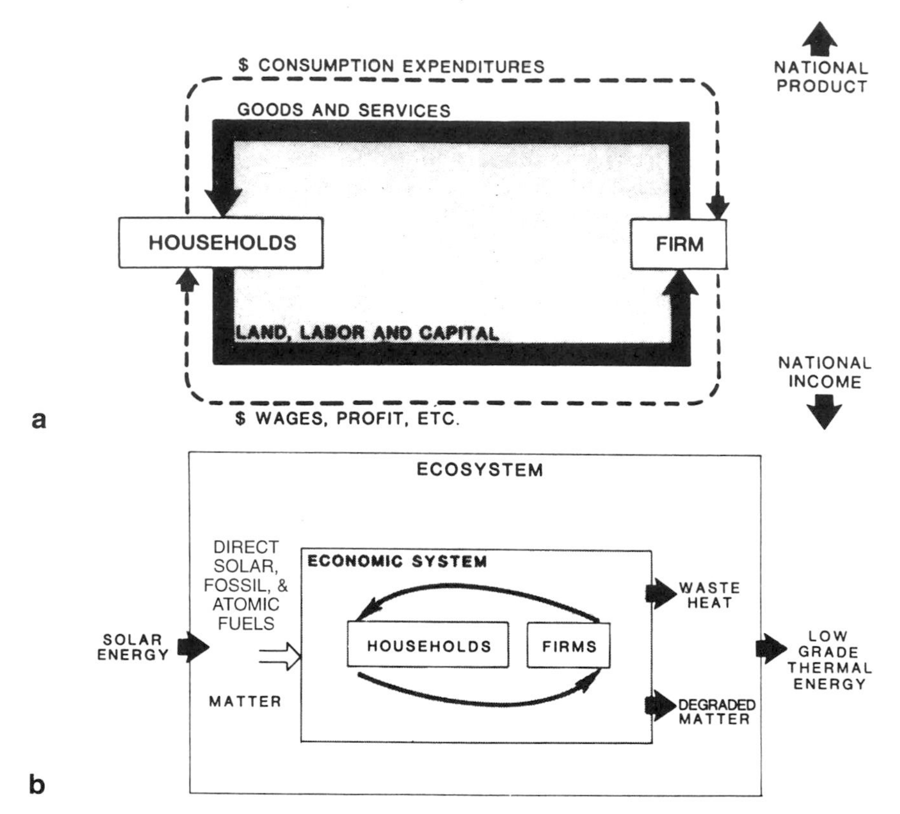


FIGURE 3-2 (a) Neoclassical view of the circular flow of economic production. Households sell or rent land, natural resources, labor, and capital to firms in return for rent, wages, and profit. Firms combine the factors of production and produce goods and services in return for consumption expenditures, investment, government expenditures, and net exports. (From Hall *et al.*, 1986, as modified from Heilbroner and Thurow, 1981.) (b) Economic production from a biophysical perspective. A continuous input of high-quality, low-entropy fuels enter the economic system, and the economy uses these fuels to upgrade natural resources, driving its circular flow between households and firms in the process. The fuel is degraded and put back into the environment as low-quality, high-entropy heat. (From Hall *et al.*, 1986, as modified from Daly, 1977.) (c) The steps of economic production. Natural energies drive geological, biological, and chemical cycles that produce natural resources and public service functions. Extractive sectors use economic energies to exploit natural resources and convert them to raw materials. Raw materials are used by manufacturing and other intermediate sectors to produce final goods and services. These final goods and services are distributed by the commercial sector to final demand. Eventually, the goods and energy return to the environment as waste products (original drawing by Charles Hall and Doug Bogen, reprinted with permission).

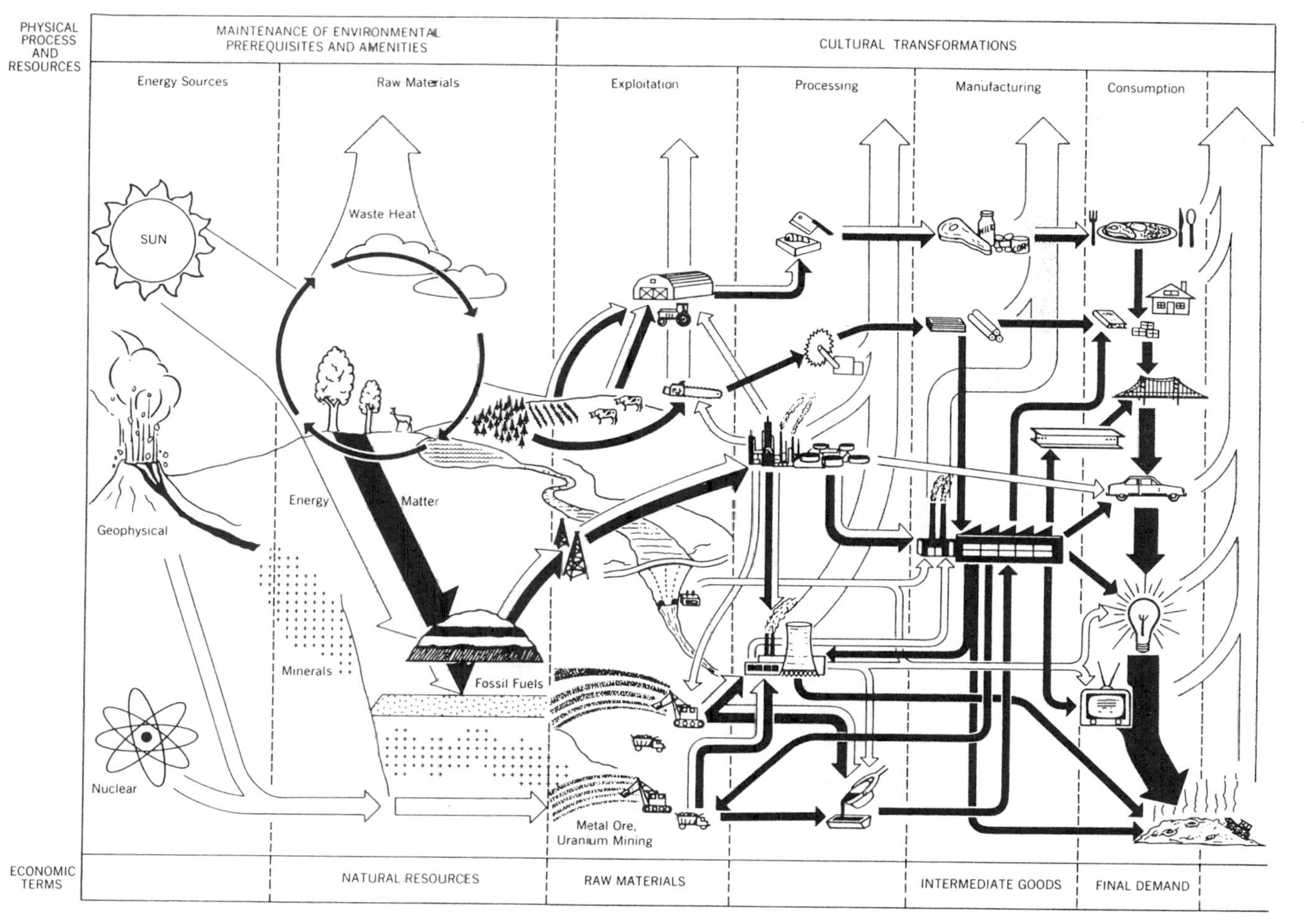


FIGURE 3-2 (*Continued*)

production uses and depends on the availability of free energy. The laws of energy and matter control the availability, rate, and efficiency of energy and matter use in the economy. They are, therefore, essential to a comprehensive and accurate analysis of economic production.

Changes in natural resource quality affect the availability and cost of fuel and matter throughput in human economies because lower quality resources nearly always require more work directly and indirectly to upgrade them into goods and services. Technological advances can counter changes in natural resource quality to varying degrees, but, historically, many technical advances that have lowered unit labor costs have been realized only by increasing the quantity of fuel used directly and indirectly to perform a specific task (Hall *et al.*, 1986). The degree to which technological change can or cannot offset declining resource quality as some basic natural resources are depleted (for example, fuel and metal ores) and/or mismanaged (some agricultural and fishery resources) is an empirical question, and cannot be predicted easily. Nevertheless such resource changes have important implications for what is and is not possible in the economy.

The importance of resources was illustrated dramatically in the years following the 1973 and 1979 oil price increases, when the price of virtually all commodities increased dramatically, as did unemployment, adverse balance of payments, and many other economic problems. These problems continued to linger to some degree in the United States through the 1980s, although the effects were much more severe in many less-developed countries such as Costa Rica. The most obvious sign of this was the increase in debt. Thus economic theory and policy must incorporate the importance of potentially rapid changes in the price and physical availability of resources if economic predictions are to be accurate and economic policies effective.

## B. Production Functions

The economic process is depicted generally in basic texts as a closed system in which the flow of output is "circular, self-feeding, and self-renewing" (Heilbroner and Thurow, 1981). But, as we have indicated, this model is incomplete. In reality, the human economy is an open system embedded in a global environment that depends on the continuous throughput of solar energy. The global system produces environmental services, foodstuffs, fossil and atomic fuels derived from solar and radiation energies, and other resources. The human economy uses fossil and other fuels to support and empower labor and to produce capital. Fuel, capital, and labor are then combined to upgrade natural resources to useful goods and services. Therefore, economic production can be viewed as the process of upgrading matter into highly ordered physical and informational (and thermodynamically improbable) structures. Where one speaks of "adding value" at successive stages of production, this cannot happen without "adding order" to matter through the use of free energy (Roberts, 1982). Odum (1988, 1995) gives a procedure for accounting for all environmental energies that are used to produce a product. He calls the final total "emergy" or energy memory (see Chapter 25).

Standard economic production functions do not account for the important physical interdependence between energy and all other factors; the availability of all factors created by humans depends on the existence and availability of free energy in the natural environment. Capital and labor are combined to extract energy from the environment, but they cannot create, in a physical sense, the free energy and matter from which they are derived. Thus, elasticities of substitution among natural resources, capital, and labor calculated at the level of an industry do not necessarily reflect true substitution possibilities over the economy as a whole. For example, if the price of energy increases, the standard economic models would increasingly replace energy with capital and labor. But producing capital and labor requires energy, both directly and indirectly, which reduces the degree to which capital and labor can substitute for fuel in production. In general we have little knowledge about the magnitude of these costs because they have not been calculated recently or comprehensively (but see Hannon, 1982).

## 1. Empirical Testing of the Alternate Economic View

Empirical testing of economic theories is a difficult but essential procedure which is too frequently ignored. Nevertheless it is essential for developing theories that we can trust. Thus we have examined the historical record of the last 90 years to test the hypotheses generated by our model. The formal development and testing of these hypotheses are given in Cleveland *et al.* (1984).

Fuel use and economic output in the United States had an extremely high correlation from 1890 to 1980 (Cleveland *et al.*, 1984). More recently the relation shows the effect of improved efficiency following each of the oil price shocks of 1973 and 1979. Subsequently, however, a linear relation was again reestablished between economic output and energy use. The high coefficient of determination of Figure 3-3 is consistent with the hypothesis that, at least in the past, economic output and fuel use have been closely linked. While a causal relation from fuel use to GNP or vice versa cannot be verified, a strong contemporaneous link between the two variables is supported (Akarca and Long, 1980). Similar relations have been shown for Argentina (Paruelo *et al.*,

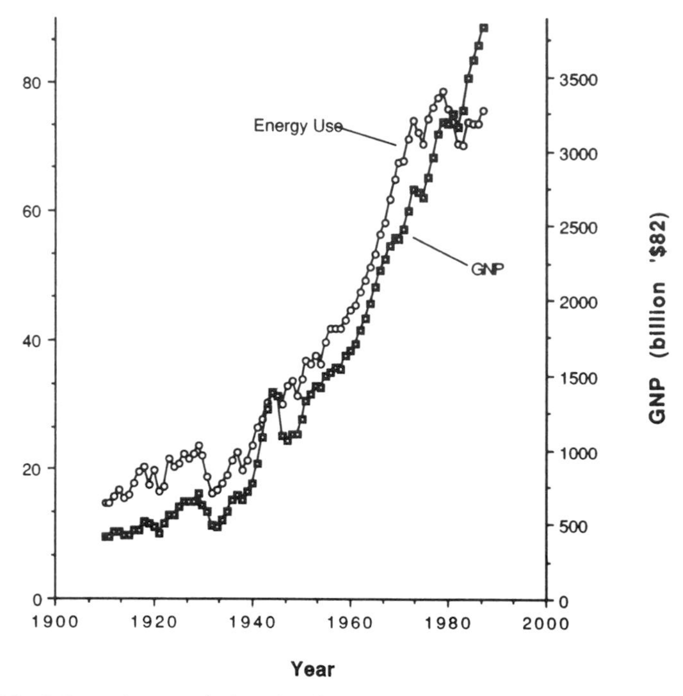


FIGURE 3-3 Relation between fuel used and economic output in the United States. Shown here is fuel use and real GNP per year from 1910 to 1987.

1987), for five other industrial countries (Kaufmann, 1992), and for four developing nations (Ko *et al.*, 1998).

It has often been argued by economists (and others) that since there sometimes has been an increase in efficiency as measured by the GNP produced per unit of fuel energy used, human ingenuity can be substituted for physical resources. But a more careful analysis showed that most of the apparent improvement in efficiency was due to increasing use of high-quality fuels and decreasing direct consumer fuel use, factors that have little to do with what most people view as efficiency. Many studies found very ambiguous results from many nations as to whether or not the efficiency of turning energy into wealth was improving (Cleveland *et al.*, 1984; Nilsson, 1993; Hall and Hall, 1992; and Ko *et al.*, 1998). Nilsson concluded that "there is some level of energy intensity, from 0.25 to 0.5 tonne of oil per 1000 (1980) international dollars, to which most countries are converging."

The ability of biophysical approaches to predict (in a statistical sense) economic phenomena also can be compared directly to economic analyses. Kaufmann (1988) analyzed available petroleum production data for non-OPEC oil-producing nations using both standard economic and biophysical procedures. For virtually all regions, he found that a biophysical analysis was a much better predictor of production level over time than an economic analysis alone, although the inclusion of price information increased somewhat the quality of the results of the physical analysis.

In summary, in all the cases that we have examined—including the Argentine, French, Korean, Mexican, and U.S. national economies, agriculture in the United States, Costa Rica, China, and elsewhere, and labor productivity over time in the United States and Europe—there have been few technical or economic advances without a concomitant increase in the use of industrial energy, principally energy derived from fossil fuels. In other words, little development has taken place without requiring a continuous input of industrial energy and hence a continuous degradation of the nation's resource base (and of the earth's atmosphere). Cottrell (1955) shows elegantly how the same relation has held throughout human history. This is not to say that increases in efficiency are not possible, but just that they have occurred only rarely.

In general, biophysical procedures offer hope for avoiding some of the major faults of conventional economic analysis, although it is also clear that the issues are not clear-cut and that some of the basic assumptions of each approach are correct at least some of the time. There is need now for insightful reviews, using comprehensive techniques, of whether or how often: (1) there has been development without a concomitant increase in energy use, (2) there has been increased efficiency with which energy is being used to generate material well-being in the developing world, (3) the neoclassical assumptions are accurate where the resource base is declining substantially in quality, (4) the explanatory or predictive ability of the neoclassical approach is exceeded by that of the biophysical approach, and (5) the biophysical approaches do or do not incorporate total costs and benefits more inclusively than the neoclassical alternative.

All of this is not to say that these biophysical flows take place in a social vacuum. Economies are also social institutions devised by humans to coordinate the complicated and complex interactions of groups of human beings as they exploit the biophysical world and each other for their own advantage. Any complete analysis of economic activity must include an analysis of the social systems and institutions involved, and the ways they are shaped by tradition, culture, history, and political power. Many social scientists believe that the neoclassical model, by focusing on individual human "rational" behavior, leaves out the important human and institutional aspects of economies just as thoroughly as it ignores the biophysical matrix (Makgetla and Seidman, 1989; O'Hara, 1995). An example of a biophysical approach to agriculture is given in the next chapter.

## C. The Hidden Connections

Although the connection between energy and development is obvious to trained individuals, it appears not so obvious to the vast majority of planners and farmers who view fertilizers as fertilizers, not as the product of petroleum wells which will be depleted within their or their children's lifetime. But is it impossible to compete in the marketplace today if one does not use fertilizers? Does the development of market economies inevitably result in the destruction of alternate nonindustrial agricultures which may be more sustainable? Are population levels too high to have nonindustrialized agriculture?

Although I am a critic of many aspects of neoclassical economics in both theory and application, my own perspective is that it has its utility too, and that probably the best way to think about the role of different approaches to economics is by thinking about the parable of the seven blindfolded Hindu men and the elephant. The man feeling the elephant's side thought it was a wall, the one feeling his leg thought it was a tree, the one feeling his trunk thought it a snake, and so on. Economies exist, and we can examine them from many different perspectives—as long as we remember that each is a perspective and not an absolute truth. This is the view that we use in this book, although the analytical focus is on a biophysical approach and its relation to the generation of wealth. After all, our perspective is that the reality, that is, the elephant, is a biophysical phenomenon, however we may wish to characterize it. Probably over the next decade or two these issues will be dealt with much more explicitly. In the past economists have shown a great unwillingness, an arrogant unwillingness in my opinion, to answer the criticisms aimed at them from other scientists. But as of this writing there is for the first time a dialogue, even if not a particularly satisfying one, on these particular issues [*Ecological Economics*, 22(3), 1997]. We suggest that the reader examine the various sides and make up his or her own mind. Perhaps what would be best would be to generally undertake analyses from both a neoclassical economic and a biophysical perspective; where they agree the path is clear, and where they do not the points of contention are clarified.

A real problem with the biophysical approach is that there is not a clear and universally agreed-upon set of procedures for calculating biophysical possibilities and costs and gains, as there is in conventional economics. We try to show how this might be done for each resource chapter by chapter, but a full formalization awaits further development. In the meantime this book in no way advocates an "energy theory of value" or decision making based entirely on a biophysical approach. What we do advocate is a careful assessment of the biophysical basis for economic activity, however that is measured, and a very cautious use of any economic procedures that stem simply from economic theory without examining the biophysical basis.

## D. Application of a Biophysical Perspective to Sustainability in Costa Rica

A main issue we address is that although there are many "new" ideas about how the Costa Rican economy can be "saved," we perceive essentially all of them as a continuation of the same general process: cranking up the rate of exploitation of the limited resource base with increasing investments in fossil fuel use. For example, tourism is expanded to make debt service payments and satisfy the demand for foreign exchange. With few exceptions this means, inevitably, an increasing use of more fossil fuel—for airplanes, microbuses, construction, and so on. So calling the development process by its rightful name it is really industrialization, and the principal issue that determines the future of the Costa Rican economy is the price of oil and oil-derived inputs to the economy relative to the price of coffee and bananas and the incomes from potential tourists. Whether we are talking about increasing revenues through transporting coffee or bananas to northern consumers, increasing agricultural production on increasingly worn-out land, shipping flowers by jet to Amsterdam, or bringing tourists by jet from New York or Minneapolis to bask on the beach or look at the marvels of such nature as remains, it is all highly, and increasingly, dependent upon fossil fuels.

If we ignore the question of fossil fuels, or possibly conjure up substitutes such as massive development of hydropower or some other solar power, then there are other definitions of "sustainable development" that may be possible: soils can be saved, forests can be protected, biodiversity can be encouraged, carbon can be sequestered in biomass, and, possibly, economies can continue to operate, at least while there is cheap energy. This book examines the possibilities and the actualities. Then policy makers might wish to ask the question, "are there policy approaches that can turn this situation into one that is sustainable, and if so, what are the limits and opportunities of such an alternative path?"

But in the meantime it is clear that one must use the words "sustainable development" very, very carefully. And we see from the preceding discussion the need for an explicit biophysical analysis to supplement conventional economic analysis and to examine constraints for development (Table 3-2). This can be done for each major resource in terms of its potential for contribution

TABLE 3-2 Defining Sustainability from a Biophysical Perspective

There are certainly many definitions of "sustainable" that one might choose to use when attempting to determine what policies might or might not be sustainable (Table 3-1). Turning these rather abstract definitions into some minimal criteria about sustainability we can, following the lead of Ayres (1996), at least begin to define what is NOT sustainable.

Nonsustainable criteria include:

1. Economic criteria

- An economy based on increasing debt, especially international debt, because that debt requires the generation of foreign exchange to pay it, and the generation of foreign exchange almost always implies the export of real resources (timber, agricultural products) which are generated through depletion of soils, forests, and imported fossil fuels.
- An economy that imports more goods than it can pay for (which of course generates the debt mentioned above).

2. Maintaining capital equipment criteria

- An economy that is based on exploiting capital rather than living on interest (Ehrenfeld, 1996). Thus agriculture that does not maintain the soil resource or fisheries that deplete the stocks of reproducing fish would not be sustainable from this perspective.
- An economy that imports increasing amounts of food. This may or may not be desirable, but it hardly seems sustainable based on what must be traded for that food.
- An economy that has no fossil fuels or significant industrial production and cannot easily pay for them from domestic resources.
- An economy that is not increasing the efficiency with which it turns inputs (land and energy, and perhaps other inputs) into wealth.

3. Social criteria

- An economy that does not generate and maintain basic government services such as health care, roads, and education.
- An economy with a large and/or increasing discrepancy between the rich and the poor.
- An economy with little stability for its workers.

to the economy and the resources needed for their mobilization. Our principal tool here is empirical analysis of the land, energy, and other inputs required for a unit of production. This can be supplemented by explicit geographical considerations, for the possibilities of each resource are extremely dependent upon the rate of exploitation. If the rate of exploitation is low, then the best land can be used for that process. But if the rate of exploitation of that resource is high, then the efficiency, and hence profitability, of the process declines, and the possibility for genuine profit to the nation from that resource also declines. And in all development options, there is a critical issue of the large and increasing connection with the industrial world. Thus it is important to understand each resource from both an economic and a biophysical perspective, both over time and spatially. We think these issues may become more important in the not too far future. A recent high-profile and apparently rigorous analysis suggests that global petroleum production may peak by about 2008 (Campbell and Laherre, 1998). This analysis comes to essentially the exact same conclusion as Hubbert's analysis of 1968.

## ACKNOWLEDGMENTS

Important parts of this chapter were contributed by Dawn Montanye and Sabine O'Hara. We appreciate critical comments from Bernardo Aguilar and Andrea Baranzini. Parts of this chapter are modified from Hall (1991 and 1992).

## REFERENCES

Aguilar, B. 1997. *Paradigmas Económicos y Desarrollo Sostenible. La Economía al Servicio de la Conservación.* Editorial UNED, San Jose, Costa Rica.

Akarca, A. T., and T. V. Long. 1980. Dynamic modeling using advanced time series techniques: Energy GNP and energy-employment interactions. In F. S. Roberts (Ed.), *Energy Modeling,* Vol. 11, pp. 501–516. Institute of Gas Technology, Chicago, IL.

Atkinson, L. C. 1982. *Economics.* Irwin, Homewood, IL.

Ausubel, J. 1996. Can technology spare the earth? *American Scientist* 84:166–178.

Ayres, R. U., and I. Nair. 1984. Thermodynamics and economics. *Physics Today,* Nov., 62–71.

Ayres, R. U. 1995. Economic growth: Politically necessary but not environmentally friendly. *Ecol. Econ.* 15:97–99.

Ayres, R. U. 1996. Limits to the growth paradigm. *Ecol. Econ.* 19:117–134.

Bailey, M. N. 1982. Economic models under challenge. *Science* 216:859.

Banks, F. E. 1988. Truth and economics. *International Association of Energy Economics Newsletter* 1:9.

Barnett, H. J., and C. Morse. 1963. *Scarcity and Growth: The Economics of Natural Resources Availability.* Johns Hopkins University Press, Baltimore, MD.

Bello, W., D. Kinley, and E. Elinson. 1982. *Development Debacle: The World Bank in the Philippines.* Transnational Institute, San Francisco, CA.

Braidotti, R., E. Charkiewicz, S. Hausler, and S. Wieringa. 1994. *Women, the Environment and Sustainable Development.* Zed Books, London.

Brown, L. R. 1996. The acceleration of history. In *State of the World,* pp. 3–20. W. W. Norton, New York.

Campbell, C. J., and J. H. Laherrere. 1998. The end of cheap oil. *Scientific American* 278:78–83.

Canadian Broadcasting Corporation. 1985. *Cut and Run: The Assault on Canada's Forests.* CBC Transcripts, Montreal, Quebec, Canada.

Clark, C. 1976. *Mathematical Bioeconomics: The Optimal Management of Renewable Resources.* Wiley–Interscience, New York.

Cleveland, C., R. Costanza, C. Hall, and R. Kaufmann. 1984. Energy and the United States economy: A biophysical perspective. *Science* 225:890–897.

Cleveland, C. 1987. Biophysical economics: Historical perspective and current research trends. *Ecological Modelling* 38:47–74.

Cleveland, C. 1991. Natural resource scarcity and economic growth revisited: Economic and biophysical perspectives. In R. Costanza (Ed.), *Ecological Economics: The Science and Management of Sustainability*, pp. 289–317. Columbia University Press, New York.

Cleveland, C. J., and M. Ruth. 1997. When, where and by how much do biophysical limits constrain the economic process? A survey of Nicholas Georgescu-Roegen's contribution to ecology. *Ecological Economics* 22:203–224.

Costanza, R. 1980. Embodied energy and economic evaluation. *Science* 210:1219–1224.

Cottrell, F. 1955. *Energy and Society*. McGraw–Hill, New York. Reprint 1970, Greenwood Press, Westport, CT.

Daly, H. E. 1977. *Steady-State Economics*. W. H. Freeman, San Francisco, CA.

Daly, H. E. 1979. Energy, growth and the political economy of scarcity. In V. K. Smith (Ed.), *Scarcity and Growth Reconsidered*, pp. 67–94. Johns Hopkins University Press, Baltimore, MD.

Dohan, M. 1977. Economic values and natural ecosystems. In C. A. S. Hall and J. Day (Eds.), *Ecosystem Modeling in Theory and Practice*, pp. 134–171. Wiley–Interscience, New York.

Dung, T. H. 1992. Consumption, production, and technological progress: A unified entropic approach. *Ecological Economics* 6:95–210.

Ehrlich, P., A. Ehrlich, and J. P. Holdren. 1985. *Ecoscience: Population, Resource, Environment*. W. H. Freeman, San Francisco, CA.

Esty, D. C. 1994. *Greening the GATT: Trade, Environment, and the Future*. Institute for International Economics, Washington, DC.

Figueres-Olsen, J. M. 1996. Sustainable development: A new challenge for Costa Rica. *SAIS Review* 16:187–202.

Fisher, A. 1981. *Resources and Environmental Economics*. Cambridge University Press, Cambridge, UK.

Friedman, M. 1953. *Essays in Positive Economics*. University of Chicago Press, Chicago, IL.

Galbraith, J. K. 1973. *Economics and the Public Purpose*. Houghton Mifflin, New York.

Gell-Mann, M. 1994. *The Quark and the Jaguar. Adventures in the Simple and the Complex*. W. H. Freeman, New York.

Georgescu-Roegen, N. 1971. *The Entropy Law and the Economic Process*. Harvard University Press, Cambridge, MA.

Ghai, D. 1994. *Development and Environment: Sustaining People and Nature*. Blackwell, Cambridge, UK.

Gillis, M., D. H. Perkins, M. Roemer, and D. R. Snodgrass. 1996. *Economics of Development*, 4th ed. W. W. Norton, New York.

Goldemberg, J. 1995. Energy needs in developing countries and sustainability. *Science* 269:1058–1059.

Goodland, R., and G. Ledec. 1987. Neoclassical economics and the principle of sustainable development. *Ecol. Modelling* 38:19–46.

Goodland, R., and H. Daly. 1996. Environmental sustainability: Universal and non-negotiable. *Ecological Applications* 6:1002–1017.

Gowdy, J., and S. O'Hara. 1996. *Economic Theory for Environmentalists*. St. Lucie Press, Baton Racon, Florida.

Hall, C. A. S. 1975. The biosphere, the industriosphere and their interactions. *Bulletin of Atomic Scientists* 31:11–21.

Hall, C. A. S., and J. Day. 1977. *Ecosystem Modeling in Theory and Practice*. Wiley–Interscience, New York. Reprinted 1990, University Press of Colorado, Boulder, CO.

Hall, C. A. S., M. Lavine, and J. Sloane. 1979. Efficiency of energy delivery systems. 1. An economic and energy analysis. *Environmental Management* 3:493–504.

Hall, C. A. S., C. J. Cleveland, and R. K. Kaufmann. 1986. *Energy and Resource Quality: The Ecology of the Economic Process*. Wiley– Interscience, New York. Reprinted 1992, University Press of Colorado, Boulder, CO.

Hall, C. A. S. 1988. An assessment of several of the historically most influential theoretical models in ecology and of the data provided in their support. *Ecological Modelling* 43:5–31.

Hall, C. A. S. 1990. Sanctioning resource depletion: Economic development and neoclassical economies. *Ecologist* 20:99–104.

Hall, C. A. S. 1991. An idiosyncratic assessment of the role of mathematical models in environmental sciences. *Environment International* 17:507–517.

Hall, C. A. S. 1992. Economic development or developing economics: What are our priorities? In M. Wali (Ed.), *Ecosystem Rehabilitation,* Vol. I. *Policy Issues,* pp. 101–126. SPB Publishing, The Hague, The Netherlands.

Hall, C. A. S., and M. Hall. 1993. The efficiency of land and energy use in tropical economies and agriculture. *Agriculture, Ecosystems and Environment* 46:1–30.

Hall, C. A. S., R. G. Pontius, J. Y. Ko, and L. Coleman. 1994. The environmental impact of having a baby in the United States. *Population and Environment* 15:505–524.

Hall, C. A. S., H. Tian, Y. Qi, G. Pontius, J. Cornell, and J. Uhlig. 1995. Spatially explicit models of land use change and their application to tropics. *DOE Research Summary* 31:1–4. [Oak Ridge National Laboratory]

Hall, C. A. S., R. Kuemmel, T. Kroeger, D. Lindenberger, and W. Eichorn. The need to bring natural sciences back into economics (manuscript).

Hall, D. C., and J. V. Hall. 1984. Concepts and measures of natural resource scarcity with a summary of recent trends. *Journal of Environmental Economics and Management* 11:363–379.

Hannon, B. 1982. Analysis of the energy costs of economic activities: 1963–2000. *Energy Systems Policy Journal* 6:249–278.

Heilbroner, R. L., and L. C. Thurow. 1981. *The Economic Problem.* Prentice-Hall, Englewood Cliffs, NJ.

Hildyard, N. 1989. Adios Amazonia? A report from the Altimira gathering. *Ecologist* 19:53–62.

Hill, P. 1986. *Development Economics on Trial.* Cambridge University Press, Cambridge.

Hunter, J. R. 1994. Is Costa Rica truly conservation-minded? *Conservation Biology* 8:592–595.

Johnson, P. M., and A. Beaulieu. 1996. *The Environment and NAFTA: Understanding and Implementing the New Continental Law.* Island Press, Washington, DC.

Kaufmann, R. K. 1987. Biophysical and Marxist economics: Learning from each other. *Ecological Modelling* 38:91–106.

Kaufmann, R. K. 1988. Higher oil prices: Can OPEC raise prices by cutting production? Ph.D. dissertation, University of Pennsylvania, Philadelphia, PA.

Kaufmann, R. K. 1992. A biophysical analysis of the energy/real GDP ratio: Implications for substitution and technical change. *Ecological Economics* 6:35–36.

Knetsch, J. L., and J. A. Sinden. 1984. Willingness to pay and compensation demand: Experimental evidence of an unexpected disparity in measures of value. *Quarterly Journal of Economics* 99:507–521.

Ko, J-Y., C. A. S. Hall, and L. G. Lopez Lemus. 1998. Resource use rates and efficiency as indicators of regional sustainability: An examination of five countries. *Environmental Monitoring and Assessment* 51:571–593.

Lekachman, R. 1986. *Economists at Bay: Why the Experts Will Never Solve Your Problems.* McGraw–Hill, New York.

Lelé, S. M. 1991. Sustainable development: A critical review. *World Development* 19:607–621.

Lelé, S., and R. Norgaard. 1996. Sustainability and the scientist's burden. *Conservation Biology* 10:354–365.

Leontief, W. 1982. Academic economics. *Science* 217:104–107.

Makgetla, N. S., and R. B. Seidman. 1989. The applicability of law and economics to policy making in the third world. *Journal of Economic Issues* 23:35–78.

Malthus, T. R. 1798/1961. *An Essay on the Principle of Population.* J. M. Dent and Sons, London.

Martinez-Alier, J., and K. Schlupmann. 1987. *Ecological Economics: Energy, Environment and Society*. Basil Blackwell, Oxford, UK.
Meier, G. M. 1994. *Leading Issues in Economic Development*. Oxford University Press, Oxford.
Merchant, C. 1983. *The Death of Nature. Women, Ecology and the Scientific Revolution*. Harper & Row, San Francisco. 348 pp.
Mies, M., and V. Shiva. 1993. *Ecofeminism*. Fernwood Publications, Halifax.
Milton, J. 1968. The careless technology. Ecological aspects of international development. In *Proceedings, Conference at Arlie House, Warrenton, Virginia, USA*.
Mirowski, P. 1988. *More Heat Than Light*. Cambridge University Press, London.
Montayne, D. R. 1998. M.S. thesis, College of Environmental Science and Forestry, State University of New York, Syracuse, NY.
Morgan, T. 1988. Theory vs. empiricism in academic economics. *Journal of Economic Perspectives* 7:159–164.
Myrnick, W. H. 1982. *The Illusions of Contemporary Economics*. West Virginia University Press, Morgantown, WV.
Nelson, R.H. 1997. In memorium: On the death of the "Market Mechanism." Ecol. Econ. **20**, 187-198.
Nilsson, L. J. 1993. Energy intensity trends in 31 industrial and developing countries 1950–1988. *Energy* 18:309–322.
Nordhaus, W., and J. Tobin. 1972. Is growth obsolete? In M. Moss (Ed.), *The Measurement of Economic and Social Performance*, Vol. 5, pp. 10–11. National Bureau of Economic Research, Columbia University Press, New York.
O'Connor, M. 1994. *Is Capitalism Sustainable? Political Economy and the Politics of Ecology*. Guilford Press, New York.
Odum, H. T. 1983. *Systems Ecology, an Introduction*. Wiley–Interscience, New York. Reprinted 1994 by the University Press of Colorado, Niwot, CO.
Odum, H. T. 1995. Self organization and maximum empower. In C. Hall (Ed.), *Maximum Power. The Ideas and Applications of H. T. Odum*, pp. 311–329. University Press of Colorado, Niwot, CO.
Odum, H. T., and E. C. Odum. 1976. *Energy Basis for Man and Nature*. McGraw–Hill, New York.
O'Hara, S. 1995. From production to sustainability—Considering the whole household. *Journal of Consumer Policy* 18(4):1–24.
O'Hara, S. 1996. Discursive ethics in ecosystems valuation and environmental policy. *Ecological Economics* 16:95–107.
Ostwald, W. 1907. Modern theory of energetics. *Monist* 17:480–515.
Paruelo, J. M., P. Aphalo, C. A. S. Hall, and D. Gibson. 1987. Energy use and economic output for Argentina. In G. Pillet (Ed.), *Environmental Economics: The Analysis of a Major Interface*, pp. 169–184. Leimgruber, Geneva, Switzerland.
Peet, J. 1992. *Energy and the Ecological Economics of Sustainability*. Island Press, Washington, DC.
Pimentel, D., *et al.* 1995. Environmental and economic costs of soil erosion and conservation benefits. *Science* 267:1117–1123.
Pimentel, D., and M. Pimentel. 1996. *Food, Energy and Society*, rev. ed. University Press of Colorado, Niwot, CO.
Preston, P. W. 1996. *Development Theory*. Blackwell, Cambridge.
Prugh, T. R. Costanza, J. H. Cumberland, H. Daly, R. Goodland, and R. B. Norgaard, 1995. Natural Capital and human economic survival. International Society for Ecological Economics. Solomons, MD. 198 pp.
Quesada-Mateo, C., and V. Solis-Rivera. 1990. Costa Rica's national strategy for sustainable development: A summary. *Futures* 22:396–416.
Reed, D. 1996. Structural adjustment, the Environment , and sustainable development. Earthscan Publications, London.

Repetto, R. 1988. *The Forest for the Trees? Government Policies and the Misuse of Forest Resource.* World Resources Institute, Washington, DC.

Roberts, P. C. 1982. Energy and value. *Energy Policy* 10:171–180.

Samuelson, P. 1976. *Economics*, 10th ed. McGraw–Hill, New York.

Schoemaker, P. J. H. 1982. The expect utility model: Its variance, purposes, evidence and limitations. *Journal of Economic Literature* 20:529–563.

Schmid, A. A. 1987. *Property, Power and Public Choice: An Inquiry into Law and Economics*, 2nd ed. Praeger Publishers, New York.

Simon, J., and H. Kahn (Eds.). 1984. *The Resourceful Earth.* Blackwell, New York.

Slade, M. E. 1982. Trends in natural resource commodity prices: An analysis of the time domain. *Journal of Environmental Economics and Management* 9:122–137.

Smith, A. 1776. An inquiry into the nature and causes of the Wealth of Nations.

Smith, V. K. (Ed.). 1979. *Scarcity and Growth Reconsidered.* Johns Hopkins University Press, Baltimore.

Smith, V. K. 1989. Theory, experiment and economics. *Journal of Economic Perspectives* 3:151–169.

Soddy, F. 1926. *Wealth, Virtual Wealth and Debt.* E. P. Dutton, New York.

Solorzano, R. R., R. de Camino, R. Woodward, J. Tosi, V. Watson, A. Vasquez, C. Villaobos, J. J. Jimenez, R. Repetto, and W. Cruz. 1991. *Accounts Overdue: Natural Resource Deletion in Costa Rica.* World Resources Institute, Washington, DC.

Solow, R. M. 1974. The economics of resources or the resources of economics. *American Economics Review* 64:1–14.

Solow, R. M. 1997. Reply. *Ecological Economics* 22:267–268.

Stewart, I. 1994. *Does God Play Dice? The Mathematics of Chaos.* Blackwell, Cambridge, MA. 348 pp.

Stiglitz, 1997. Georgescu-Roegen versus Solow/Stiglitz. *Ecological Economics* 22:269–270.

Taylor, L. 1993. The World Bank and the environment: The World Development Report 1992. *World Development* 21:869–881.

Todaro, M. P. 1994. *Economic Development*, 5th ed. Longman, New York.

U.S. Congress. House. Committee on Banking, Finance and Urban Affairs. 1994. *Oversight of the International Monetary Fund and the World Bank.* 103rd Cong., 2nd sess., 1994.

*Wall Street Journal.* 1984. Oil's Legacy. Oct. 22, p. 1.

World Bank. 1996. *World Development Report 1996. From Plan to Market.* Oxford University Press, New York.

World Commission on Environment and Development. 1987. *Our Common Future.* Oxford University Press, New York.

CHAPTER 4

# Data on Sustainability in Costa Rica

## Time Series Analysis of Population, Land Use, Economics, Energy, and Efficiency

CHARLES A. S. HALL, JUAN-RAPHAEL VARGAS, OLEGARIO SAENZ, WILL RAVENSCROFT, AND JAE-YOUNG KO



## I. INTRODUCTION

This chapter presents what we think are the basic data that are needed to understand the recent history of the Costa Rican economy. It also serves as a basis for attempting to understand the relation of the Costa Rican economy to concepts of sustainability. The data include information on human population levels and distribution, how land, especially agricultural land, is used, and various estimates of economic activity, including trade, and of energy use. Although many of these data are used and expanded upon in other chapters we think it important to include them here in one place as an introduction so that various trends and comparisons can be analyzed more explicitly.

The most important economic factor for most people is their personal purchasing power, including those government expenditures made in their behalf for roads, schools, hospitals, etc. There are many factors that contribute directly to this purchasing power: the national rate of wealth production, how that wealth is distributed, the price of the goods and services purchased (and hence the rate of inflation), and the relation of one country's monetary units to those of others. There are a series of links between the biophysical aspects of a nation and these economic factors. These include access to, and the price of, energy supplies, the quality and quantity of land and other resources, and the cost of their exploitation.

Wealth production is best measured on a per capita basis. For example, Iceland and Senegal have approximately the same gross domestic product (GDP) as Costa Rica, but Iceland is considered a wealthier country and Senegal a poorer one. In 1995 the populations of the three countries were 0.27 million for Iceland, 3.42 million for Costa Rica, and 8.31 million for Senegal (WRI, 1996). Thus per capita wealth in Iceland is more than 10 times that of Costa Rica, and Senegal's is half. Thus realized, economic wealth is mainly a reflection of a country's total economic activity as it is divided among its populace. For this reason it is necessary to examine the past, present, and future population levels of Costa Rica when considering wealth availability and production.

We also focus on energy, which is essential for our biophysical analysis. The most important energy source for Costa Rica is the sun. Solar energy runs the hydrologic cycle as well as both natural and agricultural ecosystems. Solar input is approximately constant from one year to the next, although it varies a great deal regionally. Solar energy is often thought of as free, but in reality its contribution to the economy is a function of the amount and availability of land, the market cost of land, the depth and quality of the soil, access to financing, rainfall patterns, and the type of ecosystem found at the area under study. Most of the food that is necessary to feed humans—and which is a major energy input to, and component of, the economy—is derived from land-based solar-powered agricultural activities. The solar energy falling on forested land enters the economic system indirectly as ecosystem services, such as flood protection or the provision of clean water, and as products such as sawlogs or firewood (Hall, 1975; Daily, 1997).

A more complete energy flow picture for Costa Rica would include the energy of geologic uplift that creates mountains, the energy of the winds that bring moisture to those mountains, and the energy of tides that wash the shores. Chapter 25 provides a more comprehensive energy analysis that includes all these flows.

We focus on the direct contribution of energies, including solar energy, "traditional energy" (biomass, etc.), fossil fuels, and hydropower, to the economic output as measured by GDP. Solar energy falling on the land area becomes embodied in crops, meat, timber, and other major components of the economy. Traditional energy is another important input to some segments of the economy, even though in general these two types of energy are not counted by economists in the GDP calculation, whereas, e.g., electricity production is, because these traditional energies are mainly not traded in markets. Biomass fuels, together with hydroelectric power, might be considered sustainable, although hydroelectric reservoirs tend to silt up over time and the harvest of biomass often leads to soil erosion. Since Costa Rica has no fossil fuels of its own, nor a nuclear industry, essentially all industrial energy used is either imported oil, biomass, or hydropower.

## II. METHODS

In Chapter 3 we defined economic growth as an increase in the generation of economic output as imperfectly measured in physical units (e.g., tons) or inflation-corrected dollars of GDP. We defined sustainability as not degrading or depleting the resource base that makes that economic process possible. We defined efficiency as output over input. This chapter examines time trends in economic and energy activity, and efficiencies derived from their ratios, using several published and presumably accurate international indices of economic activity[1] as well as numbers derived each year by one of our authors, Juan Vargas, an economist at the University of Costa Rica. Vargas's numbers are comprehensive and normally considered "official" for the country by national agencies and publications. We also used similar data for Costa Rica and four other countries to compare some aspects of national economies, including sustainability and efficiency (Ko *et al.*, 1998).

### A. Sources of Data: Economic Variables

To determine the *efficiency* of various processes we divided several measures of economic output by the energy (or other resource) input for the same activity or sector of the economy. For a measure of total national economic

[1] A note on the accuracy of data. All numbers used in science are inherently suspect. We have attempted to use generally accepted international values for the data we present here, as well as official data from Costa Rica itself. Nevertheless there has been considerable concern expressed about the quality of the data used in these analyses, including concern expressed by Costa Rican nationals. One person who had such reservations was such a national who was a graduate student of C.H. at the time we were generating these analyses. She felt, often with some justification, that the developed world had an inaccurate perception of Costa Rica, and frequently put its own political spin on the data used. Challenged to find errors in the data that we were using, she in fact found that her own data, gathered from Costa Rican sources, showed as much as a 10% error between two different estimates of a particular year's data point for some important variables. But the interesting thing was that the different data sets were not consistently in disagreement, but tended to "circle about" each other, so that although in any one year disagreements existed between data sources, there were not important differences that lasted over more than a year or two. The FAO data seemed a little smoother than the student's Costa Rican sources, whatever that meant. We found no reason not to use either data set, and use FAO and UN data for demographic, agricultural, energy, and land use data, and Vargas's economic data.

activity we used the GDP. Gross domestic product is similar to the more familiar gross national product (GNP) except that it does not include profits sent to, or received from, Costa Ricans living abroad. GNP measures the goods and services produced by a country's citizens for a year, irrespective of what nation they live in. GDP, more appropriate here, is the goods and services produced within the country, whether the profits went to Costa Ricans or to others. We acknowledge that GDP is a useful and standardized, but extremely incomplete, measure of material wealth, and that material wealth by any definition is a very incomplete measure of human well-being. These issues are considered in the Discussion and in Hall *et al.* (1986). Finally, even GDP per capita says nothing about the distribution of wealth among individuals or about the public vs the private sector (i.e., more schools and better roads vs more personal automobiles). These important issues are discussed in other chapters.

Having chosen GDP, of necessity, for our index of economic output, there is still confusion when selecting from among the many indicators available from various sources, each purporting to represent the rate at which the nation generated wealth. All are in constant (inflation-adjusted) currency units. The first index is GDP given in constant 1985 colónes. This reflects purchasing power for Costa Rican-produced goods and services (Figure 4-1a). The second, GDP in constant 1985 U.S. dollars, uses an internationally accepted currency exchange rate to convert colónes to U.S. currency (hereafter US$) (Figure 4-1b). This index reflects the purchasing power of Costa Rican money in terms of its ability to purchase foreign products. Two estimates are available here for the US$-based constant GDP: Vargas's and the World Bank's. Both are expressed in 1985 US$. The GDP based on "purchasing power parity" (PPP) (expressed in 1985 international dollars) is from a web site of the Penn World Tables (http://cansim.epas.utoronto.ca:5680/pwt). The purchasing power parity approach equalizes the purchasing powers of different currencies by estimating the number of units of a country's currency that would be required to buy the same amount of goods and services in the domestic market as $1 would buy in the "average" country (WRI, 1996, p. 171). As of February 1998, 1985 is the base year for the PPP-based economic data. We converted the data of Vargas and the World Bank into 1985-based data to compare the three different sources within the same year base. Vargas's original 1970 US$-based constant GDP is adjusted into values of 1985 US$ using the converter available from the web site of the consumer price index (http://www.orst.edu/dept/pol-sci/sahr/cip96.htm). The original current US$-based GDP of the World Bank is converted into the 1985 constant US$-based GDP using the same consumer price index.

The PPP approach is thought to better reflect purchasing power in practice, as many goods and services are much cheaper in Costa Rica than in the United States, and Costa Ricans buy mostly Costa Rican goods and services. Thus,

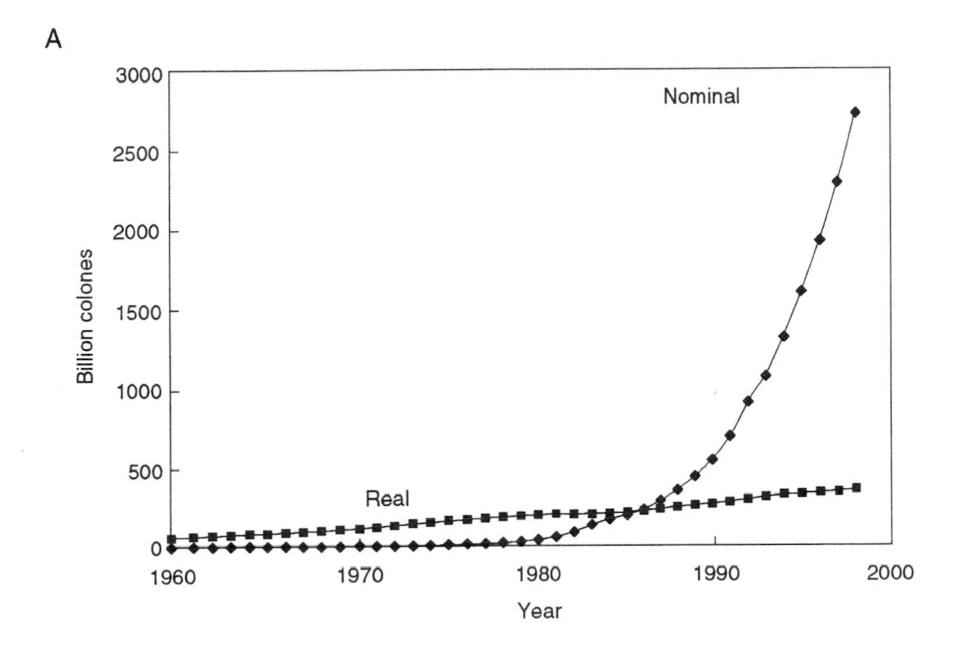


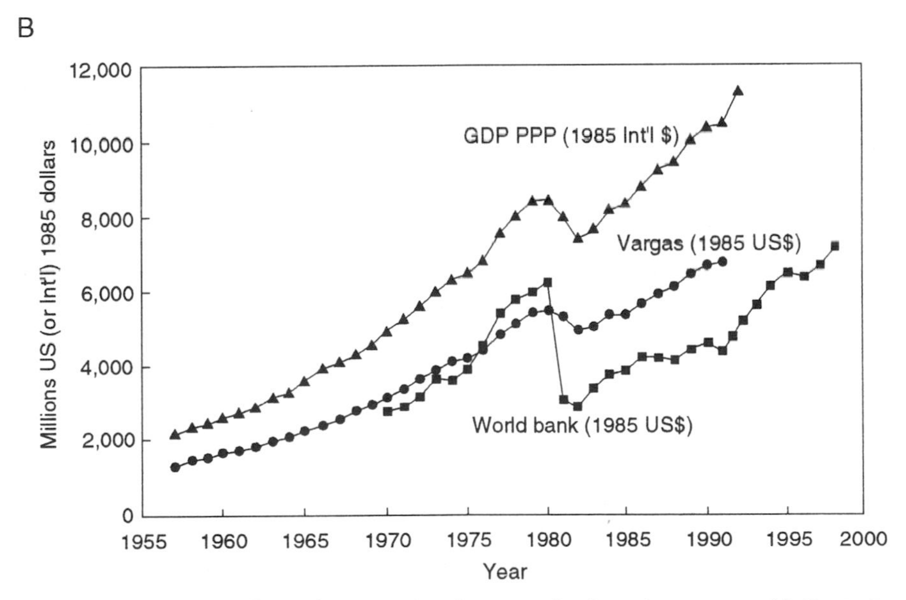


FIGURE 4-1 Measures of GDP for Costa Rica from several independent sources. (a) Nominal and constant (or real) GDP in 1985 colónes, 1957–1992. (b) Three different estimates of constant (or real) GDP, 1957–1992. The Penn World Tables data (top) indicate internal purchasing power, and the World Bank data external. 1998 GDP is 339 billion colónes, and World Bank values are 7279 in 1985 dollars.

the PPP-based GDP is higher than the other US$-based constant GDPs. Nevertheless it is unsettling that there is such a large difference among different estimates of nominally the same thing.

## B. Energy Variables

We used energy data as reported by the United Nations and which is available from the World Resources Institute (WRI, 1996). This energy is given in, or converted to, petajoules (PJ) for national-scale energy consumption (1 PJ = $10^{15}$ J = $10^6$ GJ = $10^3$ TJ; 1 kcal = 4186.8 J). Gigajoules are used for per capita consumption. One terajoule equals the energy contained in 160 barrels of oil, and is a measure of the ability of that unit of energy to do work, explicitly to heat water. It has no correction for quality, that is, the ability of different types of energy to do economic work. The "traditional" energy category is a United Nations' estimate of energy consumption by families and small consumers, mainly in the forms of firewood and charcoal, but also including agricultural wastes such as bagasse (from sugar cane) that are not accounted for in the other categories. Many of these fuels are not sold in markets, but in Costa Rica they contribute greatly to economic production, for example, on farms and in sugar mills.

For many analyses we make a correction for energy quality, reflecting the fact that not all forms of energy are equally useful to an economy per heat unit. This correction is most important for electricity produced by hydroelectric facilities. We adjust electric energy for quality by a factor of about three (Cleveland *et al.*, 1984; Hall *et al.*, 1986; Hall, 1992), except as indicated, to represent electricity's greater utility per heat unit for most economic activities. Darmstadter *et al.* (1971) argue against an across the board multiplier for assessing the contribution of electric energy to the economy, but Nilsson (1993) believes that a multiplier of about 2.6 is essential for making adequate comparisons among countries or over time. In most other national economies, where electricity is generated principally through the burning of fossil fuel at an efficiency from 30 to 40%, the use of such a conversion factor is unambiguous. In Costa Rica, where electricity is generated principally from hydropower, the use of such a conversion may be less defensible, although still important in determining the causes of any changes in efficiency, i.e., what portion of any improved efficiency might be due simply to using a fuel with more economic potency per heat unit. Since there is no unequivocal solution to this issue we present our results using both approaches.

All graphs were created using Microsoft Excel, version 7.00. The sources of the data are listed in the data bibliography and are available as Microsoft Excel worksheets from Dr. Ko.

# III. RESULTS

## A. Economic Output

Our first analyses show that when using the different GDP estimates from various sources for Costa Rica, one arrives at very different results even when presumably measuring the same information (Figure 4-1b). Since we had no criteria by which to choose one data set over another, nor could we find any explanation as to why these various indexes do not give the same results, we use mainly Vargas's official data set, believing that Costa Rican statisticians and economists should be more capable of defining their own economy than outsiders.

Our main results show that the Costa Rican economy grew fairly rapidly over the time period of our analysis as measured by GDP (Figures 4-1a,b), but the trends of GDP per capita show very different results. Figure 4-2a illustrates the rapid growth of current GDP per capita. After the current values are converted into constant 1985 colónes using Vargas's "GDP implied index," the per capita GDP remains very nearly constant, although with some ups and downs (Figures 4-2a,b). For example, GDP per capita measured in current colónes seems to increase rapidly, from 18,349 in 1980 to 280,145 in 1992. When corrected for internal inflation, however, the GDP per capita is essentially constant, from 86,905 to 86,679 in 1985 colónes over the same period (Figure 4-2a). There is virtually no change in real purchasing power. The very different growth trends between total GDP and GDP per capita are driven by the population growth in Costa Rica. The economy grew approximately in proportion to human population numbers over time.

Figure 4-2b shows the purchasing power for Costa Rican people for international goods and services. The inflation- and exchange rate-corrected GDP per capita is prepared by converting the previous 1985 constant colónes-based GDP per capita into the 1985 US$ one using the exchange rate of colónes against US$ for 1985 as a measurement of its "foreign exchange" or "external purchasing power" value. The GDP per capita reached its peak of U.S. $1625 in 1978. Overall the constant US$-based GDP per capita has been constant, which means that per capita external purchasing power has remained approximately the same during this time. The PPP-based GDP per capita, reflecting relatively more internal purchasing power, shows a higher level of GDP per capita than the US$-based indicator.

The gap between the PPP-based GDP per capita and the US$-based GDP per capita has been increasing, which means that the Costa Rican economy's external purchasing power has been decreased relative to internal purchasing power. In other words, Costa Rican people have been experiencing increased

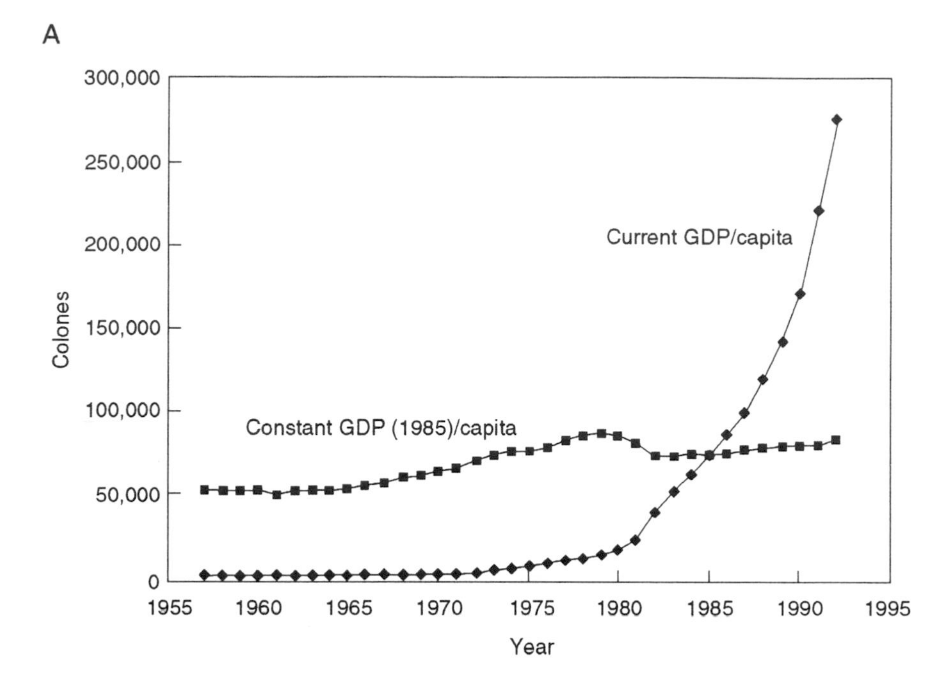


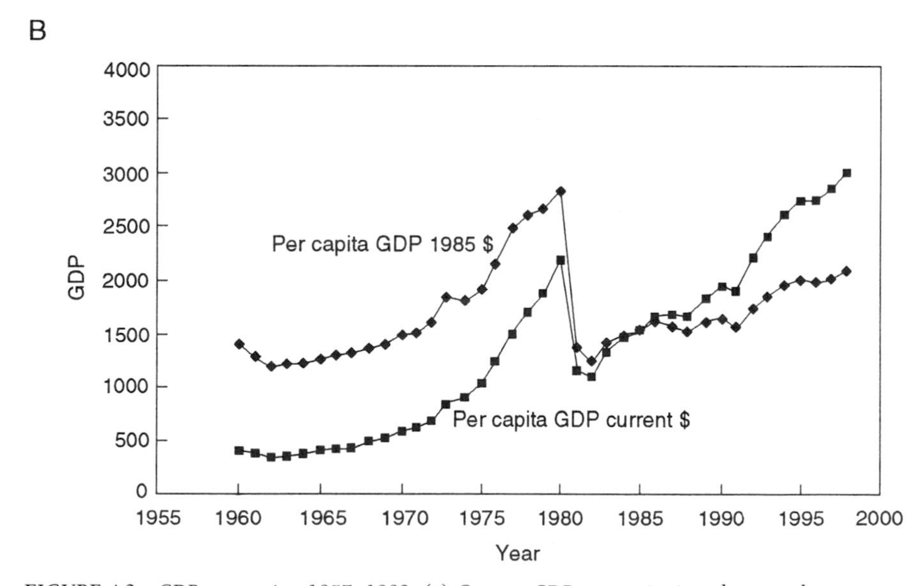


**FIGURE 4-2** GDP per capita, 1957–1992. (a) Current GDP per capita in colónes and constant GDP per capita in 1985 colónes. (b) Two estimates for constant GDP per capita, 1957–1992. Per capita GDP was about 20 percent higher in 1998.

financial challenges since 1978 when purchasing imported goods. All of these figures are complicated by the increases in debt during this period, as will be covered later.

## B. Population

There have been four distinct periods of growth in population during the last 50 years. During the first half of the century, the country had a moderately low (1.5%) population growth rate with both a high birth rate and a high mortality rate. The population of Costa Rica in 1950 was 800,875 people, one-quarter of the 1995 population.

However, since the 1950s, the population has increased dramatically (Figure 4-3a). Between 1950 and 1960 Costa Rica had a population explosion due to improved health standards and a continued high birth rate (Holl *et al.*, 1993). Costa Rica socialized its medical services in 1949 and made large investments in a potable water supply, improving health conditions in the rural areas dramatically. The result was a decrease in infant mortality and an increase in longevity (Figure 4-3b). The 3.7% rate of population growth from 1955 to 1960 was one of the highest in the world—at this rate the population would double every 19 years. As a result, Costa Rica had a population of 1,871,780 by 1975.

The decade of the 1970s brought a sharp decrease in the rate of population growth, in part due to improvements in the education system, an increase in quality of life, and a nationwide family planning program. The average number of children per family decreased from 7.3 in 1960 to 3.5 in 1985. Nevertheless the population continued to increase (Holl *et al.*, 1993). A new factor is a large influx of Nicaraguans seeking the better economic conditions in Costa Rica.

The present rate of population growth is somewhere between 2.1 and 2.6%, and may be decreasing, but we will not know for sure until the next census. The 1995 population was estimated to be 3.4 million, which translates to a population density of 67.1 inhabitants per square kilometer, much higher than that of the United States, whose population density was 27.5 for the same area (WRI, 1996). Approximately one-third of the population is concentrated in the Central Valley.

A United Nations' projection shows the growth rate decreasing to 1.27% by the 2020–2025 period and predicts a population of 5.6 million people by the year 2025, double that of 1988 (Figure 4-3a). The rural population is projected to peak in 2010 and then decline slightly. Population increases are projected to occur almost entirely in the urban environment (Figure 4-3a).

The Inter-American Development Bank (1993) shows that for the year 1990 Costa Rica had a young population, 36.4% under the age of 14, a moderate fertility rate of three children per woman, and a birth rate of about 2.6%. In 1950, "prior to development," the population distribution was relatively skewed

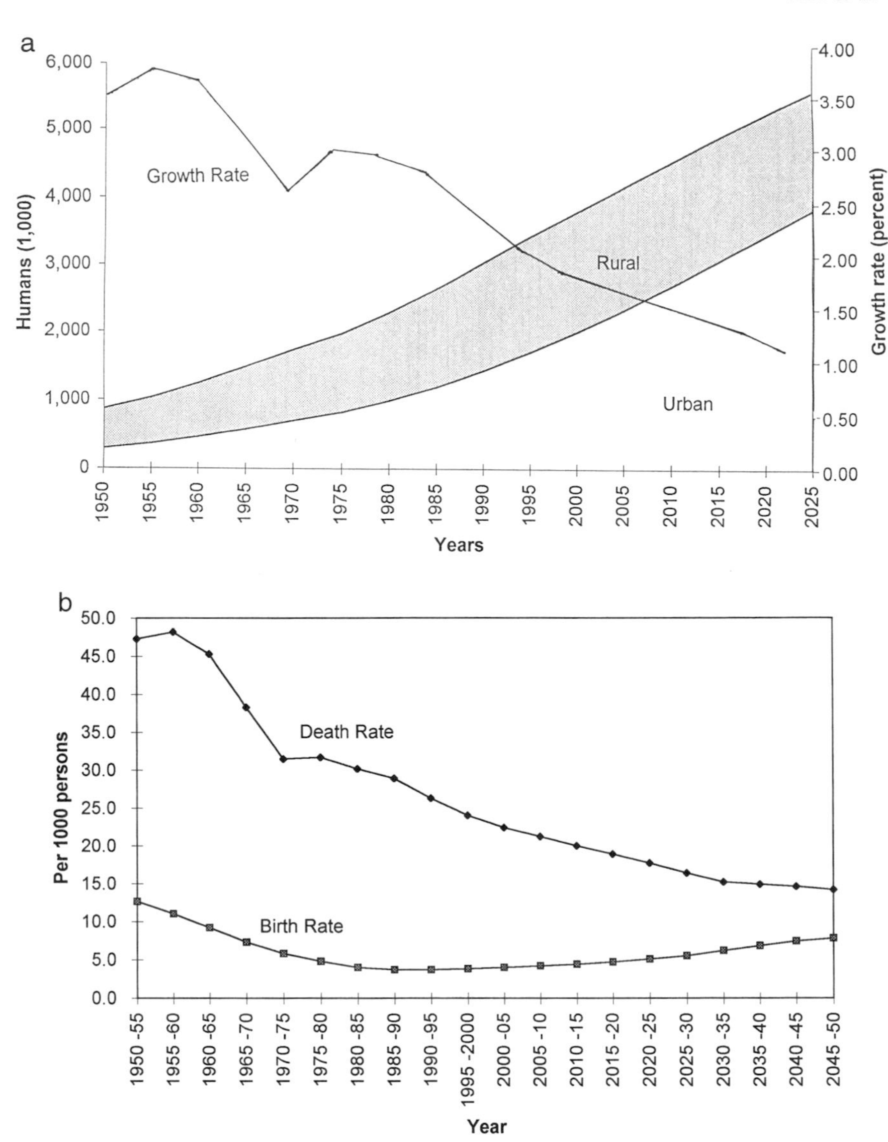


FIGURE 4-3 (a) Total population, population distribution, and growth rate. (b) Crude birth and death rates. (c) Age group distribution, 1950–1990. (Data from UN and IA Bank data.)

toward younger ages with 43.3% of the population younger than 15 years old. By 1970 that number was 36.2%, indicating increased longevity. The 1990 distribution is similar to that of 1950, and shows that the country still has a very young population (Figure 4-3c). Although the number of children born

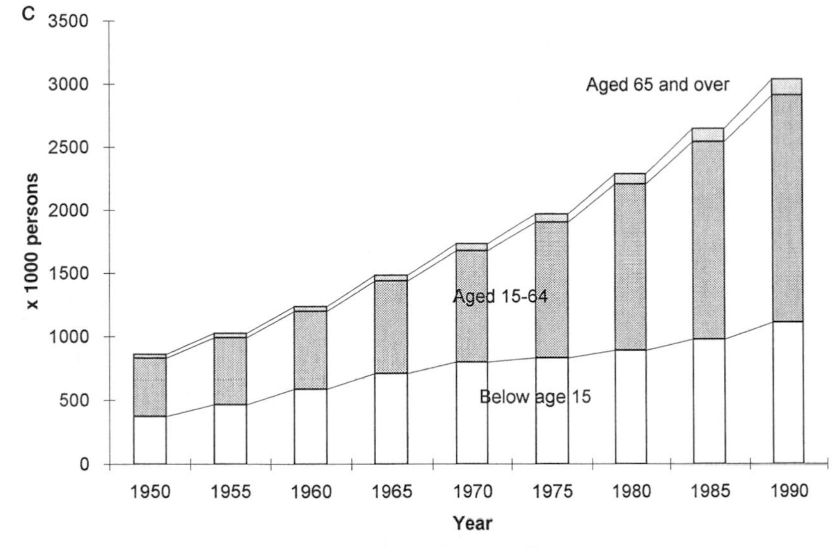


FIGURE 4-3 (*Continued*)

per family has been decreasing, the "bottom-heavy" population means that it will continue to increase for some time even if each woman has only two children. Thus it is difficult to make exact predictions until the reproductive behavior of the youngest people becomes apparent—something that we will not know for one or two decades!

There are more males born than females, but the death rate for males is higher. Thus there are more males in the early ages, no significant difference in middle ages, and fewer males in older ages. The tendency for longer life for women is similar to trends for the United States (Piel, 1992). Costa Ricans have an exceptionally long life span compared to most other nations, and especially so for a nation that is not wealthy.

## C. Land Use

Forested land in Costa Rica decreased from 3.240 million ha in 1961 to 1.638 million in 1982, a 50% decrease and a deforestation rate of 3% per year (Figure 4-4). The amount of land used for permanent and annual crops and for other uses (i.e., land used by cities and other abandoned and "unused" land) has remained almost constant since 1961. There are many reasons for deforestation (Cornell and Hall, 1994), but in Costa Rica the end result has been a net conversion of forest to pasture (see Chapters 14 and 16). Total developed land

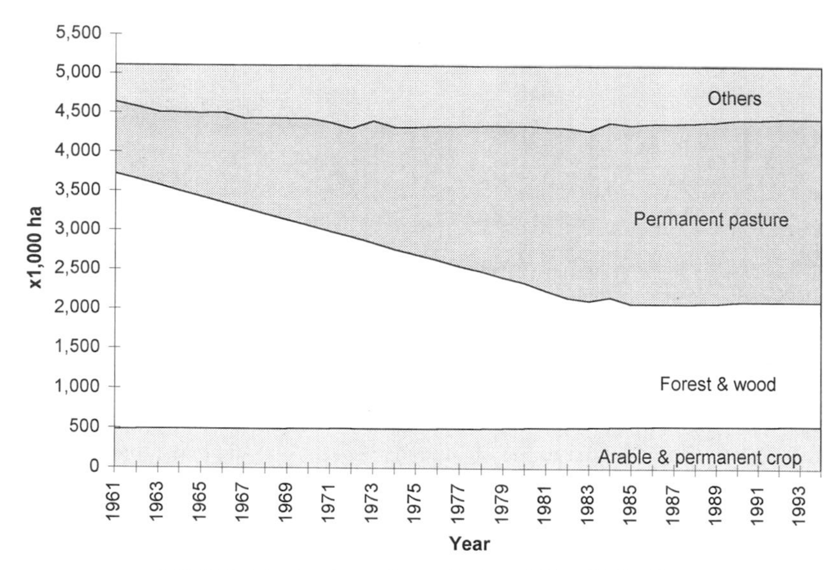


FIGURE 4-4 Land use change in Costa Rica, 1961–1994.

has increased at about the same rate as the human population, about 3% per year.

## D. Energy Inputs, Economic Outputs, and Efficiency

Solar input is assumed to be approximately constant and is estimated as roughly $190.1 \times 10^3$ PJ for the entire nation as determined by multiplying the total area of the country by the solar energy absorption constant of the tropics (Odum and Arding, 1991) (Figures 4-5a and 9-13). Figure 4-5a shows an assessment of the relative magnitude of the energy contributions to the Costa Rican economy, including solar input. Figure 4-5b shows the same data but with a quality correction. We weighted hydroelectricity by 3.0 relative to oil, and biomass 0.5 relative to oil, to give an approximate assessment of the total quality-corrected energy consumption. The quality of solar input relative to oil is corrected using a factor of 1/53,000 (Odum and Arding, 1991).

Dependence on fossil fuel has tended to increase over time, and the portion of electricity generated from hydropower has also increased. Total energy consumption per capita fluctuated between 25 and 31 GJ until 1985, when it began to increase somewhat (Figure 4-5c).

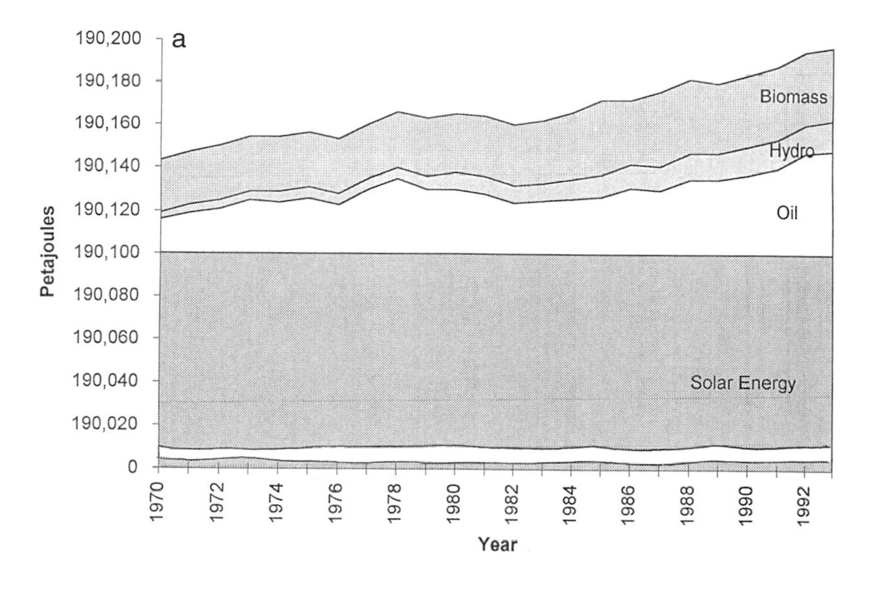


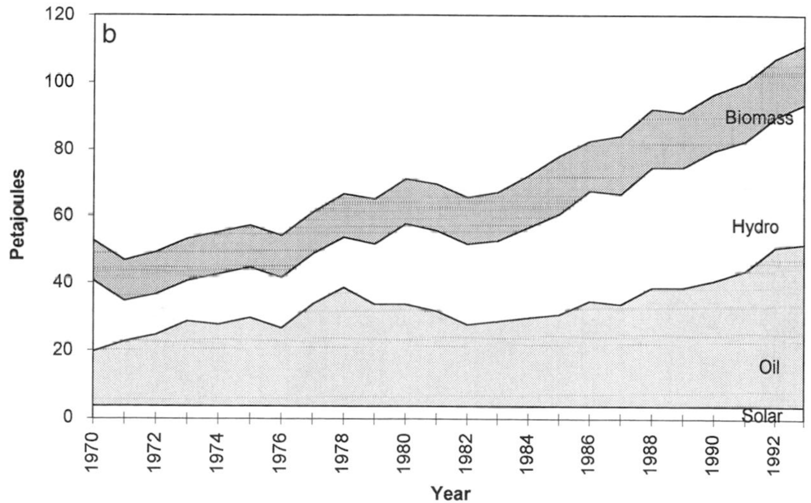


FIGURE 4-5 (a) Energy use of the Costa Rican economy, including solar energy. (b) Quality-adjusted energy use of the Costa Rican economy, including solar energy. (c) Per capita quality-adjusted energy consumption. (d) Energy use vs GDP, 1970–1991 (quality-adjusted total energy consumption with constant GDP, in 1985 US$). (e) Energy efficiency for internal purchases (top) and external purchases (bottom) to per capita GDP, 1970–1992.

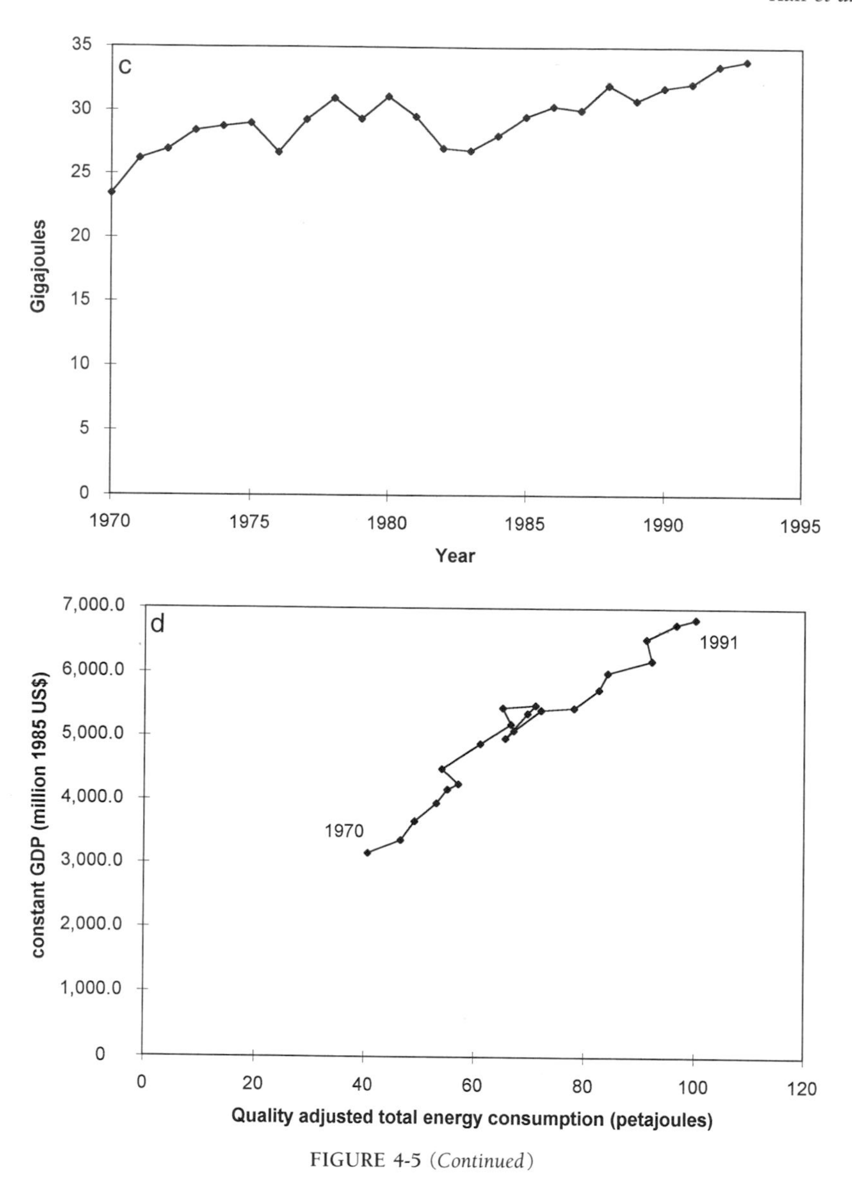


FIGURE 4-5 (*Continued*)

Economic output has increased approximately in proportion to energy use (Figure 4-5d). Whether the energy efficiency of the Costa Rican economy has changed much over the last 30 years depends upon which variables one chooses to use (Figure 4-5e). If one uses external purchasing power, which is based

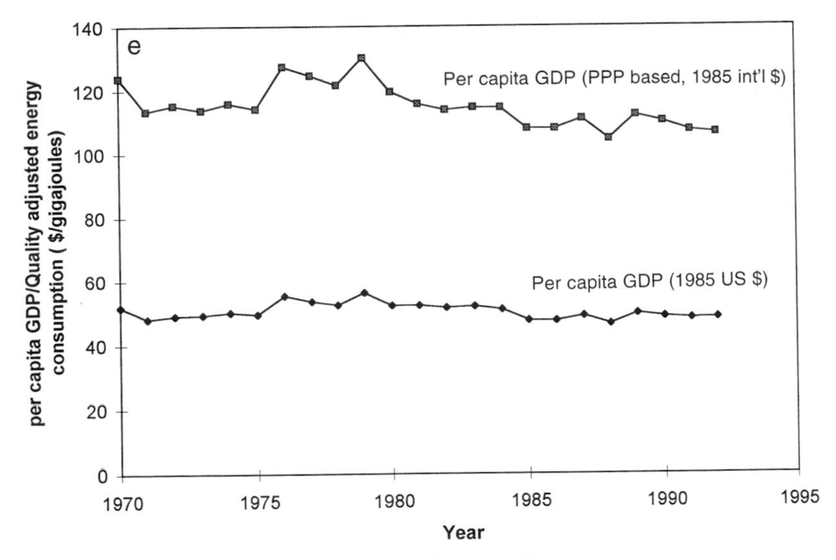


FIGURE 4-5 (*Continued*)

on US$, there is little change. But if one uses internal purchasing power, which is based on local currency (the PPP-based estimation), the efficiency appears to decline, reflecting the devaluation of the local currency. Also, the PPP-based estimation shows higher apparent energy efficiency than the (external) US$-based estimation because the economic output measured by the PPP-based GDP per capita is higher than the GDP per capita measured in 1985 US$ (Figure 4-2b). Perhaps it is more correct to say that energy use and economic production both increased roughly in proportion to the increase in the number of humans.

Fertilizer use is a form of energy input. Fertilizer consumption, which is supplied mainly by imports from foreign countries, has tended to increase over time (Figure 4-6a). Agricultural production, however, has not increased as rapidly as has fertilizer inputs. Thus, fertilizer efficiency, which is the ratio of, e.g., total cereal production to total fertilizer use on cereals, is inverse to fertilizer input (Figure 4-6b). Meanwhile food imports from foreign countries, especially cereals, have increased (Figure 4-6c). We compared these imports with the total national requirement for calories, calculated as "cereal equivalents," which we estimated by multiplying the total population of each year by the kilogram equivalent of a person's annual energy requirement (286 kg of grains a year) (Hall and Hall, 1993). In the 1990s, about one third of Costa Rica's food energy requirements appear to be supplied by imports.

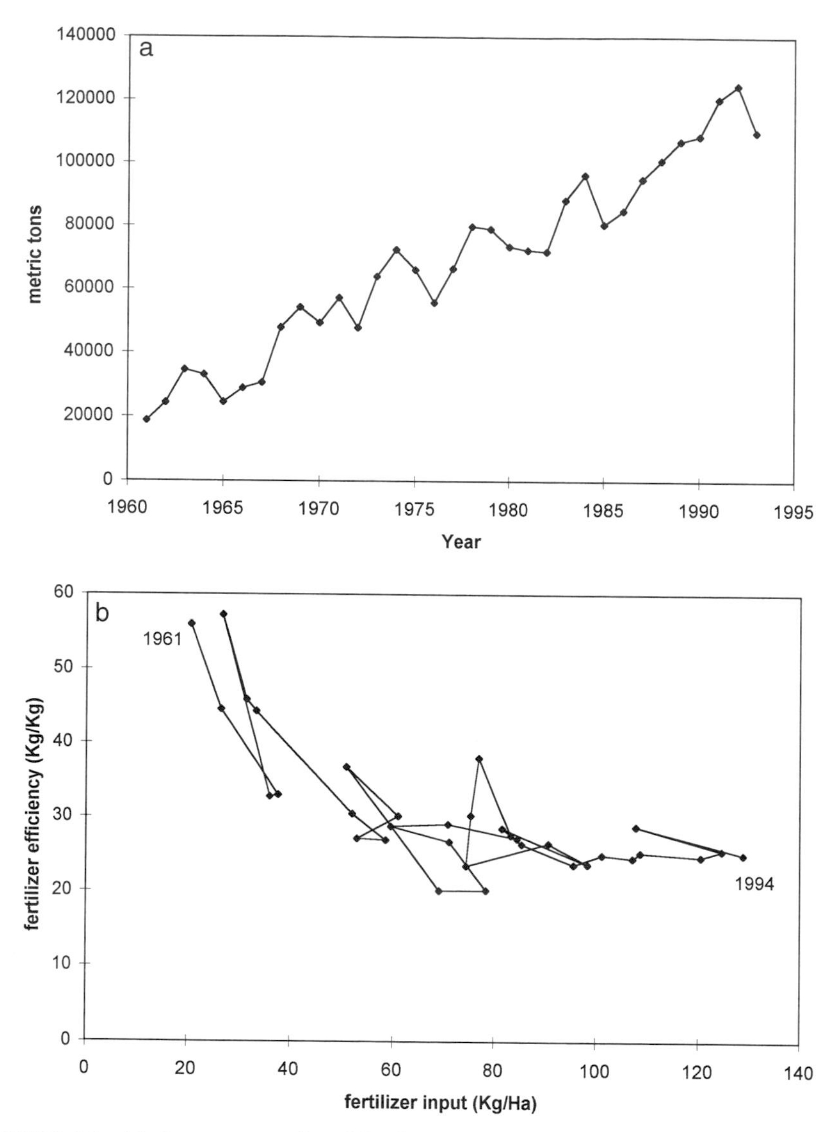


FIGURE 4-6 (a) Consumption of total fertilizer, 1961–1995. (b) Fertilizer input vs fertilizer efficiency for total cereals production in Costa Rica, 1961–1994. (c) Domestic demand and net import of cereals, 1961–1993. (Data from UN records, various years.)

## IV. DISCUSSION

A main conclusion of this analysis is that all of the basic data are interrelated, or perhaps it is more accurate to say that as the population has grown, the

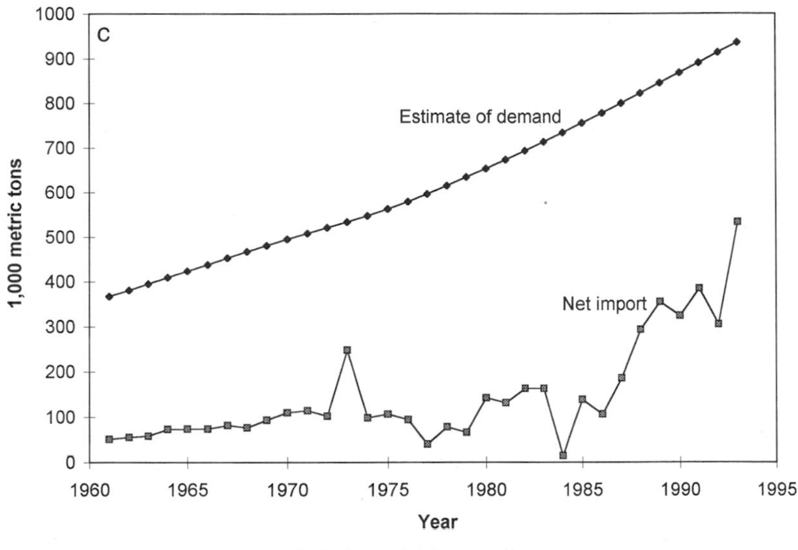


FIGURE 4-6 *(Continued)*

economy has grown more or less proportionally, along with the use of resources to support that economic growth. A second main conclusion is that from an energy perspective the national economy is strongly and increasingly based on nonrenewable resources, especially oil. Although oil dependence dropped following the oil price shocks of the 1970s, it has increased again both absolutely and proportionally since 1985 (Figure 4-7). The following sections examine these relations within the context of examining what possibilities lie ahead for the economy of Costa Rica.

## A. What Are the Projections for the Next 30 Years?

If Costa Rica continues at its present rate of population growth (2.4%) it will have a population of 4 million by the year 2000 and 5.6 million by the year 2025 (WRI, 1996). The average number of children per family, however, is expected to drop to 2.2. That would lower the projected population by the year 2025 to 5 million. The UN Food and Agriculture Organization (FAO) gives the 1990 population growth rate as 2.6%. Valdes and Gomariz (1993) estimate that it is closer to 2.3%. Therefore, it appears likely that the rate of population growth is decreasing. Immigration, however, may add substantially to total population numbers. For example, many Nicaraguans work in Costa Rican agriculture, and some of these people

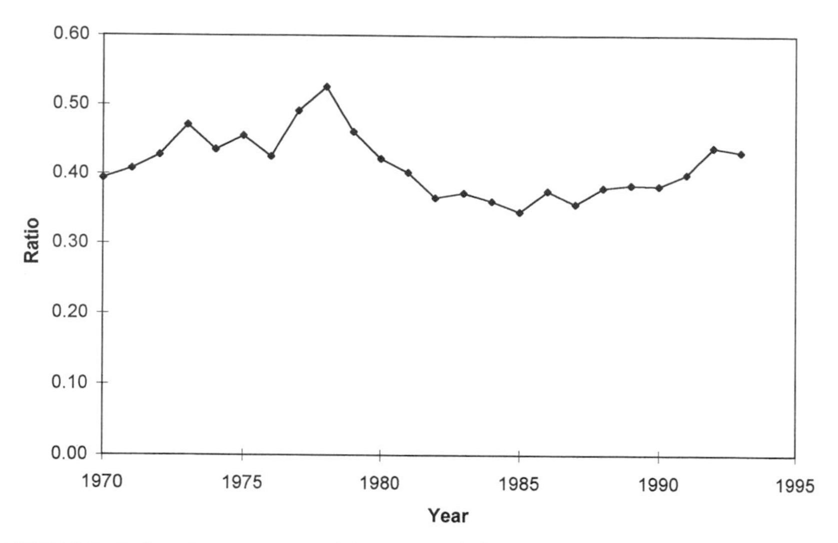


FIGURE 4-7 Index of energy sustainability. Ratio of oil use to quality-adjusted total energy consumption.

stay in Costa Rica where economic opportunities are better than at home. According to Ruben Guevara, director of CATIE, the actual population growth was 2.8% per year when corrected for immigration (personal communication).

Valdes and Gomariz (1993) show that the number of children per family decreased from 3.5 in 1985 to 3.0 in 1990. The number of children born to a family is highly correlated with level of income. Low-income families in 1985 had 5.1 children per family, medium-income families had 2.9, and high-income families 2.2. Two sources, Arcia *et al.* (1991) and Valdes and Gomariz (1993), agree that there has been a very sharp decline in births and an increase in longevity, and that these trends will likely continue. They attribute the changes to improved public services (of which health care is the most significant), a continuous emphasis on education, and an increase in the standard of living, with wide access to birth control programs and 93% of homes having electricity and potable water. Piel (1992) cites a general increase in GNP per capita since 1982 as a major cause for the population growth decline. It should be noted that 1982 represents a peculiar year to choose since per capita income had declined a great deal in the previous decade due to the oil price shocks. Since the projected population growth rate is uncertain, we provide estimates of future population levels using different assumptions (Figure 4-8).

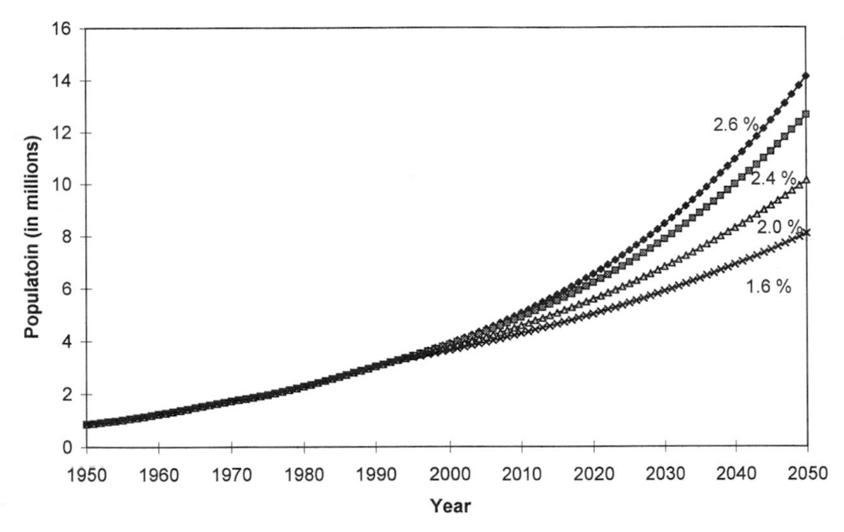


FIGURE 4-8 The population of Costa Rica from 1950 to 2050 based on empirical (1950–1993) and simulated (1994–2050) data using four different values of population growth.

An important concept in development theory is the "demographic transition," where, according to the theory, as economic development proceeds, the death rate initially falls rapidly, followed subsequently by the birth rate. The net result is that populations will eventually stabilize. This concept has been criticized heavily by Abernathy (1993), who believes that people have more or fewer children in response to their expectations of improving or declining economic conditions.

It can be concluded from this analysis that Costa Rica clearly followed the first stage of demographic transition theory, since the death rate declined remarkably, but that it has been slow in approaching the "second stage," where the birth rate falls to replacement level, being 30 to 40 years behind the United States in making that transition (Figure 4-3b). Since per capita economic conditions have neither particularly improved nor declined, it is hard to assess the movements relative to Abnernathy's hypothesis. Holl *et al.* (1993), in a particularly careful analysis, found that the Costa Rican birth rate initially fell from a very high to a moderate level in association with development, but has since remained at a relatively high "plateau" of about 3.2% (a total fertility rate, meaning approximately the number of children per family, of 3.5 to 3.8). The census initially scheduled for the mid-1990s would have allowed us a much better opportunity to make predictions about time trends in population growth, but as of 1997 it had not yet been undertaken.

## B. Environmental Impacts Associated with Urbanization

The concentration of the population in the San Jose area, less than 4% of the national territory, has produced some serious environmental problems. One of the most critical problems is air pollution. As of 1995 there were approximately 250,000 vehicles circulating in the streets of San Jose at peak hours, 88% of the total vehicle fleet in the country. A large percentage of these vehicles are operated with diesel fuel and little or no emissions control. The steep hills mean that many engines are tuned to a rich mixture of fuel to get a little more power (and a lot more pollution). About 75% of the total regional air pollution can be attributed to vehicles, and 25% to industry. Approximately 85% of the country's industry is located in San Jose. Poor air quality is a crushing experience at rush hour and is manifested in a very large increase in allergies and respiratory diseases in children and the elderly (Carlos Hernandez, personal communication).

The projected increase in population alone will double the number of vehicles by the year 2025. The country's stated goal of increasing the quality of life, generally interpreted to mean increasing the purchasing power of the population, thus means the number of cars will increase more or less rapidly depending on the success of meeting that goal. If affluence grows greatly air pollution and its associated health problems will get much worse unless there is a very different national approach to regulating vehicles. In the early 1990s there was a program to bring in less-expensive, used automobiles from the United States. Now new imported cars must meet the air standards of the United States.

The second most critical problem is solid waste. Studies show that only 46% of the waste generated in Costa Rica is collected and treated by a government agency. The remaining 56% is handled by the generators, and most is deposited in illegal dumps. San Jose has only one landfill, which reached its capacity in 1995. The operation of this landfill is deficient but has improved significantly in the last five years. It receives over 1400 tons of waste per day, which is a product of a per capita generation rate of 0.80 kg per day per person. The rate in rural areas is 0.30 kg per day per person (C. Hernandez, personal communication), compared to about 1.3 kg for the United States (Hall *et al.*, 1994).

As the urban population increases, the pressure on landfills will increase, especially if the purchasing power of the people increases and if the collection services improve. In the last five years, the government has made multiple efforts to site new landfills. But the neighbors of potential sites have reacted, first with violence and then through the legal system, and all new efforts have been stopped. The communities adjacent to the present landfill have threatened the government with closing the road to the landfill if the government does not come up with an alternate site. "Trash crises" have become a regular component of daily life.

A third problem of urbanization is the limited availability of commercial forests for lumber. Before 1994, Costa Rica had a deforestation rate of over 25,000 ha per year. This has changed significantly and it is estimated that the present rate is below 8000 ha per year. But the demand for lumber has remained constant at 0.3 $m^3$ per person per year. Prices for lumber have increased greatly as a result of scarcity. If the population increases as projected, if tree plantations do not increase greatly, and if processing efficiencies do not improve, Costa Rica will run out of commercial forests by the year 2004. This will put pressure on the national parks and protected areas. Fortunately, there has been a tendency to increase the planting of tree farms because of government incentives and market conditions which make tree farming a good business. Also, technology has improved, and harvesting and processing efficiency are on the rise.

A fourth problem of importance is the continuing loss of high-quality land on which to grow food. San Jose and its environs have become a major city of more than a million people and taken over the best and most fertile soils of the country. What were separate towns a few decades back are now a continuous metropolitan area (see "Land Use" on CD). For every 1000 additional inhabitants, San Jose has grown by 9 ha. Therefore if present population rates continue, San Jose, already sitting on much of the best agricultural land in the nation, will have invaded an additional 27,000 ha of high-quality food and coffee growing land by the year 2025, approximately 3% of the best farmland.

## C. What Is the Outlook for the Future?

There are many officials in Costa Rica, as in many other countries of the world, who believe that the nation has to take steps to lower the size of the average family. Costa Rica, however, is a deeply Catholic country. Programs to reduce the increase in population have to respect most Costa Ricans' belief that life begins at the moment of fertilization and that sex is a sacred act that must be treated with utmost care and sensitivity.

Many officials also believe it is necessary to take steps to make staying in the rural areas attractive and stop migration into the city. However, each day the farmer receives less for his or her crops, and at the same time pays more for imported inputs and technology. This discourages small farmers and there is a tendency to sell the land to the large national or transnational corporations. Industrialized farming generally brings down the economic cost of food, but at the expense of jobs and the environment.

Overall there is a tendency toward a reduction in birth rates and an increase in life expectancy. The net result appears to be a decrease in the population growth rate, although the bottom-heavy population pyramid will reduce the effects of this decrease for some time. Immigration is an unassessed factor. One important question concerns when Costa Rica will reach a zero growth rate. According to many observers it will depend on the ability of the country to sustain its social programs. Unfortunately, there has been a reduction of national investments in social programs and infrastructure. Costa Rica's foreign debt is staggering, the debtors expect payment, and a large share of government revenue is dedicated to servicing this debt. Therefore, it is very important to understand that Costa Rica's population future will be influenced heavily by how the foreign debt is handled. One disturbing scenario goes like this: if governmental health and education programs are sacrificed in order to pay the debt, then the less-educated women with less access to, or interest in, birth control information will have far more children, creating a need for more governmental services and requiring more land to feed, directly or indirectly, those new Costa Ricans, a positive feedback which in turn exacerbates the debt problem.

## D. Economics

Per capita wealth in Costa Rica is as much a function of population levels as it is of the rate of wealth generation, and it is projected that population growth will continue. Increasing or even sustaining per capita wealth would almost certainly mean increased usage of land, fossil fuels, and other inputs to the economic process. This growth will make it even more difficult to generate a sustainable society.

As we have seen there is no single best measure of economic activity or well-being. We were forced to use gross domestic product, despite its deficiencies, because it is readily available, has been calculated over a long time period, and captures many important aspects of the economy. Some of the reasons why GNP and GDP are inadequate measurements of human wealth are given in Table 4-1.

New measurements of well-being are being developed to compensate for some of the already mentioned weaknesses of GNP and GDP. The Index of Social Economic Welfare (ISEW) (Daly and Cobb, 1989) and the Human Development Index (HDI) (UNDP, 1993) both attempt to remedy inadequacies 1 and 3 in Table 4-1 while the Net Domestic Product (NDP) (Repetto *et al.*, 1989) introduces methods to account for resource depletion (number 2). Unfortunately most of these indices are not available as time series for Costa Rica, although a partial assessment of the NDP shows a decline in national

TABLE 4-1 A Number of Reasons that GNP and GDP Are Inadequate Measures of Human Material Well-Being

1. They are only a partial measure of those conditions that contribute to human happiness and well-being. For example, they do not measure access to education and health care, the richness of a country's cultural identity, or the availability of clean air and water.

2. They do not account for the depletion of natural resources that will be unavailable to future generations. Overgrazing and the susbsequent deterioration of soil quality is an example in Costa Rica.

3. They do not measure the distribution of wealth by distinguishing either the disparity between the rich and the poor, or between production of basic needs and that of luxury goods. Thus a wealthy nation can have poverty while a poorer one might not. This issue may be increasingly important in Costa Rica as the distance increases between the rich and the poor.

4. They are poor measures of production, especially in newly developing nations, because much of the production does not pass through markets but is consumed directly by families.

5. They do not measure nonmarket transactions such as environmental services and subsistence agriculture. Remediation of environmental destruction actually increases GNP and GDP of those wealthier countries that can afford these activities.

6. They do not measure existing wealth but rather the flow of new wealth into a society. Thus existing infrastructure, like roads and bridges, and the soil inventory are not accounted for.

7. Their use for measuring well-being encourages planned obsolescence, such as the production of poor-quality goods with short life spans, because the index is increased by a rapid turnover rate.

*Source:* Hall (1992) and Hall *et al.* (1986, Chap. 2).

wealth over the past several decades due to environmental degradation (Repetto *et al.*, 1988).

What is the meaning of the difference between estimating economic growth from internal purchasing power, which has recorded a slow increase, and estimating it from external purchasing power, which has been constant? It means that the Costa Rican currency has been decreasing in purchasing power relative to the dollar over time (Figure 4-2b). The widening difference between the internal and the external purchasing powers can be explained in part by the rapidly worsening exchange rate of colónes against the U.S. dollar (Figure 4-9). Costa Ricans must forego a greater amount of domestically produced goods in order to purchase one dollar's worth of foreign goods; i.e., imports are increasingly costly compared to domestic goods. Thus the increase in the standard of living seen with respect to domestic purchases is not reflected in the average person's ability to purchase imported goods. This means that Costa Ricans receive low prices for their exports (because the price they receive is based on the exchange rate instead of purchasing power). The relative advantage

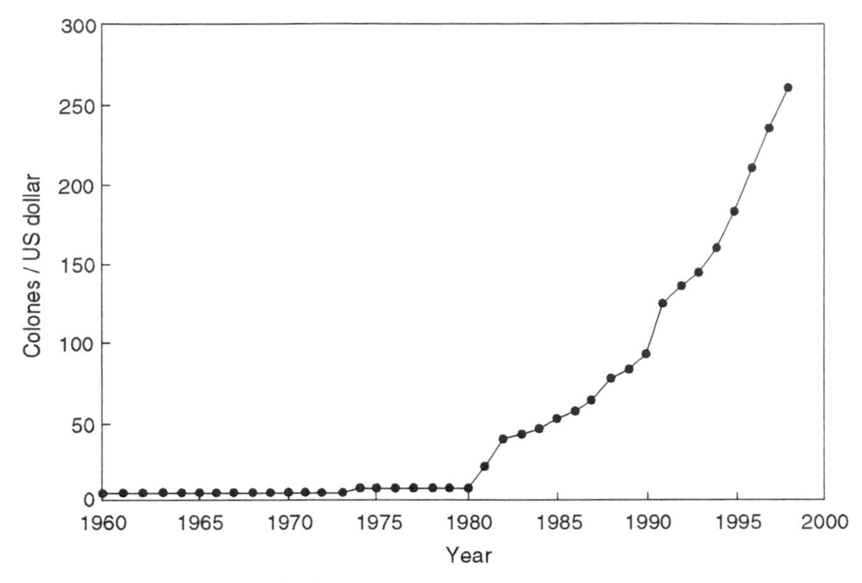


**FIGURE 4-9** Exchange rate of colónes per U.S. dollar, 1950–1992. The value for 1998 was 258 per dollar.

over time has gone to the industrial countries. But if this had not occurred, the downside is that presumably exports would be reduced if wealthy nations had to pay more for Costa Rican exports while simultaneously receiving less for goods sent to Costa Rica. The rapid devaluation of colónes against U.S. dollars has provided serious challenges for the ongoing policy of protrade economics.

## E. Has Development been Successful?

Brian Atwood, former director of the U.S. AID mission in Costa Rica, said on the disbanding of that mission in 1996 that aid to Costa Rica had been successful in that

> Costa Rica's per capita income of nearly $2,500 compares to only $1,000 in 1950. Whether this increase is because of, in spite of, or independent of the aid is debatable, but it certainly seems unlikely to have occurred without any investment. And $2,500 is a much higher annual GDP than any other Central American country except Panama.

Our own analyses give a somewhat more complex answer than Mr. Atwood's. While national GDP has indeed increased a great deal since 1950 (Figures 4-1a,b), this increase has been very little more than population growth, so that per capita domestic purchasing power income has increased only a small amount, and the per capita GDP, based on foreign currency exchange rates, has remained constant (Figure 4-2b). Meanwhile total per capita debt has increased from near zero to nearly half the difference between the 1950 income and the 1990 income ($2500).

## F. Debt

Costa Rica's foreign debt is the largest in Latin America in relation to its income, and a large share of both national revenue and national foreign exchange is assigned to the service of this debt (Figure 4-10a). The foreign debt of Costa Rica, as of March 1995, was $3417 million. There were approximately 3.4 million citizens of Costa Rica at that time, so that every citizen had a debt with a foreign country or institution of approximately $1000. Additionally, the Central Government had an internal debt that exceeded $1400 million. This means that every citizen had an additional debt with national entities of $411 (Monge, 1995).

One way of looking at this debt is in relation to foreign exchange. When the foreign exchange needed to pay the annual interest, and occasionally principal payments, is plotted along with the total amount of foreign exchange gained from the sale of all bananas, coffee, beef, textiles, tourist services, and so on, the debt consumes roughly 15% of these precious export earnings (Figure 4-10b). Costa Rica has avoided the full implications of paying off this debt by simultaneously borrowing more, so that the debt and interest paid have been roughly balanced by new borrowing, at least to 1992 (Figure 4-10c). This is a treadmill that has serious implications for the nation, but which is political suicide for any one politician to resolve. This problem is not confined to Costa Rica of course, and there is a nearly exact parallel with the United States. Costa Rica has made a little more progress since 1992, in part due to low oil prices.

Unfortunately, one result of the debt has been a reduction in national investments, especially those that affect social programs and infrastructure. As we mentioned in the population section, many Costa Ricans who have thought carefully about this issue have concluded that it is these very health and welfare issues that are responsible for the reduction in birth rates that have occurred so far. Ironically, the debt payments can be thought of as contributing to higher birth rates than would otherwise be the case, and consequently to the need for increased foreign exchange, perhaps generated through additional debt, to feed (through direct imports or fertilizers) and otherwise care for the additional children. Therefore, it is very important to understand the implications of Costa Rica's foreign debt for its development potential.

A

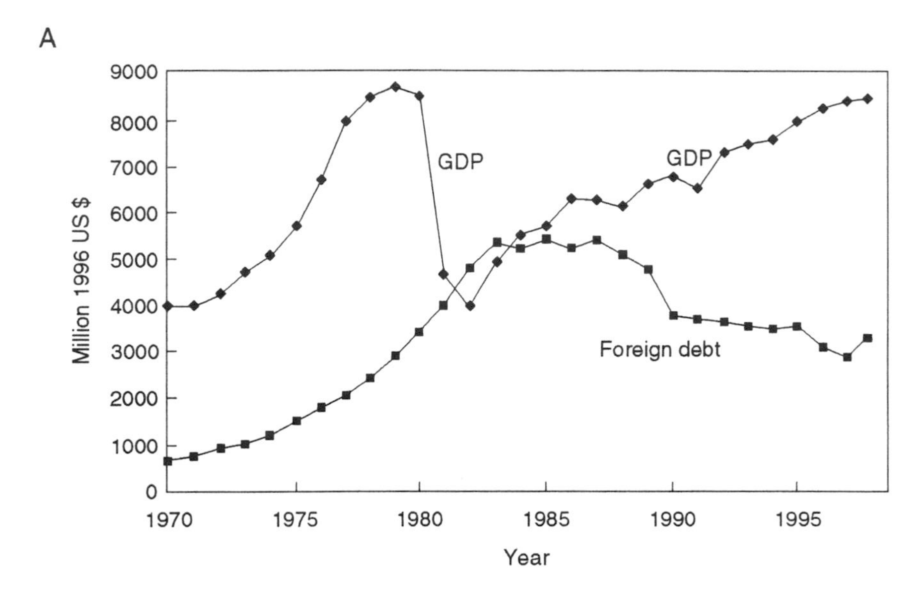


B

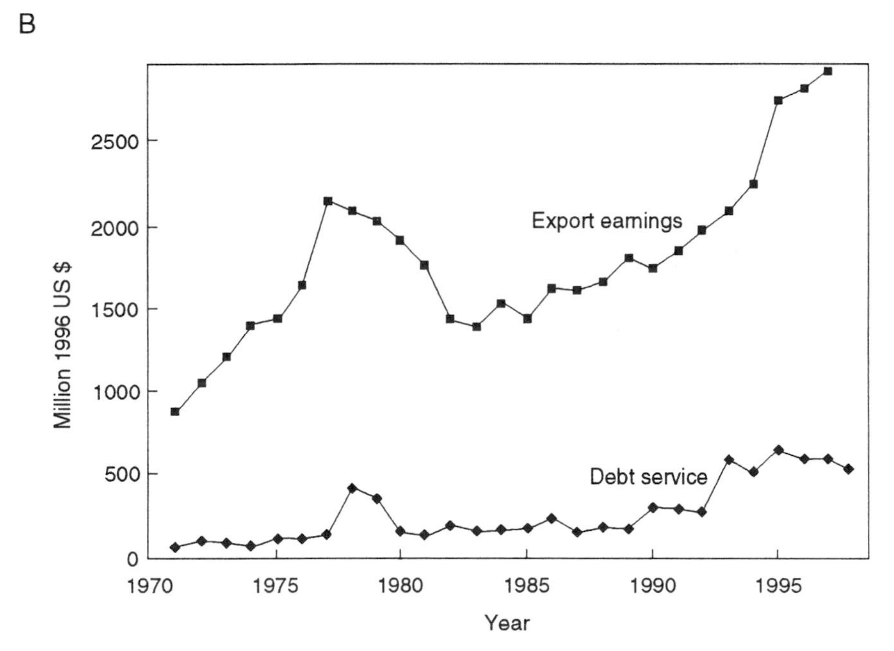


FIGURE 4-10 (a) Foreign debt and GDP (World Bank estimate). Foreign debt in 1998 was $2872 million. (b) Export earnings and debt service. (c) External borrowing and debt paid (national data). Debt service for 1998 was $405 million.

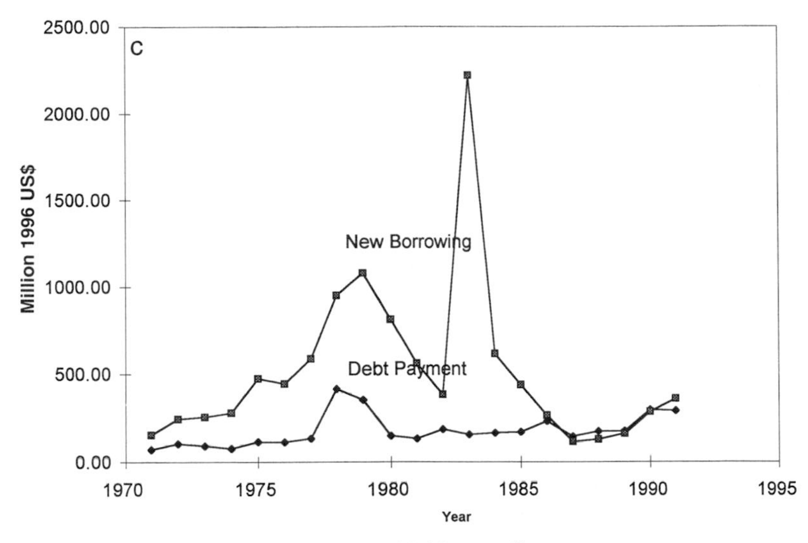


FIGURE 4-10 (*Continued*)

## 1. An Interesting Way to Reduce Debt

One interesting and useful form of debt reduction is debt-for-nature swaps in which a government or organization, such as a private conservation foundation, buys a portion of a country's debt at a substantial discount from the financial organization that holds that debt. The debtor country then issues an agreed-upon value of local currency for conservation projects within the country. One of the largest such swaps was the Dutch government's purchase of $33 million (face value) of Costa Rica's debt in exchange for an investment of $10 million in reforestation, watershed management, and soil conservation in 1990 (Postel and Flavin, 1991).

The problem with the debt-for-nature swaps is that although $33 million sounds impressive, this was only 3 to 4% of the total debt in 1990 (Patterson, 1990). Expansion of debt-for-nature swap programs requires the following factors: (1) a debt-holding country's willingness to participate; (2) a commercial bank's willingness to sell its loan at a low price; (3) a nonprofit organization with the money and desire; (4) a local nonprofit organization in the debtor country to implement the program; and (5) a financial intermediary to set up contracts and provide professional expertise (Prestemon and Lampman, 1990).

However, a favorable environment for such communication among these actors may require a serious economic crisis, because without a serious doubt about the repayment of its loan to Costa Rica, a foreign commercial bank will not agree on a huge discount (e.g., 12 cents paid on the dollar owed) (Prestemon and Lampman, 1990). Secondly, the new inflow of foreign money into the Costa Rican economy may provoke inflation when converted into local currency. For example, $50 million introduced through the swap program raised the inflation rate from 16.45 to 16.62% in 1987 (Prestemon and Lampman, 1990). Additionally, expansion of swap programs depends on a nonprofit organization's ability to raise money in Washington, DC (Patterson, 1990).

### G. Energy

At this time energy is rarely thought about by most citizens in the developed world. The price shocks of the 1970s, if they are remembered at all, generally are considered as resolved through market forces, substitutions, or improved efficiency. Energy is a little more in the public eye in Costa Rica because of the high price of imported oil compared to relatively low salaries and export prices for coffee, bananas, and so on. Energy is also important as the country continues to modernize. Finally, although the energy crises of the 1970s are left far behind in the economies of the developed world, their impact remains in Costa Rica as debt and its interest. Costa Rica's energy future is analyzed in the last chapter.

## V. CONCLUSIONS

This analysis has shown that with respect to population, land use, and energy use, Costa Rica is not increasing the sustainability of its economy and, in fact, is often doing the converse. Nor is there any indication to date that it is moving in that direction. Specifically,

1. Human populations are continuing to grow and are likely to do so for at least decades.
2. Natural areas continue to decline (Chapters 16 and 17).
3. The economy is increasingly dependent, in both absolute and proportional terms, on imported nonrenewable petroleum.
4. An increasing proportion of food calories are imported.
5. The population per unit of good land is increasing.
6. Energy efficiency in agriculture is decreasing.
7. All economic growth to date has caused little or no increase in per capita wealth.
8. External debt is large and continues to drain a great deal of internal production—much of it from social programs that have appeared effective in reducing early pregnancies.
9. There is no evidence for increased efficiency of energy use in either agricultural production or national economic production. If anything the efficiency with which energy is converted into wealth is declining.

On the other hand there are many more optimistic aspects of Costa Rica. First, and perhaps most important, Costa Rica has produced an extremely positive society on a relatively modest resource and economic base. Costa Rica is among the highest of any country in the world in terms of life span, literacy, health standards and health care, freedom of the press, lack of military or police brutality, and equitable distribution of income. The Costa Rican people have been able to generate a good quality of life with relatively little in terms of economic quantity. This might be considered to be the very best way to generate sustainable development, and could serve as a valuable guide to other nations. Also, they have a relatively large internal source of possible energy due to hydropower and biomass potential. Birth rates have fallen, and perhaps the population will become stable within the limits of available resources. The last chapter discusses the possibilities of generating a stable economy.

Nevertheless, due to imported energy dependency and balance of trade problems, as well as other problems mentioned later, sustainability, at least according to the criteria of this chapter, may not be possible or even desirable. The argument for sustainable development may be an oxymoron which is more sentimental than practical. Without a serious change in our idea of economic development and serious thoughts about population, economic growth, and resource depletion, we may never be able to implement sustainable development in any meaningful way. While it is in many ways very desirable to protect species and other biotic resources, forestlands, and so on, this is not the same as saying that the net result leads by itself to a sustainable economy if the general economy is becoming increasingly dependent upon nonrenewable fuels, imported goods and foods, eroding soils, and contaminating lands. On the other hand, perhaps we should not try to achieve the unachievable, but instead ask how we can use our remaining fossil fuels, soils, and other resources as wisely as possible. The rest of this book examines some possibilities.

## ACKNOWLEDGMENTS

Carlos Hernandez provided important information about demographic and environmental conditions in Costa Rica. Suni Edson and Luis Lopez helped in data analysis, and Chuck Levitan with graphics.

## REFERENCES

Abernathy, V. 1993. The demographic transition revisited: Lessons for foreign aid and U.S. immigration policy. *Ecological Economics* 8:235–252.

Arcia, G., L. Merino, A. Mata, and B. O'Hanlon. 1991. *Modelo Interactivo de Poblacion y Medio Ambiente en Costa Rica 1990.* Asociacion Demografic Costarricense.

Cleveland, C., R. Costanza, C. Hall, and R. Kaufmann. 1984. Energy and the United States economy: A biophysical perspective. *Science* 225:890–897.

Cornell, J., and C. A. S. Hall. 1994. A systems approach to assessing the forces that generate tropical land use change. In J. O'Hara, M. Endara, T. Wong, C. Hopkins, and P. Maykish (Eds.), *Timber Certification: Implications for Tropical Forest Management,* pp. 35–43. School of Forestry and Environmental Studies, Yale University, New Haven, CT.

Daily, G. (ed.) 1997. Nature's Services: Societal dependence on Natural Ecosystems. Island Press, Covello, Cal.

Daly, H. E., and J. B. Cobb, Jr. 1989. *For the Common Good, Redirecting the Economy toward Community, the Environment, and a Sustainable Future.* Beacon Press, Boston, MA.

Darmstadter, J., P. D. Teitelbaum, and J. G. Polach. 1971. *Energy in the World Economy, a Statistical Review of Trends in Output, Trade, and Consumption Since 1925.* Johns Hopkins University Press, Baltimore, MD.

Hall, C. A. S. 1975. The biosphere, the industriosphere and their interactions. *Bulletin of Atomic Scientists* 31:11–21.

Hall, C. A. S., C. J. Cleveland, and R. K. Kaufmann. 1986. *Energy and Resource Quality: The Ecology of the Economic Process.* Wiley–Interscience, New York. Reprinted 1992, University Press of Colorado, Boulder, CO.

Hall, C. A. S. 1992. Economic development or developing economics: What are our priorities? In M. Wali (Ed.), *Ecosystem Rehabilitation,* Vol. I. *Policy Issues,* pp. 101–126. SPB Publishing, The Hague, The Netherlands.

Hall, C. A. S., and M. H. Hall. 1993. The efficiency of land and energy use in tropical economies and agriculture. *Agriculture, Ecosystems and Environment* 46:1–30.

Hall, C. A. S., R. G. Pontius, J. Y. Ko, and L. Coleman. 1994. The environmental impact of having a baby in the United States. *Population and Environment* 15:505–524.

Holl, K. D., G. C. Daily, and P. R. Ehrlich. 1993. The fertility plateau in Costa Rica: a review of causes and remedies. *Environmental Conservation* 20:317–323.

Inter-American Development Bank. 1993. *Latin America in Graphs: Demographic and Economic Trends 1972–1992.* Distributed by Johns Hopkins Univ. Press, Washington, DC.

Ko, J-Y., C. A. S. Hall, and L. G. Lopez Lemus. 1998. Resource use rates and efficiency as indicators of regional sustainability: An examination of five countries. *Environmental Monitoring and Assessment* 51:571–593.

Monge, C. 1995. Costa Rica se Ahoga en Deudas. *La Pensa Libre,* July 11, p. 3.

Nilsson, L. J. 1993. Energy intensity trends in 31 industrial and developing countries 1950–1988. *Energy* 18:309–322.

Odum, H. T., and J. E. Arding. 1991. *Emergy Analysis of Shrimp Mariculture in Ecuador.* Center for Wetlands, University of Florida, Gainesville, FL. 114 pp.

Patterson, A. 1990. Debt for nature swaps and the need for alternatives. *Environment* 32:4–13.

Piel, G. 1992. *Only One World: Our Own to Make and Keep.* W. H. Freeman, New York.

Postel, S., and C. Flavin. 1991. Reshaping the global economy. In *State of the World.* Worldwatch Institute, Washington, DC.

Prestemon, J. P., and S. E. Lampman. 1990. Third world debt: Are there opportunities for forestry? *Journal of Forestry* 88:12–16.

Repetto, R., W. Magreth, M. Wells, C. Beer and F. Rossini. 1989. Wasting assets: Natural resources in the national income accounts. World Resource Institute, Washington, D.C.

UNDP. 1993. *Human Development Report 1993. United Nations Development Programme.* Oxford University Press, London.

Valdes, T., and E. Gomariz. 1993. *Mujer Latinoanericana en Cifras: Costa Rica.* Instituto de la Mujer, Ministerio de Asuntos Sociales de Espana y Facultad Latinoamericana de Ciencias Sociales, Santiago, Chile.

World Resources Institute. 1996. *World Resources 1994–97: A Guide to the Global Environment.* Oxford University Press, New York. 400 pp. with data diskette.

CHAPTER 5

# Land, Energy, and Agricultural Production in Costa Rica

CHARLES A. S. HALL, LOIS LEVITAN, AND TOMÁS SCHLICHTER



## I. INTRODUCTION

### A. THE GENERAL PROBLEM

If human systems are to be sustained then agriculture must be sustainable in some way since the provision of food is essential for humans. But what does sustainability mean with respect to agriculture? Globally, Malthusian concerns about the world's carrying capacity for human beings continue to be assuaged by increases in agricultural production. Since 1984, however, the worldwide rate of increase in agricultural productivity has slowed to 1% per year, while the rate of world population growth remains nearly twice that (Figures 1-1 and 1-2). In most of the developing world people are fed increasingly from the surplus of a few relatively uncrowded industrial nations. Presently the United States and Canada export grains—the main foodstuffs of the world—to over 100 countries (Brown, 1991, 1996).

For centuries agricultural yields per hectare were, such as we can tell, relatively low and unchanging except as land was eroded by use (Table 1-1).

*Quantifying Sustainable Development*

The increases in production required to feed increasing human populations were brought about principally by the expansion of area farmed. Starting about 1950 agricultural yields per hectare began to increase in many parts of the world, brought about by the expansion of industrialized agriculture and the ready availability of cheap fertilizers from munitions plants left over from World War II. By about 1980 nearly all of the world's land most suitable for agriculture had been developed for crops or pasture. Only 4% of additional land was brought into agricultural production during the 1980s and, following a pattern noted nearly two centuries ago by the economist David Ricardo (1817), this land was, on average, less productive than existing farmland. Nearly 90% of the increase in agricultural productivity in the 1980s has been the result not of an increase in the area cultivated, as had been the case in earlier decades, but of higher yields per hectare on previously developed lands (Brown, 1991, 1996). These increased yields were due to the increased use of industrially derived inputs such as fertilizers and genetic improvements in crop plants that allowed the plants to utilize these increased inputs. For example, global fertilizer use increased by 50% during the 1980s. Since the petroleum-derived basis of fertilizer and most other agricultural inputs is finite and likely to eventually become much more expensive, and since the supply of good agricultural land is limited, it is critically important to examine carefully the possible constraints on continued agricultural expansion as the world population continues to grow, especially in the tropics. For the reasons given in Chapter 3 we use a biophysical approach for this analysis.

The basic fact about agriculture almost everywhere in the world is that the vast majority of increasing yields that are occurring are attributable to the industrialization of agriculture, that is, to the processes of accelerating and focusing the photosynthesis of cultivars through the application of fossil-fueled technologies (e.g., Odum, 1967; Pimentel and Pimentel, 1979; Brown, 1981; Hall *et al.*, 1986; Odum, 1989). One way to think about modern agriculture is given in Figure 5-1, which plots isopleths of yield as a function of both intrinsic site quality and fossil-fuel-derived inputs. In general, the trajectory of a particular site, or nation, over time is from the lower right toward the upper left. Yields may increase even as the site is degraded by use because of increased inputs, but if the fossil-fuel-derived inputs are removed, yields tend to fall to levels below their original value because site quality has declined. Although the intrinsic quality of most of our major agricultural soils has declined substantially (Pimentel *et al.*, 1987; Pimentel, 1995), this is not reflected in yields because of increasing inputs. For example, the United States and Canada are losing considerable quantities of topsoil due to the intensification of agriculture, but crop yields are increasing because of the increased use of more fertilizers and plant genomes bred to convert this fertilizer into more of the edible part of the plant. Thus conventional economics may show an

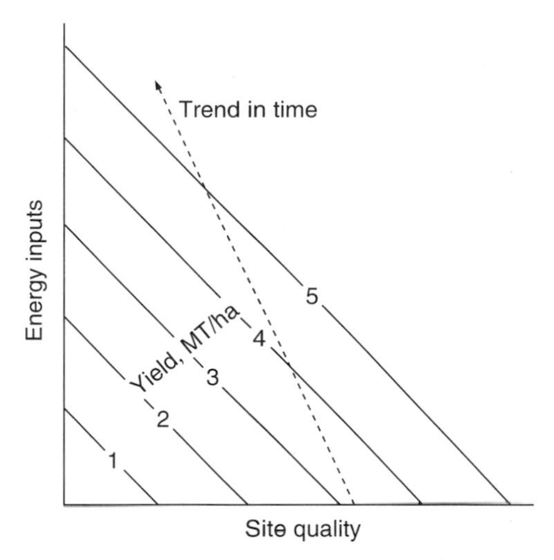


FIGURE 5-1 Isopleths of agricultural yield as a theoretical function of site quality and fossil-energy-derived inputs. For each group, sustained high yields can be a function of high site quality (in general riparian, volcanic, or glaciated soils) or as a function of inputs, which are derived largely through the use of fossil fuels. Over time, yields tend to decrease as site quality degrades as a function of agricultural activity unless inputs are increasing so that yields increase. Thus over time the trend on these axes tends to be upward and toward the left. If inputs are diminished then the yields would fall vertically, resulting in lower yields than earlier because of a decline in site quality.

increase in the value of agricultural production while the resource base for agriculture is undergoing serious degradation. For long periods this degradation is barely, or not at all, reflected in the market price.

Is it possible to increase agricultural yield per hectare without an increase in the direct or indirect use of energy? In countries where historical data are available, there is no evidence at all of increased agricultural production at the national level without an at least concomitant increase in the use of industrial energy. This is as true for China as it is for the United States (Hall *et al.*, 1986) or for the extremely intensive cultivation that occurs in Japan, Israel, Korea, or the Netherlands (Odum, 1967; Ko *et al.*, 1998). It is not clear whether new genetic engineering technologies or new types of "organic" farming will change this pattern if they are implemented on a large scale. Certainly it is a most important goal.

One way of thinking about this issue is to consider the basic things that a crop plant must do. It must capture solar energy and invest this energy in sequestering nutrients, assimilating carbon, defending itself against insects, and capturing water. In a wild plant, these processes require substantial amounts of energy, in fact, most of the energy that the plant captures through photosynthesis. With modern agricultural technology, we have increased the edible yield of plants but not the total biomass production (Gifford *et al.*, 1984). One hypothesis about the overall effect of plant breeding is that we increasingly subsidize the non-seed-producing functions of the plant with external energies; i.e., we now use industrial energy, rather than the cultivar's own energy, for water supply, nutrient supply, and pest defense. As a rule, the cultivars use less energy for root growth, secondary chemicals, and so forth, and the proportion left for production of seeds increases.

## B. Energy, Agriculture, and the Costa Rican Economy

We focus on agriculture because Costa Rica has traditionally had an agricultural-based economy, because agriculture is still the base of nearly half of the economy, and because food availability is the most critical issue for humans. Agriculture is also important in that it traditionally and currently generates more than half of the national foreign exchange, which is essential for purchasing petroleum and other industrial products, none of which are derived from Costa Rican raw materials. The basic data that we use here for analysis are presented in more detail in Chapters 4 and 12.

As mentioned previously, the economy of Costa Rica, including the agricultural sector, is in many ways extremely heavily industrialized, due in part to critical needs for increasing production and to deliberate governmental policies started in the 1960s (Chapter 2). The rapid industrialization of agriculture was facilitated during the 1970s by an overvalued currency and by an infusion of loans from international sources (Annis, 1990; Hanson-Kuhn, 1993) (see Chapters 23 and 24). During the 1970s agricultural productivity increased more rapidly than did the agricultural workforce, as productivity increases were fueled by energy subsidies—in the forms of agrochemicals, fossil fuels, and machinery—that enhanced and substituted for labor. These imported energy subsidies doubled per agricultural worker during the 1970s (Levitan, 1988). Nationally, energy consumption increased by 8.8% annually in the years between 1965 and 1980. Although the growth rate declined to 0.6% for a number of years after the energy price increases in 1979, Costa Rica still imported 700,000 tons of petroleum and 80,000 tons of energy-intensive fertilizers in the mid-1980s, and twice that by the mid-1990s. Although Costa Rica generates one-quarter of its energy from hydropower, as a nation without fossil energy resources it must import all the petroleum and petroleum products it uses. The energy price increases of the 1970s were especially hard on Costa

Rican agriculture, although the relaxation of energy prices since then has made energy less of a consideration at the present.

## C. Agricultural Technology and Its Relation to Energy Use

The ability of agriculture to sustain humans depends on many things. Among the most important are the quantity of good agricultural land available, the ratio of the area of that land to the number of people, and the productivity of that land. As we have noted, the productivity of agricultural land has been greatly increased in recent decades through, essentially, the industrialization of agriculture. Yet there are concerns about the environmental consequences of this industrialization, and indeed about the sustainability of the industrial fuel base (e.g., Clark, 1995; Campbell, 1997a). Thus it is a critical issue to Costa Rica whether its agricultural yields can be increased further.

There is considerable hope among many who study the global food situation that agricultural technology (either industrially based "hard" or environmentally sensitive "soft" technology, depending on one's view) will continue to expand agricultural production significantly (e.g., Mellor and Gavian, 1987). Unfortunately the empirical data in support of this perspective are not especially promising because of: (1) the asymptotic response of agricultural yield to inputs (Figure 5-2); (2) lower production and cultivar response to inputs as area in production is expanded, necessitating moving crops to increasingly marginal and lower yielding land; and (3) the additional fossil energy subsidies needed to compensate for the loss of soil quality on land where soil has been eroded or compacted, or where nutrients are depleted (Larson *et al.*, 1983; Hall and Hall, 1992; Pimentel *et al.*, 1995; Alfsen *et al.*, 1996). In earlier studies we found no empirical evidence that national agricultural production in any developing country studied was becoming more efficient per unit of energy (or fertilizer) input except when the area cropped was reduced (Cleveland, 1995; Ko *et al.*, 1998)—in fact, fertilizer efficiency has declined in most countries (Hall *et al.*, 1986; Hall and Hall, 1992). We also found this true for Costa Rica (Chapter 12). Since Costa Rica is now importing nearly a third of its food calories, the issue is far from academic.

### 1. Exports

Prior to the debt explosion in the 1980s, more than half of the foreign exchange needed to pay for industrial inputs to agriculture (such as fertilizers, pesticides, and trucks) as well as other imported goods, and to service the debt, was

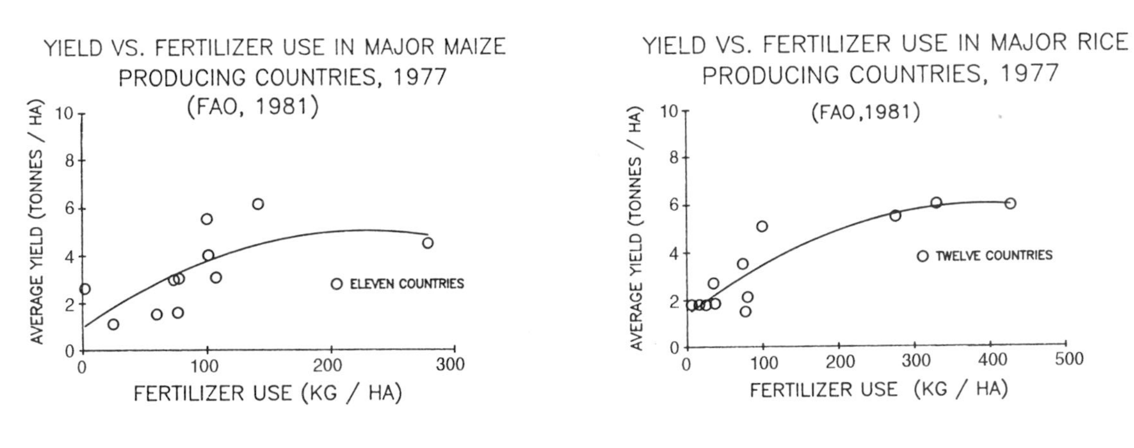


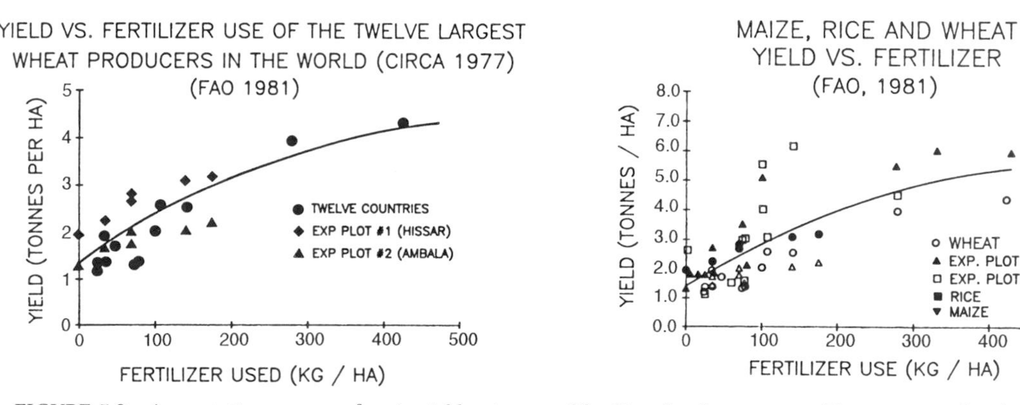


**FIGURE 5-2** Asymptotic response of grain yield to inputs of fertilizer for the major world grains considered separately and together. (From FAO, 1983.)

earned from the export of only a few agricultural commodities—principally coffee and bananas—which consequently played a pivotal role in the domestic economy. But despite the importance of coffee to the domestic economy, Costa Rica has a limited influence in setting the world trade price for coffee because it produces only about 2% of the coffee traded internationally. Coffee is grown commercially in 53 countries and is, after oil, the second most heavily traded commodity on the world market in terms of dollars (Levitan, 1988). Thus Costa Rican agricultural production is extremely sensitive to the prices of coffee and bananas relative to those of petroleum and petroleum-derived products.

## 2. The Petroleum Dependence of Agricultural Fertility and the Impact of Energy Price Increases

Traditionally in Costa Rica and, indeed, most of the tropics, agricultural fertility was maintained through shifting cultivation and other extensive fallow systems (Conklin, 1961; Watters, 1971; Lal and Cummings, 1979) (see Chapters 2 and 12). Shifting cultivation is a truly sustainable system, perhaps the only one for humans. It can maintain agricultural production indefinitely with little reliance on external inputs when population levels are low—roughly no more than one person per two hectares of agriculture land.

But the human population in the 1990s was about 10 persons per hectare of active crop land. Thus the population density in 1995 was roughly 20 times what might be supported by "zero input" shifting agriculture on all land presently used for agriculture. In fact if we assume that 2 ha is needed to support one person in a shifting cultivation system (including fallow), the present Costa Rican population of 3.5 million people would require 7 million ha to be fed by shifting cultivation. This is 40% more land area than exists in the country, even assuming that all land could be cultivated and none used for any other purpose. Thus it is clear that current population levels are far too high to return to the traditional low-input, extensively fallow systems. Presently cropland is now left fallow at most just one year out of three, and fertility is maintained instead with fertilizers (Levitan, 1988).

Consequently, agricultural production in Costa Rica, as in many other tropical countries, has increasingly utilized, and then become dependent upon, high levels of fossil energy. This energy is used directly to drive irrigation pumps and transport goods to market, and, more importantly, is embodied in farm machinery, fertilizers, and pesticides. Most U.S. citizens would be surprised to learn that even as early as 1985, on average, Costa Rican farmers used about 40% more chemical fertilizer per hectare of land in annual and permanent crops (133 kg/ha) than was used in the United States (94 kg/ha)

(FAO, 1981, various years). According to the FAO (1992) and Levitan (1988), about 80% of this fertilizer was used for crops grown for the export market. The use of fertilizers in Costa Rica has continued to increase, and by the mid-1990s its use was about two times as high as even the 1985 figure (FAO, various years; Bleggi and Gonzales, 1994). In developing countries as a whole, although not in Costa Rica, about 70% of the fossil energy used directly in agriculture is embodied in fertilizer manufacture and distribution (Mudabar and Hignett, 1982), particularly in the manufacture of energy-intensive nitrogen fertilizer. Currently in both Costa Rica and worldwide, the input of nutrients in chemical fertilizers is about the same as the quantity removed in harvests (FAO, 1981). Thus there is no escaping the fact that at today's population levels the production of crops in Costa Rica is completely dependent upon nonrenewable resources, such as petroleum, and the industrialized nations. Consequently true sustainability is impossible without a great reduction in the human population.

The amount of income from the export sale of coffee does not reflect costs of production and profitability within Costa Rica as much as it does the political, economic, and climatic conditions elsewhere in the world. Therefore, it was not possible for growers in Costa Rica to increase the price of exported coffee commensurate with the tremendous increases in production costs which resulted from increased costs of inputs during the early 1980s. The monetary cost of the fertilizer used to produce coffee, for example, increased from 5 to 35% of profits in the 30-year period after 1953 (Levitan, 1988). Although that proportion has shrunk somewhat since 1983 as petroleum prices relaxed, coffee production remains highly subsidized by fossil fuels.

Another problem is that most of the land with ideal conditions for coffee is already growing coffee or some other equally valuable product. To a lesser degree this is also true for bananas. Many of the best lands are already used and abandoned (see Figure 2-1 and Chapters 12 and 20). If fossil energy does again become more scarce and costly, we expect that the cost of fertilizer will rise as it did in the past (Figure 5-3), and that a greater proportion of the earnings from export crops will be needed to purchase industrial inputs for agriculture (Figure 5-4). Even if this does not occur the generation of foreign exchange remains the most critical element facing the debt-ridden Costa Rican economy at this time, and agriculture tends to remain the main generator of foreign exchange. Therefore we have concluded that the relation of energy and its products to agricultural yield, as well as the relation between imports for agriculture and the generation of foreign exchange, is of fundamental importance to understanding the future potential of the Costa Rican economy.

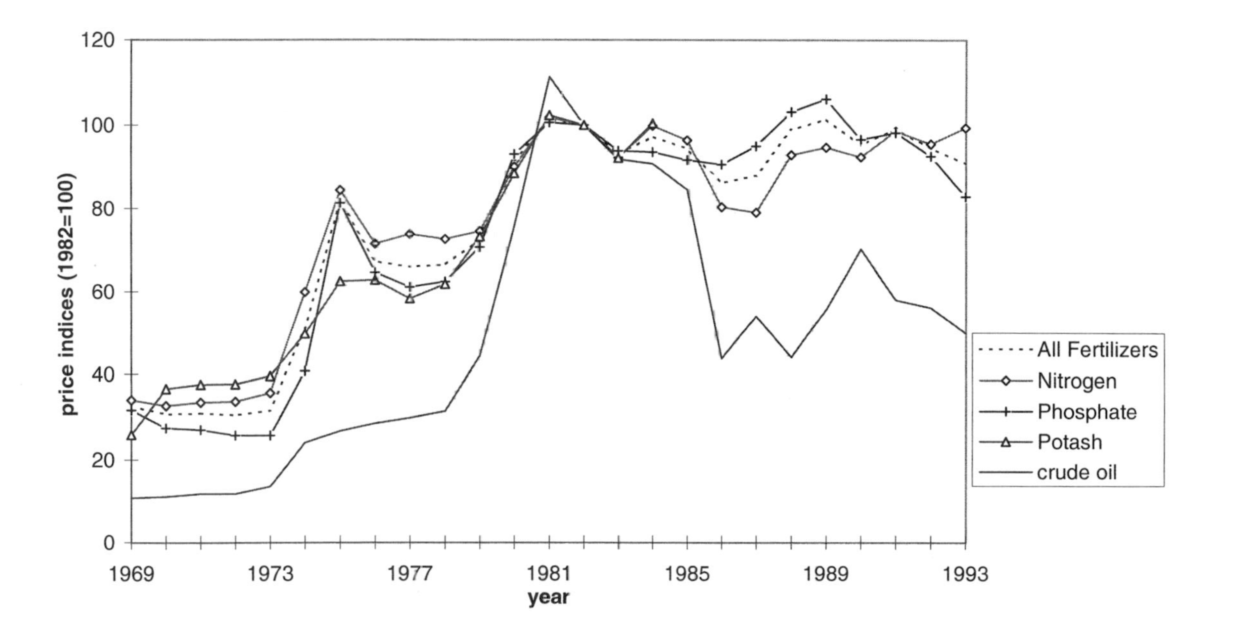


FIGURE 5-3 Price indexes of crude oil and selected fertilizers (producer prices, nominal) in the United States, 1969–1993. Oil prices are based on nominal price of domestic crude oil, first purchase price per barrel, in the United States. No numbers were available for potash after 1984. (Source: H. H. Taylor. *Fertilizer Use and Price Statistics, 1960–1993*. USDA Economic Research Service.)

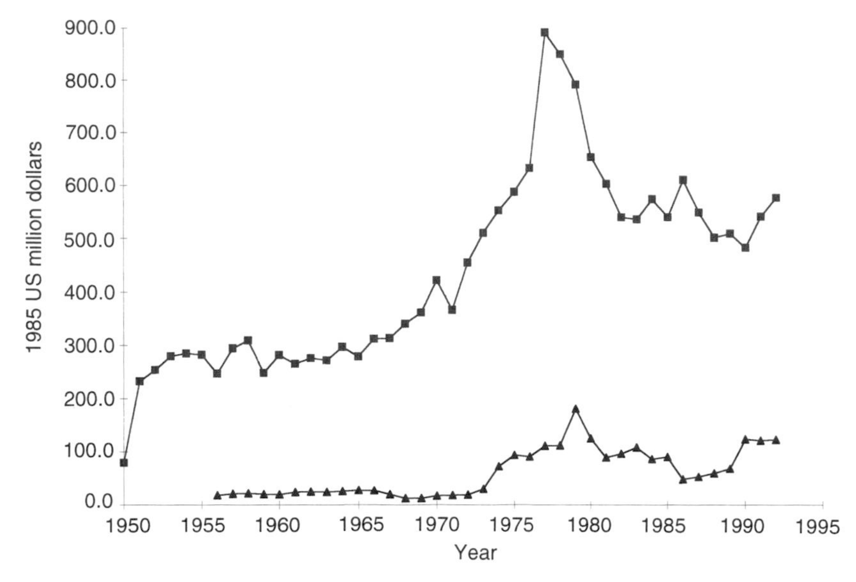


**FIGURE 5-4** Value of imported petroleum and petroleum products (△) and value of all agricultural exports (□), 1940–1995. (Prices in 1990 dollars.)

The objective of this chapter[1] is to examine past and present patterns of land use in Costa Rica with the objective of examining the biophysical limits to crop production and developing a simulation model of food and coffee production sensitive to land quality and quantity and to inputs, especially fertilizers. Then we will use this model to project production and its implications into the future as a function of different assumptions about population growth, land use, energy prices, and policies. Later chapters will examine these same questions with greater spatial and economic sophistication.

## II. ASSESSING THE FOOD PRODUCTION CAPACITY OF COSTA RICA

Over time we have constructed a set of four increasingly detailed and sophisticated analyses to assess the possible level of future food production in Costa

[1] This chapter, written initially in 1990 and revised for this volume, is based on a national-level overview of the food production system in Costa Rica. It left us with a number of questions, the most important of which is whether a more spatially explicit analysis would generate a different, perhaps more optimistic, scenario. Charles Hall spent his 1995 sabbatical in Costa Rica in order to pursue an answer to this question. Much of the rest of this book is the result.

Rica as a function of land capacities, land use, and the availability of fertilizer and other agricultural inputs. These models then compare that simulated food production to the food requirements of the growing population based on the number of people and the Costa Rican diet (Table 5-1). The first analysis can be found in Hall and Cornell (1990), the second in Levitan (1988), and some important elements of the third are published in Hall and Hall (1993). This chapter represents a fourth level, and includes a comprehensive simulation of the question of food production and foreign exchange at the level of the entire country of Costa Rica.

The essence of this fourth-level simulation model is a time loop that reads and plots many types of empirical data from 1940 to 1990, and then allows the user to make a series of decisions for the nation as a whole (or its farmers collectively). The future values for the variables are then projected through simulation, depending upon the decisions made. For example, the model predicts human population as a function of a user-defined growth rate from 1990 forward. The principal decisions allowed are (1) how much fertilizer to use, (2) which crops to use it on, (3) what crops to grow and how much area to grow them on, and (4) the population growth rate.

This model is included in the enclosed CD and can be downloaded by the reader should he or she desire. For example, the user may decide to increase the imports of fertilizers greatly or to change their allocation to crops for internal consumption vs export, here characterized by coffee. The model will then compute crop production for the nation as a whole using generalized

**TABLE 5-1 Costa Rican Diet**

| | Calories (%) | Protein (%) |
|---|---|---|
| Cereals | 37 | 30 |
| Rice | 22 | |
| Maize | 5 | |
| Wheat | 10 | |
| Beans | 9 | 22 |
| Animal products | 19 | 41 |
| Dairy | 10 | 17 |
| Meat | 7 | 20 |
| Eggs | 2 | 4 |
| Fats and oils | 14 | 0 |
| Sugar | 13 | 0 |
| Fruits, vegetables | 8 | 0 |
| Total | 100% | 93 |

*Source:* Levitan (1988).

production equations. The model predicts the quantity of food calories required by the population for each year, the quantity of calories produced by the chosen agricultural strategy, the amount of money (1990 dollars) required to pay for the fertilizer and other imports required for that agricultural production, the amount of money gained from export crops, and the area of land required to maintain that intensity of land use.

The model calculates agricultural production as a function of land area, land quality, and fertilizer input. Land area in each use is modeled as a function of total area dedicated to each crop type and a sophisticated function that assigns land uses to the land available as a function of land quality.

This chapter uses our "version 17" simulation model called "Fancy plot" that contains about 2000 lines of commented FORTRAN code linked with dynamic color displays and synthetic tabular output (see enclosed compact disc for the actual FORTRAN code). The model is driven by empirical data (mostly from FAO data—see file CRINT93.DAT on CD) from 1940 to 1990, and then allows the user to make either extrapolations or projections based on assumptions of his or her choice (Table 5-2). The setup table that allows the user to specify various aspects of the simulation is given with base assumptions in Table 5-3, and directions for its use are in Table 5-4.

The main results given here, and that are the base case scenario found on the CD, are derived from an extrapolation of present conditions and our decision to attempt to maximize agricultural production. For example, it is assumed that the population will continue to grow at 2.4% per year and that increased land and fertilizer will continue to be devoted to producing both subsistence and export crops. We also make the assumption for our base case that the number of cows will not be increased beyond the number existing in 1990.

The basic model structure is a main program that handles the time loop and calls a series of subprograms for each of the major components of the society, with an emphasis on agriculture (demography, food production, fertilizers, foreign trade, etc.), as well as several graphics programs (Table 5-5). Most of the simulation procedures are straightforward, atheoretical, and traditional, although their simultaneous linkage has rarely been undertaken. This model reflects our philosophy that real resource-related problems are associated with a complex suite of basic costs and benefits. In our view it is better to show simultaneously the many individual processes occurring in both the real world and its simulation rather than trying to aggregate all factors into one number, whether that be a dollar cost–benefit ratio, an environmental index, an energy summary, or whatever.

We have developed a complex but user-friendly output format that allows both the novice and the expert to comprehend the many complex processes and trade-offs that occur over time. Two particular advantages of this approach

TABLE 5-2 Basic Assumptions for Model FPLTCR (Fancy Plot Costa Rica)

1. A person requires about 2650 kcal per day, or 1 million kcal per year. This is the average diet for 1977–1985, and is slightly higher than the UN standard of 2600 kcal (Collins, 1982; Wilkie and Perkal, 1985; Wilkie, 1989).

2. A ton of grain contains about 3.5 million kcal and hence can support about 3.5 people for a year at minimum diet and with small postharvest losses.

3. Pastures support one cow per 2 hectares and produce roughly 200 kg (or about 700,000 kcal)/ha/yr (live weight), which would feed a person for about 3/4 of a year.

4. Population will increase at 2.4% per annum, slightly lower than the 1985–1990 rate of 2.6% per year (Comision Economica, 1989).

5. The types of food eaten in Costa Rica, determined from a study of the Costa Rican diet by the Ministerio de Salud (1979), will remain about the same (Table 5-1).

6. Agriculture growth parameters are reasonably well specified by the parameters in the setup file, Table 5-3. In this table YnoFert = yield if no fertilizer is applied, SfR is scope for response from adding fertilizer, 1/2 sat is the half-saturation response defining fertilizer response (see Figure 12-8), Degr is the decrease in yield due to erosion (% per year), and % Fert is the proportion of national fertilizer use allocated to that crop (from FAO, 1992b; FAO, various years).

7. Agricultural yield declines as a function of land quality. This is represented by the Relative Productivity table in Table 5-3. In this table, for example, the agricultural yield of annual crops on class II (second best) land is assumed to be (over time) 75% of that on class I land, while perennial crops has the same yield. And so on.

8. Values for extrapolating numbers of humans and the quantity of each crop type used after 1990 are given in the fourth line of the table in CRINT93.DAT. These also may be changed by the user with PEDIT or another ASCII editing program once the model and data are downloaded to the user's own computer.

are that it allows complex phenomena occurring over time to be comprehended readily (in one sense obviating the need for a discount rate) and shows how a particular policy effects various parameters simultaneously. When this model was shown to Oscar Arias, former president of Costa Rica, he said, "I like it. It forces the decision maker to face up to the various consequences of his or her decisions." Likewise the enclosed model allows the reader to face and understand the various effects of his or her assumptions and decisions. To the best of our ability it reflects the essence of the Costa Rican basic agricultural economy.

## A. Simulating Crop Yields

Since we were unable to find much specific yield-response data for particular crops that are generally applicable to Costa Rica, we derived agricultural yield

TABLE 5.3 Setup File for Program Fancyplot CR[a]

```
 INPUT FILE FOR CONTROLLING PROGRAM 'FNCYPLT'...                DEFAULT VALUE
 MAIN TITLE       ="COSTA RICA                           "      TITLE1
 SUBTITLE =       ="BASE PARAMETER RUN +                 "      TITLE2
 TIME START       = 1940
 TIME END         = 2040
 LAST TIME DATA   = 1991
 PLOT WANTED?     =    2            (MOVE PLT STUFF DOWN)     0 = NO, 2 = YES
 # BOXES          =   10                                      10
 # PARAMETERS     =   25            # empirical parameters read in....
 PRINT CONTROL    =    2   1  20    Print: 2=CROPS;3=COFFEE;4=GRAZING;5=PRICES

++++Parameters to determine agricultural productivity:
CROP Ynofert 1/2sat  SfR rDegr %fert
grain   0.30  90.   3.00  .005 0.18
Coffe   0.40  90.   1.60  .000 0.30
Banan   0.70  90.   2.00  .000 0.30
P.Food  0.60  90.   2.00  .000 0.10
Cattle  0.30.0015                      30% on Land Class I
Fallow  284. 4.79                      This is 1000's ha
++++ Parameters to extrapolate after last yr of data
grain   0.30  90.   3.00  .005 0.18
Coffe   0.40  90.   1.60  .000  .30
Banan   0.70  90.   2.00  .000  .30
P.Food  0.60  90.   2.00  .000  .10
Cattle  0.20 .00                       0.20   .00        This is %
Fallow  100.   0.                                        This is 1000's ha
TITLE: PRESENT LAND USE PATTERN
RELATIVE PRODUCTIVITY:
CLASS >     I     II    III     IV      V    Notes:
A.CROP    1.0    .75    .50    0.1    0.0    estimates from Jamie Echeverria
PERREN    1.0    1.0    0.6    0.3    0.1    and Joseph Tosi,
 COFFE    1.0    1.0    0.6    0.3    0.1    Tropical Science Center, assuming
PAST.     1.0    0.8    0.4    0.1    0.0    one or more decades of erosion...
OTHER     0.0    0.0    0.0    0.1    0.0
OTHUR     0.0    0.0    0.0    0.2    0.0
FOREST    1.0    1.0    1.0    1.0    1.0
>>>>>>>>>>>>>>>>>>>>>>>>>>>>>>>>>>>>
```

TABLE 5-4 How to Run and Modify the Model "Fancyplot"

Model output is easily seen by going to the CD and pushing the "synthesis" button on the screen. More detailed output is given in a series of ".dmp" files generated for land use and for agricultural production. These are easily read by typing "PE," followed by pressing the "F5" button and typing "*.dmp." For example, select "landuse.dmp" and type "1." The output file should appear on your screen for the year 1940. Pushing the "+" button on the right side of your keyboard will allow you to scan through the output very conveniently and interestingly. The user can modify the model setup file "crica.set" to change the parameters for each model run. An example for yield levels for crops in Costa Rica (ton/ha) follows.

| Yield scenario | Rice | Maize | Beans | Coffee |
|---|---|---|---|---|
| Current | 2.44 | 1.44 | 0.54 | 1.4 |
| Optimistic | 3.00 | 1.80 | 0.80 | 2.0 |

Yield levels were set as a result of personal communication with Dr. Margaret Smith, Department of Plant Breeding, Cornell University, Ithaca, New York, on January 14, 1988.

relationships from more general cultivar responses to fertilizers (Figure 5-2; see also Figure 12-4), adjusted downward slightly to match the lower yields observed in the tropics. We started with assumptions about what yields would be in the absence of fertilizers (Table 1-1). For example, it is possible to estimate from data given in Seligson (1980) that coffee yields prior to the use of fertilizers were roughly 0.3 tons per hectare. Higher yields occur with more inputs up to a little more than 1.5 tons per hectare. Such values are summarized in Figure 12-4c. These results are tested against empirical national-level land and fertilizer use rates and yields (FAO, various years). Although it is difficult to predict accurately what particular yields might be in any one place because of the variable nature of agricultural response to all conditions, this is less important for a large region such as an entire nation because different conditions will even out from one region to another. For example, deriving the impact of erosion, a particular problem in large parts of Costa Rica, is difficult, since in general the impacts of erosion are compensated for by farmers increasing their application rates of fertilizer. Rates of yield decrease of about 1 to 2% per year from erosion—in the absence of compensating fertilizer—seem most defensible (Pimentel *et al.*, 1987, 1995; Alfsen *et al.*, 1996). Thus although we cannot predict the yield in one particular place as erosion occurs over time, we can predict national-level yields. We use in our base case a conservative 0.5% per year. A more sophisticated model would make this a function of land quality.

The model uses FAO and other data up until 1990, and then allows the user to extrapolate into the future using whatever assumptions he or she wants

TABLE 5-5 Synopsis of Main Program of Fancyplot

```
        PROGRAM MAIN
C       Note: a "C" in the first column means that what follows is
C         a comment rather than functioning code.
C       Add declaration code
        INCLUDE DECCR

C       Simulate for 100 years, 1940 > 2040
        DO YEAR = 1,100
C          Determine real year:
           RYR  = 1939 + YEAR
C          Call subroutines where all the detailed calculations are made:
           CALL DEMOGRAPHY
           CALL EXTRAPOLATION
           CALL LAND USE
           CALL FERTILIZER
           CALL AGPROD
           CALL CATTLE
           CALL COFFEE
           CALL ECONOMIC
           CALL CARBON
           CALL DISPLAY
       ENDDO
       END

C      Example subroutine
       Subroutine DEMOGRAPHY
C      Simple exponential growth of human population as a function
C        of growth rate (gr) which is roughly 2.4 % per year:
         Hpop(t  ) =   HPop(t-1) * (1+gr)

C        Food energy required for a person for a year:
         EReg      =   HPop * 2650 * 365

C        Tons of grain required per capita (in Kcal/Ton Grain)
         TGrreq    =   EReq / 3.64 e6

       RETURN
       END

       Subroutine AGPROD
C      Definitions
C      COMMON statements

C      Grain production = Basic (i.e. no fertilizer) production times an
C      erosion factor, plus fertilizer effects times scope for response
C      to erosion.
       GYpHSb = (NFYpHa + ScfRes * (TFpHaG /(TFpHaG + tGhsat)) * erofac)

C      Total grain production is yield/Ha on best land times
C      area times a land quality factor
       GrProd(t) =   GrArea   * GYpHa * LQFac

       RETURN
       END
```

to about future patterns and rates of change of land allocation, fertilizer use, or types of crops grown (Table 5-3). The importance of poorly known parameters was investigated through sensitivity analyses.

## B. Predicting the Location of Land Use

As noted by the early economist David Ricardo (1817), farmers tend to use their best (generally meaning flattest, most fertile, and of medium moisture) land first. In addition people tend to use the very best agricultural land (i.e., Land Class I) for annual crops such as grains and vegetables, and land of progressively poorer quality for perennial (tree) crops and for pastures. This is clearly the case in Costa Rica. Land of the poorest quality from the perspective of human use (i.e., the steepest land with shallow, infertile soils) tends to be left in forests. Thus we can rank soil use as a gradient of desirability for human use, with the first use getting the best soils, as annual crops > perennial crops > pastures > forests. There are several exceptions to this pattern. First, over time and with population expansion people tend to build urban areas on the best, meaning flat and fertile, lands. Their grandparents or great grandparents built their few houses on that land because it was next to the land they thought best and used for farming, and their children and their children's children settled there too (see Land use on CD). Second, wealthy people often own the best land, which they frequently put to extensive use, such as pastures.

There are five basic land quality types for Costa Rica (Table 5-6). Class I land, about 20% of the total area, is generally flat and fertile. Class V land should be used only for forests and watershed protection, while intermediate

**TABLE 5-6 Land Use and Land Use Potential Based upon Soil Type**

| | Land use potential | | Land use, 1987 | |
|---|---|---|---|---|
| Land class | 1000 ha | % of total land | 1000 ha | % of total land |
| I. Clear tillage | 944 | 18.6 | 325 | 6.4 |
| II. Pasture | 466 | 9.2 | 2300 | 45.0 |
| III. Permanent crops | 816 | 16.2 | 207 | 4.1 |
| IV. Forest | 1609 | 31.7 | 2104 | 41.2 |
| V. Protected areas | 1249 | 24.6 | | |
| Other | 149 | 3.4 | | |
| Total | 5084 | 100.0 | 5084 | 100.0 |

*Source:* Costa Rica Atlas Estadistico (1981). Land use potential from Hartshorn *et al.* (1982). Land use from Levitan (1988).

lands (II–IV) are variously suited for grazing and tree crops. Class I land can be put to any use, but higher numbered classes have increasing restrictions. Each land class is assigned a different production potential for each use type, which generally means less production in higher land classes. When there is little land used by humans (i.e., in 1940) all uses (urban, annual crops, perennial crops, and grazing) can occur on Class I land. But as time goes on, and total land use increases, there is increasingly not enough land for all uses (i.e., crops grown) to occur on land of the highest quality. Our model compares total land needed (as determined from year by year FAO estimates of land use) and allocates whatever land is available to the "highest priority" use [in the order of urban, pastures on the best land (assumed to be one-quarter of total pastureland), annual crops, perennial crops, general grazing land, and forest]. In other words, urban areas will be placed entirely on Class I land, and then 30% of pastures and all annual crops will also be placed there if enough Class I land is available. As total land use increases, the lower priority land uses (perennial crops and the rest of pastures) are pushed increasingly onto land of progressively lower quality. Perhaps the greatest uncertainty in constructing this model is deciding how much yields decline as lower quality land is used. Our best guess of land productivity as a function of land quality, generated with the assistance of scientists at the Tropical Science Center, San Pedro, is given in Table 5-3. These values reflect the losses in yield expected after one or two decades of use of lower quality land compared to yields expected from Class I land, other things being equal.

## C. Foreign Exchange

The model also recreates (until 1990), and then simulates (after 1990), the foreign exchange derived from the external sale of coffee and bananas and the foreign exchange required for all externally derived inputs to the agricultural sector, such as fertilizers and trucks. We used historical international prices (corrected to 1990 dollars). We then used 1990 prices for base case extrapolations, i.e., coffee at $2655 per ton; bananas at $155 per ton; wheat at $230 per ton; beans at $702 per ton; and nitrogen fertilizer at $333 per ton. We assume the cost of freight and handling to be 30% on all commodities (Levitan, 1988).

## D. Model Scenarios

We ran various scenarios representing possible policies and physical circumstances for the future of Costa Rica. Our base scenario, represented by the

TABLE 5-7 Examples of Food Production Scenarios for Costa Rica

Scenario I. Base case. Continue the present pattern but maximize production by increasing crop area and increasing imported agricultural inputs such as fertilizers at the same rate as population growth. Domestically produced annual crops are estimated to provide about 50% of the calories in the current diet, with the remainder produced abroad (20–30%) or derived from perennial (22%) and pasture (8%) crops grown in Costa Rica (Levitan, 1988). Based on the Ministerio de Salud survey (1979); Atlas Estadistica (1981); Guillen and Gutierrez (1981); Levitan (1988); UN (1985 and earlier editions); and unpublished data compiled by Rafeal Statler (CATIE). Industrial inputs to agriculture, in dollars, assumed to be a conservative 4 times the cost of fertilizer (Levitan, 1988).

Scenario II. Same as Scenario I but do not increase fertilizer inputs after 1990.

Scenario III. Maximize food self-sufficiency. Assume that the percentage of rice and maize in the diet would increase and that only beans, equal to 5% of calories consumed, would continue to be imported, since there is not sufficient suitable land in Costa Rica to grow enough beans to satisfy demand. In this case 80% of fertilizers are used on annual crops, and 20% on domestic perennial crops and pasture. No animal feed stuffs are imported and 13% of the calories from domestically produced grain are fed to animals.

Scenario IV. Maximize export crop production and food imports. Assume that 20% of calories would be derived from domestically produced annual crops, 30% from domestic perennial crops and pasture, and 50% purchased from abroad. This scenario assumes that only fresh fruits and vegetables and half the required grain are produced domestically. Only 2% of the calories derived from domestically produced grain are fed to animals.

Scenario V. Same as Scenario I but assume a doubling of oil prices relative to export crops.

Scenario VI. Same as Scenario I but assuming a doubled crop response to fertilizer inputs.

The results of these scenarios follows.

| | Millions of people fed domestically (with imports)[b] | | | Forest area ($10^6$ ha) | | | Cost of agric. inputs (millions 1990 dollars) | | |
|---|---|---|---|---|---|---|---|---|---|
| Scenario[a] | 2000 | 2020 | 2040 | 2000 | 2020 | 2040 | 2000 | 2020 | 2040 |
| I | 2.0(4.1) | 2.4(4.5) | 2.6(4.9) | 1.3 | 0.8 | 0.0 | 77 | 124 | 201 |
| II | 1.9(3.8) | 1.8(3.8) | 1.6(3.6) | 1.3 | 0.8 | 0.0 | 78 | 78 | 78 |
| III[c] | 3.0(3.3) | 3.6(3.9) | 4.1(4.1) | 1.3 | 0.8 | 0.0 | 100 | 161 | 261[c] |
| IV | 1.9(4.2) | 2.3(4.6) | 2.6(5.0) | 1.3 | 0.8 | 0.0 | 100 | 161 | 261 |
| V | 2.0(4.1) | 2.4(4.5) | 2.6(4.9) | 1.3 | 0.8 | 0.0 | 154 | 248 | 402 |
| VI | 3.3(7.0) | 3.9(7.9) | 4.2(8.6) | 1.3 | 0.8 | 0.0 | 100 | 161 | 261 |
| Projected population | 3.8 | 6.4 | 10.8 | | | | | | |

[a] These numbers are a reasonable but very approximate estimate of the possibilities.
[b] The first numbers are number of people fed from domestic land area. The number in parentheses includes people fed from trade (estimated using coffee as example for all export crops). The number fed is derived as the net annual profit divided by the price of wheat per ton times three people fed for one year per ton of wheat.
[c] Scenario III is unfeasible as it generates almost no foreign exchange to pay for fertilizers, etc., so is just for the "what if" calculation.

conditions in the base setup file (Table 5-3), assumes that current rates of population growth, prices, and agricultural response to land quality and inputs remain at present levels. Obviously there are many other possible futures for Costa Rica. Our model is meant to allow the easy exploration of those possibilities. The directions for doing this are given as Table 5-4.

# III. SIMULATION RESULTS

## A. Model Predictions

We find that even with what we perceive to be optimistic assumptions about land utilization, fuel and fertilizer prices, and yield increases, Costa Rica as a nation will be extremely unlikely to feed an exponentially growing population from domestic food production ever again (Table 5-7, Figure 5-5). The problem

**FIGURE 5-5** Output for computer data analysis and simulation program for various entities for Costa Rica. The program is found on the accompanying CD, and can be downloaded, run, and modified by the reader. This output summarizes a comprehensive computer analysis representing the process of development, and its consequences, in Costa Rica from 1940 to 2040. The central image, based on data in Sader and Joyce (1988), represents the extent of basically undisturbed forest in the country at various times from 1940 to 1983. The 10 small graphs are time series of FAO and other data from 1951 to 1990. Numbers from 1940 to 1950 are educated guesses, and post-1990 values are the results of extrapolations (population, fertilizers), user-supplied assumptions, or simulations. (**a**) Projected conditions near the beginning of the simulation in 1943 (year of first author's birth). (**b**) Conditions in 1990. All graphs and images represent good data from FAO and Sader and Joyce (1988) except for simulated food production in the fifth top graphlet, which shows close agreement with actual FAO production data, shown in black. (**c**) Simulation to 2040 assuming a 2.4% per year population growth and an emphasis on agricultural expansion by increasing land area in both annual and permanent crops and fertilizer input by 2.4% per year. (**d**) Same, but assuming no erosion effects.

The first "graphlet" is the number of humans and cows. The second represents the land area in agriculture. The horizontal lines are the (cumulative) amounts of land in Land Classes I (best), II, and III from the bottom upward. The lower black area is urban areas, the second is the area of pastures on the good lands, the next is the area of perennial crops, and the fourth is the area of permanent (tree) crops. The third graphlet represents fertilizer used (gray, rising toward the right), the yield per unit fertilizer (black), and the quantity used on grain and coffee. The fourth graphlet is yield per hectare for grain and coffee. Note the saturation of yield in the 1990s, perhaps a consequence of fertilizer saturation, cumulative erosion, or production forced onto lower quality land. The fifth graphlet is the food calories required to feed the population (top), and a simulation of national food production based on area cropped, land quality, and fertilizers. Empirical levels of food production are represented in black. In the bottom row of graphlets, the first is total area of forests, second is the carbon released (or taken up) by land use change, third is the total national area divided into land usages, fourth is the soil lost by erosion (based on Repetto *et al.*, 1989), and fifth is foreign exchange (billions of dollars) gained from exports of coffee and the quantity required to pay for the fertilizers and other inputs to the agricultural sector.

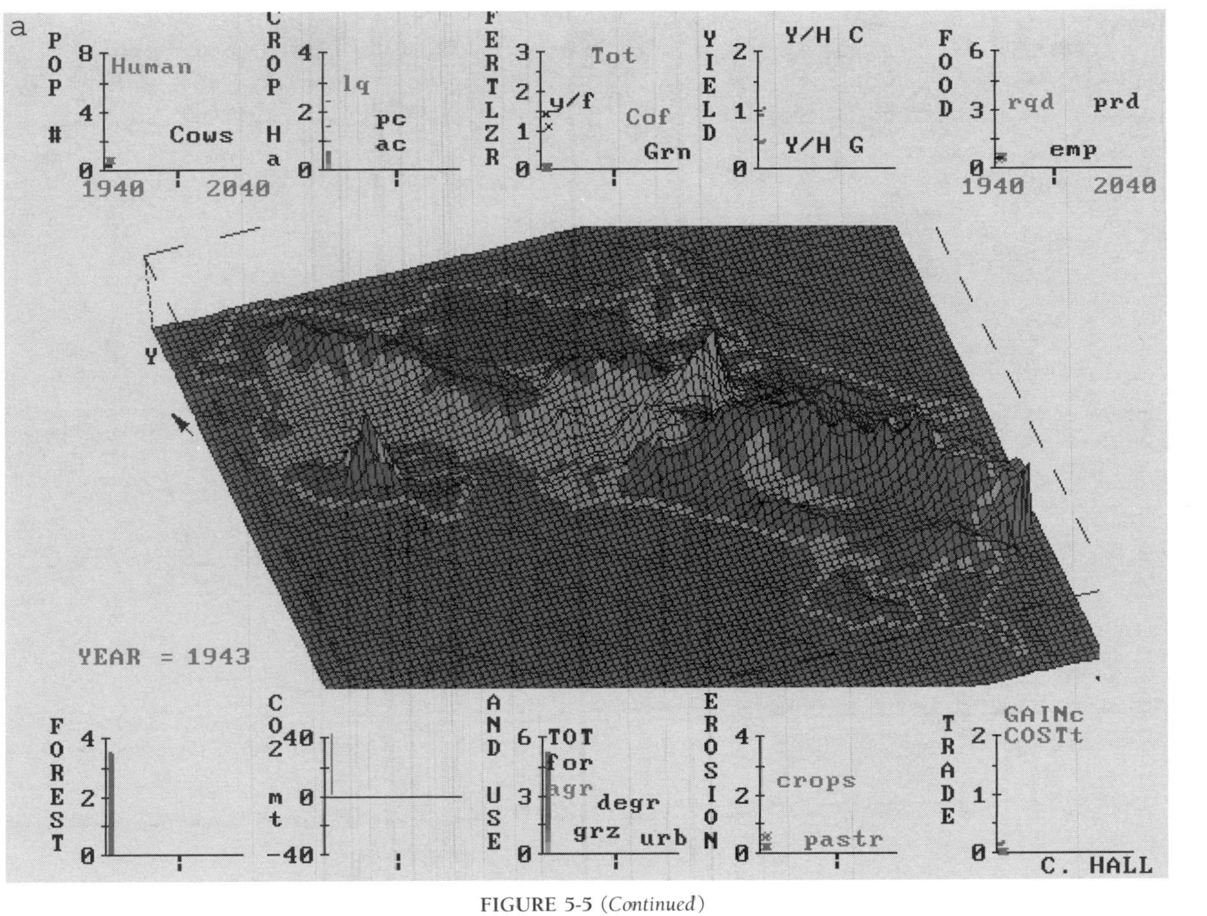


FIGURE 5-5 (*Continued*)

FIGURE 5-5 (*Continued*)

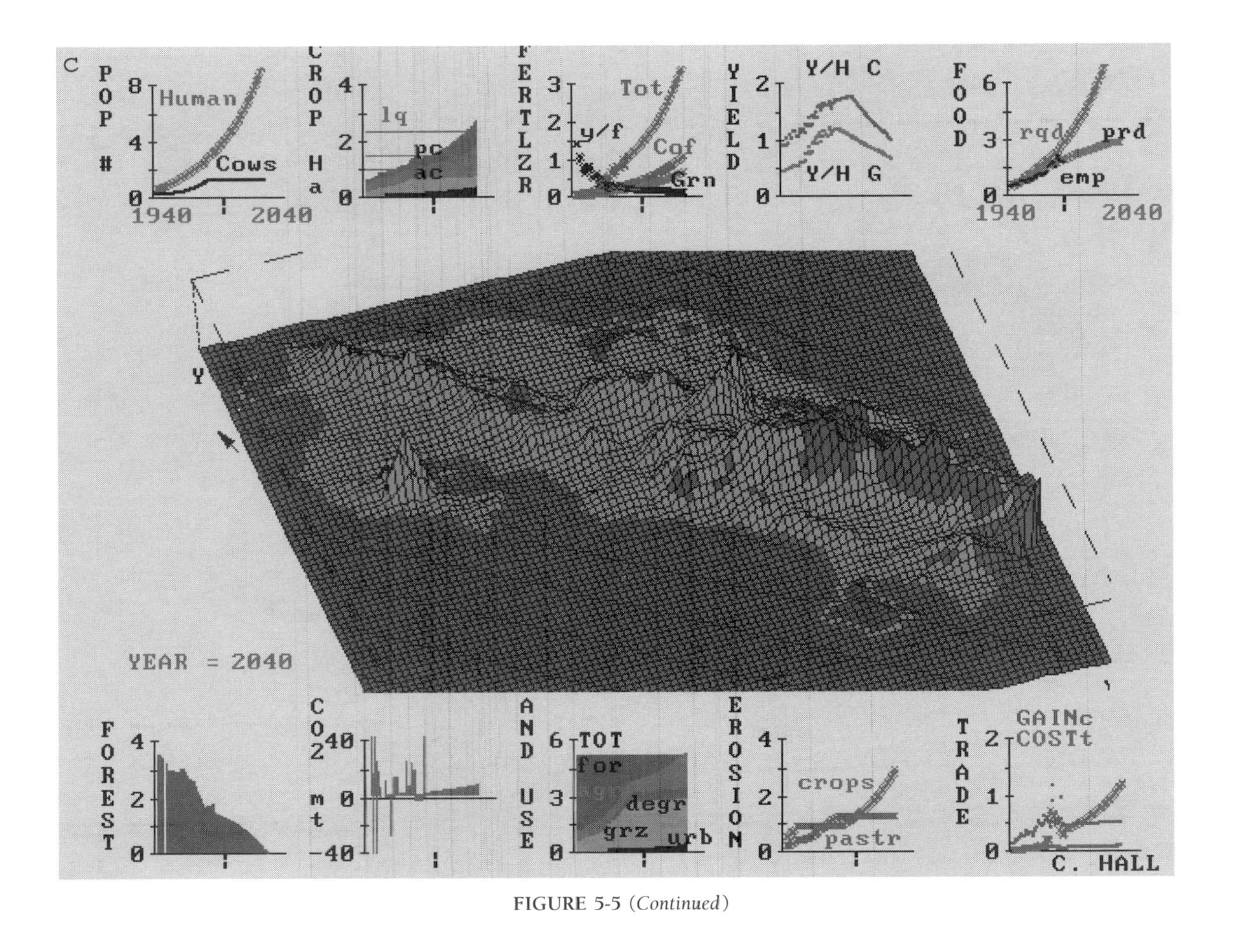


FIGURE 5-5 (*Continued*)

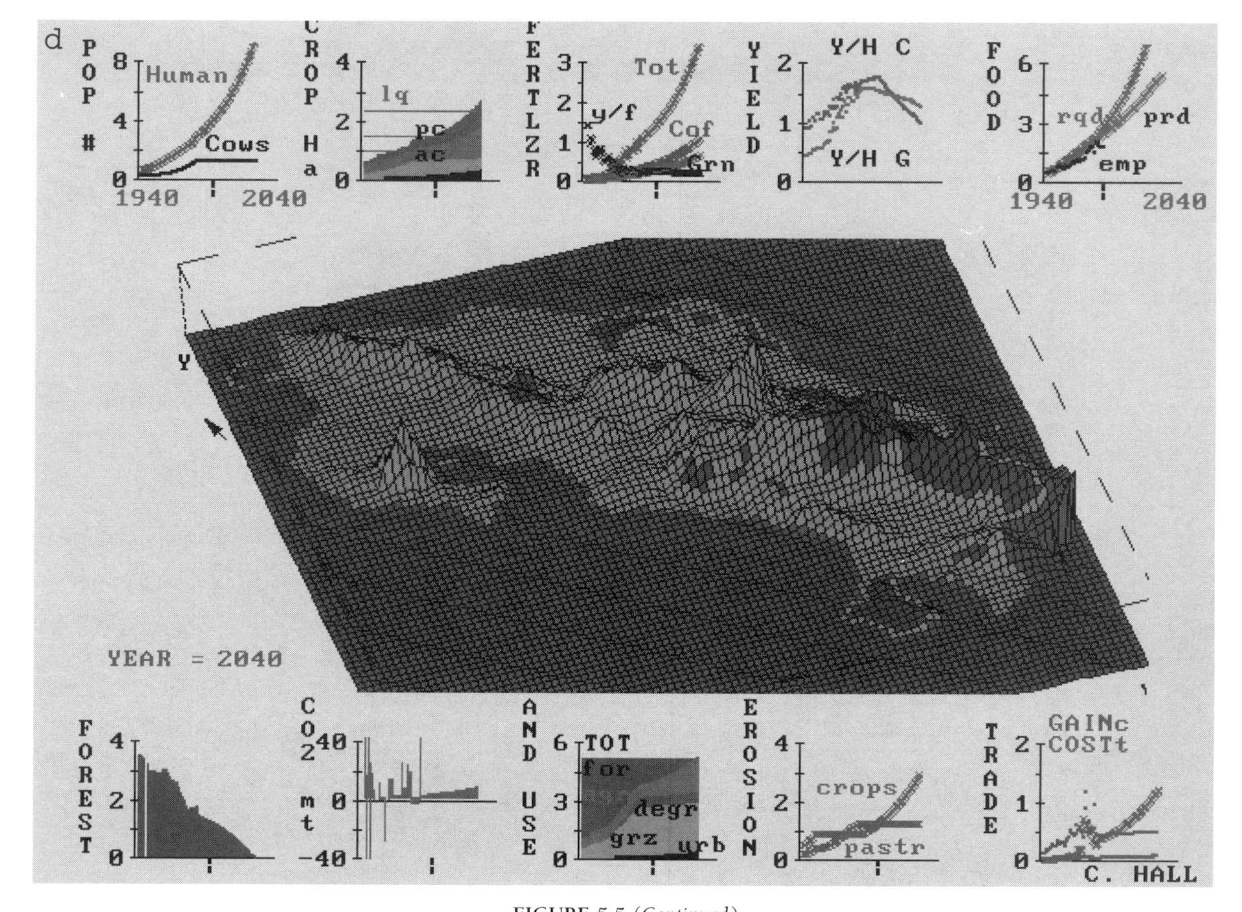


FIGURE 5-5 (*Continued*)

becomes considerably more difficult if the full effects of soil erosion or potential oil price increases are included in the model. If the population is stabilized it becomes possible, although difficult, to feed the population from indigenous production early in the next century. The model used here gives essentially the same qualitative and quantitative results that we have obtained with earlier analyses based on a spreadsheet analysis (Levitan, 1988) and on simpler simulations (Hall and Cornell, 1990). Thus we believe our results robust, at least as long as our assumptions are met.

One of the principal reasons it will be difficult to feed all Costa Ricans from domestic resources in the future is that increasingly there will not be enough land available of sufficient quality to produce high yields of food crops. Thus until about 1964, but not after, it was possible to both feed the country on domestic harvests and put all perennial crops (export tree crops) into Class I land. Subsequently perennial crops were of necessity allocated to land of ever lower quality (Figure 5-5). The use of high-quality lands for pasture is of special note. If all cows could be moved off of Class I land, all perennial crops currently grown could still be accommodated on Class I land, and all perrenials on Class II land, until about the year 2000.

One national-level strategy for maximizing domestic production of food calories that we can explore in the model is moving cattle off of the best land, maximizing fertilizer and other inputs on that land, and growing grain. Another possible strategy is to maximize the production of export crops such as coffee and trade them for grain. About six times more people can be fed per hectare growing coffee than can be fed growing food crops. We think this is consistent with the fact that Costa Rica is actually importing an increasing proportion of its food calories—about 30% in 1995—and paying for it in a net sense with increased exports. We find no evidence in the data up until the present that technology has improved the efficiency with which fossil energy is turned into either crops or general economic welfare—in fact, the efficiency appears to be declining (Figures 5-5 and 12-4) (Ko *et al.*, 1998). The most important results of other assumptions derived through sensitivity analyses are given in Table 5-7.

Assuming (1) an international demand for those export crops that physically can be grown (such as coffee) and (2) reasonable prices for imported grains, then it would be possible to feed Costa Ricans indirectly from their own land well into the next century as long as sufficient quantities of imported fertilizers and other inputs are available—although it would be extremely difficult to pay for the inputs (Table 5-7, Figure 5-5). Twenty to thirty times as much nitrogen fertilizer would be required as is now used, primarily on crops such as coffee, and a large commitment of land. Since much of this coffee would need to be grown on land marginal for coffee, we assume that in fact other high-value export crops would be substituted for coffee, and we also assume that these crops would be similarly land and input requiring. If the demand for coffee and bananas does not continue to grow, or the fertilizers are not affordable, then Costa Rica, traditionally one of the world's great agricultural countries, will become overwhelmingly dependent on food imports from other countries unless a completely different basis for agriculture or the economy is forthcoming and viable. No matter what happens, Costa Rica's increasing human population seems almost certain to become increasingly vulnerable to changes in the availability or price of fossil fuel and fossil-fuel-derived fertilizer, weather in the prairie states of North America, and North American export policies, economics, and breakfast preferences.

### B. Model Validation and Sensitivity Analysis

The output of our model is complex, and hence full validation is not possible. Many parts of it can be subjected to validation, however, and those that cannot are so straightforward that it is difficult to conceive where the predictions can run astray. First of all, essentially all of the data presented in Figure 5-5 are empirical up to 1990. The exception is the total food production graphlet (the fifth from the left on the top). In this case food production is simulated as a function of land area, fertilizer used, and land quality. The results of this simulation up to 1990 are almost identical (on average) to total empirical food production derived from FAO statistics (darkest line), resulting in a noncircular, nontrivial validation of the most important modeled component. Simulated yield per hectare increases and then reaches an asymptote (fourth graphlet on the top), just as is occurring in Costa Rica (Figure 12-4). Whether mean yield per hectare will decrease in the future as it becomes increasingly necessary to use lower quality land and as the cumulative impact of erosion continues is not yet known. One aspect of the model that is uncertain is that it assumes that the impact of erosion on crop yields accumulates year after year. What may happen instead is that farmers simply may use more and more fertilizer to compensate. Thus the impact of erosion may be evidenced initially more in farmer costs than in yield declines.

The principal sensitivity analysis we have undertaken is in different land use scenarios. Specific model response to uncertainties in, e.g., possible future agricultural technologies, is given in Table 5-7.

## IV. WHAT THIS MEANS FOR THE FUTURE

### A. Alternative Ways to Feed the Growing Population

There will not be a specific "moment in time" when Costa Rica no longer has adequate land on which to grow food. In fact, the case can be made that it has already occurred, since Costa Rica imports from a quarter to a third of its food calories and is already using more than two-thirds of its land area for agricultural production, including exports. On the other hand Costa Rica exports nearly the same quantity of calories as beef, which could be consumed at home were it not so valuable as a generator of foreign exchange. Given the finite amount of arable land, the people of Costa Rica have several options for meeting food demand: (1) eliminating export crops and raising crops only for domestic consumption, (2) redistributing land with the assumption that each hectare would be farmed more intensively, (3) importing more food, (4) reducing production and consumption of animal products, (5) farming more intensively, or (6) stabilizing population. It is important to examine each of these in turn to see what is possible and desirable.

### 1. Food Self-Sufficiency (Scenario III)

For political and economic reasons a policy of national food self-sufficiency might be considered desirable. The main argument for this is that while from 15 (Chapter 10) to 19% (Hartshorn *et al.*, 1982) of the 5.1 million ha of land in Costa Rica are considered suitable for cultivation, only about one-third of this land, or 6.4% of the national land area, is used for annual crops or their fallow years (Chapter 10). This 6.4% of the land base provides about 50% of the calories now consumed by Costa Ricans (Levitan, 1988). The remaining two-thirds of Class I land is used primarily as pasture, and produces only an estimated 3–4% of the food calories consumed (Table 5-8). Thus one potentially viable strategy would be to move all cattle off of Class I land and use this land for food crops. Unfortunately much of the land would need to be used for cash crops to generate the foreign exchange required to purchase the inputs required on the annual crops for good yields. So if this was done, yields per hectare probably would fall. Two other problems with this approach are that (1) row crops require much more fertilizer than pastures, so that more export crops would be needed, and (2) tilled land has much higher erosion rates than pasture land. Nevertheless it clearly could increase food yields.

Even if all the Class I land were used for the cultivation of food crops, this would be a short-lived strategy, viable for no longer than 30 or 40 years assuming the present fertilizer use, diet, and population growth rate (Figure 5-5). By the year 2040 all the Class I land—that considered suitable for growing annual crops—will be required for grain production without allowing for any fallow periods, pastures, or export crop production. If yields were to increase to more optimistic levels this land could feed Costa Ricans until about 2060. Unfortunately yields per hectare seem to be leveling off despite increased uses of fertilizer (Figure 12-4).

TABLE 5-8 Land Use in Costa Rica in More Detail (in Thousands of Hectares and Percent)

| Description of land use | 1973 | 1987 | Change |
|---|---|---|---|
| Total land area | 5110 | 5110 | |
| 1. Nonfarm, nonforest | 79 (1.5%) | 100 (2.0%) | |
| 2. Forest, nonfarm | 1908 (37.3%) | 1742 (34.2%) | −9% |
| 3. Farm | 3123 (61.1%) | 3268 (64.0%) | +5% |
| 3a. Cultivated land | 283 (5.5%) | 325 (6.4%) | +15% |
| i. Annual crops | 141 | 225 | |
| ii. Fallow | 125 | 100 | |
| iii. Other | 17 | — | |
| 3b. Permanent crops | 207 (4.1%) | 207 (4.1%) | 0% |
| 3c. Pasture | 1558 (30.5%) | 2300 (45.0%) | +48% |
| 1. Cultivated and improved (cut) | 92 | | |
| ii. Cultivated and improved (not cut) | 641 | | |
| iii. Natural range | 826 | | |
| 3d. Forest, scrub and wild, within farms | 1001 (19.6%) | 362 (7.1%) | −64% |
| 3e. Other farmland | 74 (1.4%) | 74 (1.4%) | |
| Total forest, wild and scrub[a] | 2909 (56.9%) | 2104 (41.2%) | −28% |

*Sources:* 1973 data compiled from Atlas Estadistico (1981); 1987 data are estimates projected by Levitan (1988).

[a] Sum of lines 2 and 3d.

The ramifications of "self-sufficiency" programs must be looked at carefully. For example, in the 1980s Costa Rica, which had begun to import milk, undertook a policy of milk self-sufficiency. This was in fact achieved, although it is less well known that this has required the import of huge amounts of animal feeds from the United States to feed the dairy cows! These imports are more or less the largest dollar item of all agricultural imports (Levitan, 1988).

## 2. Redistributing Land and Cultivating Land More Intensively

This scenario is related to the preceding section. Costa Rica is traditionally a nation of small to medium farmers, although the trend is toward greater discrepancies between the rich and the poor. Presently much of the best land is in the hands of wealthy people, many of whom use the land for extensive uses such as grazing. It is commonly argued that redistributing land from these fewer large landowners to larger numbers of small farmers who will farm the land intensively will increase yields. Our sensitivity analysis found that eliminating cattle on the best land would allow Costa Rica to continue to feed itself for about an additional 20 years, other things being equal.

### 3. Maximizing Food Imports (Scenario IV)

A third strategy would be to maximize food imports. Wheat imported from the United States has become an important staple of the Costa Rican diet, particularly in urban areas, in large part as a consequence of its relatively low price and a subsidy by U.S. foreign aid. In the 1970s and through the mid-1980s, it was twice as expensive to buy rice grown in Costa Rica as wheat imported from the United States, even though the price for wheat included freight and handling charges equal to about 30% of the international trade price (Levitan, 1988). The per capita cost of imported food rose from 11% of wages in the 1970s to more than 20% in the early 1980s. The direct impact on workers has not been quite as large yet because the cost of food and other imported purchases has been absorbed by the growing debt.

We examined the possibility of maximizing food imports by increasing the land area devoted to export crops so that only perishable fruits and vegetables, eggs, and half of the basic grains would be produced domestically. Then more land would be available to produce export crops, and the demand for cultivated land would be far less pressing. Using current yield levels (Table 5-4), the Class I land already under cultivation would suffice until 2020 for growing both the basic food crops required to supply 20% of calories and the coffee needed to pay for importing the remaining 80% of calories. With this reduced demand for domestically produced food, there would be sufficient arable Class I land—assuming it was all used for grain crops or coffee—beyond the year 2050 (Figure 5-5). The requirements for fertilizers and other inputs at this level of production, however, would be extremely difficult or impossible financially, and make the nation far more vulnerable to potential future energy shocks.

#### *a. Problems with Growing Coffee to Pay for Imported Food*

While this trade-intensive scenario does reduce the demand for good agricultural land, it has other costs. It assumes that a reasonable price and market for coffee or comparably high-valued export crops will exist, and that inputs will be available and affordable. In the mid-1980s, the net income from nearly one-third of the coffee exported was required just to generate sufficient foreign exchange to buy the imported 20% of food calories (Levitan, 1988). Projecting from current crop yields and relative prices of coffee and imported food crops, after the year 2030 more land would be needed to grow the amount of coffee needed to pay for imported food than is now used to grow the entire coffee crop (Figure 5-5). If a larger proportion of calories were to be imported, as suggested by the import-maximization strategy, and yields do not increase, income from virtually all coffee exports would be needed to pay for food by as soon as the year 2000. This would not allow for the purchase of fertilizer for any crop, including coffee (Table 5-7). Furthermore, the 100,000-plus ha currently used to grow coffee is greater than the optimal area available nationally for coffee (54,000 ha), and a large proportion of the optimal plus suboptimal area of 184,000 ha (Levitan, 1988; Costa Rica, 1993) (see Chapter 12). Thus, the export-dependent scenario projected beyond the year 2000 (Table 5-7) may in fact require the cultivation and marketing of other high-value export crops that can be grown in other climatic and soil zones, such as flowers and macadamia nuts (Levitan, 1988). The successes and failures of these alternative crops are discussed in Chapter 23.

#### *b. Advantage of Growing Coffee Instead of Domestic Food Crops*

Coffee sales generate six to seven times as much money per hectare as does the sale of basic foods. In 1984, for example, 25,000 ha, or 0.05 ha per person, was required to grow the coffee that generated sufficient income to purchase imported food sufficient for half a million people (roughly 20% of calories consumed nationally). If these calories had been produced domestically, they would have required 165,000 ha, nearly seven times more land. Thus coffee is a good way to feed increasing numbers of people from a finite land base. The downside is that coffee requires about eight times as much nitrogen fertilizer per ton produced as do the basic food crops. Thus approximately the same amount of nitrogen fertilizer would be required whether the food was produced directly or via coffee grown as an export crop; the difference is that a much smaller land area is needed per unit food produced through coffee production and trade. In the 1990s coffee prices rose somewhat, giving coffee slightly greater economic leverage when traded for food.

### 4. Reducing Consumption of Animal Products

About two-thirds of Costa Rica is in pasture. Turning this land into cropland and farming it intensively might seem to provide far more land than is needed to feed any foreseeable population. The problem is that most of this land is of very poor quality for agriculture, being inherently infertile, too dry, too wet, or very steep. Cattle provide a means of using this land even though the yield per hectare is low (Chapter 14). In addition, cattle production requires very little industrial input unless the pastures are improved, and erosion is much less than from annual food crops (Chapter 14). We do not find the arguments for using this land for annual crops compelling, although we think that if Costa Ricans are desperate for food they will indeed utilize this land for annual crops.

### 5. Farming More Intensively: Technology as a Means of Improving Yields

Many observers hold a great deal of hope for continued technological progress in increasing agricultural yields. Unfortunately the track record to date leads one to conclude that progress has come essentially only through increased reliance on fossil fuels and fossil-fuel-based technology (Chapter 12), which is expensive and potentially unstable due to uncertain future fuel prices. Therefore the application of increasing technology, at least as it has come to pass thus far, is severely limited, first by the ability to pay for fertilizers and other fossil-energy-based inputs, second by the ultimate limit of cultivar response to increased fertilizer application (Figures 5-2 and 12-4), and in the longer run by the availability of the fossil fuels needed to produce fertilizers and other production inputs.

There are many who advocate investment in knowledge about low-input agricultural and environmental interactions as a means to most efficiently utilize fossil-energy-based inputs. It is possible that labor-intensive technologies, such as those used to grow rice in many parts of Asia, could expand yields with little additional use of fossil fuels. That may also apply to agroforestry, in which food and tree crops are combined. Perhaps the greatest return on investment in agriculture would be from those practices which enhance soil and water conservation. Such investments in turn would make Costa Rica relatively less dependent upon uncertain supplies of fossil energy in the future. But Costa Rican society is changing rapidly in ways that make this labor-intensive approach difficult. Already it is frequently difficult to get enough Costa Rican field workers, and farm laborers are often imported from Nicaragua. In general, industrial inputs to the land have been used more intensively over time, not less, as pressures for food and foreign exchange mount.

### 6. Reducing Population Growth

This is covered under Conclusions.

## B. Environmental Considerations

The scenarios just presented all have devastating impacts on Costa Rica's remaining forests and soils unless virtually all of the land presently in pastures is released from that use. Otherwise, as the remaining forests are used, carbon will continue to be released to the atmosphere until there are no more forests—about 2040 in most of our simulations. By this model's calculations, $CO_2$ will be released at roughly 10 million tons per year until that time (Chapters 15–16). Producing enough food for an increasing Costa Rican population beyond about 2030 is constrained seriously by both land and energy, regardless of the production strategy employed. According to evaluations of land use potential, no more than 44% of the land in Costa Rica should be used for agriculture, including rangeland, and already more than 55% is so used—45% for rangeland alone (Hartshorn *et al.*, 1982).

## C. Economic Considerations

Our analysis could be used to justify increasing reliance on export crops. But there are important environmental and other reservations about the increasing use of export crops to generate foreign exchange. Commercial banana production, for example, is extremely dependent upon a whole pharmacopeia of chemicals, many of which are environmentally harmful. Another ramification of increasing exports and using the money to increase food imports is that this increases the cycle of dependency, whereby "less-developed" countries must produce increasing quantities of export products in order to earn foreign exchange to pay for food, as well as servicing their debt. Nevertheless, within the already established context of the need to generate foreign exchange, and given the constraints imposed by a growing population, agricultural export crops may be the only viable means of feeding Costa Rica. Fossil-fuel-derived inputs used in agricultural production appear to have a considerably greater financial return on investment than do inputs into most manufacturing processes (Levitan, 1988) (Chapter 23). Specifically, in the 1980s agricultural exports returned five times the dollar value of production inputs imported for use compared to manufactured items in Costa Rica.

If export crops are to be emphasized, there are arguments for developing other high-value agricultural export crops in addition to coffee and bananas. In ideal circumstances coffee is a good means of generating foreign exchange, but relying on its exchange-generating capabilities is precarious because of the frequent glut of coffee on the world market, the increasing cost of agrochemical inputs, which make production risky for the individual farmer, and the relatively long time it takes to add or subtract coffee trees from production. In fact, as of 1990, Costa Rica was generating more import income from nontraditional crops, such as macadamia nuts, than from coffee. Dependence on export agriculture, however, is a very risky strategy. For example, in 1995 the European Union (EU) made a political decision to tax only those bananas not grown in their former colonies. This political decision suddenly eliminated a major portion of Costa Rica's market for export bananas.

Given its relative lack of control over export income, Costa Rica faces major economic constraints in generating the hard currency to continue importing either production inputs or food to feed its growing population. As of 1996, the government of Costa Rica decided to again emphasize traditional export crops such as coffee since the nontraditional crops were expensive to grow and were not generating as much foreign exchange as once hoped (D. Kass, personal communication).

There is probably no realistic alternative for Costa Rica besides a continued investment of additional energy into agriculture. Even with two children per family (a zero net reproduction rate), the size of the Costa Rican population will continue to increase for at least 20 years as the more numerous younger aged cohorts reach reproductive maturity. How else will these people be fed? Such dependency, however, must be done with an awareness of the increasing vulnerability to petroleum price changes and potential social and environmental consequences of the concentration of resources.

Finally, our analysis is based on the assumption of the need for all of the industrial inputs presently being used to support agriculture. Fertilizer, the most critical component, is only 5 to 10% of the total dollar value of the imports required for agriculture (Chapter 23). It is not clear how much of the other inputs are actually required or are in some sense superfluous. The issue of animal feeds is one example. Or what about trucks? They constitute a large proportion of inputs, but if food was consumed near where it was grown, might the inputs for trucks and fuels be reduced substanitally? Such an analysis might indicate some quite useful policies for reducing the inputs required for agriculture.

## V. CONCLUSIONS

The fundamental conclusion of this analysis is that human population levels and growth matter very much, but in ways more subtle than previously recognized. It is only through the interplay of land area, land quality, and the generation and use of foreign exchange that one can understand adequately the effects of population increase. The basic mechanism is this: An increasing population requires, obviously, more food. As more food is required there is not enough land area of even moderate quality to generate that food from more extensive systems (Table 5-7). Traditional low-input, extensive systems are replaced by more intensive, industrially assisted systems which generate more food per hectare, but require foreign exchange to operate. Increasing amounts of land are needed to generate the foreign exchange to purchase the inputs required for all agricultural production or imported food. In addition, of course, additional land is needed to generate foreign exchange for all other imports.

In Costa Rica coffee and other export crops are grown increasingly instead of food crops. Many people criticize this, but it is one viable strategy in a crowded country without fossil fuels or significant other raw industrial resources. Coffee yields per hectare are only a little less than those for most grains, but coffee is worth about 10 times what grain is, or six or seven times if transport is included. Thus even given the additional cost of growing coffee and the cost for transport, more people can be fed on a hectare by trading coffee for wheat than by growing grain. People who argue for food self-sufficiency as part of a "greening" of Costa Rica need to examine where the inputs will come from for that policy. The emphasis on export crops makes sense (such as anything makes sense), and it is essentially what had to happen given the ratio of people to good land. Now Costa Rica imports some 30% of its food calories and pays for them with exports or sometimes debt.

Together the increased land requirements for more local food and more export crops are pushing agriculture into increasingly marginal land, where either yields are lower or more inputs are required per unit of production (Chapter 12). Meanwhile urban areas increasingly cover the best farmland. While there is a great deal of agricultural land left, its average quality is much lower than that of the land developed initially, and the inputs required to utilize them are large. Additionally, the longer a given hectare has been in intensive agriculture, the more the cumulative effects of erosion. Thus even though Costa Rica has been increasing its inputs over the past decade, mean yields per hectare have been stable or declining (Chapter 12). At the same time there has been an increase in many environmental problems (Chapters 7, 14, 15, and 16), many related to the continued growth of agriculture. But without that industrialization of agriculture the standard of living would be much lower and, probably, people would be starving, for there just is not enough good land to feed 3.5 million people without the inputs or the trade (Table 5-6).

Thus with respect to feeding Costa Ricans there seems to be no possible conclusion about sustainability except that it can no longer in any sense be based on renewable resources alone. There are far too many people for shifting cultivation to feed, and almost not enough land to grow the food on Costa Rican soil if we include the land needed to pay for agricultural inputs, let alone other imported items. An additional perspective on sustainability is the increasing vulnerability to food shortages faced by Costa Ricans due to the vagaries of the international market, climate, and weather. Agriculture can be made relatively more sustainable mostly by protecting soil resources. This would decrease the industrial inputs required in the future and maintain the most important component of agricultural production, the soil itself. If the Costa Rican population were 500,000 none of these problems, economic, environmental, foreign dependence, or food production, would exist. If the future population is doubled, the problems will be more than twice as difficult, for all agricultural activities would be pushed even more into submarginal lands. That is the bottom line of our analysis.

## ACKNOWLEDGMENTS

This work was supported initially by a grant for Charles Hall to study with Tomas Schlichter at CATIE in 1984 as part of Schlichter's program on the relation of energy to Central American agriculture, as well as by a SUNY-sponsored sabbatical for Charles Hall at CATIE in the spring of 1995. Other funding was in part from the U.S. Department of Energy's Carbon Cycle Program. Much of the detailed programming was done while using the facilities of the University of Montana's Biological Station. We thank the staff at CATIE, especially Alfredo Bonano and Viviana Palmira, and ESF graduate student Ruth Tiffer for help in uncovering and analyzing data, Joe Cornell for help with early simulations, Margaret Smith and Donald Kass for insight into farming systems in Costa Rica, and Cutler Cleveland for critical review.

## REFERENCES

Alfsen, K. H., M. A. DeFranco, S. Glomsrod, and T. Johnson. 1996. The cost of soil erosion in Nicaragua. *Ecological Economics* 16:129–146.

Annis, S. 1990. Debt and wrong-way resource flows in Costa Rica. *Ethics and International Affairs* 4:107–121.

Atlas Estadistico de Costa Rica No. 2. 1981. *Statistical Abstract of Latin America*, Vol. 23. UCLA, Center Publication, Los Angeles, CA.

Brown, L. 1981. World population growth, soil erosion and food security. *Science* 214:955–1002.

Brown, L. 1996. The acceleration of history. In *Journal State of the World*, pp. 3–20. W. W. Norton, New York.

Campbell, C. J., and J. H. Laherrere. 1998. The end of cheap oil. *Scientific American* 278:78–83.

Cleveland, C. J. 1995. Resource degradation, technical change, and the productivity of energy use in U.S. agriculture. *Ecological Economics* 13:185–201.

Collins, J. 1982. *What Difference Would a Revolution Make? Food and Farming in the New Nicaragua.* Institute for Food and Development Policy, San Francisco, CA. 185 pp.

Comision Economica Para America Latina y el Caribe. 1989. *Annuario Estadistico de America Latina y el Caribe. Naciones Unidas.* HA 755 A634.

Conklin, H. C. 1961. The study of shifting cultivation. *Current Anthropology* 2:27–61.

Costa Rica. 1993. *Informacion Basica del sector Agropecuario,* Vol. 7, Cuarto 49 y 50.

FAO. Each year. *Production Yearbook.* FAO, Rome.

FAO. 1981. Crop production levels and fertilizer use. *FAO Fertilizer and Plant Nutrition Bulletin.* FAO, Rome.

FAO. 1992. *Fertilizer Use by Crop.* ESS/MISC/1992/3, FAO, Rome.

Gifford, R. M., J. H. Thorne, W. D. Hitz, and R. T. Giaquinta. 1984. Crop productivity and photoassimilate partitioning. *Science* 225:801–808.

Guillen, R., and D. Gutierrez. 1981. Situacion de los fertilizantes en Costa Rica [Fertilizer situation in Costa Rica]. Presented at FAO/FIAC Seminar on the Politics of Prices and Subsidies of Fertilizers, Lima, Peru, October 21–23, 1981. SEPSA, San Jose, Costa Rica.

Hall, C. A. S., C. J. Cleveland, and R. K. Kaufmann. 1986. *Energy and Resource Quality: The Ecology of the Economic Process*. Wiley–Interscience, New York. Reprinted 1992, University Press of Colorado, Boulder, CO.

Hall, C. A. S., and J. D. Cornell. 1990. Simulating Costa Rican agricultural production. Presentation at First Meeting of International Society of Ecological Economics, Washington, DC.

Hall, C. A. S., and M. H. Hall. 1993. The efficiency of land and energy use in tropical economies and agriculture. *Agriculture, Ecosystems and Environment* 46:1–30.

Hartshorn, G. S., L. Hartshorn, A. Atmella, L. D. Gomez, A. Mata, R. Morales, R. Ocampo, D. Pool, C. Quesada, C. Solera, R. Solarzano, G. Stiles, J. Tosi, A. Umaña, C. Villalobos, and R. Wells. 1982. *Costa Rica: Country Environmental Profile: A Field Study*. Tropical Science Center, San Jose, Costa Rica.

Hansen-Kuhn, K. 1993. Sapping the economy. Structural adjustment policies in Costa Rica. *The Ecologist* 23:179–184.

Ko, J-Y., C. A. S. Hall, and L. G. Lopez Lemus. 1998. Resource use rates and efficiency as indicators of regional sustainability: An examination of five countries. *Environmental Monitoring and Assessment* 51:571–593.

Lal, R., and D. J. Cummings. 1979. Clearing a tropical forest. I. Effects on soil and micro-climate. *Field Crops Research* 2:91–107.

Lal, R. 1987. Effects of soil erosion on crop productivity. *Critical Reviews in Plant Science* 5:303–369.

Larson, W. E., F. J. Pierce, and R. H. Dandy. 1983. The threat of soil erosion to long-term crop productivity. *Science* 219:458–465.

Levitan, L. C. 1988. Land and energy constraints in the development of Costa Rican agriculture. Unpublished M.S. thesis, Cornell University, Ithaca, NY.

Mellor, J. W., and S. Gavian. 1987. Famine: Causes, prevention, and relief. *Science* 235:539–595.

Ministry of Health (Ministerio de Salud). 1979. *Encuesta Nacional de Nutricion: Evaluacion Dietetica Ano 1978 (National Nutrition Study: Dietetic Evaluation for the Year 1978)*. Ministerio de Salud, San Jose, Costa Rica.

Mudabar, M. S., and T. P. Hignett. 1982. *Energy and Fertilizer: Policy Implications and Options for Developing Countries*. T-20, International Fertilizer Development Center, Muscle Shoals, AL.

Odum, E. P. 1989. Input management of production systems. *Science* 243:177–182.

Odum, H. T. 1967. Energetics of world food production. In *Problems of World Food Supply*, pp. 55–94. *President's Science Advisory Committee Report,* Vol. 3. White House, Washington, DC.

Pimentel, D., and M. Pimentel. 1979. Food, energy and society. Wiley, N.Y.

Pimentel, D. 1987. Soil erosion effects on farm economics. In J. M. Harlin and G. M. Berardi (Eds.), *Agricultural Soil Loss: Processes, Policies, and Prospects,* pp. 217–241. Westview Special Studies in Agriculture, Science and Study, Boulder, CO.

Pimentel, D., C. Harvey, P. Resosudamo, K. Sinclair, D. Kurz, M. McNair, S. Crist, L. Shpritz, L. Fitton, R. Saffouri, and R. Blair. 1995. Environmental and economic costs of soil erosion and conservation benefits. *Science* 267:1117–1123.

Repetto, R., W. M. Magrath, C. B. Wells, and F. Rossini. 1989. *Wasting Assets, Natural Resources in the National Income Accounts*. World Resources Institute, Washington, DC.

Sader, S. A., and A. T. Joyce. 1988. Deforestation rates and trends in Costa Rica 1940 to 1983. *Biotropica* 20:11–14.

Seligson, M. A. 1980. *Peasants of Costa Rica and the Development of Agrarian Capitalism*. University of Wisconsin Press, Madison, WI.

Wilkie, J. W., and A. Perkal (Eds.). 1985. *Statistical Abstract of Latin America,* Vol. 23. UCLA Latin American Center, Los Angeles, CA.

# SECTION III

# *Adding a Spatial Dimension*

## TOOLS FOR DYNAMIC GEOGRAPHICAL ANALYSIS

Neither standard economic models nor even most previous biophysical models include space explicitly. Some economists argue that this is not a problem, since all factors, including whatever factors space might entail, will be included in present market prices. This is of course foolish; for example, if agricultural response was strongly a function of the quality of the soils developed, if that quality varied a great deal over the landscape, and if a cost–benefit analysis was made on the basis of existing costs, then costs of future expanded production might be seriously underestimated because of the higher production costs on the new, lower quality soils.

This is but one example of the many ways that space matters. Agronomists and others have been struggling with space for decades. But now the job has become far easier with the development of computerized GIS (geographical information system) technologies. Although the gathering of the information is not made easier with GIS's (although it is with remote sensing—see Chapter 11), the comparison, manipulation, and application of that information have become enormously easier. This section gives a philosophical and theoretical overview of the issue of making resource science and analyses spatial. It includes a less technical introductory chapter that explains many of the difficulties in building national-level GIS data bases and provides some common sense advice for any new attempts to construct such national-level information bases. This is followed by a more technical chapter on how ecological perspectives of how plants and animals respond to their environment can be integrated with GIS's and computerized approaches. The ideas in both chapters form the conceptual basis for the next section on building a geographical data base for Costa Rica.

CHAPTER 6

# The Derivation and Analysis of National-Level Geographical Information

## A New Model for Accessibility and an Easy-to-Use Micro GIS Program

CHARLES A. S. HALL, GREGOIRE LECLERC, AND PATRICK VAN LAAKE



## I. INTRODUCTION

Everywhere, in nearly all scientific disciplines related to the earth, investigators are turning to Geographical Information Systems (GIS's) in order to maintain information about the earth in a spatially referenced and organized manner, generate maps of landscapes as they exist or might become, manage natural resources, and make decisions for the future. People from many countries are learning the intricacies of these systems, and ambitious government and private organization plans for "digitizing" whatever information available will allow future decisions to be as informed as possible. The availability of such systems also gives great hope for synthesizing disparate information sets. Conservationists can interact with foresters' data, soil scientists with geologists' maps, and so on. For those of us interested in a systems approach to environmental and other problems the methodological future looks bright indeed. The ideal is that the computer will be a willing tool which investigators can use to get directly to their problem without worrying very much about data availability, compatibility, spatial fidelity, and so on.

The reality that both young and experienced practitioners are soon confronted with, however, falls far short of the ideal. The computer systems are idiosyncratic and difficult to learn, maps are generated only very, very slowly, and almost no one's data files are compatible with anyone else's. And these are not the principal problem, which is that "existing" data simply are not available. Ministry X cannot let out its data because it is "being revised" or it is "politically sensitive"; Ministry Y is happy to give you its digital database but "everyone is too busy" to make it available; private environmental firm Z does not want to let out its data because it was very expensive to collect, and they have not yet derived their full value from it. Finally, of course, is the problem that the data itself is often organized for one purpose in such a way that makes it incompatible for another, including the one for which you wish to use it. The net effect is that too often we have not gone very far from where things were before GIS's, and any kind of integrative systems-oriented science is very, very difficult to get off the ground. Thus we found it remarkably difficult to do what should have been the easiest part of this book—getting the data. This turned out to be even more difficult than doing what we thought would be the hard part—interpreting the data and writing it up. Although our experience and examples are related to the "developing" nations, we think they are probably often equally applicable to the "developed" world.

Other scenarios came into play that we thought you might enjoy reading, although they were not much fun while they happened. At CATIE, an agricultural research station with a long history of excellent but widely dispersed data gathering, we learned by accident of an incredible data base (digital!) that would be extremely useful to our analysis. To begin this saga, Charles Hall was vaguely aware of the agricultural data that currently forms the basis for Chapter 12 from his previous work in Costa Rica in 1983. However, he could not find anyone at CATIE who could tell him how to find that data, or indeed whether it existed any longer. The logical step of asking a librarian was not even taken for months because a number of knowledgeable researchers said that no such national surveys had ever been taken.

Finally, however, someone acknowledged that indeed such a survey had been done, and it was on computer tape at the Centro de Computo (where we all worked), but that the lead investigator had gone to work for a computer firm in San Jose. After considerable effort, requiring weeks, we did indeed locate this person, and he said that such a tape did exist, but "why were we asking him?" The key person, he said, was Gustavo. And it turned out that Gustavo had the office directly across the corridor from Charles Hall. Gustavo did indeed have the data, and was very willing to cooperate, but gave us the bad news that the data was for only a relatively few plots, and they were all from the eastern zone. That did not help us very much for our gradient analysis since the eastern zone is relatively homogeneous in terms of climate compared to the rest of Costa Rica.

Fortunately, on Charles Hall's first visit to the university office of Arturo Sanchez, one of our contributors, he noticed an undergraduate typing in numbers from a thick book into a computer. Upon closer examination the numbers turned out to be exactly the data for which he had been looking for over five months. (Of course even though the tables in that book were obviously print outs from a computerized data set it was necessary for Arturo to hire someone to type them into the computer, as the computer tapes at the Ministry of Agriculture and Livestock had been lost, at least for practical purposes.) Upon determining the title and author of the books, it was easy to find them at the CATIE Library. But about five months had been wasted.

Other such stories are widespread. A favorite of ours is that a key agricultural data base was stored at the Centro de Computo at CATIE. We had the tape in hand, and a tape driver was all set up to read it and transfer it to diskette. But we could not access the tape. The data had been recorded with a required password. And we could not get that password because the only man who knew it had died! We still have not read that tape.

We have experienced all of this and much more with the development of this book. Perhaps the most common problem was that people who initially promised to generate data failed to deliver. Another problem was incompatibility of data types. But we also found a lot that was good. The files of the Instituto Meteorológico Nacional, for example, were extremely well organized and put into a digital format in an up-to-date fashion, and the head of the unit was very helpful in assigning someone to download the data into a useful format. Carlos Elizondo, the subdirector of the Instituto Geographico Nacional, was very helpful in aiding us to derive a consistent format for all data. It was also critical for our work that there had been a special project jointly sponsored by the Costa Rican Ministerio de Agricultura y Ganaderia (MAG) and the United Nations Food and Agriculture Organization (FAO). Out of this effort came Patrick Van Laake's high-resolution digital elevation map for Costa Rica (Chapter 8).

We have also been impressed in the past by the work done by Scott Madrey and his group at the Geospatial Laboratory of Cook College, Rutgers University. Scott started with extensive U.S. Army data sets and, in general, has acquired as many maps as possible for geographical and environmental information around the world. He has entered these into a completely consistent, rasterized format that allows any user to overlay his map "A" of Nigeria, for example, a soil map, perfectly with his Map "B" of Nigeria, for example, a map of the border of the nation. We had previously found this approach to be critical when we undertook complex land use modeling for Africa (e.g., Pontius, 1994;

Hall *et al.*, 1995a,b). It is especially important to have a reliable base for boundaries and coastlines.

Unfortunately this procedure had not been done yet for the various Costa Rica maps, even when they had been digitized, so we had to do this ourselves. The problems that we commonly experienced are summarized under the next sections.

## A. Inadequate Training and Experience

In many organizations we saw in Costa Rica, the GIS was run by subject matter specialists (e.g., agronomists, foresters, or engineers) rather than by geographers or cartographers. Although a profound knowledge of the subject one is trying to model in a GIS is indispensable, so is a thorough understanding of the implications of whatever operation one is performing in the GIS. This is as true for a *vector* data (line-based) GIS, with its intricate relationships between elements in the data set, as it is for a *raster* data (matrix-based) GIS (Figure 6-1). Most of our raster maps are derived from vector-based maps first, except for satellite images, which are raster to start with. Apart from these rather technical observations, there are also the more mundane little inconveniences such as the inability of many professionals in other countries to read manuals written in English. In one organization the manuals remained unread—and the GIS an inaccessible black box—until somebody on a temporary contract, who could not read English either, took the bold step of translat-

```
ELEVTU.SQR             4/28/98          Page 1
 2696 2739 2593 2557 2387 2197 1981 1810 1759 1579 1376 1283 1176 1006  946  823  729  586  474  496  416  300  331  487
 2633 2519 2500 2329 2154 2062 1822 1636 1616 1441 1342 1200 1085 1000  900  800  762  700  513  452  392  324  353  414
 2492 2368 2329 2169 1965 1825 1745 1567 1400 1298 1243 1105 1010  955  900  822  793  667  532  410  365  407  446  523
 2501 2222 2125 1953 1772 1605 1543 1504 1414 1298 1108 1079  949  835  788  746  752  674  538  400  400  431  542  585
 2277 2038 1817 1664 1600 1489 1375 1380 1294 1200 1050  963  942  826  708  645  616  562  500  400  500  554  600  626
 2077 1784 1641 1541 1576 1314 1219 1161 1234 1200 1193  989  832  783  707  661  508  466  400  427  545  631  681  700
 1899 1713 1700 1577 1413 1270 1123 1075 1000  987 1072 1173 1081  886  729  600  530  417  500  508  539  673  780  796
 1951 1757 1700 1600 1280 1184 1092 1000  956  911  800  838  905  890  813  609  499  472  659  639  617  712  802  877
 1661 1643 1600 1728 1485 1252 1115 1068  983  831  795  721  693  717  835  646  500  599  700  748  771  800  884  861
 1593 1597 1497 1477 1386 1300 1239 1121 1032  900  735  696  647  651  689  568  568  600  714  836  858  936 1008  900
 1487 1473 1458 1384 1298 1300 1203 1200 1034  901  738  677  620  600  600  500  746  771  792  878 1101 1000 1105  957
 1304 1350 1299 1290 1198 1211 1160 1200 1128 1013  883  719  647  600  600  506  776  806  869  968 1175 1120 1100  930
 1273 1244 1200 1270 1137 1146 1107 1070 1191 1126  982  788  680  641  600  589  657  840  996 1000 1156 1200  955  845
 1199 1167  971 1005 1008 1093 1031  905 1075 1156 1039  868  729  665  592  600  597  700  895 1000 1192 1200 1079  936
 1035  918  840  919  774  858  883  880  863 1092 1072  914  799  705  600  595  588  599  736  898 1004 1200 1200 1194
 1066  996 1182 1097  890  821  719  765  700  801  853  756  655  710  620  592  586  606  663  741  801 1023 1187 1202
 1220 1339 1358 1256 1148  915  926  782  708  705  729  694  642  603  600  589  599  648  600  651  736  831 1009 1046
 1253 1283 1384 1457 1200 1118 1077  896  800  751  801  722  605  659  677  597  709  771  774  700  700  692  776  792
 1333 1450 1500 1510 1200 1058 1018  857  801  713  711  700  618  699  693  624  817  975  800  851  838  796  786  784
 1189 1300 1538 1589 1247 1200 1134 1236 1012  789  780  757  690  643  647  725  964 1001 1076 1088  943  900  950  943
 1383 1597 1700 1484 1400 1300 1000  891  919  696  663  692  816  920  716  703 1041 1188 1246 1175 1117 1116 1105  946
 1674 1782 1700 1458 1362 1300 1222  900  810  697  689  699  747 1018  922  798  900 1283 1400 1253 1290 1300 1082 1100
```

**FIGURE 6-1** A basic "square" file, representing elevation values for a portion of Costa Rica near Turrialba at a resolution of 1 km. This is an example of a raster-based map.

ing all the manuals, page by page, into Spanish. This process took several months of meticulously looking up every technical (and many nontechnical) terms in a dictionary. The transcripts, in awkward but useful Spanish, were much sought after by all the GIS users in that organization.

## B. Inappropriate Procedures and (Lack of) Standards

The lack of clear and concise standards is another cause for incorrect or unreliable results. In Costa Rica there have been a few modest attempts to draw up a set of guidelines and standards for spatial data, but they have mostly remained confined to the GIS of the group that defined them. In 1995 the Instituto Geográfico Nacional developed some standards related to data quality (precision and accuracy) and fidelity. These were basically a translation of the standards for converting analog maps to a digital environment, but were never published on a broad scale, let alone formalized or mandated. The proliferation of GIS's in Costa Rica has continued, but as far as we know without any national-level standards. Instead each group uses ad hoc standards, the default standards of the system, or even none at all, or just use the available data sets. Even the names of districts vary from one ministry or organization to another.

## C. Technology Not Understood by Decision Makers

Applying GIS technology to solve a problem involves more than buying a powerful computer, installing expensive software, and training a few operators to perform the magic that is usually expected in return for the investment made. The changes that need to be made in the organization or the work procedures that would serve to integrate the GIS into ongoing work and take advantage of its full potential are seldom actually made.

## D. Insufficient Resources

Many organizations do not have the resources to acquire the computers and software they actually need for the tasks they have at hand. They are even less able to hire experienced people to operate the system or to build and maintain a library of standard procedures and functions relating to metadata (discussed later), spatial data, attribute data, data formats, export formats, etc., all of which would allow them to cooperate much more easily with the subject matter specialists who require this data in order to analyze their region of interest within a GIS.

### E. Lack of Local or Long-Term Support for GIS Software

Often the resources to acquire and use GIS's come from international aid agencies, who see a GIS as an ancillary tool to achieve their stated objectives, as it should be. But they cannot hire experienced staff on a long-term basis. Furthermore, the recipient organization may not make the necessary provisions (i.e., include funds in the budget!) to maintain the GIS system once the international aid dries up. An interesting detail about GIS acquisitions using international aid is the position of the local vendors of GIS's. Quite often aid agencies bring their own software, or impose a particular software on the recipient organization, because the international experts know the software, because they have developed applications elsewhere, or because they may simply be advocating a particular system because of other interests. The market for GIS's is not very large in Costa Rica, and probably even smaller in many other developing countries, and the local vendors become marginalized. Also, there may be no local service available for the software and upgrades may have to come from abroad, to be paid for in scarce dollars. The good news is that computers powerful enough for GIS work are becoming very cheap.

## II. A PLAN TO GENERATE NATIONAL-LEVEL DATA BASES

We would like to recommend a solution that we believe would overcome some of the problems we encountered. It is simply to have the formatting, archiving, and description of any government or national-level data have more priority than the gathering of the data itself. Although this requires very little money, the more difficult thing to achieve is a change in the mentality of those involved. Anyone involved with research knows that, in general, research proceeds project by project, or more accurately, contract (or grant) by contract. Scientists in general are constantly concerned about where their next contract will come from, and money is rarely given to rework or archive old data, for the emphasis in science is on the new. But as data has accumulated over recent decades we are much less data limited than in the past and more limited in our ability or efforts to synthesize and integrate existing data. Thus most researchers do an insufficient amount of archiving of the data they gathered and of discussing its characteristics. An exception, and a model, is the Carbon Dioxide Information Center at Oak Ridge, Tennessee, which has as its priority the archiving of all data and models pertaining to climate measurements and potential climate change. Twenty percent of the research program was so earmarked, and in our opinion it is an excellent example to follow.

Such a system can be achieved where a great deal of attention is given to *metadata,* which means data about data. Thus our prescription is simple: no GIS-type research should be done unless the results can be made generally available in a PC (or Macintosh)-type format in accordance with a set of criteria laid down by the national equivalent of the Instituto Geographico Nacional or (in the United States) the USGS. And it is the responsibility of those agencies to make their standards well known, easily accessible, easily understood, and, in general, available in ASCII (which stands for "American Standard for Computer Information Interchange" and means simply that the data is given in a "what you see is what you get" format) as well as other formats. Each nation should budget a geographical agency to set standards, exert quality control, and archive all nonproprietary digital geographical data obtained in the nation.

This is not to say that these would be the only formats that could be used for this data; obviously many formats can exist to fulfill the different requirements of different projects. But the results should also be made in the standard national (or perhaps even international) format. Our preference is for unformatted (ASCII, digits and decimals under each other and right justified) data. Such data can be examined and read by everyone else. We also like storing map data in what we call "square" files. A ".SQR" file is a matrix of cells of raster data that has the spatial properties of a map, that is, north is to the top of the file and east to the left in an intuitive manner (Figure 6-1). Then if there is a problem with the data one can go in with an editing program and check it out easily! Our suggestions also include improving the accessibility of the data. For example, data stored in one of the best known GIS's is extremely difficult to access for use in a computer model. The reason is that the data structure is "hidden," that is, it is in its own special format that is not accessible to standard (ASCII or other) reading procedures. For this reason we recommend other GIS's that allow investigators to use the data freely with other computer programs, that is, to read and write from and to the GIS data files. Unfortunately our experience in July 1999 suggests that neither the new "Standard U.S." SDTS format nor the software vendors response has resolved this issue.

It is our hope that among the readers of this book will be people with administrative power in developing countries who find these suggestions useful and who would like to generate a similar analysis for their own country or region. They probably would like to know how difficult an endeavor such as the analyses in this book might be for their nation. Our answer is "fairly difficult," but if one does it correctly from the start one should be able to reduce the effort per area of analysis to about one-quarter of that which we encountered. In other words, about three-quarters of the effort in this book went into sleuthing out data sets and making them compatible with each other.

## III. OUR PLAN TO PROVIDE A UNIFORM FORMAT FOR COSTA RICAN GEOGRAPHICAL DATA

Once we had decided to undertake this book, one of the first things that the editors did was to agree that we needed to decide upon a consistent format for the data that we were going to include. A problem that we immediately encountered is the evolving standards for what were thought to be standardized data formats. For example, we wanted to print our GEOPLOT square files of rainfall. We had an expensive programmer change our GEOPLOT program to give output as printable "TIFF" files. It then turned out that our printing program (HIJACK) could not read these TIFF files—for while we had worked on them, another standard for TIFF formats had been developed by the powers that decide these things. It was incompatible with the previous standard that we had paid for! Consequently we have converted, where possible, our files to the ASCII format, the most general, readable, and stable format.

The principal disadvantage of the ASCII format, other than that not all software uses it, is that it uses more memory space to save a given amount of information. We felt this was not a problem for a country the size of Costa Rica and the resolution of the data we had available. In our case each digital map requires a little more than half a megabyte to store; i.e., about 2 maps can fit onto a standard 3.5-in. high-density diskette. (One hundred or more different digital maps fit comfortably on a PC hard drive when we started our analysis in 1995.) Thus all the computer programs used here, such as GEOPLOT, read from ASCII-formatted data, which can be examined easily with any text editor, such as PEDIT. Both programs are on the enclosed CD. Our approach to spatial data has been to generate ".SQR" files, which to us were the most intuitive choice. The incredible storage and processing power of new computers and the availability of 100-MB and GB removable disks largely make data compression or ultraefficient storage procedures moot, although for very large data sets an argument also can be made for a binary form.

Our next question concerned which cell resolution to use for our raster data. In the international and scientific environment of CATIE there was no question that the unit of measurement would be metric. That, at least, most of the world (but not the United States) has come to agree upon. We also agreed upon a raster (matrix or grid) rather than a vector (line type—see the next chapter) mapping and data storage format since this allowed us to interact easily with conventional FORTRAN, the computer language in which we write our models.

But what should be the resolution of the individual cells? In other words, how much ground area in real-world measurements should each cell cover? Normally this is determined by the scale of the base maps one has available, and cannot be finer than that which is readable on one's smallest scale map. Most of our maps, DEMs (digital elevation maps), and soil data were 1 : 250,000 in scale. According to Forbes *et al.* (1987) the minimum legible area on a map of this scale is 250 ha (2.5 $km^2$). Normally the resolution for a grid map is one-fourth the minimum legible area. We discussed the possibility of everything from cells of 100 m on a side (1 ha) to ones of 10,000 m on a side (100 $km^2$). Fortunately many things argued for cells of 1000 m on a side (1-$km^2$ cells). First, this would require a map of roughly 400 by 400 cells to cover all of Costa Rica, and this was a convenient size to display on most computer screens. Second, the horizontal resolution of the digital elevation map, the basis of many analyses, was about 20 m, so that we would be "pushing it" to go to a 1-ha level (see Chapter 8). Third, given the speed of the current generation of desktop computers, manipulating 400-by-400-element arrays is much more reasonable than manipulating 4000-by-4000-element arrays. Finally, most of the spatial data was known to a resolution of 1 $km^2$ (if we were lucky) so that there was no reason to go to a scale finer than our source data.

Thus we settled for a 1-by-1-km resolution (approximately), and all of our data is placed accordingly. For the kind of analysis that we were contemplating—modeling natural resources and semidetailed agricultural production and potential for an entire country—a 1-by-1-km resolution was just fine. Obviously, designing a rural road network in a canton to market the agricultural production that we might forecast would require a finer resolution in order to capture more of the variation in soils and topography, and probably even a completely different GIS, i.e., one that is vector based and much more appropriate for working with linear elements. Some applications such as erosion modeling, where the slope length is a required input factor, or non-point source pollution modeling, in which the slope of the terrain plays a fundamental role, require a map resolution of much finer scale (30, 60, or 90 m) in order to come close to mimicking reality (for a discussion of this see Vieux *et al.*, 1988). Thus another important issue related to spatial resolution is the availability of data at a scale appropriate to the application, and if not, the cost of acquiring it. Cutting the cell dimension in half will quadruple the amount of data (or money, time, computer storage, computer processing, analytical algorithms, and so forth) that is required to maintain the same level of informational integrity.

A second major problem that we encountered was map projection. Many different map projections are available, all of them using slightly different mathematical transformations for representing a round earth on a flat piece of paper (good discussions of this problem are found in Dent, 1985). After much discussion we finally agreed that a geographic "projection," i.e., using latitude and longitude instead of the Lambert Conformal Conic (north) projection that is traditionally used throughout Costa Rica, would be best for a variety of reasons. First, much information is available with geographic coordinates of degrees longitude and latitude attached, including that reported using Global Positioning Systems (GPS's). Second, almost all GIS's are capable of changing geographically referenced maps to other projections.

However, in a meeting of all Costa Rican authors that took place at CATIE in May of 1995, Carlos Elizondo, the subdirector of the Instituto Geográfico Nacional, argued that instead we should use the Universal Transverse Mercator (UTM) projection because this is going to be the new standard coordinate system for Costa Rica (as it is in the surrounding countries). However, since the conversion from the current Lambert Azimuthal projection to UTM is going to take several years, and since many users of GIS's store their Costa Rican data in Lambert Azimuthal form, it seemed more appropriate to use this existing system, rather than to anticipate the new standard. Thus we have done that.

The procedures for converting various data sets into this national-level grid system are given in the individual chapters of Section III. We will give one technique here, however, because of its importance. This is converting political-level data into a geographical map. Most of the data in the world is collected by political or administrative units. This reflects the nature of bureaucracies: they are responsible for their bureaucratic territory, which normally is defined by arbitrary or historical processes more or less unrelated to basic physical/ecological characteristics, such as elevation, watershed, or location on the earth's surface. So, for example, all agricultural data from Costa Rica is defined at the canton or district level. But we wanted to assess the response of crops to climate. Each political unit has more or less different meteorological and other characteristics compared to any other political unit. How can the physical characteristics of political units be quantified?

Our approach was first to construct a digital political map so that each of the 1-by-1-km cells of Costa Rica had a district number assigned to it. Here the problems proliferated rapidly! First of all we found three "official" lists of the districts in Costa Rica! Second, the names used for many districts in the official lists were different from those used in the agricultural surveys, some of which were more than 40 years old, requiring us to eliminate that district or guess that "tigre" and "el tigre" were the same district. (That one was easy—how about "San Francisco" and "Agua Caliente" being the same names for the same place, or "Jaco" being listed in one canton in the national name list and in another in the agricultural survey?)

Once the first of these problems was resolved, the boundaries of all districts were digitized and fit to the standard 1-by-1-km format. The second problem was resolved by writing a FORTRAN program that would read in each name from the agricultural file and then read the entire official "master" name list. When the names agreed, and they were in the same canton, we accepted that as the same district, and were thus able to relate the agricultural data of that district to the average of the environmental gradient values found in all cells located in that district. When no matching name could be found the agricultural political unit name was changed, where possible and upon consultation with Costa Ricans, to the appropriate currently accepted name for the canton or district. All of this was an extremely time-consuming process, requiring a great deal of difficult programming. It would have been unnecessary if there had been a completely consistent terminology for cantons and districts which was used consistently in the Ministry of Agriculture. That is, of course, a lot to ask for data that goes back to 1950!

## IV. DEVELOPMENT OF MEAN ENVIRONMENTAL CHARACTERISTICS FOR EACH POLITICAL UNIT

Ideally we would like to examine the relation of a given entity at each square kilometer to other parameters measured at that exact square kilometer. Since there are about 51,000 $km^2$ in Costa Rica it is just wishful thinking to expect that number of weather stations, soil samples, or agricultural yield measurements. As already mentioned, the agricultural yield information has been gathered at the district or canton political level. But if we wish to examine the behavior of agricultural crops vs climatic or edaphic (soil) factors it is necessary to generate mean values for these parameters for each political district. This was done by taking the average of all cells for that political unit. The computer program ASSGNDIS was written to scan all cells for their political identification, and simultaneously compute the average rainfall, temperature, etc., for all corresponding cells found in the maps of environmental or physical variables.

For example, when we wanted to determine the mean annual temperature for the district of Turrialba we took the average of all temperature values for each of the 1-km grids that were found in that district (Figure 6-2). This was done by using the CR_DIST.SQR file, which is a square file of all districts in Costa Rica. Turrialba is given the number 30501 in the official University of Costa Rica list of political entities, and is number 276 in our map of districts (see the NewNam.DAT and CR_DIST.SQR files on the CD). The computer program ASSGNDIST.FOR searches row by row and column by column through the CR_DIST.SQR file, and every time it encounters the number 276 it goes to file TEMPMEAN.SQR and reads the mean annual temperature for the same cell and sums that with the temperatures of other cells in that district. It also sums the number of times that district is found. When the search of the district file is complete the program divides the sum of the temperatures by the number of occurrences to get the mean (and standard error) of the temperature for that district. A similar search was undertaken for canton-level temperatures.

Likewise a similar analysis was undertaken to generate mean values of temperature month by month, as well as rainfall, relative humidity, soil moisture, soil fertility, and so on. For example, a temperature and rainfall summary by canton is given in Table 6-1, and in more detail by district on the CD. A summary of the programs developed for this process is given in Table 6-2, and a flow chart for the programs written for these analyses is summarized in Figure 6-3. The programs are included on the CD under "PROGRAMS" or "DATA." Once this task was undertaken it was possible to examine, for example, the production of coffee for each district vs the environmental conditions for

```
TDIUTU.SQR          4/28/98        Page 1
  12  12  13  13  14  15  16  17  17  18  19  20  20  21  21  22  22  23  24  24  24  25  25  24
  12  13  13  14  15  15  17  18  18  19  19  20  21  21  22  22  22  23  24  24  24  25  24  24
  13  14  14  15  16  17  17  18  19  19  20  20  21  21  22  22  22  23  24  24  24  24  24  24
  13  15  15  16  17  18  18  18  19  19  20  21  21  22  22  22  22  23  23  24  24  24  23  23
  14  16  17  18  18  18  19  19  20  20  21  21  21  22  23  23  23  23  24  24  24  23  23  23
  15  17  18  18  18  19  20  20  20  20  20  21  22  22  23  23  24  24  24  24  23  23  23  23
  16  17  17  18  19  20  20  21  21  21  21  20  21  22  22  23  24  24  24  24  23  23  22  22
  16  17  17  18  20  20  21  21  21  22  22  22  22  22  22  23  24  24  23  23  23  23  22  22
  18  18  18  17  18  20  20  21  21  22  22  23  23  23  22  23  24  23  23  22  22  22  22  22
  18  18  18  19  19  19  20  20  21  22  22  23  23  23  23  23  23  23  23  22  22  21  21  22
  18  19  19  19  19  19  20  20  21  22  22  23  23  23  23  24  22  22  22  22  21  21  20  21
  19  19  19  20  20  20  20  20  20  21  22  23  23  23  23  24  22  22  22  21  20  20  21  21
  20  20  20  20  20  20  20  21  20  20  21  22  23  23  23  23  23  22  21  21  20  20  21  22
  20  20  21  21  21  21  21  22  21  20  21  22  22  23  23  23  23  23  22  21  20  20  21  21
  21  21  22  21  22  22  22  22  22  21  21  22  22  23  23  23  23  23  22  22  21  20  20  20
  21  21  20  21  22  22  23  22  23  22  22  22  23  23  23  23  23  23  23  22  22  21  20  20
  20  19  19  20  20  21  21  22  23  23  22  23  23  23  23  23  23  23  23  23  22  22  21  21
  20  20  19  19  20  20  21  22  22  22  22  23  23  23  23  23  23  22  22  23  23  23  22  22
  19  19  18  18  20  21  21  22  22  23  23  23  23  23  23  23  22  21  22  22  22  22  22  22
  20  19  18  18  20  20  20  20  21  22  22  22  23  23  23  23  21  21  21  21  21  22  21  21
  19  18  17  18  19  19  21  22  21  23  23  23  22  21  23  23  21  20  20  20  20  20  20  21
  17  17  17  19  19  19  20  22  22  23  23  23  22  21  21  22  22  20  19  20  20  19  21  21
```

FIGURE 6-2 Mean temperature values at a 1-by-1-km resolution for the vicinity of Turrialba. Derived from the elevations of Figure 6-1.

TABLE 6-1 Examples of Geographical and Gradient Parameter Summaries for 20 Cantons[a]

| Code no. | Political entity (San Josè Province) | Area | Mean | | Gradient | |
|---|---|---|---|---|---|---|
| | | | Lat. | Long. | Temp. °C | Rain (M) |
| 10100 | CANTON CENTRAL | 5470 | 9.86 | 84.29 | 23.43 | 1.70 |
| 10200 | ESCAZU | 860 | 9.85 | 84.20 | 20.31 | 1.56 |
| 10300 | DESAMPARADOS | 6360 | 9.68 | 83.33 | 23.32 | 2.27 |
| 10400 | PURISCAL | 4520 | 9.67 | 83.60 | 22.85 | 2.66 |
| 10500 | TARRAZU | 300 | 9.53 | 84.10 | 21.16 | 2.26 |
| 10600 | ASERRI | 5130 | 9.69 | 84.19 | 19.46 | 1.91 |
| 10700 | MORA | 3230 | 9.82 | 84.25 | 22.54 | 1.96 |
| 10800 | GOICOECHEA | 4300 | 9.86 | 85.31 | 25.35 | 1.65 |
| 10900 | SANTA ANA | 4190 | 9.77 | 84.94 | 26.21 | 1.71 |
| 11000 | ALAJUELITA | 1480 | 9.82 | 84.38 | 23.53 | 2.26 |
| 11100 | CORONADO | 1760 | 9.86 | 83.67 | 20.42 | 2.85 |
| 11200 | ACOSTA | 4230 | 9.68 | 84.29 | 22.76 | 2.32 |
| 11300 | TIBAS | 1940 | 9.89 | 84.14 | 21.43 | 1.41 |
| 11400 | MORAVIA | 1580 | 9.95 | 84.07 | 19.90 | 2.33 |
| 11500 | MONTES DE OCA | 3230 | 9.87 | 84.15 | 20.44 | 1.67 |
| 11600 | TURRUBARES | 1550 | 9.69 | 84.56 | 25.96 | 2.38 |
| 11700 | DOTA | 230 | 9.53 | 83.97 | 15.73 | 2.26 |
| 11800 | CURRIDABAT | 1380 | 9.82 | 84.47 | 25.54 | 1.65 |
| 11900 | PEREZ ZELEDON | 2640 | 9.31 | 83.73 | 20.52 | 2.76 |
| 12000 | LEON CORTEZ | 180 | 9.64 | 84.13 | 17.61 | 1.80 |

[a] Derived by averaging all parameter values for all 1-km cells that lie within that canton.

TABLE 6-2 Principal Computer Programs[a] Used to Sort Data into a Geographical Context

| Name | Function | Input | Output |
|---|---|---|---|
| ASSGNDIST | Assigns physical parameters to political entities | 1. Digital map of location of political entities<br>2. Digital map of values for parameters | Table of political entities, each with assigned parameter values |
| MERGE6 | Determines parameter values for crop yields<br>--and--<br>Plot yields, etc., on national map and in gradient space | Output from above<br><br>Output from above | <br><br>National and gradient maps |

[a] These programs are contained in the "PROGRAMS" part of enclosed CD.

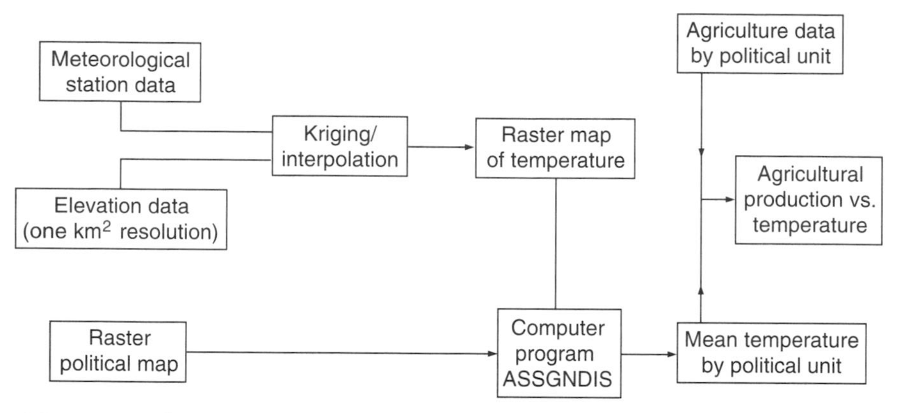


FIGURE 6-3 Flow chart of the derivation of agricultural production as a function of mean temperatures for a district.

that district. Although we believe that even better analyses could be generated by starting with explicitly spatial data rather than that which is aggregated by political unit, nonetheless our approach is much better than the nongeographical analysis undertaken previously, and greatly helps us understand the possibilities and limitations for managing resources, including forests, agriculture, and biodiversity. The results of this approach to understanding and formalizing the geography of Costa Rica are summarized in the various chapters of Section III.

## V. A FREE MICRO GIS PROGRAM

We found a need for displaying data easily in both geographical and gradient space (see next chapter). This was the impetus for developing the computer program MERGE.FOR. The program, available on the CD, will do the following: (1) read tabular data on, e.g., agricultural output (yield per hectare for a particular crop) and plot it on a map template provided by the user that has corresponding map coordinates for the map corners, and (2) plot the same data as a function of environmental parameters (gradients—see the next chapter). To use this program for a new area the user has to prepare files similar to those given in Table 6-3. Basically this requires preparing a list for each political unit considered that includes the geographic (longitude and latitude) coordinates of the center of that unit and the mean values for, e.g., temperature and rainfall, or any other physical or climatic variables for a region of interest. Once this is prepared go to the CD and download MERGEG.FOR. With patience

TABLE 6-3a Basic Data[a] Sets Required to Use the Gradients Program "Merge.FOR"

| | | | Mean | | Gradient | |
|---|---|---|---|---|---|---|
| Code no. | Political entity | Map no. | Lat. | Long. | Temp. °C | Rain (M) |
| 10100 | SAN JOSE PROV | | | | | |
| 10100 | CANTON CENTRAL | | 9.86 | 84.29 | 23.43 | 1.70 |
| 10101 | Carmen | 294 | 9.87 | 84.12 | 21.00 | 1.47 |
| 10102 | Merced | 287 | 9.88 | 84.14 | 21.50 | 1.45 |
| 10103 | Hospital | 297 | 9.87 | 84.15 | 22.00 | 1.49 |
| 10104 | Catedral | 302 | 9.86 | 84.13 | 21.00 | 1.49 |
| 10105 | Zapote | 311 | 9.86 | 84.11 | 21.00 | 1.47 |
| 10106 | San Fco de Dos | 321 | 9.85 | 84.11 | 21.00 | 1.46 |
| 10107 | Uruca | 259 | 9.90 | 84.18 | 22.13 | 1.34 |
| 10108 | Mata Redonda | 286 | 9.88 | 84.16 | 22.00 | 1.44 |
| 10109 | Pavas | 269 | 9.89 | 84.06 | 19.38 | 1.87 |
| 10110 | Hatillo | 305 | 9.84 | 84.58 | 27.30 | 1.99 |
| 10111 | San Sebastián | 318 | 9.85 | 84.14 | 21.60 | 1.53 |
| 10200 | ESCAZU | | 9.85 | 84.20 | 20.31 | 1.56 |
| 10201 | Escazù | 310 | 9.86 | 84.20 | 21.17 | 1.50 |
| 10202 | San Antonio | 319 | 9.83 | 84.19 | 18.71 | 1.71 |
| 10203 | San Rafael | 257 | 9.88 | 84.20 | 22.17 | 1.37 |
| 10300 | DESAMPARADOS | | 9.68 | 83.33 | 23.32 | 2.27 |
| 10301 | Desamparados | 329 | 9.83 | 84.12 | 21.00 | 1.48 |
| 10302 | San Miguel | 353 | 9.78 | 84.10 | 18.56 | 1.34 |
| 10303 | San Juan de Dio | 500 | 9.82 | 84.14 | 21.00 | 1.60 |
| 10304 | San Rafael | 345 | 9.81 | 84.13 | 21.00 | 1.52 |
| 10305 | San Antonio | 332 | 9.84 | 84.10 | 21.00 | 1.45 |
| 10306 | Frailes | 397 | 9.63 | 82.91 | 25.61 | 2.70 |
| 10307 | Patarrá | 352 | 9.81 | 84.08 | 19.78 | 1.39 |
| 10308 | San Cristobal | 391 | 9.70 | 84.05 | 16.92 | 1.54 |
| 10309 | Rosario | 379 | 9.74 | 84.14 | 19.38 | 1.48 |
| 10310 | Damas | 340 | 9.83 | 84.16 | 21.00 | 1.68 |
| 10311 | San Rafael | 333 | 9.83 | 84.14 | 21.00 | 1.57 |
| 10312 | Gravilias | 338 | 9.82 | 84.39 | 23.80 | 2.19 |
| 10400 | PURISCAL | | 9.67 | 83.60 | 22.85 | 2.66 |
| 10401 | Santiago | 356 | 9.76 | 84.36 | 23.27 | 2.26 |
| 10402 | Mercedes Sur | 374 | 9.72 | 84.46 | 23.77 | 2.64 |
| 10403 | Barbacoas | 360 | 9.80 | 84.41 | 22.41 | 2.35 |
| 10404 | Grifo Alto | 324 | 9.82 | 84.45 | 23.85 | 2.39 |
| 10405 | San Rafael | 372 | 9.76 | 84.33 | 23.67 | 2.33 |
| 10406 | Candelarita | 380 | 9.73 | 84.39 | 24.32 | 2.27 |
| 10407 | Desamparaditos | 360 | 9.66 | 83.23 | 22.00 | 2.76 |
| 10408 | San Antonio | 490 | 9.81 | 84.36 | 23.07 | 2.30 |
| 10409 | Chires | 10 | 9.60 | 84.45 | 26.31 | 2.38 |

[a] Climate data

TABLE 6-3b Costa Rica Coffee Production for 1973

| | District | Production | Yield |
|---|---|---|---|
| 457 | SAMPLES IN THIS FILE | | |
| 100 | COSTA RICA | 77918.8 | 369205120 |
| 100 | PROVINCE SAN JOSÉ | 24509.8 | 94022131 |
| 101 | CANTON CENTRAL | 500.2 | 3230766 |
| | Zapote | 83.8 | 1051500 |
| | San Fco de Dos Ríos | 83.2 | 669831 |
| | Uruca | 209.0 | 1195328 |
| | Mata Redonda | 3.5 | 11591 |
| | Pavas | 3.9 | 16405 |
| | Hatillo | 96.6 | 142112 |
| | San Sebastián | 19.9 | 143919 |
| 102 | ESCAZU | 465.0 | 1761096 |
| | Escazú | 24.7 | 77683 |
| | San Antonio | 283.4 | 965531 |
| | San Rafael | 156.8 | 717882 |
| 103 | DESAMPARADOS | 2726.7 | 10229578 |
| | Desamparados | 144.6 | 445368 |
| | San Miguel | 710.3 | 2369248 |
| | San Juan de Dios | 146.5 | 572040 |
| | San Rafael | 338.9 | 1103460 |
| | San Antonio | 103.8 | 517616 |
| | Frailes | 467.6 | 2563751 |
| | Patarrá | 255.9 | 686359 |
| | San Cristobal | 154.3 | 661163 |
| | Rosario | 286.5 | 942585 |
| | Damas | 27.7 | 77276 |
| | San Rafael | 90.2 | 290704 |
| 104 | PURISCAL | 1466.6 | 5320445 |
| | Santiago | 358.9 | 1485244 |
| | Mercedes Sur | 268.6 | 894773 |
| | Barbacoas | 313.2 | 1166400 |
| | Grifo Alto | 116.1 | 426797 |
| | San Rafael | 160.0 | 614990 |
| | Candelarita | 109.4 | 271342 |
| | Desamparaditos | 35.5 | 117095 |
| | San Antonio | 104.6 | 343804 |

TABLE 6-3c ASCII Template for Geographical Output

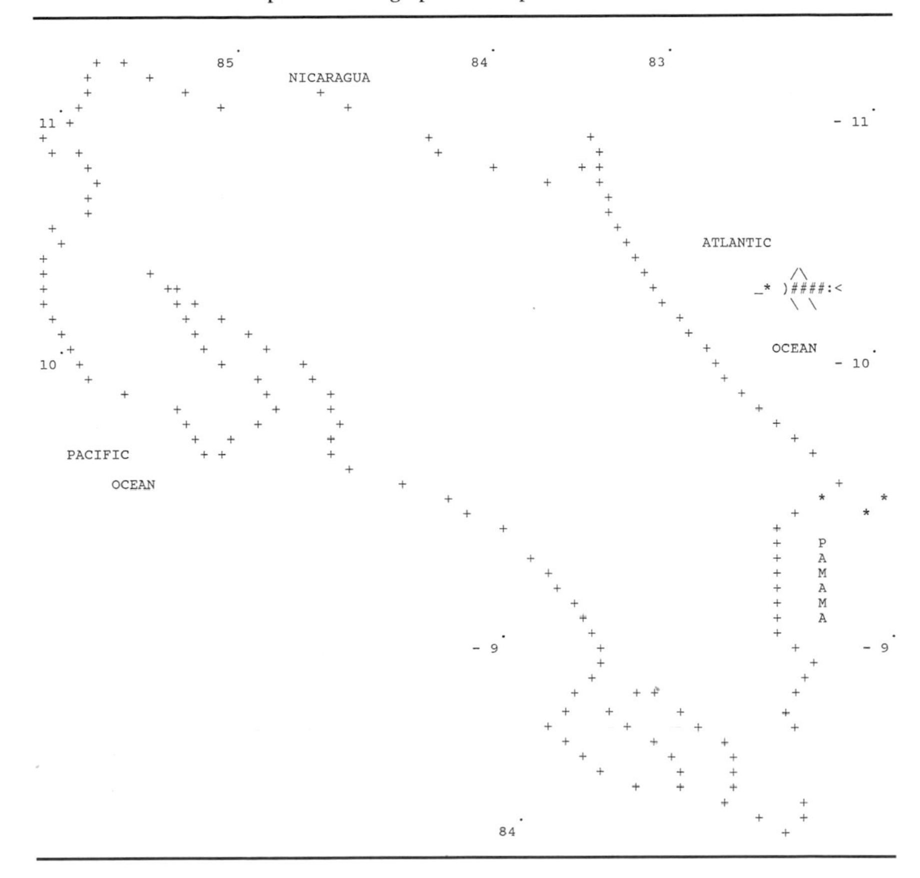


and some programming skills you should be able to get as output figures similar to Figures 12-12 and 12-13 which are readable and printable with PEDIT and any printer.

## REFERENCES

Dent, B. D. 1985. *Cartography: Thematic Map Design*. 2nd Ed. Brown, Dubuque, IA.

Dikau, R., E. E. Brabb, and R. Mark. 1991. *Landform Classification of New Mexico by Computer*, pp. 91–634. United States Geological Survey Open File Report.

Forbes, T., D. Rossiter, and A. Van Wambeke. 1987. Map scale and map texture. In *Guidelines for Evaluating the Adequacy of Soil Inventories*, pp. 2–5. Cornell University Press, Ithaca, NY.

Hall, C. A. S., H. Tian, Y. Qi, G. Pontius, and J. Cornell. 1995a. Modeling spatial and temporal patterns of tropical land use change. *Journal of Biogeography* 22:753–757.
Hall, C. A. S., H. Tian, Y. Qi, G. Pontius, J. Cornell and J. Uhlig. 1995b. Spatially explicit models of land use change and their application to the tropics. DOE Research Summary, No. 31, Oak Ridge National Laboratory.
Pontius, R. G. 1994. Modeling tropical land-use change and assessing policies to reduce carbon dioxide release from Africa. Ph.D. thesis, State University of New York, College of Environmental Science and Forestry, New York.
Vieux, B. E., V. F. Bralts, and L. J. Segerlind. 1988. Finite element analysis of hydrologic response areas using geographic information systems. In *Modeling Agricultural, Forest, and Rangeland Hydrology*, pp. 436–437. Publication 07-88, Am. Soc. Agr. Engs., St. Joseph, MI.

CHAPTER 7

# Geographical Modeling

## The Synthesis of a GIS and Simulation Modeling

MYRNA HALL, CHARLES A. S. HALL, AND MARSHALL TAYLOR



## I. THE CONCEPTS BEHIND LAND USE DECISION MAKING

### A. A BRIEF CONCEPTUAL HISTORY OF LAND USE DECISION MAKING

Land evaluation has become an important component of sustainability analysis for developing countries (FAO, 1976, 1983; Beek, 1978; McCrae and Burnham, 1981). When land was abundant relative to human populations, natural resource availability and environmental degradation, including the availability and maintenance of good land for agriculture, were rarely issues. Increasingly, however, in most parts of the world, population pressures are coming up against limited good land. Hence land use and land availability play an ever more important role in determining if a country's economic development strategy will be sustainable over time. It is likely this decrease in land per capita contributes to increasing intergroup tensions and even wars in various parts of the world (Homer-Dixon *et al.*, 1993). The issue has relatively little

*Quantifying Sustainable Development*

to do with absolute area of land, but rather is related more to both quantitative and qualitative aspects of land availability. As formalized by the English classical economist Ricardo in 1817, farmers tend to use the most productive land first. "Productive" generally refers to the flat, frequently riparian land at low elevations which reaps the largest difference between economic costs, including labor, and economic returns. Historically, farmers have cultivated fertile river valleys first. As populations have increased over time, they have had to develop new lands upstream and upslope.

These concepts pertaining to productivity can be understood through the use of Howard Odum's "maximum power" principle. The principle states that in general organisms have been selected for behaviors and other adaptations that operate to obtain the largest net power flow possible given the environmental conditions at their disposal (Odum, 1983; Hall *et al.*, 1995). We have formalized this concept slightly differently (by ignoring the time rate) as maximizing the energy return on energy investment (EROI), again within the constraints of the environment (Hall *et al.*, 1986, 1992a). Logically, maximizing EROI should lead to maximizing economic return on investment if a very comprehensive perspective is used.

A problem facing those responsible for managing the resources of a nation is that what may appear to be logical investment decisions for individual farmers is often less logical from the perspective of optimizing a nation's resource use. For example, certain crops may give high short-term economic returns to the farmer but only at the expense of large longer term soil erosion. Putting very high-quality land (i.e., where EROI is high) into pastures may give the owner a relatively substantial profit because expenses are low, but may work against feeding a hungry nation. Heavy use of fertilizer may give a farmer high profits but erode a nation's foreign exchange. It is not easy to determine in any of these situations which is the better strategy. What is right for one location, time, and soil type is almost certainly not the best solution for another location, time, and soil type.

Until relatively recently it was extremely difficult to put all the information required to make such decisions into the hands of decision makers, be they farmers, extension agents, corporate managers, government agencies, or national presidents. One result has been that decisions were often made on the basis of blind ideology (e.g., "free market" or "collective welfare") or local custom, with relatively little exploration of the possibilities of alternative decisions or the long-term economic and environmental consequences of each possibility.

Cheap microcomputers and recent software development are changing all that. Specifically, the relatively recent development of Geographic Information Systems (GIS's) has allowed the generation and manipulation of extraordinary quantities of complex geographically referenced information that for the first time allows the generation of integrated spatially specific models of agricultural production, climate, hydrology, erosion, and related processes. Up until recently, however, there was neither a conceptual framework nor an adequate data base that would allow for the development of such a model that integrates both biophysical parameters and economic costs and gains at the scale of, e.g., a small country. We develop here a technique that we think resolves this problem by deriving an explicit, quantitatively testable approach.

The method is based on what we call *geographical modeling,* and on the ecological concept of *gradient analysis.* We use this method principally for analyses of biological species and agronomic cultivars. The concept, however, is in principle applicable to more or less any problem that has geographically referenced environmental variables. The methodology is an important step in the evolution of sustainability analysis from concepts to actual applications.

## B. A Systems Approach

Human-induced development and its associated social and technological activities throughout the world are contributing to rapid and often stressful changes in the earth's environment. Human land use practices in forestry, agriculture, industry, transportation, and residential development have and will continue to alter our terrestrial and marine ecosystems significantly (Wheeler, 1993; Rapport, 1997). The overexploitation of soils, fisheries, and forests, the proliferation of human pollution and wastes, the potential implications of the many changes to the global atmosphere, and the loss of habitat for many species, including humans, are increasingly driving science today. More than ever many scientists are grappling with the "what-ifs" of global change. Ironically most other segments of society are pushing for the acceleration of these changes through arguments and programs for increasing economic activity. It is becoming increasingly evident that the problems and the potential solutions are connected in ways not previously understood. Thus the most important aspect of all environmental analysis is the need for a systems approach, meaning a formalized set of procedures for attempting to understand not simply the phenomenon of interest but the ways in which that phenomenon is linked to other processes at similar and different scales. An easily accessible and very useful introduction to systems thinking is given in Peet (1992).

A heightened concern about potential global change has been a large impetus for the development of what we call geographic modeling. In order to understand the interactions of terrestrial, aquatic, and atmospheric systems at a regional or global scale, modelers of earth systems have required new models linked to geographically referenced data that cover broad areas. Over the past 15 years we have acquired a new suite of tools in the form of GIS's that allow us to analyze whole landscape interactions, greatly increasing our power relative to the site-specific ecosystem models of the past. Now we can predict many aspects of environmental change at local, regional, and global levels by combining the mapped data of a GIS with environmental simulation models. Such models can incorporate descriptions of key physical processes that modulate the behavior of many aspects of the earth's systems and aid our understanding of these processes.

## II. TOOLS FOR ANALYSIS

### A. Simulation Modeling

Models are simplified representations of real objects or processes. More formally, models can be defined as: (1) a simplification, (2) devices for predicting complex, poorly understood phenomena from simpler components that are understood, and (3) (our favorite) a formalization of our assumptions about a system (Hall and Day, 1977). Although we tend to think of models as some sort of computer-based phenomena, in fact there are many types of models, conveniently classified as conceptual, diagrammatic, quantitative, and computer. These four types of models represent a series of steps for increasingly formalizing our understanding of the system of interest. Models are most powerful when combined with empirical (field) studies. Good field data of course lead to good models. What is less often appreciated is that good models can lead to better field sampling by helping to define what data are important for the question being asked.

Models use two general approaches to solve equation sets: (1) analytic or closed form and (2) simulation or computer. Analytic models are frequently mathematically complex and require a great deal of training for their use. Most are based on a series of differential equations, Markov chains, or Leslie matrices, often with explicit steady-state assumptions and without spatial considerations (Hunsaker *et al.*, 1993). Generally they do not give solutions over many points in time easily. Simulation models use the iterative (repeating) power of computers to simplify the mathematics and to add time as a factor, thus changing the output from static to dynamic. With the advent of digital mapping, simulation models are increasingly adding space as a factor as well. Combining the dynamics of space and time is essentially impossible analytically but essential for representing the real world.

Environmental models, at least of the computerized sort, have been around for considerably longer than GIS's. The original concept can be traced back to Howard Odum's 1960 paper, "Analog Circuits in Ecology" (for a review see Kangas, 1995). The early computer models were viewed with great enthusiasm by many in the environmental science community in the 1960s (i.e., Watt, 1968; Hall and Day, 1977; Hagen, 1992) but the enthusiasm became more muted in the 1970s and 1980s when the original, sometimes inflated, pronouncements of the virtues and power of models were met only partially. The most conspicuous of the environmental models, the "Club of Rome" models (Forrester, 1968; Meadows *et al.*, 1974), often have been castigated because the environmental and economic collapse that they predicted did not come to pass. Most critics, however, do not seem to recognize or perhaps acknowledge that the predicted apocalypse, likely or unlikely, was not predicted to occur until 2020 or so. The model predictions are basically "on track" as of 1999.

Far less ambitious predictive or optimization models are used routinely now in resource management agencies to determine the best management approaches and to allocate increasingly scarce resources. It is difficult to find major resource management agencies in the developed world whose management policies are not based substantially on the use of models and/or GIS's. Some examples include fishery population models (Karl English, personal communication; Jowett 1988), wildlife management (e.g., Nielsen *et al.*, 1997), hydrology models (Devantier and Feldman, 1993; Arnold *et al.*, 1990; Meyer *et al.*, 1993; Maidment *et al.*, 1996), water quality models (Huber and Dickinson, 1988; Al-Abed and Whiteley, 1995; DeVries and Hromadka, 1993), and reservoir management models (HEC, 1976, 1982; Hall *et al.*, 1989; Jourdonnais *et al.*, 1990; Loucks *et al.*, 1995). Assessment models are a little different, and are used to estimate the effects of some activity not yet performed (such as development), the results of which are difficult to observe or estimate in advance (Hall and Day, 1977). A modeling framework, by providing a basis for formalizing and evaluating alternative scenarios, can help slow or alter the negative impact humans have had on the environment in recent decades.

The three most important roles of models, therefore, are probably (1) to help formalize and understand complex systems, including the organization and synthesis of data, (2) to understand the importance of poorly measured or poorly understood components of a system through the process of sensitivity analysis, and (3) to test whether or not one's understanding of the system is accurate through the process of validation, that is, comparing the output of the model to the behavior of the real system.

Historically ecologists have focused on changes in state variables at one location over time. Today, largely due to the power of storing spatially referenced data in a GIS, resource researchers are moving toward incorporating spatial pattern and influence into their models. They are able to apply their models to ever larger geographic areas such as Yellowstone Park or entire countries, as we do here. Hunsaker *et al.* (1993) point out that not all ecological topics require spatially explicit models. To determine whether spatial position is important in a model, one should ask, (1) what are the spatial units and their interactions, and (2) how do the relative sizes and locations of these spatial units affect state variables such as biomass and population size, and such ecosystem functions as energy flow and biological productivity? "The null hypothesis is that the future state of a landscape unit is independent of adjacent units" (Hunsaker *et al.*, 1993, p. 248). Most ecologists generally find that the answer to this question is "yes, space matters" (Graham *et al.*, 1990; Turner *et al.*, 1997; Hall *et al.*, 1995; Tian *et al.*, 1988).

## B. Geographical Information Systems

"Geographic" (or "spatial") implies that attribute information such as soil texture, millimeters rainfall, elevation, and vegetation type varies over, and can be located on, the earth's surface using earth system coordinates such as longitude, latitude, UTM meters, or state plane English units. Geographical information systems have existed since the dawn of our species in the form of maps drawn in the dirt or on parchment, but the present day use of the term as a computerized entity is only a few decades old. The old and the new are linked in the name Idrisi, originally a great 12th century Sicilian-Arabic cartographer (who wrote Kitab Rujjar, or book of Roger, in 1154) and now the name of a very useful GIS program (Eastman, 1997).

The popular definition of a GIS was coined by Burrough (1986, p. 6): A computerized system for "collecting, storing, retrieving at will, transforming, and displaying spatial data from the real world for a particular set of purposes." Spatial or geographic data can describe renewable and nonrenewable natural resources as well as cultural and human resources. Typically the data appear as maps that tell us the location, magnitude, or characteristics of a resource at each location on a map. Each *data layer,* or individual map, in a GIS contains one and only one kind of information or attribute. Typical data of geographic interest include elevation, percent slope, aspect (compass direction that a land slope faces), soil characteristics, vegetation type, land use, or average annual rainfall (Figure 7-1). This information is gathered from aerial photographs, satellite images, resource sampling using global positioning systems (GPS), surveying, or by digitizing or scanning already created paper maps.

The computer stores this information as numeric values in one of two fundamental approaches. In the *vector* format a series of Cartesian coordinates record the position of a point, a line, or aerial data taken from a map (e.g., a soil sample site, an elevation contour line, or an area of 300-year-old Redwood trees; Figure 7-2). A relational data base stores information (e.g., the soil pH, depth, texture, etc. at a sample site; the elevation in meters for the contour line; or the age leaf area index, growth, etc., of a forest) about each of these map

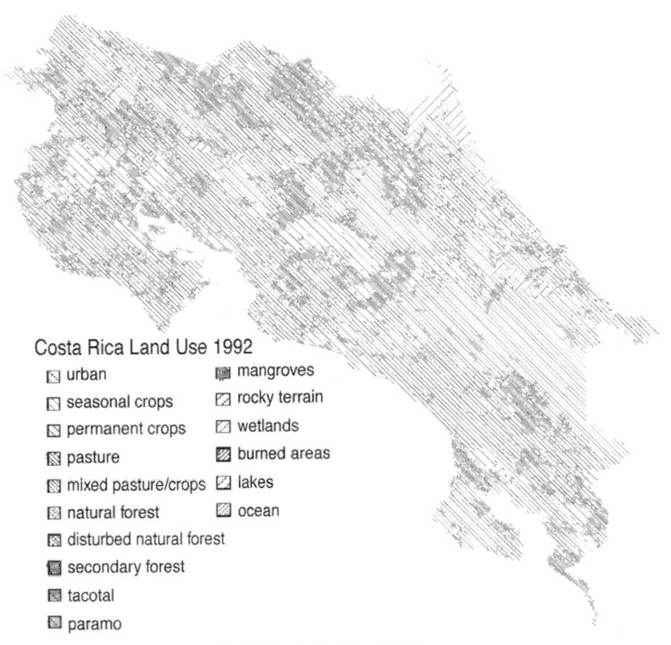


FIGURE 7-1 Spatial data.

features for each position. In the *raster* format each grid cell (i.e., row–column position) of a map matrix holds a unique numeric data value for the attribute recorded. You can think of it as a digital checkerboard, with a unique value stored in each square representing either data that vary continuously across the landscape such as soil permeability or elevation, or discrete values representing such information as forest type. These values may belong to one of four scales of measurement (Griffith and Amrhein, 1991): (1) the nominal scale where the number simply represents the name of an attribute, e.g., where 1 = forest, 2 = agriculture, etc.; (2) the ordinal (or ranked) scale, which indicates comparability (e.g., 3 = good, 2 = better, or 1 = best soil for sorghum); (3) interval scale data, which include numerical data from a continuum without an absolute zero, such as temperature in degrees Celsius, or elevation in meters; and (4) the ratio scale, which is used where an absolute zero value exists, as with distance from a fixed reference point such as sea level, zero income, or zero degrees Kelvin.

Geographical information systems were designed originally as a tool to support decision making for land use planning. In the 1960s, Canada was experiencing increasing competition among potential uses of land within its commercially accessible land zones. In response the government initiated a

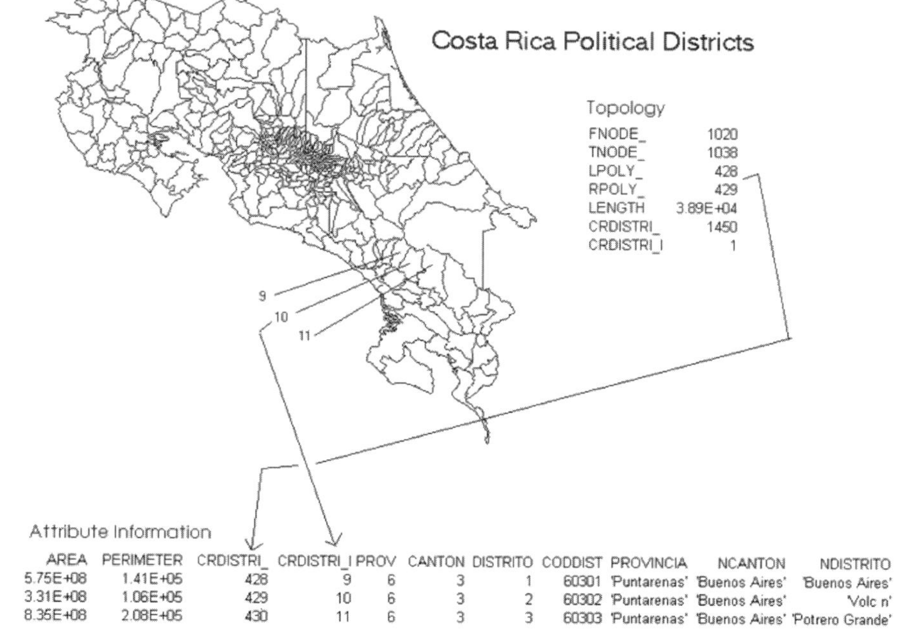


| AREA | PERIMETER | CRDISTRI_ | CRDISTRI_I | PROV | CANTON | DISTRITO | CODDIST | PROVINCIA | NCANTON | NDISTRITO |
|---|---|---|---|---|---|---|---|---|---|---|
| 5.75E+08 | 1.41E+05 | 428 | 9 | 6 | 3 | 1 | 60301 | 'Puntarenas' | 'Buenos Aires' | 'Buenos Aires' |
| 3.31E+08 | 1.06E+05 | 429 | 10 | 6 | 3 | 2 | 60302 | 'Puntarenas' | 'Buenos Aires' | 'Volc n' |
| 8.35E+08 | 2.08E+05 | 430 | 11 | 6 | 3 | 3 | 60303 | 'Puntarenas' | 'Buenos Aires' | 'Potrero Grande' |

FIGURE 7-2 An example of a vector format GIS map (ESRI) showing the outlines of political districts in Costa Rica. Attribute information for each district polygon is stored in a database file and referenced by, in this case, two identification numbers (crdistri_ and crdistri_i). Topology, or the position of each line segment with reference to its neighboring (left and right) polygons), is also stored.

country-wide land inventory to map the current uses of its land and the capability of each spatial unit of the land for agriculture, forestry, wildlife, and recreation. Government perceived that it had an increasing role to play in making decisions about land management and planning the utilization of natural resources because conflicts were increasing dramatically.

Credit for the first functional computerized GIS goes to Roger Tomlinson, who at the time of the Canadian government initiative was working for an Ottawa aerial survey company charged with the task of inventorying all of Canada's forest resources. Using manual map analysis techniques he estimated that his assignment would require 500 technicians and three years to complete. At just about this same time transistors replaced vacuum tube computer technology. Suddenly computers could calculate faster, were more reliable and cheaper, and most importantly had memory, or the ability to store information. Tomlinson conceived the idea that many maps could be put into numeric form and linked together to form a complete picture of natural resources throughout Canada.

Since that time there has been dramatic growth in the industry. Development of the technology has been driven primarily by demands for information management, such as Tomlinson faced, rather than by research needs for spatial analysis and modeling functions. The major success of the GIS is due to its mapping and query capabilities, i.e., its ability to answer the questions, "What lies where?" and "Where lies what?" Commercial development, because of financial necessity, supported the demand of the majority of users, which was to answer these two questions.

What a GIS provides, in addition to excellent mapping capabilities, is the spatial overlay function, an idea developed in the 1960s by landscape architect Ian McHarg, who envisioned in his book, *Design with Nature,* the power for informed land use decision making by overlaying transparent acetate maps. When this concept was transferred to the computer environment it became possible to integrate various sources of data, represented by multiple digital data layers, within a single system and single format. Such spatially referenced data sets now can be manipulated and queried to produce new information that was not known or available previously. No current GIS, however, has both the capability for manipulating and representing spatial and temporal data and the algorithmic flexibility to build and test complex process models internally (Nyerges, 1993). Models provide solutions over time and the resolution of complex processes in a way that a GIS does only superficially (Hall and Day, 1977). Dangermond (1993), the founder of ESRI, the manufacturer of the ARC/INFO GIS, the dominant GIS on the market today, agrees that "while GIS technology has been used extensively in both the environmental field and in modeling, it has not been used very often for modeling in the environmental field." In addition, neither GIS's nor modeling has evolved to represent the dynamic display of ecosystem processes.

## C. The Synthesis

### 1. What Is Geographic Modeling?

To fill this void we (and some others) have developed *geographic modeling* by integrating the geographic or spatial data base structure of GIS raster format maps into an environmental modeling perspective. Models that once addressed ecosystem processes at only one location can now be run for multiple locations, in the form of landscape units (polygons or grid cells), using data from multiple mapped data layers as model inputs. This allows us to predict and analyze outcomes across a spatially varied landscape, and to incorporate the influence of spatially adjacent landscape units, which was not possible in the models of the past.

What led us to this synthesis was that as systems thinkers we were becoming more and more interested in the interaction of human and terrestrial systems. We wanted, therefore, to be able to analyze what was going on in a system, whether a river, a lake, a watershed, or a swidden field, as a function of other spatially varying parameters, like rainfall, insolation, soil quality, or human population density. As these parameters vary across a landscape, so do the ecological outcomes predicted by our models. Similar approaches have been undertaken to model, e.g., land changes in southern Louisiana as a function of Mississippi River sedimentation (e.g., Costanza *et al.*, 1990), the interactions of land and water management in the Florida Everglades (e.g., Wu *et al.*, 1996), and forest productivity in Montana as a function of climate and soil parameters (e.g., Running *et al.*, 1989).

## 2. How Do Geographic Models Work?

Geographic modeling treats the landscape as a matrix of geographically referenced landscape units or cells (Figure 7-3). Although both vector-based and raster-based GIS maps can be used to do geographic modeling, we prefer the latter for two reasons. (1) It allows us to do much higher resolution distributed modeling as opposed to the lumped parameter modeling commonly encountered with environmental models based on vector GIS polygon input maps. (2) Raster maps interface well with the ability of FORTRAN and other programming languages to handle matrices of numbers. We can predict changes in these matrix-referenced variables, i.e., their row–column cartographic position, across hours, days, months, or years by using iterative "DO" loops. For example, in our TOPOCLIM model (Everham *et al.*, 1991) we predict the solar insolation on a plot of land as a function of the sun's position relative to that place on earth at a given time of year and time of day and the plot's slope and solar aspect. Additional corrections are made for topographical features that block the sunlight to that cell.

Geographic modeling requires transdisciplinary skills in the complex process of modeling, in GIS technology, and in the subject matter being represented. Inexperience in any can lead to misuse of the GIS technology and error propagation in models (Moore *et al.*, 1993). To date there is no commercial software package available that allows a completely smooth interface between models, their mapped data inputs, and their graphic output. We, however, have created three software packages, FNCYPLOT, GEOPLOT, and ECOPLOT, that allow an experienced programmer to do this quite easily by incorporating Resources Planning Associates' (RPA) graphic interface programmer libraries (CAPLIB) with our ecosystem models. Each provides a dynamic graphic display of simulation output. Both GEOPLOT and ECOPLOT are included on the enclosed CD. GEOPLOT allows a time series display of up to nine maps on the screen at a time (e.g., Hall *et al.*, 1992b) (Figure 7-4). FNCYPLOT and ECOPLOT display a changing 3D landscape with system variables graphed at the perimeter of the screen (Hall and Hall, 1993; M. Hall, 1994) (Figure 5-5). The simulation model of Chapter 5 and the meteorological displays of Chapter 9 (and the CD) are examples of their use.

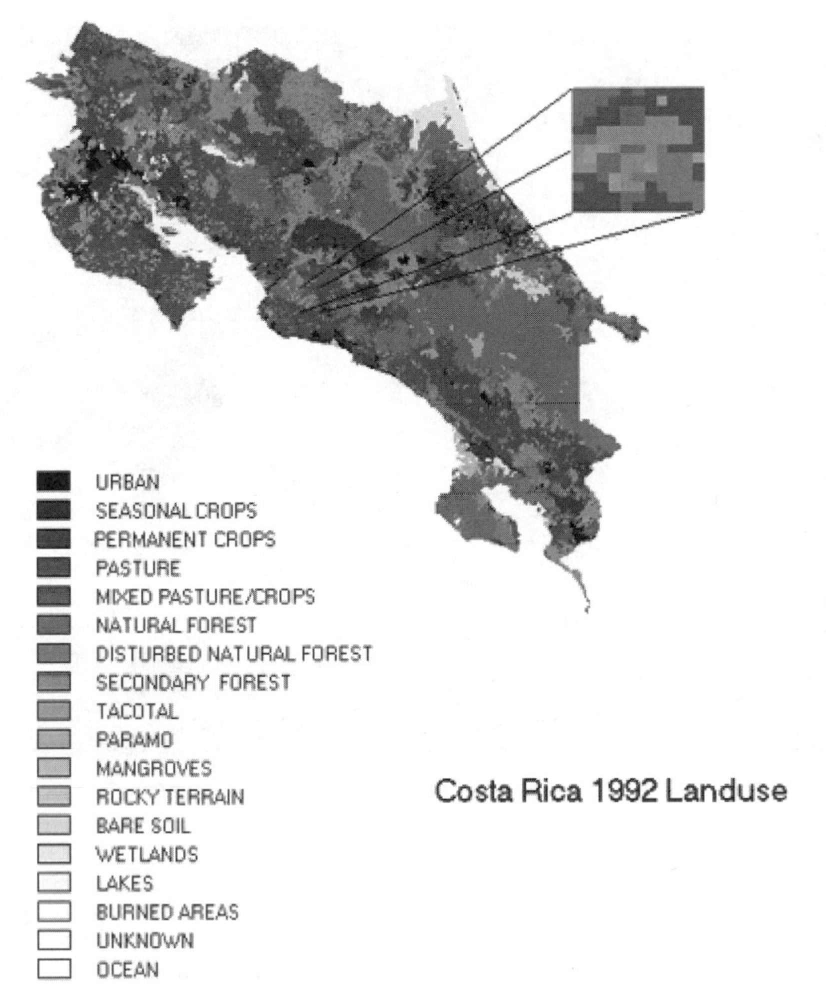


FIGURE 7-3 An example of a raster (i.e., grid or matrix) format map, where each cell stores, in this case, a value indicating type of land use.

## D. Gradient Analysis, an Approach to Extrapolating Spatial Data

### 1. Background

The basic concept behind gradient analysis is that each species responds to the environmental conditions of a given site independently, so that communities at any one location are a function of the collective sets of species that grow well

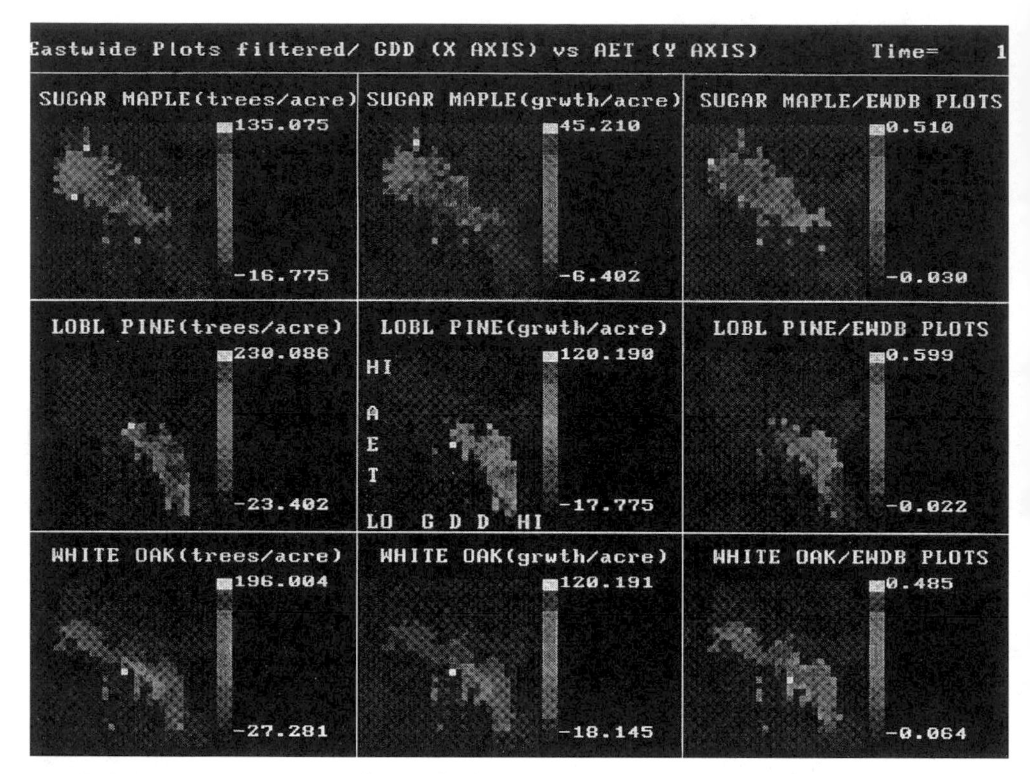


**FIGURE 7-4** GEOPLOT visualization displaying run-time model output on nine maps simultaneously. GDD, growing degree days; AET, actual evapotranspiration. (Original in color.)

there. Changes in space imply changes in environmental conditions. Thus gradient analysis is an approach to examining the spatial distribution and production of plants and animals as a function of environmental characteristics, which together are perceived as a location along a gradual change in conditions (hence the word "gradient"). The essence of the approach was first written about by Aristotle's student Theophrastus (Kormandy, 1969), and more fully developed by the Russian L. G. Remensky (1924).

Since essentially all environmental conditions can be analyzed as gradients, it is possible to determine the conditions empirically in which each species (or cultivar) does well or poorly. Each species tends to be distributed more or less normally, or skewed normally, over each important gradient. In general, physical gradients (e.g., light intensity, temperature, and moisture) are the most important determinants of species distribution. Within that context interbiota factors are sometimes important, and occasionally dominant. Two exceptions to that rule appear to be browsing intensity and pathogen load, although of course these too can be considered gradients. In all cases disturbance history can be critical, and that too can be analyzed as a gradient (Kessell, 1977, 1979).

The history of the use of gradient analysis is inextricably tied up with the history of major ideas in ecology, especially the ideas by which ecologists have attempted to summarize complex spatial relations of species and communities. In the English-speaking literature, the principal early geographic focus was on the distribution of plant "communities," which were perceived as more or less self-generating entities (e.g., Clements, 1931). The second, or "individualistic," view of plant community development proposes that plant distribution and growth result from the interaction of individuals and species responding to geographic variability in the conditions of soil and climate (Gleason, 1926). The two have been resolved in large part in favor of the individualistic concept by Whittaker's (1956) studies in the Great Smoky Mountains.

In this early work Whittaker examined the distribution of tree species as a function of location along elevational gradients (Figure 7-5). Later, he and his student Kessell (1977, 1979) extended the concept to the examination of tree species distribution and abundance as a function of two gradients: soil

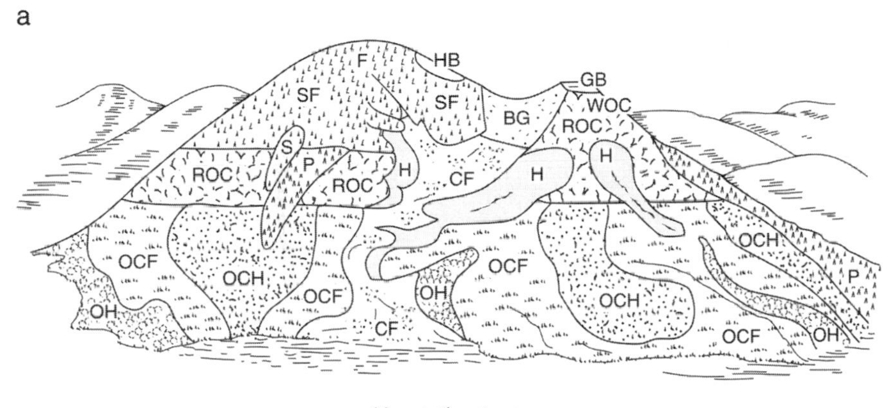


FIGURE 7-5 (a) The distribution of tree species as a function of geographical space and elevation in the Great Smoky Mountains of the eastern United States. (Source: Whittaker, 1956.) (b) Distribution of tree communities in Figure 7-5a as a function of two gradients: elevation and topographic soil moisture index. (Source: Kessell, 1979.)

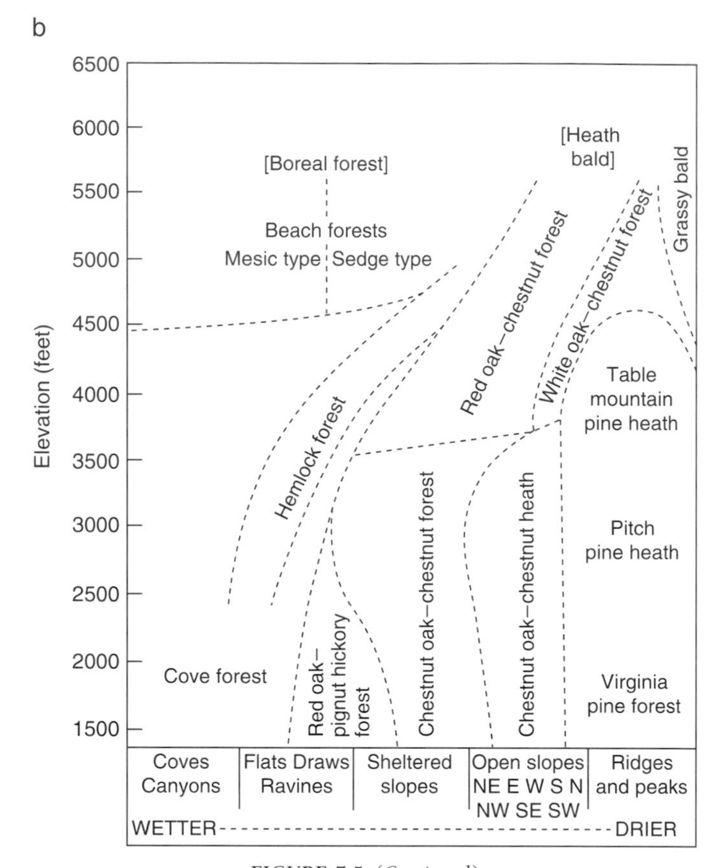


FIGURE 7-5 *(Continued)*

moisture and elevation (Figure 7-6). The empirical record from examining the distribution of species over elevational gradients appears to give rather strong support to the power of gradients for examining spatial properties of organisms. Unfortunately, this method has not been applied especially widely by other than Whittaker and his students, so that we do not understand well the power of this approach other than the obvious changes in species abundance and distribution found as one progresses up or around a mountain. [See Austin (1987), Prentice *et al.* (1990, 1992), and ter Braak and Prentice (1988) for later perspectives on gradients.]

Recently Hall *et al.* (1992a) derived a mechanism for the gradient pattern of distribution based on energy costs and gains (Figure 7-7). The derivation of this approach requires more space than is appropriate here, but is found

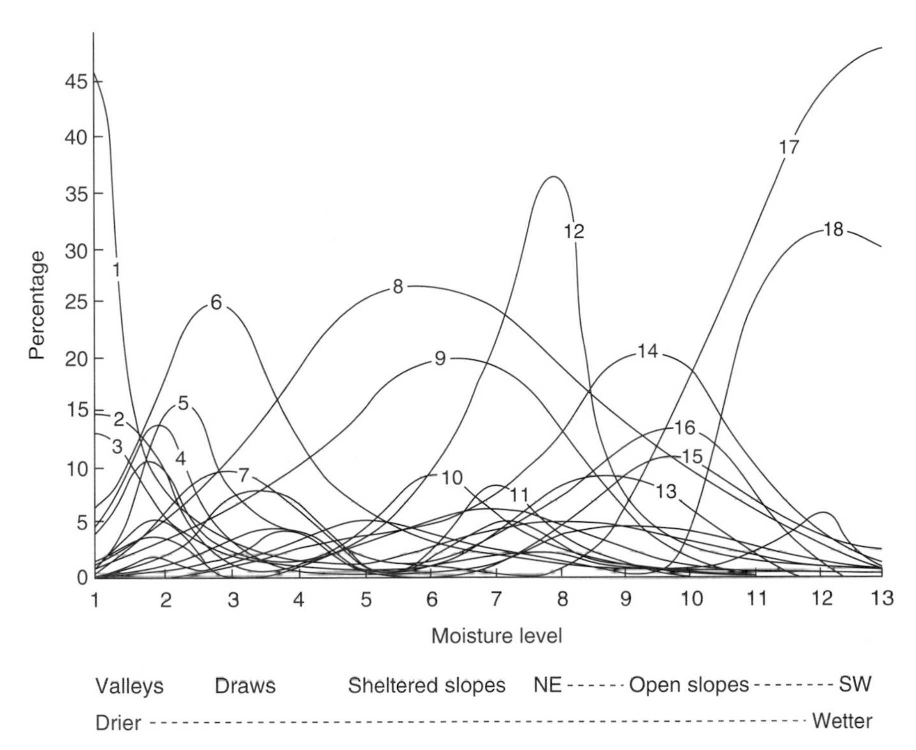


FIGURE 7-6 Distribution of 18 tree species as a function of soil moisture. (Source: Kessell, 1979.)

in the cited publication. Briefly, the fundamental concept is that plant growth and reproductive potential are a function of environmental gradients, which reflect the plant's energy balance between costs and gains. An important question is, "how many gradients are necessary?" Hilborn and Stearns (1982), among others, have shown the power of examining data as a function of several independent variables when one variable alone shows little promise. Whittaker, and later Kessell (e.g., 1977, 1979), used two gradient dimensions (generally elevation and moisture) and plotted both community type and isopleths of abundance for species (Figure 7-6b). Hall *et al.* (1992a) presented a concept for drawing up to nine gradients in two dimensions, but its implementation has been difficult.

Many bioclimatic classification schemes have been devised previously to associate vegetation types and climate regimes. These include Griesbach (1838), von Humboldt (1867), Koppen (1900), Thornthwaite (1931, 1948), Holdridge (1947, 1967), Troll and Paffen (1964), Box (1978, 1995), Solomon *et al.* (1981) and Nielson *et al.* (1992). All of these systems are based in some way on the

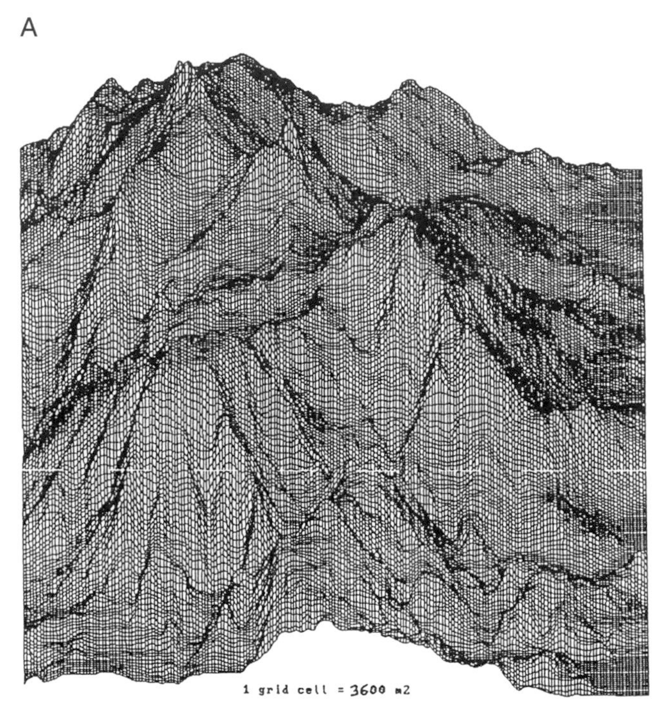


FIGURE 7-7 The relation of geographical space (A) to gradient space (B, facing page). Each spot in geographical space can be represented by a location in gradient space which represents the environmental conditions at that spot. A species may be found over an entire landscape (A) but will tend to be concentrated near the center of its gradient space, that is, toward the center of the range of environmental conditions in which it is found. For example, temperature tends to decrease going up a mountain, and rainfall to increase. Each are also affected by, e.g., aspect. Thus each spot on the mountain will have its particular mean temperature and rainfall conditions. Each species will tend to have a particular location in gradient space where it grows optimally and has a high-energy profit and hence potential reproductive output.

relation of plant species or types to climate. But the development of modern fast computers allows for a more explicit and quantitative application of the gradients approach. A fundamental problem with spatial data is that the cost of obtaining that information is often much, much larger than that of nonspatial information. A gradients approach can help, especially with biological problems, in estimating species abundance and distribution for unsampled locations. It can also assist in interpreting previously obtained spatial information.

New graphical computer techniques offer the possibility of doing much more interesting and complex gradient analyses, but have not been developed previously for this problem. The availability of extensive data sets (i.e., NASA digital climate records on compact discs or USFS "Eastwide" forest inventory)

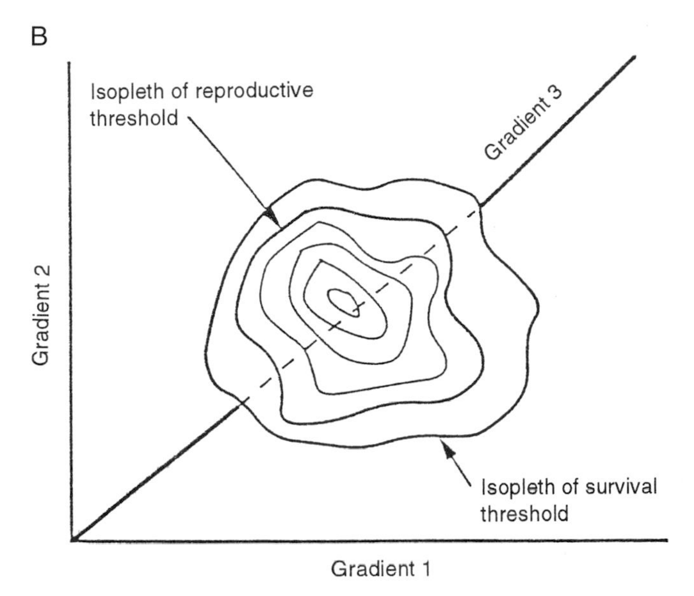


FIGURE 7-7 (*Continued*)

offers the possibility of examining various aspects of species and ecosystems in new, data-intensive ways. Furthermore, the availability of computer predictions of possible future climates at various locations and scales offers the possibility of examining the potentially changing relation between geographical space and what we call gradient space, both for present conditions and under various climatic scenarios. For example, once we have derived the initial analysis of species distribution patterns we can return to the landscape and predict which species will be found where following disturbances such as fire, grazing, road building, or climate change. We believe that this approach has enormous power for synthesizing existing plant distribution and growth patterns and projecting what they are likely to be in the future. The next section develops procedures for using the gradients idea within the context of more recent theoretical and computational developments.

## 2. Integrating Geographical and Gradient Space

Species are most often mapped as to where they live in geographical space. For example, maps of species' geographic distribution are given in various publications of the U.S. Forest Service (USDA Forest Service, 1913; Little, 1971; Burns and Honkala, 1990). Maps of general vegetation types based on climate zones (life zones) exist for Costa Rica as well (Holdridge, 1967). This is, in a sense, paradoxical, because, as already explained, the vegetation is not responding to that particular piece of geography but rather physiologically to the location on gradients of temperature, water availability, nutrients, and so forth that they find there, as well as the history of that site. We have devised a technique that allows us to record and summarize plant response at any geographic location as a function of the gradient space found there.

Some definitions are in order: *geographical space,* at least as used here, is the same as what is represented on maps, that is, two-dimensional space on the earth's surface (i.e., in terms of Cartesian coordinates such as UTM meters, state plane English units, or degrees longitude and latitude). But most organisms do not occupy geographical space randomly; they are found in patches and corridors as a function of the changing microclimatic, edaphic, and historical conditions imposed upon a region by patterns of climate and disturbance as influenced by topography and geology. Examples are all around us: willows and alders are found along streams; ponderosa pines are found in mesic western United States; brook trout are found in cool headwater streams of the northeast; and other such commonly perceived and obvious relations of organisms to their environment that can be represented on maps. A particularly well-developed gradient approach is that of Holdridge (1947, 1967). His "life zone" approach uses a clever schematic to examine the distribution of vegetation types as a function of temperature (as it varies with latitude or altitude, and represented on the vertical scale) and moisture, based on the interaction of rainfall and evapotranspiration and represented as the interaction of two diagonal sets of lines (Figure 17-2). A special feature of the Holdridge diagram is that it represents each independent variable as a geometric progression, which seems to sort out vegetation well. Holdridge's system was inspired by, and worked out in, the complex environment of Costa Rica. Brown and Lugo (1984) and Helmer (Chapter 17) have synthesized biomass using this approach, and it works well.

*Ecological* or *gradient space* refers to an organism's location along Whittaker's gradients or, by extension, Hutchinson's "*n*-dimensional hyperspace" (Figure 7-7). Thus, sugar maples are found where winter temperatures are no lower than about 40°C and where the soil is wet or moist for most months of the year, and ponderosa pines are found in normally unsaturated soils between winter temperatures of minus 40 and 0°C. Brook trout are found where summer water temperatures are between about 8 and 20°C, and where the pH is no less than about 5.5 and no more than about 8. Thus, the ecological space inhabited by brook trout, for example, is described by the various ("*n*" in the mathematical sense) physical and chemical environmental conditions that the organisms can tolerate (see Hutchinson, 1965; Day *et al.*, 1989; Hall *et al.*, 1992a).

There are at least two approaches to modeling vegetation distribution. Gates (1993) differentiates them as empirical models versus physiological models, although there is overlap. In a sense, all are gradient models. We prefer to call them direct versus indirect response models since both are in essence based on gradient analysis. The first methodology, the direct response models, derives taxon response to resource conditions directly and empirically. Then a climate is generated for some time, past or future, and vegetation distribution is predicted as a function of the conditions created using previously measured empirical relations of taxa to climate. Examples include Emanuel *et al.* (1985), Bartlein *et al.* (1986), Webb (1986), Overpeck and Bartlein (1989), Zabinski and Davis (1989), Prentice (1990), Graham *et al.* (1990), Prentice *et al.* (1992), Lenihan (1993), Gates (1993), and Monserud *et al.* (1993).

Indirect response models introduce a time (and hence successional) element. Often succession is mimicked on disturbed sites, called forest gaps, and the number and biomass of individual species growing there are predicted using the empirical physiological response of the taxon taken from field observations. Botkin *et al.* (1972), Shugart and West (1977), and Solomon *et al.* (1981) similarly create environmental conditions and then "grow" each taxon to the conditions created.

We developed the model GEO2ECO, a direct response model, to determine the relationship between a species' geographic location and its ecological or gradient location (M. Hall, 1994). GEO2ECO (one version is called "MERGE") calculates the relative frequency of occurrence for each species, expressed as a percentage, as a function of both one and two environmental gradients (Figure 7-8). The relative frequency, therefore, represents the proportion of the number of times a species occurs under a given set of environmental conditions to the number of times that given set of environmental conditions occurs within the geographic area being analyzed. This procedure corrects for the tendency of absolute abundance values giving high values simply because those environmental conditions are common on that particular landscape. It has the disadvantage of giving very high values to low absolute, but high relative frequency occurrences of vegetation on rarely occurring gradient conditions.

We synthesize the preceding arguments by hypothesizing that: (1) species (including agricultural cultivars) will tend to have the highest net energy return, and hence the highest growth, toward the middle of their environmental or ecological range, which may or may not be the middle of their geographical range (Figure 7-7); (2) species will also tend to be most abundant in the middle of their gradient space, for they will have had the highest net energy gains here, allowing maximum successful reproduction; and (3) species will be found in a larger number of microsites in the middle of their gradient space.

## III. APPLYING GRADIENT ANALYSIS THROUGH GEOGRAPHICAL MODELING TO COSTA RICA

The steps required to accomplish the goal of building a data network for Costa Rica in both geographical and gradient space are as follows:

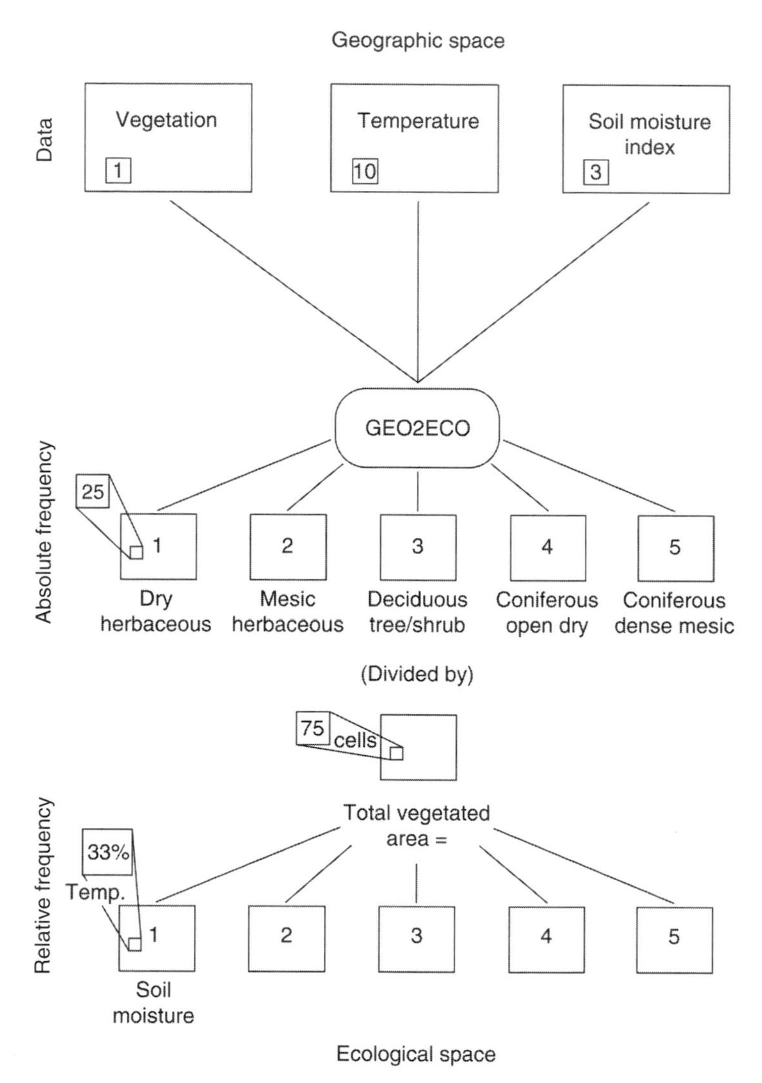


FIGURE 7-8 Overview of GEO2ECO program showing steps in converting a species' distribution over geographical space into distribution as a function of gradients. The numbers 1–5 refer to vegetation types. The axes of the lowest set of boxes are temperature and soil moisture.

1. Obtain high-quality digital elevation information for the country of Costa Rica (Chapter 8). From that derive slope and aspect.

2. Obtain mean monthly meteorological records for principal weather stations in Costa Rica (Chapter 9). Use either interpolation (Kriging) or our

correlation program TOPOCLIM to extend this information to the entire country of Costa Rica at a resolution of 1 $km^2$. If possible obtain some independent information to validate the extrapolation.

3. Obtain detailed geographically based soil descriptive information for Costa Rica (Chapter 10). Attempt to interpret the soil information in a gradients context.

4. From the preceding data sources derive for each square kilometer of Costa Rica an estimate of its location along major environmental gradients. The gradients include temperature, rainfall, evapotranspiration, solar radiation, soil texture, soil pH (if possible), soil depth, and so on. Finer resolution will be undertaken if and as data quality and computer processing capacity allow.

5. Analyze location-specific crop production, etc., in terms of gradient space using GEO2ECO. These data are derived from existing large data sets with district- and canton-scale resolution (Chapters 6 and 12).

6. Examine the statistical relation between agricultural yield and the gradients listed in step 5. Fit, if possible, normal curves or other statistical fits to this data set (Chapters 12 and 13).

7. Generate a graphical display of data and analysis. Predict optimal production for various crops. Compare with realized yields under different past fertilizer application levels (Chapters 12 and 13).

8. Add an erosion component as a function of climatic and soil variables. Predict future yields as a function of expected future erosion scenarios and management practices (Chapter 15).

## IV. MODELS AND POLICYMAKING

Following the steps above one can present to decision makers a realistic view of the potential future outcomes of the various economic, political, social, and environmental decisions they are having to make now. Many of the rest of the chapters in this book use a similar approach and show the implications for decision making.

As a final note, there is one aspect about geographical modeling that has an interesting political twist. As stated at the beginning of this chapter, a good model can predict probable effects of a decision at future times. According to studies undertaken by King and Kraemer (1993) models tend to have three singular and powerful influences on policy process. (1) They clarify issues in a debate and provide systematic arguments for and against various biases, including the biases incumbent in the model itself. (2) They enforce a discipline of analysis and discourse. (3) They provide an interesting and powerful form of advice on what not to do. By illustrating the possible disastrous outcomes of a particular proposed policy, models can be highly valuable in the policymaking process.

The sort of geographical models that we provide allow one to view, in an easily comprehendible manner, the future results of present day decision making on various factors, including the economy and the environment. Taking this concept to its logical extreme, we envision the kind of models presented in Chapter 5 as a means of democratizing decision making. If someone has an idea about how to develop a region let it be done first in a computer using a comprehensive systems approach, and let the results be projected into the future. Then let the people whose lives will be affected have an opportunity to see the consequences of those development decisions. Public feedback could allow decision makers to weigh alternative strategies that impact ecosystems and human populations. We think that this process could democratize the process of development choice and political choice in general, and give all people a greater ability to anticipate and hence control processes that are affecting their lives. Perhaps this concept is idealistic, but why not? Compare that to how development decisions are made now (e.g., Chapter 24).

## REFERENCES


Al-Abed, N., and H. R. Whiteley. 1995. Modeling water quality and quantity in the lower portion of the Grand River watershed, Ontario. *Proceedings, ASAE International Symposium on Water Quality Modeling,* 213–223.

Arnold, J. E., J. R. Williams, A. D. Nicks, and N. B. Simmons. 1990. *SWRBB—A Basin Scale Model for Soil and Water Resource Management.* Texas Agricultural and Mechanical University Press, College Station, TX.

Austin, M. P. 1987. Models for the analysis of species' response to environmental gradients. *Vegetatio* 69:33–45.

Bartlein, P. J, I. C. Prentice, and T. Webb. 1986. Climatic response surfaces from pollen data for some eastern North American taxa. *Journal of Biogeography* 13:35–57.

Beek, K. J. 1978. Land Evaluation for Agricultural Development: Some Explorations of Land-use Systems with Particular Reference to Latin America, ILRI, Pub. 23, Wageningen.

Botkin, D. B., J. F. Janak, and J. R. Wallis. 1972. Rationale, limitations, and assumptions of a northeastern forest growth simulator, *IBM J. Res. Dev.* 16:101–116.

Box, E. O. 1978. Ecoclimatic determination of terrestrial vegetation physiognomy. Ph.D. dissertation, University of North Carolina, Chapel Hill, North Carolina.

Box, E. O. 1995. Factors determining distributions of tree species and plant functional types. *Vegetatio* 121:101–116.

Brown, S., and A. Lugo. 1984. Biomass of tropical plants: A new estimate based on forest volumes. *Science* 223:1290–1293.

Burns, R. M., and B. H. Honkala. 1990. *Silvics of North America.* USDA Agriculture Handbook no. 654, USDA Forest Service, Washington, DC.

Burrough, P. A. 1986. Principles of Geographical Information Systems for Land Resources Assessment. Oxford, New York.

Clements, F. E. 1931. Nature and structure of the climax. *Journal of Ecology* 24:252–284.

Costanza, R. Sklar, F. H., and M. L. White. 1990. Modeling coastal landscape dynamics. *Bioscience* 40, 91–107.
Dangermond, J. 1993. The role of software vendors in integrating GIS and environmental modeling. In M. Goodchild, B. Parks, and L. Steyaer (Eds.), *Environmental Modeling with Geographic Information Systems*. Oxford University Press, New York.
Day, J. W., Hall, C. A. S., Kemp, M., and Yanez Arencibia, A. 1991. Estuarine Ecology, Wiley, New York.
Devantier, B. A., and A. D. Feldman. 1993. Review of GIS applications in hydrological modeling. *Journal of Water Resources Planning and Management* 119(2):246–261.
Devries, J. J., and T. V. Hromadka. 1993. Computer models for surface water. In D. R. Maidment (Ed.), *Handbook of Hydrology*. McGraw–Hill, New York.
Eastman, R., *et al.*, 1997. Idrisi software package.
Emanuel, W. R., H. H. Shugart, and M. P. Stevenson. 1985. Climatic change and the broad-scale distribution of terrestrial ecosystem complexes. *Climatic Change* 7:29–43.
Everham, E. M., III, K. B. Wooster, and C. A. S. Hall. 1991. Forest landscape climate modeling. In *Proceedings of the 1991 Symposium on Systems Analysis in Forest Resources*. GTR SE-74, USDA Forest Service, Washington, DC.
FAO. 1976. A framework for land evaluation. *Soils Bulletin*, no. 32. FAO, Rome; ILRI Publication no. 22, University of Wageningen, Wageningen, The Netherlands.
FAO. 1983. Guidelines for land evaluation for rainfed agriculture. *FAO Soils Bulletin*, no. 52. FAO, Rome.
Forrester, J. W. 1968. *World Dynamics*. Wright-Allen, Cambridge, MA.
Gates, D. M. 1993. *Climate Change and Its Biological Consequences*. Sinauer, Sunderland, MA.
Gleason, H. A. 1926. The individualistic concept of the plant association. *Bulletin of the Torrey Botanical Club* 53:7–26.
Graham, R. L., M. G. Turner, and V. H. Dale. 1990. Increasing atmospheric $CO_2$ and climate change: Effects on forests. *Bioscience* 40:575–587.
Griesbach, A. 1838. Ueber den Einfluss des Climas auf die Begrenzung der natuerlichen Floren. Linnea 12:159–200.
Griffith, D. A., and C. G. Amrhein. 1991. *Statistical Analysis for Geographers*. Prentice Hall, Englewood Cliffs, NJ.
Hagen, J. B. 1992. *An Entangled Bank: The Origins of Ecosystem Ecology*. Rutgers University Press, New Brunswick, NJ.
Hall, C. A. S. 1992. Economic development or developing economics: What are our priorities? In M. K. Wali (Ed.), *Ecosystem Rehabilitation*. Volume I. pp. 101–126.
Hall, C. A. S., and J. Day. 1977. *Ecosystem Modeling in Theory and Practice*. Wiley–Interscience, New York. Reprinted 1990, University Press of Colorado, Boulder, CO.
Hall, C. A. S., C. J. Cleveland, and R. K. Kaufmann. 1986. *Energy and Resource Quality: The Ecology of the Economic Process*. Wiley–Interscience, New York. Reprinted 1992, University Press of Colorado, Boulder, CO.
Hall, C. A. S., J. H. Jourdonnais, and J. A. Stanford. 1989. Assessing the impacts of stream regulation in the Flathead River Basin, Montana, USA. I. Simulation modeling of system water balance. *Regulated Rivers: Research and Management* 3:61–77.
Hall, C. A. S., J. A. Stanford, and R. Hauer. 1992a. The distribution and abundance of organisms as a consequence of energy balances along multiple environmental gradients. *Oikos* 65:377–390.
Hall, C. A. S., M. Taylor, and E. Everham. 1992b. A geographically-based ecosystem model and its application to the carbon balance of the Luquillo Forest, Puerto Rico. *Journal of Water, Air and Soil Pollution* 64:385–404.
Hall, C. A. S., and M. H. Hall. 1993. The efficiency of land and energy use in tropical economies and agriculture. *Agriculture, Ecosystems and Environment* 46:1–30.

Hall, C. A. S., H. Tian, Y. Qi, G. Pontius, and J. Cornell. 1995. Modeling spatial and temporal patterns of tropical land use change. *J. Biogeogr.* 22:753–757.

Hall, M. H. P. 1994. Predicting the impact of climate change on glacier and vegetation distribution in Glacier National Park to the year 2100. M.S. thesis, State University of New York, College of Environmental Science and Forestry, Syracuse.

HEC (Hydrologic Engineering Code). 1976. HEC-3 Reservoir System Analysis for Conservation. U.S. Army Corps of Engineering, Vicksburg, MS.

HEC (Hydrologic Engineering Code). 1982. HEC-5 Simulation of Flood Control and Conservation Systems. U.S. Army Corps of Engineering, Vicksburg, MS.

Hilborn, R., and C. Stearns. 1982. On inference and evolutionary biology: The problem of multiple causes. *Acta Biotheor.* 31:145–164.

Holdridge, L. R. 1947. Determination of world plant formation from simple climatic data. *Science* 105:367–368.

Holdridge, L. R. 1967. *Life Zone Ecology*. Tropical Science Center, San Jose, Costa Rica.

Homer-Dixon, T. F., J. H. Boutwell, and G. W. Rathjens. 1993. Environmental change and violent conflict: Growing scarcities of renewable resources can contribute to social instability and civil strife. *Scientific American* 268:38–45.

Huber, W. C., and R. E. Dickinson. 1988. *Storm Water Management Model Version 4 User's Manual*. ERL USEPA, Athens, GA.

Humboldt, A. von. 1867. Idee zu einer Geographie der Pflanzen nebst einem Naturgemaelde der Tropenlaender, Tuebingen, FRG.

Hunsaker, C. T., R. A. Nisbet, D. C. L. Lam, J. A. Browder, W. L. Baker, M. G. Turner, and D. B. Botkin. 1993. Spatial models of ecological systems and processes: The role of GIS. In M. F. Goodchild, B. O. Parks, and L. T. Steyaert (Eds.), *Environmental Modeling with GIS*. Oxford, New York.

Hutchinson, G. E. 1965. The niche: An abstractly inhabited hypervolume. In *The Ecological Theory and the Evolutionary Play*, pp. 26–78. Yale University Press, New Haven, CT.

Jourdonnais, J., J. A. S. Stanford, F. R. Hauer, and C. A. S. Hall. 1990. Assessing options for stream regulations using hydrologic simulations and cumulative impact analysis: Flathead River Basin, USA. *Regulated Rivers* 5:279–293.

Jowett, I. G. 1988. *River Hydraulics and Habitat Simulation, RHYHABSIM. User's Manual*. Freshwater Fisheries Centre, Riccarton, New Zealand.

Kangas, P. G. 1995. Teaching about ecological systems: The HTO influence. In C. A. S. Hall (Ed.), *Maximum Power, the Ideas and Applications of H. T. Odum*, pp. 11–18. University Press of Colorado, Niwot, CO.

Kessell, S. R. 1977. Gradient modeling: A new approach to fire modeling and resource management. In C. A. S. Hall and J. W. Day (Eds.), *Ecosystem Modeling in Theory and Practice*, pp. 575–605. Wiley, New York.

Kessell, S. R. 1979. *Gradient Modeling*. Springer-Verlag, New York.

King, J. L., and K. L. Kraemer. 1993. Models, facts, and the policy process: The political ecology of estimated truth. In M. F. Goodchild, B. O. Parks, and L. T. Steyaert (Eds.), *Environmental Modeling with GIS*. Oxford, New York.

Koppen, W. 1900. Versueh einer klassification der klimate, vor zugsweise nach ihren Beziehungen zur pflanzenwelt. *Geog. Z.* 6:593–611.

Kormandy, E. J. 1969. *Concepts of Ecology*. Prentice Hall, Englewood Cliffs, NJ.

Lenihan, J. W. 1993. Ecological response surfaces for North American boreal tree species and their use in forest classification. *Journal of Vegetation Science* 4:667–680.

Little, E. L., Jr. 1971. *Atlas of United States Trees*, Vol. 1, *Conifers and Important Hardwoods*. Miscellaneous Publication no. 1146, USDA Forest Service, Washington, DC. [Maps]

Loucks, D. P., P. N. French, and M. R. Taylor. 1995. *Interactive River-Aquifer Simulation Model (IRAS)*, Program Description and Operation Manual. Resources Planning Associates, Ithaca, NY.

Maidment, D. R., F. Olivera, Z. Ye, S. Reed, and D. C. McKinney. 1996. Water balance of the Niger River. In *Proceedings, ASCE–North American Water and Environment Congress '96 (NAWEC '96)*.

McCrae, S. G., and C. P. Burnham. 1981. Land evaluation. Clarendon Press, Oxford.

Meadows, D. H., D. L. Meadows, J. Randers, and W. W. Behrens III. 1974. *The Limits to Growth: A Report for the club of Rome's Project on the Predicament of Mankind.* Universe Books, New York.

Meyer, S. P., T. H. Salem, and J. W. Labadie. 1993. Geographic information systems in urban storm-water management. *Journal of Water Resources Planning and Management* 119:206–228.

Monserud, R. A., N. M. Tchebakova, and R. Leemans. 1993. Global vegetation change predicted by the modified Budyko model. *Climatic Change* 25:59–83.

Moore, I. D., A. K. Turner, J. P. Wilson, S. K. Jenson, and L. E. Band. 1993. GIS and land-surface-subsurface modeling. In M. F. Goodchild, B. O. Parks, and L. T. Steyaert (Eds.), *Environmental Modeling with GIS.* Oxford, New York.

Nielsen, C. K., W. F. Porter, and H. B. Underwood. 1997. An adaptive management approach to controlling suburban deer. *Wildlife Society Bulletin* 25:470–477.

Nielson, R., G. A. King, and G. Koerper. 1992. Toward a rule-based biome model. *Landscape Ecology* 7:27–43.

Nyerges, T. L. 1993. Understanding the scope of GIS: Its relationship to environmental modeling. In M. F. Goodchild, B. O. Parks, and L. T. Steyaert (Eds.), *Environmental Modeling with GIS.* Oxford University Press, New York.

Odum, H. T. 1983. *Systems Ecology, an Introduction.* Wiley–Interscience, New York. Reprinted 1994 by the University Press of Colorado, Niwot, CO.

Overpeck, J. T., and P. J. Bartlein. 1989. Assessing the response of vegetation to future climate change: Ecological response surfaces and paleoecological model validation. In J. B. Smith and D. A. Tirpak (Eds.), *The Potential Effects of Global Climate Change on the United States,* Appendix D, pp. 1-1 to 1-32. U.S. Environmental Protection Agency, Washington, DC.

Peet, J. 1992. *Energy and the Ecological Economics of Sustainability.* Island Press, Washington, DC.

Prentice, I. C. 1990. Bioclimatic distribution of vegetation for general circulation model studies. *Journal of Geophysical Research* 95:11811–11830.

Prentice, I. C., W. Cramer, S. P. Harrison, R. Leemans, R. A. Monserud, and A. M. Solomon. 1992. A global biome model based on plant physiology and dominance, soil properties and climate. *Journal of Biogeography* 19:117–134.

Rapport, D. J. 1997. Is economic development compatible with ecosystem health? *Ecosystem Health* 3:94–106.

Remensky, L. G. 1924. *Basic Regularities of Vegetation Covers and Their Study* [In Russian], pp. 37–73. Vestnik Opytnogo de la Stredne-Chernoz, Ob.-Voronezh.

Running, S. W., R. R. Nemani, D. L. Peterson, L. E. Band, D. F. Potts, L. L. Pierce, and M. A. Spanner. 1989. Mapping regional forest evapotranspiration and photosynthesis by coupling satellite data with ecosystem simulation. *Ecology* 70(4):1090–1101.

Shugart, H. H., and D. C. West. 1977. Development of an Appalachian deciduous forest succession model and its application to assessment of the impact of the chestnut blight. *Journal of Environmental Management* 5:161–179.

Solomon, A. M., D. C. West, and J. A. Solomon. 1981. Simulating the role of climate change and species immigration in forest succession. In D. C. West, H. H. Shugart, and D. B. Botkin (Eds.), *Forest Succession Concepts and Applications,* pp. 154–177. Springer-Verlag, New York.

Ter Braak, C. J. F., and J. C. Prentice. 1988. A theory of gradient analysis. *Advances in Ecological Research* 18:271–317.

Thornthwaite, C. W. 1931. The climates of North America according to a new classification. *Geog. Rev.* 21:633–655.

Thornthwaite, C. W. 1948. An approach toward rationale classification of climate. *Geog. Rev.* 38:55–94.

Tian, H., C. A. S. Hall, and Y. Qi. 1998. Modeling primary productivity of the terrestrial biosphere in changing environments: Toward a dynamic biosphere model. *Critical Reviews in Plant Science* 15:541–557.

Troll, C., and K. H. Paffen. 1964. Karte der Jahreszeiten klimate der Erde. *Erkund. Arch. Wiss. Geog.* 18:5–28.

Turner, B. 1997. Spirals, bridges and tunnels: engaging human-environment perspectives in geography. *Ecumene* 4:196–217.

U.S. Department of Agriculture (USDA) Forest Service. 1913. *Forest Atlas. Geographic Distribution of North American Trees. Part 1: Pines.* By George B. Sudworth. USGS, Washington, DC.

Watt, K. E. F. 1968. *Ecology and Resource Management: A Quantitative Approach.* McGraw–Hill, New York.

Webb, T., III. 1986. Is vegetation in equilibrium with climate? How to interpret late-Quaternary pollen data. *Vegetatio* 67:75–91.

Wheeler, D. J. 1993. Commentary: Linking environmental models with geographic information systems for global change research. *Photogrammetric Engineering and Remote Sensing* 59(10):1497–1501.

Whittaker, R. H. 1956. Vegetation of the Great Smoky Mountains. *Ecol. Monogr.* 26:1–80.

Wu, Y., F. H. Sklar, K. Gopu, and K. Rutchey. 1996. Five simulations in the Everglades using parallel programming. *Ecological Modeling* 93:113–124.

Zabinski, C., and M. B. Davis. 1989. Hard times ahead for Great Lakes forests: A climate threshold model predicts responses to $CO_2$-induced climate change. In J. B. Smith and D. A. Tirpak (Eds.), *The Potential Effects of Global Climate Change on the United States,* Appendix D, pp. 5-1 to 5-19. U.S. Environmental Protection Agency, Washington, DC.

# SECTION IV

# *Building a Geographical Database for Costa Rica*

This section discusses the construction of a geographical database for Costa Rica that can help serve as a base for the biophysical approach introduced in Chapter 3 and the analysis of the major issues of Costa Rica as introduced in Chapters 1, 2, and 5. The first chapter in this section discusses how we built the most important component of a national-level geographical database, a digital elevation map (DEM), for Costa Rica. The second and third develop a spatial database for Costa Rica for meteorological and soils information. The meteorological chapter focuses especially on the statistical problems in extrapolating relatively few weather stations to an entire nation, while the soils chapter discusses in some detail the nature of the very varied soils in Costa Rica and how that variation effects decisions about how soils can and should be used.

The final chapter in this section discusses how remote sensing applications can greatly enhance our ability to assess and understand the landscape about us, and can greatly reduce the cost of managing agricultural resources.

CHAPTER 8

# Developing a Nationwide Topographical Database

PATRICK VAN LAAKE AND GLENN HYMAN



## I. INTRODUCTION

### A. COSTA RICAN PHYSICAL GEOGRAPHY

Costa Rican topography is as diverse as one can find in almost any part of the world. The forces of orogeny and denudation (erosion) are both very strong due to the country's unique location on the border of the Cocos and Nazca tectonic plates and in the humid equatorial zone. Thus rates of mountain building are rapid, as evidenced by frequent volcanic and earthquake activity. But the high rainfall of the humid tropics works in an opposite direction to create mass movements and landslides in the mountainous areas, and a great abundance and variety of fluvial landforms throughout the country. Recent research has emphasized the importance of the relatively high temperatures of the tropics for maintaining large mountain masses at high elevations. This is because these mountains are not subject to the more intensive long-term erosion characteristic of regions with glaciers, which occur at lower elevations at higher latitudes (Brozovic *et al.*, 1997).

This nearly unique physical geography generates a wide range of both problems and opportunities for the country's agriculture, economy, and environment. From a human perspective the bad includes earthquakes, volcanic eruptions, floods, and other disasters sometimes brought by nature's forces, and the good includes a diversity in local climates and resources that allows for a multitude of economic activities. But each of these activities is in turn limited by, and effects, the environment. Prosperous long-term development requires a rational evaluation of the merits and the risks related to each activity at each location, and an acknowledgment of the overall limitations inherent in a complex terrain of generally suboptimal topographical resources, at least from an agricultural perspective. In this chapter we discuss the development and application of numerical terrain analysis for natural resource development and management, including especially the computerization of that effort.

There has been very little digital terrain mapping in Costa Rica to date. Most efforts have been aimed at digitizing available hand-derived maps of geology, geomorphology, and soils, and have usually been focused on specific study areas and interests. We develop in this chapter a detailed (200-by-200-m resolution) digital topographical database for the entire nation, derived from the 1 : 200,000 scale topographical base maps from the Instituto Geográfico Nacional (IGN). This is the starting point for the other digital products discussed in this chapter. This chapter also develops a slope map and a landform map derived from topography.

## B. Geomorphology and Landforms

### 1. Background

Geomorphology means the shape of the land. The term "landform" also refers to the shape of the land but in reference to the formative geologic processes, and includes, for example, alluvial fans, river-cut terraces, and volcanic craters. Landform analysis is important for deriving proper land use because it not only describes the current situation, but it also tells us how the current landform came to be. For example, areas where landslides occur are not suitable for habitation because of potential loss of life and property, or for agriculture because it would increase the instability of slopes by removing the natural vegetation. Instead they should be used as protective regions with natural vegetation only. Old river terraces are generally suitable for economic activities and infrastructure as they indicate geological stability and usually possess stable and relatively flat soils. Young alluvial plains have high agricultural potential because they tend to be rich in mineral components, although they might have problems with flooding. The most common type of broad-scale mapping has been the slope map and some of its derivatives. Many slope diagrams have been generated manually, and these allow the analyst to determine units of similar slope. Traditionally, landform mapping has been very time consuming and largely has depended on the subjective judgment of the analyst and on field knowledge to delineate a landform based on interpretation of aerial photographs or topographic maps. Hammonds' (1964) work was especially important in quantitatively characterizing land surface forms from topographic maps. He mapped the contiguous United States from 1 : 250,000 scale maps. The process took years to complete because it required the analyst to convert contour lines to areas of similar slope, to determine relief for sections of each map, and to classify land profile types. That process could be done today with computers in a fraction of the time.

## 2. Historical Morphological Studies

To date most studies of Costa Rican land surface form have been conducted manually on small areas to support applied environmental analyses. A notable exception is the series of 1 : 50,000 scale landform maps produced by the Costa Rican National Geographic Institute (IGN) (Bergoeing and Malavassi, 1981). These maps cover the entire Central Valley of Costa Rica and are exceptional because very little mapping of landforms has been done at this scale anywhere in the world. Another major publication on landforms by the same authors is the series of 1 : 100,000 scale maps of the northwestern Nicoya Peninsula and Guanacaste Province.

To our knowledge only two geomorphological maps of the entire country exist. Bergoeing and Brenes' (1978) 1 : 1,000,000 map delineates landform units based on age of the surface, lithology, and morphology. While the geomorphological units make reference to landform, this map is primarily a genetic classification. Madrigal and Rojas (1980) mapped the landforms of all of Costa Rica using the 1 : 200,000 IGN series as their base map, and their results are presented as nine unconnected map sheets. They describe nine general morphographic regions and numerous individual features as well, such as fault lines, escarpments, and cones.

The two maps previously mentioned contain a wealth of information on the geomorphology of Costa Rica, but are less helpful for national-scale analysis of land use, because the classification is too general, resulting in large units being homogeneous only in broad morphogenetic characteristics. For land use and agricultural or ecological modeling the analyst needs data on surface and regional topography, such as slope, aspect, and relief. A combination of morphogenetic and morphometric data would yield a powerful data set capable of answering many queries regarding territorial planning in terms of the essential resource: the land. It would answer queries on liabilities such as tectonic instability (volcanoes, earthquakes), slope instability (erosion, landslides), etc.

In this chapter we will present a methodology to derive a morphometric map on the basis of topography. Unfortunately, as previously outlined, the map by Madrigal and Rojas (1980) is too general and the IGN maps cover only part of the country.

### 3. Digital Mapping of Landforms

A digital geographic model of landforms has several advantages. First, it is based on topographic mapping covering the entire country and the derived output will represent quantitatively the entire surface form to the degree that the original source maps are of good quality. Second, the digital geographic model allows easy replication of the results and ease of experimentation with methods. Classification schemes can be altered quickly and adapted to the question at hand. Third, the derivation of the map can be completed quickly if a digital elevation model (DEM) is available. And finally a quantitative classification of land surface form is directly and readily applicable to modeling.

Today the numerical analysis of land surface form is progressing rapidly and a useful body of literature is developing (see Pike, 1993, for review). A large portion of research in this field has focused on the development of DEMs and the derivation of land surface characteristics from these models (e.g., Jensen and Domingue, 1988; Papo and Gelbman, 1984). Some of the land surface characteristics commonly derived with computers are slope, aspect, relief, drainage network, watershed boundary, and slope curvature. The chief beneficiaries of the basic research on DEMs and their derivatives are modelers of surface and groundwater hydrology, slope stability, terrain characteristics, and engineering schemes (see Moore *et al.*, 1991 and 1993 for review). One area of research that has received less attention is the classification of land units according to morphologic types (e.g., Dikau *et al.*, 1991; Dymond *et al.*, 1995; Pike, 1993), although with the increasing availability of DEMs the quantity and quality of research in numerical analysis of land surface morphology are expanding rapidly. Geographic modeling with GIS technology now allows the modeler to put more effort into the model (of, e.g., erosion) relative to the preprocessing stage of database development.

## II. CONSTRUCTION OF THE TOPOGRAPHICAL DATABASE

We derived the base information for the topographical database from the nine official national 1 : 200,000 scale topographical maps generated by IGN. We scanned and vectorized the original contour line color separates that were used to print the maps, and subsequently corrected, annotated, and merged the nine resulting coverages into one nationwide coverage using ArcInfo GIS software, as detailed below. Essentially this meant putting each map into a large flatbed scanner and generating a digital image of that map. Then the resultant digital image was corrected by removing all information that did not consist of contour lines (e.g., elevation labels) using commercial software. This generated a vector map of contour lines. These images were then georeferenced, that is, related to a standard coordinate system for the earth itself, as discussed below.

The RMS (root mean square) error of registering the individual scanned images to the geographical coordinate system was typically about 7 m, and never more than 18 m. The process of joining the nine separate digital images, however, revealed a considerable matching error between sheets. This error amounted typically to a horizontal shift of about 45 m, although values as high as 127 m were recorded. There may be several causes for this error, including manuscript errors (carrier material deformation by temperature and humidity), scanning deformation (representing a portion of the round earth on a flat piece of paper), and registration error. In all cases the deformation was greatest on the edges and it thus affects only a small area of the complete coverage.

Therefore the contour lines will not be reliable locally (e.g., where the San Jose and Limon sheets meet, at 83°; 45′; west longitude, between 9°; 40′; north and 10°; 30′ north latitude, and to a lesser extent between some other sheets), and the degree to which they are unreliable cannot be specified exactly. However, the contour lines are not used but for the generation of the DEM. The error will of course propagate into the DEM, but since the DEM is an approximation of elevation over a relatively large area (relative with respect to the magnitude of error in the contour lines) the error in the DEM will not be so pronounced. Furthermore, the DEM is not likely to be used just to query the elevation on a point-by-point basis; instead it will be used to do regionalized analysis. This chapter gives two such examples. A slope map is derived using a square analysis window of 600 by 600 m, and a landform map is derived using a circular analysis window of approximately 10 km in diameter. In both cases a local error will smooth out over the entire analysis window. The user of the DEM should, however, not be oblivious to this error, particularly when a small area is being studied.

The official national coordinate system of Costa Rica is based on the Lambert Azimuthal projection. There are two distinct sets of projection parameters yielding two distinct coordinate systems, officially called Costa Rica Norte and Costa Rica Sur. The two coordinate systems cover the areas north and south of 9°; 40′; north latitude, respectively. This division was made to keep distortions due to the projection within a certain limit. About two-thirds of the national landmass falls within the Costa Rica Norte coordinate system. In order to facilitate geographical analysis we decided to project the entire contour line coverage to the Costa Rica Norte coordinate system, thereby sacrificing some geometric precision. However, since the aim is to produce a DEM at a 200-m resolution, this sacrifice is negligible. Two benefits of projecting the coverage to the Costa Rica Norte coordinate system are that GIS software will report length and area in metric units (as opposed to decimal degrees and square decimal degrees!) and that, for the northern two-thirds of the country, the coverage coordinates translate directly to the coordinates on the topographic sheets produced by IGN. [The parameters for converting from geographic coordinates (latitude and longitude) to the Costa Rica Norte and Costa Rica Sur projections are included on the CD-ROM.]

In order to create a digital elevation model, i.e., a surface in which each grid cell holds an elevation value, we had to convert the scanned elevation contour values from the vector-based GIS (see Figure 8-1) to a grid or raster-based format. We used Delauney Triangulation to create a triangulated irregular network (TIN). The triangulation approach is more accurate than linear interpolation because the size of each triangle is inversely proportional to the density of the contour lines. In other words, where the slope is steep there are many contour lines and thus small triangles. Each triangle has an equal quality since each is composed of three observed elevations; the distinguishing factor is its size. Also the triangulation will follow the longitudinal curvature of the contour lines and it will cover the entire land area exhaustively. Thus the quality of triangular interpolation is intrinsically better than that of linear interpolation. In the triangulation all nodes and vertices from the vector coverage were included as triangle points. Due to the nature of contour lines this resulted initially in many flat areas that may in fact be very steep; for example, mountain peaks were "cut off" at their nearest contour interval. To avoid this problem and improve the quality of the data set we added to the TIN a total of 7272 georeferenced spot elevations from the more detailed 1 : 50,000 topographical base maps of IGN. These spot elevations represent mountain peaks, saddles, depressions, and other landmarks. In a final step the TIN was converted to a DEM using a 200-m cell dimension (equal to 1 mm on the source maps) by interpolating the $z$ (elevation) values on the TIN surface.

Figure 8-2 shows a shaded relief map generated from the DEM. The major geomorphologic features of Costa Rica are identified easily. Figure 8-3 is a detail (close to Limon) of the shaded relief map where the effect of adding spot elevation to hilltops can be appreciated. Note also the triangular artifacts which are a result of spot elevations at greater distances in a relatively flat area (i.e., where there are no contour lines). Where there are no spot elevations in low-lying interior areas there is no apparent relief on the DEM, since these areas are bounded on all sides by 100-m elevation contours.

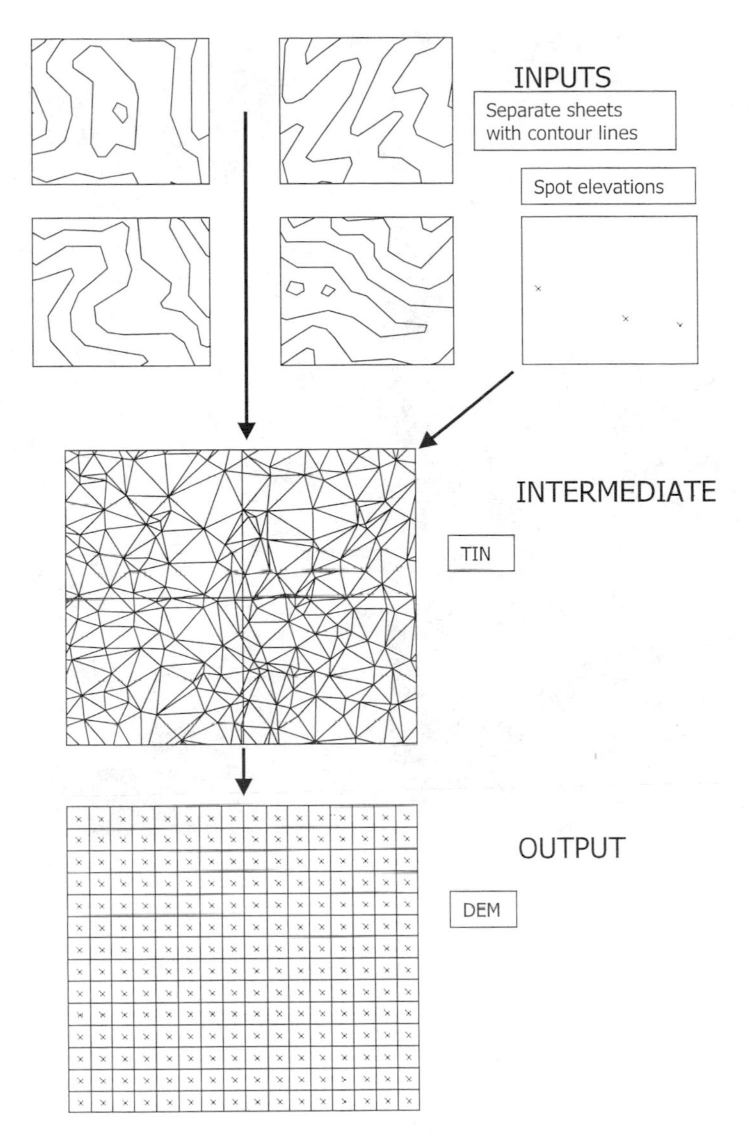


FIGURE 8-1 The difference between a vector map (top), in which many coordinates defining a line (such as a contour line) are stored in the computer, vs a raster map (bottom), which is a matrix in which each spatial (matrix) element receives a numerical value (represented by crosses and often called a $z$ value) representing that value for that attribute at that location.

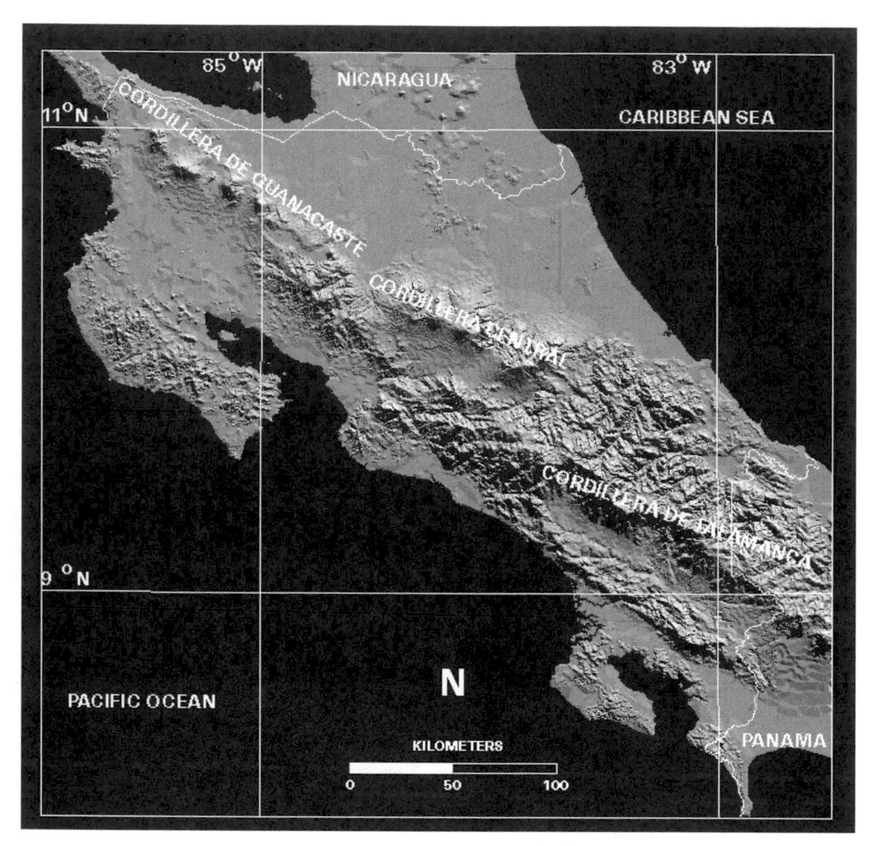


FIGURE 8-2 Shaded relief map of the digital elevation map of Costa Rica prepared in this study.

The quality of this DEM is probably not quite as good as one that might be made (at a much greater cost) from stereoscopic analysis of very high-resolution satellite images (e.g., SPOT images) (Bolstead and Stowe, 1994). Although the DEM is essentially as accurate as the contour lines on the 1 : 200,000 maps, it may not describe accurately all of the terrain now. Costa Rica is a tectonically active country and some major earthquakes and many landslides have occurred since the preparation of the maps. Also, in certain steep and dissected areas the resolution of the DEM does not allow for the accurate representation of the terrain. To investigate potential inaccuracies of our national-level approach we also prepared a similarly derived DEM from a portion of a 1 : 50,000 topographic map with a 50-m cell resolution. The national-level map shows, not surprisingly, a considerable generalization of the relief at a 50-m resolution (Figure 8-4). The DEM can be used with confidence, however, to analyze spatial relationships and trends that are significantly greater than its resolution of 200 by 200 m. For example, most of the analyses used in other chapters are based on a 1-by-1-km

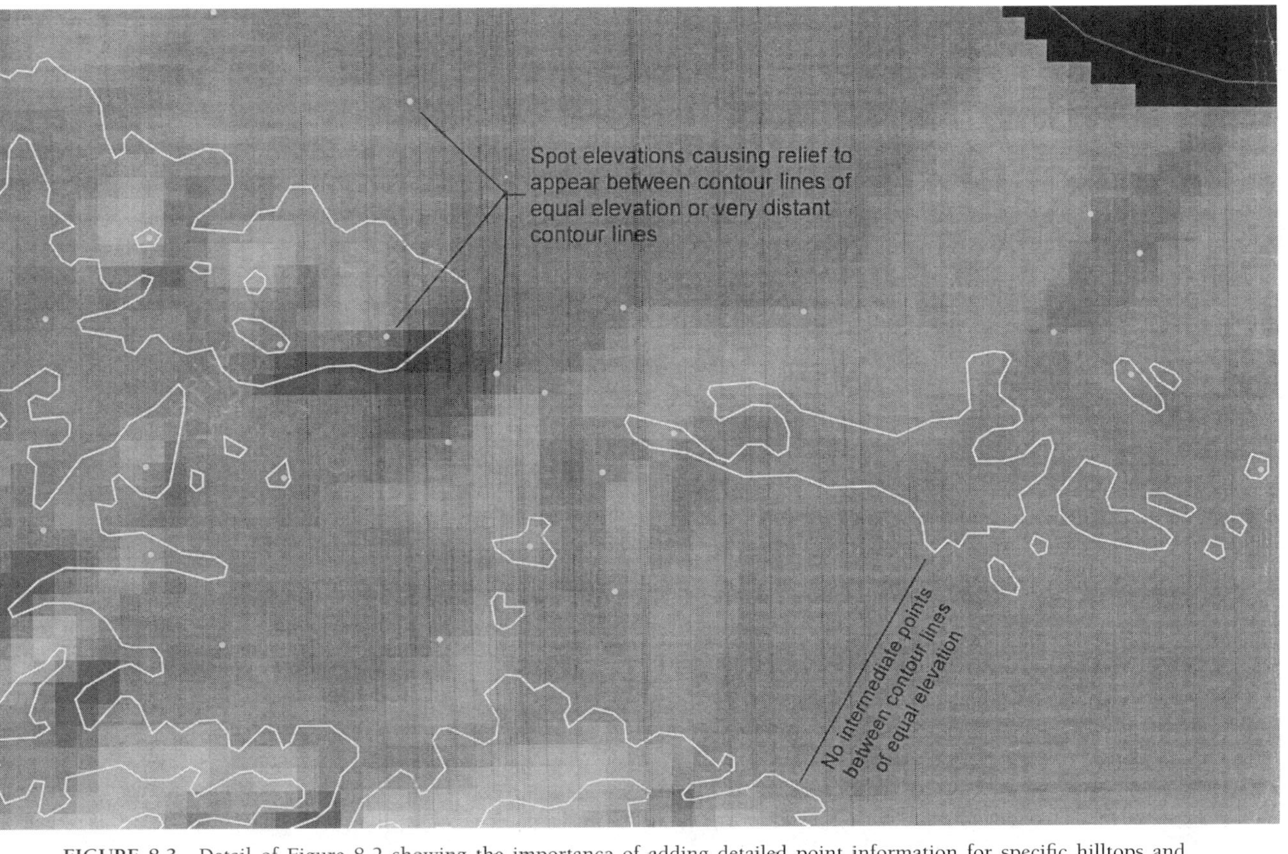


FIGURE 8-3 Detail of Figure 8-2 showing the importance of adding detailed point information for specific hilltops and valley bottoms.

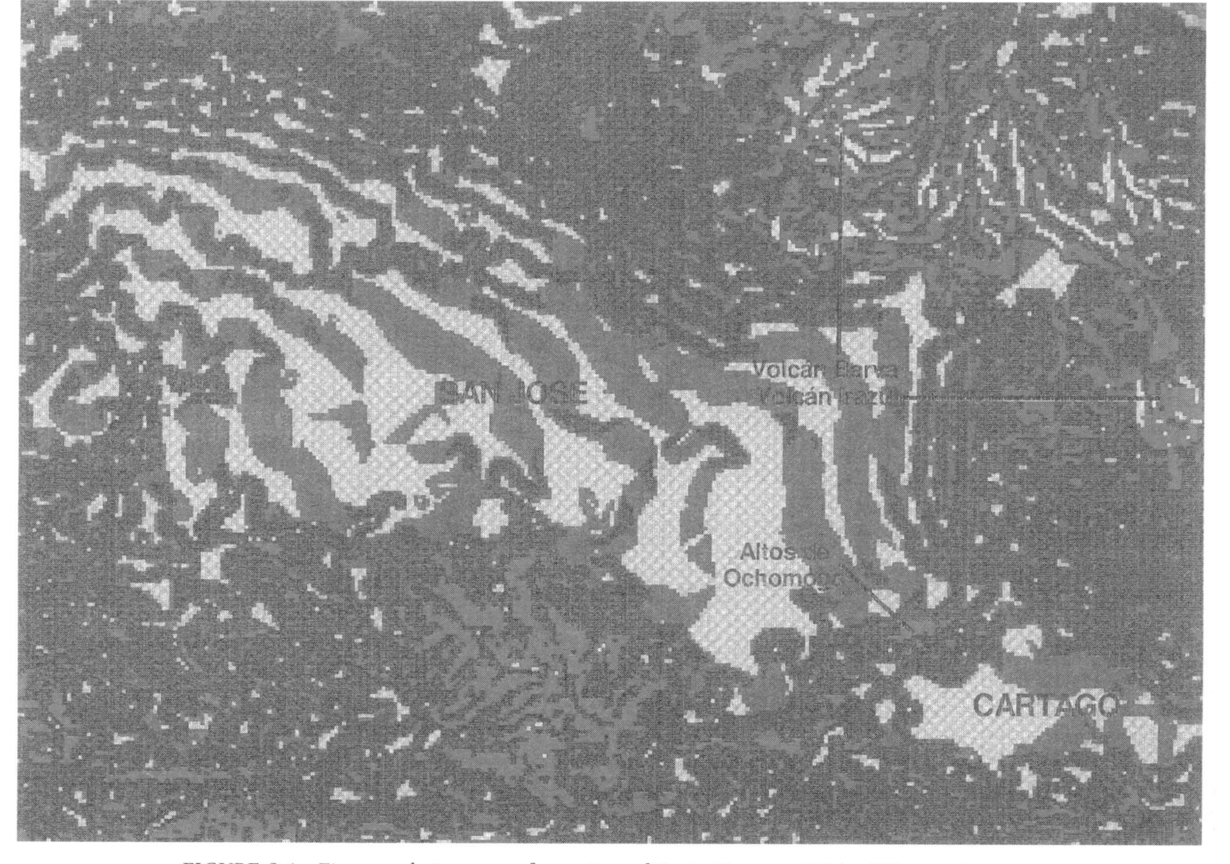


FIGURE 8-4 Finer resolution map of a portion of Costa Rica at a 200-by-200-m scale.

resolution DEM, meaning a resolution which dissolves any error present in the contour line coverage or the 200-m resolution DEM.

## III. GENERATING DERIVATIVES OF THE DEM

A number of other terrain parameters can be derived from the DEM which have their basis in their value and position relative to other such values within the environment. This general process is called terrain analysis. Many such analyses are important for human activities, because the land surface geometry dictates to some degree the type of activities that can be carried out at a given location, and because most locations present problems to be overcome for optimum land use. Terrain analysis can help identify limitations for certain activities and suggest where modifications of the activity are necessary in order to sustain the natural resource base. To this end geographic modeling of landforms and surface geometry provides a basis for wise planning of agricultural and economic activities, and helps to ascertain which kinds of land use are more or less sustainable.

### A. Digital Landform Maps for Costa Rica

#### 1. Creation of the Morphometric Classification of Landforms in Costa Rica

With this in mind we set out to produce a morphometric classification of the landforms of Costa Rica based on the DEM described earlier. We chose the classification system proposed by Hammond (1964a) and developed for DEMs by Dikau *et al.* (1991). After experimenting with several different procedures we have modified the classification system some to adapt it to conditions found in Costa Rica.

We passed a circular moving window over the DEM and other raster map layers derived from the DEM to calculate parameter values at the center of the window or processing cell. In contrast, Hammond (1964b) used rectangular non-overlapping windows for his map of the contiguous United States and Dikau *et al.* (1991) uses a rectangular moving window to map New Mexico. After some experimentation we found a circular moving window more appropriate for Costa Rica because in this rugged terrain landform units such as lone hills and ridgetops appeared blocky with the rectangular moving window. The circular moving window yielded shapes that approximated more closely the land units as they appear on the ground. Figure 8-5 shows the countrywide landform map produced from the 200-m-pixel-resolution raster DEM described earlier using the circular moving window.

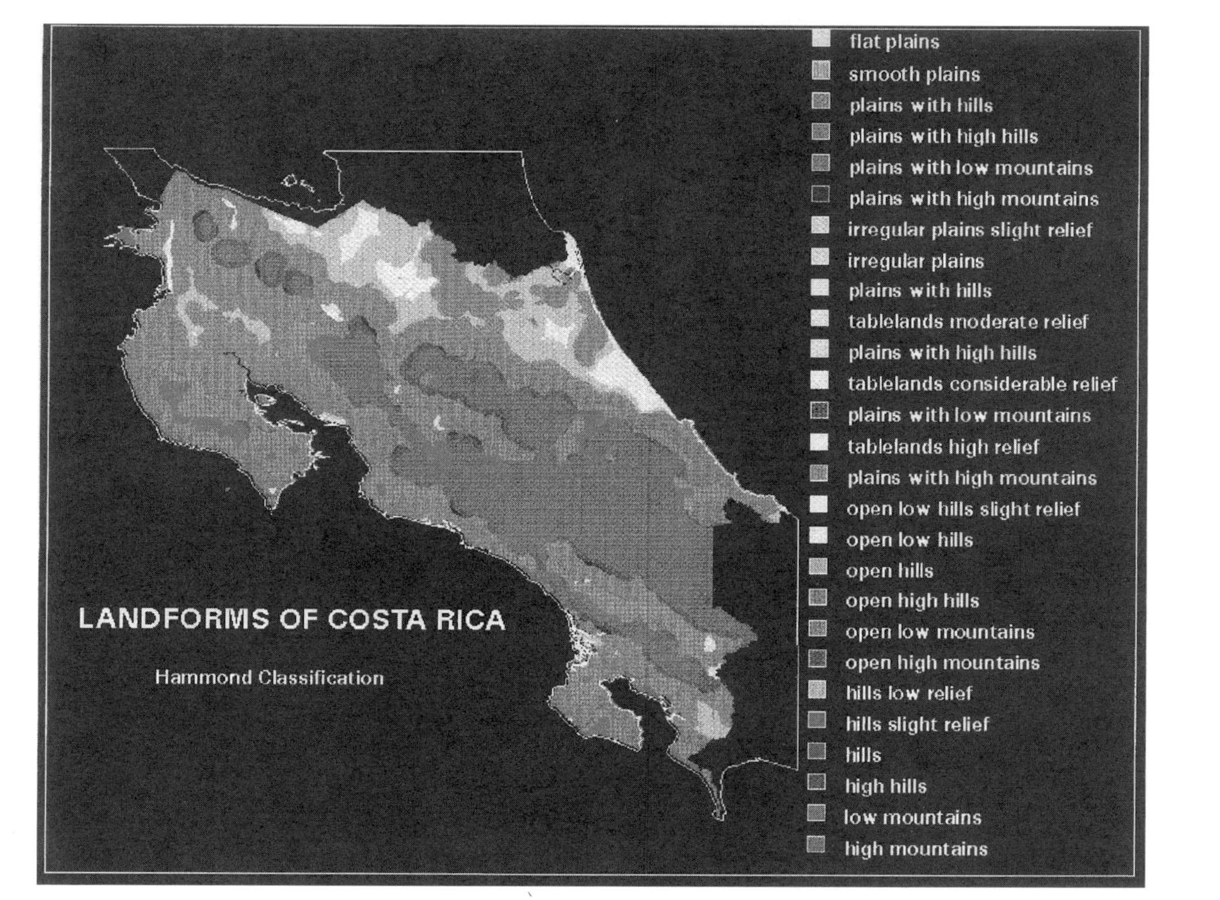


FIGURE 8-5 Digital landform map for all of Costa Rica prepared in this study.

Landform unit types are based on local relief, percentage of gently sloping land (less than 8%), and a profile type parameter (Hammond, 1964a). A local relief map is produced by identifying the lowest and highest elevation within the moving window and returning the difference to the processing cell. The relief values are then classified into Hammond's six relief types (Table 8-1).

The percentage of land that was sloping gently was determined by creating a binary map that stored 1 for all pixels with slope values less than 8% and 0 for all slope values greater than 8%. For each cell the percentage of gently sloping land was calculated by summing the values of the binary map within the moving window and then dividing by the area of the window. Then each cell was classified into Hammond's four categories. Areas with more than 80% gently sloping land are considered plains. Irregular plains, tablelands, and plains with hills and mountains are defined as those areas with 50 to 80% of the area sloping gently. Open hills and mountains have 20 to 50% of their area sloping gently. Those areas with less than 20% are classified as hills and mountains.

TABLE 8-1 Hammond Classification Scheme (Modified after Dikau *et al.*, 1991)

| Class | Definition |
|---|---|
| | Slope |
| A | >80% of neighborhood gently sloping |
| B | 50–80% of neighborhood gently sloping |
| C | 20–50% of neighborhood gently sloping |
| D | <20% of neighborhood gently sloping |
| | Relief |
| 1 | 0–29 m |
| 2 | 30–91 m |
| 3 | 91–152 m |
| 4 | 152–305 m |
| 5 | 305–915 m |
| 6 | >915 m |
| | Profile |
| a | >75% of gentle slope in lowland |
| b | 50–75% of gentle slope in lowland |
| c | 50–75% of gentle slope in upland |
| d | >75% of gentle slope in upland |

We also determined profile type, which is a measure of the gentleness of slope in relation to the adjacent lowland or upland environments. The profile type distinguishes between gently sloping areas on highland plateaus and those in lowlands. The measure is calculated by subtracting the elevation at the processing cell from the maximum elevation in the moving window. If this value is less than one-half of the local relief for that cell, the pixel is classified as upland. If the value is greater than one-half of the local relief, the pixel is lowland. Hammond distinguished four profile types according to the percentage of gently sloping land found in lowland or upland situations. But ultimately the profile type parameter is used to distinguish between tablelands and plains with hills or mountains.

### 2. Analysis of the Morphometric Map

The quality of any landform map depends greatly on the source topographic maps used to make the DEM and on the interpolation process. Areas of great relief are more reliable than lowlands, because the high frequency of contour lines in the highlands makes the interpolation of elevation more accurate in this zone. As noted earlier we countered that problem by adding a large number of spot heights from 1 : 50,000 scale maps. The map suggests that Costa Rica could be divided into four broad landform regions: 30% of Costa Rica is class D, mountainous with less than 20% of the surrounding area characterized by gently sloping terrain; 18% consists of class C open hills and mountains; 27% of the land area is class B gently sloping plains with minor relief; and 21% is class A characterized by plains adjacent to hills and mountains.

The relative geological youth of Costa Rica and its location in a tectonically active area are reflected in this distribution. The gently sloping plains are recently uplifted portions of land that were under water in the recent geological past and have had little time to experience downcutting by river networks. The older mountain areas in the Talamancas and the volcanic residues in the Central and Guanacaste cordilleras reflect rapid uplift and subsequent channel incision. The remaining landscapes are the boundaries between mountains and plains (plains with hills or mountains category) or open areas surrounded by hills and mountains. One example of the latter category is the large crater near Moravia de Chirripo. Tablelands are rare which probably reflects the tectonic complexity of the region and rapid dissection of landscapes by rivers.

## B. Slope Map

Information on slopes is needed in virtually every kind of land use planning, and therefore slope maps have been prepared on the basis of semiautomatic and visual interpretation of aerial photographs. The IGN has published a series of slope maps at a 1 : 200,000 scale, where the slope is classified according to the seven categories as they are defined in the "Metodología para la determinación de la capacidad de uso de las tierras de Costa Rica" (MAG/MIRENEM, 1991). Although this is a very useful map the generalization which is inherent to the visual interpretation of topographic information on base maps led to relatively large mapping units. These fail to reveal even medium-scale features, such as the terracing that occurs on some slopes of the major volcanoes.

We derived slope relatively easily using a standard algorithm (i.e., the difference in elevation divided by cell edge size) on a 3 by 3 moving cell. We then reclassified the resulting slope image into six classes which correspond to the aforementioned MAG/MIRENEM classification, except that we merged the last two classes (60–75% and over 75%) into one (Table 8-2). It must be noted that most steps in the preparation of the DEM and the subsequent generation of the slope map tend to smooth the terrain, leading to an underestimation of the area falling into the steeper categories.

The relatively flat lands constitute 40% of the national area and are concentrated in the north (Huetar Norte topographic map), the northeast (Llanura Atlantica and the coastal area south of Limon), the Tempisque Basin, and the area around the Golfo de Golfito. Although the low slope of these regions would seem to make them ideal for agriculture, the economic possibilities of all these areas are restricted by hydrology (an excess of surface water on the Atlantic side, and a seasonal shortage in the Tempisque Basin) and peripheral location (their relative inaccessibility). Classes 2 and 3 offer much better opportunities for agriculture, as they are mostly on the footslopes of the volcanic range or the Central Valley, but the area falling into these classes is fairly limited (27% of national area) and most is already cultivated intensively. Classes 4 and 5 are present throughout the volcanic range on the middle and higher reaches of the slopes and on the Nicoya Peninsula. They are cultivated extensively (mostly for coffee) but degrade easily under improper management.

TABLE 8-2 Percentage of Costa Rican Territory by Slope Class

| Class | Slope interval (%) | % of territory |
|---|---|---|
| 1 | 0–3 | 40.1 |
| 2 | 3–8 | 14.0 |
| 3 | 8–15 | 13.4 |
| 4 | 15–30 | 21.0 |
| 5 | 30–60 | 10.9 |
| 6 | over 60 | 0.6 |

Class 6 is present only in high mountainous areas and usually falls into protected zones.

Figure 8-6 shows a detail of the slope class map covering part of the Central Valley. The banding that occurs on the slopes of the volcanoes has its origin in the lava flows that formed the slopes (perhaps enhanced by the relatively coarse resolution of the DEM and its subsequent processing into this slope class map).

## IV. POTENTIAL USES IN SUSTAINABILITY ASSESSMENT

The DEM created in this analysis is now ready to be applied to many other issues in Costa Rica. For example, we used this information to derive temperatures as

FIGURE 8-6 Detail of slope map for Central Valley region.

a function of elevation in Chapter 10, to estimate the agricultural potential using a gradients approach in Chapter 12, and to estimate erosion potential in Chapter 14.

## REFERENCES


Bergoeing, J. P., and L. G. Brenes. 1978. *Mapa Geomorfológico de Costa Rica 1 : 1,000,000.* Instituto Geografico Nacional, San Jose, Costa Rica.

Bergoeing, J. P., and E. Malavassi. 1981. *Geomorfologia de la Valle Central de Costa Rica (Explicacion de La Carta Geomorfológica 1 : 50,000).* Universidad de Costa Rica, San Jose, Costa Rica.

Bolstad, P. V., and T. Stowe. 1994. An evaluation of DEM accuracy: Elevation, Slope and Aspect. *Photogrammetic Engineering end Remote Sensing* 60:1327–1332.

Brozovic, N., D. W. Burbank, and A. J. Miegs. 1997. Climatic limits on landscape development in the northwestern Himalayas. *Science* 276:571–574.

Dent, B. D. 1985. *Cartography: Thematic Map Design,* 2nd ed. W. C. Brown, Dubuque, Iowa.

Dikau, R., E. F. Brabb, and R. Mark. 1991. Landform Classification of New Mexico by Computer. USGS open file report 91–634.

Dymond, J. R., R. C. Derose, and G. R. Harmsworth. 1995. Automated mapping of land components from digital elevation data. *Earth Surface Processes and Landforms* 20:131–137.

Hammond, E. H. 1954. Small scale continental landform maps. *Annals of the Association of American Geographers* 54:33–42.

Hammond, E. H. 1964a. Analysis of properties in landform geography: An application to broad-scale landform mapping. *Annals of the Association of American Geographers,* March, 11–19.

Hammond, E. H. 1964b. Classes of land surface form in the forty-eight states, U.S.A. *Annals of the Association of American Geographers* 54. Map Supplement Series no. 4. [Scale 1 : 5,000,000]

Jensen, S. K., and J. O. Domingue. 1988. Extracting topographic structure from digital elevation data for geographic information system analysis. *Photogrammetric Engineering and Remote Sensing* 54:1593–1600.

Madrigal, R., and E. Rojas. 1980. *Manual descriptivo del mapa geomorfologico de Costa Rica.* Imprenta Nacional, San Jose, Costa Rica.

MAG/MIRENEM. 1991. *Metodologia para la determinacion de la capacidad de uso de las tierras de Costa Rica.* MAG, San Jose, Costa Rica.

Moore, I. D., R. B. Grayson, and A. R. Ladson. 1991. Digital terrain modeling: A review of hydrological, geomorphological, and biological applications. *Hydrological Processes* 5:3–30.

Moore, I. D., A. K. Turner, J. P. Wilson, S. K. Jenson, and L. E. Band. 1993. GIS and land-surface-subsurface modeling. In M. F. Goodchild, B. O. Parks, and L. T. Steyaert (Eds.), *Environmental Modeling with GIS.* Oxford University Press, New York.

Papo, H. B., and E. Gelbman. 1984. Digital terrain models for slopes and curvatures. *Photogrammetric Engineering and Remote Sensing* 50:695–701.

Pike, R. J. 1992. Machine visualization of synoptic topography by digital image processing. In *Selected Papers in Applied Computer Sciences.* United States Geological Survey, Reston, VA.

Pike, R. J. 1993. *A Bibliography of Geomorphometry with a Key to the Literature and an Introduction to the Numerical Characterization of Topographic Form.* United States Geological Survey Open-File Report 93-262-C, Menlo Park, CA.

CHAPTER 9

# Synthesis of Costa Rican Meteorological Information in a Geographical Context

GREGOIRE LECLERC, CECILIA REYES, AND CHARLES A. S. HALL



## I. INTRODUCTION

Climate and weather affect most human activity, and are especially important in cultures such as Costa Rica that depend heavily on agriculture. A study of climate is an essential component of understanding the biophysical basis for a region's economy, including the growth and productivity of agricultural plants and fruit trees, erosion, other forms of environmental degradation, land use, land use change, and many other subjects of critical importance to Costa Rica. More specifically, temperature, moisture relations, and solar radiation determine potential agricultural crop production; rainfall is important for estimating the risks of losses in crop production due to an excess or deficiency of water; relative humidity is associated with the incidence of diseases in plants; and evapotranspiration, which is a function of several weather variables, determines the amount of water required by the crop, whether it is supplied by rainfall or by irrigation. Consequently many human farm operation decisions are made in consideration of climate. In addition, there is a growing scientific interest in the effects of climate and possible climate changes on agriculture,

*Quantifying Sustainable Development*

biodiversity, and the behavior of natural plants and animals. Economic decisions made in the absence of climatic information can lead to disastrous investments, for example, expanding the production of a crop when the only land available is outside the optimal climate range for that crop, or constructing a reservoir in a region where severe storms will cause excessive siltation. Finally any consideration of sustainability has to ask whether a strategy selected would hold if the climate were to change.

Climate also effects us indirectly. For example, Jones *et al.* (1997) demonstrated that there is an intimate link between climate and the origin of cassava and beans, which evolved from wild relatives of these species in response to climate. The extensive biodiversity database being built by INBio (in Heredia, Costa Rica) will probably lead to important discoveries when analyzed in a climatic context. These and many more examples show the necessity of understanding climate patterns in their relation to issues of sustainability, and provide a strong incentive for estimating different weather parameters accurately.

Unfortunately deriving any large-scale—such as a national—synthesis of climate is very difficult, even when the nation in question is small and contains a relatively sophisticated scientific system. Meteorological stations are rarely dense enough anywhere to give a true geographical picture of climate. This is a problem that is not restricted to the developing world. For example, Nemani *et al.* (1993) state that for the western United States there is only one meteorological station for every 10,000 $km^2$. In addition, these are found generally only at airports and in heavily settled areas, which tend to have very special characteristics (such as being on flat land and in valleys) that may not characterize the general landscape. And it is often misleading to extrapolate between weather stations. For example, if two cities are located in adjacent valleys it would be very misleading to interpolate through the higher land located between the cities. Thus it is fair to say that we do not have a comprehensive view of climate hardly anywhere because rarely have we approached the issue of climate in a geographically sensitive way.

## A. Objective

The objective of this chapter is to construct a comprehensive spatial view of meteorology in Costa Rica by synthesizing the monthly climate data gathered by different institutions. Subsequently in Chapters 12–15 and 18 we use that information to help understand the limits of agricultural production, land use, hydroelectric potential, and other aspects of sustainability in Costa Rica.

We use well-documented interpolation techniques to estimate values for those areas not provided for in the data, and develop 1-km-resolution national-level maps of mean monthly values for key climate factors, including rainfall, temperature, solar radiation, and evapotranspiration. These maps supplement the excellent but increasingly outdated weather maps made manually (at a 1 : 1,000,000 scale) that are available from the Costa Rican Instituto de Metero-logía Nacional (IMN), such as the *Atlas Climatológico de Costa Rica* (IMN, 1986). In addition, since our maps are in a digital format they provide for the first time very convenient information for undertaking geographical analyses, such as gradient analyses of vegetation (Chapter 18) or agricultural production (Chapter 12). The maps that we produced in this study are summarized in this chapter and given in full on the CD-ROM accompanying this book, along with files of the original data and the programs used to produce them.

Costa Rica is a nation with exceptionally diverse climate within a very small area. For example, anyone used to temperate climates might be surprised to see ice within 10° of the equator! The very large mountains in close proximity to two very different oceans and the diversity of landform types (i.e., rugged mountains in the center and the south, and two more or less flat plains to the northeast and northwest), plus the diversity of physical forces impinging upon the country, make for perhaps as diverse a climate as can be found in such a small area anywhere in the world. One can find in several places a rain-fall gradient of 50 mm/month/km, which means that one farm will receive 250 mm (more than a foot) more rain in a month than another farm only 5 km (2 miles) away. Along the Pacific coast it rains 20 mm in the north during December but 200 mm in the south! This results in very large differences in runoff, soil genesis, moisture, and erosion, as well as solar radiation, plant disease, growth rate, etc., within only a few kilometers. A profile for December rainfall and maximum temperature along a west–east–north–south transect demonstrates the very large changes over relatively limited space (Figure 9-1).

## B. The Tropical Setting and the Global Heat Engine

All processes require energy for their operation, and the earth's weather system is no exception. This energy, like most of the other energy that runs ecosystems, comes from the sun. But the distribution of this energy on the earth's surface is uneven. The reason is not because the poles are further than the equator from the sun (the relation of 6400 km to 240 million km is trivial). Rather it is because at the equator the angle of the land to the incoming photons is about 90°, and becomes progressively less as one moves toward the poles (Figure 9-2). Thus the flux of solar energy per square meter of the earth's surface, and the resultant heating, is greatest at the equator and significantly less at the poles.

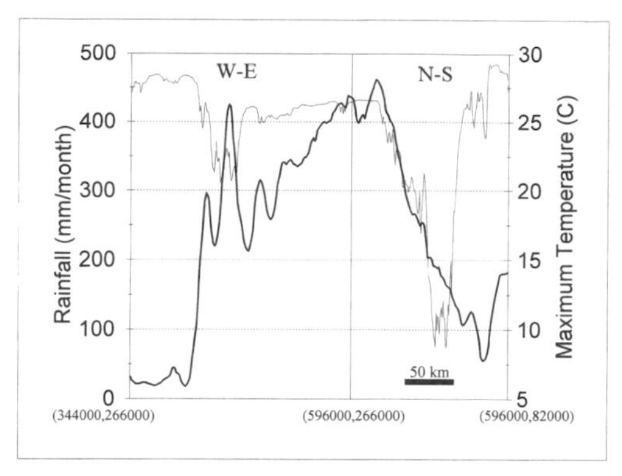


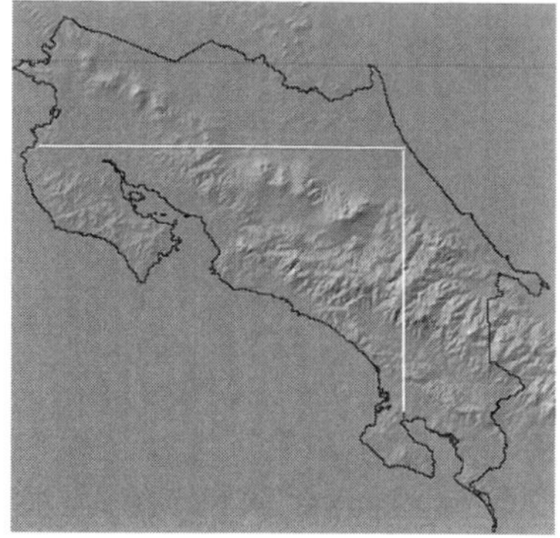

**FIGURE 9-1** Profile of December rainfall (thick line) and maximum temperature along a transect from west to east and then north to south. This transect is plotted on top of the elevation map.

One effect of greater radiative flux at the equator is that the atmosphere warmed there becomes buoyant and rises. As it rises it cools, and since cooler air (which has less energy) can hold less water vapor, clouds form and precipitation occurs. The air masses gain even more energy from the condensation of the rain, and rise even further. This process tends to make it very rainy on or near the equator. The location of the solar equator, and hence (more or less) the most intensely rising air masses, changes from 23.5° north (the tropic of Cancer) to 23.5° south (the tropic of Capricorn) over the year as the sun moves relative to the earth's surface from one tropic to the other. If you are planning a trip to the tropics and want to know whether to bring an umbrella, you can generally count on the rainy season being the time of year when the sun is directly overhead for that particular location. Thus it rains the most in Central America in the boreal summer, and in southern Brazil in the boreal winter. It rains on the equator twice a year—in roughly September and March.

As the air rises it piles up over the equator. This relatively high pressure tends to push the air north and south at about a 15-km altitude (Figure 9-3). At about 30° north and south this relatively dry air is no longer more buoyant than the air below it. Here it tends to sink, producing high-pressure air at the earth's surface. The sinking air warms. Warm air can store more water (if any is available) because warm air has more energy, and more energetic water molecules do not condense easily. When these descending air masses hit the surface of the earth they generate a region of high pressure. This high pressure forces that air back toward the lower pressure areas on the equator (and also poleward). The air's potential to hold more water pulls moisture from the

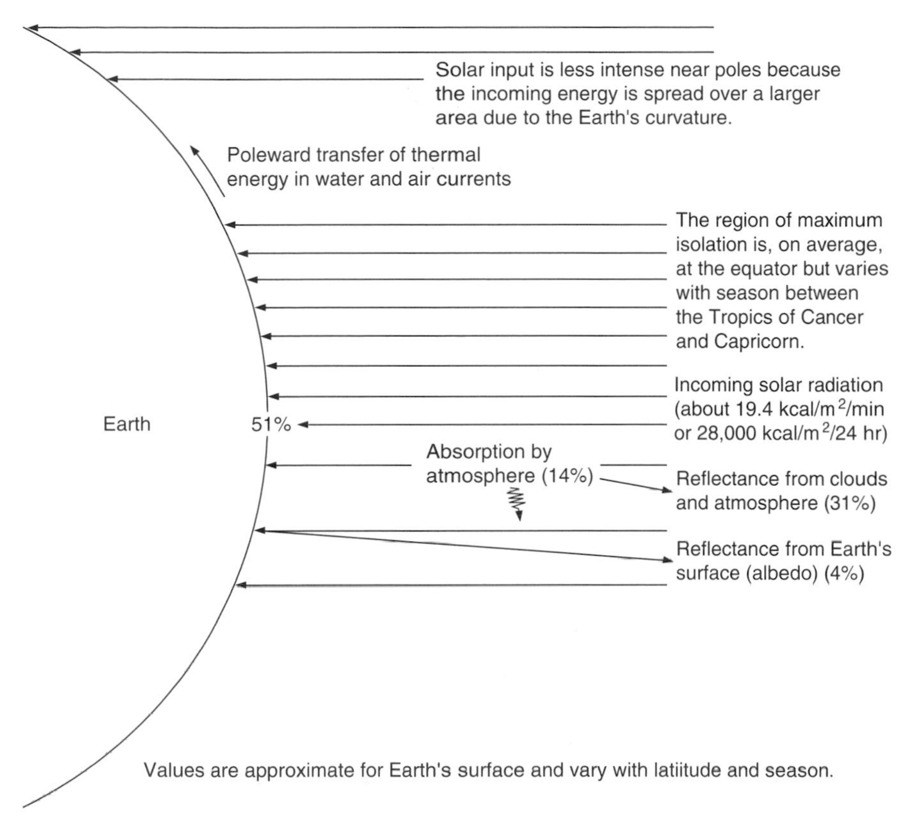


FIGURE 9-2 Incoming solar radiation in relation to the earth's surface at different latitudes. The tropics are warmer not because they are closer to the sun, but because each square meter receives a higher density of photon flux.

land, and creates the great deserts of 30° north and south. Over the oceans it absorbs large quantities of water. The return air flows toward the equator and creates the *trade winds* that complete the cycle. The entire process generates the circular air movements that dominate the climatic conditions in the tropics and subtropics, and controls the overall climate of Costa Rica.

This basic north–south pattern is modified by the Coriolis force, generated by the spin of the earth. It describes the fact (one supposedly first observed with long-range artillery) that a moving mass in the northern hemisphere will be deflected to the right, but to the left in the southern hemisphere. Thus the surface winds flowing toward the equator are deflected toward the west, so that to an observer on the earth's surface tropical winds appear to be coming

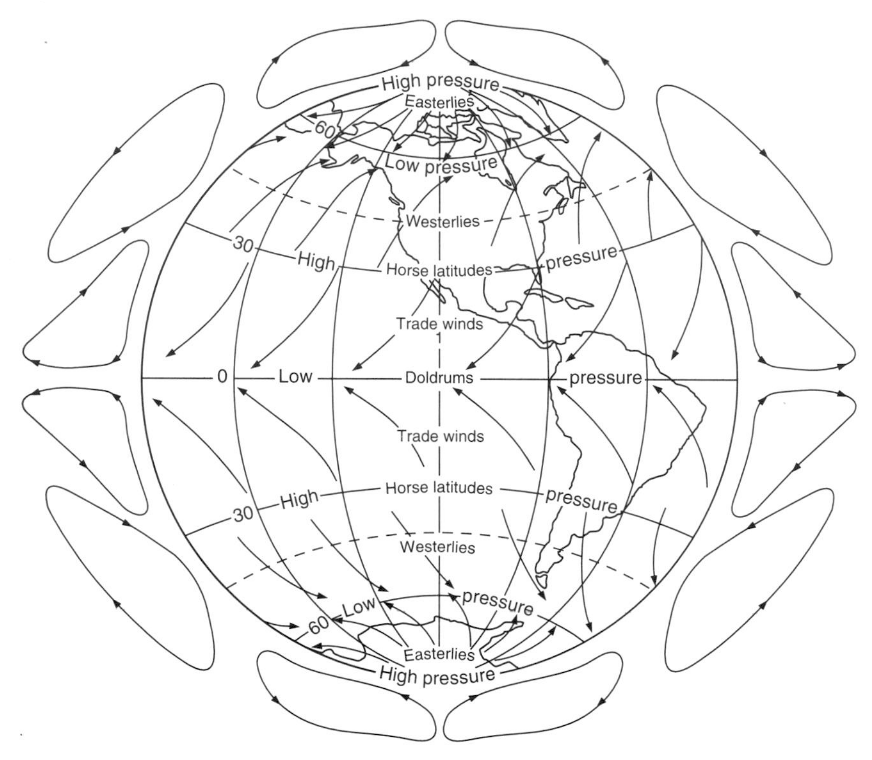


**FIGURE 9-3** Diagrammatic representation of Hadley cell circulation. In general the intense heating of the earth's surface at the solar equator causes air masses to rise at the equator. The high pressure formed at altitude pushes the air masses north and south until they descend at about 30° north and south. Hadley cells are most explicitly the atmospheric cells between the equator and 30° N and 30° S. (Modified from Rumney, 1968.)

from the northeast north of the equator, or the southeast south of the equator. These elliptical cycles of winds, going on average from the equator to 30° north and 30° south, are called *Hadley cells* after the early English meteorologist George Hadley. These cells, and the wind patterns that they generate, move north and south with the seasons. The location on the earth's surface where the winds converge (at the thermal equator, depending on the season) is called the *Intertropical Convergence Zone* (ITCZ). These wind patterns, and especially their upward movement, produce very wet conditions at the equator. Regions to the north and south of the equatorial regions are seasonally wet and seasonally dry, based on the seasonal movements of the ITCZ. Very dry conditions are found at 30° north and south, although these zones move north and south too with the seasons.

The basic wind and precipitation patterns due to the Hadley cells are modified by their relation to oceans and mountains. Moisture enters the atmosphere principally over warm tropical oceans. As already described, when moisture-laden air is lifted, it is cooled, causing it to lose energy and hence hold less moisture in suspension. Thus when air that has been over the sea hits the land the moisture begins to be squeezed out, especially if the air mass is forced up mountains. Hence there tends to be considerable rain on the windward side of mountains. As the air mass descends on the other side and becomes warmer a dry area is formed, often called a "rain shadow." This pattern is called the *orographic* effect, orographic meaning pertaining to mountains. Costa Rica has large mountains right in the path of the trade winds. The Atlantic side is normally very wet in more or less all months of the year, and it is very dry for about half the year on the Pacific side, especially in the north, because it is in the rain shadow of the mountains (Figure 9-4). All this produces a very diverse and variable climate over short distances (Figure 9-1).

## C. The Regional Climate Setting

Costa Ricans commonly talk about two seasons: summer (December–May) and winter (May–December). This "opposite" terminology for someone from North America or Europe may seem confusing. But in Costa Rica the temperature does not change much over the year, while the rainfall changes a great deal. There is a precipitation maximum in June–September, which brings some cooling, and a minimum in Febuary–April. In some areas (such as Turrialba) there is also a second rainfall peak in November–December.

According to Hastenrath (1976) and Waylen *et al.* (1996), surface winds in Costa Rica are dominated by two basic patterns: cold continental outbreaks ("nortes") that move from North America across the Gulf of Mexico and the Caribbean, and the northeast trades flowing in from the Caribbean ("alisios"). The trade winds vary in direction and intensity, being more intense during the boreal summer and less in winter, with a greater meridional component causing short dry periods called "veranillos" during July–August. "Nortes" get warmer and acquire moisture over the Gulf of Mexico and the Caribbean Sea, and precipitate that water especially on the Atlantic coast. In the very northeast of Costa Rica, close to the Nicaraguan boundary, very high rainfall (up to 7000 mm—7 m or 25 ft—annually) results from the effect of the "nortes." The orientation of the coastline relative to the pressure gradients promotes a pile up of moisture along the coast which generates increased coastal precipitation following the passage of cold fronts. Although few actually reach Costa Rica, cyclones also occasionally influence climate—curiously not on the side of the mountain ranges ("cordilleras") where the cyclone is moving (the eastern

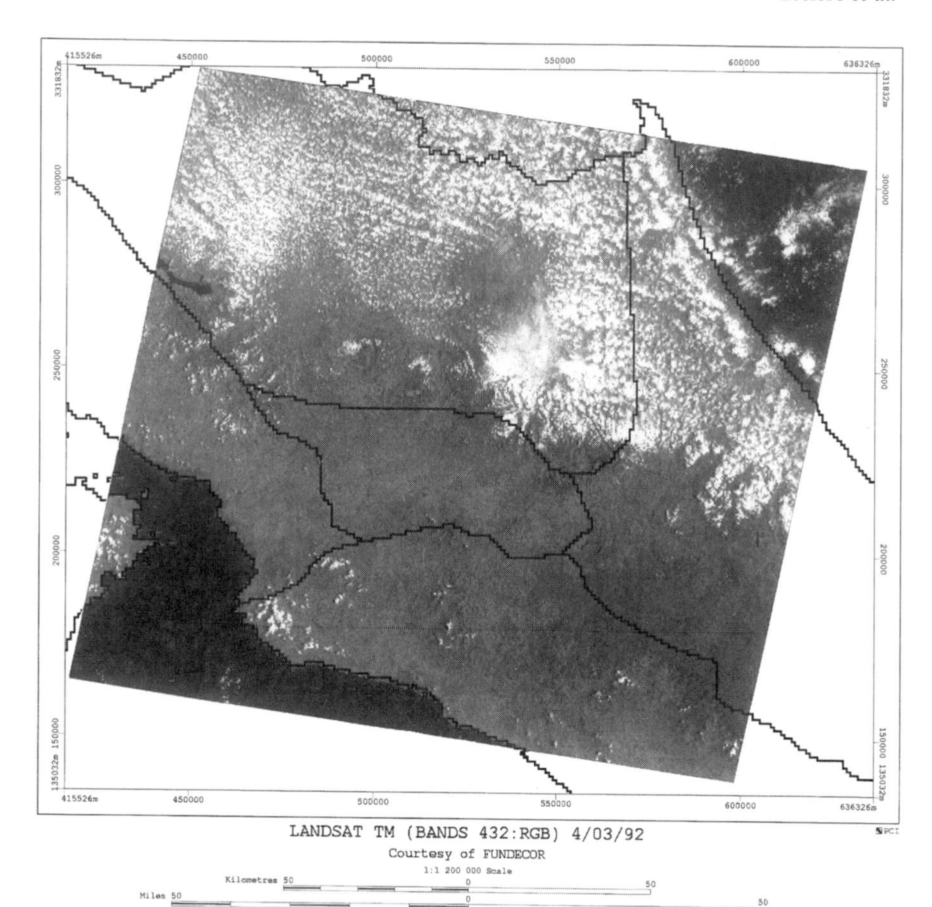


FIGURE 9-4 Landsat TM image taken over central Costa Rica. This image shows clearly the effect of the cordillera in defining separate climate patterns. A close examination of this image shows clouds literally draped around the watersheds.

front) but on the opposite side. Thus cyclones and other tropical storms in the Caribbean periodically cause very heavy rainfall ("temporales") along the Pacific coast, followed by deadly and destructive floods.

Because the mountains are oriented perpendicular to the trade winds coming in from the Atlantic there is a very strong orographic division in rainfall between east and west, and a somewhat less-pronounced division between north and south. Although the Central and Talamanca Cordilleras cut Costa Rica into two well-defined zones (the Atlantic and Pacific zones), the exact role of elevation and the orientation of the cordilleras in determining precipitation is still unclear (see figures under "Climate" on CD). Toward the northwest, there is also a chain of volcanoes, but these consist mainly of isolated peaks and they do not mark a clear transition as is the case in the central and southern regions. In the central and southern part of the cordilleras, pressure gradients caused by the air moving in this complex landform cause many microclimates.

Along the Pacific coast the climate is dominated by the seasonal migration of the Intertropical Convergence Zone north and south, and also by convective events inland (thunderstorms from ground-heating-induced rising of air masses) in May–November. These processes produce rainfall along the Pacific coast at that time equivalent to that of the Atlantic zone, with occasional periods of extremely heavy precipitation. There is also a slight monsoon (land-warming-induced sea breeze) effect which generates weak equatorial "westerlies" that penetrate from the Pacific coast. These are sometimes enhanced by strong daily sea breezes during the summer, which makes the Pacific coast very attractive to surfers.

## D. Migration of the ITCZ to the South

During February and March two forces combine to generate very dry conditions in northwestern Costa Rica. First, the Hadley cells and the ITCZ are relatively far to the south, so the trade winds are weaker during these months. Second, the orographic effect of the cordilleras is to dump any water in the trade winds on the Atlantic side. Thus rainfall is typically very low in February–March in the central and northern part of the Pacific coast, which makes the seasons very pronounced in these areas. On the Atlantic side of the cordilleras the seasons are much less marked, although there is still some reduction in the rainfall during February and March. Figure 9-4 is a Landsat TM (thematic mapping) image taken over central Costa Rica which shows clearly the effect of the cordilleras in defining separate climate patterns. A close examination of this image shows clouds literally draped around the watersheds. Migration of the ITCZ to the south during the boreal winter (November–March) produces westerlies in southwest Costa Rica and an orographic effect on the Caribbean side of the Talamanca Cordillera (in the southeast of Costa Rica) as well as in the Central Valley.

## E. Climate Change

Throughout the world there is considerable concern that the activities of humans, principally as a consequence of the burning of fossil fuel and hence the injection of $CO_2$ and other greenhouse gases into the atmosphere, may be producing large-scale climate change (see Chapters 16 and 17). These concerns have generated one of the largest focused interdisciplinary scientific investigations ever undertaken, and as a consequence we know much more about climate and the interaction of climates and ecosystems than ever before. One of the results of this analysis is that we now know much better that indeed climate change exists—that there always has been climate change and there almost certainly always will be. The important question is whether humans are causing climate to change more, or more unpredictably, than in the past, and to what degree humans can adapt to climate change and climate unpredictability, whether that is caused by humans or whether it is natural. The greatest concern for tropical regions is not so much temperature changes but possible soil moisture changes. Even though warmed air can hold more moisture it also causes greater evapotranspiration. Thus the net result of a warmer earth is thought to be a drying of tropical soils (Rind *et al.*, 1990), something that would have critical consequences for most of the tropics, including much of the central and Pacific portions of Costa Rica.

One natural source of climate variability is the El Nino Southern Oscillation (ENSO) (the equatorial wind–ocean interactions that cause, for example, the warming of oceans and the failure of fisheries near Peru). This factor may be changing due to human-induced or other factors, or it may be just continuing its largely unknown long-term periodicity. Costa Rica is located near the origin of the largest known single cause of the ENSO. Costa Rica also appears to be experiencing a changing climate, including increases in both droughts and excess precipitation, and changes in magnitudes and durations of rainfall events. Waylen *et al.* (1996) examined the relation of regional events to local climate patterns by analyzing correlations of rainfall with the Southern Oscillation index using 105 rainfall stations with more than 20 years of records. They also considered the relation between Central America and Caribbean precipitation totals and the ocean–atmosphere circulation in the tropical Pacific. They found that the Pacific coast experiences longer dry periods just before the onset of an ENSO event, while the Atlantic coast receives increased summer and winter rainfall during ENSO periods. The interior regions showed mixed responses to such events. In Costa Rica, it is believed that droughts are associated with ENSO events and that heavier rainfall is experienced on the Pacific coast in the year following an ENSO event (Hastenrath, 1976; IMN, 1988). But Waylen's analysis showed that the issue was more complex—specifically, that the response to a "warm-phase" ENSO was opposite to that of a "cold-phase" ENSO, and that each region responded independently. Nevertheless it is obvious that there are strong links between ENSO patterns and climate in Costa Rica, and if the ENSO is changing, so must the climate of

Costa Rica. The very strong El Nino of 1997–1998 was accompanied by many extreme climatic events.

Another important issue pertaining to climate change is whether the deforestation that is occurring in Costa Rica (see Chapters 16 and 17) is causing changes in climate. Vegetation affects climate in many ways. First, an important cause of rain is vertical advection of air parcels following the heating of the earth's surface. This is a familiar pattern in many parts of the world as it causes warm-season late afternoon thunderstorms. The degree to which the air masses are lifted by the absorption of heat is obviously a function of the heating of the earth's surface, and that heating in turn is a function of the albedo, or reflectivity, of the surface. Satellite pictures show us that the surface of the desert areas of the earth is very bright, indicating much reflected light (high albedo), and that the area of forests is very dark, indicating absorption of light. It is this absorbed light that causes the heating of the atmosphere and advective rain. Thus it would seem that as deforestation takes place the earth's surface would absorb less light and generate less vertical heat movement and, consequently, less rain. If this were true it could cause severe problems for a nation heavily dependent upon agriculture and hydroelectricity. Vargas and Trejos (1994) report that over 75% of Costa Rica experienced diminishing precipitation in the period 1960–1990. However, Waylen *et al.* (1996) applied simple regressions on precipitation time series from 105 stations and obtained significant linear trends (positive and negative) for only 13 stations. Clearly the strong effect of ENSO and its coupling with the land prevent the detection of most of the trends if any.

A second way that vegetation affects climate is through effects on evapotranspiration. Obviously the more vegetation, the more the transpiration. This effect may be small in Costa Rica, however, since transpiration is a function of leaf density (or LAI, leaf area index). In the favorable environment for growth found in Costa Rica, vegetation, once disturbed, quickly recovers to its original leaf density of about 6 $m^2$ of leaves per square meter of ground even if biomass is much less. Even crop areas tend to have a leaf density similar to natural forests. Therefore in Costa Rica this is probably not an important issue except in the dry northwest where vegetation can be quite limited. Whether the climate is changing due to long-term natural forcing processes, or whether indeed human-induced climate change is occurring, Costa Rica has been subject to unprecedented severe weather in recent years: droughts in the northwest, especially in 1993–1995, and floods in the south in 1996. These events are extremely deleterious to an already fragile economy. For example, the floods of 1996 destroyed a large section of the Cartago–Osa Peninsula highway, requiring the borrowing of many millions of dollars of very expensive foreign capital to finance the repairs. These changes have not been considered to date in government planning. It is our hope that eventually geographically based long-range meteorological forecasting will allow for better predictions of what can be expected in different parts of the country and improve long-range agricultural and environmental planning. As a first step in this process we develop here what we believe is the first digital geographical database for the entire country.

## II. METHODOLOGY—AN OVERVIEW

Our overall approach was to gather as much digital meteorological data as possible for Costa Rica and to synthesize them into a digital and spatial format at a resolution of 1 $km^2$. This required some rather complicated procedures for interpolating weather parameters between or among weather stations located within a very complex terrain. The resulting climate maps allow for presumably much more complex analysis of the relations between climate, plant growth, erosion, hydrological potential, possible climate change, and so on. There was enough empirical information to generate maps of varying degrees of accuracy, for rainfall, temperature, hours of sunlight, solar radiation, and evapotranspiration. Because of data and time limitations, other variables such as relative humidity, vapor pressure, and wind were not included in the study. Also, no attempt was made either to develop a process-based model of the complex patterns of the weather variables or to include the effects of greenhouse gases, deforestation, or the ENSO.

### A. Gathering and Processing of Data

The Instituto Meteorológico Nacional of Costa Rica is the official compiler of meteorological data. As of the 1990s it operated 105 stations itself, and receives and archives data from many other stations. It publishes monthly data of standard weather parameters from its own stations, and summaries of daily data are available for purchase at a cost of approximately $10,000. An inventory of climatic stations for Costa Rica shows that to date about 817 weather stations have been operated by various entities. Data are available from IMN, other national institutions such as the Instituto Costarricense de Electricidad (ICE), banana growers (Chiquita Brand, Standard Fruit Corporation, CORBANA), international organizations such as CATIE, and specific projects in certain regions (e.g., the Arenal Reservoir). Many stations gather data on rainfall and/or temperature only, while fewer stations gather data on solar radiation, evaporation, humidity, wind, and other parameters. The meteorological data for each station are summarized routinely, and some are entered into a computer, although unfortunately with no standard format. Only minimum and maximum temperature are recorded, and the other values such as mean, daily, or diurnal temperature are derived from the former.

IMN could provide us data only from its own stations, so we asked other institutions such as SENARA, ICE, PINDECO, and the Atlantic Zone Project at CATIE for their weather records. They provided us with data which were primarily precipitation and mean temperature.

Because of our own time limitations we used only data already in digital format. These were checked against the IMN inventory file NOMINA.INV for correctness of coordinates (latitude and longitude) and elevation of the location. When a difference in elevation was found, the location data were compared with our 200-m-resolution digital elevation model (DEM). Because the computer program we used for spatial analysis does not allow duplicate coordinates, we chose the record with a longer time record for those cases where two stations had the same coordinates but different elevations. From 350 station records, we ended up with 256 stations with a minimum of 10 years of data starting in 1960. As can be seen in Figure 9-5, the locations of the stations are mostly concentrated in the Central Valley and the Guanacaste region, principally because that is where the most important hydroelectric plants in Costa Rica (Cachi and Arenal) are located. For each location, the overall monthly and yearly averages were calculated. Rainfall data as early as 1880, 1905, and 1927 have been recorded from 3 stations. Since 1935 more stations were opened to collect data on rainfall and/or temperature. Data on other weather variables started to be recorded in 1937. Costa Rica is therefore a good source of data for studying long-term climate changes in this part of the world.

The Costa Rican national meteorological institute (IMN) has subdivided the geographical climate setting based mostly on relief. IMN constructed these zones manually by considering geophysical factors such as mountain ranges and wind directions (Alfaro, personal communication). These five zones (IMN zones) are shown in Figure 9-5 and serve as a base for analysis by several authors, including ourselves. For example, we used the results of their regression of temperature vs elevation for each IMN zone to produce our temperature maps.

We synthesized precipitation maps ourselves from the raw data. We derived most of the other variables presented here by extrapolating the results of a previous statistical study (Rojas, 1985) to a spatial framework. Rojas's study was an agroclimatic characterization of Costa Rica. He used the divisions derived by IMN to group weather stations into the five different fundamental climatic zones of Costa Rica (Figure 9-5). Rojas used a data bank with 72 stations of daily rainfall, 54 stations with daily maximum and minimum temperature records, and 22 stations of sunshine hours to produce regression coefficients with elevation as the independent variable.

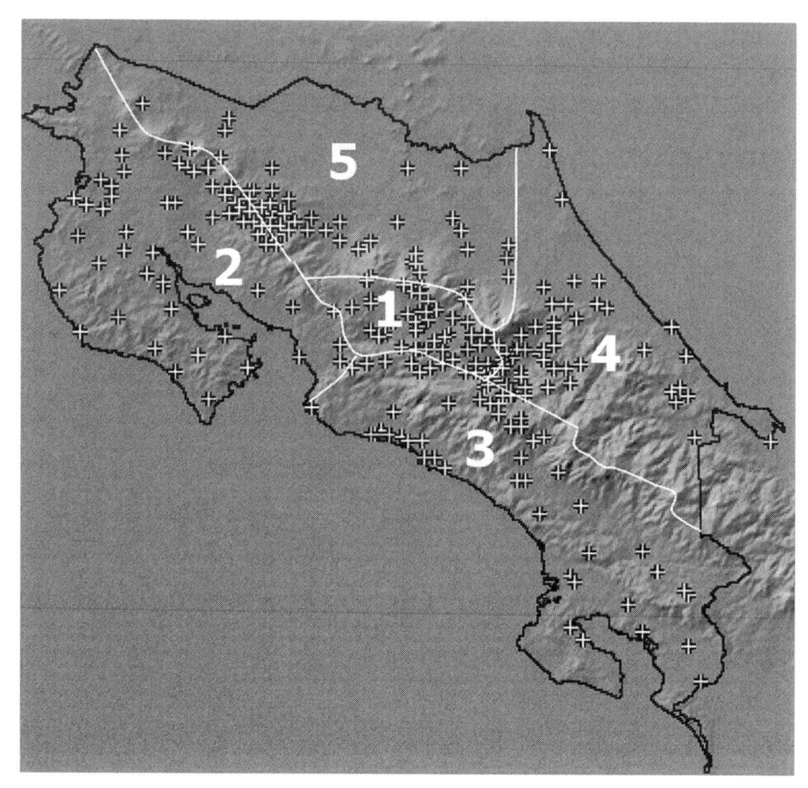


FIGURE 9-5 Location of meteorological stations in Costa Rica, and the five general zones used for meteorological analysis.

## B. Deriving Spatial Information

The basic problem that we faced as we attempted to derive spatial representations of climate, one common in geographical modeling, was how to generate accurate maps from relatively little sampling (i.e., sparse point-based information). We have already mentioned the problem with predicting climate on a mountain from relatively close weather stations that tend to be located at airports in valley bottoms. The sections that follow review some possible procedures for dealing with this problem and develop the particular procedures we used to generate climate maps for each square kilometer of Costa Rica.

## C. Generating Spatial Maps from Point Data

Once we had derived the best data possible from the existing weather station records we were confronted with an enormous technical problem: how do we translate that data into a map that has any hope of capturing the effect of the very large impact of regional spatial and topographical variation on the predicted climate? There are two general approaches that can be used: *regressions* and *interpolations*. The following section may be skipped by those uninterested in spatial analysis techniques.

### 1. Regressions

Regressions use the site-by-site relation of the dependent values of interest (such as temperature) to independent predictors (such as elevation) that are much more abundant. Simple arithmetic is used to generate a landscape of the dependent values. Interpolation, including *Kriging*, is used when there are no such relations. There is also an approach that combines the two approaches called *co-Kriging*.

Regressions, either single or multivariate, are a generally effective way to predict the relation between a dependent variable and one or a set of independent variables based on a suite of measurements of each. The mathematical relation so derived can then be used to predict the expected value at an unsampled location (or time). For example, an analysis of temperature vs elevation records always shows linear relationships with high correlation coefficients. The slope of the regression line (called the "lapse rate") is variable, however, and depends on the time of year and whether minimum, mean, or maximum temperature is considered. Under very special conditions the relation can even change sign: colder air settles in valleys and can drag hot air downslope (Hutchinson, 1991). In the case of rainfall, regression analysis generally performs poorly given the complexity of the mechanisms involved.

Regression results can be improved when "local windows" are considered around each station, meaning that mathematical relations are derived that emphasize local relations among measured variables. Such an approach is used in the PRISM model (Daly *et al.*, 1994). This model: (1) estimates elevation for each station using a digital elevation model; (2) assigns each DEM grid cell to a topographic "facet" or directional face by assessing slope orientation; (3) develops a weighted localized rainfall–DEM elevation regression function using slope orientations of nearby stations to weigh stations; and (4) predicts rainfall at each grid cell using the local regression function. To compute the regression function, greater weight is given to stations with similar location, elevation, and aspect. Software to implement the PRISM model was not publicly available at press time.

### 2. Interpolation

In cases where the variability in the weather parameters is not well explained by the regression equations, that is, where there is no obvious relation between meteorological variables and, e.g., elevation, interpolation is normally done. Simple interpolation uses the spatial relation among the relatively few dependent variables (such as temperature) to predict values over the landscape, and areas that are intermediate spatially to existing data are given intermediate values. For example, if one weather station had a temperature at some time of 10°C and another 20°C, then the site half way in between would get a value of 15°C. Software is available to do this in two dimensions. This procedure obviously does not deal well with our mountain-between-the-airports problem.

A more sophisticated interpolation approach uses *Kriging* or *co-Kriging* (Bogaert *et al.*, 1995; Staritsky, 1989). These are well-documented and often successful techniques that were developed originally to determine the best place to mine gold based on test drillings. Kriging is a spatial interpolation technique that recognizes that there may be no smooth statistical relation among the spatially separated values and hence uses the theory of random functions. The interpolations are generated by first exploring the statistical variability of the known data points and then using that information to fill in the regions between measured values (Burrough, 1986). Weighted averages of parameter values are calculated by taking into account spatial dependence and the distance of the observations to the predicted location of a series of individual points over the map surface (Delfiner and Delhomme, 1975; Stein, 1991). The variance of pairs of observation points is examined in relation to their distance to each other, called the *lag*. The resulting relation of variance vs lag is used to obtain the *semivariogram* fitting function, whose coefficients are used in the interpolation process (Figure 9-6). An idealized semivariogram shows that the variables are dependent upon each other up to a certain distance. Above that distance they are not. The equation for the semivariogram $\gamma$ is given by

$$(\gamma h) = \frac{1}{2n} \sum_{i=1}^{n} [Z(x_i) - Z(x_i + h)]^2 \qquad (1)$$

where $h$ is the "lag" (distance between one station and a group of other stations), $n$ is the number of pairs considered in the calculation, $Z(x_i)$ and $Z(x_i + h)$ are the parameter measurements at point $x_i$ and $x_i + h$, respectively. The semivariogram is nothing more than a characterization of the difference in the measured variable, $Z$, at one point and $Z$ at a distance $h$ from this point, for different values of $h$. For spatial 2D interpolation, space is searched around $x_i$ for all measurements $Z$ within a ring or annulus at a distance between $h$ − step/2 and $h$ + step/2, averaging $Z$ to improve statistics. The Kriging method depends critically on the step size, on the range over which measurements will be considered (maximum value of $h$, or $A_0$), and on estimating a semivariogram function that fits well and which will be used to compute the weights to assign to neighboring stations for averaging. The method can be hampered by ad hoc assumptions on the form the semivariogram should take and the difficulties in assessing the merit of different functional forms (Dubrule, 1983).

*Thin plate splines* are different mathematical procedures that can be used to perform optimal interpolation. They are easier to use than Kriging and also allow the use of user-defined submodels, but do not help to understand spatial correlations as Kriging does. Hutchinson (1991) used this method successfully to interpolate mean monthly climate surfaces in Australia.

These spatial interpolation procedures, including Kriging, are available through commercial Kriging software such as GS+ (Gamma Design Software) or Surfer (Golden Software). The thin plate spline software is available commercially from ANU at http://cres20/anu.edu.au/software/anusplin.html. Public domain software also is available, such as GEOEAS. A set of programs (which includes co-Kriging) has been developed by the FAO (Bogaert *et al.*, 1995) especially to handle climate interpolations. See Varekamp *et al.* (1996) for an assessment of popular public domain geostatistical software.

## III. METHODS WE USED TO COMPUTE COSTA RICAN DIGITAL CLIMATE MAPS

### A. Rainfall

We assumed originally that there was a clear relation of rainfall to elevation. Consequently we used the stepwise regression analysis PROCEDURE REG (SAS Institute Inc., 1988) on the yearly and monthly rainfall averages for the 256 stations that met our criteria. We examined latitude, longitude, elevation, and their interactions and squares as independent variables. Distance to the sea was not included in the list of independent variables to prevent multicollinearity problems since elevation and distance to the sea were highly correlated. We used a significance level of 15% as a threshold criterion for the variables to enter and stay in the model. Unfortunately the original regressions did not give acceptable $R^2$ values so we were forced to redo the analysis on subsets, as defined by the clustering, of the original data.

We also assessed the potential of co-Kriging for improving our results by seeking better regression coefficients within smaller groups of stations. Co-Kriging is an approach that can be superior to standard Kriging because it corrects for the effects of highly correlated independent variables (such as elevation and temperature). We divided the entire set of observations into different clusters based on the average linkage method of the PROCEDURE CLUSTER in SAS. We chose the number of clusters based on standard criteria: the cubic clustering criterion, pseudo-*F*, a coefficient of determination

($R^2$) higher than 0.6, and no more than 10 clusters to facilitate analysis. We checked the grouping of the clusters visually by mapping them on the IMN zones map (Figure 9-5), which helped to identify atypical stations (i.e., those having a cluster number different from the surrounding stations). These were excluded from the analysis. We then performed regressions and generated a semivariogram within each valid cluster to see if the results improved.

We used a step size of 0.0167° (1.8 km) for all interpolations to facilitate the synthesis of the results. The range of values used for modeling the semivariogram was limited to include points shortly to the right of the beginning of the sill (the asymptote of the semivariogram). The interpolation grid interval was 0.02°. The interpolation radius, which defines the length of the distance within which observation points will be used in the estimation, was limited to the point where the sill started. Finally, we used a Gaussian (normal distribution) model to derive the semivariogram as this allows a better description of semivariograms where there exists a trend in the data (universal Kriging, which includes the effect of a trend, was not available readily to, nor tested by, us). We used all the stations for the estimation to avoid any visual artifacts in regions where the density of stations is low. We subjected our predictions to a validation procedure by using a "jacknife" analysis, where an observed value from the set of observations is temporarily deleted and the Kriging estimate is compared to that actual value. This analysis was readily performed with GS+.

## B. Temperature

We derived the monthly parameters for each square kilometer from Rojas's coefficients and our elevation map (Chapter 6) as summarized in Table 9-1. Diurnal and daily mean temperatures were calculated from the maximum and minimum temperatures using

$$
\begin{aligned}
T_{di} &= \frac{2T_{max} + T_{min}}{3} \\
T_{mn} &= \frac{T_{max} + T_{min}}{2}
\end{aligned}
\tag{2}
$$

where $T_{max}$ and $T_{min}$ refer to the maximum and minimum temperatures, and $T_{di}$ and $T_{mn}$ are the diurnal (daytime) and daily mean temperatures, respectively. Rojas derived averages of the diurnal and daily temperatures for each 10-day interval (i.e., three times a month) and ran a regression analysis of these values against elevation. In general this method was a good predictor of temperatures as judged by $R^2$ values of 0.85 to 0.95 (Table 9-2). An exception is the North Pacific region, which had low $R^2$ values for the months April to December. Except for these low values, it can be said that elevation is sufficient to explain temperature. The lapse rate, however, is not constant, but ranges from 4 to 10°C/km depending on place and season.

We found that applying the regression equations by IMN zone produced a temperature map with two problems: (1) The temperature predicted at a location corresponding to the site of a meteorological station is the value from the regression equation and only rarely the exact temperature measured at that station, and (2) there are sharp changes in temperature at the boundaries of the IMN zones since the lapse rate is different.

There are technical solutions to these problems but we could not implement them because we could not obtain enough raw temperature data. The first problem can be avoided by first computing the sea-level temperature for each station using the lapse rate of Table 9-1 [i.e., $b = T(0) = T(H) - a$, where $T(H)$ is the temperature measured for the station]; second, Krig the zero-elevation temperature to produce a continuous surface; third, build the continuous temperature map [$b = T(H) = T(0) + a \times H$], where $H$ is the DEM elevation and $T(0)$ the zero-elevation temperature surface.

The second problem can be eliminated by using the same lapse rate function for the entire country. This can be derived from the data of Table 9-1, or we could use a "universal" regional value. Jones (1991) used a relationship based on mean tropical atmosphere night measurements for the West Indies to derive universal Latin American temperature maps. It can be expressed as

$$T_{la} = T(0) + H\left(5.8 + \frac{1}{0.25 + 0.3505\,H^2}\right)$$

where $T_{la}$ is the temperature (in degrees Celsius) at an elevation $H$ (expressed in kilometers).

The best approach which can be adopted is to produce a continuous surface by Kriging the lapse rate from Table 9-1 [$a(x, y)$] and a zero-elevation temperature map [$b(x, y)$]. Then the minimum and maximum temperature maps can be computed using the digital elevation map [or map of $H(x, y)$]:

$$T(x, y) = [b(x, y) + a(x, y)] \times H(x, y).$$

## C. Hours of Sunlight

Rojas estimated values for hours of sunlight every 10 days for each of the IMN geophysical zones. But because of the scarcity of data for the Atlantic and North Atlantic zones, it was necessary to make a general model for all of Costa

TABLE 9-1 **Monthly Daily Maximum and Daily Minimum Temperature Regression Coefficients ($T = a \times \text{elev.} + b$)**

| | Maximum temperature | | | | | | | | | |
|---|---|---|---|---|---|---|---|---|---|---|
| | Central Valley | | North Pacific | | South Pacific | | Atlantic | | North Atlantic | |
| | *b* | *a* | *b* | *a* | *b* | *a* | *b* | *a* | *b* | *a* |
| January | 32.04 | −0.00902 | 29.27 | −0.00875 | 29.79 | −0.00684 | 26.20 | −0.00517 | 26.50 | −0.00534 |
| February | 33.40 | −0.00994 | 30.13 | −0.00948 | 30.30 | −0.00686 | 26.28 | −0.00506 | 26.86 | −0.00543 |
| March | 33.71 | −0.00938 | 31.05 | −0.00882 | 30.70 | −0.00684 | 27.04 | −0.00503 | 27.80 | −0.00547 |
| April | 32.70 | −0.00820 | 31.31 | −0.00837 | 30.47 | −0.00667 | 27.57 | −0.00510 | 28.46 | −0.00567 |
| May | 29.84 | −0.00566 | 30.04 | −0.00632 | 29.77 | −0.00653 | 28.36 | −0.00540 | 28.64 | −0.00557 |
| June | 28.84 | −0.00534 | 28.77 | −0.00592 | 29.27 | −0.00640 | 28.02 | −0.00540 | 28.38 | −0.00557 |
| July | 29.45 | −0.00595 | 28.97 | −0.00668 | 29.22 | −0.00640 | 27.57 | −0.00539 | 27.94 | −0.00550 |
| August | 28.81 | −0.00539 | 28.87 | −0.00653 | 29.01 | −0.00631 | 27.73 | −0.00542 | 28.00 | −0.00555 |
| September | 27.69 | −0.00455 | 28.34 | −0.00586 | 28.95 | −0.00647 | 28.01 | −0.00550 | 28.15 | −0.00570 |
| October | 27.80 | −0.00476 | 28.10 | −0.00548 | 28.75 | −0.00645 | 27.89 | −0.00547 | 27.84 | −0.00555 |
| November | 28.93 | −0.00598 | 28.18 | −0.00573 | 28.90 | −0.00656 | 27.38 | −0.00545 | 27.10 | −0.00532 |
| December | 30.93 | −0.00817 | 28.47 | −0.00687 | 29.30 | −0.00675 | 26.63 | −0.00530 | 26.55 | −0.00533 |

| | Minimum temperature | | | | | | | | | |
|---|---|---|---|---|---|---|---|---|---|---|
| | Central Valley | | North Pacific | | South Pacific | | Atlantic | | North Atlantic | |
| | *b* | *a* | *b* | *a* | *b* | *a* | *b* | *a* | *b* | *a* |
| January | 29.14 | −0.00789 | 27.25 | −0.00722 | 27.47 | −0.00640 | 24.47 | −0.00507 | 24.88 | −0.00507 |
| February | 30.33 | −0.00871 | 28.08 | −0.00814 | 27.86 | −0.00646 | 24.50 | −0.00496 | 25.13 | −0.00516 |
| March | 30.61 | −0.00822 | 28.98 | −0.00775 | 28.32 | −0.00644 | 25.27 | −0.00500 | 25.93 | −0.00517 |
| April | 29.80 | −0.00710 | 29.32 | −0.00763 | 28.34 | −0.00637 | 25.84 | −0.00507 | 26.56 | −0.00533 |
| May | 27.52 | −0.00500 | 28.27 | −0.00580 | 27.90 | −0.00629 | 26.63 | −0.00530 | 26.79 | −0.00520 |
| June | 26.74 | −0.00480 | 27.22 | −0.00546 | 27.48 | −0.00610 | 26.37 | −0.00530 | 26.68 | −0.00517 |
| July | 27.22 | −0.00518 | 27.35 | −0.00580 | 27.39 | −0.00613 | 26.02 | −0.00529 | 26.41 | −0.00517 |
| August | 26.51 | −0.00465 | 27.19 | −0.00565 | 27.17 | −0.00598 | 26.09 | −0.00532 | 26.36 | −0.00524 |
| September | 25.71 | −0.00409 | 26.76 | −0.00524 | 27.18 | −0.00613 | 26.28 | −0.00532 | 26.40 | −0.00530 |
| October | 25.93 | −0.00438 | 26.59 | −0.00497 | 26.99 | −0.00611 | 26.18 | −0.00530 | 26.13 | −0.00520 |
| November | 26.68 | −0.00528 | 26.54 | −0.00493 | 27.11 | −0.00623 | 25.72 | −0.00530 | 25.54 | −0.00499 |
| December | 28.19 | −0.00703 | 26.61 | −0.00551 | 27.30 | −0.00640 | 24.93 | −0.00520 | 24.96 | −0.00503 |

Based on decade regressions (Rojas, 1985).
$b$ is in degrees C; $a$ is in $C \cdot m^{-1}$.

TABLE 9-2 Range of $R^2$ Values for the Linear Relation of Temperature and Elevation

| Zone | Diurnal temperature | Daily temperature |
|---|---|---|
| Valle Central | 0.84–0.95 | 0.80–0.94 |
| Pacifico Norte | 0.45–0.87 | 0.39–0.87 |
| Pacifico Sud | 0.95–0.98 | 0.95–0.97 |
| Atlantico | 0.97–0.99 | 0.98–0.99 |
| Subvertiente Norte | 0.94–0.98 | 0.93–0.99 |

Rica. Rojas examined the relation of hours of sunlight to various variables (including mean diurnal and daily temperature, mean precipitation, elevation, latitude, and longitude). He found high correlation coefficients between hours of sunlight and precipitation when he used quadratic equations and the data for all years. The equations with their respective correlation coefficients are given in Table 9-3. We used the equation for Costa Rica with the interpolated values of rainfall to prepare the national maps of hours of sunlight.

Another method for calculating hours of sunshine involves first calculating clear-sky sunlight intensity using the solar constant corrected for the optical density the light must pass through (a function of latitude), and then adjusting for cloud cover and correcting for the slope and aspect of the ground (Everham *et al.*, 1991). However, because of scarcity of data on cloud cover, we could not employ this method. Such data might be obtained from satellites, however, and many such assessments could be used to derive an independent estimate of hours of sunshine.

## D. Solar Radiation

Total solar radiation is an important variable for predicting the agricultural potential of a region, for predicting evapotranspiration and hence water bud-

TABLE 9-3 Relation between Daily Mean Values of Sunshine Hours (ssh) and Daily Rainfall ($p^2$) Calculated over All Years, and Their Correlation Coefficients $(R)^2$

| Zone | Regression equation | $R^2$ |
|---|---|---|
| Valle Central | $ssh = 8.4799 - 0.077975p + 0.00039392p^2$ | 0.72 |
| Pacifico Norte | $ssh = 8.9524 - 0.068464p + 0.00027891p^2$ | 0.78 |
| Pacifico Sur | $ssh = 9.1104 - 0.056360p + 0.00016309p^2$ | 0.71 |
| Costa Rica | $ssh = 8.6388 - 0.063440p + 0.00022611p^2$ | 0.71 |

gets, and for assessing the relation of natural vegetation types to climate. Maps of solar intensity have been produced manually by the IMN. We developed a set of digital surfaces based on the same database.

There are two methods available to derive solar radiation: an analysis of existing empirical solar input data (derived from instruments called heliographs and actinographs) and through the use of an empirical equation, the Black–Prescott equation [Equation (3)], which gives solar radiation based on the regional relation between incoming solar radiation, which can be predicted easily from the time of year and latitude, and regional cloudiness. A comparison of the two can then be used to determine the likelihood that the estimate for a given area is reasonable. The Black–Prescott (1940) equation is

$$R_g/R_{go} = b + a - (n/N), \tag{3}$$

where $R_g$ is daily global radiation at the earth's surface for that location, $R_{go}$ is solar radiation on top of the atmosphere, $n$ is sunlight hours (actual day length), $N$ is the astronomical day length, and $a$ and $b$ are statistically derived empirical constants that vary from station to station. The sum $a + b = 1$ corresponds to a perfectly transparent atmosphere for which, if $n = N$, we obtain $R_g = R_{go}$.

Our estimates of the empirical intensity of global radiation were based on Castro (1987), which includes a listing of all the daily raw and corrected values from Costa Rican heliographs and actinographs since 1982. Castro corrected the raw data set by inspecting the record for obvious errors, both temporal (abrupt changes, missing data) and spatial (unusual changes over space).

Castro corrected the data further based on (1) historical records (date when the instrument was changed or recalibrated, and comments of the observers); (2) any discrepancies with the climatic context of the region, for example, when one value seems very different from other values in that region; (3) the correlation between sunshine hours and solar radiation for that location; and (4) independent measurements made by ICE in 1982 to evaluate the state of actinographs for several stations. Therefore, the raw solar radiation data can be classified into (a) data with no or acceptable errors (within 10–15% of the value from a calibrated instrument); (b) data with a systematic error greater than 15% with respect to the Black–Prescott relation [Equation (3)] or with respect to the data from other instrument; and (c) data whose quality could not be determined due to a lack of independent measurements.

Only two stations (numbers 84023 and 98022) had heliographic (solar) records for more than 10 years. For these stations Rojas's estimates of the constants $a$ and $b$ of the Black–Prescott equation are given in Table 9-4. Since there were no other stations with adequate records, the constants derived for station 84023 have been used for all stations above 500 m, and the ones from

TABLE 9-4 Regression Coefficients for the Black–Prescott Equation for the Two Most Reliable Stations

| Station | Elevation | Period | *a* | *b* | $R^2$ |
|---|---|---|---|---|---|
| 84023 | 840 | 70–80 | 0.303 | 0.438 | 0.992 |
| 98022 | 350 | 70–80 | 0.278 | 0.414 | 0.919 |

station 98022 for all stations below 500 m. There is some evidence that the use of these constants overestimates global radiation for South Pacific lowlands, but this remains unclear. Also, this approach underestimates global radiation for the Irazu Volcano (and probably for all the mountains above 2000 m), and overestimates it in the very cloudy valleys of the upper watershed of the Reventazon River (Turrialba and Orosi). $R_{go}$ and $N$ can be computed using the subroutine SUASTC in Penning de Vries *et al.* (1989). For 10° north latitude, these values used by Castro are given in Table 9-5.

We interpolated digital maps from the corrected values provided by Rojas by using Kriging. We also used a Gaussian (normal curve) function in generating the semivariogram. Step size and maximum search radius were 0.06° and 1.4°, respectively. Generally the semivariogram shows an important "nugget," which indicates that the difference in values between stations in close proximity is large and that there are still substantial errors in the data. Also the points in the semivariogram are considerably scattered, which again demonstrates the relatively poor quality of the national radiation data.

## E. Evapotranspiration

Potential evapotranspiration is important for predicting agricultural potential and water budgets. We derived this for each square kilometer using the Priestley–Taylor equation:

$$\mathrm{PET} = 1.26 \frac{\delta}{(\delta + \gamma)} (\mathrm{R_n} - \mathrm{G}) \tag{4}$$

TABLE 9-5 Monthly Values for Constants in the Black–Prescott Equation, as Used by Castro (1987)

| | Jan. | Feb. | Mar. | Apr. | May | June | July | Aug. | Sep. | Oct. | Nov. | Dec. |
|---|---|---|---|---|---|---|---|---|---|---|---|---|
| $R_{go}$ | 31.4 | 33.9 | 36.8 | 37.6 | 37.2 | 36.4 | 36.8 | 37.2 | 36.4 | 34.7 | 32.6 | 30.5 |
| $N$ | 11.5 | 11.7 | 12.0 | 12.2 | 12.4 | 12.5 | 12.4 | 12.3 | 12.0 | 11.8 | 11.6 | 11.5 |

where PET refers to the potential evapotranspiration, $\delta$ is the slope of the curve of the maximum vapor pressure and temperature, $\gamma$ is the psychrometric constant, $R_n$ is the daily net radiation, and $G$ is the flow of heat at ground level, which is approximately 5% of $R_n$ (Rojas, 1985). The term $\delta/(\delta + \gamma)$ is dimensionless and varies with temperature. For the range of temperatures 10–30°C, which includes most of Costa Rica, this term can be estimated as $0.430 + 0.012T_{di}$, where $T_{di}$ is the daily diurnal temperature.

There are other methods for calculating potential evapotranspiration (Penning de Vries *et al.*, 1989). They involve weather variables, however, such as vapor pressure and wind, which have not been measured systematically over Costa Rica. One implication of this is that for dry periods with strong winds the Priestley–Taylor equation may strongly underestimate the value of evapotranspiration. The resulting digital maps were prepared with standard algebra [Equations (4) and (5)] based on global radiation and temperature maps.

## IV. RESULTS

We were able to produce reasonable-appearing monthly maps for all of Costa Rica for rainfall, temperature, hours of sunlight, solar intensity, and evapotranspiration. All of these maps are presented as 400-by-400-element grid files on the enclosed CD, and most may be viewed easily in the "data" directory. A discussion of the characteristics and limitations of these data follows. Unfortunately one principal conclusion is that the weather stations are not sufficiently dense (or perhaps the data not sufficiently comprehensive) to provide precise results when these values are extrapolated to the entire country. In a nation as topographically complex as Costa Rica, truly accurate climatology maps may continue to elude us without a greater investment in meteorological stations.

### A. Rainfall

The results of the stepwise regression analysis using the yearly and monthly rainfall averages as the dependent variable and latitude, longitude, elevation, and their interactions and squares as independent variables are given in Table 9-6. The values of the coefficient of determination ($R^2$) for the unweighted values are low, indicating that the independent variables do not explain the variability in rainfall sufficiently. Two other sets of regression runs made using the averaged values but weighted according to the proportion of the number of years on record and the inverse of the variance gave even lower $R^2$ values in most cases. Therefore, Kriging was used to interpolate the values.

TABLE 9-6 $R^2$ Values of the Stepwise Regression Runs for Yearly and Monthly Rainfall Averages Using Latitude, Longitude, Elevation, and Their Combinations and Squares as Independent Variables

| Period | Unweighted | Weighted according to *N* | 1/var |
|---|---|---|---|
| Year | 23.63 | 16.78 | 22.37 |
| January | 39.74 | 38.40 | 39.04 |
| February | 37.94 | 37.16 | 9.23 |
| March | 34.14 | 34.22 | 40.23 |
| April | 44.85 | 43.11 | 51.57 |
| May | 21.64 | 20.76 | 19.05 |
| June | 8.54 | 8.25 | 5.69 |
| July | 31.12 | 27.49 | 27.34 |
| August | 11.73 | 10.54 | 9.21 |
| September | 24.08 | 22.53 | 24.68 |
| October | 29.12 | 24.65 | 23.22 |
| November | 36.94 | 34.17 | 31.03 |
| December | 43.49 | 41.73 | 41.39 |

*Note.* *N*, proportion of the number of years on record. 1/var, reciprocal of the variance.

In general, the scatter graph of the covariance of pairs of observations against distance (i.e., the semivariogram) increases up to a certain distance and becomes more or less stable, and then increases again. The initial increasing trend is as according to expectations; i.e., the farther apart the measurements, the higher the covariance (i.e., lower correlation) between pairs of observations. The more or less stable trend is the sill, where the semivariance becomes constant (i.e., rainfall values become independent when the distance is large). It starts at a distance of about 0.13° (1° is about 108 km) and ends at a distance of about 2°, which is about half the distance separating the furthest points in the set of observations (Figure 9-6). The final trend (not shown) is the increasing covariance when the pairs of observations are very far apart.

We investigated these spatial correlation trends using a linear isotropic model with the "nugget" and sill to fit the semivariogram. The formula for this model is

$$\begin{aligned} \gamma(\mathrm{h}) &= C_O + [\mathrm{h}(C/A_0)] \qquad \text{for h} < \text{Ao} \\ \gamma(\mathrm{h}) &= C_O + C \qquad \text{for h} > \text{Ao} \end{aligned} \tag{6}$$

where $h$ is the lag, $C_0$ (the nugget) is the variance when the lag h is zero, $C$ is the structural variance ($C > C_0$), and $A_0$ is the range or the point where the sill starts. A step size of 0.0167° (1.8 km) was used in the analysis, and the maximum lag was set at 0.30° (32.4 km). The values of the parameters and

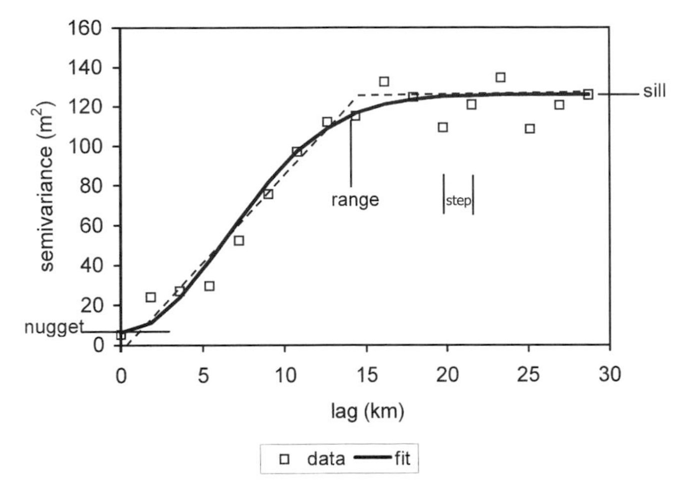


FIGURE 9-6 Semivariogram for January rainfall (step size = 1.8 km). The "nugget" has been exaggerated for illustration. The fit corresponds to the Gaussian function used to produce interpolated surfaces. Dotted line fit with $\gamma(h) = Co + C\,[1 - \exp(h(C/A_0))]$ sill model of equation 6.

the $R^2$ values for the semivariogram and the cross-validation analysis are given in Table 9-7. Fortunately the $R^2$ values of the semivariograms are high and those of the cross validations are acceptable, so we could use this method (Figure 9-9).

**TABLE 9-7 Parameter Values and Results of Kriging of Rainfall**

| | | | | $R^2$ | |
|---|---|---|---|---|---|
| Period | $C_0$ | $C_0 + C$ | $A_0$ | Kriging | Validation |
| Year | 10 | 1.214E4 | 0.14 | 95.4 | 77.0 |
| January | 20 | 1.211E4 | 0.13 | 92.2 | 73.8 |
| February | 10 | 6.264E3 | 0.12 | 95.1 | 73.6 |
| March | 10 | 5.631E3 | 0.13 | 93.9 | 65.6 |
| April | 10 | 6.392E3 | 0.12 | 90.1 | 73.3 |
| May | 10 | 1.278E4 | 0.14 | 88.8 | 67.5 |
| June | 10 | 1.578E4 | 0.14 | 92.4 | 63.2 |
| July | 10 | 2.035E4 | 0.15 | 97.4 | 71.5 |
| August | 10 | 1.931E4 | 0.15 | 95.2 | 66.7 |
| September | 10 | 1.460E4 | 0.15 | 89.8 | 53.4 |
| October | 10 | 1.897E4 | 0.13 | 87.7 | 63.3 |
| November | 10 | 1.802E4 | 0.12 | 96.1 | 72.9 |
| December | 10 | 1.928E4 | 0.13 | 91.1 | 74.0 |

As a rule of thumb, $C$ gives an indication of the variability of the regional rainfall, $C_0$ is an indication of the intrinsic error in the data, and $A_0$ is the distance over which there is spatial correlation (i.e., rainfall at a given location is related in some way to rainfall in another location). Theoretically, when we estimate a value at a given point, we should not have to consider stations that fall outside a radius of $A_0$, but in practice this would lead to holes in the interpolated surface. From Figure 9-7 one can see that rainfall is less spatially correlated during the dry season than the rainy season, and that rainfall is more variable during the rainy season. We also found that the nugget ($C_0$) is generally very small, indicating that the original data are reliable.

We considered using co-Kriging in order to improve the relationships in those months with an $R^2$ value less than 0.70. It was thought that elevation might be used as a covariable of rainfall. However, the correlation coefficients of the rainfall averages and elevation (Table 9-8) are not high and mostly not very different from zero. Therefore, co-Kriging was not continued. The correlation results do indicate that in Costa Rica many other factors other than elevation determine rainfall patterns, including wind, low pressure, sea surface temperature, and episodic cyclones. But these variables have not been studied in any systematic, quantitative way, so cannot be used to drive better rainfall analyses.

We also examined the results of both the regression and the Kriging by grouping observations via statistical clustering to see if we could improve our estimates by dealing with regional rather than countrywide patterns. Five sets

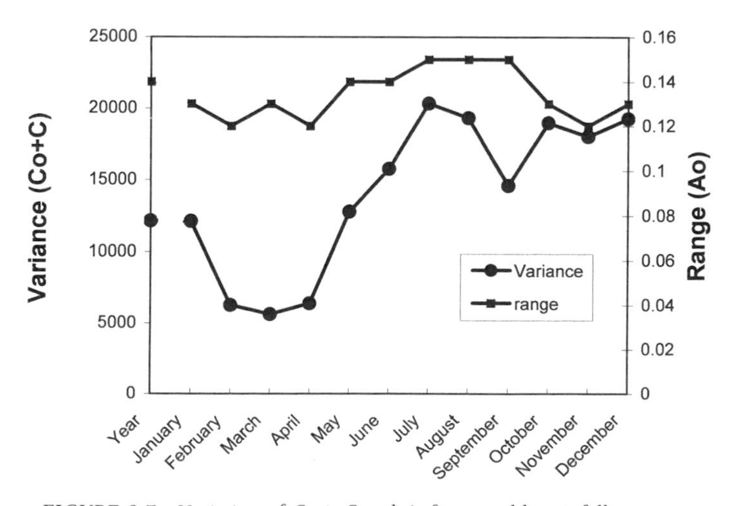


FIGURE 9-7 Variation of $C_0 + C$ and $A_0$ for monthly rainfall averages.

TABLE 9-8 Correlation Coefficients of Rainfall and Elevation and Their Tests of Significance

| Period | Correlation coefficient | Prob > $\|R\|$ under $H_0$: $\rho = 0$ | Significance |
|---|---|---|---|
| Year | 0.061860 | 3242 | ns[a] |
| January | 0.022120 | 7262 | ns |
| February | 0.018620 | 7682 | ns |
| March | 0.030320 | 6319 | ns |
| April | 0.029710 | 6381 | ns |
| May | 0.17923 | 0.0042 | ** |
| June | 0.15263 | 0.0149 | * |
| July | −0.06640 | 0.2918 | ns |
| August | 0.05024 | 0.4253 | ns |
| September | 0.12788 | 0.0425 | * |
| October | 0.11784 | 0.0613 | ns |
| November | 0.04659 | 0.4616 | ns |
| December | 0.03624 | 0.5669 | ns |

[a] ns, not significant.
* significant at 95 percent level.
** significant at 99 percent level.

of cluster analysis were performed to group the observations. The following groups of independent variables were used:

1. Latitude, longitude, and rainfall pattern
2. Latitude, longitude, rainfall pattern, and their variance
3. Latitude, longitude, rainfall pattern, and their coefficient of variation
4. Latitude, longitude, and elevation
5. Latitude, longitude, and rainfall averages

The rainfall data used were calculated by dividing the average monthly by the average yearly rainfall value. It was interesting to see how the clusters developed when we used the different independent variables, especially when we examined monthly means. Then the effects of the seasonal movement of low pressure, wind, and topographic factors can be observed through the migration of the clusters from one zone to another (Figure 9-8). The first three groups of variables were not studied in much detail because the results for a preliminary small number of clusters did not divide Costa Rica as well as the last two. Five clusters for the fourth group were considered, and five to seven clusters for the fifth group. The divisions set by the fourth and fifth group are shown in Figure 9-4. This result is the same if distance to the sea is included in the list of variables for clustering. The results of the stepwise regression procedure, using the same list

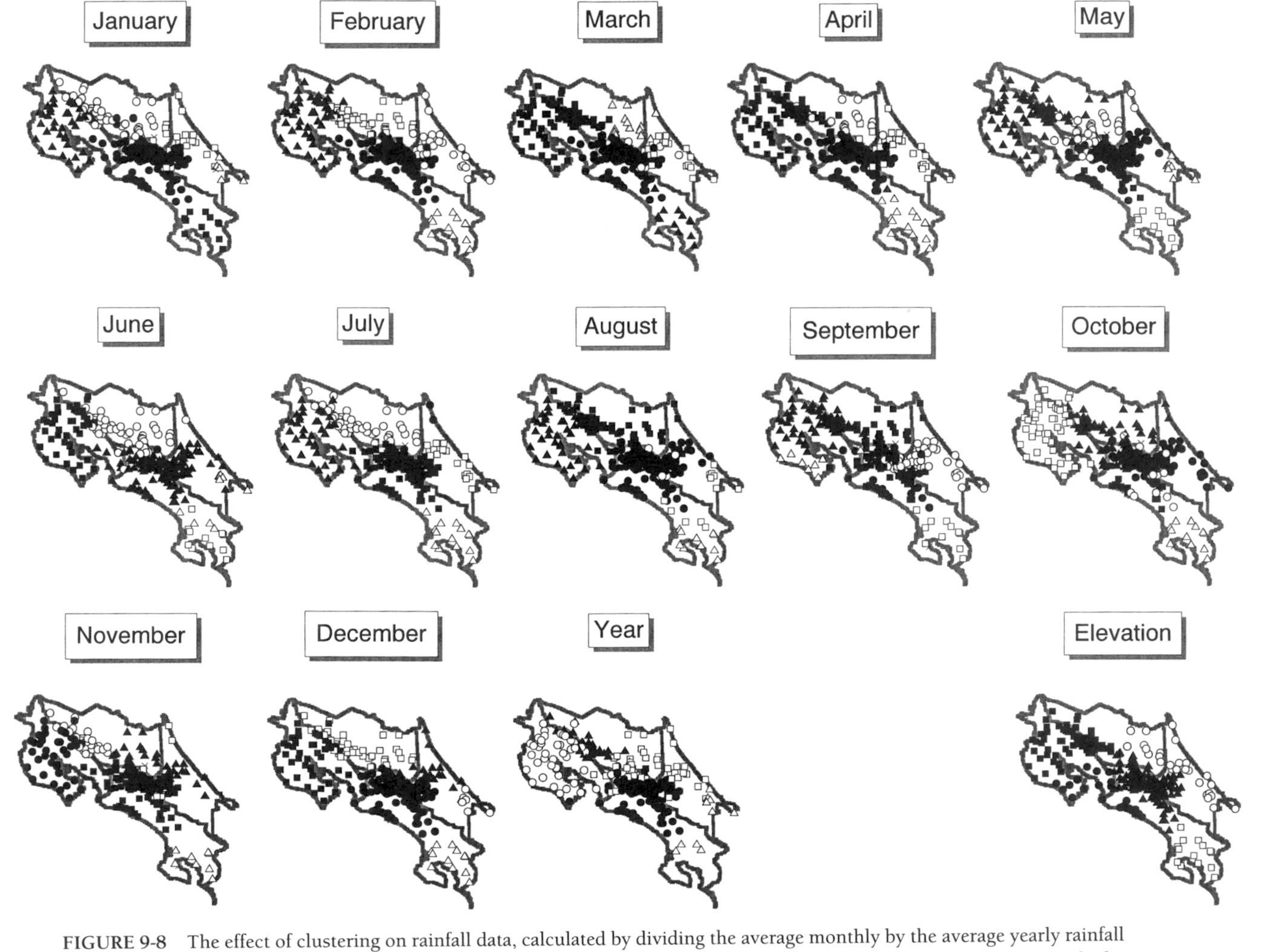


FIGURE 9-8 The effect of clustering on rainfall data, calculated by dividing the average monthly by the average yearly rainfall value. The effects of the seasonal movement of low pressure, wind, and topographic factors can be observed through the

of independent variables as in the runs where all the observations were considered, are given in Tables 9-9 and 9-10.

In some cases, none of the independent variables met the requirement of 15% significance, so were not included in the model. In Table 9-9, it can be observed that, in general, the $R^2$ values are high for the North Pacific, upper South Pacific,

**TABLE 9-9** $R^2$ **Values for the Different Stepwise Regression Runs Made with the Clusters Formed by the Variables Latitude, Longitude, and Elevation between Rainfall and Latitude, Longitude, and Elevation**

| Period | Cluster | $R^2$ | Period | Cluster | $R^2$ |
|---|---|---|---|---|---|
| Year | 1 | 88.47 | | | |
| | 2 | 83.11 | | | |
| | 3 | 13.51 | | | |
| | 4 | 76.26 | | | |
| | 5 | 20.15 | | | |
| January | 1 | 82.27 | July | 1 | 88.97 |
| | 2 | 85.08 | | 2 | 81.32 |
| | 3 | 28.65 | | 3 | 15.25 |
| | 4 | — | | 4 | 73.25 |
| | 5 | 26.99 | | 5 | 32.53 |
| February | 1 | 73.82 | August | 1 | 87.02 |
| | 2 | 83.40 | | 2 | 79.98 |
| | 3 | 22.18 | | 3 | — |
| | 4 | — | | 4 | 63.79 |
| | 5 | 18.32 | | 5 | 19.13 |
| March | 1 | 86.80 | September | 1 | 74.64 |
| | 2 | 84.88 | | 2 | 42.62 |
| | 3 | 18.26 | | 3 | — |
| | 4 | 34.33 | | 4 | 80.42 |
| | 5 | — | | 5 | — |
| April | 1 | 90.27 | October | 1 | 73.61 |
| | 2 | 73.98 | | 2 | 70.03 |
| | 3 | 14.47 | | 3 | — |
| | 4 | — | | 4 | 67.34 |
| | 5 | — | | 5 | — |
| May | 1 | 82.38 | November | 1 | 91.31 |
| | 2 | 34.02 | | 2 | 87.16 |
| | 3 | 5.52 | | 3 | 21.28 |
| | 4 | 23.67 | | 4 | 72.48 |
| | 5 | — | | 5 | 35.89 |
| June | 1 | 76.56 | December | 1 | 83.36 |
| | 2 | 67.33 | | 2 | 83.50 |
| | 3 | — | | 3 | 32.60 |
| | 4 | 60.23 | | 4 | 70.41 |
| | 5 | — | | 5 | 33.91 |

TABLE 9-10 $R^2$ Values for the Different Stepwise Regression Runs Made with the Clusters Formed by the Variables Latitude, Longitude, and Rainfall Averages between Rainfall and Latitude, Longitude, and Elevation

| Period | Cluster | $R^2$ | Period | Cluster | $R^2$ |
|---|---|---|---|---|---|
| Year | 1 | 42.55 | | | |
| | 2 | — | | | |
| | 3 | 77.27 | | | |
| | 4 | 7.61 | | | |
| | 5 | 28.85 | | | |
| | 6 | 66.54 | | | |
| | 7 | — | | | |
| January | 1 | 46.89 | July | 1 | 75.86 |
| | 2 | 71.87 | | 2 | 28.72 |
| | 3 | 87.75 | | 3 | 75.45 |
| | 4 | 57.30 | | 4 | 7.75 |
| | 5 | 60.13 | | 5 | 41.01 |
| | 6 | 65.73 | | 6 | 49.40 |
| February | 1 | 24.93 | August | 1 | 37.27 |
| | 2 | 26.43 | | 2 | 5.72 |
| | 3 | 71.76 | | 3 | 51.44 |
| | 4 | 68.93 | | 4 | — |
| | 5 | 9.11 | | 5 | 22.31 |
| | 6 | — | | 6 | — |
| March | 1 | 23.36 | September | 1 | 39.91 |
| | 2 | 61.70 | | 2 | 16.23 |
| | 3 | — | | 3 | 61.72 |
| | 4 | — | | 4 | 46.13 |
| | 5 | — | | 5 | 42.45 |
| | 6 | 36.26 | | 6 | 33.02 |
| April | 1 | 29.37 | October | 1 | 20.60 |
| | 2 | 53.36 | | 2 | 36.97 |
| | 3 | 86.19 | | 3 | 8.70 |
| | 4 | 42.43 | | 4 | 16.10 |
| | 5 | 26.56 | | 5 | 48.23 |
| | 6 | — | | 6 | 86.32 |
| May | 1 | 38.26 | November | 1 | 44.13 |
| | 2 | 32.57 | | 2 | 36.20 |
| | 3 | 20.98 | | 3 | — |
| | 4 | — | | 4 | 71.44 |
| | 5 | — | | 5 | — |
| | 6 | 41.15 | | 6 | — |
| | 7 | 54.40 | | | |
| June | 1 | 40.16 | December | 1 | 44.98 |
| | 2 | 17.40 | | 2 | 72.33 |
| | 3 | 25.72 | | 3 | 41.29 |
| | 4 | — | | 4 | — |
| | 5 | — | | 5 | 14.37 |
| | 6 | — | | 6 | 67.02 |

North Atlantic, and Atlantic regions. For these cases the results of the regression analysis can be used to estimate or predict the rainfall values for areas not covered by the observations. The results of the stepwise regression analysis between rainfall and latitude/longitude/elevation for clusters formed from latitude/longitude/ rainfall are more difficult to generalize; i.e., sometimes the $R^2$ for the different regions are high, and sometimes they are low (Table 9-10).

For those individual cases where $R^2$ values are high, it is possible to use the regression equations with the parameters of Table 9-10 for generating maps. But since the results for the different clusters were uneven, Kriging was appropriate. The results of the semivariogram and cross-validation analysis on clusters produced from the fifth group of variables are given in Table 9-11, along with the $R^2$ of the validation when all the observations were taken together (from Table 9-9). In some cases, especially when the cluster covers the Central Valley, the value of the $R^2$ improved compared with the results for the entire country. But the results for the other clusters, especially the southern part of the side near the Pacific Ocean (Cluster 5), were not encouraging. This was mostly because of the difficulty in obtaining a good semivariogram model when the stations are quite far apart. Similar results were obtained (not shown here) when the clusters were formed using the latitude, longitude, and rainfall averages. We tried to improve the statistics and resulting semivariogram estimates by aggregating clusters, but ended with the same conclusions. One solution to this problem was to take a combination of the results of the regression and the Kriging. For example, for group four, the best results were to use the regression results for the North Pacific, upper South Pacific, North Atlantic, and Atlantic regions, and the Kriging results for the Central Valley. But the lower side of the South Pacific remained a problem. Because of this, we decided to use the Kriging results for all the data of Costa Rica to prepare the maps.

We used a latitude–longitude grid with a 0.02 pixel size for interpolations, and then projected these results to Lambert Conformal Conic Costa Rica Norte with a 1-km pixel size and bilinear interpolation. The maps of the interpolated values are given in Figures 9-9a and 9-10a. The maximum search radius was set to 0.5° to include all stations for interpolation. If we used instead the maximum radius suggested by the saturation of the semivariogram ($A_0$) the interpolated area would not completely cover the whole of Costa Rica: some stations or groups of stations would be completely isolated, such as in the North Atlantic and in the south. The data suggest that in order to make a complete map of Costa Rica with statistical rigor, all rainfall stations should be within no more than 13 km ($A_0 = 0.12°$) of each other.

This analysis of rainfall could be improved with a denser network of stations (all 817 stations in Costa Rica and those from neighboring countries), and if

TABLE 9-11 Kriging Parameters and Results for the Clusters Obtained from the Variables Latitude, Longitude, and Elevation

| Period | Cluster | Step size | Max lag distance | Model type[a] | $R^2$ model | Radius $R^2$ value | $R^2$ | Overall $R^2$ |
|---|---|---|---|---|---|---|---|---|
| Year | 1 | 0.0249 | 0.36 | l/s | 0.822 | 0.22 | 0.541 | 0.770 |
| | 2 | 0.02499 | 0.33 | l/s | 0.830 | 0.16 | 0.673 | |
| | 3 | 0.01666 | 0.30 | l/s | 0.949 | 0.14 | 0.833[b] | |
| | 4 | 0.03332 | 0.66 | g | 0.835 | 0.66 | 0.675 | |
| | 5 | 0.06664 | 0.53 | g | 0.909 | 0.40 | 0.405 | |
| January | 1 | 0.03332 | 0.32 | g | 0.716 | 0.42 | 0.878[b] | 0.738 |
| | 2 | 0.02499 | 0.27 | s | 0.895 | 0.18 | 0.715 | |
| | 3 | 0.01666 | 0.28 | l/s | 0.949 | 0.13 | 0.710 | |
| | 4 | 0.05831 | 0.55 | g | 0.825 | 0.55 | 0.513 | |
| | 5 | 0.06664 | 0.53 | l/s | 0.840 | 0.28 | 0.374 | |
| February | 1 | 0.02499 | 0.30 | g | 0.139 | 0.28 | 0.562 | 0.736 |
| | 2 | 0.02499 | 0.41 | l/s | 0.781 | 0.16 | 0.711 | |
| | 3 | 0.01666 | 0.28 | l/s | 0.962 | 0.12 | 0.689 | |
| | 4 | 0.05831 | 0.81 | g | 0.944 | 0.81 | 0.627 | |
| | 5 | 0.07497 | 0.50 | l/s | 0.687 | 0.26 | 0.597 | |
| March | 1 | 0.02499 | 0.30 | s | 0.715 | 0.21 | 0.795[b] | 0.656 |
| | 2 | 0.02499 | 0.42 | l/s | 0.836 | 0.16 | 0.752[b] | |
| | 3 | 0.01666 | 0.26 | l/s | 0.949 | 0.13 | 0.571 | |
| | 4 | 0.05831 | 0.58 | l/s | 0.718 | 0.34 | 0.078 | |
| | 5 | 0.06664 | 0.53 | l/s | 0.813 | 0.30 | 0.478 | |
| April | 1 | 0.02499 | 0.25 | l/s | 0.541 | 0.11 | 0.549[b] | 0.733 |
| | 2 | 0.02499 | 0.23 | l/s | 0.884 | 0.16 | 0.619 | |
| | 3 | 0.01666 | 0.19 | l/s | 0.975 | 0.14 | 0.676 | |
| | 4 | 0.05831 | 1.10 | l/s | 0.573 | 1.01 | 0.217 | |
| | 5 | 0.06664 | 0.39 | l/s | 0.926 | 0.30 | 0.067 | |
| May | 1 | 0.02499 | 0.30 | l/s | 0.755 | 0.29 | 0.522 | 0.675 |
| | 2 | 0.02499 | 0.24 | l/s | 0.545 | 0.16 | 0.217 | |
| | 3 | 0.01666 | 0.19 | l/s | 0.987 | 0.16 | 0.681[b] | |
| | 4 | 0.05831 | 0.58 | l/s | 0.859 | 0.38 | 0.472 | |
| | 5 | 0.07497 | 0.52 | g | 0.825 | 0.35 | 0.062 | |
| June | 1 | 0.02499 | 0.30 | l/s | 0.608 | 0.18 | 0.358 | 0.632 |
| | 2 | 0.02499 | 0.24 | l/s | 0.911 | 0.16 | 0.532 | |

| | | | | | | | | |
|---|---|---|---|---|---|---|---|---|
| | 3 | 0.01666 | 0.19 | l/s | 0.973 | 0.16 | 0.705[b] | |
| | 4 | 0.05831 | 0.81 | l/s | 0.747 | 0.67 | 0.606 | |
| | 5 | 0.04998 | 0.50 | l/s | 0.826 | 0.24 | 0.303 | |
| July | 1 | 0.02499 | 0.41 | l/s | 0.696 | 0.22 | 0.559 | 0.715 |
| | 2 | 0.02499 | 0.50 | s | 0.829 | 0.38 | 0.738[b] | |
| | 3 | 0.01666 | 0.24 | l/s | 0.974 | 0.15 | 0.717[b] | |
| | 4 | 0.04165 | 0.62 | g | 0.725 | 0.62 | 0.569 | |
| | 5 | 0.06664 | 0.53 | s | 0.989 | 0.54 | 0.479 | |
| August | 1 | 0.03332 | 0.37 | l/s | 0.609 | 0.22 | 0.340 | 0.667 |
| | 2 | 0.02499 | 0.34 | e | 0.894 | 0.34 | 0.698[b] | |
| | 3 | 0.01666 | 0.30 | l/s | 0.945 | 0.14 | 0.693[b] | |
| | 4 | 0.05831 | 0.81 | l/s | 0.778 | 0.73 | 0.659 | |
| | 5 | 0.07497 | 0.52 | l/s | 0.961 | 0.17 | 0.487 | |
| September | 1 | 0.02499 | 0.30 | l/s | 0.692 | 0.18 | 0.117 | 0.534 |
| | 2 | 0.02499 | 0.38 | g | 0.757 | 0.26 | 0.401 | |
| | 3 | 0.01666 | 0.30 | l/s | 0.903 | 0.14 | 0.562[b] | |
| | 4 | 0.05831 | 1.63 | l/s | 0.891 | 1.27 | 0.717[b] | |
| | 5 | 0.04998 | 0.49 | l/s | 0.710 | 0.42 | 0.005 | |
| October | 1 | 0.02499 | 0.36 | g | 0.812 | 0.29 | 0.436 | 0.633 |
| | 2 | 0.02499 | 0.38 | e | 0.862 | 0.38 | 0.520 | |
| | 3 | 0.01666 | 0.19 | l/s | 0.975 | 0.16 | 0.652[b] | |
| | 4 | 0.05831 | 0.81 | l/s | 0.920 | 0.73 | 0.767[b] | |
| | 5 | 0.04998 | 0.57 | l/s | 0.323 | 0.13 | 0.530 | |
| November | 1 | 0.02499 | 0.38 | l/s | 0.804 | 0.36 | 0.709 | 0.729 |
| | 2 | 0.02499 | 0.33 | e | 0.861 | 0.33 | 0.718 | |
| | 3 | 0.01666 | 0.30 | l/s | 0.943 | 0.12 | 0.757[b] | |
| | 4 | 0.05831 | 0.81 | g | 0.681 | 0.81 | 0.631 | |
| | 5 | 0.06664 | 0.46 | g | 0.895 | 0.31 | 0.626 | |
| December | 1 | 0.02499 | 0.33 | g | 0.519 | 0.29 | 0.825[b] | 0.740 |
| | 2 | 0.02499 | 0.41 | e | 0.834 | 0.41 | 0.738 | |
| | 3 | 0.01666 | 0.24 | l/s | 0.938 | 0.12 | 0.679 | |
| | 4 | 0.05831 | 0.70 | g | 0.796 | 0.63 | 0.508 | |
| | 5 | 0.06664 | 0.50 | l/s | 0.831 | 0.30 | 0.696 | |

[a] Model types are l/s, linear with sill; s, spherical; e, exponential; and g, Gaussian.

[b] $R^2$ of the validation for the cluster is greater than the overall $R^2$.

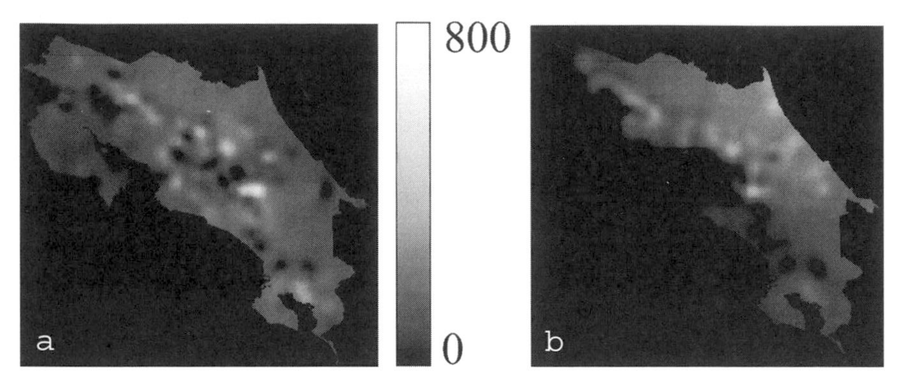


FIGURE 9-9 Map of interpolated mean rainfall for Costa Rica for June (a) and December (b). See "view" on CD for color rendition.

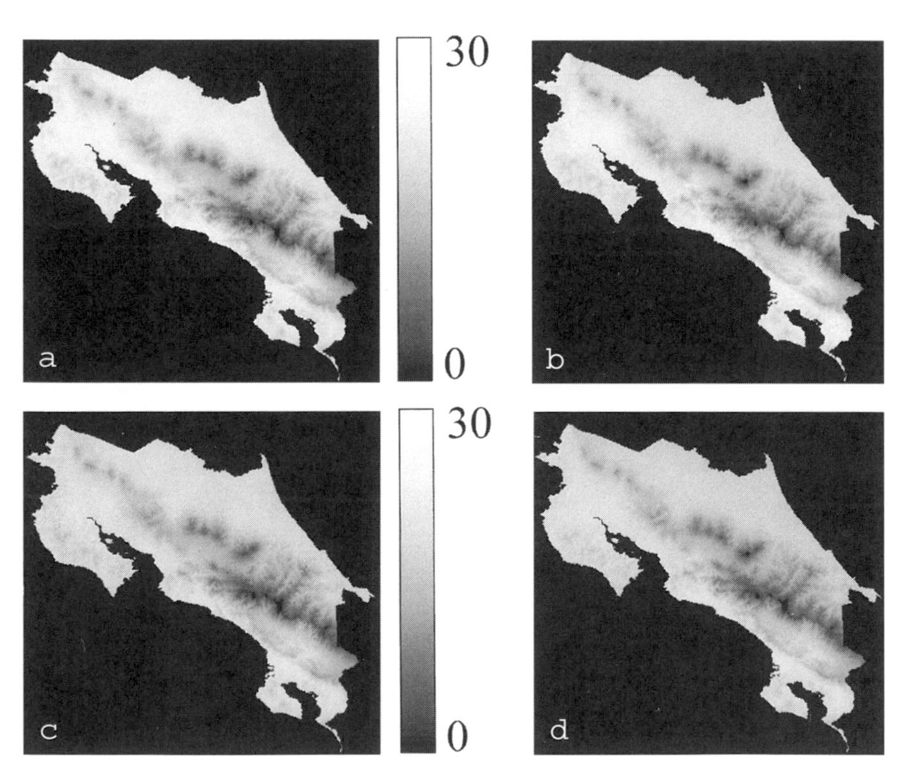


FIGURE 9-10 Average daily temperatures: (a) June maximum, (b) December maximum, (c) June minimum, (d) December minimum. See "view" on CD for color rendition.

other weather variables such as temperature and wind were found to be correlated with rainfall. Use of thin plate splines or co-Kriging, which allows weighted regression submodels, or the PRISM model, could be tried. A significant correlation must be found with some independent variables having a higher spatial resolution in order to apply such models. Results from Tables 9-9, 9-10, and 9-11 can be examined for this purpose.

## B. Temperature

Maps of minimum, maximum, daily, and diurnal temperatures are shown in Figure 9-10. A problem of the regression approach we used is that the boundaries between IMN zones show small discontinuities. These are unavoidable if we do not consider a constant lapse rate throughout Costa Rica. Particular care should be observed when using the results of the North Pacific region for the months April to December, for the $R^2$ values were low. Additional studies could be done to try to improve the regressions for this region, and we are currently working on a different approach, in many aspects similar to PRISM. Regressions can be performed for each station with its neighbors with respect to elevation (which is known to correlate well with temperature), and maps of regression coefficients can be interpolated with Kriging. This allows one to identify local variations of the lapse rate and possible unknown factors that contribute to local climate.

Another approach would be co-Kriging or thin plate splines, which allow the use of an obvious covariable, elevation. Although extremely computer intensive (to have the 1-km grid resolution we would need 50,000 data points of the independent variable), it could improve accuracy.

## C. Solar Radiation

Our estimates of average incoming solar radiation range from 10 to 25 MJ $m^{-2}$ $day^{-1}$, depending on the site and time of year. The highest values are found in the North Pacific region and on some mountain tops. Lower values are located in the lowlands, on the windward mountain slopes, and in other regions with zones of persistent cloudiness.

Our predicted maps of the average hours of sunshine per day, for each month, are given in Figure 9-11. Our estimates of monthly averages of net solar radiation, $R_n$, from global radiation, $R_g$ (using $R_n/R_g = 0.60$ for Costa Rica), were calculated from the Black–Prescott equation and from the values of $a$, $b$, and $N$ in Tables 9-4 and 9-5. Our results could be improved if regressions for these parameters for each region could be made, but this requires more

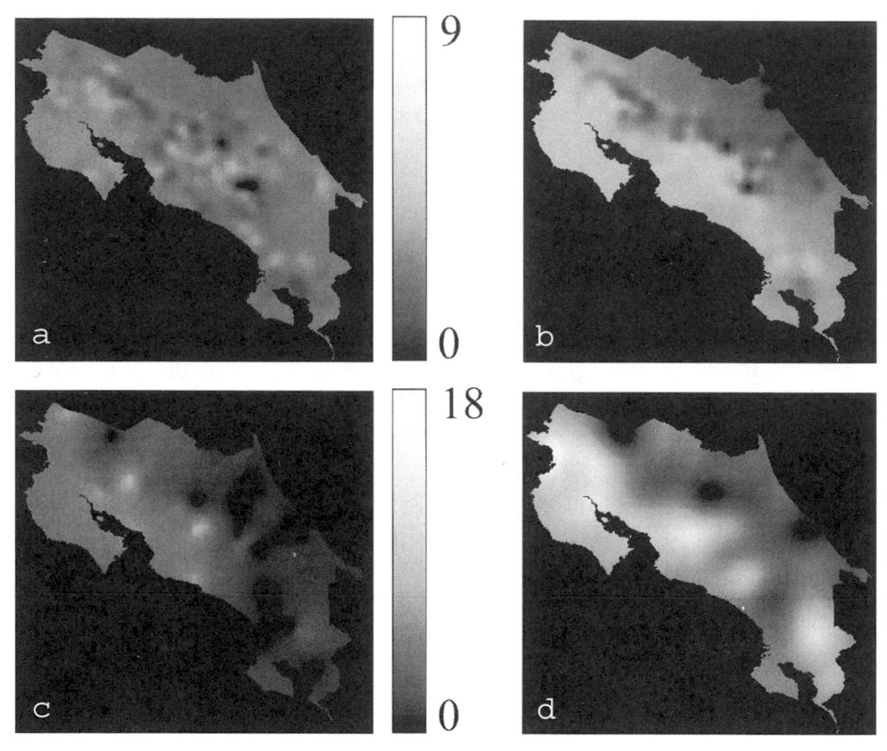


**FIGURE 9-11** June (a) and December (b) maps of our predicted average hours of sunshine per day, for each month. June (c) and December (d) maps of estimated monthly averages of net solar radiation, $R_n$.

data than are available presently. Since the hours of sunshine each day were estimated from rainfall predictions, all limitations of the rainfall results are reflected in the estimated hours of sunshine.

## D. Evapotranspiration

A map of potential evapotranspiration values is given in Figure 9-12. The accuracy of the evapotranspiration results is affected by the values used for the slope of the curve of the maximum vapor pressure and temperature, the daily net radiation, the flow of heat at soil level, and temperature. Perhaps more important is the absence of wind corrections, the use of which might give quite different results. Jimenez (personal communication) has computed evapotranspiration from CATIE (station 73010) data using the Penman, Tuck, Garcia-Lopez, Thornthwaite, Hargreaves, and Linacre methods. He found that

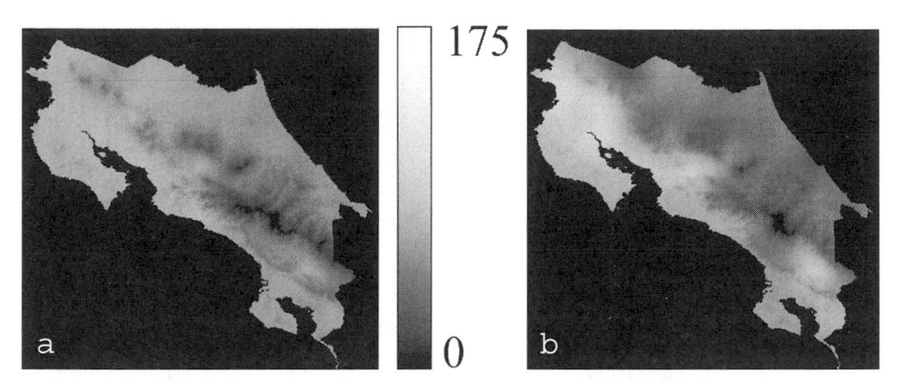


FIGURE 9-12 Map of potential evapotranspiration for Costa Rica. (a) June, (b) December.

estimated values of evapotranspiration differed by as much as 30%, which is perhaps not too bad considering the large possibility of error.

## V. DISCUSSION—VALIDATING THE ACCURACY OF THE GENERATED MAPS

We chose to check the validity of our spatial predictions by examining the output of the model relative to one particularly well-studied weather station. The Centro Agronómico Tropical de Investigación y Enseñanza (CATIE) has maintained a fully equipped and professionally staffed meteorological station with one of the longest time sequences of data for rainfall (55 years), temperature (38 years), solar radiation (32 years), sunshine (35 years), relative humidity (38 years), potential evapotranspiration (38 years), and wind (12 years). These data are of variable quality. For example, we recently found, using a global positioning system, that the reported official coordinates of the station are off by 2.4 km! Moreover, Castro (1987) had to discard solar radiation data from this station for having an important systematic error. This is, however, a good case with which to illustrate that Costa Rican climate data are far from perfect, being sometimes incorrectly calibrated, or having missing values or erroneous coordinates or elevations. And this is for one of the best meteorological networks in the developing world! In addition, the length of the period from which the various averages were computed may affect the resulting value. Thus although we did not have a "perfect" station against which we could check our maps, we decided to use CATIE's station mostly as a case-study validation data set.

There are many factors that could contribute to differences between the values we calculated at CATIE's coordinates from our various regression and interpolation techniques and CATIE's data. These include errors in the data used for interpolation (in estimates of coordinates or elevation, or bad values for the parameters), incorrect or incomplete interpolation algorithms (i.e., wrong semivariogram fitting), and the coarse pixel size. The error is lower in areas of uniform relief since the spatial gradients were smaller. But Turrialba is located on the slopes of Turrialba Volcano in the corridor that links the

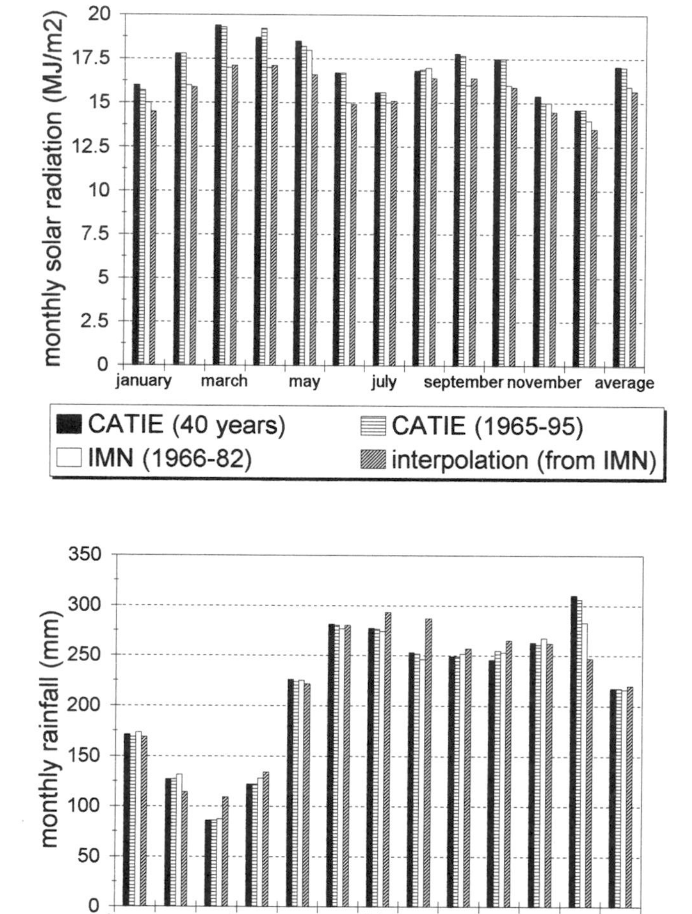


FIGURE 9-13 Comparison of our interpolated results (rainfall and solar radiation) with data from CATIE's meteorological station. It is obvious that there is a fairly close agreement over the year between model predictions and measured values.

Atlantic zone with the Central Valley, which means it is in an area of strong meteorological gradients. Thus we might expect that the agreement between model and data might not be too strong.

# VI. CONCLUSIONS

Despite all of the potential errors and problems we were able to predict most meteorological information to within 10%, giving us a certain amount of confidence in our predictions in general. Probably some of our success comes simply from the overall averaging effect of integrating over large amounts of space and time. For example, from Figure 9-13 one can see that the data coming from our interpolated maps exhibit smoother behavior than individual data, which denotes the smoothing effect caused by deriving our values over wider areas. Since this was an unfunded study we believe that a focused well-funded study could generate spatial weather maps of very high quality, especially if a greater effort was made to increase the density of weather stations or, more importantly, the number of parameters measured at existing stations. All programs (except those protected by commercial license), results, and data used are included on the compact disc, and other investigators are encouraged to improve upon our analysis. In the meantime we believe that we have created a very useful, digitally accessible, and reasonably accurate series of climate maps at the national scale.

# ACKNOWLEDGMENTS

We thank Rosario Alfaro from IMN for providing us with IMN data and contact persons for obtaining data from other institutions. Special thanks to Francisco Jimenez from CATIE for his help in many aspects of this work. We acknowledge the support of Lic. Sadi Laporte (ICE), Ing. German Matamoros (SENARA), Sr. Freddy Vargas, and Ir. Donatus Jansen (Atlantic Zone Project of CATIE).

# REFERENCES

Bogaert, P., P. Mahau, and F. Beckers. 1995. *The Spatial Interpolation of Agroclimatic Data: Cokriging Software and Source Code, User's Manual*, Version 1.0b. Agrometeorology Series Working Paper no. 12, Environment Information Management Service, FAO, Rome, Italy.

Burrough, P. A. 1986. *Principles of Geographical Information Systems for Land Resources Assessment*. Oxford, New York.

Castro, V. 1987. *Radiacion solar global en Costa Rica*. Nota de investigacion numero 6, Instituto Meteorologico Nacional and Ministerio de Agricultura y Ganaderia, San Jose, Costa Rica. 31 pp.

Daly, C., R. P. Nielson, and D. L. Phillips. 1994. A statistical-topographic model for mapping climatological precipitation over mountainous terrain. *J. Appl. Meteor.* **33**, 140–158.

Delfiner, P., and J. P. Delhomme. 1975. Optimum interpolation by Kriging. In J. C. Davis and M. J. McCullagh (Eds.), *Display and Analysis of Spatial Data,* pp. 96–114. Wiley, New York.

Dubrule, O. 1983. Cross validation of Kriging in a unique neighbourhood. *Journal of Mathematical Geology* 15:687–699.

Everham, E. M., III, K. B. Wooster, and C. A. S. Hall. 1991. Forest landscape climate modeling. In *Proceedings of the 1991 Symposium on Systems Analysis in Forest Resources.* GTR SE-74, USDA Forest Service, Washington, DC.

Hasteonath, S. L. 1976. Variations in low latitude circulation and extreme climatic events in the tropical America's. *J. Atmos. Sci.* **33**, 202–215.

Hutchinson, M. F. 1991. The application of thin plate smoothing splines to continent-wide data assimilation. In J. D. Jasper (Ed.), *Data Assimilation Systems,* pp. 104–113. BMRC Research Report no. 27, Bureau of Meteorology, Melbourne.

IMN (Instituto Meteorológico Nacional). 1986. *Atlas Climatológico de Costa Rica.* IMN, San Jose, Costa Rica.

IMN. 1988. *Droughts and El Nino.* IMN, San Jose, Costa Rica.

Jones, P. G. 1991. *A Climate Database for South and Central America. Machine Readable Data Version 3.70.* CIAT, Cali, Colombia.

Jones, P. J., Galwey, N., Beebe, S. E., and Tohme, J. 1997. The use of geographical information systems in biodiversity exploration and conservation. *Biodivers. Conserv.* **6**, 947–958.

Nemani, R., S. W. Running, L. E. Band, and D. Peterson. 1993. Regional hydroecological simulation system: An illustration of the integration of ecosystem models in GIS. In M. F. Goodchild, B. O. Parks, and L. T. Steyaert (Eds.), *Environmental Modeling with GIS,* pp. 296–304. Oxford University Press, Oxford.

Penning de Vries, F. W. T., D. M. Jansen, H. F. M. Ten Berge, and A. Bakema. 1989. *Simulation of Ecophysiological Processes of Growth in Several Annual Crops.* PUDOC, Wageningen, The Netherlands.

Rind, D., R. Goldberg, J. Hansen, C. Rosensweig, and R. Ruedy. 1990. Potential evapotranspiration and the likelihood of future drought. *Journal of Geophysical Research* 95:9983–10004.

Rojas, O. E. 1985. *Estudio agroclimático de Costa Rica.* Instituto Interamericano de Cooperación para la Agricultura, San Jose, Costa Rica.

Rumney, J. R. 1968. Climatology and the world's climate. McMillan Co.

Staritsky, I. G. 1989. *Manual for the Geostatistical Programs SPATANAL, CROSS, MAPIT.* Agricultural University, Wageningen, The Netherlands.

Stein, A. 1991. Spatial interpolation. Ph.D. thesis, Agricultural University, Wageningen, The Netherlands.

Taylor, L. 1993. The World Bank and the environment: The World Development Report 1992. *World Development* 21:869–881.

Varekamp, C., Skidmore, A. K., Burrough, P. A. B. 1996. Using public domain geostatistical software for spatial interpolation. *Photogram. Eng. Remote Sensing,* **62**(7), 845–854.

Vargas, A. B., and Trejos, V. F. S. 1994. Changes in the general circulation and its influence on precipitation trends in Central America: Costa Rica. *Ambio,* **23**, 87–90.

Waylen, P., M. Quesada, and C. N. Caviedes. 1996. Temporal and spatial variability of annual precipitation in Costa Rica and the southern oscillation. *International Journal of Climatology* 16:173–193.

CHAPTER 10

# Properties, Geographic Distribution, and Management of Major Soil Orders of Costa Rica

FLORIA BERTSCH, ALFREDO ALVARADO, CARLOS HENRIQUEZ, AND RAFAEL MATA



## I. INTRODUCTION

Soils are the most important of the earth's resources for humans. That tiny layer of the earth's surface, usually from about 1/10 to 1 m deep in temperate regions, may penetrate to more than 2 m in some tropical environments. Or, if abused, may cease to exist in some locations altogether. Any consideration of sustainability has to start with soils, for just as agriculture is essential for any concept of human sustainability, so soils are essential for agroecological systems. And whether soils are to be sustained or not depends upon what natural processes occur and what we as humans do with them. And what we do with soils depends upon how we view them. To the noninitiate soils are just "dirt," relatively boring and homogeneous. In reality a scoop of soil is a living system—complicated, often heterogeneous, its nature completely dependent upon its physical, chemical, and biological properties, filled with thousands of species and millions of individuals (Dindal, 1992). The soils of an entire nation are proportionately more complex, and need to be understood in their complexity if we are to understand their sustainability.

*Quantifying Sustainable Development*

The Costa Rican economy has been dependent historically on its agricultural soils. Since before colonial times natural vegetation growing in fertile lands nourished the local population, and subsequently sustained the European settlers. Cocoa, tobacco, and corn were adopted by Spaniards from among the many local crops that the indigenous population consumed on a daily basis. Cocoa and tobacco were used for trade with Europe and formed the basis for much of the early economies. Later, introduced coffee, bananas, and sugarcane were planted in areas thought adequate for their cultivation. The original planters were correct, for soon yields obtained in Costa Rica for coffee and bananas were the highest worldwide. The resulting dual production system of traditional low-input farming for local consumption and more intensively managed coffee, bananas, and sugarcane for exports lasted until the early 1970s. Subsequently the production of all crops was greatly intensified and industrialized. At present, the agriculturally based economy has expanded to include many nontraditional crop production systems and even government-sponsored reforestation. These are all part of a more diversified economy that includes the expansion of industry and tourism and were possible thanks to the richness of Costa Rica's soils, which generally have provided nutrients and maintained their physical properties through time.

Costa Rican soils have been an important contributor to social sustainability in the past. For example, the potential of the soils for coffee production allowed small farmers with only 5 or 10 ha of land to maintain a decent standard of living for many generations. This pattern of land use changed, however, with the appearance of large plantations and the intensification of beef production, leading to many problems as chronicled in Chapters 2 and 16. Celis and Alvarado (1994) describe recent governmental efforts to reverse this pattern by making landownership more dependent on sustainable uses. Indicators of sustainability for the agricultural and forestry sectors of the country were presented in the 1994 National Agronomy Meetings, demonstrating the concerns of professionals for preserving the Costa Rican soil environment (Alvarado *et al.*, 1993; Torres, 1993).

The rest of this chapter is divided into three sections. The first gives an overview of soils at the national level, including their types, patterns of use, and conservation status. A large middle section examines particular Costa Rican soils in much more detail, including their origin, physical and chemical nature, and management needs. A short final section synthesizes this information from the perspective of sustainability.

## II. HISTORICAL PERSPECTIVE

According to Alvarado (1996), soil studies in Costa Rica started at the beginning of this century when the first soil maps of the coastal regions of the country were drawn in order to establish banana and rubber plantations (Prescott, 1918; Bennett, 1926). With the expansion of high-altitude crops (coffee, sugarcane, and vegetables) in the 1950s, soil studies were carried out in the Central Valley (Dondoli and Torres, 1954; Costa Rica MAI, 1958), and with this small amount of information the first Costa Rican soil map was published at a scale of 1 : 5,000,000 (Quiros, 1954). The scientific research associated with the government-sponsored colonization programs for the northern region in the 1950s and 1960s led to a very large increase in the understanding of this region's natural resources (Costa Rica ITCO, 1964; Sandner *et al.*, 1966). These studies also allowed the drafting of the first maps of potential land use for Costa Rica, at a scale of 1 : 750,000 (Plath and van der Sluis, 1965; Coto and Torres, 1970), as well as the first semidetailed taxonomic soil map of the country (USAID, 1965).

Subsequently, soil maps of the entire country were prepared at a more detailed scale (1 : 200,000) by Perez *et al.* (1978), Vasquez (1979), and SEPSA (1991). These documents incorporated detailed studies prepared for two large hydroelectric development projects in mountainous regions, irrigation projects in Guanacaste, banana expansion in the Atlantic region, and rural development in the border region with Panama. Thus, most early soil analyses were associated with plans for development.

Currently, maps for the country are being refined to a more detailed scale of 1 : 50,000, which is sufficiently detailed to allow precise taxonomic mapping to be used and will aid in developing programs for sustainable development. Several studies have already integrated geographic information systems with concepts of sustainable development (Stoorvogel and Eppink, 1995; Arroyo, 1996; Ugalde, 1996).

## III. CLASSIFYING COSTA RICA'S DIVERSE SOILS

Costa Rica has extraordinary soil diversity in a very limited area, greatly enhancing vegetation diversity and thus possibilities for any kind of agricultural operations. The reason for this diversity is the highly variable parental material, a heterogeneous relief, and the action of a greatly variable climate and biota (Alvarado, 1985; Bergoeing, 1998). Most of the soils are relatively young, reflecting the relatively recent formation of the Costa Rican landmass. Some shallow volcanic soils in mountainous regions are less than 400 years old, formed from ashes deposited by geologically recent eruptions (Harris, 1971), and others have developed A horizons from ashes deposited in 1963–1965 (Chaverri and Alvarado, 1979). The age of the rich organic soils in the alluvial plains has been reported to be less than 3400 years old (Cohen *et al.*, 1986).

These are remarkably young soils by world standards, and their youth has important implications for their potential for human use.

The parental material of Costa Rican soils is dominated by Quaternary alluvial sediments and igneous rocks, with some additional influence of Holocene volcanic ashes and limited regions of older (Cretaceous and Tertiary) sedimentary formations that include limestone (Sandoval *et al.*, 1982). In general the terrain is very irregular, so that the main economical activities are carried out in the flatter narrow alluvial coastal plains and the intermountain Central Valley.

Precipitation, evapotranspiration, and temperature variations, associated principally with changes in altitude, have produced 13 life zones with different biotic conditions (Holdridge, 1979) (see Chapters 9 and 17). These different climatic conditions have been one important determinant of the soil types found in Costa Rica. Variations in precipitation patterns associated with the Intertropical Convergence Zone dominate regional differences in climate. The northern and central Pacific regions receive less than 1.8 m per year of rain and have a distinctive dry season which varies between 3 and 6 months (Figure 9-12). The dry season lasts less than 3 months in the rest of the country, and the rainfall varies from 1.8 to over 4.8 m per year (Figure 9-12). In some locations, precipitation can increase up to as much as 7 or 8 m per year at relatively high elevations, although it drops to lower values at the Paramo, which is above 3200 m elevation.

It is possible to find in Costa Rica all 12 major soil orders recognized by soil taxonomists except desert soils (Aridisols) and frost soils (Gelisols) (Soil Survey Staff 1996, 1998). Six of these 10 have major agricultural relevance: Inceptisols, Ultisols, Andisols, Entisols, Alfisols, and Vertisols (Table 10-1).

**TABLE 10-1 Extension of Soil Orders in Costa Rica**

| Order | Approximate meaning | $km^2$ | % |
|---|---|---|---|
| Inceptisols | New soils | 15,642 | 38.6 |
| Ultisols | Relatively new soils | 8,402 | 21.0 |
| Andisols | Volcanic Ash soils | 5,874 | 14.4 |
| Entisols | Depositional soils | 4,963 | 12.4 |
| Alfisols | Old, highly weathered soils | 3,857 | 9.6 |
| Vertisols | Shrink and swell soils | 621 | 1.6 |
| Mollisols | Black Prairie soils | 546 | 1.4 |
| Spodosols | Ashlike color soils | 62 | 0.2 |
| Oxisols | High sesquioxides soils | 60 | 0.2 |
| Histosols | Swamp soils | 390 | 0.2 |

*Source:* Mata (1991).

*Note:* There have been recent changes in soil taxonomy and nomenclature and we incorporate these changes.

## A. Nutritional Situation

The soils of Costa Rica are, in overview, moderately fertile by global standards, but very fertile when compared to other tropical areas. The old soils of much of Africa and the Amazonia present far more severe problems, but the new soils of northern glaciated areas in the U.S. Midwest are much more fertile. An overview of the nutritional status of the country's soils can be seen in Figure 10-1 (Bertsch, 1986).

According to Bertsch (1986), the most important chemical qualities of soils with respect to their nutritional value for plants relate to the abundance and availability of the major plant nutrients, nitrogen, phosphorus, and potassium (NPK). By these criteria specific tracts of land in Costa Rica present often severe problems for economic agricultural activities. As in other tropical areas, 100% of the Costa Rican soils are deficient in N. In addition, 74% are deficient in P, and 22% are deficient in K. Calcium and magnesium are low in 35% of the soils, a problem considered more relevant than Al toxicity (20–30%) of samples analyzed. There is less information on micronutrients. Boron deficiency is probably most important, followed by Zn (26%) and Mn (23%). Very small regions of the country present Fe deficiencies (6%), and a few areas show Cu toxicity problems (Cordero and Ramirez, 1979). All of these nutritionally related problems can be overcome easily with the use of soil amendments such as lime and fertilizers.

More recent literature seems to be showing that the physical properties of soils can also restrict agriculture at a national level. Most important are soil shallowness, steep relief, erosion susceptibility, flooding risk, drought-

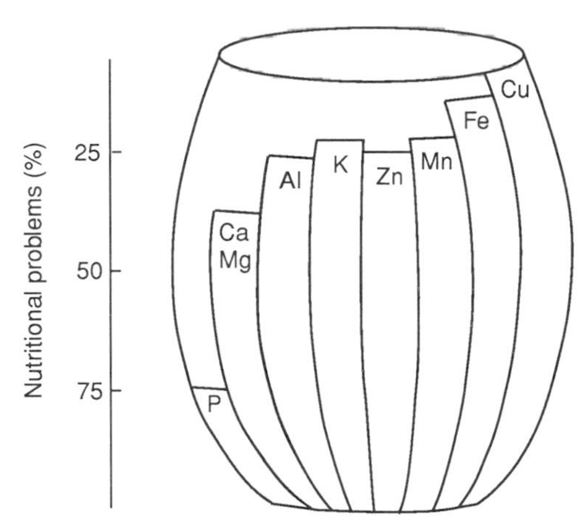


FIGURE 10-1 Nutritional deficiencies of Costa Rican soils.

inducing sandy textures or heavy textures (generated by swamps), and compacted soils (Beets, 1990). In this context, Higgins *et al.* (1984) estimated that under no-input agriculture the country would be able to support a population of 7.9 million inhabitants, which could increase to 20.5 million with intermediate inputs and 51.6 million with high inputs. The last figures, and perhaps the first as well, seem unrealistic considering the present yields, the situation of food imports, and the need to preserve the environment for future generations.

Animal as well as human nutrition can be influenced by the soils via mineral concentrations in feed material (Lag, 1990). The quantity of most nutrients in pasture soils and hence forages is adequate (Bertsch, 1986), except that 63% of pasture leaf samples had insufficient levels of Co or Cu for ruminant growth (Vargas *et al.*, 1992; Vargas and McDowell, 1993). In past years Costa Ricans have suffered from iodine deficiency, a typical problem related to volcanic-ash-derived soils. When salt was iodized this deficiency disappeared except for some remote coastal regions where people consume noniodized salt (Ascencio, 1999).

## B. Land Use Capacity

Approximately 15% of Costa Rica's soils have agricultural potential or capacity (e.g., Classes I to IV in Table 10-2) following the Klingebiel and Montgomery (1961) definition. At the present time, 13% of the national territory is in agriculture (Costa Rica MAG, 1996), and large areas of degraded grasslands which should never have been cleared are on secondary forest regrowth or being planted with forest species.

**TABLE 10-2 Land Use Capacity in Costa Rica, 1992 (Hectares by Soil Class)**

| Soil class | ha |
|---|---|
| I | 16,637 |
| II | 415,452 |
| III | 660,596 |
| IV | 853,001 |
| V | 68,167 |
| VI | 805,370 |
| VII | 829,948 |
| VIII | 1,478,628 |
| Total | 5,097,804 |

*Source:* Costa Rica MAG (1996) and Costa Rica MIDEPLAN (1996).

The land area used for agricultural activities in the country was more or less stable at around 10% from about 1940 to 1979 (FAO, various years). The land area in crops, especially perennial crops, may be increasing for the first time in half a century; the latest survey, in 1979, shows 13% of the land in crops (Table 10-3). It is important to note the loss of some fertile soils in the Central Valley region over the past 50 years due to urban expansion. In contrast to the stable situation with crops, large areas of forests were converted to grasslands for pastures (Sader and Joyce, 1988), a trend recently reversed by official efforts to increase reforestation and the preservation of forested areas as well as natural forest regrowth.

Major agricultural areas of the country are planted with perennial or semi-perennial crops (coffee, bananas, sugarcane, grassland, etc.). This situation is well suited to the agricultural potential of the land. However, in many circumstances land is overused, meaning that it is used too intensively for crops or grazing. Also, some land is often misused, that is, it is used for crops that are inappropriate for its qualities, such as maize on steep slopes. Other land is insufficiently used, for example, when good agricultural land is used for grazing or forest plantation. Rodriguez (1999) arrives at similar conclusions when looking at Costa Rica scenarios for the year 2025. The final problem, often related to each of the preceding, is an inequitable tenure pattern of the land.

TABLE 10-3 Land Use by Activity in Costa Rica in 1979 and 1992

| | 1979 | | 1992 | |
|---|---|---|---|---|
| Land use | ha | % | ha | % |
| Pasture | 820,557 | 16.1 | 1,565,076 | 30.7 |
| Pasture in agriculture | 66,430 | 1.3 | 101,460 | 1.9 |
| Annual crops | 101,355 | 2.0 | 132,955 | 2.6 |
| Perennial crops | 246,279 | 4.8 | 369,210 | 7.2 |
| Natural forest | 2,085,906 | 40.7 | 1,286,456 | 25.2 |
| Selected logging forest | 367,090 | 7.2 | 484,071 | 9.5 |
| Secondary forest | 882,164 | 17.3 | 695,903 | 13.6 |
| Abandoned grassland | 292,287 | 5.7 | 228,445 | 4.5 |
| Paramo vegetation | 19,625 | 0.4 | 13,495 | 0.3 |
| Mangroves | 66,523 | 1.3 | 49,374 | 0.9 |
| Rocky lands | 15,292 | 0.3 | 8,567 | 0.2 |
| Plowed land | 0 | 0.0 | 26,469 | 0.5 |
| Wetlands | 113,267 | 2.2 | 106,058 | 2.1 |
| Lakes and dams | 18,432 | 0.4 | 9,797 | 0.2 |
| Burnt lands | 0 | 0.0 | 10,063 | 0.2 |
| Urban | 14,792 | 0.3 | 22,599 | 0.4 |
| Total | 5,110,000 | | 5,110,000 | |

*Source:* Costa Rica MAG (1996) and Costa Rica MIDEPLAN (1996).

## C. Land Ownership Patterns

Land ownership in Costa Rica is highly skewed (Martinez *et al.*, 1983; Leonard, 1986; FAO, 1990). Most of the land is in the hands of a relative few and, in addition, the best land belongs to the same people. Distorted incentive policies have favored over- and underutilization of this land, as is reflected in low productivity, high rates of deforestation, and to a lesser degree illegal invasions of private land (Strasna and Celis, 1992). Land-related problems seem to be associated more with farmers without land than to the latifundia system, although both factors are closely linked (Rodriguez, 1989). The net effect is to push small and medium farmers onto the slopes and low-quality soils, which are better suited for forestry (Leonard, 1986). This process generates problems ranging from land invasion to the use of Class VIII land, which in turn leads to a great deal of erosion.

Other factors that affect land ownership in Costa Rica are urban development (with associated migration patterns) and colonization or resettlement programs. A former government office known as ITCO and currently as IDA (National Institute for Agrarian Development) granted a total of 1,375,095 ha to 56,668 families between 1963 and 1986. This program is still active and benefits an average of 250 families per year. Land ownership slightly improved from 1950 to 1984 in Costa Rica as shown by a slight decrease in the Gini Coefficient (a measure of ownership concentration—the higher the index, the higher the concentration) from 0.76 to 0.73 (Gonzalez, 1993).

As a result of perceived land ownership distortions, recent debates cover subjects such as incentive regulations (Lutz and Daly, 1991), land use planning (Lucke, 1993; Vasquez, 1993; Barahona, 1993), and land taxation (Celis and Alvarado, 1994). This pattern has contributed to the rural–rural, and to a lesser degree rural–urban, migration (Kleysen, 1990). It also influences urban–rural migration and reduces population outmigration in low-income regions since farmers' family members move to work in the increasing tourism activities.

In general, the development of agriculture, pastures, forestry, tourism, and urban areas has been disorganized and has not been related to logical patterns of land use that would subject land use to constraints such as agroecological potentialities and land fragility (Vasquez, 1996). Thus much of the potential for using the land and the soil sustainably and in a pattern that would provide the greatest economic and social welfare has been lost. In recent years the "efficiency" claimed under globalized economics is contributing to the enlargement of farm size for orange groves, palm hearts, and forestry plantations.

## D. Erosion Situation

When heavy rainfall exceeds the soil infiltration rate, overland runoff ensues. This runoff is accelerated if the soil is lacking in vegetative cover. This is the primary cause of natural soil erosion under tropical humid conditions. Other common causes are catastrophic climatic events, including tropical storms, hurricanes, floods, and drought, and geologic phenomena, including volcanic ash deposition and earthquakes. Overall, however, it is important to emphasize that many of these processes also form soils. The problem of erosion occurs when the natural factors interact with loss of vegetation cover. In particular, the level of erosion is high in areas with greater road and human settlement density. This can be seen in a comparison of the Pacific slope, which was settled first, and the Atlantic slope (Table 10-4). Even though the Atlantic slope gets more rain, the Pacific slope has greater erosion. This is enhanced by the occurrence of highly erosive geologic formations along the Puriscal and Aquacate mountains.

Erosion in Costa Rica, according to the CCT (Centro Cientifico Tropical/WRI, 1991), is larger in areas planted with annual crops than in grasslands, and relatively small when the land is planted with perennial crops. Although estimated erosion values for the country increased steadily from 1973 to 1983, losses have been stable since then at an estimated value of 190,000 MT/year. The CCT (1991) estimated the replacement of the amount of NPK loss by erosion to be around 2.6 million 1984 colónes, equal to roughly U.S. $60,000 (Chapter 14).

## E. Description of Major Soil Orders

Different soils have very different potentials for human use, depending on their inherent physical and chemical properties. This section describes the utility and importance of various soils for agriculture as a function of their taxonomy, that is, as a function of the human-derived classification scheme imposed on the more or less infinitely varying properties of soils. The analysis includes a summary of the type of areas which each major soil type occupies,

TABLE 10-4 Area Affected by Hydric Erosion for 1981

| | Region | | | | | |
|---|---|---|---|---|---|---|
| | Nation | | Atlantic | | Pacific | |
| Degree of erosion | km$^2$ | % | km$^2$ | % | km$^2$ | % |
| Tolerable | 29,436 | 58 | 18,256 | 77 | 11,180 | 41 |
| Slight to moderate | 12,405 | 25 | 4,366 | 18 | 8,039 | 30 |
| Severe | 7,301 | 14 | 981 | 4 | 6,320 | 23 |
| Very severe | 1,689 | 3 | 121 | 1 | 1,568 | 6 |
| | 50,831 | | 23,724 | | 27,107 | |

*Source:* Hartshorn *et al.* (1983). Costa Rica MIDEPLAN (1996).

the crops each is used for, its geographic distribution in the country and position in the landscape, its probable means of origin, the principal mineralogical, physical, and nutritional characteristics of each group, and the management practices that could be applied to each to achieve the best expression of its productive potential (Bertsch *et al.*, 1994).

### 1. Entisols

#### *a. Distribution, Area, Use, and Origin*

Entisols are very new soils that exhibit little development, so that it is not possible to distinguish a defined horizon sequence in the profile. The most common Entisol suborders in the country are fluvents, aquents, orthents, and psamments (i.e., soils derived from recent alluvial deposits, soils formed under stagnant water, shallow soils on hard rock, and sandy textured materials, respectively). The presence of psamments in coastal beachfronts or elevated coastal structures is a consequence of continental uplift due to plate tectonic activity.

Recent alluvial deposits lead to the formation of fluvents in areas where frequent flooding does not allow soils to remain undisturbed long enough to permit the development of horizons. Under these conditions, a sequence of layers of contrasting particle sizes occurs. In the same geomorphic surfaces, these soils turn into aquents when the water table remains near or above the soil surface and restricts soil development for long periods of time.

Orthents are the most common type of Entisols that are abundant on rocky hillsides where there is low temperature and/or very erosive rains, recent volcanic depositions such as ash or lava, and the presence of parental material resistant to weathering. Other orthents are found in low-relief portions of regions of riolithic origin in the dry Pacific areas where rainfall is less than 1500 mm per year.

Entisols cover 12% of the country, of which the orthents are the most abundant. Most Entisols are of limited agricultural potential due to their high flood risk, restricted soil rooting depth, and low fertility status or because they are on steep slopes. Their use by humans should be restricted to forestry or conservation activities. Nevertheless, in Costa Rica these soils are frequently used for annual crops and extensive cattle raising in both flat and steep lands.

#### *b. Mineralogical, Physical, and Nutritional Properties*

Due to their minimal development, these soils reflect the properties of the parent material from which they were formed, which is why they have very varied mineralogy. In general, they are not good agricultural soils because of shallow rooting depth, reduced conditions, and, as noted, frequent flooding and high susceptibility to hydric and eolian (wind) erosion.

### *c. Management*

In wetlands, aquents and fluvents are associated with aquepts (Inceptisols) covered by a useful natural vegetation community of yolillo (*Raphia* spp.) and cativo (*Prioria copaifera*). When drained, these soils have been used to plant bananas, cocoa, and oil palm. Orthents form on thin volcanic ash deposits over lava flows [e.g., in Cervantes, Cartago Province (see picture in "Land Use" on CD), and Paso Canoas, Puntarenas Province]. Although they are not very productive soils, in some cases they are heavily fertilized and planted with vegetables to take advantage of closeness to markets. In other regions, where the rock exposed is not very hard (e.g., riolithic materials in Guanacaste), they are used for ranching activities and, recently (and with very little success), for forestry production. On the hillsides around the southern and central Pacific regions the orthents are commonly used for low-technology bean planting ("Frijol Tapado").

## 2. Inceptisols

### *a. Distribution, Area, Use, and Origin*

Inceptisols are relatively new soils that are not greatly changed from their parent material. They are widely distributed in Costa Rica. They are common on hillsides where they are called tropepts, and where erosion due to a combination of earthquakes and heavy rainstorms induces landslides, which limit soil formation to a profile with a very weak horizon development. Under these conditions, a typical toposequence includes inceptisols with high organic matter content in the upper positions, and inceptisols with low base status in the medium and lower positions of the landscape.

Another group are found in rolling and flat geomorphic surfaces. Among these, dystrustepts (long dry season), eutrudepts (high base saturation), and distrudepts, formed from the weathering of relatively old alluvial and/or coalluvial fans, are common. In the same environment, Inceptisols classified as aquepts are found when there is a perched water table. These are the most important soils for agriculture of the lowlands less than 100 m above sea level. A sulfidic horizon forms where there is brackish water and mangrove vegetation along coastal backswamps. Those soils are classified as sulfaquepts.

The most important inceptisols for humans are found in alluvial valleys in the coastal plains. These soils have the highest agricultural potential of the country and can be found along the valleys of the Tempisque, Bebedero,

Tarcoles, Parrita, Terraba, Sierpe, and Coto rivers on the Pacific side, and the Matina, Reventazon, Parismina, Pacuare, Estrella, and Sixaola rivers on the Caribbean side (Figure 10-2). In many cases these soils develop from basic parent materials such as limestone, from which they inherit their high base saturation, adequate texture, and moisture retention.

Because Costa Rica is geologically and geomorphologically young, Inceptisols cover about 39% of the national territory. The youth of the soils also means that the soils reflect strongly their parent material, including lithic (rock), fluventic (riverine), andic (volcanic), vertic (clay mineralogy), or oxic (trivalent cation accumulation) properties.

Inceptisols (except those with poor drainage) generally have good characteristics for management, since they do not possess the properties of more developed soils, such as cation depletion, that affect management adversely. For this reason, they can be used for a wide range of agricultural and livestock production activities, including bananas, oil palm, sugarcane, cocoa, coffee, staples, livestock, forestry, and, recently, nontraditional crops such as mango, avocado, cantaloupe, pepper, roots and tubers, and tropical flowers. Even the sulfaquepts along the coast are important for mangrove forestry, shrimp aquaculture, and extraction of salt.

FIGURE 10-2 Geographical distribution of Inceptisols in Costa Rica.

### b. *Mineralogical, Physical, and Nutritional Properties*

Chemical and mineralogical properties of Inceptisols vary according to the origin of these soils. Therefore, the range of chemical characteristics is wide. Each soil tends to include many clay mineral types mixed together, including smectites, allophane, kaolinites, and organic and oxidic coatings. When there is a preponderance of volcanic ash materials some amorphous clays develop. In the alluvial valleys of both the Caribbean and the Pacific sides, montmorillonite is found. The extreme weathering conditions in tropical environments of the El General River Valley result in the formation of 1 : 1 clays and oxides in red soils with very high acidity values and cation depletion. These are the most infertile Inceptisols of the country.

### c. *Management*

Those Inceptisols used for commercial plantations in the poorly drained lowlands require drainage. For example, as eastern banana plantations spread from the slopes toward Limon, extensive networks of 1- to 2-m-deep ditches are required. Such ditches are economically viable only when the flood frequency remains low.

The fertility of Inceptisols in the North Atlantic zone is much higher than that in the South Atlantic region because the former soils are developed from volcanic materials spread downslope by rivers. In the South Atlantic region Inceptisols were formed from much less productive calcareous materials, and also are subject to a much greater frequency of flooding (Stoorvogel and Eppink, 1995).

The fertility of Inceptisols in the Guanacaste lowlands can be greatly enhanced with applications of S and Zn, especially in rice plantations (Bornemisza, 1990; Cordero, 1994). Moisture availability is critical to these Inceptisols in ustic (long dry season) environments. These properties have been mapped and used for categories of crop insurance (IICA, 1979).

During the early days, these lowlands were planted with bananas and cocoa without any fertilizer application; but the banana plantations were subject to massive applications of copper-containing Bordeaux fungicides. Rice cultivars on Inceptisols in the South Pacific valleys were subject to Cu toxicity in the 1940s and 1950s (Cordero and Ramirez, 1979), and much of this very fertile land had to be abandoned. Due to natural rejuvenation by silt deposition from river flooding they have been rehabilitated for annual crops. At present, improved varieties and higher productivity of oil palm and banana plantations require large amounts of complete fertilizer formulas as well as drainage systems.

Small farmers living in government settlements, particularly in the northern and Atlantic regions of the country, plant cereals and roots and tubers on these soils using low levels of inputs. Because of the very wet environment predominant in these regions, the traditional slash and burn system is not practiced since the biomass chopped down does not dry and thus cannot be burned. This problem does not allow the beneficial effect of liming and fertilizing with the added ashes. Instead, the biomass slowly decomposes with time, releasing nutrients gradually (Bertsch and Vega, 1991).

### 3. Andisols

#### *a. Distribution, Area, and Use*

Andisols are formed from volcanic ash deposits and occupy: (a) the Central Valley and surrounding mountains; (b) hillsides of the Guanacaste Mountain Range, (c) the region between Coto Brus and the border with Panama influenced by the Baru Volcano's ashes; and (d) some regions of the northern and Atlantic zones where fluviovolcanic depositions occur (Figure 10-3).

Although they cover only 14% of the national territory, major agricultural activities (such as coffee, sugarcane, and vegetables), a few nontraditional export crops (flowers, ferns, strawberries), and dairy production are carried out on them. Part of the latest large banana boom of the 1990s was carried out on volcanic soils of the northern zone and parts of the Atlantic region. In

FIGURE 10-3 Geographical distribution of Andisols in Costa Rica.

the lowlands, Andisols can produce very good yields of nontraditional crops, roots and tubers, heart of palm, and a huge range of tropical ornamental plants.

### *b. Origin*

The frequent rejuvenation of these soils by andesitic volcanic ash additions has constantly enriched the environment with nutrients. Different volcanic and environmental conditions generate different soil types. Large depositions of debris, particularly near the craters, allow the formation of vitrands, while udands form under repetitive deposition of thin volcanic layers in the middle positions of the landscape in udic (seasonally dry) environments. At the bottom of the landscape, where a distinctive dry season occurs, ustands are predominant.

Andisols of lighter color are found along the Guanacaste Mountain Range and in the north of the country, and originated from the deposition of riolithic/dasitic ashes. Darker andesitic-basaltic ashes predominate in the central and southern parts of the country, and give rise to dark-colored soils.

The effective depth of the top soil layer of Andisols generally depends on the magnitude of the volcanic deposition that formed that layer. Deep soils tend to be formed from the deposition of many small layers of ash, while thin Andisols are formed by one event, which can be of small or large magnitude. It is possible to observe the ash deposition frequency and magnitude in deep road cuts, as well as the presence of Paleosols with a different degree of weathering.

### *c. Mineralogical Properties*

Soil particles are generated and distributed initially by the nature of the original volcanic activity, and then sorted by prevailing winds according to particle size and density, creating a textural gradient along the hillsides of volcanic craters. Coarser material is deposited in the vicinity of craters, resulting in sandy to sandy loam materials. Further away silty loam or loam textures are predominant. Even finer textures are found farther away from the volcano, particularly in the B horizons of well-developed soils. This textural gradient notoriously affects nutrient availability and irrigation needs.

Once the original deposition has taken place, climatic forces predominate. For example, in a moist and cold environment the weathering process of the volcanic glass is weak, releasing small amounts of Si, Al, and Fe oxides and hydroxides. If long periods of volcanic inactivity follow, the translocation of the oxides will form a cemented layer (called a placic horizon) wherever an abrupt textural change is present.

Farther from the crater, allophane becomes predominant. This type of clay is an amorphous and hydrated colloid, which forms organomineral compounds and represents the required product of volcanic ash decomposition in humid zones. Allophane is an unstable clay-size particle of high reactivity that gives a peculiar behavior to these soils. Secondary, organomineral compounds possess a very large hydration capacity that produces an enlarged total surface, therefore increasing the capacity of the clay to retain or exchange cations and anions.

In the Central Valley, further away from the volcano, the rainfall decreases under Costa Rican conditions, and the long dry period causes the formation of a 1 : 1 crystalline clay named halloisite. This type of clay has "shrink and swell" and low water retention characteristics and less nutrient retention than allophanes. Halloisite is predominant in the brown-yellowish soils of the coffee and sugarcane plantations of the Central Valley. Each mineral type gives a characteristic color to the soils that are formed from them. Dark-colored Andisols are associated with a high allophane content; brown-yellowish Andisols are dominated by halloisite; and brownish Andisols are related to kaolinite. White-colored Andisols are associated with the presence of gibbsite (Colmet-Daage *et al.*, 1973; Besoain, 1985).

Volcanoes are still quite active in Costa Rica, and their activity influences agricultural potential directly as well as indirectly through soil building. The emissions of acidic clouds from volcanoes become acid rain in nearby zones, which leads to an intensive weathering of the system, enhancing basic cation leaching and causing considerable loss of crop yields.

### *d. Physical Properties*

Due to the presence of highly stable organomineral compounds, especially in the A horizon, Andisols tend to be very well structured. This results in a high infiltration capacity leading to both good drainage and good moisture retention characteristics. One unfortunate consequence of these properties is that these soils promote the leaching of nitrate from agricultural systems down to underground waters, contaminating groundwaters and reducing their value for human use (Reynolds, 1991). This problem may have a variety of solutions (Radulovich *et al.*, 1992), which are currently being studied.

These soils have low bulk density and low resistance to tangential forces, making them easy to plow. In Costa Rica this task should be done by animal traction in order to prevent erosion, instead of by heavy machinery that tends to compact the soil (RELACO, 1996). Overgrazing causes similar compaction.

During periods of high volcanic activity, large amounts of very unstable ashes are deposited as blankets that cover the landscape. This material partially dissolves when subject to alternating dry and moist periods, inducing redistribution of soluble elements at the surface, cementing small pores, and reducing infiltration by crusting. This phenomenon develops into massive erosion, which encourages the formation of coluvio-alluvial fans at the bottom of the landscape. This is the main factor generating catastrophic events when deposited as "lahares" in populated areas. Also, these soils are used intensively for agricultural activities that greatly encourage their erosion and the silting of hydroelectric dams.

### e. *Nutritional Properties*

Most Andisols have a moderate fertility status depending on the composition of the parent material. In general, soils formed from the ashes of the Irazu Volcano are richer in bases than those formed from Poas Volcano materials (Alvarado, 1975); Andisols around the Baru Volcano in the southern region are even poorer than those of the Central Valley. Nutrient leaching in volcanic areas is counterbalanced by new additions of volcanic ash; this process has enabled nature to maintain the base saturation of the ecosystem.

Generally, Andisols have pH values near neutral except in agricultural areas with poor management or where they are gradually acidified by the decomposition of the abundant organic matter content of the Andisol soils, particularly when large amounts of N are applied. When this happens they do respond to liming with calcitic (Ca carbonate) or dolomitic (Ca and Mg carbonates) product amendments.

The soil fertility potential of Andisols can be estimated by a sum of cations (Ca, Mg, K, and Na); higher values indicate a better condition for crop development and imply that other nutrients are also abundant. In Andisols of the southern region, the predominance of plagioclases over orthoclases creates a pronounced K deficiency (Molina *et al.*, 1986; Henriquez and Bertsch, 1994).

In recent volcanic ashes, N is the most limiting factor for crop production. But P, although abundant in total, creates difficulties for farmers too. Phosphorus is held tightly by the clay lattices of Andisols so that it is not available to plants. Retention is generally over 70%, which is very high, and it can easily reach values of 95%. This problem constitutes by far the major limitation for crop development on these soils (Alvarado, 1982; Canessa *et al.*, 1987). In addition, B and S can also be held tightly as anions. The application of these two elements is essential for coffee production all over Costa Rica (ICAFE, 1989).

Andisols formed in the lowlands of fluviovolcanic origin of the northern zone and part of the Atlantic region, along the Sarapiqui, Sucio, Chirripo, Tortuguero, and Destierro rivers, are poorly understood. Under very high temperature and rainfall conditions, they seem to weather to form soils with more nutritional problems than those of the highlands. In addition, the low relief of these areas enables water to accumulate on the surface, enhancing soil compaction, particularly in pasturelands.

### *f. Management*

Due to the high P retention of Andisols, most crops require large applications of soluble fertilizer P. The exact location and granule size of the fertilizer are important, which should be applied along with light applications of lime that increases the availability of the P held in organic materials.

Nitrogen is also a limiting factor for crop production, except when legume species are planted to fix N, such as when white clover is planted with kikuyo grass. Large applications of ammonium N result in the release of hydrogen ions, enhancing acidification of extensive areas, particularly in grasslands and coffee plantations. To correct for this condition, frequent applications of lime are required to get good yields, as has been done for sugarcane (Chaves and Alvarado, 1994).

Other elements, such as Mg, can limit crop performance if the parent material is low in Mg (Poas slopes, primarily) or when large K applications induce nutrient antagonisms. As with B and Zn, foliar and soil analyses should be run on a regular base to correct for deficiencies.

Tree-shaded coffee plantations require smaller additions of fertilizer than full-sun plantations since shading reduces photosynthesis of the coffee plants. If the shading tree is a legume species, such as *Erythrina* or *Inga,* biologically fixed N is added to the system. Coffee plantations also are associated with contour planting, the use of tills, windbreak barriers, and hedge rows, practices which are necessary to reduce erosion, particularly during crop establishment.

The use of agrochemical products on these types of soils has different effects over long periods. In the case of potato fields many years of fertilization have generated P accumulation. In areas where potatoes have been cropped for more than 25 years, concentrations of more than 80 ppm of available P have been found so that sometimes no further fertilization is required. In the case of cupric fungicides, used as disease protectors in intensively managed coffee plantations, however, Cu accumulates at a rate of approximately 1 ppm/year. This could become a long-term problem because plant toxicity occurs beginning at 100 ppm (Cabalceta *et al.*, 1996).

## 4. Vertisols

### *a. Distribution, Area, and Use*

Vertisols are depositional soils found mainly in areas of the dry northwest Pacific region of Costa Rica, on either plains or depressions, where the dry season extends from 4 to 6 months, and are often associated with small patches of similar Mollisols. Although most Vertisols have a neutral or basic status, a few of them located near the border of Nicaragua are acid.

Vertisols occupy only 2% of the country's area, and are principally located in depressional areas on the most important alluvial valleys of the dry Pacific region and to similar locations on the western part of the Central Valley (Santa Ana, Pozos, Lindora, and Ciruelas) (Figure 10-4).

Vertisols are used intensively for both agricultural and, in the Central Valley, urban development. The main crop on these soils during the rainy season is rice, either flooded or rain-fed. With irrigation and adequate soil water management, sugarcane, soybean, melon, and cotton, or even hot chili peppers and sauce tomatoes, can be grown. Trees grow poorly on these soils, due to root damage caused by alternate seasonal periods of dryness and water excess. Thus, commercial forests are neither abundant nor recommended on Vertisols. Even though pastures are found there, their management is very difficult and beef production very poor.

### *b. Origin*

Vertisols in Costa Rica originate mostly from riolithic tuffs high in biotitic micas, with some recent additions of very fine volcanic ash. The simultaneous

FIGURE 10-4 Geographical distribution of Vertisols in Costa Rica.

occurrence of several factors is necessary for Vertisols to form: the presence of a depressional zone which prevents good drainage, the occurrence of materials rich in Si, Ca, and Mg that accumulate in alluvial and/or fluviolacustrine deposits, and a well-defined dry season. An exception occurs in the Central Valley, where climatic and tectonic dynamics created and then eliminated lakes, which served the same functions as depressional areas.

### c. *Mineralogical Properties*

The conditions necessary for Vertisols also favor the formation of 2:1 montmorillonite-type clays, which have very high Si content and the strongest colloidal properties of all clays. This generates Si films interlayered between, but very poorly bonded with, montmorillonite particles. The result is very small, highly individual particles with unlimited and reversible water absorption capacity. The resulting soil is highly expandable, and has high specific surface activity, cohesivity, adhesivity, plasticity, and water retention capacity. What this means is that these soils can get both very wet and very dry, change their shape greatly, crack when they get dry, and become extremely slippery and sticky when wet. Even with their large water holding capacity, the difference between their field capacity and the permanent wilting point is rather low, so from a plant's perspective they dry out easily. Overall Vertisols are poorly suited for most agricultural and engineering operations due to their contractions and expansions in response to seasonal fluctuations of rainfall.

### d. *Physical Properties*

Most Vertisols (usually usterts) are less than 1 m deep and dark-colored, and have little horizon differentiation and a clayish texture. Shallower Vertisols tend to lay over low-permeability tuffs and become water saturated and anoxic during the rainy season. Because of the reduced conditions, grayish subsurface horizons (aquerts) are present.

When the dry season arrives, Vertisols dry very drastically, forming massive blocks with open cracks that affect irrigation canals, electric poles, and engineering operations. At the onset of the rains, a vertical flow of water takes place down the cracks, causing the subsoil clays to expand rapidly. This effectively seals the whole system, which floods heavily as the rains increase. Therefore the use of mechanized cultivation on these soils is difficult and expensive.

### e. *Nutritional Properties*

Vertisols are potentially quite fertile soils, with a high pH and high Ca and Mg content. When organic matter is provided, favorable conditions for nutrient release may result. Thus, the constraints for high productivity on Vertisols are mainly physical rather than nutritional. Nevertheless there are many nutritional problems that get in the way of the high potential fertility being expressed as high plant growth. Organic matter additions under flooding conditions may induce the reduction of Fe and Mn to toxic levels for most crops.

But the high Ca and Mg concentrations generate additional problems leading to difficult uptake of other nutrients by plants, and hence poor plant growth, especially when the K content is low. Even though calcium–phosphate complexes are the most soluble among all phosphates, a plant's ability to use P is limited due to its binding to Ca in the soil. Additionally, the content of minor cations is low in response to the high pH. All of these lead to serious limitations on plant growth.

#### *f. Management*

The basic nutrient management strategy for Vertisols is maintenance fertilization with particular consideration of the levels of K and Zn (Sancho *et al.*, 1984). Sulfur fertilization may also be useful (Bornemisza, 1990). Clays of 2 : 1 composition display a high cation retention capacity, especially for K and $NH_4$, on both the external and the internal surfaces, resulting in peculiar behaviors of these cations. To reduce K-induced deficiencies, potassium needs to be applied, particularly for annual crops. The use of pesticides must be planned carefully when crop rotation is practiced on Vertisols because the active ingredients can be trapped on the clay particle during the first culture cycle, only to be released later when irrigation is applied during the second cycle.

Irrigation of Vertisols in Costa Rica will be possible soon because of the hydropower plant projects in Guanacaste Province. Significant investments in infrastructure, such as canals, need to be done in order to achieve a sustainable and profitable use of this irrigation. Research and technology adaptation programs are needed to ensure, for example, that the expandable clays will not destroy the new infrastructure.

### 5. Alfisols and Ultisols

The oldest and most weathered soils of Costa Rica belong to these orders, the differences being the chemical properties of the subhorizon. These are the typical "tropical" soils, red or yellow in color, slippery when wet, and generally infertile. Alfisols have more basic subhorizons and, particularly in Costa Rica, occur in drier environments. In agronomic terms, both types of soils have a very similar "plow layer." The real differences arise after intensive use, when Ultisols start exhibiting more marked fertility problems.

*a. Distribution, Area, and Use*

In Costa Rica, these soils occupy a large area, about 31% of the territory (21% Ultisols, 10% Alfisols). In earlier times, and in other regions of the tropics today, the prevalent land use for these soils was for slash and burn agriculture. This is not relevant in Costa Rica today because of the prevailing high-input agricultural system, the wet climatic conditions where natural vegetation cannot be burned, and the high quality requirements for agricultural products. In general, they are considered marginal for contemporary agriculture because of their low and rapidly declining fertility, and only some of them are in use, principally for root, tuber, and pineapple production. During the beef cattle boom of the 1970s, these soils were the most preferred for grazing. The cattle degraded these soils rapidly, however, and most such pastures were abandoned, leading eventually to establishment of secondary forest.

These soils, however, can be productive when properly managed. Virtually all the pineapple produced in Costa Rica is grown on these soils, as well as significant amounts of citrus, mangoes, avocados, palmito (heart of palm), sugarcane, roots and tubers, etc. In the southern Pacific region, large coffee plantations and *Gmelina arborea* plantations for pulp production are being established, although both face severe nutritional constraints. The acidity problems of many Ultisols might be reduced by liming, which decreases the Al availability and increases fertility, or by the selection of species, varieties, or strains tolerant to acid soils and low P contents (Acuna and Uribe, 1996; Uribe, 1994).

Ultisols are found in the northern zone of Costa Rica in the Sarapiqui, San Carlos, and Cutris districts, in areas of the southern portion of the country in the Perez Zeledon and Buenos Aires districts, and in proximity to the border of Panama, as well as the foothills (Atlantic and Pacific) of the Talamanca Mountain Range (Figure 10-5). The main areas of Alfisols are located on the Nicoya Peninsula and, associated with Vertisols, the floodplains of the Tempisque River. In these areas, commercial forests of *Tectona grandis, Bombacopsis quinatum,* and *Gmelina arborea* have been successfully established, along with small coffee plantations. Alfisols also occur in the central Pacific zone in the Grecia, Atenas, Orotina, and San Mateo districts, where small-scale fruit plantations (mangoes, tamarind, cashew, and caimito) and recreational villas are the main forms of land use. Wherever they are found, these soils occupy the highest positions in the watersheds and slopes. Thus Alfisols are not subject to frequent additions of fresh materials and are dominated instead by mild leaching conditions resulting in base accumulation in the subsoil.

*b. Origin*

These soils originate from the downward flow of water through the profile over long periods of time, under high-temperature conditions, and from practi-

FIGURE 10-5 Geographical distribution of Ultisols in Costa Rica.

cally any parent material. Their main feature is the presence of an argillic clay horizon formed from the water-borne migration of clays from the superficial horizons to the deepest layer of the soil. For this movement to occur, precipitation must be greater than potential evapotranspiration under free-drainage conditions; that is, the water table must extend very deep into the soil and be separated from the surface. This process causes the loss of Na, K, Ca, and Mg from the soil profile, leaving behind high concentrations of Al, Fe, and Si (tri- and tetravalent cations)—to a greater extent in Ultisols than in Alfisols.

The high concentration of hydrated iron accounts for the reddish color of these soils. More specifically, they are brownish-red to reddish in the concave parts of the relief and brownish-yellow to yellow in the convex depressions when Fe is bound with water molecules. The most relevant criterion considered for classification of Ultisols and Alfisols is the presence of an argillic and/or kandic subsurface horizon, which is acidic in Ultisols (humid tropics) and neutral or basic in Alfisols (dry/humid tropics).

### *c. Mineralogical Properties*

Clays of 1 : 1 compositions (mainly kaolinites) as well as Fe and Al oxides predominate in these soils. Even when composed of fine materials, the formation of H bonds in 1 : 1 clays fosters particle aggregation and, therefore, a more developed structure. As these aggregates get coated by oxides a larger particle, known as "pseudosand," is formed. In some regions, Fe and Al accumulation is so high as to allow their exploitation as bauxite (aluminum ore). Such deposits, also known as plinthite, display a white mottled surface over a red matrix in which gibbsite can be found.

#### *d. Physical Properties*

The presence of stable aggregates in granular structures gives these soils excellent physical properties for agriculture, especially with respect to drainage. Overgrazing and intensive mechanization, however, can cause their favorable physical properties to deteriorate irreversibly. Liming improves fertility, but when excessive, increases erosion by favoring clay deflocculation. Such actions affect productivity much more drastically in Ultisols, because of their low fertility.

#### *e. Nutritional Properties*

Unfortunately the good aggregation qualities of these soils also result in ideal conditions for nutrient losses, especially bases (Ca, Mg, K). This in turn brings about severe acidity problems, including toxicity caused by Al and, to a lesser extent, Mn, and P availability problems due to its fixation on Fe and Al oxides and hydroxide surfaces. The rapid leaching also results in poor effective cation exchange capacity (ECEC) due to a restricted specific surface on the clay particle aggregations. As no favorable conditions for organic matter accumulation are present, nitrates are lost easily by leaching and N availability is always limited. Leaching of micronutrients due to this acidity results in deficiencies more commonly observed in even older soils highly exposed to runoff. All of these properties collectively account for the low fertility of Alfisols and especially Ultisols.

#### *f. Management*

The priority in managing these soils is replacing the lost Ca and Mg by liming, along with the selection of acid-tolerant cultivars. Agriculture in these soils is possible with an intense and well-balanced NPK fertilization program if an adequate supply of minor elements is included. The use of organic fertilizers, along with liming, can be an important source of nutrients while at the same time improving the physical properties altered by mismanagement.

# IV. CONCLUSIONS

Remarkably, the soil diversity in Costa Rica has allowed the country to diversify its economy, thus helping to attain its sustainability to date. The many different environments available, mostly determined by soil and climatic factors, are a significant factor in defining living conditions, biodiversity, and other related socioeconomic aspects of development. In addition, the basic information is available now in a digital form, and new methods of combining GIS's with modeling allow some very interesting cost-effective possibilities for extending analyses (Chapter 13).

To summarize, the country lowlands with fertile soils developed on recent alluvial deposits are planted with bananas and oil palm plantations to generate export income. The hillsides covered by recent volcanic ash deposits are mainly planted with coffee, sugarcane, and vegetable plantations where large numbers of Costa Ricans work on a daily basis. The most abrupt mountains are protected as national parks and used to generate income in ecotourism. In this way, good and poor soils are used more or less properly to satisfy the needs of a growing population.

Sound agronomic practices have been developed through time by Costa Rican scholars in order to overcome nutrient deficiencies, and thanks to extension and the high literacy rates of Costa Ricans they tend to be implemented. In recent times, more environmentally friendly practices such as composting, cover crops, and green manuring have been implemented, and these have supplemented, but not replaced, the high use of chemical fertilizers that are necessary on most of these soils. Another beneficial trend, although not related to soil fertility, is that research on organic pesticides has helped to ameliorate pollution problems.

Practices of the past, when uncultivated land was abundant, still exist and present a challenge for scientists to work on. The rate of erosion is effected by climatic factors, slopes, and land use, and it remains high. Substantially larger amounts of soil are lost from annual crops than from pastures or perennial vegetation (Table 10-5). The dependency on imported fertilizers to fulfill the nutrient requirements of the crops is being reevaluated and new approaches, such as a greater use of biological nitrogen fixation, are being implemented. The use of very large amounts of fertilizers might be beginning to affect groundwater quality, especially in dairy areas where urine nitrogen is also added to the system. $N^{15}$ studies are being undertaken to elucidate the N sources. In high-input systems, such as banana plantations and ornamentals in confined environments, the very intensive use of agrochemicals continues to be a problem (Chapter 20). This situation is heavily criticized by environmental groups and the companies are implementing more sound alternatives.

The agricultural sector of the country has been impacted greatly by structural adjustment and globalization through the increasing exports of nontraditional

TABLE 10-5 Costa Rica Soil Losses by Erosion in 1984 by Crop Type, ha, Percent, and T/ha/year

| Type of erosion | Units | Total | Crops | | Pasture (cattle) |
|---|---|---|---|---|---|
| | | | Annual | Perennial | |
| Total area in production | $10^3$ ha | 2435 | 413 | 252 | 1770 |
| Total erosion | $10^6$ T | 224 | 126 | 14 | 84 |
| | Percent | 100 | 56 | 6 | 38 |
| | T/ha/year | 92 | 304 | 56 | 48 |
| No sustained erosion | $10^6$ T | 189 | 119 | 9 | 60 |
| | Percent | 100 | 63 | 5 | 32 |
| | T/ha/year | 77 | 289 | 37 | 34 |

*Source:* Centro Cientifico Tropical and World Resources Institute (1991). Costa Rica MIDEPLAN (1996).

products and a reduction in the area planted in staple crops and vegetables, mainly because of lower production costs in neighboring countries. Meanwhile new crops (e.g., cantaloupe, watermelon, mangoes, ferns, pineapples, and heart of palm) have expanded. In this way, the agricultural sector in Costa Rica maintains the same economic contribution, in absolute terms, as years before. The diversity and basic good qualities of the soils of Costa Rica have helped the nation adjust to the changing economic environment imposed by the rest of the world. A more difficult question is whether this system, sustainable so far—in the sense that it still feeds directly or indirectly the Costa Rican population—can continue that sustainability in terms of the increased costs of inputs, the cumulative long-term effects of erosion, and the increased pressure placed on the soils by a continually growing population. These questions are explored in other chapters, especially in Chapters 15 and 23.

## ACKNOWLEDGMENTS

We thank Donald Kass and Scott Barr for contributions to this chapter.

## REFERENCES

Acuna, O., and L. Uribe. 1996. Inoculacion con 3 cepas seleccionadas de *Rhizobium leguminosarum* bv. phaseoli. *Agronomia Mesoamericana* 7:35–40.

Alvarado, A. 1975. Fertilidad de algunos Andepts dedicados a protreros en Costa Rica. *Turrialba* 25:265–270.

Alvarado, A. 1982. Phosphate retention in andepts from Guatemala and Costa Rica as related to other soil properties. Ph.D. Thesis, North Carolina State University, Raleigh. 82 pp.

Alvarado, A. 1985. *El origen de los suelos.* CATIE, Turrialba, Costa Rica.

Alvarado, A., E. Gutierrez, M. Baldares, and L. G. Brenes. 1993. Indicadores de sostenibilidad para los sectores agricola y de recursos naturales en Costa Rica. Presented at Congreso Nacional Agropecuario y de Recursos Naturales (9, 1993, San Jose), Colegio de Ingenieros Agronomos, San Jose, Conf. 1.

Alvarado, A. 1996. El papel del estudio del suelo en la agricultura costarricense. In F. Bertsch *et al.* (Eds.), *Congreso Nacional Agronomico y de Recursos Naturales (10, 1996, San Jose)*: Memorias, pp. 1–5. EUNED/EUNA, San Jose, Costa Rica.

Arroyo, L. A. 1996. Metodo de evaluacion de tierras para cultivos anuales, por medio del sistema de informacion geografica: Estudio de caso, distrito de Upala. In F. Bertsch *et al.* (Eds.), *Congreso Nacional Agronomico y de Recursos Naturales (10, 1996, San Jose)*. Memorias, pp. 29–37. EUNED/EUNA, San Jose, Costa Rica.

Ascencio, M. 1999. Logros del programa de control de losdesordenes causados por deficiencia de yodo. UNICEF, Costa Rica 3(1):92–97.

Barahona, R. 1993. Anteproyecto de ley de ordenamiento territorial (OTI). Presented at Congreso Nacional Agropecuario y de Recursos Naturales (9, 1993, San Jose), Colegio de Ingenieros Agronomos, San Jose, Conf. 14.

Beets, W. C. 1990. *Raising and Sustaining Productivity of Smallholder Farming Systems in the Tropics.* AgBe Publishing, Holland. 738 pp.

Bennett, H. H. 1926. General soil regions of eastern and northern Costa Rica. In J. C. Treadvell (Ed.), *Possibilities of Rubber Production in Northern Tropical America,* pp. 66–83. Bureau of Foreign and Domestic Commerce, Washington, DC.

Bergoeing, J. P. 1998. Geomorfologia de Costa Rica. San José. Instituto Geografico Nacional. 409 pp.

Bertsch, F. 1986. *Manual para interpretar la fertilidad de los suelos de Costa Rica.* Universidad de Costa Rica, Oficina de Publicaciones, San Jose, Costa Rica. 81 pp.

Bertsch, F., and V. Vega. 1991. Dinamica de nutrimentos en un sistema de produccion con bajos insumos en un Typic Dystropept del tropico muy humedo, Rio Frio, Heredia, Costa Rica. In T. J. Smyth, W. R. Raun, and F. Bertsch (Eds.), *Manejo de suelos tropicales en Latinoamerica,* pp. 28–32. North Carolina State University, Raleigh.

Bertsch, F., R. Mata, and C. Henriquez. 1994. Caracteristicas de los principales ordenes de suelos presentes en Costa Rica. Presented at Congreso Nacional Agropecuario y de Recursos Naturales (9, 1993, San Jose), Colegio de Ingenieros Agronomos, San Jose, Conf. 15.

Besoain, E. 1985. *Mineralogia de arcillas de suelos.* Vol. 60, *Serie de libros y materiales educativos.* IICA, San Jose, Costa Rica. 1205 pp.

Bornemisza, E. 1990. *Problemas del azufre en suelos y cultivos de Mesoamerica.* Editorial de la Universidad de Costa Rica, San Jose. 101 pp.

Cabalceta, G., A. D'Ambrosio, and E. Bornemisza. 1996. Evaluacion de cobre disponible en Andisoles e Inceptisoles de Costa Rica plantados de cafe. *Agronomia Costarricense* 20(1), 125–134.

Canessa, J., F. Sancho, and A. Alvarado. 1987. Retencion de fosfatos en Andepts de Costa Rica. II. Respuesta a la fertilizacion fosforica. *Turrialba* 37:211–218.

Celis, R., and A. Alvarado. 1994. Land taxation for sustainable development in Central America: The role of soil and social scientists. In *Proceedings, World Congress of Soil Science (15, Mexico, 1994)*, Vol. 9 (supplement, transactions), pp. 104–105. ISSS, Acapulco, Mexico.

Centro Cientifico Tropical and WRI. 1991. *La depreciacion de los recursos naturales en Costa Rica y su relacion con el sistema de cuentas nacionales.* CCT/WRI, San Jose, Costa Rica.

Chaverri, D., and A. Alvarado. 1979. Cambios quimicos de importancia sufridos por las cenizas del Volcan Irazu en 15 anos. *Agronomia Costarricense* 3:181–182.

Chaves, M., and A. Alvarado. 1994. Manejo de la fertilizacion en plantaciones de cana de azucar (Saccharum spp.) en Andisoles de ladera de Costa Rica. In *Proceedings, World Congress of Soil*

*Science (15, Mexico, 1994)*, Vol. 7a, pp. 353–372. Commission VI Symposia. ISSS, Acapulco, Mexico.

Cohen, A., R. Raymond, S. Mora, A. Alvarado, and L. Malavassi. 1986. Caracteristicas geologicas de los depositos de turba de Costa Rica. *Revista Geologica de America Central* 4:47–67.

Colmet-Daage, F., *et al.* 1973. *Caracteristiques de quelques sols derives de cendre volcanique de la Cordillera Central du Costa Rica.* Prov. Office de la Recherche Scientifique et Technique Outre Mer, Centre des Antilles, Bureau del Sols, Guadaloupe. 32 pp.

Cordero, A., and G. F. Ramirez. 1979. Acumulamiento de cobre en los suelos del Pacifico Sur de Costa Rica y sus efectos detrimentales en la agricultura. *Agronomia Costarricense* 3:63–78.

Cordero, A. 1994. *Fertilizacion y nutricion mineral del arroz.* Editorial UCR, San Jose, Costa Rica.

Costa Rica ITCO (Instituto de Tierras y Colonizacion). 1964. *Estudio de la region de Upala.* ITCO, San Jose, Costa Rica.

Costa Rica MAG (Ministerio de Agricultura e Ganaderia). 1996. *Informe del Departamento de Suelos y Evaluacion de Tierras.* MAG, Direccion de Investigaciones Agropecuarias, San Jose, Costa Rica.

Costa Rica MAI (Ministerio de Agricultura e Industria). 1958. *Estudio preliminar de suelos de la region occidental de la Meseta Central.* Vol. 22, *Boletin Tecnio.* MAI, San Jose, Costa Rica. 64 pp.

Costa Rica MIDEPLAN (Ministerio de Planificacion Nacional y Politica Economica). 1996. *Principales indicadores ambientales de Costa Rica.* MIDEPLAN, San Jose, Costa Rica. 122 pp.

Coto, J. A., and J. E. Torres. 1970. *Mapa de uso potencial de la tierra de Costa Rica.* Ministerio de Agricultura y Ganaderia, San Jose, Costa Rica. [Scale 1 : 750,000]

Dindal, D. L. 1990. *Soil Biology Guide.* Wiley, New York. 1349 pp.

Dondoli, C., and J. A. Torres. 1954. *Estudio geoagronomico de la region oriental de la Meseta Central.* Ministerio de Agricultura e Industrias, San Jose, Costa Rica. 180 pp.

FAO. 1990. *Forest Resources Assessment.* FAO, Rome.

Gonzalez, R. 1993. *El regimen de tenencia de la tierra en Costa Rica.* EUNA, Heredia, Costa Rica. 178 pp.

Harris, S. A. 1971. Podsol development on volcanic ash deposits in the Talamanca range, Costa Rica. In D. H. Yaalon (Ed.), *Paleopedology*, pp. 191–209. ISSS, Jerusalem, Israel.

Hartshorn, G. S., L. Hartshorn, A. Atmella, L. D. Gomez, A. Mata, R. Morales, R. Ocampo, D. Pool, C. Quesada, C. Solera, R. Solarzano, G. Stiles, J. Tosi, A. Umaña, C. Villalobos, and R. Wells. 1982. *Costa Rica: Country Environmental Profile: A Field Study.* Tropical Science Center, San Jose, Costa Rica.

Henriquez, C., and F. Bertsch. 1994. Efecto de la aplicacion fraccionada del fertilizante potasico en un Andisol bajo cultivo de maiz y frijol en Coto Brus, Costa Rica. *Agronomia Costarricense* 18:53–59.

Higgins, G. M., *et al.* 1984. *Capacidades potenciales de la carga demografica del mundo en desarrollo.* FAO/FNUM/IISA, Rome.

Holdridge, L. R. 1979. *Ecologia basada en zonas de vida.* Serie de libros y materiales educativos no. 34. IICA, San Jose, Costa Rica.

ICAFE (Instituto de Café de Costa Rica). 1989. *Manual de recomendaciones para el cultivo del cafe.* Programa Cooperativo ICAFE–MAG, San Jose, Costa Rica. 122 pp.

IICA (Instituto Interamericano de Ciencias Agricolas). 1979. *Estudio de planificacion agricola del Pacifico Seco.* IICA, San Jose, Costa Rica. [Scale 1 : 200.000, 1 h.]

Kleysen, B. J. 1990. A flows-counter flows matrix accounting method for the analysis of internal migration: Application to Costa Rica 1973–1984. Ph.D. thesis, Cornell University, Ithaca, NY. 130 pp.

Klingebiel, A. A., and P. H. Montgomery. 1961. *Land Capability Classification.* Agricultural Handbook 210, Department of Agriculture, Washington, DC.

Lag, J. (Ed.). 1990. *Geomedicine.* CRC Press, Boca Raton, FL.

Leonard, H. J. 1986. *Recursos naturales y desarrollo en America Central: Un perfil ambiental regional.* Trad. Por G. Budowski y T. Maldonado. CATIE, Turrialba. 267 pp.

Lucke, O. 1993. Bases de un marco conceptual y lineamientos generales para el diseno de un sistema de planificacion ambiental y ordenamiento territorial en Costa Rica. Presented at Congreso Nacional Agropecuario y de Recursos Naturales (9, 1993, San Jose), Colegio de Ingenieros Agronomos, San Jose, Costa Rica, Conf. 6.

Lutz, E., and H. Daly. 1991. Incentives, regulations, and sustainable land use in Costa Rica. *Environment and Resources Economics* 1:179–194.

Martinez, R., E. Liboreiro, and L. Flores. 1983. Tenencia de la tierra y reforma agraria en Centroamerica. Mimeograph. IICA, San Jose, Costa Rica. 18 pp.

Molina, E., A. Cordero, and F. Bertsch. 1986. Potasio en Andepts de Costa Rica. II. Respuesta a la fertilizacion con P y K en invernadero. *Turrialba* 36:289–298.

Perez, S., A. Alvarado, and E. Ramirez. 1978. *Asociaciones de subgrupos de suelos de Costa Rica (mapa preliminar).* OPSA, San Jose, Costa Rica. [Scale 1 : 200.000, 9 h.]

Plath, C. V., and A. J. van der Sluis. 1965. *Mapa de uso potencial de la tierra.* IICA, Turrialba, Costa Rica.

Prescott, S. C. 1918. *Examination of Tropical Soils.* Bulletin no. 3, United Fruit Company, Research Laboratory. La Lima, Honduras. 594 pp.

Quiros, T. 1954. *Geografia de Costa Rica.* Instituto Geografico de Costa Rica, San Jose, Costa Rica. 191 pp.

Radulovich, R., P. Sollins, P. Baveye, and E. Solorzano. 1992. Bypass water flow through unsaturated microaggregated tropical soil. *Soil Science Society of America Journal* 56:721–726.

RELACO. 1996. El uso sostenible del suelo en zonas de ladera: El papel esencial de los sistemas de labranza conservacionista. In F. Bertsch and C. Monreal (Eds.), *Reunion Bienal de la Red Latinoamericana de Labranza Conservacionista (3, San Jose, 1995). Memorias.* ACCS, San Jose, Costa Rica. 307 pp.

Reynolds, J. N. 1991. Soil nitrogen dynamics in relation to groundwater contamination in the Valle Central, Costa Rica. Ph.D. thesis, University of Michigan, Ann Arbor, MI.

Rodriguez, A. (Ed.). 1999. *Escenarios de uso del territoria para Costa Rica en el año.* 2025. MIDEPLAN, San José, Costa Rica, 108 pp.

Rodriguez, C. R. 1989. Concentracion de la tierra en Guanacaste, 1950–1970. *Ciencias Sociales* 43:73–80.

Sader, S. A., and A. T. Joyce. 1988. Deforestation rates and trends in Costa Rica 1940 to 1983. *Biotropica* 20:11–14.

Sancho, F., A. Cordero, and A. Alvarado. 1984. Fertilidad actual de los suelos de tres toposecuencias en el Pacifico Seco de Costa Rica. *Agronomia Costarricense* 8:9–16.

Sandner, G., *et al.* 1966. *Estudio geografico regional de la Zona Norte de Costa Rica.* Programa Zona Atlantica (CATIE/UAV/MAG), Guapiles, Costa Rica.

Sandoval, L. F., *et al.* 1982. *Mapa geologico de Costa Rica.* MINEREM Direccion de Geologia, Minas y Petroleo, San Jose, Costa Rica. [Scale 1 : 200.000]

SEPSA. 1991. *Estudio para aumentar el nivel de detalle del mapa de suelos de Costa Rica a escala 1 : 200.000.* Ministerio de Agricultura y Ganaderia, Convenio MAG/SEPSA–MIDEPLAN, San Jose, Costa Rica. 367 pp.

Soil Survey Staff. 1998. Keys to soil taxonomy. NRCS, USDA 8th edition. 325 pp.

Stoorvogel, J. J., and G. P. Eppink. 1995. *Atlas de la Zona Atlantica Norte de Costa Rica.* Programa Zona Atlantica (CATIE/UAV/MAG), Guapiles, Costa Rica.

Strasna, J. D., and R. Celis. 1992. Land taxation, the poor and sustainable development. In A. Sheldon (Ed.), *Poverty and Natural Resources in Public Policy in Central America,* pp. 143–169. New Transactions Publications, Brunswick.

Torres, J. 1993. El agro costarricense: De la explotacion a la sostenibilidad. Presented at Congreso Nacional Agropecuario y de Recursos Naturales (9, 1993, San Jose), Colegio de Ingenieros Agronomos, San Jose, Costa Rica, Conf. 2.

Ugalde, M. A. 1996. Evaluación de suelos y tierras por medio de modelos y sistemas de información geográfica. In F. Bertsch *et al.* (Eds.), *Memorias del Congreso Nacional Agronómico y de Recursos Naturales (10, 1996, San Jose)*, pp. 3–28. EUNED/EUNA, San Jose, Costa Rica.

Uribe, L. 1994. Formacion de nodulos de Rhizobium: Factores que puede conferir ventaja competitiva. *Agronomia Costarricense* 18(1):121–131.

USAID (U.S. Agency for International Development). 1965. *Costa Rica, analisis regional de recursos fisicos*. USAID, Washington, DC. [Scale 1 : 750.000]

Vargas, E., R. Solis, M. Torres, and L. McDowell. 1992. Selenio y cobalto en algunos forrajes de Costa Rica; efecto de la epoca climatica y el estado vegetativo. *Agronomia Costarricense* 16:171–176.

Vargas, E., and L. McDowell. 1993. Cobre, azufre y molibdeno en forrajes de diferentes zonas de Costa Rica; efecto de la epoca climatica y el estado vegetativo. *Agronomia Costarricense* 17:55–59.

Vasquez, A. 1979. *Mapa generalizado de suelos*. Unidad de Suelos, Direccion de Investigaciones Agricolas, MAG, San Jose, Costa Rica. (1):87–94.

Vasquez, A. 1993. Situacion actual del ordenamiento territorial en Costa Rica. In *Congreso Nacional Agropecuario y de Recursos Naturales (9, San Jose, 1993)*. Memorias, Vol. 1, no. 7. Colegio de Ingenieros Agronomos, San Jose, Costa Rica. 13 pp.

Vasquez, A. 1996. El ordenamiento territorial y los cambios en el uso de la tierra en Costa Rica. *Agronomia Costarricense* 20.

CHAPTER 11

# Remote Sensing and Land Use Analysis for Agriculture in Costa Rica

GREGOIRE LECLERC AND CHARLES A. S. HALL



## I. INTRODUCTION

In most developing countries, such as Costa Rica, the word "development" in the popular pronouncements and political statements is no longer synonymous with "green revolution" but with "sustainability." In the minds of many people and many programs, sustainability means somehow that *land* will be used sustainably, meaning that human economic activity will not automatically degrade it. But different land uses have different impacts and contribute differently to sustainability or the lack thereof. Land use analysis, which involves the quantification, often over time, of natural areas and the areas of land used for agronomy and forestry, plus ecology, economics, and marketing, must be interdisciplinary, quantitative, and objective, while still representing accurately an extremely complex problem. Remote sensing (RS) can play an active role in the information technology packages aimed at assisting our multidisciplinary teams in this task. Remote sensing, with its many platforms, from satellite to airborne, its many sensors, from photographic to hyperspectral, and its wide range of time coverage, can provide an assortment of imagery needed to assess land cover and land use, but only if the correct tools are used for the correct questions.

This chapter gives an introduction to those tools and their uses. It is divided into three main parts: the first reviews the historical development of RS and categorizes the tools available and some of their characteristics. The second section examines the historical applications of remote sensing to the assessment of forest cover and land use for all of Costa Rica. The third section examines specific applications of remote sensing to Costa Rican agriculture.

## A. Historical Development of Remote Sensing

Early efforts to map land use and land cover were based on special observation and drawing skills, and a capacity to estimate distances. The first known land use map (Figure 11-1) shows almost all the elements needed for a good map, except for a scale or the use of universal cartographic elements. The accuracy of the map is also doubtful and, although it certainly filled all the needs at the time, it would not be considered seriously in a modern geography environment except for the truism that having any map is better than none at all, and that even a crude map is the start for most development projects.

World War I greatly enhanced the infant discipline of photogrammetry. It did not take the combatants long to realize that they could measure distances, areas, and elevation, identify targets, and produce precise maps of all of these

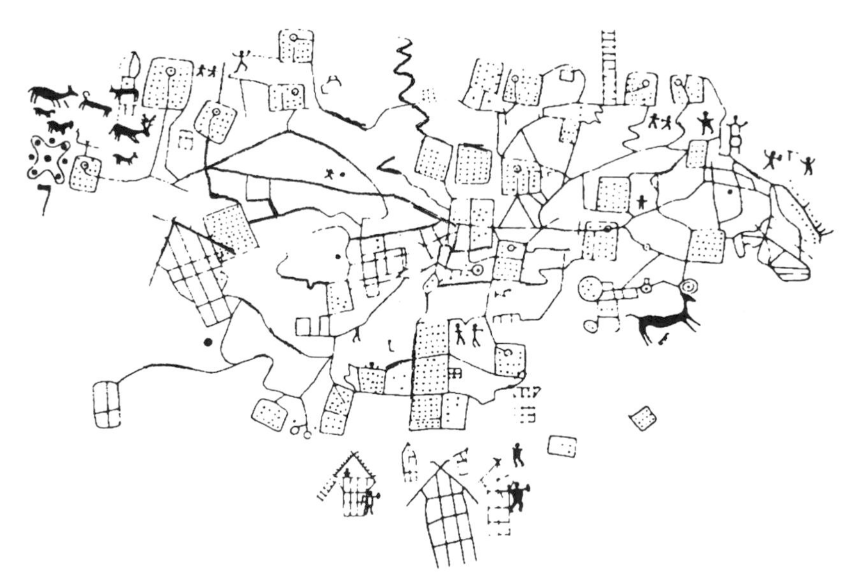

FIGURE 11-1 The earliest known map of an inhabited site: the Bedolina map from northern Italy, circa 200 B.C. From Eric Dudley, Map Maker Pro User's Manual.

using photographs taken with good large-format cameras on airborne platforms. The advantages of air photos compared to ground-level observations are (1) wider coverage, which led to better regional interpretation, (2) less obstruction (allowing for the first time a resolution to the age-old problem of "you can't see the forest for the trees"), and (3) a large reduction in perspective (therefore distance and area estimates are easier and more accurate). A look at an aerial photograph allows us to determine not only what is on the ground (*land cover*), but the way it is organized (*land patterns*), and for what purpose it is used by humans (*land use*). When the scale is large enough, meaning when the spatial resolution is fine enough, it is often possible to identify farming practices, for example, sun vs shade coffee plantations, associated cultures, or drainage patterns in banana plantations. More recently, multispectral imagery has opened the way to determine, from the sky, the developmental and health state of plants and trees and of soil characteristics. Now for the first time we can fully understand and easily quantify how farmers and others are collectively using the land over a region, and from this we can draw many conclusions about whether what we are doing is or is not sustainable and by what criteria.

## B. Satellite Sensor and Image Types

The past 25 years have seen the birth of satellite RS and its evolution into increasingly sophisticated technology designed to help us monitor the earth and the environment. The satellite programs include the U.S. NASA Landsat series, the French SPOT series, the U.S. NOAA TIROS weather satellites and later the ESA ERS-1 and ERS-2, the Japanese NASDA JERS-1, the Indian IRS-1, the Canadian RADARSAT, and RESURS from the former USSR (Table 11-1). Early satellites collected imagery designed to determine land surface areas precisely and give accurate pest monitoring and forest inventories. But with their coarse resolution and uneven radiometric quality they fell far short of delivering the results promised. With the deployment of more precise and versatile sensors, and the development of better software running on faster machines, so finally the early promises of remote sensing are starting to be realized.

## C. Passive Sensors

The source of the energy that allows our eyes and most satellite sensors to characterize the material world around us is, of course, the incident photons from the sun (Figure 11-2). Most satellite sensors operate by measuring the characteristics of the solar radiation that is reflected back from the earth's surface, in other words, the incoming radiation that is not absorbed. Different

TABLE 11-1 Current and Planned (before Year 2000) High-Medium-Resolution RS Satellites

| Country | Owner / OBJ | Program | Launch Actual/ Scheduled Date | Instrument Type | Resolution [m] | | | # Multi-spectral Bands, radar band | Stereo Type |
|---|---|---|---|---|---|---|---|---|---|
| | | | | | P | M | R | | |
| CANADA | G/O | Radarsat | 1995 | R | - | - | 10 | C | C/T,I |
| CHINA/ BRASIL | G/O | CBERS-1 | 1999 | M | - | 20,80,160 | - | 3,5,1 | |
| ESA | G/O | ERS-1 | 1991 | R | - | - | 30 | C | F/A,I,C/T |
| ESA | G/O | ERS-2 | 1995 | R | - | - | 30 | C | F/A,I,C/T |
| ESA | G/O | ENVISAT | 1998 | R | - | - | 30 | ? | F/A,I,C/T |
| FRANCE | G/O | SPOT-2 | 1990 | P,M | 10 | 20 | - | 3 | C/T |
| FRANCE | G/O | SPOT-4 | 1998 | P,M | 10 | 20 | - | 4 | C/T,F/A |
| INDIA | G/O | IRS-1 A | 1988 | M | - | 36.25 | - | 4 | |
| INDIA | G/O | IRS-1 B | 1991 | M | - | 36.25 | - | 4 | |
| INDIA | G/O | IRS-P2 | 1994 | M | - | 32x37 | - | 4 | |
| INDIA | G/O | IRS-1 C | 1995 | P,M | 5.8 | 25 | - | 4 | |
| INDIA | G/O | IRS-P3 | 1996 | M | - | 32x37 | - | 4 | |
| INDIA | G/O | IRS-1 D | 1997 | P,M | 10 | 20 | - | 4 | |
| INDIA | G/O | IRS-P5 | 1999 | P | 2.5 | - | - | - | |
| Israel/US | | WIS Eros-A | 1999 | P | 1.5 | - | - | - | |
| Israel/US | | WIS Eros-B1 | 1999 | P | 0.82 | - | - | - | |
| JAPAN | G/O | JERS-1 | 1992 | R,M | - | 18.3 | 18 | 3,L | C/T,I |
| JAPAN | G/O | ADEOS | 1996 | M | 8 | 16 | - | 4 | C/T |
| RUSSIA | G/O | Resurs-01-2 | 1988 | M | - | 45,160,600 | - | 3,4,1 | |
| RUSSIA | G/O | Resurs-01-3 | 1994 | M | - | 45,160,600 | - | 3,4,1 | E |
| RUSSIA | G/O | SPIN-2 | 1990 | P,M | 2 | 10 | - | | C/T |
| RUSSIA | G/O | Resurs-F1 | 1996 | P,M | 6-8 | 15-30 | - | 3 | C/T |
| RUSSIA | G/O | Resurs-F2 | 1996 | P,M | - | 6-8 | - | 6 | C/T |
| RUSSIA | G/O | Resurs-F3 | 1996 | P | 2 | - | - | ? | C/T |
| RUSSIA | G/O | Resurs-01#4 | 1998 | M | - | 30,160,600 | - | 3,4,1 | |
| RUSSIA | G/O | Resurs-01#5 | 1999 | M | - | 30,160,600 | - | 3,4,1 | |
| U.S. | G/O | NMP/EO1 | 1999 | M | | 10 | | 315 | |
| U.S. | G/O | Landsat 5 | 1984 | M | - | 30,80,120 | - | 6,4,1 | |
| U.S. | G/E | CTA Clark | 1998 | P,M | 3 | 15 | - | 3 | F/A |
| U.S. | C/O | Space Imaging IKONOS-1 | 1999 lost | P,M | 1 | 4 | - | 4 | F/A |
| U.S. | C/O | Space Imaging IKONOS-2 | 1999 | P,M | 1 | 4 | - | 4 | F/A |
| U.S. | G/O | TERRA | 1999 | M | - | 275 | - | 4 | F/A |
| U.S. | G/O | Landsat-7 | 1999 | P,M | 15 | 30,60 | - | 6,1 | |
| U.S. | C/O | EarthWatch Quickbird-1 | 1999 | P,M | 0.8 | 3.2 | - | 4 | F/A |
| U.S. | G/O | EO-1 | 1999 | P,M | 10 | 30,250 | - | 548 | F/A |
| U.S. | C/O | Orbview-3 | 1999 | P,M | 1-2 | 4-8 | - | - | |

M=Multispectral R=Radar P=Panchromatic
O=Operational C=Commercially Funded G=Government Funded E=Experimental
F/A=Fore&Aft Stereo I=Interferometry C/T=Cross-Track Stereo

materials absorb and reflect different wavelengths, so that the material on the earth's surface can be characterized by the difference between the intensity of particular incoming and outgoing wavelengths. For example, trees appear green to our eyes because chlorophyll absorbs red and blue, but not green, light. But in addition to a reflection peak around 550 nm (green), vegetation reflects near- and medium-infrared radiation very strongly (above 700 nm), except for absorption bands in the medium infrared corresponding to its water content (at 1450 and 1950 nm). The ability of different wavelengths to discriminate among materials on the ground (i.e., "themes") is the basis of thematic mapping satellite systems. Table 11-2 summarizes the interpretation

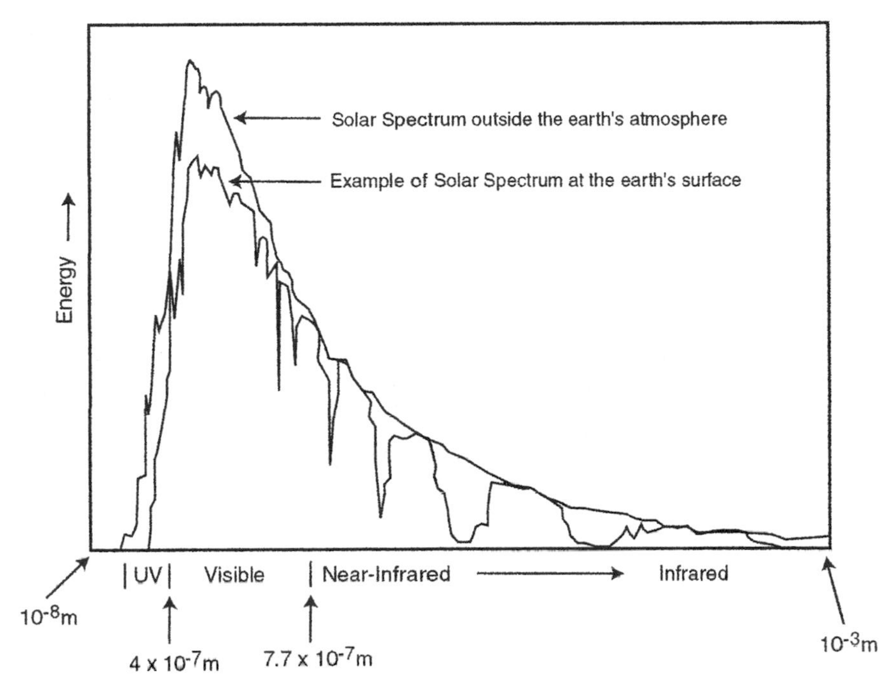


FIGURE 11-2 Summary of basic properties of incident light. Source: Epply Company.

**TABLE 11-2 Information Obtained from Thematic Mapper Satellites by Interpretation of TM Bands from Landsat Satellite Imagery**

| Band | Information represented |
|---|---|
| TM1, TM2, TM3 | Information about the chlorophyll and pigment concentration of the first layer of leaves near the top of the canopy. |
| TM4 | Detects reflection by multiple leaf layers such that reflectance increases nonlinearly with the number of leaf layers. A full canopy, often containing about six layers of leaves, reflects 85% more light than a single leaf layer. |
| TM5 | Sensitive to both leaf area and leaf water content. Detects interaction of light with more than one leaf layer but not as deep in the canopy as detected by TM4. |
| TM6 | Thermal IR band, sensitive to surface temperature (120-m resolution). |
| TM7 | Also sensitive to both leaf area and leaf water content. Detects reflectance from the first layer of leaves. |

of TM (thematic mapper, on board Landsat 5) wavelength bands with respect to light–canopy interactions.

The TM (thematic mapper, on board Landsat 5) is probably the sensor that has been used the most in general and in Costa Rica. Remotely sensed data are downloaded from satellites to ground stations where they are processed and transformed, and made available through various agencies. For example, Landsat imagery for Costa Rica is available from various agencies, including the U.S. EROS data center in South Dakota where it costs $450–600 for the new Landsat 7 TM imagery (resolution 15 m pan-chromatic, 30–60 m multispectral) or $425 for Landsat 5 TM imagery older than 10 years, and $200 for multispectral scanning (MSS), which has a resolution of 80 m. Landsat 5 TM imagery can be obtained through EOSAT (a private company) for a cost of approximately $4400 per scene for imagery less than 5 years old, and $2200 for additional scenes 5–10 years old. Six scenes, each covering 180 × 180 $km^2$, are needed to cover all of Costa Rica. The CLIRSEN agency, based in Cotopaxi, Ecuador, is another distributor of Landsat images for Costa Rica, but, unfortunately, periodic failure of the equipment, and some disorder in the archiving system, makes it sometimes difficult to obtain imagery. Costs are sometimes lower than those of EOSAT. SPOT imagery is available mostly through SPOT IMAGE, at a cost of FF 11,700 (about $2800 per scene). It has a much better spatial resolution than TM (20 m for multispectral and 10 m for panchromatic images), but covers only 60 × 60 $km^2$.

## D. Active Systems, Including Radar

Passive systems have a number of inherent limitations, including especially their inability to see through clouds. This is a particular problem in much of the tropics because many areas are cloudy almost all of the time (Figure 9-4). A study conducted by Maraux and Garcia using long time series of GOES (a weather satellite) imagery showed that for the SPOT and Landsat satellites the probabilities of getting clear imagery for Central America are very low, in particular for the Atlantic coast of Nicaragua, Costa Rica, and Panama (Maraux and Garcia, 1990). Another problem with passive systems is that they cannot detect soil moisture accurately, which is very important because crop production is often highly correlated with soil moisture. Finally, we may need to know the bare soil areas during the rainy season to predict high erosion areas, but then clouds make it impossible to use satellite information.

One answer to these problems is the use of active systems in which the platform with the sensors also sends out the signal, the characteristics of which are designed for the purposes of the survey (including the penetration of clouds). The most familiar, and the most useful, of these techniques is radar.

Basically a microwave signal is sent out and the reflected microwave signal is analyzed for the spatial patterns of backscatter intensity, commonly called "reflectivity" or "brightness." Most of the studies using radar imagery carried out in tropical environments have focused on geology or forestry research. It might seem logical to use radar to assess other biological information, including crop coverage. This type of study has been performed in temperate regions. In effect, it has become possible to discriminate crop types and growth stages in flat areas with radar imagery (Brown *et al.*, 1993; Wooding, 1988) as well as soil moisture (Dubois *et al.*, 1995). But in the tropics the crops are more difficult to differentiate by radar, the fields being rarely large, flat, uniform, or of regular shape, characteristics that make radar useful in the developed countries. Thus there is not much known about the potential of SAR (synthetic-aperture radar) images for the inventorying and monitoring of non-export tropical crops. The lack of this type of research in Central America can be understood easily, since the combination of microparcelation and a pronounced topography complicates the matter significantly. In addition, topography causes particular geometrical distortions in radar images, including foreshortening, layover, and shadowing, that must be corrected for in order to map features and measure areas. Also, the slopes facing the antenna appear very bright and the ones facing away appear darker. These effects impede visual interpretation of vegetation as well as quantitative analysis. Fortunately simple image processing methods have been developed to alleviate these problems (Leclerc *et al.*, 1996; Baudoin *et al.*, 1994), but they have yet to be applied on a very large scale.

Costa Rica has been surveyed a few times with airborne radar sensors, such as Intera X-SAR, SEASAT, ERS-1/2, and JERS-1, or RADARSAT. The most complete high-resolution coverage was performed in April 1992 by the Canada Center for Remote Sensing using the C-SAR airborne radar sensor. These images are used in the following analysis so we describe them in detail here. The SAREX images are two bands and "seven-look" (which means that each pixel is the result of the averaging of seven measurements, and therefore very smooth) that have been acquired in two modes of polarization, horizontal emitted–horizontal received (HH), and vertical emitted–vertical received (VV). The signal is expressed as the "digital number" (DN), which is proportional to the square root of the received radar power. The images were acquired with different slant range resolutions (6 or 12 m) with a pixel size of 4 or 15 m in the range direction (perpendicular to flight direction) and 4.31 or 6.9 m in the azimuth direction (along the flight direction). Incidence angles (in the range direction) are from 0 to 75° and the swath width is approximately 18 km. The SAREX mission covered 60% of Costa Rica (2700 km$^2$), in flat areas as well as mountainous areas (Figure 11-3). The images are available on CD-ROM from IGN and the Canadian Center for Remote Sensing (CCRS).

FIGURE 11-3 Position of flight strips from SAREX '92 radar flights. 8 = flight lines 8.1, 8.2, 8.3; 7 = flight lines 7.1, 7.2; 4 = flight lines 4.1.

Acquisition of imagery was not the only purpose of the SAREX program. It also included several national efforts for interpretation of the imagery and application to specific areas, such as coastal areas mapping, agriculture, geology, and forestry, as well as several workshops. The results are reported in an extensive document available at IGN. Our contribution is summarized in section III of this chapter.

## E. Land Cover and Land Use Maps

Once the RS data are obtained the problems are far from over. Given the wide range of possibilities that remote sensing offers, with its multiple sensors and diverse spatial and spectral resolutions, and given the frequency with which a site can be revisited from RS platforms, codification and classification of data are critical. Land use maps are a lot more complicated to generate than land cover maps because we are dealing with the interaction of humans with land in all its infinite variations. This need originally resulted in a seemingly infinite number of classification systems developed for particular purposes. For example, the Land Use Coding of the U.S. Geological Survey (USGS), which is applied in the United States, is a useful hierarchical system that has barely been applied in Latin America. The FAO, with decades of experience working in developing countries, devised a land use coding system that may be best, but that has severe deficiencies when linked to information models (Stoorvogel, 1995). A third approach is reflected in the International Institute for Aerospace Survey and Earth Sciences (ITC) system that makes a clear distinction between land use and land cover, and establishes well-defined links between the two. In effect, one type of land cover can be classified as one or more land use types and vice versa. For example, a forest can be recreational or productive. Unfortunately the utility of this very flexible system is limited because it is not hierarchical.

In Latin America, hierarchical systems such as the one developed by Centro Interamericano de Fotointerpretacion (CIAF) (Vargas, 1994) are the most widely used, perhaps because CIAF has trained many remote sensing specialists in Latin America. Generally four levels are adopted:

1. Exploratory (identifying land cover types)
2. Recognition (determining the quality or condition of level 1)
3. Semidetailed (purpose, management, or quality of level 2)
4. Detailed (function, or quality of level 3)

Table 11-3 shows an example of the CIAF classification scheme.

## F. Classification Methods

The production of a map from satellite imagery is usually called either an *interpretation* (when the image is interpreted visually, and map elements are drawn by hand), or a *classification* (when computer algorithms convert a measured reflectance or combination of reflectances into a category or feature).

For interpretation of multispectral imagery, usually spectral bands are combined, by means of specialized software, into a *color composite* that allows the display of three times as much information as each band taken alone. For example, a "TM 453 RGB color composite" is a color image where Landsat TM band 4 is displayed in the red channel of the color monitor, band 5 in the green, and band 3 in the blue. On this image, a pixel with high intensity

TABLE 11-3 CIAF Land Use Classification System (translated from Vargas, 1994)

| LEVEL 1 Exploratory | LEVEL 2 Reconnaissance | LEVEL 3 Semi-detailed | LEVEL 4 Detailed |
|---|---|---|---|
| 1. Constructions | a. Urban | 1. Residential | Single-family, multi-family, hotels |
| | | 2. Industrial | Textile, transportation, factory, crafts |
| | | 3. Commercial | Supermarkets, open areas, closed areas |
| | | 4. Educational | Schools, universities |
| | | 5. Recreational | parks, theaters, clubs |
| | b. Rural | 1. Dense | villages, industries, parks |
| | | 2. Dispersed | Housings, corrals, cellars, greenhouses |
| 2. Crops and parcels | a. Perennial and semi-perennial | 1. Irrigated | Fruits, sugarcane, commercial crops |
| | | 2. Non-irrigated | Coffee, banana, oil palm, fruits, sugarcane |
| | b. Temporary | 1. Irrigated | Forages, beans, tomatos, flowers, commercial crops |
| | | 2. Non-irrigated | Cotton |
| | c. Confined | 1. Horticulture | Lettuce, cabbage, radish |
| | | 2. Flowers | Roses, carnations, ferns |
| | | 3. Nursery | Pine, coffee |
| 3. Herbaceous vegetation | a. Natural pasture | 1. Herbaceous | Type, protection, shepherding |
| | | 2. Shrubby | Type physionomic characteristic, floristic structure |
| | b. Pasture ground | 1. Irrigated | Leguminous, grasses, species |
| | | 2. Non-irrigated | Mixed, species |
| | c. Tundra | 1. Herbaceous | Type, species |
| | | 2. Shrubby | Protection shepherding |

| | | | |
|---|---|---|---|
| 4. Forest | a. Natural | 1. Broadleaf | Protection, commercial, species, density, structure |
| | | 2. Coniferous | Protection, commercial, species, density |
| | | 3. Bushes | Protection, commercial, specie, density |
| | b. Planted | 1. Broadleaf | Protection, commercial, specie, density |
| | | 2. Coniferous | Protection, commercial, specie, density |
| | | 3. Reforestation | Protection, commercial, specie, density |
| 5. Water bodies | a. Open areas | 1. Natural | Lake, snow, ice |
| | | 2. Artificial | Dam, reservoir |
| | b. Marsh | 1. Permanent | Herbs, eutrophic, shrubs |
| | | 2. Temporary | Herbs, shrubs |
| 6. Non-tilled Land | a. Exposed rocks | 1. Massive | Scarp, inselbergs |
| | | 2. Fragments | Debris |
| | b. Bare ground | 1. Induced erosion | Furrow, cavity |
| | | 2. Natural erosion | Mass movements, gullies |
| | | 3. Quarries, mines | Sand, limestone |
| | | 4. Riversides and beaches | Gravel, sand, tourism |
| | | 5. Dunes | |

in bands 4 and 5, but low in band 3, will appear yellow on the screen. Healthy vegetation will produce high intensity in band 4 (essentially because of large total leaf area), low intensity in band 5 (because of a high water content), and low intensity in band 3 (lots of chlorophyll), resulting in a red-orange color on the screen. Dry bare soil will show a large reflectivity in all three bands (especially band 3), which produces a white or bluish color on the screen. Bands can therefore be combined according to the user's needs, based on the spectral characteristics of the sensor and of the targets of interest, to produce a high-contrast color image. Interpreters learn to identify land use classes from tone, saturation, texture, ancillary data, and contextual information. Most of the remote sensing work done in Costa Rica has been based on visual interpretation of high-quality color composite prints obtained from imagery distributors. This is now changing as software and hardware have become more accessible. Now color composites can be produced in-house from digital imagery, and people are trained to produce numerical classifications from raw imagery.

Multispectral images can be interpreted visually, although RS software and statistical analysis permit a more objective analysis. Thanks to these considerable and still ongoing efforts one can now find a large range of shape recognition algorithms as well as classifiers, going from per pixel to contextual, using statistical clustering algorithms, knowledge-based Bayesian techniques, or neural networks to partition spectral space into interpreted categories. We describe briefly here some of the best known classification algorithms.

The Maximum Likelihood Classifier (MLC) has been around since the mid 1970s and is one of the most widely used supervised classification techniques (Swain and Davis, 1978). Interestingly, more sophisticated techniques have led to only marginal increases in classification accuracy while sometimes requiring huge amounts of disk space or computer time. The MLC is fundamentally simple: for ground truth training sites (areas on an image where the interpreter knows what is on the ground), the computer determines basic statistics for each band (such as mean and standard deviation) as well as the correlation matrix between bands. The assumption is that a given class has a normal probability distribution in the *N*-dimensional space of surface reflectivity and that this probability distribution can be computed from the training sites statistics. For example, bare soil has high reflectivity in all bands, so the probability that a pixel is bare soil is high if the intensity values for all bands for this pixel are high. This is very similar to the gradient analysis described in Chapter 7. All the manual work is focused on delineating training sites of adequate meaning and size. The algorithm then scans all pixels of the image, retrieves the intensity values for all bands, computes for each class the probability of membership based on training site statistics, and assigns the pixel to the class with the highest probability of membership (Figure 11-4a).

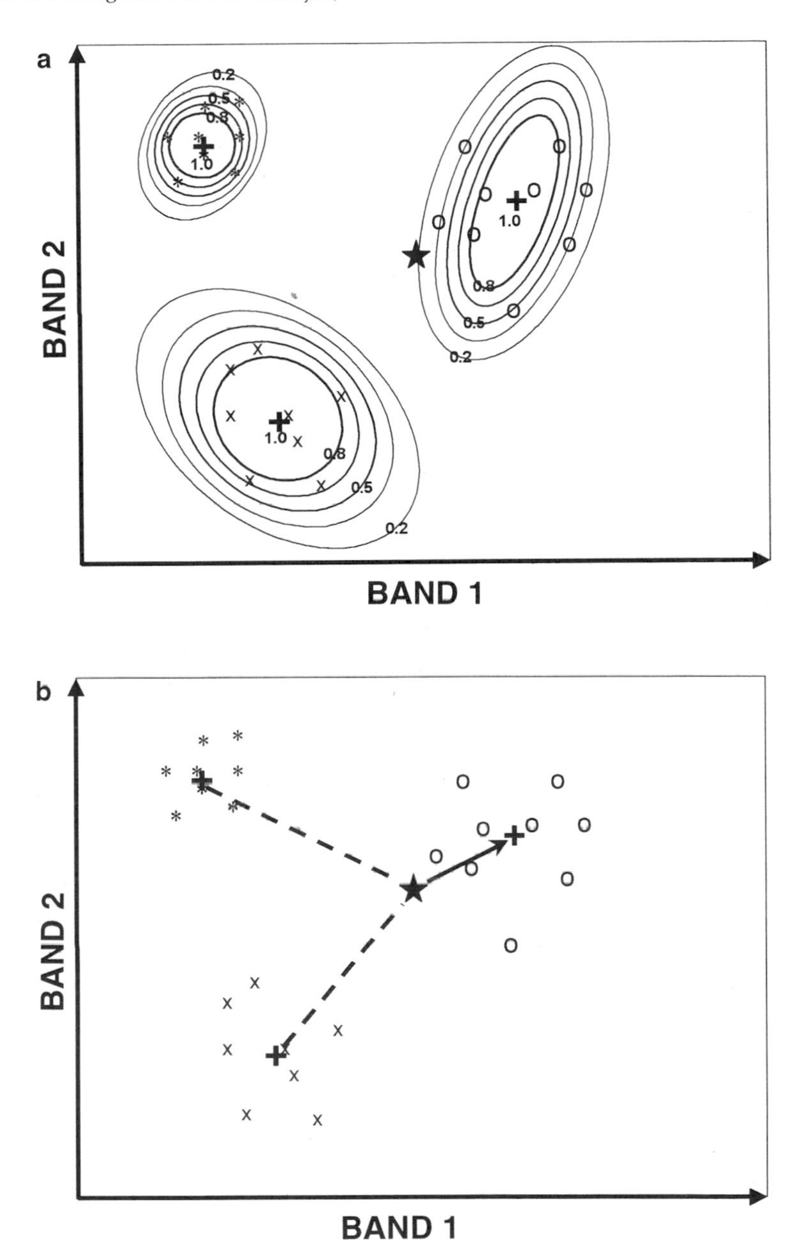


FIGURE 11-4 (a) "Maximum Likelihood" classifier. The pixel represented by a star is assigned to cluster "o" since it has a probability of 0.2 of being a member of cluster "o," which is higher than that for the other clusters.(b) "K-means" classifier. The pixel represented by the star in the two-dimensional space determined by bands 1 and 2 is assigned to cluster "o" since the distance to that cluster center ("+") is the shortest.

The K-means classifier (KMC) is one of the most widely used unsupervised classifiers. This method takes an approach somewhat the reverse of that of the MLC. It is basically a clustering method that examines the distribution of intensities present in each band of the image, and finds natural groupings. These groupings have no meaning until an interpreter labels them. Unlike the MLC, the ground truth is used not for training but for labeling. The user determines what land cover constitutes a cluster. The method first determines initial cluster centers (i.e., a set of characteristic intensities for each band, for each cluster). It next samples a subset of the image and assigns each pixel to the nearest cluster center. It iteratively recalculates the cluster center based on the new pixel's intensity values, then resamples the image, and so on until there is no significant change in cluster centers (Figure 11-4b).

The MLC and KMC are two "per-pixel" classifiers that assign land cover or land use classes based on the intensity values for individual pixels. Contextual classifiers, on the other hand, measure the neighborhood characteristics around each pixel. Generally, contextual classification starts by producing new bands containing regional characteristics computed around each pixel. The contextual bands are used with supervised or unsupervised classifiers instead of spectral bands. Popular contextual measures include "homogeneity," "coefficient of variation," and "entropy." Multiscale decompositions, such as the wavelet transform, are another group of popular contextual measures.

For crop mapping digital "per-field" classifications (which are based on averages for each agricultural field) have more potential than per-pixel classification (Ban *et al.*, 1993) because they tend to be less sensitive to variability in a crop's reflectivity. However, they require previous knowledge of field boundaries, which are difficult to obtain in the tropics (if we exclude export agriculture). In Central America, information on field boundaries is very scarce to say the least, and radar images are more likely to be of help in this respect since the fields are often separated by live fences which are detected by radar.

## G. Deriving Land Use Change Analyses from Remote Sensing

Techniques for land use change analysis can be classified into two broad groups:

1. Map to Map (GIS). This approach consists of comparing a map for one point in time with a map for another. A frequently used approach is to generate a table of land areas in the form [(land use X at time 1) → (land use X at time 2)]. Analysis of the matrix of the difference between two maps can help quantify the degree of change. The problem with this approach comes from the fact that if the classification system and mapping techniques are not exactly the same there is an additional error introduced in change estimates.

2. Image to Image (Remote Sensing). This approach uses one of two tools. The first is image differencing—the entire image from one date is subtracted from the same image at a second date. Sophisticated change detection algorithms based on image differencing have been developed for radar. For optical sensors, the NDVI (Normalized Difference Vegetation Index, a ratio of spectral reflectance images) is often used instead of raw imagery and is subject to standardized principal component analysis (see Eastman and Folk, 1993, for an interesting analysis of Africa).

It is important to emphasize the necessity of perfect geometric matching of the images for these analyses. Suppose we compare a map with the same map using a matrix of the differences for each pixel: we will find 100% accuracy. Now shift one of the maps by one pixel; the accuracy will decline suddenly (the greater the number of classes and fragmentation, the greater the decline). It can easily decline to 70% or less just by being off one pixel. When poor registration of maps is suspected, pixel by pixel analysis in error-prone and zonal statistics are preferred. Case studies for automated land use change analysis in the United States can be found in Green *et al.* (1994).

## H. Finding the Drivers for Land Use Change

There has not been a consistent and well-tested theory as to the cause of land use change (other than that hungry people will try to eat), and hence it has not been easy to make assessments of where and why land use change occurred or where it is likely to occur next. A summary of the most frequently mentioned causative agents of deforestation is given in Cornell and Hall (1994). Essentially all direct agents are exacerbated by population growth. Predicting the place where deforestation will occur, or is likely to occur in the future, is a more interesting question. Some previous analyses have simply taken a multivariate statistical approach.

New approaches based on temporal probabilities and spatial statistics are leading to substantial advances in our understanding of the mechanisms of change. In particular, the derivation of land use change indicators based on scale-consistent landscape indices and Markov chain coefficients are promising for better integration of the variables involved. Another approach is the use of semiempirical demand–pressure response spatial models such as the one described by Hall *et al.* (1995) or in the CLUES model (Veldkamp and Fresco, 1997). Hall *et al.* assumed that the principal aspect of the landscape, and hence maps, that determined where subsequent deforestation would take place was the expected energy return on energy invested by the farmer. Based on this, principle parcels are likely to be cleared that were adjacent to existing agriculture, or that had favorable physical characteristics based on an examination of the characteristics of areas already cleared. Hall *et al.* were able to predict the state (forested or nonforested) of more than 90% of a 1983 Costa Rican national map from a 1940 map (kappa statistic of about 0.50) based on an application of these principles (Figure 11-5). The validity of the results of such complex models relies essentially on calibration with actual land use, which will certainly benefit from more systematic input from RS.

## II. APPLICATION OF REMOTE SENSING TO ASSESSING FOREST COVER AND LAND USE CHANGE IN COSTA RICA

### A. How Land Cover Is Distributed in Costa Rica

There has been a great deal of international attention focused on the changes in the extent of tropical forests, including those of Costa Rica, because of

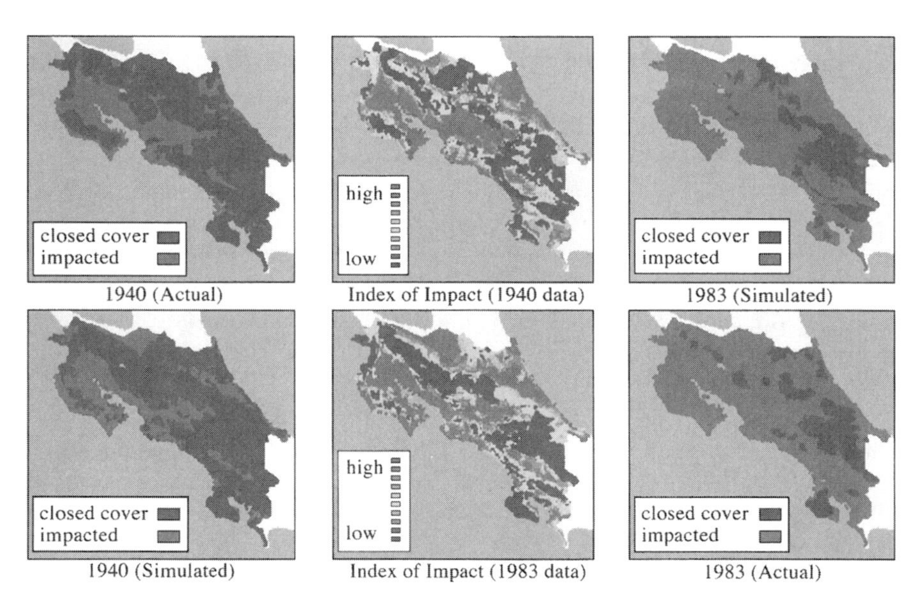


FIGURE 11-5 Land use change in Costa Rica from 1940 to 1983. (Top) Forward simulation starts with actual (left) to simulated (right) data. The quality of the 1983 simulation results can be determined visually by comparing the top right panel with the actual 1983 map below it. (Bottom) The backward simulation starts at the lower right and goes to the lower left—the effectiveness of this simulation can be seen by comparing the simulated vs measured maps on the left (1940) and on the right (1983), respectively. Middle maps indicate national probability of development based on the land already developed at the initialization of the simulation.

concerns about deforestation and the impacts such deforestation might have on the carbon dioxide concentration of the atmosphere and hence global climate. But for development and planning issues, it is as important to assess how other land use areas are distributed (such as pasture, urban and suburban, or agriculture), and how rapidly areas and patterns are changing. A variety of assessments, using a variety of methods, have been used as reviewed below and in Chapter 16.

## B. Early Land Use Assessments

Keogh (1984) produced a rough land cover (LC) map for the year 1800, when there were an estimated 52,000 Costa Ricans, based on historical records. This map illustrates the crudeness of our efforts before remote sensing, and shows that if remote sensing did not exist now our knowledge of the spatial distribution of LC would resemble that of 1800 [Figure 11-6a (color version on CD); note that on this map the contour of Costa Rica comes from modern maps]. The earliest contemporary assessment of forest cover in Costa Rica was in Zon and Sparhawk's 1923 assessment of the entire tropics. They suggested that at that time Costa Rica was estimated to be about 75% forested. Since their methods were not clearly ascribed, nor the results mapped, it is difficult to interpret the validity of their efforts. Nevertheless all of these early maps suggest that Costa Rica was largely forested before 1940 but even then human impacts were far from trivial.

From 1950 through 1984 land use and land use change were typically quantified in Costa Rica through the use of agricultural census data, which are based on a verbal on-the-ground sampling scheme, have the spatial resolution of the administrative unit, and can contain substantial errors. This is true especially for forest cover, which is considered unproductive, resulting in its undervaluing and incorrect areal estimates (Sanchez-Azofeifa, 1996). Such a national agricultural census was carried out in 1950, 1955, 1963, and 1973.

## C. The Advent and Use of Remote Sensing

Since the introduction of air photo interpretation in the 1940s, a considerable amount of reasonably reliable land use data have been produced for Costa Rica. The first was done in 1943 by the U.S. Department of Agriculture, in part to ensure sufficient coffee for the U.S. Navy! Little additional work was done until forest cover maps were produced to supplement the agricultural census of 1961 (IGN, 1967) and 1977 (MAG, 1977). Perez and Protti (1978) made a survey of the forest sector for the period 1950–1977. The Central Bank

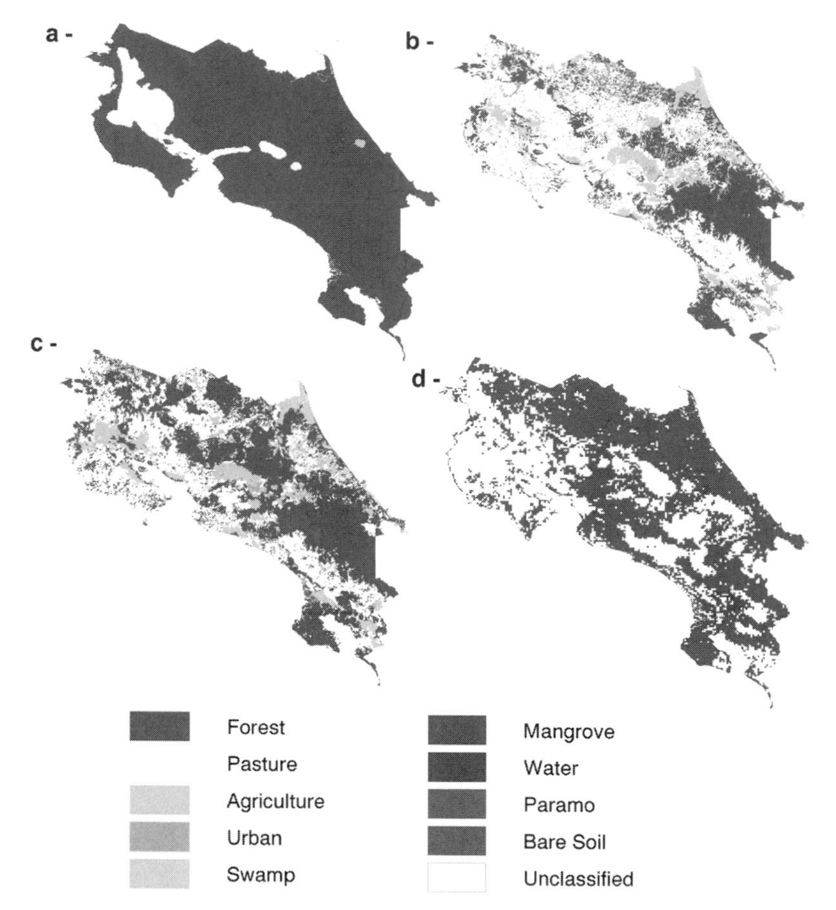


FIGURE 11-6 (a) Land cover in 1800 as derived by Keough (1984), based on data in Hall (1978). (b) LC1, land cover in 1992 from UNA–Clemson University. (c) LC2, land cover in 1992 from IMN. (d) LC3, land cover in 1992 from USGS. (Color version on CD.)

of Costa Rica also has produced relevant data. In 1978 Sylvander produced a forest map of Costa Rica based on satellite image interpretation. Most of the national-level RS efforts in Costa Rica are listed in Table 11-4. The products of those studies generally are available from the respective institutions.

There have been few if any purely Costa Rican RS efforts, mostly because of the cost of imagery, software, and equipment, the emigration of trained manpower, and the lack of a proper organizational context. In the past, chronic violation of satellite imagery copyrights helped Costa Rican (and U.S.) universities and government organizations develop a wide range of RS applications for the country. But today such copyright violations are fewer and more dangerous to those doing it, necessitating considerable efforts to obtain funds to purchase the imagery, to maintain newly installed facilities, and to keep trained personnel in the country. Financial restrictions have also limited participation of Costa Rican nationals in international conferences. For example, the 1993 International Geographical and Remote Sensing Conference in Tokyo, whose theme was "Better Understanding of Earth Environment," hosted five times more participants from developed countries than from developing ones (Perera and Tatieshi, 1995).

Thus most of the remote assessments of land use in Costa Rica have been international efforts. These assessments include TREES, FAO's world forest assessment, NASA's Pathfinder tropical deforestation project, and the USGS Global Land Cover Project to use satellite imagery to map and monitor tropical forests and estimate deforestation rates and forest biomass. The result of these efforts is that Costa Rican forests have been mapped relatively frequently, beginning in 1943 (USDA, 1943.) These maps, made at different scales by different techniques and at different dates, have allowed a relatively complete time series of forest–nonforest maps to be made (Sader and Joyce, 1988).

## D. Estimates of Land Use and Land Cover

Relatively simple land use trend analyses based on RS have been applied to Costa Rica with uneven success. The best known is that of Sader and Joyce (1988), who produced consistent forest cover maps based on various types of imagery for 1940, 1950, 1961, 1977, and 1983. These maps showed clearly the extremely high deforestation rate that occurred in Costa Rica during that time period and identified where it occurred. More recent work by Sanchez that builds upon and critiques some aspects of Sader and Joyce's study is given in Chapter 16. Both Sanchez-Azofeifa (1996) and Leclerc (1996) have shown that contemporary conditions are such that although net deforestation continues, there are considerable areas of secondary forests developing in some regions where pastures have been abandoned.

Changes in forest estimates between 1986 and 1992 in the Central Cordillera Conservation Area have been analyzed in relation to spatial pressure factors (Leclerc and Chacon, 1998). The simple weighted multicriterion model based on local population density, terrain slope, access to transportation network, forest management plans, and data from the Instituto de Desarollo Agrario (the national organization in charge of the agrarian reform) predicted deforestation correctly for a wide range of land change pressures. Stoorvogel (1995) analyzed land use change for a 3000-$km^2$ area in the Atlantic zone using three methods: (1) single time analysis of spatial patterns, based on a qualitative knowledge

TABLE 11-4 Large Coverage and National-Level RS Efforts in Costa Rica

| Project | Coordinating institution | Coverage | Period | Source | Products | Type of activity (b : baseline; r : research) |
|---|---|---|---|---|---|---|
| Aerial photos | IGN | Costa Rica | 1945–1989 | | Aerial photos | b |
| Base maps | IGN | Costa Rica | 1945–1989 | Aerial photos | Base maps 1 : 10,000; 1 : 25,000; 1 : 50,000; 1 : 200,000; 1 : 250,000 contour, infrastructure, land cover | b |
| Global rain forest mapping | NASDA, JPL, JRC | Global | 1995–1996 | JERS-1 | JERS-1 100m resolution digital mosaics; 12.5m resolution digital scenes | b, r |
| Global land cover/ global AVHRR | USGS/NASA | Global | 1991 | AVHRR | Land cover digital data, 1.1 km, maximum ndvi avhrr data | b, r |
| Forest cover (Sader) | USAID, Costa Rica government | Costa Rica | 1940, 1950, 1976, 1977, 1983, 1984, 1986 | Aerial photos, MSS | Land cover maps | b, r |
| LC change 1979, 1992 | IMN/MAG/IGN/ DGF | Costa Rica | 1979, 1990–1992 | Aerial photos, MSS, TM | LC 1979, 1992, LC changes 79-92, 1 : 300,000 map; 1 : 200,000 sheets digital | b |
| North American Landscape Characterization Program—pathfinder | EPA, EDC | Mexico, Central America, Caribbean | 1971, 1986, 1991 | MSS, TM | Land cover maps and digital data, 3 dates | |
| Forest 1961 | IGN | Costa Rica | 1961 | Aerial photos, MSS, TM | 1 : 750,000 scale map | b |
| Land use 84 | IGN | Costa Rica | 1984 | Landsat TM prints, b/w and IR aerial photos | LC maps, 1 : 200,000 scale (unpublished) | b |

| | | | | | | |
|---|---|---|---|---|---|---|
| Soil mapping | SEPSA, MAG | Costa Rica | 1979 | Aerial photos | 1 : 250,000 soils maps | b |
| Tropical deforestation and habitat fragmentation | UCR, CIEDES, FONAFIFO | Costa Rica | 1976, 1986, 1991, 1996 | Aerial photos, TM, MSS | 1 : 250,000 scale land cover, digital | r |
| COSEFORMA | DGF, GTZ | Costa Rica-North Atlantic | 1994 | Landsat TM | Forest cover photomaps, 1 : 50,000 | b |
| GAP | UNA, Clemson | Costa Rica | 1984, 1991–1992 | Landsat TM 1984 land cover map | Habitats, LU 1992, 1987, digital files | r |
| SAREX-92 | IGN, CCRS | Costa Rica (60%) | 1992–1994 | C-band HH/VV high-resolution radar, aerial photographs | Applied research in geology, agriculture, forestry, natural resources; digital imagery and recent area photos | r |
| Cobertura vegetal y de suelo | CCAD B2PROARCA | Central America | 1992, 1993, 1996 | AVHRR | Mapa de cobertura vegetal y de suelo 1 : 2,000,000 | b |
| TREES | JRC (EC) | Tropics | 1990 | Landsat TM AVHRR | 1992 LC map (focus on forest), 1.1 km forest cover | r |
| LC mapping | FUNDECOR | ACCVC | 1986, 1992, 1996 | Landsat TM | Lc and LC changes maps (focus on forest) | b, r |
| ADRO (RADARSAT) | CIAT | Central America | 1996–1997 | RADARSAT (all modes) | Evaluation of RADARSAT modes for land applications | r |
| Digital cartography | Comisión Terra | Costa Rica | 1997–1998 | Aerial photos | Digital cartography 1 : 25,000 | b |

of temporal patterns of land use evolution; (2) a Markov chain approach describing the probabilities of land use changes; and (3) a new adaptation of Markov chains that includes the influence of shape and size of polygons and neighbor associations. The third approach was based on two hypotheses: (1) that land cover conversion is influenced by land cover or land use in adjacent areas, and (2) that the shapes of adjacent areas are related with respect to both land type and rate of change. The validity of both hypotheses, although theoretically sound, has not been supported by the data.

## E. How Accurate Are the Surveys Based on Remote Sensing?

In this section we compare three 1992 land cover maps for Costa Rica produced independently by three different groups. The degree to which these maps do or do not agree should give us some confidence in the degree to which we are able to assess land use change, or at least classify land use consistently. The studies were as follows:

1. Map LC1 (Figure 11-6b), developed by Universidad Nacional Autonoma of Heredia, Archbold Tropical Research Center, and Clemson University, is based on GPS field work, digital processing of 1991–1992 Landsat TM imagery, and a comparison with unpublished 1:200,000 scale 1984 land use maps (IGN, 1967). The final product was a digital vector map with 10 land cover classes at a 200-m resolution in Arc/Info EXPORT format. The study was funded by USAID Grant HRN-5600-G-00-2008-00 USAID under the program, "Gap Analysis of Biological Resource Diversity Mapping."

2. Map LC2 (Figure 11-6c), developed by Instituto de Minas y Recursos Naturales of MINAE (Ministerio del Ambiente y Energia), was based on a visual interpretation of 1 : 60,000 scale aerial photographs from IGN and 1 : 100,000 scale Landsat TM false-color printouts from Dirección General Forestal (DGF), land use maps (1990, 1 : 10,000 scale) from IGN, land use maps (1992, 1 : 100,000) from the WAU-CATIE Atlantic Zone Project, and forest cover maps for the North Atlantic zone (1994, 1 : 200,000) from COSEFORMA. The final product was a digital Arc/Info vector map with 16 classes and a printed atlas at a 1 : 300,000 scale. The project was funded by UNEP Climate Unit Project GF/4102-92-42.

3. Map LC3 (Figure 11-6d), "USGS Global Land Cover Characterization" (http://edcwww.cr.usgs.gov/landdaac/glcc/glcc.html, USGS EROS Data Center, University of Nebraska-Lincoln, JRC, Version 1), is based on long time series of 1.1-km-resolution AVHRR NDVI (to get rid of clouds, the maximum value of NDVI for tens of daily images was used); this map has been validated from

high-resolution data sets. The final product is a digital raster map coded according to several classification schemes.

The three maps were available in digital format. We converted LC1 and LC2 from a vector format to a uniform 200-m-resolution grid raster format which corresponds to the minimum mapping unit for a 1 : 250,000 scale. LC2 and LC3 were reclassified into the same categories as LC1. Doubtful categories, such as "tacotal" or "mixed crops/pasture," were assigned to the class that produced the least error when compared to LC1.

Even for the broad land cover categories considered, the match between the three maps is extremely poor. The two that were the most similar are compared in Table 11-5.

For the Forest class, there are additional independent assessments for about the same time period. Sanchez-Azofeifa (1996) estimated from 1992 Landsat imagery covering 93% of the country that "closed forests" covered 1.36 million ha (Chapter 16), which is 70% less than the 1.70 million ha given by LC1. The FAO forest resources assessment for 1990 estimated the forest cover of Costa Rica as 1.43 million ha. Thus, these three estimates of forest area are reasonably close, and we conclude that the 2.46 million ha of LC2 is a substantial overestimate. The major difference between LC1 and LC2 may lie in the definition of "forest." Most analyses of the forest in Costa Rica assume that what defines a forest is a crown closure of greater than 80%. LC2 appears to have grouped together forest patches and included the land between the patches as forests. We see that despite many problems with the implementation of forest assessments using remote sensing, there is a fair agreement that there existed about 1.4 million ha of forest area remaining in 1990, and that the rate of deforestation was very rapid in the relatively recent past. There is some argument as to whether net deforestation is continuing through the 1990s or whether the high rate of deforestation is beginning to be balanced by the abandonment of pastures. The principal problem with giving a uniform assessment appears to be inconsistent definitions and standards applied by different projects.

Considering the 10 land cover classes together, and LC1 as a standard, LC2 and LC3 gave kappa values of 0.47 and 0.0, respectively. LC3 is therefore equivalent to a map drawn at random in comparison to LC1. Note that a value of kappa around 0.60 is generally considered good. We would like to warn the reader about using a land cover map with no accuracy assessment, and in general land cover areas should be considered with extreme caution.

Agreement between LC1 and LC2 is good for other land use classes except for Bare Soil. In Figure 11-6c one can see that there are large bare soil areas toward the Atlantic. These are thousands of landslides triggered by a 1991 earthquake (7.1 Richter scale) which had its epicenter near Puerto Limón on the Atlantic coast of Costa Rica. They are mapped in LC2 but not in LC1. If

TABLE 11-5a Areas ($km^2$) Obtained for LC1 and LC2

| | LC2 | LC1 | % diff. |
|---|---|---|---|
| Forest | 24591 | 16996 | 36.5 |
| Pasture | 18997 | 24161 | 23.9 |
| Agriculture | 5043 | 4236 | 17.4 |
| Urban | 226 | 203 | 10.6 |
| Swamp | 1059 | 1147 | 8.0 |
| Mangrove | 494 | 375 | 27.3 |
| Water | 98 | 100 | 2.7 |
| Paramo | 134 | 135 | 0.5 |
| Bare Soil | 350 | 111 | 104.0 |
| Unclassified | 3 | 3569 | |

TABLE 11-5b LC2 Accuracy vs LC1

| | LC2 | | |
|---|---|---|---|
| LC1 reference | User accuracy | Producer accuracy | Kappa |
| Forest | 0.80 | 0.60 | 0.52 |
| Pasture | 0.58 | 0.79 | 0.50 |
| Agriculture | 0.60 | 0.53 | 0.39 |
| Urban | 0.49 | 0.44 | 0.31 |
| Swamp | 0.72 | 0.81 | 0.62 |
| Mangrove | 0.69 | 0.59 | 0.47 |
| Water | 0.68 | 0.70 | 0.53 |
| Paramo | 0.63 | 0.67 | 0.48 |
| Bare Soil | 0.19 | 0.05 | 0.04 |

*Note.* For these nine categories (unclassified pixels were masked out), we found the following LC2 accuracies: overall, 0.67; kappa, 0.47; and tau, 0.57. User accuracy is 1 − (comission error); producer accuracy is 1 − (comission error).

it was not for remote sensing, the extent of the affected area would have been undetected given the extreme inaccessibility of the area.

## III. ANALYSIS OF CROPS

In this section we look at how remote sensing has been used to analyze agricultural issues in Costa Rica, and what role it might play in the future. The extremely wide range of possibilities makes an exhaustive review difficult. We will have a bias toward radar partly because that is where we have more experience and because we believe in the strong potential of this technology in Costa Rica where cloud cover seriously impedes timely and systematic use of optical sensors.

## A. General Overview of the Use of RS in Agriculture

The ability to identify crops, their health, and their areal extent is an important tool for predicting inventories, yield, and price, and for assessing taxes by governments. Although the main users of yield forecasts are governments and agribusiness, the individual farmer can benefit from them indirectly in terms of improved price stability and improvements in the quality of advice offered by agricultural support services. Also, RS techniques are extremely useful for identifying areas that are most susceptible to soil erosion, as well as monitoring with great sensitivity the implementation or nonimplementation of soil conservation practices such as agroforestry or the leaving of postharvest agricultural residues (Beaulieu, 1998).

A number of agricultural RS applications have become routinely operational in developed countries. CORINE is a land cover mapping system for all of Europe. Earlier European projects were AGRESTE, DUTA, and AGRIT. Similar projects for the United States include LACIE, CITARS, and AgRISTARS. Remote sensing has been used routinely for many years to monitor yields and pests in sugar beet farms and vineyards. Throughout the developed world an agricultural census is now a hybrid of conventional field techniques and RS image interpretation. Precision agriculture, whose goal is to improve yields and reduce environmental costs by monitoring the cropping system with high accuracy, also relies in part on airborne or satellite RS, which gives key information on spatial and temporal aspects of crop health, the efficiency of the irrigation system, and soil quality.

## B. The Use of RS for Agriculture in the Tropics

In the tropics RS use in agriculture has been limited mostly to examining the effects of particular crop stressors and to predict crop yields. When extreme cold struck the Brazilian coffee plantations in 1994, for example, speculators created a panic and international prices rose instantly. Researchers, however, determined by analyzing AVHRR images that the damage was much less than speculated, and the panic subsided.

## C. The Use of RS for Agriculture in Costa Rica

Only a subset of RS applications in Costa Rica have been systematic enough to allow conclusions to be drawn and RS operationalization achieved. Such applications as have occurred have been used for rice and coffee (shade and no shade), banana, pineapple, agroforestry, African palm, sugarcane, and pasture.

Figure 11-7 (on the CD) shows one agricultural area of the Central Valley, as seen by different sensors: TM, SPOT, and radar (SAREX '92). This area corresponds to one 1:10,000 scale map sheet which has an estimated planimetric precision of 3 m. Land uses comprise coffee, coffee with pine, urban areas, plantain, pine reforestation, greenhouses, pasture (wet and dry), etc. More examples of how major crops in Costa Rica appear when examined remotely under ideal conditions are given in Figure 11-8 (on the CD). Figure 11-9 (on the CD) shows another agricultural area of the Central Valley, near the Carchi dam, taken in (a) April and (b) June. The overall reddish, darker tone of the June image is indicative of more healthy vegetation cover during the rainy season. Next we summarize what we believe are the best examples of RS use for agriculture in Costa Rica.

### 1. Example: Synthetic Studies of the Atlantic Lowlands

The Wageningen Agricultural University of the Netherlands, in collaboration with CATIE and MAG, has worked in the Atlantic zone of Costa Rica since the late eighties on the development of a model of sustainable land management at the farm and regional levels. It is this team, perhaps, that has formulated the most comprehensive approach to this important issue in Costa Rica. The methodology they developed is built on modern information technology, i.e., synthesis of simulation of crop growth, GIS, RS, linear programming, and econometrics, and serves as a prototype of a process-oriented system that integrates these aspects. One of the results of this experiment is the definition of a new and perhaps better unit of analysis, the LUST (Land Use System and Technology), a static and descriptive formulation of land use systems which combines land use and land use technology based on the quantification of land unit inputs and outputs. The full results of this study were not available at the time of the writing of our chapter, but a good preliminary perspective is given in Veldkamp and Fresco (1997).

### 2. Example: Identifying Crop Type in the Central Valley with Radar

The most obvious agricultural feature apparent on the SAREX radar images of the Central Valley is the clear division of fields that are separated by tree

a

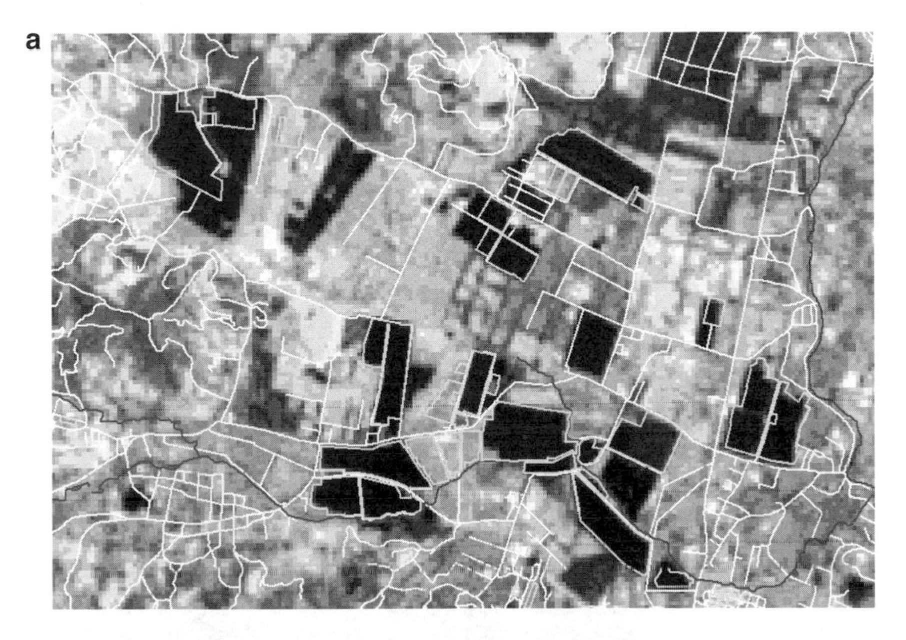

b

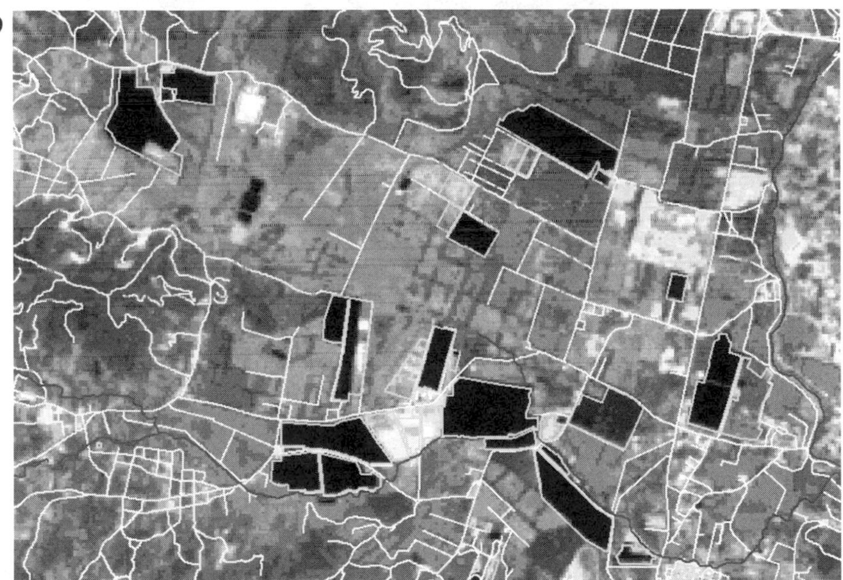

FIGURE 11-7 Area covered by 1:10,000 scale map of Purires, Central Valley, as seen by (a) Landsat TM 453, (b) SPOT XS 321, and (c) SAREX '92 C-HH high-resolution SAR. (On CD.)

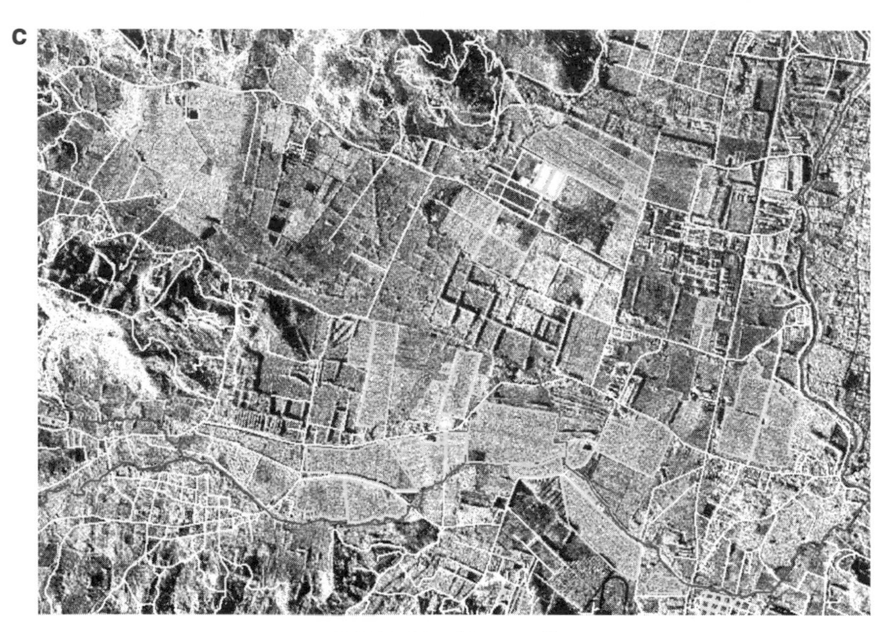

FIGURE 11-7 *(Continued)*

lines, fences, or small roads. Agroforestry, the use of living fences, "alley" cropping, windbreaks, and the presence of dispersed trees in pastures are very apparent in these images, even though identifying these features is often a problem with optical satellite imagery. Figure 11-10 shows the mottled texture corresponding to the mosaic of pasture and potato fields with dispersed live trees found near the summit of the Irazu Volcano.

We could distinguish the following land use classes from these images using tone, texture, and context in visual interpretation: forest, subsistence agriculture, export crops, and urban extensions. The shape and size of the fields and the way in which they are divided are elements that can help greatly in identifying crop types. For example, the export crops banana and sugarcane are often cultivated in large fields of regular shapes. In the Central Valley, the vegetable crops are cultivated in smaller fields, often separated by tree lines, and a large variation in color (i.e., on HH-VV color composites) could be seen from one field to another. In the windy region north of Cartago, the pasture fields also are separated by tree lines used as windbreaks. Pastures have very low tone variations between parcels so they could be distinguished readily from crops. In other parts of the images, the pastures covered extensive areas uninterrupted by fences or tree lines but had a much smoother texture than the forest areas.

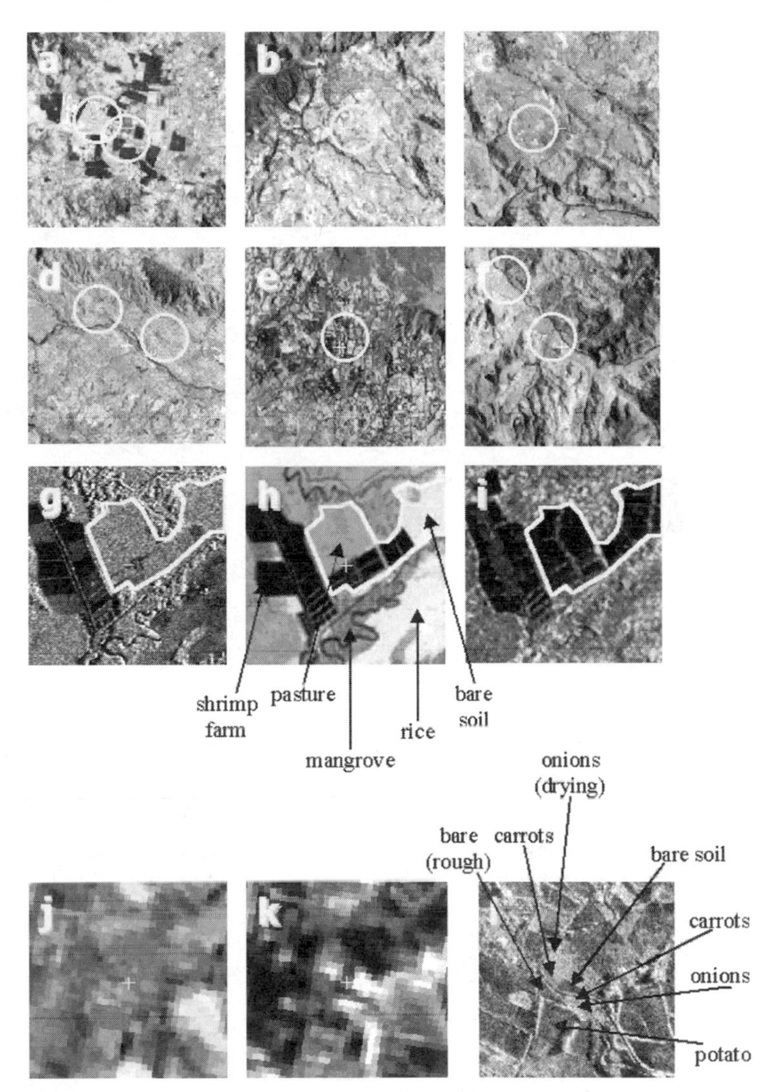


FIGURE 11-8 (a–f) Highlands agriculture: Landsat TM, 4-5-3, red-green-blue. (a) Greenhouses: decorative ferns for export; (b) pasture and potato area; (c) sugarcane (blue, bare soil); (d) coffee (with shade); (e) pasture and tree lines as wind breakers; (f) coffee (with and without shade). (g–i) Coastal agriculture. The yellow line shows the expansion of aquaculture. (g) SAREX'92, HH-VV, red-cyan; (h) Landsat TM, 4-5-3, red-green-blue; (i) Radarsat S1-F5, red-cyan. (j–l) Central valley. (j) Landsat TM, 4-5-3, red-green-blue; (k) Landsat TM, NDVI (4 − 3/4 + 3), Brighter, more vegetation; (l) SAREX'92, HH-VV, red-cyan.

a

b

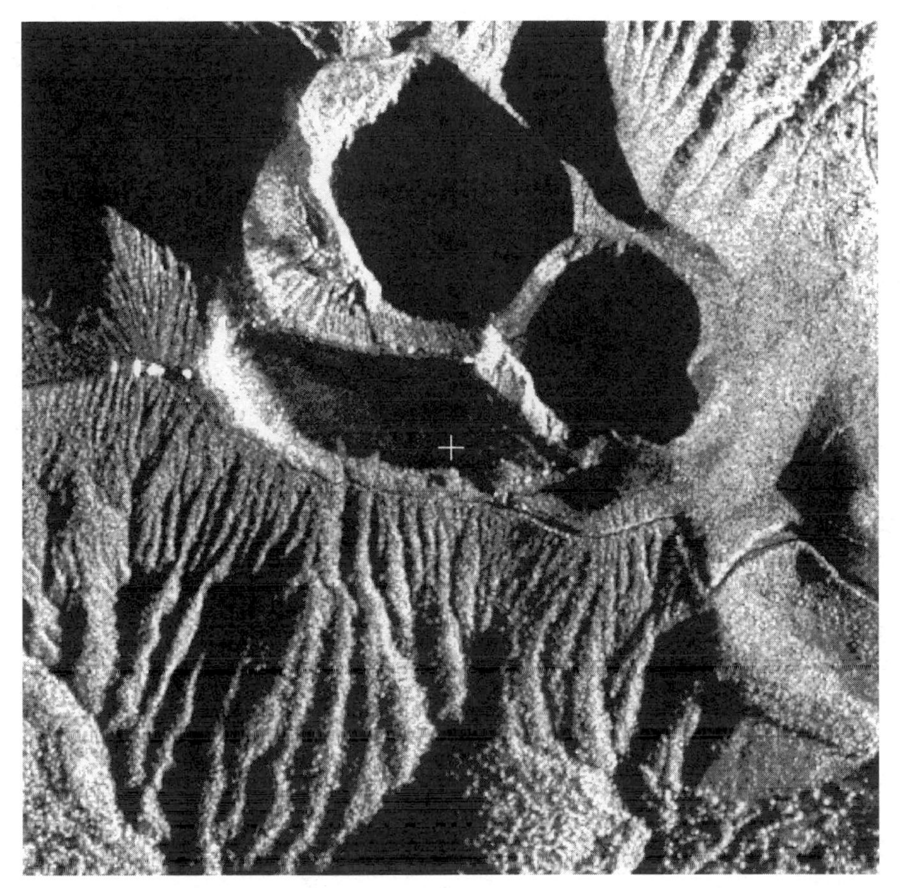

FIGURE 11-10 High-resolution SAREX '92 radar image subset of potato and pasture fields with dispersed trees, a few kilometers from the crater of the Irazu Volcano (top of image), showing a characteristic mottled texture (bottom of image).

In the case of coffee, very little variation was observed in the images among fields even though on the ground there were significant differences in plant height or state of health. In certain cases, stripes could be observed in fields

---

FIGURE 11-9 Landsat TM 453 RGB color composite of an agricultural area near Cachi dam. Crops are coffee with and without shade, chayote, pasture, and pine reforestation. (a) April (dry season) image; (b) June (wet season) image. The June color composite is almost entirely reddish, which indicates healthy vegetation. Rain forest areas (bottom right) maintain the same coloration throughout the year. (On CD.)

that had been submitted to a practice, called "Rock and Roll," that involves the total trimming of one out of every three rows, and partial trimming of the next row, while leaving the third one at its full height. In one case where an entire field had been submitted to a total trimming and was invaded by weeds, the derived color composites displayed a reddish color. This reddish color was also apparent in overgrown pastures which were adjacent to one of the plantations, and was even more intense in a large marshy high-grass patch occupying a depression in one of the coffee farms. The reddish color can be explained by the fact that the grass and the weeds have a greater number of vertical constituents than the coffee bushes. This effect should be visible for incidence angles greater than 40° in the C-band (which is the case for the SAREX images), as has been observed with prairie grass (Bakhtiari and Zoughi, 1991). What this means is that radar can be used to differentiate many, but not all, types of land use, and for the first time allows us to get, in some cases, a detailed assessment of land use crop by crop and by ecosystem type.

Although radar mapping of some land use types is promising, our first objective was to identify bare soil areas, which are enormously more susceptible to erosion than vegetated areas. Bare soil areas appear white on aerial photos, and clear on TM 453 false-color composite (see Figure 11-8). In radar images their appearance depends on if the soil is wet or dry, and smooth or rough. In wet conditions (soil water content on the order of 0.5 $g/cm^3$), bare soil areas were indistinguishable from vegetation, while they appeared very dark in dry conditions (soil water content on the order of 0.1 $g/cm^3$). Our qualitative observations suggest that separability of bare and vegetated soil increases with incidence angle, and decreases with higher soil roughness and water content. If the soils drain well, then mapping with radar is possible even in the rainy season.

In the case of vegetable crops in dry or well-drained areas, radar images can show a clear distinction among smooth bare soil (dark), young plants (intermediate brightness), and fully developed plants (bright) (Figures 11-8l and 11-11 on CD). In other sites where there were rough, recently plowed bare soil fields, the difference in intensity between them and full vegetation was smaller, but still obvious in most cases. Thus in some cases ground truthing is still required.

Recently harvested fields still covered with sugarcane residues appear bright due to the long sugarcane residues laid horizontally on the ground. Recently plowed fields also have a very high reflectance in both HH and VV. Despite their similarity, we found that it was possible to identify these similar-appearing ecosystems unambiguously, even in areas with a rolling topography, by correcting the images computationally for topographic effects.

FIGURE 11-11 High-resolution SAREX '92 radar image of potato, carrot, and onion fields. Smooth dry soils appear dark, fields with young plants have an intermediate brightness, and fields with well-developed plants appear the brightest.

## 3. Example: Determining Growth Stage in Sugarcane

Sugarcane is one of the most important crops in Costa Rica and the Caribbean. We did a quantitative study on two sugarcane sites with the objective of observing the effect of the management and growth stages on both radar backscatter intensities and Landsat TM imagery. We undertook this analysis at both the relatively wet Juan Viñas site and the relatively dry Taboga site.

*a. The Juan Viñas Site*

The company Hacienda Juan Viñas, S.A., manages over 1500 ha of sugarcane (in 130 plots) at Juan Viñas in the Cartego province. The company maintains weekly records of the activity carried on in each field, and this inventory and monitoring are needed for management and economic predictions. The economic effectiveness of this information could be greatly enhanced with remote sensing of the state of the crop in different regions. The heavy cloud cover present in these regions makes radar an attractive tool. This area coincides with the southeastern limit of coverage of the Costa Rican territory by 1 : 10,000 maps published in 1992 and produced from aerial photos acquired in 1989 (IGN and JICA, 1992). The portion of this area that is cultivated with sugarcane is either flat or extremely abrupt, making it an excellent site to test our radiometric correction methodology. Radiometric corrections of SAR imagery, both for topography and antenna pattern, were necessary in order to make any quantitative measurement from the radar signal, especially when comparing the intensities from different fields. Figure 11-12a (color version on the CD) shows a photograph of a portion of the site taken from a small airplane at low altitude. The light-colored strips in the front are residues. Figure 11-12b shows a close up of 1-month-old sugarcane. The varieties planted here attain maturity after 2 years. We constructed a detailed map of the fields and integrated that into a GIS for the purpose of extracting image parameters and to ensure that the samples of radar backscatter intensity were not derived subjectively. This map was established from the 1:10,000 topographic maps and from a map that Hacienda Juan Viñas, S.A., had developed from 1979 aerial photography. It was updated with the help of the agricultural engineer of the company.

We next constructed a georeferenced data base which contained the information on the various fields, including date of last cut, date of last intervention (burning, plowing, planting of new cane), age of sugarcane, state of the field, observations from oblique photographs taken at the time of radar imaging, and age at the time of acquisition of the Landsat TM image (i.e., 3 weeks before the radar ones). The fields that had portions in cane of different ages were divided into homogeneous polygons when their limits were identifiable features on the topographic map (roads, streams, etc.). Once this map was prepared an 8-m buffer zone was removed from the perimeter of each polygon to avoid edge effects or the inclusion of roads or fences into the samples. This gave us a total of 140 polygons that we used for the extraction of mean image

---

FIGURE 11-12 (a) Photograph of the sugarcane fields in the region of Juan Viñas taken from a small airplane. (b) Close up of 1-month-old sugarcane (from Beaulieu, 1998). (Color version on CD.)

a

b

intensity; 130 were cultivated with sugarcane. The others were soccer fields, patches of forest, and pasture. The SAREX radar images and the Landsat TM image, both taken in April 1992, were aligned with the map using image processing software. The Juan Viñas site received 207 mm of rain during the 3 days preceding the radar acquisition, and 41 mm the same day.

We analyzed various radar images (flight lines 8.2 and 8.3 in Figure 11-6), as well as a Landsat TM image taken 3 weeks before the radar ones. In addition, we derived from each radar image a ratio that highlights the geometric characteristics of the vegetation, the Normalized Difference Polarization Index [NDPI = (HH − VV)/(HH + VV)] (Beaulieu *et al.*, 1994). From the TM image we computed the NDVI ((TM4 − TM3)/(TM4 + TM3)), which is sensitive to the greenness of the vegetation. We extracted the average value for each field from each image and compared that to the state or age of the sugarcane. The fields undergoing renovation were covered with rough bare soil and given an "age" value of −5, whereas the recently harvested fields, which were still covered with residues, were given a value of −2 to distinguish them from the fields which had been submitted to burning of residues (age = 0). At the time of the radar flight, none of the fields in the data base were covered with burnt residues.

Unfortunately, the scattergraph of average DNs for each parcel as a function of age showed a large dispersion of values (Figures 11-13b and 11-13d). When the fields were grouped by age ranges, as shown in Figures 11-13a and 11-13c, there was a slight tendency for an increase in backscattering intensity in the first months of growth, and then a decrease with maturing of the cane. The only parameter that shows promise for distinguishing one class from the others is the NDPI, which was almost invariably higher in the fields covered with residues. The great dispersion encountered in the radar backscatter intensities could be due to an imperfect radiometric correction, to topographic features finer than what the DEM can account for, or to differences in water content in the soil and vegetation. Indeed, the images had been acquired after a rainy weekend, a condition that emphasizes differences in drainage and water retention capacities.

The scattergraph of the NDVI vs sugarcane age showed a much more significant trend, as can be seen in Figures 11-13e and 11-13f. However, determination of the age of sugarcane based on NDVI alone is possible only in the 0 to 6-month age classes. Above this age there is a confusion between younger and older sugarcane since the NDVI saturates around an age of 6 months and then decreases. This can be explained by the fact that the leaves of mature sugarcane become drier and therefore have a lower NDVI. The scatter of NDVI values for the same-age crop can be attributed to the absence of radiometric correction for topographic effects, variations in fertilization or plant density, and variations in leaf yellowing age between varieties.

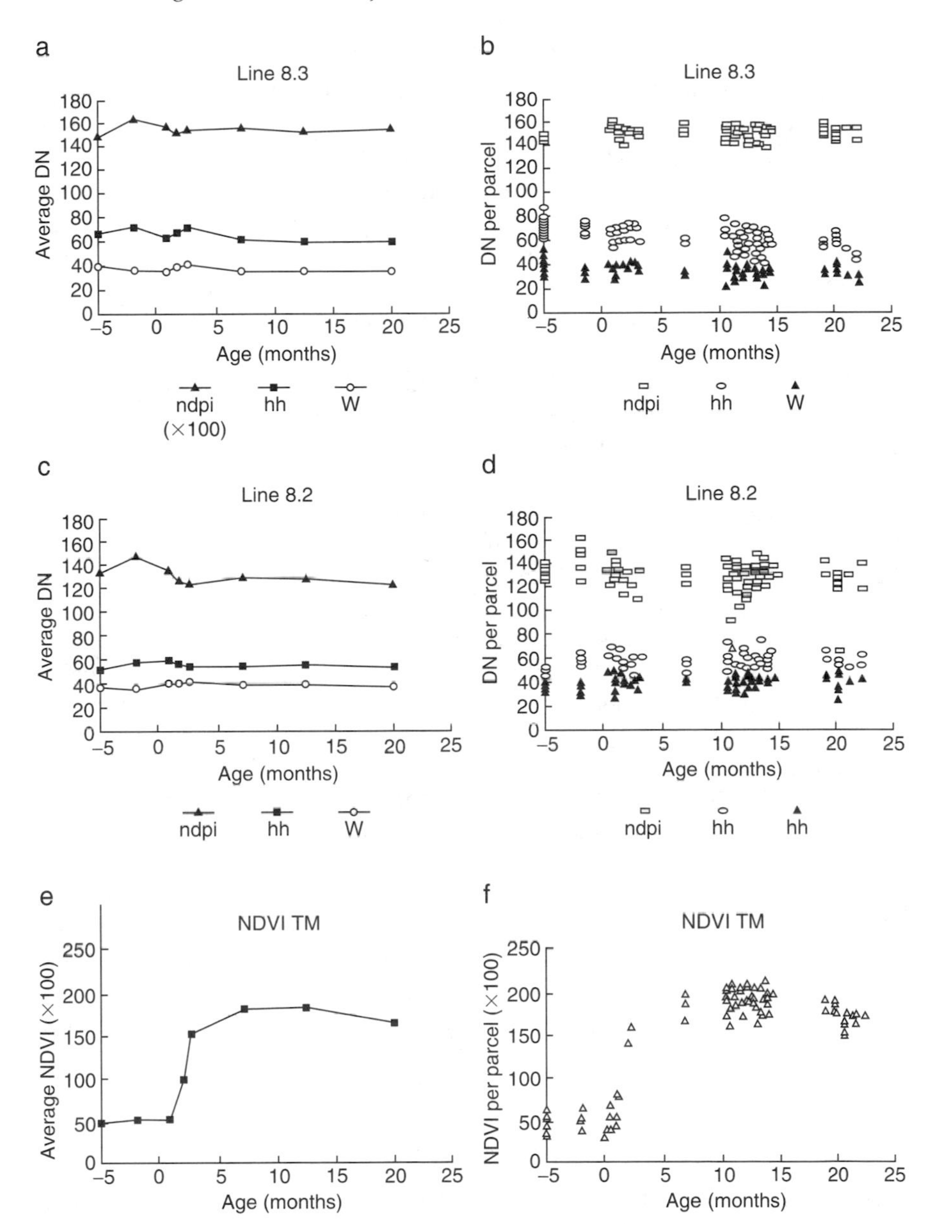


FIGURE 11-13 (b, d) Scattergraph of average C-SAR backscatter intensity for each parcel for different growth stages shows a large dispersion of values. When the values for all the fields in each growth stage are averaged, as shown in Figure 11-5a, c, the spread is much less. (f) Scattergraph of average NDVI from Landsat TM imagery, as a function of growth stage. (e) NDVI for all fields with the same growth stage averaged. Age is −5 for moist rough bare soils and −2 for recently harvested fields covered with residues.

### *b. The Taboga Site*

Hacienda Taboga, S.A., gave us access to agronomic data for its 223 sugarcane plots in the Taboga area, near Cañas (northwest Costa Rica). This is a much drier area than Juan Viñas, and cane needs irrigation during the summer. Availability of clean optical imagery during the dry season is generally not a problem; however, we decided to explore the possibilities of radar without the limitations of topography (the area is flat) and soil water content (soils were dry when the radar imagery was acquired). The varieties planted here attain maturity after 1 year. The site received no rainfall during the 6 days preceding the acquisition of the radar images. Flight lines 7.1, 7.2, and 4.1 of the SAREX mission covered the site (Figure 11-3). Calibration data were available for these images, which allowed us to derive backscattering coefficients. Figure 11-14 (on the CD) shows a color composite of the three radar images.

When applied to the Taboga site, the same approach and analysis used at the Juan Viñas site give slightly different results (Beaulieu, 1998). First, it becomes possible to reliably distinguish between bare soils and >2-month-old sugarcane. Second, by dividing an image taken at one radar incidence angle by another one, we obtain a linear relationship with sugarcane age (Figure 11-15); therefore the combination of radar imagery taken at two angles of incidence can help determine the growth stage of the sugarcane fields.

Thus overall we have to conclude that there is no "silver bullet" offering a simple unambiguous approach to identifying sugarcane or its developmental stage using remote sensing. Techniques that were successful in one region often cannot be applied to another or to the entire country. Results depended especially upon whether the crop was produced in a dry (irrigated) area or in a humid one (Beaulieu, 1998). Nevertheless if data are gathered from a sequence of days, if the days upon which the data are gathered are chosen carefully to avoid times when the soils are too wet, and if a judicious use is made of several types of remote sensing data, then nearly any type of crop data is obtainable remotely. The question then becomes one of cost: when are the data worth the expense of getting sufficient images to justify the expense?

## 4. Example: Mapping of Sweet Bananas and Banana Plantations

Sweet banana (*Musa AAA*, referred to as "banana") is the most important fruit in the world in terms of volume of production, which was over 70 million tons in 1990 (Loeillet, 1992). There are more than 100 producing countries in five continents, all located in tropical and subtropical regions. Bananas, exported to developed countries as well as consumed by locals, play an important role in the nutritional balance and in the economy of the producing

FIGURE 11-14 Color composite of high-resolution SAR imagery from SAREX '92 over the Taboga site, taken at incidence angles of 13–63°, 62–73°, and 70–79° (in RGB, in that order). The very dark fields are rice paddies (from Beaulieu, 1998). (On CD.)

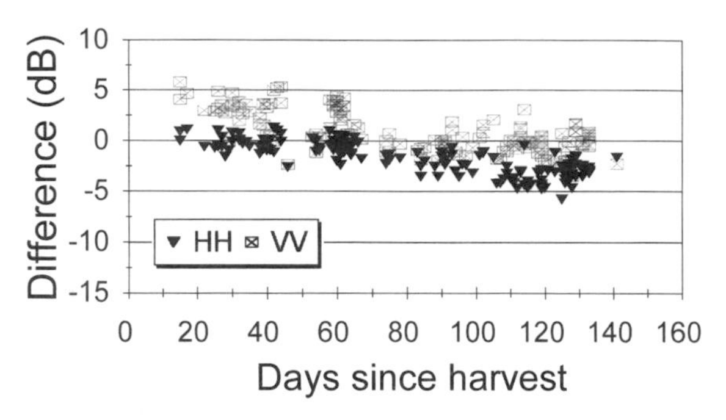


FIGURE 11-15 Scattergraph of the difference of C-SAR backscatter power (in dB, averaged over entire fields) from two SAREX '92 images taken at incidence angles 70–79° and 62–73°, as a function of sugarcane growth stage, for the Taboga site (from Beaulieu, 1998).

countries. In Costa Rica, banana exports are the largest source of foreign income. A second type of banana, the plantain (*Musa AAB*), is consumed almost exclusively locally, and is an important part of the Costa Rican diet. The importance of monitoring these crops at the national level is unquestionable.

Although the extent of the areas cultivated with banana is supposedly relatively well known, the surveys are not brought up to date regularly. With the continuous expansion of the cultivated areas and abandonment of other areas by the banana companies, the maps produced and areas calculated quickly become out of date. In addition, we found that the three existing summaries of areal data provided for individual plantations rarely agree. These summaries are: (1) the areas calculated from digitized and georeferenced banana company plan maps, (2) the areas reported in the legends of these plans, and (3) the areas provided by the managers of the farms. These disagreements showed the need to develop a uniform method for the estimation of production areas, a need that is not unique to Costa Rica (Valdivieso, 1993). Thus in spite of the importance of bananas in the country's economy, Costa Rica has not yet instituted a mechanism to map and follow the change in location of production zones at the national level. For plantain bananas, the case is much worse. There is very little knowledge of the area planted with this crop, which generally is cultivated in an artisanal manner.

### *a. Optical Estimates from Landsat TM*

Yield cannot be measured directly based on spectral information, but biomass and leaf area can be estimated. Thus there is a possibility that yield can be predicted if there is a correlation with these variables. Until recently no data have been published to test the validity of such an assumption. Veldkamp *et al.* (1997), however, studied a 370-ha banana plantation in the Atlantic zone of Costa Rica, and showed that yields could be predicted well with TM imagery combined with soils data. Forty-six percent of the variation in gross production (kg/ha/week), number of bunches, and bunch weight within the plantation could be predicted from TM band 4 and the "greenness index." Sixty-seven percent of the variance of gross production was explained by soil units alone, which emphasizes again the importance of soils in banana cropping. Accuracy in that prediction was not increased by adding TM imagery analysis. Since the canopy of a banana plantation is not completely closed, an important part of the TM4 reflectance originates from the soils and this may explain the poor correlation between TM4 and plant parameters. Spectral unmixing techniques could resolve the ambiguity.

Valdivieso (1993) have stated the necessity of determining precisely banana production areas for the banana growing areas of Ecuador. Their attempt to use satellite imagery to do so has resulted in the development of a combination of color indices that readily allow visual discrimination of all production stages (Figure 11-16 on the CD). However, satellite-derived production areas were not consistent with official surveys on the ground, and the question remains as to which ones are wrong. Our experience with Costa Rica (see next section) indicates that RS is providing more consistent areal estimates.

### *b. Using Radar to Estimate Banana Areas and Production*

Anyone seeing a radar image of a banana-producing region for the first time is surprised by the very intense brightness of the crop, resulting mainly from the large leaf area of the plant (Figure 11-17). One can imagine that radar imagery could become a very useful tool for mapping plantations because of this capability of distinguishing banana plantations from surrounding vegetation independent of atmospheric conditions. But at the time we began our analysis questions remained as to whether precise plantation boundaries could be established, whether it is possible to distinguish different stages of production, or whether bananas can be distinguished from plantains or pineapples, which exhibit similar reflectance.

With SAREX imagery, we found that the banana plantations can be identified easily because of their brightness and geometric shape. The roads and drainage channels within the plantations also appear very clear. We could also identify preproduction zones, containing young plants 0.5 to 1.5 m tall which have not yet attained maturity, because of their lower brightness. Darker areas seem to correspond to lower plant density, due to inadequate soils or drainage, lower fertilizer use, higher incidence of black leaf streak, or bad management. Surprisingly, plantain, which is planted with a higher average density than

**FIGURE 11-16** Color composite obtained from TM imagery of the Atlantic zone of Costa Rica, using the method developed by Valdivieso to discriminate banana growth stages. (On CD.)

banana (typically 1800 to 1900 vs 1700 to 1800 plants/ha), appears darker. This can lead to confusion between plantain and homogeneous forest, although their textures are different. In addition, the delineation of plantains is difficult, especially when they are surrounded by open homogeneous forest, because they are planted in an almost random manner and have irregular limits. Radar images need a geometric correction to project the image from slant range to ground range. Once this is done, the borders of banana field limits can be delineated on the screen, which allowed us to obtain a precise and georeferenced map of the production zones of individual plantations quickly and easily with low-cost software such as Map Maker.

We evaluated the area of bananas in production for nine plantations from boundaries visually delineated from SAPEX imagery . Then we compared these areas with (1) the areas calculated from digitized and georeferenced plans of the respective farms; (2) the areas appearing in the legends of these plans; and (3) the areas provided by the managers of the farms. Figure 11-18a shows the area from method 2 vs 3 ($y$ axis) with respect to method 1 ($x$ axis). Note

**FIGURE 11-17** SAREX '92 C-SAR imagery of a banana production area in the Atlantic zone of Costa Rica, showing the great potential of this all-weather imaging technology for mapping banana plantations.

the significant deviation in the *y* axis (which shows inconsistencies between methods 2 and 3), and that a line of slope 1 passes exactly through the data points (which shows that radar estimates are consistent). For one farm we knew exactly the location of three production stages. Figure 11-18b shows how the per-field average of the SAR backscatter intensity changes for each production stage, which indicates that C-SAR can successfully discriminate preparation, preproduction, and production areas.

To study the possibilities of using a more computer algorithm-based, or "automated," mapping technique, we ran various supervised and unsupervised

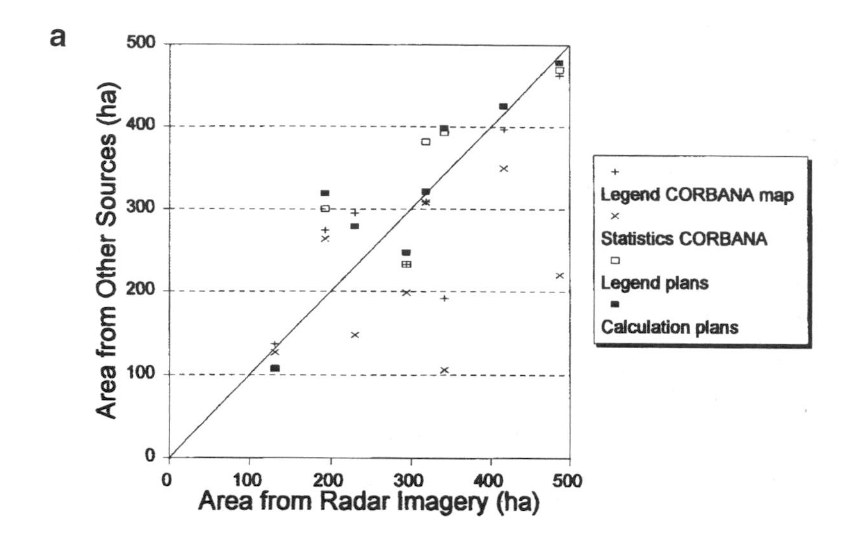


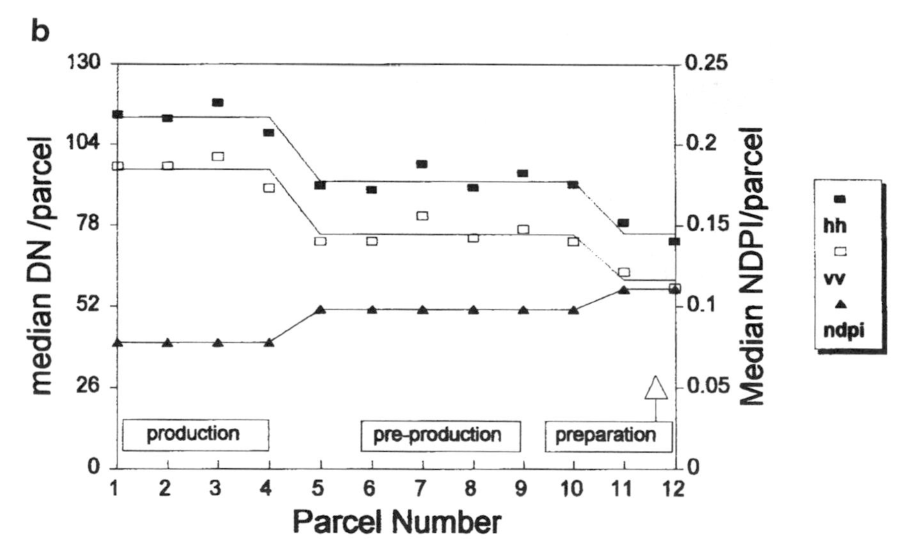


FIGURE 11-18 (a) Graph of area determined from official figure, plans, or maps, vs area determined from high-resolution C-SAR imagery. The line represents the one-to-one relationship. Note the high dispersion of values along the *y* axis, which shows the inconsistencies in official figures. (b) Graph of median digital number in individual parcels, as a function of production stage for the Turqueza Dorada banana plantation.

classifications on image subsets, using both vertical and horizontal polarizations. The original classifications obtained from the unfiltered data had a salt and pepper appearance, with a great number of dots associated with different classes both inside the plantations and out. This generated problems for the automated classification algorithms, which were resolved by first applying median or Lee filters to the images. In this way, image radiometry was homogenized inside the banana plantations, although two different radiometric classes still remained within them. Bananas were obvious by their brightness. The fields with lower brightness were either plantain plantations, preproduction banana fields, or areas with low plant density. An easy way to improve uniformity of the automated classification is to contract the image by a factor of 8 by averaging groups of 8 × 8 pixels. Then the automated classifications provided unambiguous data, showing that good results can be expected with filtered low-resolution satellite imagery (Figure 11-19). All this indicates that an automated classification of production zones in banana regions is feasible based on radar imagery.

### c. *Diagnosis of Black Leaf Streak Infections*

The main disease affecting banana and plantain crops is black leaf streak (*Micosphaerella fijensis*), a fungus that produces severe leaf necrosis, slows the growth of the fruit, and reduces yield significantly. This is an especially important issue because there is no genetic variation in a banana plantation, which makes it very sensitive to resistant diseases. A national commission was created by order of the Costa Rican government to look at black leaf streak. The commission has recognized the necessity of the government's involvement through a Declaration of National Emergency (CORBANA, 1993).

Currently, the infestations of black leaf streak are not monitored at the national level because of a lack of resources. Radar, because of its sensitivity to water content and to the geometry of targets (in this case the leaves of the plants), was thought to be a significant tool for this monitoring. We investigated this potential in a preliminary study using the preceding data.

We extracted the average of horizontal and vertical polarity radar intensity in three to eight samples of 100 pixels in six banana plantations. The severity of infestations also was expressed in field inspections by a parameter called PPI (Promedio Ponderado de Infección), a measure of the proportion of leaf area affected by the fungus. This parameter was provided to us by the farm managers for each field for April 1992. We found a strong negative correlation between average brightness and PPI, particularly in the VV polarization (Figure 11-20). More details on this banana and plantain study can be found in Pigeonnat (1993). Nevertheless our preliminary study suggests that black leaf streak hot spots can be mapped, treatment costs can be reduced (by treating

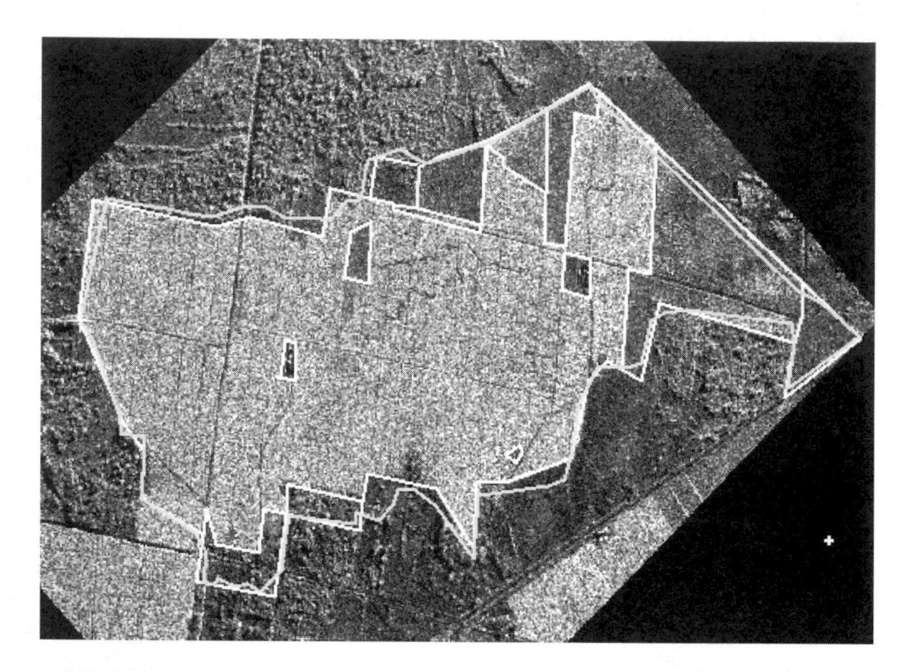

FIGURE 11-19 Raw radar image subset of La Guaria banana plantation. Areas in production appear very bright. Banana plantations usually have a geometrical shape that can be delimited easily. Radar image and farm map have been georeferenced independently, but the farm map is the one with the most intrinsic error. High-production areas are obtained automatically from smoothed C-SAR imagery.

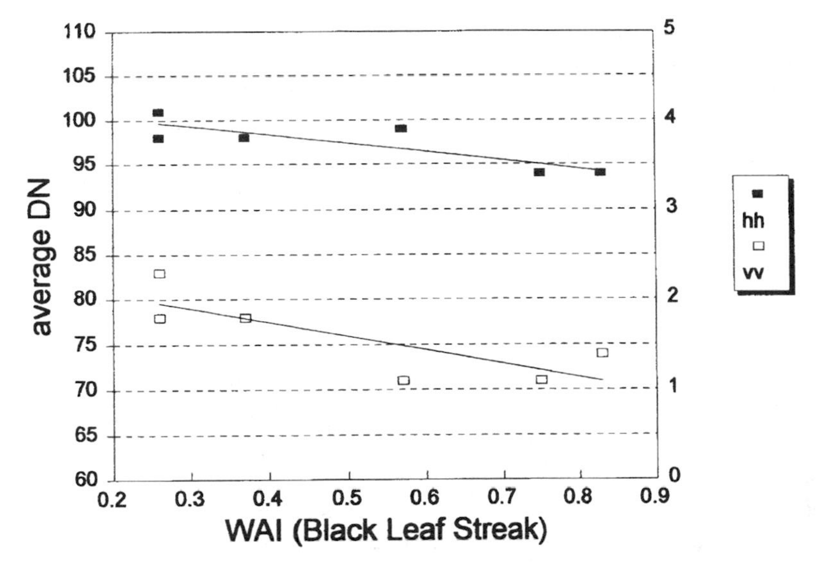


FIGURE 11-20 Average intensity (for 3–5 sampling sites per plantation) of SAREX '92 C-SAR imagery as a function of an on-the-ground indicator of the level of infection by black leaf streak for several plantations. The higher the infection, the lower the backscatter intensity.

only those fields that need it, when they need it), and production improved with regular radar monitoring and an automatic analysis of the data so obtained.

## IV. THE FUTURE OF COSTA RICAN INSTITUTIONS DEVELOPING REMOTE SENSING APPLICATIONS

Remote sensing, if it is to be as useful as its potential suggests—indeed if it is to be useful at all—requires a strong institutional base and support to standardize operations and keep users up to date with the most recent technical changes. With that in mind it is important to look at the present RS capacity in Costa Rica.

### A. IGN

The Instituto Geográphico Nacional (IGN), although the pioneer in RS in Costa Rica and a participant in major RS conferences since 1980, has not been able to build a solid GIS/RS laboratory to date using the latest digital technology. It is partly because IGN direction is now highly politicized, and thus dependent on what government is in place. The results of the 1998 elections could lead to major changes, probably to an all-digital approach. The Japanese International Cooperation Agency (JICA) has been IGN's main partner during the last 10 years, and is likely to continue to provide high-quality air surveys and cartography to Costa Rica.

## B. CATIE

CATIE, a regional institution working on matters related to natural resources in the American tropics, has played a leading role in the development of remote sensing and geographic information systems in Central America. Since the initialization of image processing capacity in 1986, CATIE has conducted numerous projects based on the use of optical remote sensing applied to agriculture and land use mapping in Guatemala, El Salvador, Honduras, Nicaragua, and Costa Rica. Researchers at CATIE were faced with the difficulties of obtaining reasonably cloud-free optical satellite images of most parts of the region, and consequently developed many of the radar approaches reported upon in this chapter. CATIE has been pioneering digital RS in Central America and is one of the few consistently to do training and research in GIS/RS, which puts the institution in a favorable position in Central America.

## C. National Universities (UCR, UNA)

The lower cost of hardware and software, the new Landsat 7 TM pricing and copyright policy, and better communications with the exterior (the Internet, for instance) are giving national universities a chance to produce good remote sensing research. Still, obtaining free imagery is an ongoing tradition, and this is done through space agency announcements of opportunities or collaborations.

## D. IPGH

Although not Costa Rican, the Instituto Panamericano de Geografía y Historia (IPGH) has played an important role in the development of government remote sensing in Costa Rica. Through IPGH-sponsored regular meetings and training, and a Spanish language monthly technical journal, IGN personnel have been participating in regional capacity building in RS. This includes aerial photo processing and interpretation, cartography, digital image processing, and land use analysis. Its role in defining standards together with USGS is vital for the usability and quality of the information generated in Central America.

## V. CONCLUSION

This chapter has shown that despite the great potential of remote sensing to aid the assessment of the resource base of Costa Rica, the actual delivery has not always been at the level of the expectations. Probably the greatest problems have been (1) the lack of a consistent classification scheme that would allow for the gathering of data in a way that would allow for a clear comparison of different studies; (2) the difficulties of working in an area frequently covered with clouds; and (3) the lack of a consistent institutional and financial base within the country. Studies reported in this chapter have shown that one of the most promising ways to improve the second problem is using radar imagery. This is true especially for the mapping of banana plantations because of the very high brightness exhibited by this crop. High-resolution radar imagery has a definite potential for monitoring soil conservation practices such as agroforestry and the leaving of residues on the ground after harvest. Radiometric corrections developed to compensate for the effects of topography improve significantly the possibility of monitoring crops in mountainous regions, and could be of benefit to many areas other than agriculture. As the technology matures it should be possible to combine optical and radar images to give very sensitive tools for the analysis of land use and land use changes.

If all of these procedures pan out we will indeed be in a better situation to understand the opportunities and limitations of the Costa Rican landscape for improving, if possible, its potential for a more sustainable future. It should be clear, however, that even the imperfect studies reported here show a landscape that is (1) severely deforested; (2) pushed very hard toward its limits for economic production; (3) subject to substantial erosion, in part due to crop management practices brought about by a very high need for export crops; and (4) very vulnerable to crop diseases. These observations are consistent with the results of other chapters and do not lend a great deal of encouragement to the construction of a substantially more sustainable landscape, at least based on traditional land uses.

## ACKNOWLEDGMENTS

We thank the many people and institutions who supported us in these studies, assisted us in the laboratory, and helped us gather field and agronomic data. In particular we thank Alexander Salas, digitizing operator at CATIE, Jorge Antonio Salazar and Carlos Gonzalez Quesada from

Hacienda Juan Vias, S.A., Juan Carlos Guevara, Ronald Vargas, and Luis Diego Ching of Cooperación Bananera Nacional, Juan Luis Sens of the plantain cooperative COOPEPALACIOS, Jetse Stoorvogel from the University of Wageningen, and the Fundación para el Desarrollo de la Cordillera Volcanica Central, FUNDECOR, for access to their Landsat TM image. We also appreciate Keith Raney and Frank Ahern, who gave us much support and guidance; Nathalie Beaulieu, for discussions and data; and Darryl Murdoch, for revising the manuscript. We thank the Instituto Geográfico Nacional of Costa Rica for its support in the scope of Proyecto Radar/Costa Rica/ Canada and the Proyecto Regional de Manejo de Cuencas of CATIE for its logistic support. We also thank PCI, Inc., for lending us the EASI/PACE image processing software. This study would not have been possible without the financial contribution of the International Development Research Center (IDRC) of Canada, the National Science and Engineering Research Council (NSERC) of Canada, and the Consultative Group for Agricultural Research (CGIAR).

## REFERENCES

Bakhtiari, S., and R. Zoughi. 1991. A model for backscattering characteristics of tall prairie grass canopies at microwave frequencies. *Remote Sensing of the Environment* 36:137–147.

Ban, Y., P. M. Treitz, and P. J. Howarth. 1993. Improving the accuracy of synthetic aperture radar for agricultural crop classification. In *Proceedings, 16th Canadian Symposium on Remote Sensing and 8e Congrès de l'Association Québécoise de Télédétection, Canadian Remote Sensing Society, Sherbrooke, Canada,* pp. 367–370.

Baudoin, A., Deshays, M., Piet, L. Stussi, N., and Le Toan, T. 1994. Retrieval and analysis of temperate forest backscatter signatures from multitemporal ERS-1 data acquired over hilly terrain. Proceedings of the First ERS-1 Project Workshop, Toledo, Spain, 22-24 June 1994(ESA SP-365), pp. 283–289.

Beaulieu, N., Leclerc, G., Velasquez, S. M., Pigeonnat, S, Gribius, N., Escalant, J.-V., and Bonn, F. 1993. "Investigations at CATIE on the potential of high-resolution radar images for monitoring of agriculture in Central America", in: South American Radar Experiment SAREX-92 Workshop Proceedings, 6–8 de Diciembre de 1993, European Space Agency ESA WPP-76, Marzo de 1994. pp. 139–153.

Beaulieu, N., Leclerc, G., and Bonn, F. 1995. "Facteurs affectant la possibilité de distinquer les stades de culture de la canne a sucre a l'aide d'images RADAR", Proceedings of the International Symposium: Retrieval of bio- and geophysical parameters from SAR data for land applications. 283–291.

Beaulieu, N. 1998. Utilité des images de radar aeroporté en band C pour l'évaluation du stade de croissance de la canne a sucre et des cultures marchères en milieu tropical, dans une optique de conservation des sols. These, Ph.D. en télédétection, Université de Sherbrooke, Département de géographie et de télédétection, Sherbrooke, Canada.

Brown, R. J., B. Brisco, R. Leconte, D. J. Major, J. A. Fisher, K. D. Reichert, K. D. Korporal, P. R. Bullock, H. Pokprat, and J. Culley. 1993. Potential applications of Radarsat data to agriculture and hydrology. *Canadian Journal of Remote Sensing* 9(4):317–329.

CORBANA (Corporacion Bananera Nacional). 1993. Sigatoka Negra, un grave problema en los bananales de Costa Rica. *Carta Informativa,* Año 2, N9.

Cornell, J., and C. A. S. Hall. 1994. A systems approach to assessing the forces that generate tropical land use change. In J. O'Hara, M. Endara, T. Wong, C. Hopkins, and P. Maykish (Eds.), *Timber Certification: Implications for Tropical Forest Management,* pp. 35–43. School of Forestry and Environmental Studies, Yale University, New Haven, CT.

Dubois, P., van Zyl, J. and Engman, T. 1995. Measuring Soil Moisture with Imaging Radars. IEEE Transactions on Geoscience and Remote Sensing, 33(4), 915–926.

Eastman, R. J., and M. Fulk. 1993. Long sequence time series evaluation using standardized principal components. *Photogrammetric Engineering and Remote Sensing* 59(8):1307–1312.

Green, K., D. Kempka, and L. Lackey. 1994. Using remote sensing to detect and monitor land cover and land use changes. *Photogrammetric Engineering and Remote Sensing* 60:331–337.

Hall, C. 1978. El Café y el desarrollo historico-geografico de Costa Rica. Editorial Costa Rica y Universidad Nacional.

Hall, C. A. S., H. Tian, Y. Qi, G. Pontius, J. Cornell, and J. Uhlig. 1995. Spatially explicit models of land use change and their application to tropics. *DOE Research Summary* 31:1–4. [Oak Ridge National Laboratory]

IGN 1967. Cobertura de bosques de Costa Rica. San Jose, Costa Rica. Map: scale 1 : 1,000,000.

IGN. 1984. *Mapa de Uso de la Tierra.* IGN, San Jose, Costa Rica.

IGN/JICA. 1992. Capellades and Birris 1:10 000 topographic maps.

Keogh, R. M. 1984. Changes in the forest cover of Costa Rica through history. *Turrialba* 34:325–331.

Leclerc, G. 1996. "Tropical Forest Assessment in Central Costa Rica Using Landsat Thematic Mapper Imagery". Technical report. Contract EN9406951V TREES/CEE. 50p.

Leclerc, G., Beaulieu, N., and Bonn, F. 1996. "Two simple methods to correct radiometric distortions due to the slant-range projection in a radar image" in: Proceedings of the 2nd Intl Airborne Remote Sensing Conference and Exhibition. I-305-314.

Leclerc, G., and Rodriquz Ch., J. 1998. Using a GIS to determine Critical Areas in the Central Cordiliera Conservation Area, Costa Rica, In: "Conservation Policy Making using Digital Mapping Technologies: Case Studies in Costa Rica." Biology and Resource Management in the Tropics Series.(T. E. Lacher, Jr., and B. G. Savitsky, Eds). pp. 108–126. Columbia University Press.

Leclerc, G., Beaulieu N., and Bonn, F. 1999. A simple method to account for topography during the radiometric correction of radar imagery. Submitted for publication.

Loeillet, D. 1992. Eléments d'Information et de Réflexions sur la Future Organisation Commune du Marché de la Banane dans la CEE. Rapport d'activités de l'observatoire des marchés CIRAD-IRFA, CIRAD, Paris.

MAG (Costa Rican Ministerio de Agricultura e Ganaderia). 1950, 1955, 1965, 1973, 1977, 1984. *Agricultural Census.* MAG, San José, Costa Rica.

Maraux, F., and A. Garcia. 1990. *Estudio Probabilstico de la Nubosidad: Aplicación para la Toma de Imagenes Satélite en Centoamérica.* Report of the Proyecto Regional de Agrometeorología. CATIE/CIRAD/ORSTOM, Turrialba, Costa Rica. 75 pp.

Perera, L. K., and R. Tatieshi. 1995. Do remote sensing and GIS have a practical applicability in developing countries? *International Journal of Remote Sensing* 16:33–35.

Perez, S., and Protti, F. (1978). Comportamiento del sector forestal durante el periodo 1950–1977. Oficina de Panificaciun Sectorial Agropecuaria OPSA. DOC-OPSA15. San Jose, Costa Rica.

Pigeonnat, S. 1993. *Potentiel de la Tldtection radar (C-ROS) pour l'etude d'une Culture Tropicale: Le Bananier (musa AAA et AAB).* Rapport de Stage du Certificat d'études Supérieures Agronomiques, Ecole Nationale Superieure Agronomique de Montpellier.

Sader, S. A., and A. T. Joyce. 1988. Deforestation rates and trends in Costa Rica 1940 to 1983. *Biotropica* 20:11–14.

Sanchez-Azofeifa, G. A. 1996. *Assessing Land Use/Cover Change in Costa Rica.* Earth Sciences Department, University of New Hampshire, Durham, NH. 20 pp.

Stoorvogel, J. J. 1995. *Geographical Information Systems as a Tool to Explore Land Characteristics and Land Use with Reference to Costa Rica.* Agricultural University, Wageningen, The Netherlands. 151 pp.

Swain, P. H., and S. M. Davis. 1978. Remote sensing: The quantitative approach. McGraw–Hill, New York, 166–174.

Treitz, P. M., P. J. Howarth, O. R. Filho, E. D. Soulis, and N. Kouwen. 1993. Classification of agricultural crops using SAR tone and texture statistics. In *Proceedings: 16th Canadian Symposium on Remote Sensing and 8e Congres de l'Association Quebecoise de Teledetection*, Canadian Remote Sensing Society, Sherbrooke, Canada, pp. 343–347.

U.S. Department of Agriculture (USDA). 1943. The Forests of Costa Rica. Trans. to Spanish by F. Sancho. 1945. *Revista de agricultura* 17(1,2,3,4,8). 1946. *Revista de agricultura* 18(1,2,4,7).

Valdivieso, J. M. 1993. Aplicación de la informacion satelitaria Landsat TM en el inventario de las superficies de banano. Presented at VI Simposio Latinoamericano de Percepción Remota, Cartagena, Colombia.

Vargas, E.(1994). Analisis y clasificación del uso y cobertura de la tierra con interpretación de imagenes, Instituto Geográfico Augustín Codazzi, Bogota, Colombia.

Veldkamp, A., and Fresco L. O. 1997. Reconstructing land use drivers and their spatial scale dependence for Costa Rica (1973 and 1984). *Agricultural Systems* 55(1):19–43.

Veldkamp, E., E. J. Huising, A. Stein, and J. Bouma. 1997. Variation of measured banana yields in a Costa Rican plantation as explained by soil survey and thematic mapper data. *Geoderma* 47:337–348.

Wooding, M. 1988. *Imaging Radar Applications in Europe, Illustrated Experimental Results (1978–1987)*. ESA TM-01, Commission of the European Communities Joint Research Centre and European Space Agency, ESA Publications division, Noordwijk, the Netherlands.

Zon, R., and W. Sparhawk. 1923. *Forest Resources of the World*. McGraw–Hill, New York.

# SECTION V

# *Application to Sustainability Issues for Costa Rica's Natural Resources*

This section begins the process of generating a biophysical analysis of the Costa Rican economy while undertaking specific assessments of various components of that economy in terms of the degree to which it is, or is not, sustainable. The first three chapters in this section deal with agriculture: Chapter 12 in terms of its empirical behavior over the last several decades, Chapter 13 with respect to the possibilities of using spatially dependent modeling to improve our ability to assess agricultural potential and optimal land use, and Chapter 14 in terms of the importance and impact of pastures. Each of these chapters stresses the interplay of models and empirical analysis for a more comprehensive and explicit understanding of how space effects agricultural potential and production.

Chapter 15 examines the issue of erosion, an issue that is critical to any discussion of sustainability, from a spatial perspective. The final four chapters address in various ways Costa Rica's valuable and unique forests, their extent, their conservation, their destruction and its consequences, their role in the economy, and the ways they are and are not protected by government programs. Chapters 12, 16, and 17 explicitly use the gradients approach introduced in Chapter 7, and the other chapters often do so implicitly. Each of these chapters is an important and generally thorough analysis of the biophysical reality of the resource being considered, which must serve as the basis for any discussion of sustainability.

CHAPTER 12

# Temporal and Spatial Overview of Costa Rican Agricultural Production

CHARLES A. S. HALL, CARLOS LEON, WILL RAVENSCROFT, AND HONGQING WANG



## I. INTRODUCTION

If human systems are to be sustainable then agriculture must be sustainable, as the provision of food is essential for humans. Food production is an explicitly biophysical process, involving the interaction of plant cultivars, climate, soils, animals, and management. Food production replaces inherently stable systems (in the case of Costa Rica, natural forests) with inherently unstable systems, or perhaps more accurately, systems whose maintenance requires continual intervention and cultural or fossil energy input by humans. The process of agricultural production itself generally degrades the potential of that land for further production. This may not be apparent to most people because that degradation is compensated for by fertilizers, which may increase the agricultural sustainability of that site by contributing to the exhaustion of fossil fuels and fertilizer deposits elsewhere.

Costa Rica normally is considered principally an agricultural country, that is, its land area is overwhelmingly used for agriculture (including pastures). In addition its economy and even much of that portion considered "industry" traditionally has been based on agriculture, and a majority of its people based their livelihood directly or indirectly on agriculture up until about 1995. The history of Costa Rica is very much tied up with agriculture, as developed in Chapter 2, and agriculture continues to be the principal source of foreign exchange. All this is true even as Costa Rica changes to a predominately urbanized, industrialized nation.

This chapter is a summary of the most important characteristics of Costa Rican agriculture from a biophysical perspective. Section II gives a brief history of land use and agricultural yields, and examines the potential for increasing the importance of agriculture in future economies. Section III examines agriculture production from a geographical context, and Section IV synthesizes Costa Rican agricultural production using energy flow analysis diagrams.

## II. TRENDS IN LAND USE AND AGRICULTURAL YIELD

An important issue with respect to understanding the degree to which Costa Rica is or is not approaching its stated goal of sustainability is to examine the efficiency of agriculture over time to see if there are any important trends relative to criteria which would reflect sustainability. Efficiency, as will be defined, is one such criterion. Increasing efficiency should allow less resources to be used to meet whatever level of production is required or desired. It is often used in the sustainability literature.

### A. Methods

Our principal source of data on land use and on crop production was the Food and Agricultural Organization of the United Nations (FAO, various years). The FAO reports yields for certain individual crops (cultivars) as well as seven more general categories: cereals, roots and tubers, pulses (e.g., beans), oil crops, vegetables, fruit, and tree nuts. We analyzed each of these aggregate categories except tree nuts (which are unimportant in Costa Rica). We also analyze production trends for the most important individual crops, including rice, maize, beans, and coffee. Finally we examine the most important trends overall for agriculture by summing all crops. While it is not entirely accurate to examine the total agricultural output of a country by summing all crops, since different crops have different values and water contents, are variously productive, and are variously responsive to fertilizer and other inputs, a great deal can be learned about general patterns, including overall efficiency of agriculture, by examining aggregate national patterns. We solve some of the problems raised by this approach by converting crop production to calories, which generates much more consistent information than does wet weight, the original units used. Energy values were derived using the conversion factors given in Table 12-1.

We provide three methods to estimate agricultural efficiency:

1. *Land efficiency* (or productivity) is total crop production (in metric tons or energy content) divided by the number of hectares of land devoted to permanent and annual crops. Where good quality land is limited, such as in Costa Rica, this index is of critical importance. In general land efficiency is increased through agricultural technology, including an increased use of inputs from outside the agricultural system.
2. *Fertilizer efficiency* is total crop production (in metric tons or energy content) divided by total fertilizer use (in metric tons). Ideally we would like to have a more comprehensive index that included all inputs (tractor horsepower-hours, truck ton-miles transportation, other chemicals, cubic meters irrigation) but data on these factors are not maintained.
3. *Labor efficiency* is total crop production (in metric tons or energy content) divided by the number of workers in agriculture.

Unfortunately the FAO data base does not include an assessment of the quality of the land and soils (although that is treated in Chapter 10) so we cannot directly correct land area used for its quality, although the change in land quality presumably is reflected in the first of the preceding ratios.

Total national fertilizer use, the sum of nitrogen, phosphorus, and potassium (NPK), has been published for each year since 1951 by the FAO. Determining

**TABLE 12-1 Water Content and Caloric Density of Major Crops Grown in Costa Rica (Based on "Dry" Harvested Weight)**

| | |
|---|---|
| Mean caloric values[a] | |
| Vegetable matter | 4.25 kcal/g dry weight |
| Meat | 5.0 kcal/g |
| Moisture contents[b] (%) | |
| Cereals | 15 |
| Roots and tubers | 70 |
| Pulses | 11 |
| Tree nuts | 6 |
| Oil crops | 7 |
| Vegetables and fruit | 80 |
| Sugar | 80 |

[a] From Whittaker (1975).

[b] Estimated from Pimentel, D. 1980 Handbook of Energy Utilization in Agriculture. CRC Press. Boca Raton, Fla.

how much of that was used on each crop, necessary for the calculation of trends in efficiency, was more difficult. Explicit FAO values for individual crops exist for most countries, including Costa Rica, only for 1991 (FAO, 1992). We were able to find independently derived rough mean national application rates for major crops for 1970–1984 (Raphael Statler, CATIE) and for 1988–1993 (FERTICO, undated). Although neither of these publications is formal, their data seem reasonable and consistent with trends in national total use trends and each other, although an occasional shift in the decimal place is necessary. Sometimes we had to make educated guesses for conversions when the original data were in just nitrogen or where the data appeared to be for total formula weight (i.e., including nonnutritive components) as opposed to just the weight of the N, P, and K in the fertilizer salt. Fortunately, as will be derived, we had means for making several independent estimates.

We derived an independent procedure for determining year by year crop-specific rates ($Fert_{c,yr}$) with the formula (Ko *et al.*, 1998; Hall *et al.*, 1998)

$$Fert_{c,yr} = Fert_{t,yr} \times (Fert_{c,91}/Fert_{t,91}) \times (Area_{c,yr}/Area_{t,yr}),$$

where $Fert_{t,yr}$ is the total national fertilizer use in year "yr," $Fert_{c,91}$ is the fertilizer used on that crop in 1991, $Fert_{t,91}$ is the total national use in 1991, $Area_{c,yr}$ is the area used for that crop in year "yr," and $Area_{t,yr}$ is the area used for all crops in year "yr." In other words we derived crop-specific rates by multiplying trends in national fertilizer use by the changes in area devoted to that specific crop and by the per hectare fertilizer intensity in our index year 1991. In some cases the estimates of, e.g., land area in a particular crop, total national area in annual and permanent crops, or total fertilizer used, were somewhat different in the crop by crop assessment (FAO, 1992b) and the annual year by year FAO summary for 1992. When this occurred we adjusted the fertilizer intensity and land areas to be consistent with the year by year estimates so that the FAO accurately represented the *proportion* of the total 1991 fertilizer use that went on each crop. Our later sensitivity analysis shows that these small corrections had no important effect on our conclusion.

We were concerned about whether we were generating the year to year fertilizer intensities for each crop in the right way, and whether we might be missing some important real data. We called the International Fertilizer Research Institute in Mussel Shoals, Alabama, and talked to their quantitative fertilizer expert, Mr. Harris. He confirmed that indeed there were not crop-specific records kept for most nations, including Costa Rica, and told us that their own scientists had independently derived the same equation as we had for their own needs. Thus we believe that we have done as well as possible, and anyway we are able to compare these estimates with independent local estimates in our results. The formula-derived crop-specific values and the independent estimates by Costa Rican experts generally agree within a factor of roughly 50% although they are occasionally different by a factor of two. Because the independent estimates are not consistently higher or lower than the formula-derived values, the emerging trends in efficiency are essentially always the same. Hence we feel confident in our general conclusions about efficiencies in fertilizer use.

## B. Results

### 1. Patterns of Land Use Change

Forested land in Costa Rica decreased from 3.240 million ha in 1961 to 1.638 million in 1982, a 50% decrease and a deforestation rate of 3% per year (Figure 4-4). The amount of land used for permanent and annual crops and for other uses (e.g., abandoned and "unused" land) has remained roughly constant since 1960. There are many reasons for deforestation, but in Costa Rica the end result has been a net conversion of forest to pasture (see Chapters 15 and 16). Much of that net conversion, however, was through clearing for agriculture which was subsequently abandoned to pasture. Total developed land has increased at about the same rate as the human population, both about 3% per year. Urbanized land probably also has increased at about that rate, but no explicit figures are kept on that. We derive a rough estimate by multiplying the population by about 0.028 ha per person (Levitan, 1988).

### 2. Agricultural Production, General FAO Categories

Raw data for six of the FAO summary groupings—cereals, roots and tubers, pulses (e.g., beans), oil crops, vegetables, and fruit—and sugar are given in Figure 12-1. The only individual cultivar whose yield is large enough to be seen on this graph is sugarcane, the largest crop by weight, followed by fruit and cereal. Vegetables, oil crops, pulses, and roots and tubers are roughly equivalent by weight. Total production of all agricultural commodities increased from 1.748 to 4.915 million metric tons (harvested weight) between 1961 and 1990, a 280% increase.

The slight increases in area of land in cultivation do not explain this 280% increase by weight in agricultural production from 1961 to 1990. The 592% increase in fertilizer use is a much more likely explanation. The number of tractors in service increased over the 23 years of reported data from 4900 in 1968 to 6500 in 1991, although no data were found regarding the size of the tractors. This seems an unlikely major contributor to the increase in yield.

The results for the same data expressed as dry weight and as energy are shown in Figures 12-2 and 12-3. Fruit and sugarcane appear extremely

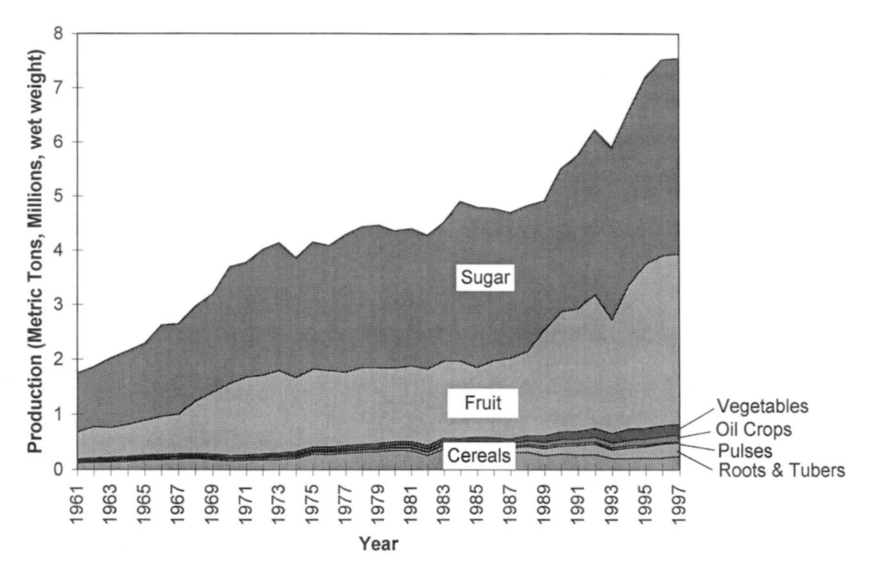


FIGURE 12-1 Total national yield for each of seven general categories of agricultural production (including sugar) expressed as wet weight. (Source for Figures 12-1 to 12-3 is FAO, Rome.)

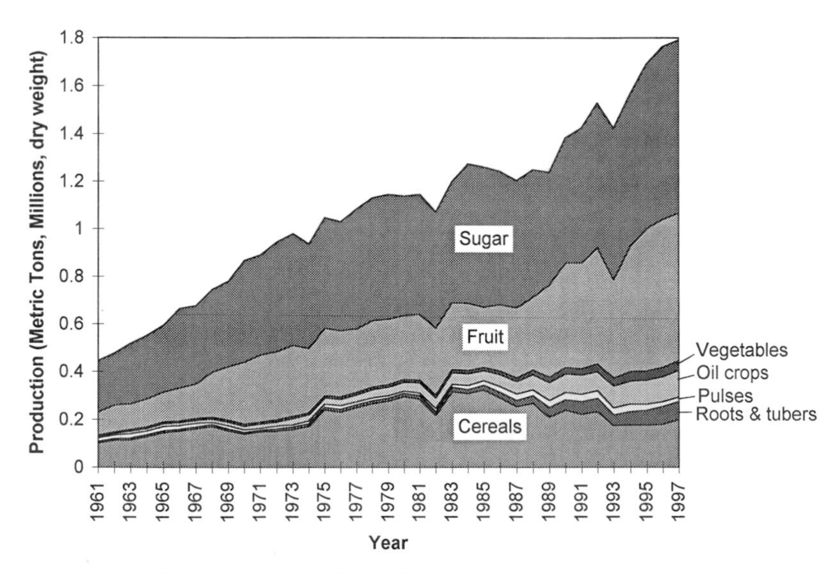


FIGURE 12-2 Same as Figure 12-1 except expressed as dry weight.

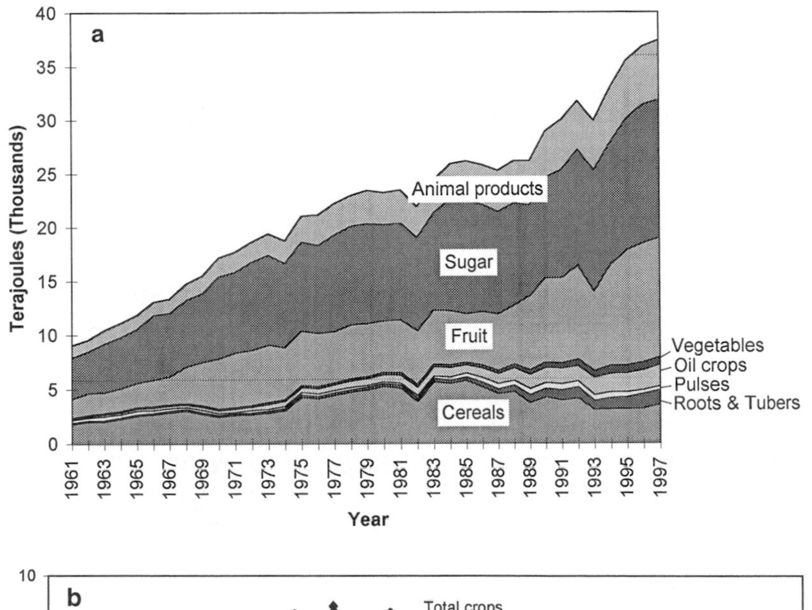


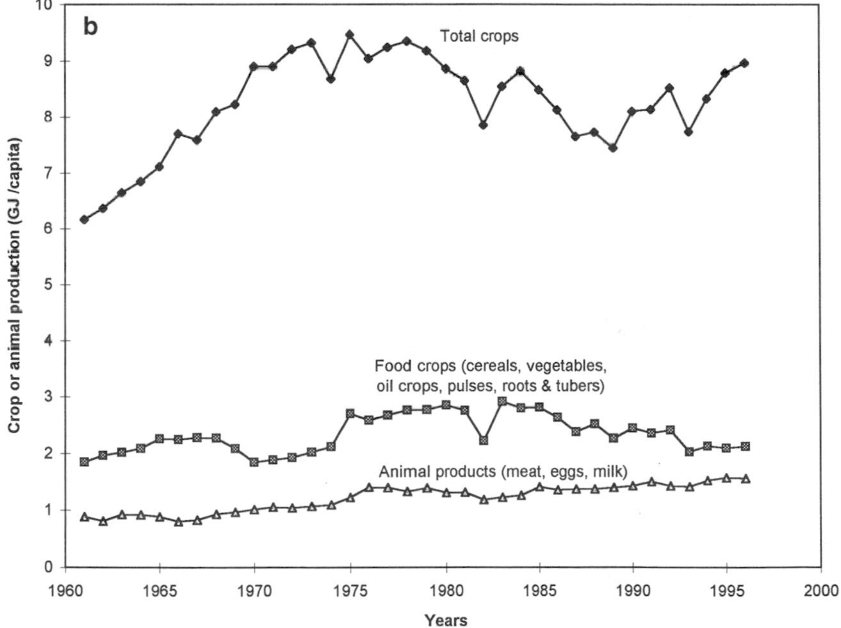


FIGURE 12-3 (a) Same as Figure 12-1 except expressed as terajoules. Energy content = [raw production (in metric tons, MT)] × (1 − proportion moisture) × ($10^6$ g/MT) × [energy content (kcal/g)] × ($4.183 \times 10^{-9}$ TJ/kcal). (b) Crop and animal production per capita in Costa Rica.

important in Figure 12-1, but that is due partly to their high water content. They are less important relatively in Figures 12-2 and 12-3. Animal production (wet weight) is almost too small to be seen, although the small energy contribution of animal products does not reflect their high dietary and economic value.

## 3. Agricultural Production, Principal Crops

Costa Rica has a very wide range of climatic conditions (Chapter 9), and as a result many different crops can be grown there. For example, although tropical bananas are grown abundantly near Siquirres, one can drive only one hour up onto the slopes of the Turrialba Volcano and see luxurious fields of crops such as potatoes and cabbages that generally North Americans associate with temperate regions. The most important crops grown in Costa Rica traditionally have been rice, maize, beans, sugarcane, coffee, and bananas. The total land area planted with these main crops (considered together) has been reasonably stable over recent decades, although the area planted to grains has decreased dramatically, while that planted to export crops has increased. Other important crops are listed in Table 12-2.

**TABLE 12-2** "Nontraditional" and Less Important Crops Planted in Costa Rica (1992–1993)

| Crop | Area planted (ha) | Production (tons) | Value of exports (1000s of dollars) |
|---|---|---|---|
| Oranges | 18,000 | 116,000 | — |
| Macadamia | 6,680 | 2,000 | 1,097 |
| Mango | 5,780 | 19,202 | 629 |
| Melon | 4,218 | 91,069 | 23,476 |
| Ornamentals | 4,280 | — | 72,958 |
| African Palm | 26,600 | 283,430 | 11,032[a] |
| Palmito | 3,822 | 19,110 | |
| Papaya | 778 | 42,762 | 2,256 |
| Pepper | 488 | 2,500 | — |
| Pineapple | 7,000 | 145,075 | 45,310 |
| Plantains | 8,300 | — | — |
| Chayote | 220 | — | — |
| Coco | 4,500 | 1,789 | 1,167 |
| Strawberries | 50 | — | 88 |

*Source:* Costa Rica (1993).
[a] Value of exported oils of all kinds.

Data are available for the area planted with, and also for the production of, some important crops over time (Figures 12-4a–f, top and third graphs). As derived in the methods, we were able to estimate annual crop-specific fertilizer data from the quantity of fertilizer used nationally (second graph from top). We are able to synthesize the important relations of efficiency with respect to land area and fertilizer use intensity in the lowest panel of Figure 12-4 as well as in in Figure 12-5. In general fertilizer use and area planted have increased over time. Fertilizer use intensity seemed sensitive to the energy crises of 1973 and 1979. Total national yields have tended to increase, especially from 1960 to 1980, as a function of both area planted and intensity of fertilizer use, but yield per capita has tended to decline, especially since about 1980, as populations have increased more rapidly than total yield (Figure 12-4, third graph). The area planted and production of maize decreased greatly after about 1987 due to the removal of government subsidies for its production.

## 4. The Inverse Relation between Intensity and Efficiency

As *total* yields have tended to increase with increased inputs of land and fertilizers, *efficiency* of land use, expressed as yield per hectare, and especially per ton of fertilizer input, has tended to stabilize or decrease (Figure 12-4, fourth graph). Thus in general it appears that major Costa Rican crops are approaching the asymptote for fertilizer response, and little increase in land efficiency has occurred for about 15 years. It also appears that fertilizer efficiency has declined progressively and continuously over time. Neither of these trends bodes well for the concept of a more sustainable agriculture. In fact the converse is true. Over time agricultural production is becoming less sustainable in terms of its relation to its use of fossil-fuel-derived inputs even while it becomes more dependent upon those fuels as increasing populations and limited high-quality lands require intensification of their use.

Another way of examining the inverse relation of intensity and efficiency is seen in the three graphs of Figure 12-5. These graphs show that while indeed total yields increase with an increasing use of fertilizer (*x* axis) and land area (*y* axis), yield per hectare (land efficiency) for a given level of fertilizer declines with increasing land use for all crops examined. For example, for rice (Figure 12-5e) at a national fertilizer use of about 6000 tons the yield is more than 3.5 tons per hectare per year when less than 50,000 ha was planted, from 2 to 3 tons per hectare per year when an intermediate area was planted, and less than 2 tons per hectare per year when more than 75,000 ha was planted. Different ways of representing efficiency are given in Figure 12-5.

These figures, which are remarkably consistent over the different crops, show that at any given level of fertilizer application the more land that is used the lower the yield will be. This again has very important implications for

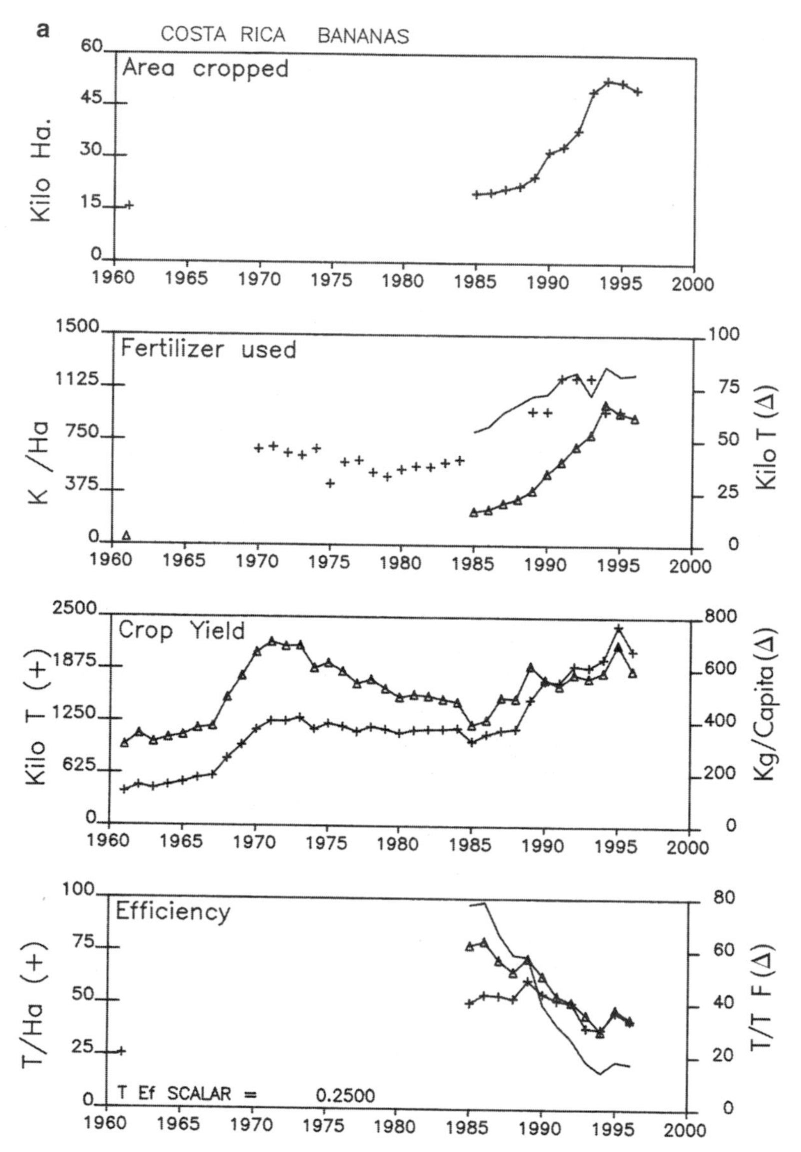


FIGURE 12-4 (a–f) Summaries of annual inputs, outputs, and efficiencies for major Costa Rican crops. The top graph represents the area planted. The second graph represents fertilizer inputs, both per hectare (left axis and line without symbols) and total (right axis and line with triangles). Crosses represent independently derived per hectare values. The third graph is production, both total (left axis and crosses) and per capita (right axis and triangles). The fourth graph is efficiency calculated as tons per hectare (left axis and crosses) and per ton of fertilizer (right axis and triangles). [Independent fertilizer values courtesy of Rafael Statler, CATIE (1970–1984), and Ministerio Agricultura and Ganaderia (1988–1994); FAO calibration values from FAO (1992).] The line is yield per ha per T fertilizer, scaled as at left.

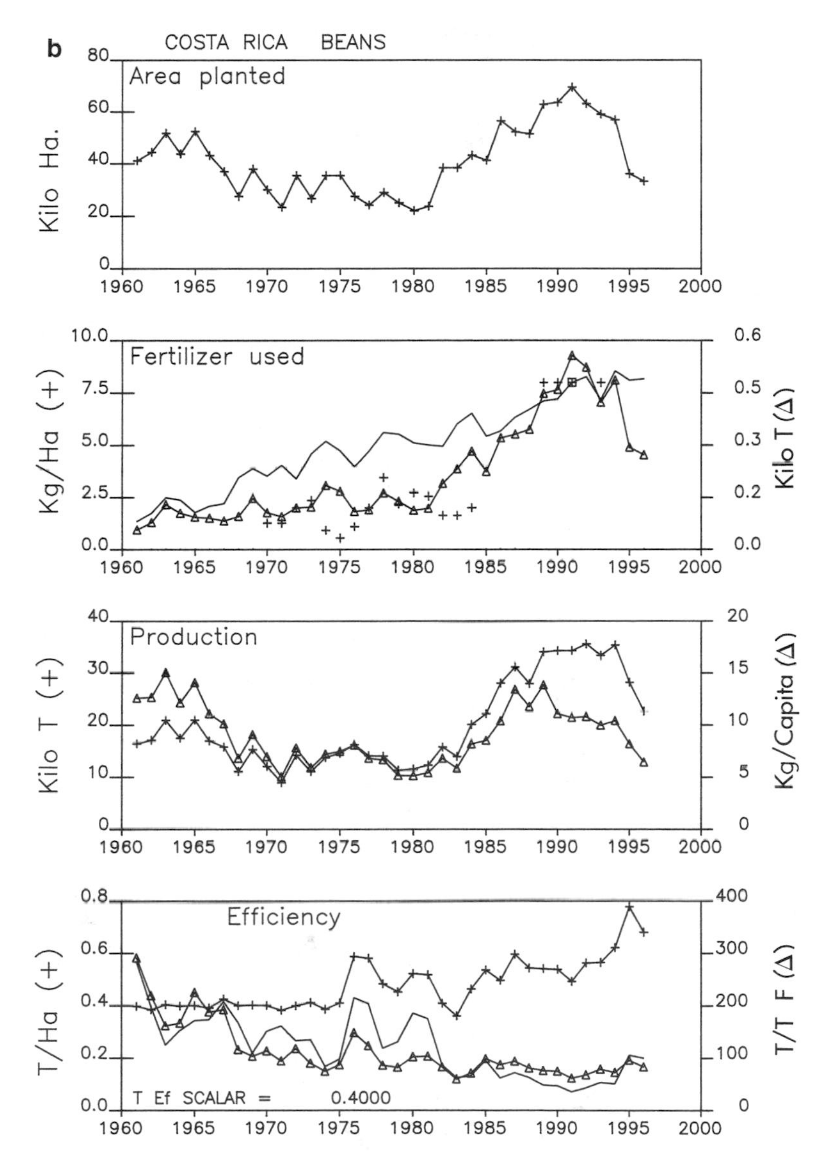


FIGURE 12-4 (*Continued*) (b) Note the lack of yield response to fertilizer input increase from 1980 to 1992, when land area was increasing, compared to post 1990.

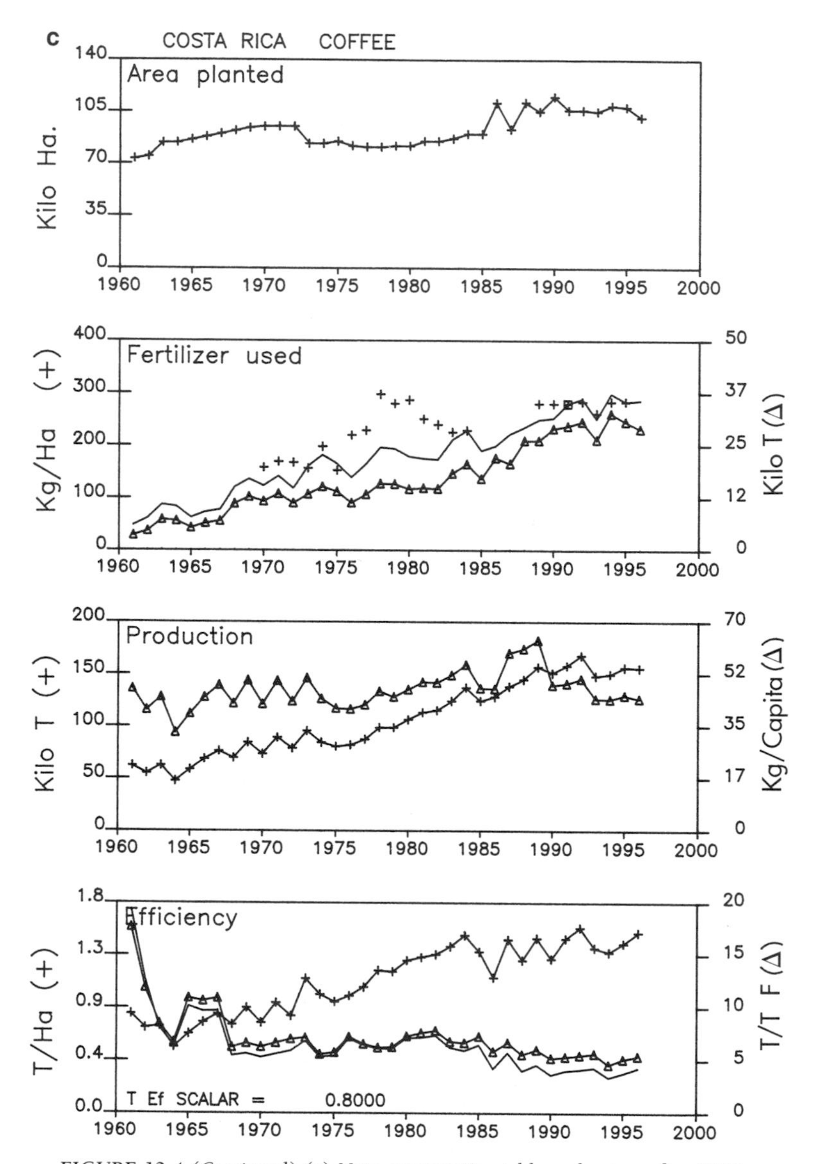


FIGURE 12-4 (*Continued*) (c) Note asymptotic yield per hectare after 1984.

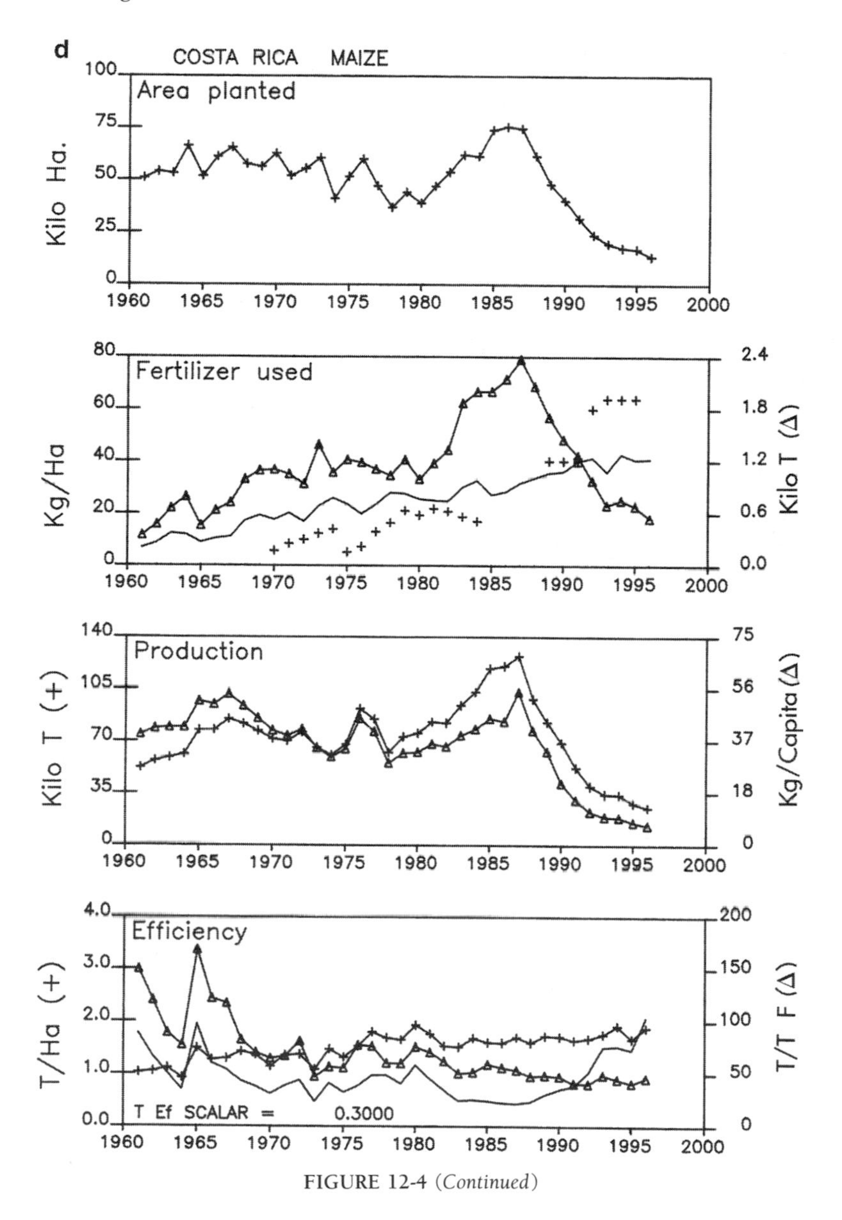


FIGURE 12-4 (*Continued*)

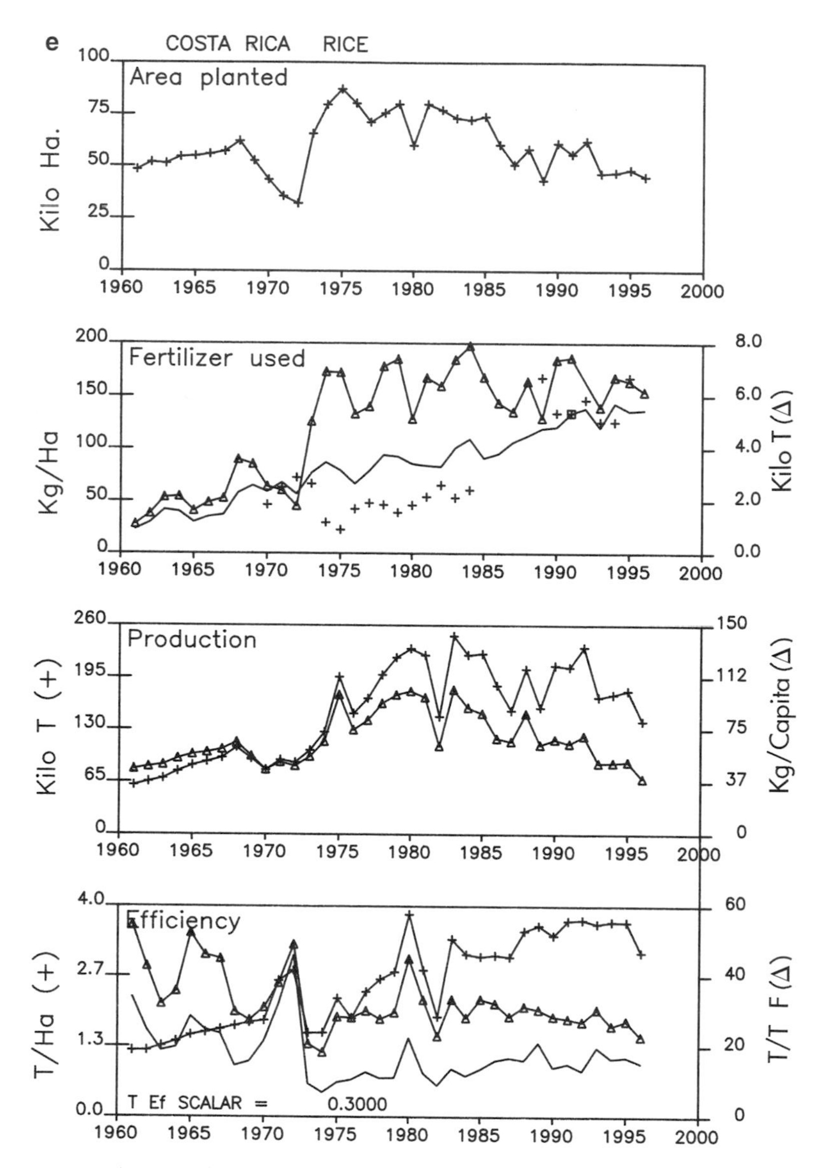


FIGURE 12-4 (*Continued*) (e) Note response in efficiency (lower panel) to decrease and then increase in area planted 1968–1975.

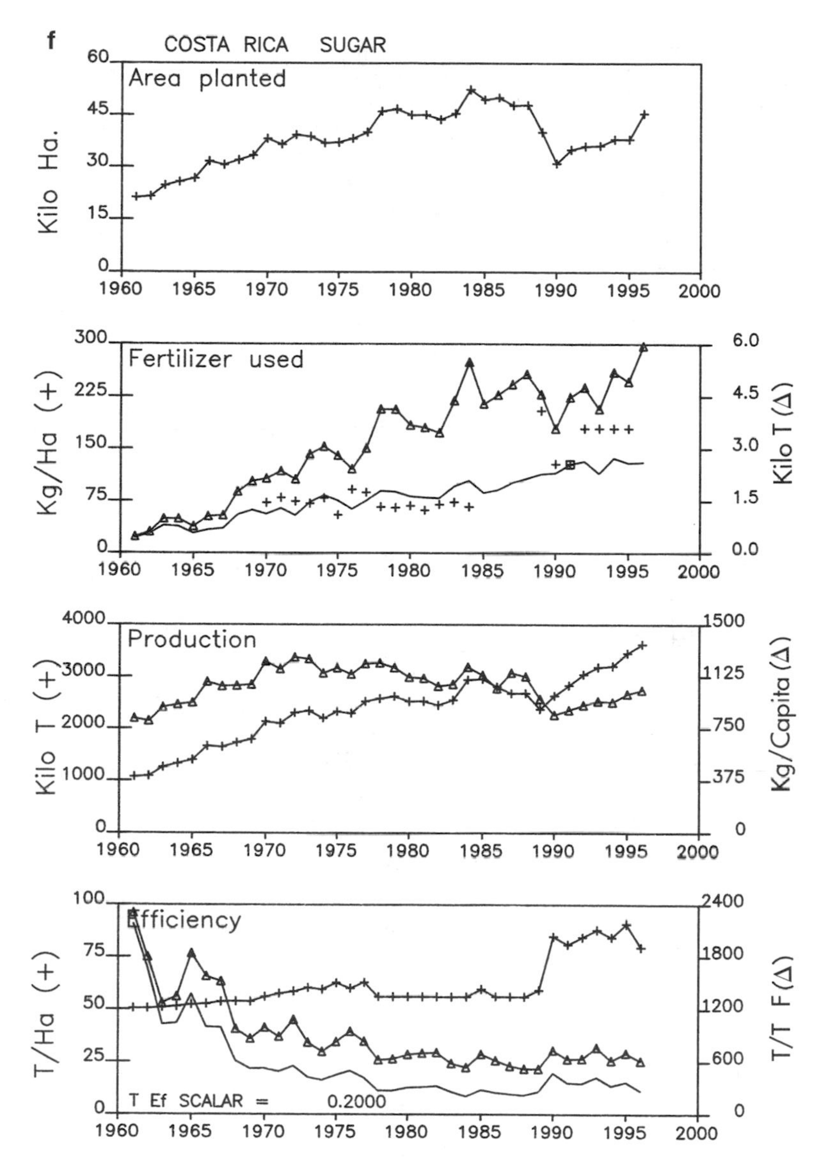


FIGURE 12-4 (*Continued*) (f) Note increase in yield per ha following reduction in area planted following 1987.

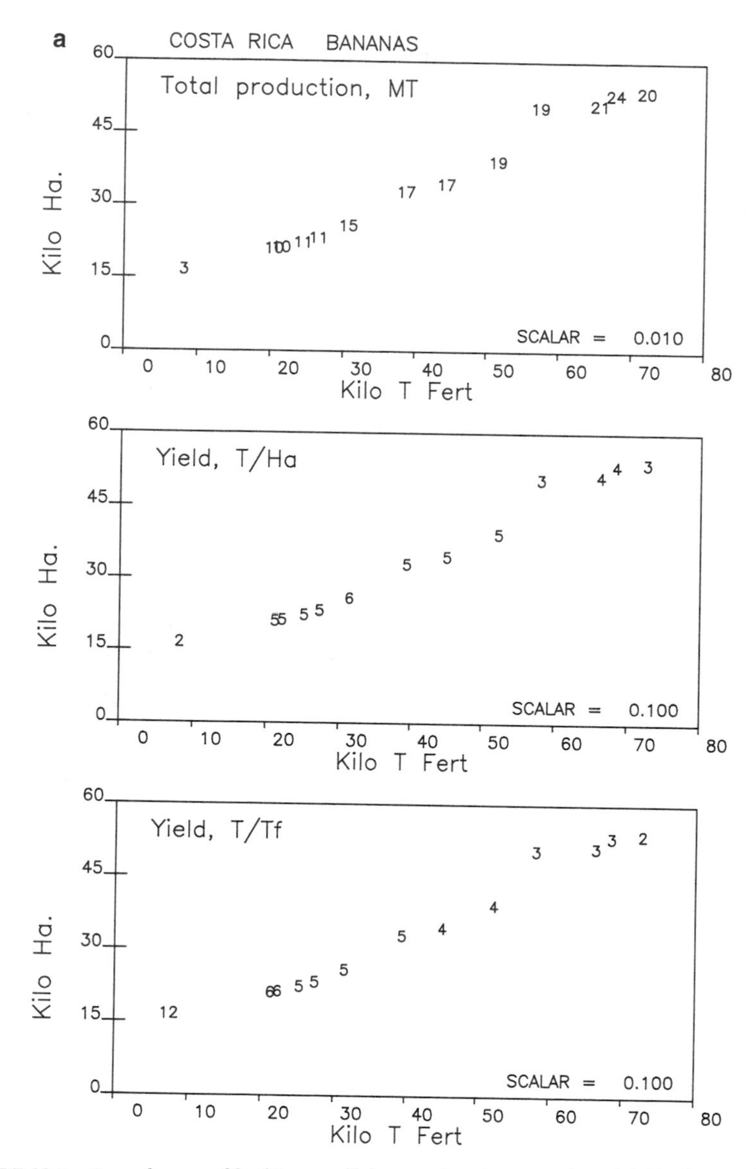


FIGURE 12-5 Several ways of looking at efficiency using the same data: (a) Total annual yield (million tons) of crop as a function of intensity of land use and fertilizer. (b) tons yield (×10) per hectare, and (c) tons yield per ton of fertilizer. Figure (a) is perhaps preferred for statistical reasons since the data plotted are completely independent of the axes, but the others are probably more useful in thinking intuitively about the issue. These figures all show in various ways that efficiency is inverse to intensity of land or fertilizer use.

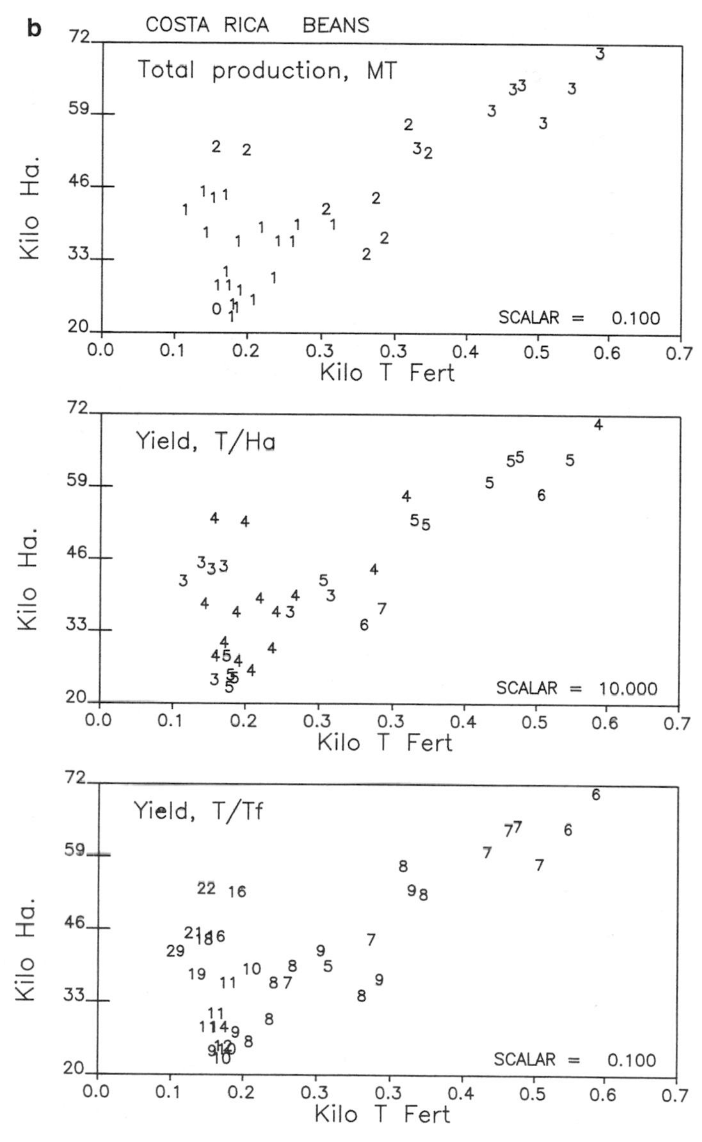


FIGURE 12-5 (*Continued*)

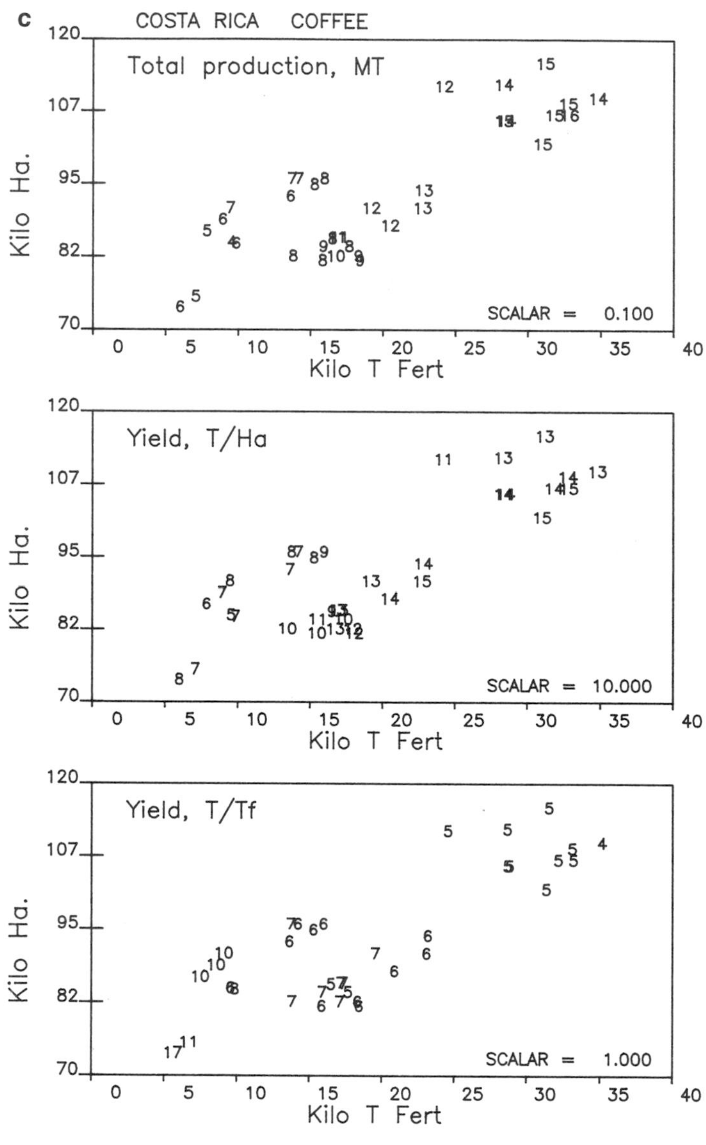


FIGURE 12-5 (*Continued*)

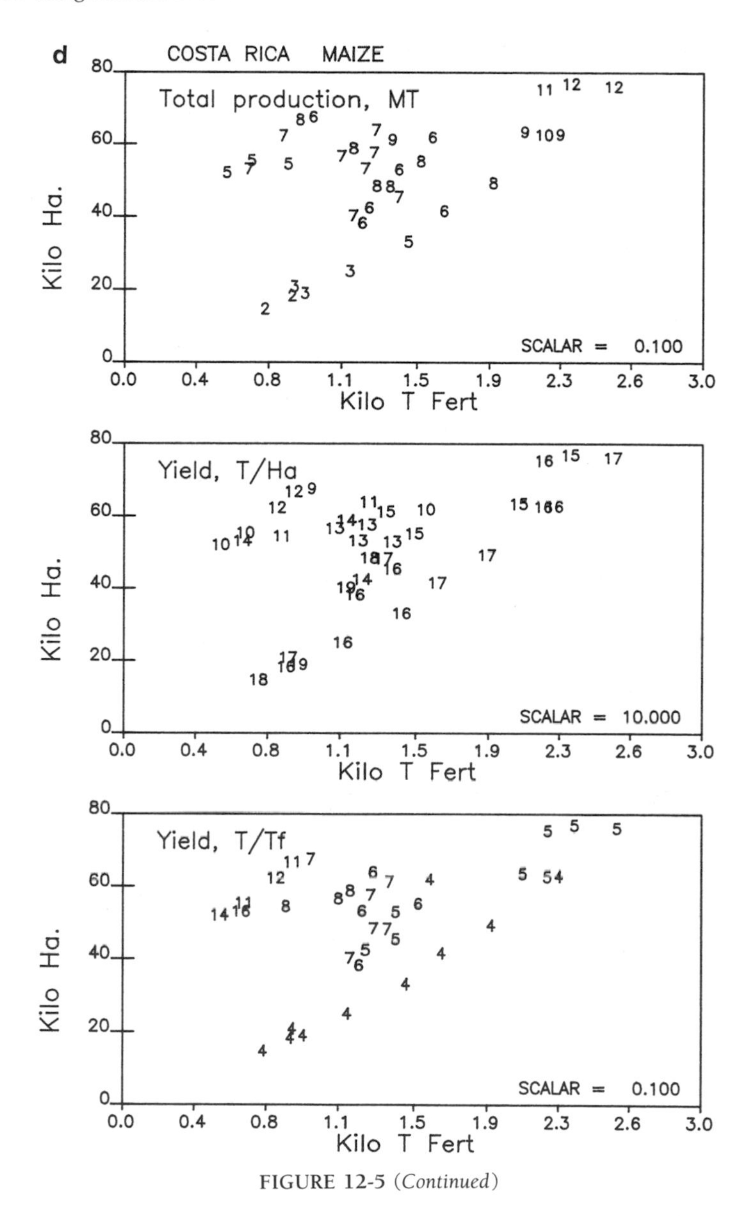


FIGURE 12-5 (*Continued*)

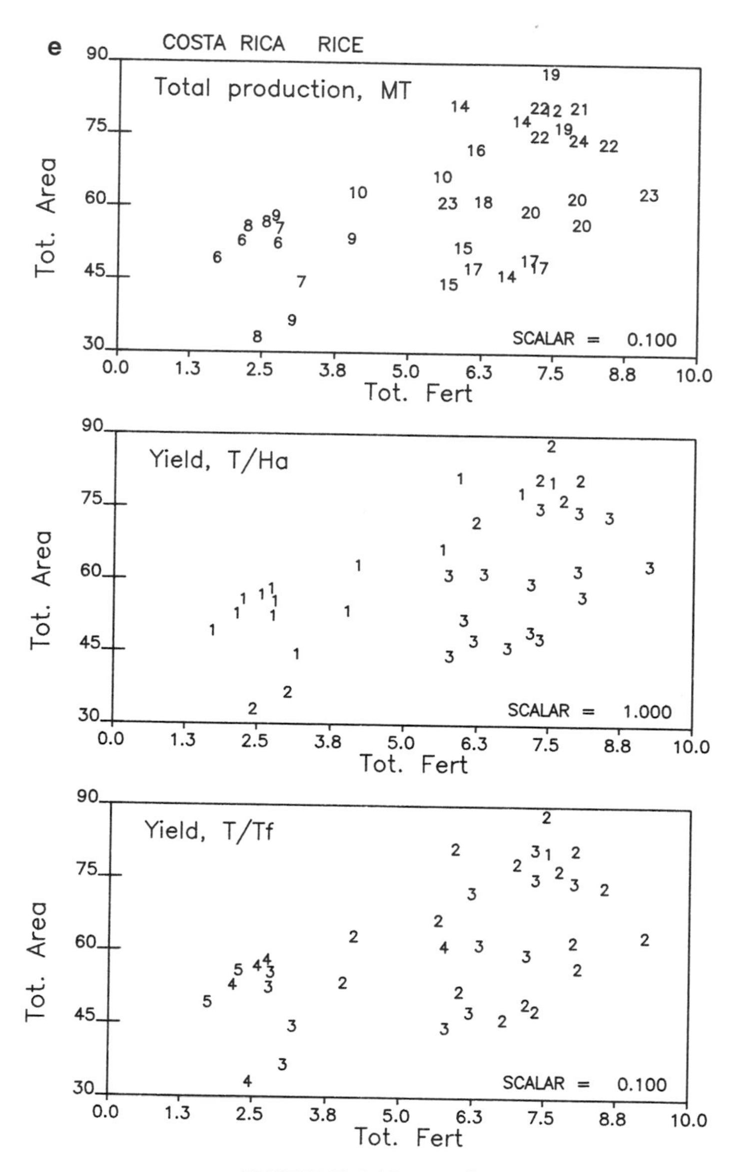


FIGURE 12-5 (*Continued*)

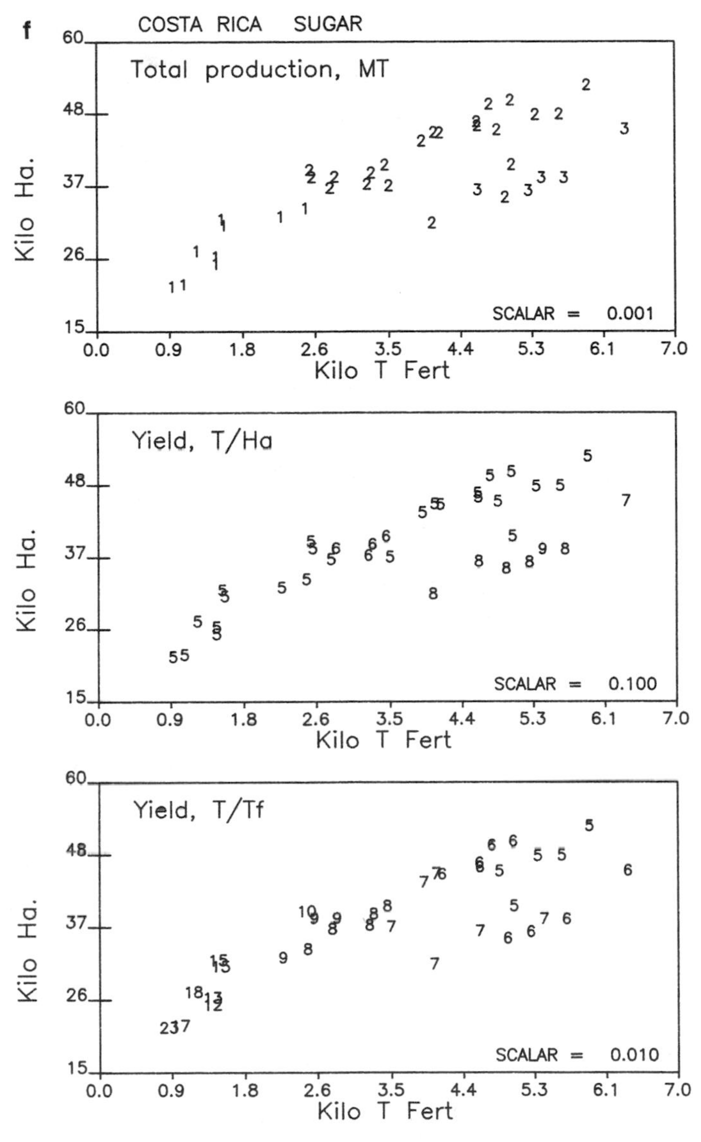


FIGURE 12-5 (*Continued*)

sustainability (see Hall *et al.*, 1999). First it implies that Costa Rican farmers are very, very clever in determining which land (on which soils and in which gradient space) is best for growing particular crops. Second, it implies that as land is taken out of production for a particular crop the poorest land is taken out progressively. Third it implies that those who wish to increase production through increasing land area in production must consider that the new land will have sufficiently lower yields to decrease the mean yield of the entire national crop significantly. This relation is seen, with a surprisingly small statistical variation, at every level of fertilizer application for every crop. Finally these analyses show that the ideas of David Ricardo, the early classical economist, are as relevant in Costa Rica today as in England in the 18th century and provide a very serious cap on our sustainability possibilities while populations are growing.

## 5. Agricultural Efficiency, General FAO Categories

Both outputs and inputs to agriculture increased during the period 1960 to 1990. For specific crops there has been a reduction in yields per unit increase in both land and fertilizer as agricultural production has been intensified and expanded (Figure 12-4). This is also true for all national crop production in the aggregate. Total agricultural land (not including pastures) has increased from 480,000 ha in 1961 to 505,000 in 1997, the number of workers in agriculture from 196,000 to 325,000 in the same period, and the use of fertilizers from 19,000 to 203,000 tons (Figure 12-6).

Both land and labor productivity have increased, the former from 3.68 tons (wet mean for all crops) per hectare in 1961 to 9.31 tons per hectare in 1990 (0.93 to 2.61 tons dry weight, and 16,481 to 46,474 MJ per hectare) (Figure 12-7). The smaller increase in resource use efficiency when the results are reported as energy units indicates the growing importance of sugarcane, a relatively wet crop. Productivity per worker has increased from 8.92 (wet) tons of agricultural product per worker in 1961 to 21.92 tons in 1995 (2.27 to 5.51 dry tons per worker). Yield per Costa Rican increased from 1.36 tons to 1.81 dry tons, although if only subsistence food crops are considered there was no change (0.093 tons per Costa Rican). The record is clear that there has been an increase in the proportion of agricultural effort that has gone into nonsubsistence crops, much of which was exported.

Fertilizer use efficiency has decreased from 93.5 tons (wet) of crops per ton of fertilizer in 1961 to 38 tons per ton in 1997, a decline to 40% of the original value. Since the energy cost of fertilizer is known (Pimentel *et al.*, 1995; Hall *et al.*, 1986), this efficiency also can be calculated as kilocalories per kilocalorie (identical to megajoules per megajoule). When expressed this way there was a similar decline by 1997 to less than half of the 1961 value. This is consistent with the fertilizer saturation phenomenon seen for all crops in Figure 12-4.

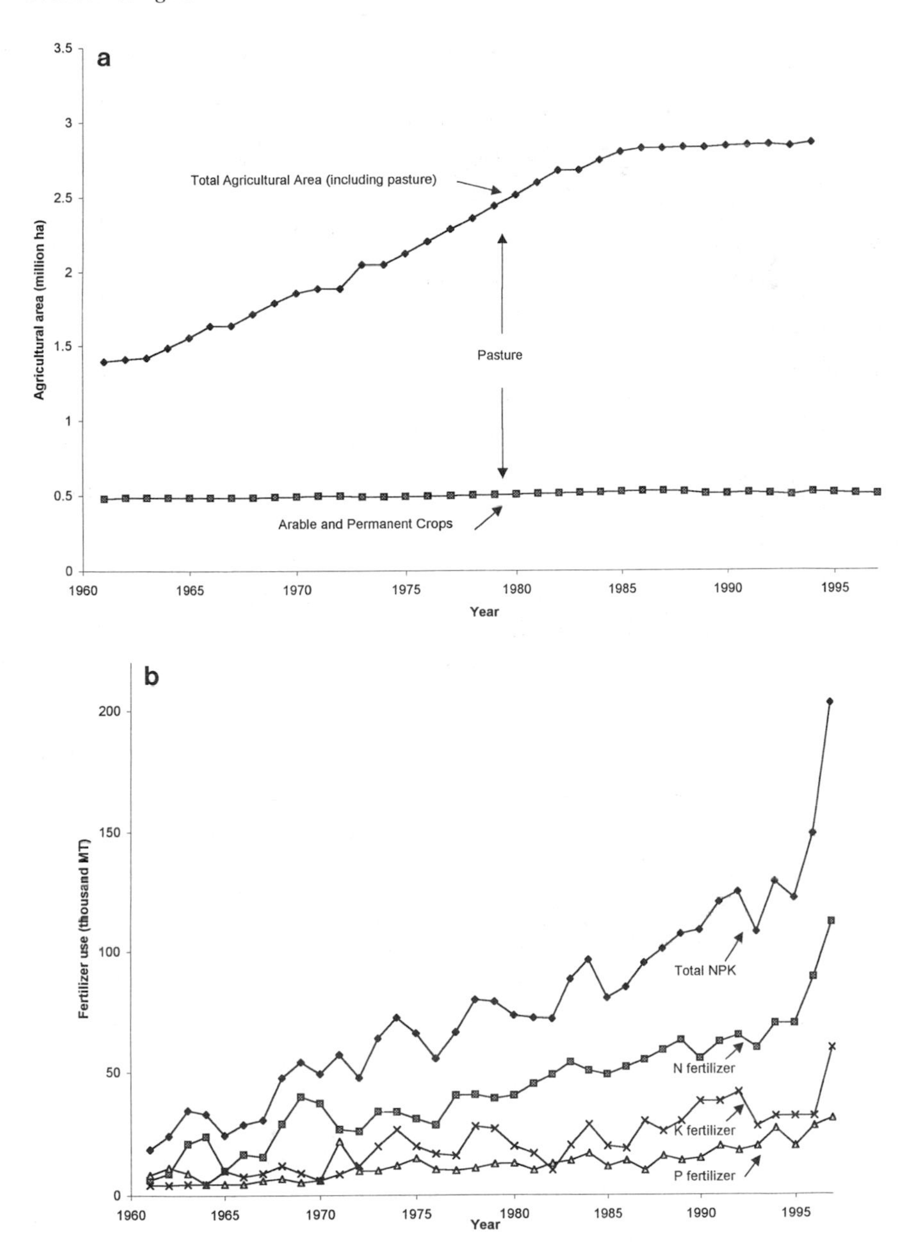


FIGURE 12-6 Total national inputs to agriculture: (a) land area, and (b) fertilizer as NPK (nitrogen, phosphorus, and potassium).

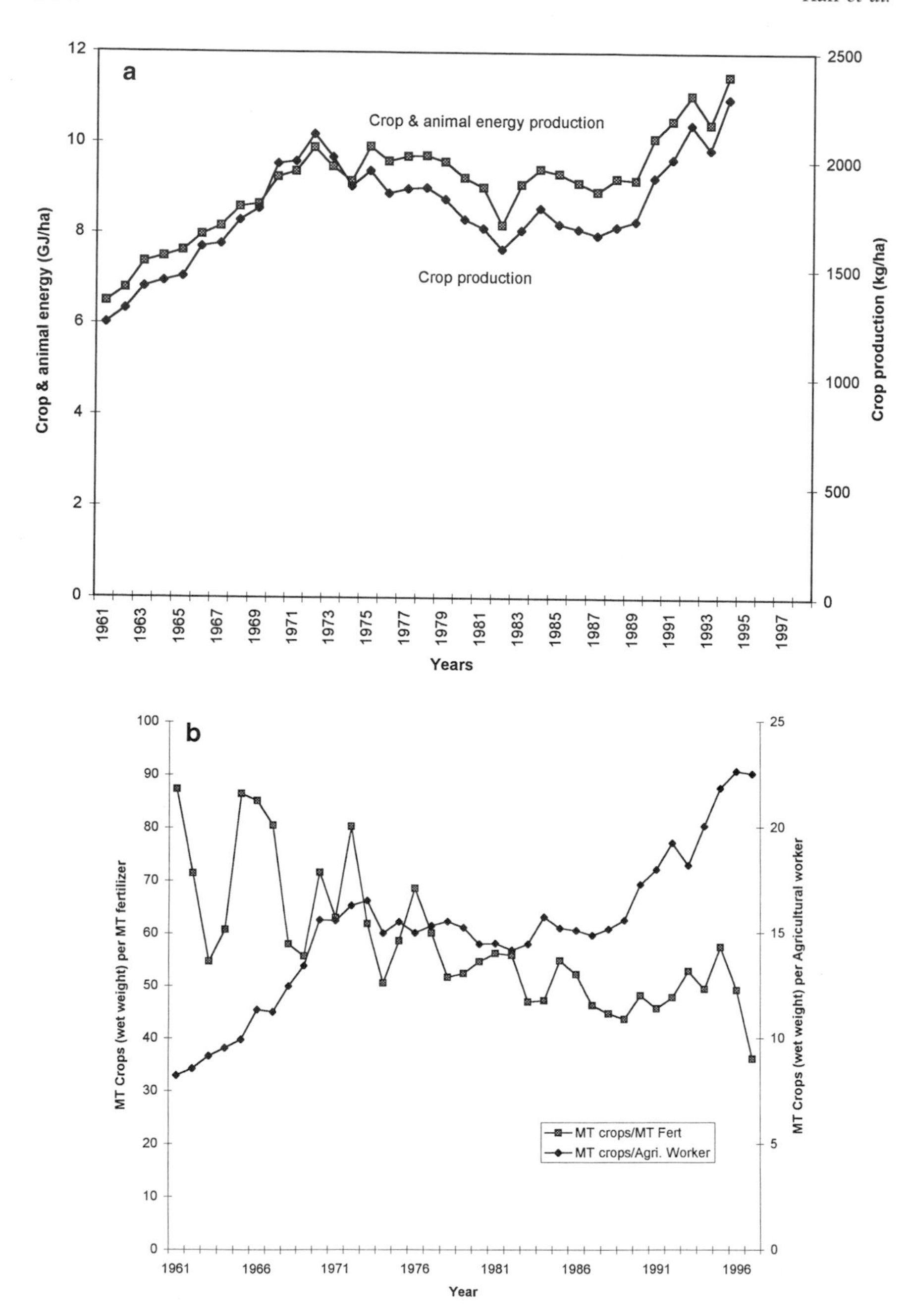


FIGURE 12-7 Total national efficiency of agriculture as defined as: (a) crop and animal production per hectare, (b) yield of all crops per agricultural worker (right axis) and per ton of fertilizer (left axis), and (c) total national yield vs total national fertilizer use.

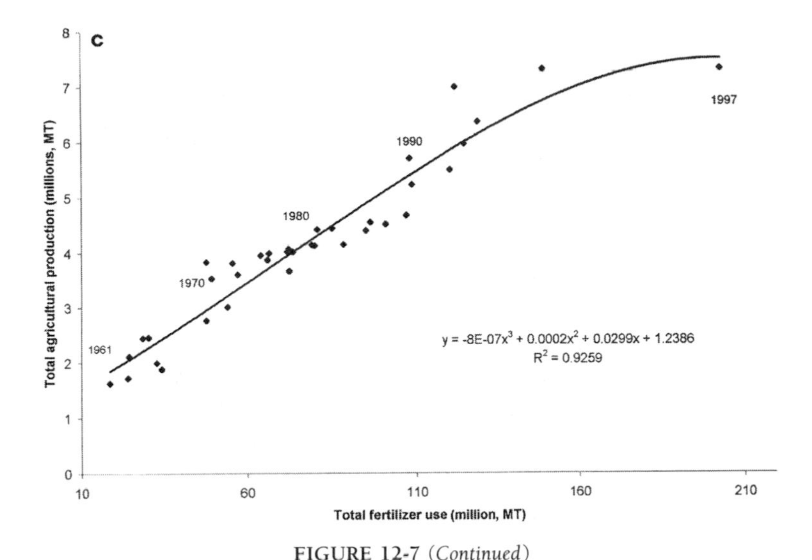


FIGURE 12-7 *(Continued)*

## 6. How Many Costa Ricans Can Be Fed?

One consequence of the continued human population increase is that more land has been needed for all human activities. At this time roughly 100,000 ha are covered by urban systems (Levitan, 1988), including much of the country's best land. Ironically, Costa Rica's increasing reliance on industrialization and fossil fuels has required an increase in the use of land and solar energy to generate the foreign exchange needed to pay for that industrialization. More than 90% of coffee and bananas, and 80% of the beef produced, is exported. Thus we can estimate that roughly 150,000 ha of cropland and 2 million ha of pastures are used to generate foreign exchange. This is nearly half of the total area of the country and more than the total area of Land Classes I, II, and III, those classes suitable in any way for agriculture or pastures.

The three main human activities that convert land from tropical forests to other uses, and thus redirect solar energy inputs directly into the economy, are logging, clearing for pasture, and clearing for farming (which includes shifting cultivation) (Cornell and Hall, 1994). While Costa Rica has a diverse and biotically rich landscape, it is only moderately fertile for agricultural production (Table 2-2, Figure 4-4). Additionally, only about one-fifth to one-quarter of the land is suitable for long-term production of the kind of crops that feed people (even this may be too high an estimate for sustained cropping; see Chapter 10). If we take the larger number, then the present 3.5 million people have about 1.4 million ha of good cropland, about 0.40 ha per person or 2.5 people per hectare.

A simple calculation can be done to examine the number of people that might be fed from existing agriculture in Costa Rica. Presently Costa Rica has about 200,000 ha in annual (food) crops. Mean yields of grains and pulses together are roughly 2.0 tons per hectare—the present yield is 1.5–2.0 dry tons per hectare for maize, 2.5–3 for rice, and about 0.6 tons per year for pulses (beans) and vegetables, so that about 400,000 tons of food are being produced on that land. Each ton can feed about 3 to 4 people for a year (Hall and Hall, 1992). Thus about 1.2 to 1.6 million people can be fed from the existing annual cropland. This is obviously not nearly enough food to feed the present population of about 3.5 million people, the difference coming from pastures, fisheries, and imported foods. How much additional food might be grown? If agricultural technology were somehow able to double yields on this land, then the number who could be fed would be 2 to 3 million people. If all of the good agricultural land (Class I, which is 1 million ha, or 20% of the national area) were put into intensively managed grains and pulses at present yields, then 1.5–2 million tons of grains would be produced and 4.5 to 8 million people could be fed. Costa Rica will probably have about 6 million people by about 2020 or 2025 if current population growth rates continue. Thus it would seem that Costa Rica should have no problem feeding its population from a strict biophysical analysis of the situation, at least for 20 years.

The main problem, however, is that much of this land is needed to grow coffee and other export crops to pay for the fertilizers and other general inputs to agriculture (not to mention everything else imported). We undertook a simple simulation to determine the maximum number of people that might be fed from an optimal allocation of land to grains vs coffee (Figure 12-8). The result of this analysis is that some 6 million people might be fed by putting all of Class I land (1.0 million ha) into grains and pulses, and all of Class II land (0.4 million ha) into coffee, assuming 1995 ratios of the price of coffee to fertilizer. This analysis assumes no other uses of that land as well as no other uses of the foreign exchange earned. In reality much of Class I land is already used for pastures, urbanization, and other uses, so that much of this proposed production would have to take place in Class III land, where yields would be much lower and erosion much greater.

It is hard to escape the fact that Costa Rica cannot possibly feed itself if the population continues to grow (indeed it does not now) without either some as yet unknown procedure to increase yields per hectare or eliminating all cattle on the best land or, if necessary, on the best three classes of land. If

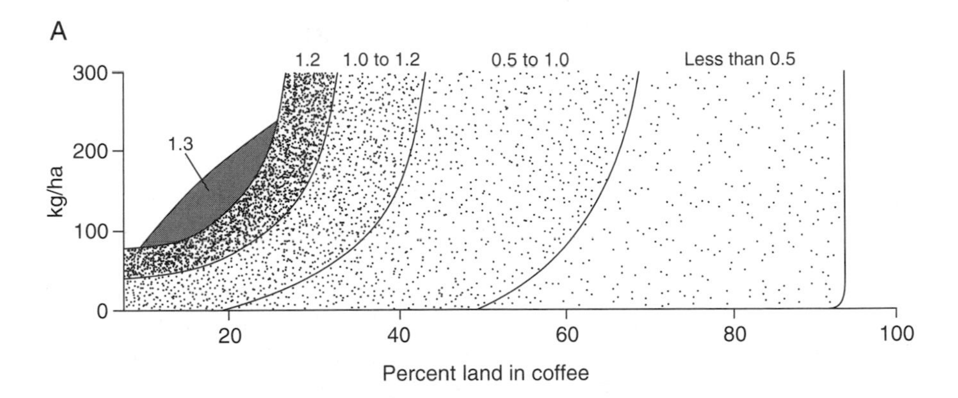


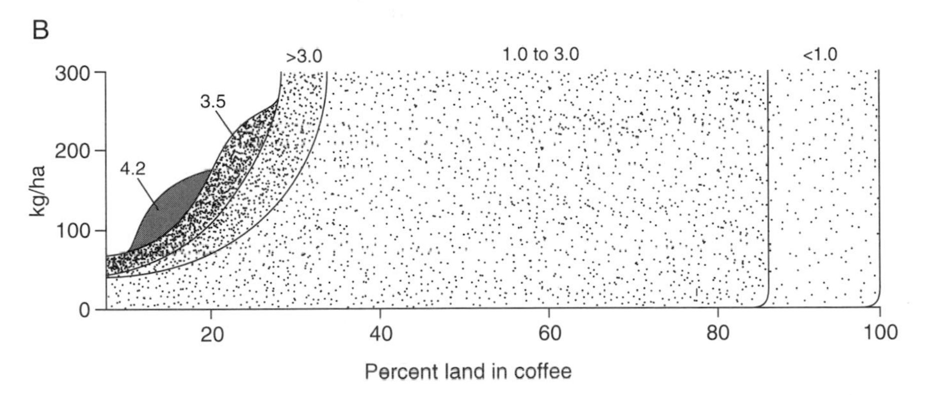


**FIGURE 12-8** Results of simulation model (LANDAPROP.FOR) for determining optimal pattern of fertilizer inputs and allocation of land between an export crop (coffee) and a subsistence crop (maize) with the objective of feeding the largest number of people by selling coffee to derive fertilizer for both coffee and maize (results are given in millions of people fed): (a) for 0.3 million ha, the present area of annual crops plus coffee, and (b) for 1 million ha, a probable maximum of land available for annual crops and coffee. In both cases the maximum number of people fed is with 15 to 20% of the land in coffee (and the rest in, e.g., maize) and a fertilizer usage of about 120 kg per hectare. The use of higher levels of fertilizer increases yields on both coffee and maize, but requires more land in coffee to pay for that fertilizer, reducing food yield. (Prepared with the assistance of Carrie Griffith and Dawn Montanye.)

that were done then it would indeed be possible to feed the present population of Costa Rica, but probably not a much larger population. In addition it would be essentially impossible to count on export crops to pay for all of the other inputs to Costa Rican society (Chapter 23).

## III. THE GEOGRAPHY OF AGRICULTURE

One possibility for the apparent cessation of increases in agricultural yields is that the good areas available for expanding production are all being used. Certainly this is obvious for some major crops. Every Costa Rican farmer knows that the perfect conditions for growing coffee are found in the Central Square of San Jose. Of course significant commercial coffee production cannot be found within many tens of kilometers of that square because of urbanization. Does that mean that all coffee is being grown in suboptimal environments? What about bananas? They do not compete with urban settlements for land, but as once-good land is destroyed by pesticides (Thrupp, 1991) the plantations must be moved to other land, which might not be quite as good. This section applies the gradients concept, introduced in Chapter 7, to these questions, which also are explored further in Chapters 20 and 21.

### A. Determining the Optimal Location for Growing Various Crops

There are other reasons that we might be interested in determining where various crops had the greatest yield, or alternately to ask what the yield potential might be at each location within a country.

First of all, any farmer would want to grow the crop that is best suited for his or her physical location. Custom or taste might dictate where a particular crop is grown although it might be that, in fact, the situation is far from optimal for that crop or that some other crop might grow better under the particular conditions of that farmer's land. But it might be that the farmer simply was not aware that the land he or she farmed was not a very good place to grow the crop, and this information was hidden by using large quantities of expensive fertilizer or through some other energy-intensive management.

Or we might want to take a larger view. If a nation were trying to feed itself then it would want to grow the highest yielding cultivars where conditions are optimum for getting maximum yields. Alternatively it may wish to take a strategy of maximizing economic return, which would also take into account the effectiveness of fertilizer and other inputs, and hence the expense of growing a crop, at each location in the country. Or a nation might attempt to grow its crops on the very best locations for each cultivar in order to reduce the need to clear new land from forests that have higher value in other uses.

All of these issues are very important in Costa Rica because so much of the best land for essentially any crop is already devoted to pastures. If grains were grown on land that could yield 2 tons per hectare, obviously far less land would be needed than if they were grown on lands that had a yield of 1 ton per hectare. But that potentially high-yielding land may already be used for pasture. Even though the yield from that pasture might be relatively good for beef (say, 0.2 kg dry weight per ha per year), the beef yield might be nearly identical on the land that gave poor grain yields. Thus total national food production could be optimized by using the land optimally. Of course it may be difficult or undesirable to institute such a plan because many people feel that the government should not interfere with a farmer's own best judgment as to how to use his or her own land.

In theory it might be possible to use data from agricultural experimental stations. But in general this information is difficult to use because such stations are few and tend to be located on the best land; thus any given station cannot represent the large set of possibilities of soil and climate type that a region might have. In addition there is a general sense that experimental yields are nearly always considerably higher than what farmers actually obtain. Thus perhaps the most useful information can be gained directly from the farmers themselves, that is, from data gathered on actual crop production. If such data are gathered in a systematic fashion over a large region (such as the country of Costa Rica) then very useful information can be gained.

We were fortunate in that Costa Rica had a number of very detailed national inventories of agricultural production. These summaries were generated by political reporting units: province (similar to a state in the United States), canton (similar to a county), and district (similar to a township). A summary of the information available is given in Table 12-3.

## 1. Determining Mean Environmental Conditions for Political Units

Since most of the information that is available for crop production in Costa Rica is summarized by political units (province, canton, and district—see Figure 12-9), it was necessary for our analysis to determine the mean environmental conditions for each of these units. Since we generated the environmental conditions at a 1-km scale (e.g. Chapter 9), we found it necessary to determine the mean of each condition considered over that political unit. For example, when we wanted to determine the mean annual temperature for the district of Turrialba we took the average of all temperature values for each of the 1-km grid cells that were found in the province of Turrialba (Figure 8-2). The computer program MEANDIS.FOR searches the DISTRICT.SQR file, which is a matrix map of Costa Rica where each square kilometer is given a value for its political district, say number 305, the district number for Turrialba. Every time the program encounters the number 305 it goes to the file TEMPMEAN.-SQR and reads the geographically corresponding temperature and sums it with other temperatures for other cells in that district. It also sums the number of

TABLE 12-3 National-Level Studies Available for Agricultural Production in Costa Rica

| Year | Agency | Crop | Level[a] |
|---|---|---|---|
| 1955 | MAG[b] | Bananas | C |
| | | Beans | C |
| | | Coffee | C |
| | | Maize | C |
| | | Rice | C |
| | | Sugarcane | C |
| | | Pineapples | C |
| | | Pasture | C |
| 1963 | MAG | Bananas | C |
| | | Beans | C |
| | | Coffee | C |
| | | Maize | C |
| | | Rice | C |
| | | Sugarcane | C |
| | | Pasture | C |
| | | Pineapples | C |
| 1973 | MAG | Bananas | D |
| | | Beans | C |
| | | Coffee | D |
| | | Maize | D |
| | | Rice | D |
| | | Sugarcane | D |
| | | Pasture | C |
| | | Pineapple | C |
| | | Pineapples | D |
| 1984 | MAG | Bananas | C |
| | | Beans | C |
| | | Coffee | C |
| | | Maize | C |
| | | Rice | C |
| | | Sugarcane | C |
| | | Pineapples | C |
| 1994[c] | MAG | Bananas | ? |
| | | Beans | ? |
| | | Coffee | ? |
| | | Maize | ? |
| | | Rice | ? |
| | | Sugarcane | ? |
| | | Pineapples | ? |

[a] D, district; C, canton.
[b] Ministerio de Agricultura y Ganaderia.
[c] Not available as of 1999. Status of report is unknown.

FIGURE 12-9 Map of provinces for Costa Rica. Several provinces mentioned in the text are identified.

times that a cell for that district is found. When the search of the entire country is complete the program divides the sum of the temperatures by the number of occurrences of cells in that political unit to get the mean (and standard error) of the temperature for that district. Canton-level temperatures were determined by the same procedure. A similar analysis was undertaken to generate mean values of temperature month by month, as well as rainfall, relative humidity, evapotranspiration, solar insolation, soil moisture, and soil fertility for each political unit. A summary of some meteorological properties by canton is given in Table 6-1, and in more detail by district in Table 6-3b.

## 2. Determining Crop Production, and the Climatic Determinants of that Production

The Costa Rica Ministry of Agriculture and Pastures data sets (MAG, 1977, 1984) for each crop type were entered into canton- and, where possible, district-level data bases. The original data were given as total hectares planted and total yield for each political unit. An index of the physical suitability of the land to generate various level yields for crops of different types was obtained by dividing the yield by area in production to get yield per hectare, and then summarizing these results by (a) political units and (b) position in gradient space based on the climatic characteristics of each political unit (see Chapter 7). A summary of the computer program that does this (and generates simple output maps) is given in Table 12-4, and the program (GRADIENT.FOR) is found on the enclosed CD under "Programs."

**TABLE 12-4 Summary of Operation and Basic Structure of Computer Program GRADIENT**

This program, developed by Charles Hall in 1995 and based on Myrna Hall's GEOTOECO, takes agricultural data generated at the Canton and District (political) level and compares the name of the input file with the names on a corresponding master file that has the mean temperature and other characteristics of that district. The master file uses Costa Rica's "official" districts. Since there is often no clear pattern of naming political districts, the input data, some of them many decades old, must be aligned with the master file (Names3.dat). The program works by progressively reading the names from the agricultural census and then spooling down the master file until a matching name is found. (Sometimes, especially with older data sets, no such file is found.) Once the corresponding political unit is found the program reads the environmental (gradient) characteristics from that file, for example, mean annual rainfall or minimum temperature. Then these values are used to locate the position of that biotic characteristic, in this case yield per hectare per year, along gradients of environmental variables. For example, the production of coffee for Puriscal District was 9.3 wet tons per hectare in 1977. The mean rainfall for Puriscal District was 2.66 m per year, and the mean temperature 23.32°C. The production value would be plotted at the coordinates along the graph represented by those environmental variables. Thus the program generates plots in gradient space as well on crude geographical maps.

Basic structure of program
(Full code found on CD.)

I. Define variables, initialize variables, and define map scalars.
II. Assign files to be read in and analyzed.
III. Read in characteristics of run: gradient parameters, size of gradient axes, etc. Generate a "blank character slate" of gradient space which will be filled in by the program.
IV. Main loop to read in agricultural input files with yields per hectare for each political unit.
V. Secondary (internal) loop to seek a matching political unit and read in gradient characteristics.
VI. Assign position in gradient space as a function of gradient values.
VII. Plot yield values as a function of each variable, as a function of two variables, and in geographical (map) space.

A difficulty with our analyses is that we have no spatial information about the use of fertilizers and other inputs. Thus if farmers in one province use much more fertilizer for the same crops than farmers elsewhere our results could be biased. On the other hand, fertilizers in Costa Rica tend to be applied according to the recommendations of the extension agencies. These tend to be more or less uniform nationally for each crop except as they are changed to overcome local nutrient deficiencies. And, if strong and consistent patterns are found as a function of gradients then we can assume that these basic physical relations are strong and that much of the remaining variance is due to variation in inputs. In future agricultural surveys it would be good to include specific information about fertilizer use per hectare.

## B. Results

Our analysis found a strong pattern in the relation of agricultural yield to spatial and gradient patterns. For example, yield per hectare values for coffee were found scattered all over the country (Figure 12-10). A more careful examination, however, finds that maximum production is centered at intermediate elevations associated with the major volcanoes (Figures 12-10a,b). The relation between yields and gradient space was even better (Figure 12-10b). Coffee, for example, did best at maximum summer temperatures from about 20 to 27°C and at rainfalls from about 1.2 to 3.2 m, although low yields are possible over a much wider climate.

We found similar results for every major crop grown in Costa Rica except for sugarcane, which did not seem to have any climatic optimum (Figures 12-11 to 12-14). Despite all the technical interventions climate and soils remain principal determinants of agricultural yield, at least where there is moderate fertilization. The "fly points" may represent locations of especially intense fertilizer use.

We made a prediction of where the best locations would be for coffee. When this is compared with where crops are actually grown it is obvious that farmers are already using more or less optimal land for their particular crops. The next chapter shows our estimate for optimal areas for major crops based on climate. Obviously many other factors might be important in determining exactly what crop should be grown at each location.

## C. Discussion

### 1. Limitations on Yields per Hectare

The data presented in Figures 12-4a to 12-4f give an overall sense that the ability of Costa Rican farmers to increase the yields per hectare for

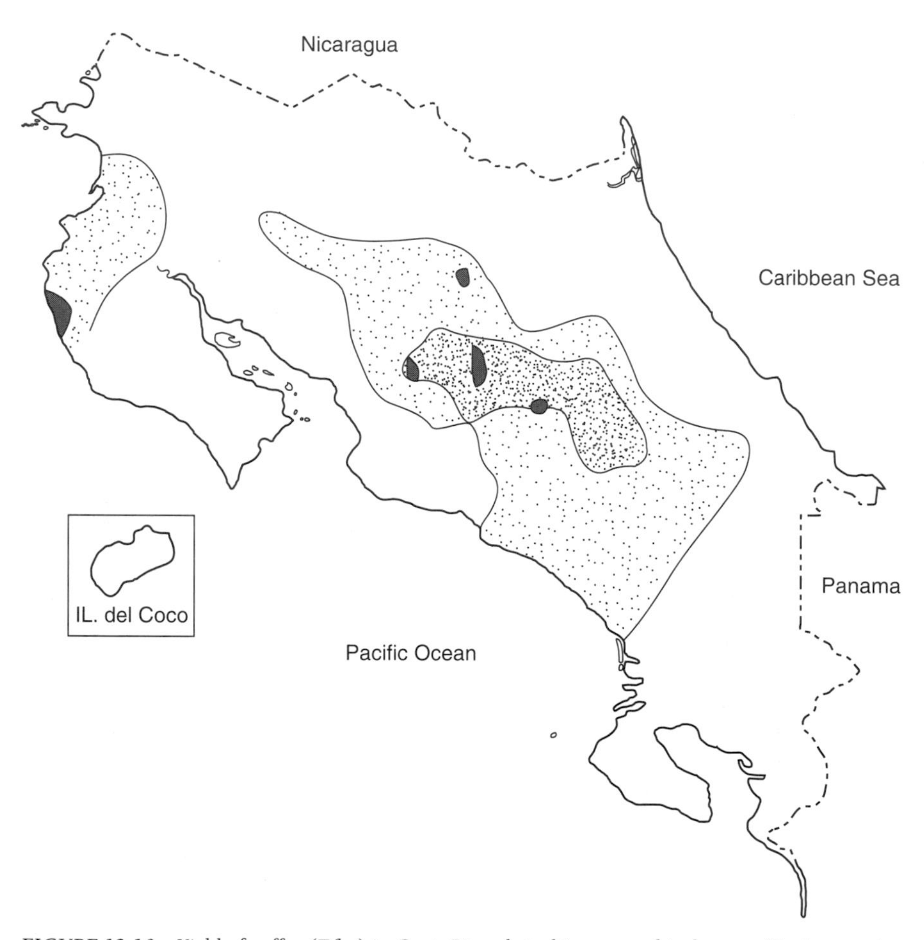


FIGURE 12-10 Yield of coffee (T/ha) in Costa Rica plotted in geographical space. (Facing page) Same data plotted as a function of geographical space, in this case as a function of mean annual temperature and precipitation. The relatively crude appearance of the output is from the simple (but free) SIMGIS GIS program included on the CD. (Black) Yields of at least 1.4 T/ha/yr are possible; (stippled) yields of at least 1.0 T/ha/yr are possible; light area = yields are less than 1.0 T/ha/yr. Dotted line delineates gradient space in Costa Rica.

these crops appears limited, as yields per hectare appear to be approaching an asymptote. In those few cases where yields per hectare are still increasing it is only because of greatly increasing inputs, so that efficiency is declining. There are several possible biophysical reasons for this, all driven by the expanding population base of Costa Rica: (1) as the best land is increasingly

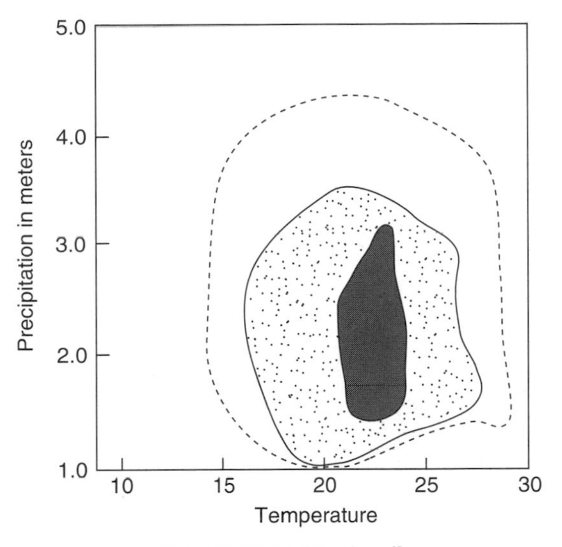


FIGURE 12-10 *(Continued)*

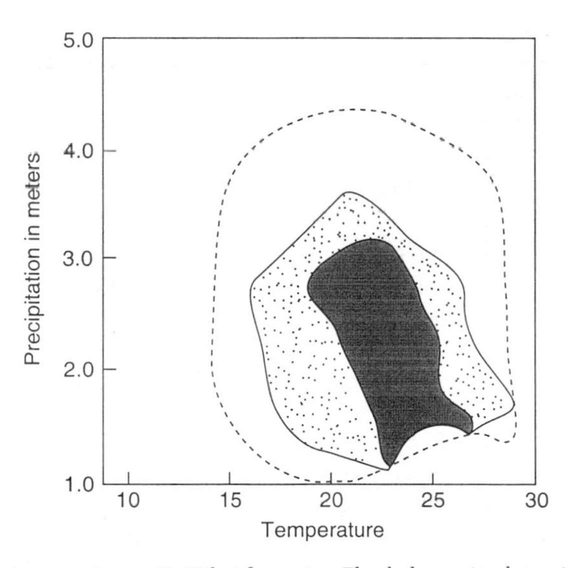


FIGURE 12-11 Same as Figure 12-10 but for maize. The dark area is where yields >1 T/ha can occur; the stippled area is where yields of >0.5 T/ha can occur.

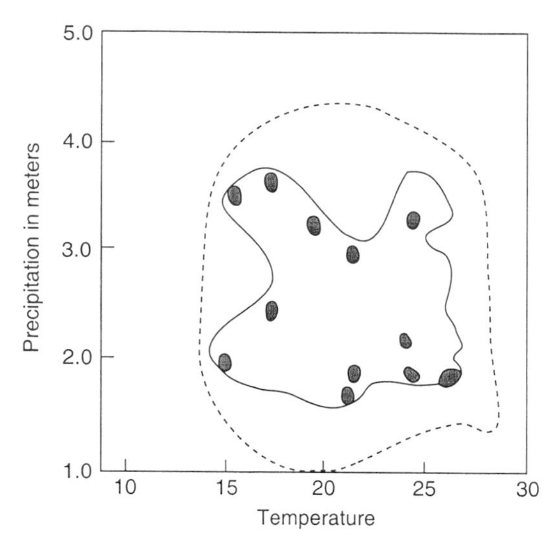


**FIGURE 12-12** Same as Figure 12-10 but for sugarcane. Cane grows reasonably well within the stippled area. High yields (>80 T(wet)/ha) occur at the dark spots.

used for urbanization, cropping is increasingly pushed into more and more marginal regions since new high-quality land is not readily available (Chapter 5; this issue is also seen in the data of Figure 12-5), and (2) the cumulative effects of erosion may be becoming increasingly important in terms of declining yields (Chapter 15). Fertilizer application may compensate only partially in each of these situations. We are not aware of any analyses that address this question directly.

An important question regards the maximum yield that one might reasonably expect in tropical agriculture. While yields of 5 to occasionally 10 tons per hectare per year are common in temperate regions, it is rare to find yields anything like that in the tropics. Whether the reason is a low degree of development or because of factors intrinsic to the tropics is an extremely important matter. The former is, of course, quite amenable to resolution, while the latter is more difficult.

The standard reference for tropical agriculture is Hans Ruthenberg's *Farming Systems in the Tropics*. This book, published in 1976, gives a taxonomy of farming systems found throughout the tropics. A particularly interesting chapter by MacArthur examines the intrinsic limitations of the tropics for agriculture. The highest yields observed by that author are not in the tropics but in the subtropics, and throughout this book the yields reported for many very different types of agricultural systems in the tropics are generally 1–2 tons/ha/yr, with yields less than 1 ton being more common than not. An exception is irrigated rice.

There is no shortage of reasons that Ruthenberg and MacArthur give as to why the tropics should have intrinsically lower yields. The factors they give include: (1) constraints due to intrinsically low soil fertility; (2) the need to avoid risks with multiple cropping strategies (vs one crop that might be ideal for the location); (3) low labor productivity ("since the work is both tedious and strenuous, so not surprisingly farmers show a marked preference for

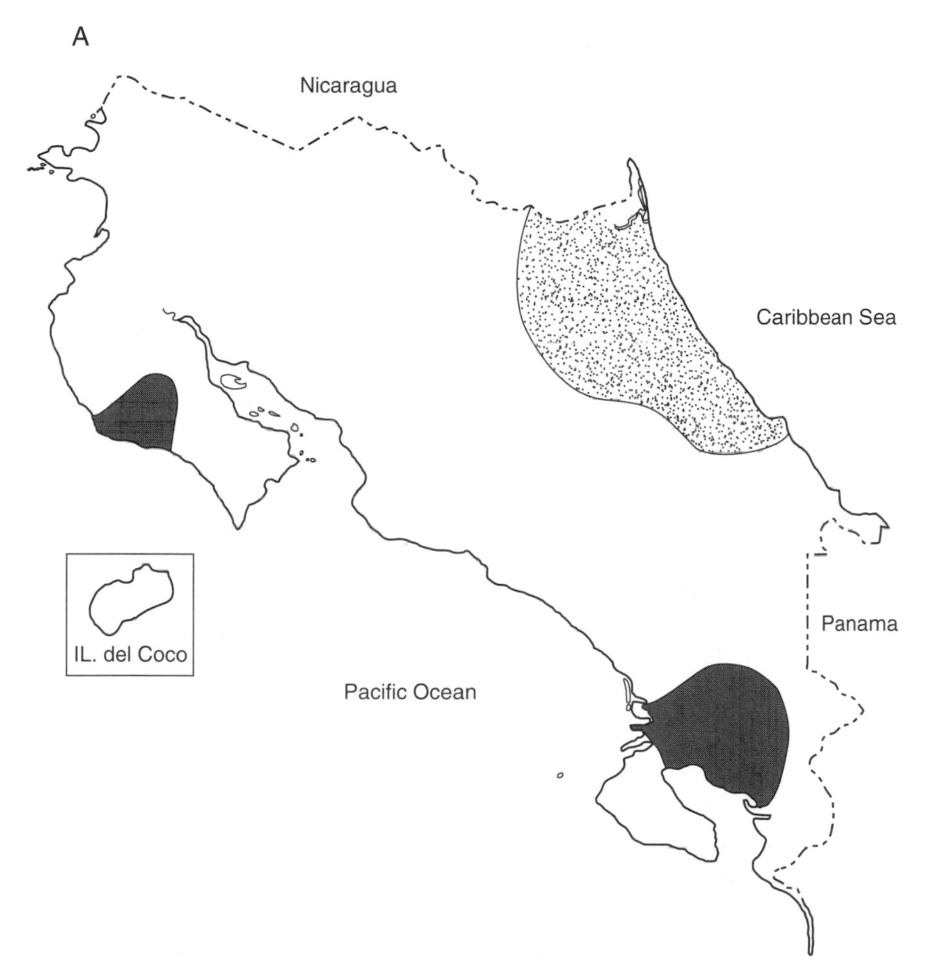


FIGURE 12-13 Same as Figure 12-10 but for pineapple.

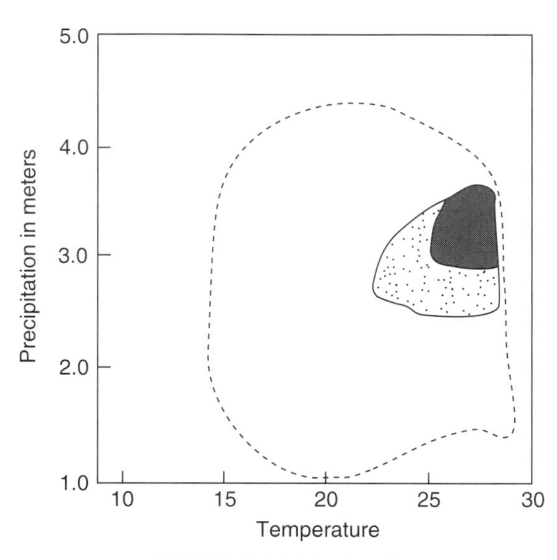


FIGURE 12-13 *(Continued)*

leisure"); and (4) problems due to seasonality (the tropics are notoriously seasonal, especially with respect to moisture, making year-round production difficult). We add two additional factors: first, the human machine simply cannot do as much work at higher temperatures (Sundberg, 1992). A second and potentially very important component has been suggested to us by Dr. Margaret Smith, a maize geneticist at Cornell University. Tropical nights are never less than about 11 hours long. During the night a plant requires considerable energy for its own respiration, that is, for maintenance metabolism. Thus during the night a plant burns up some of the day's profit. The longer the night, the more photosynthate is used and the lower the energy profit (i.e., the seeds). Dr. Smith thinks that this may be a principal reason that one almost never finds maize yields higher than 1 or 2 tons per hectare in the tropics. Smith's analysis is in agreement with the empirical analysis of Chang, 1981, who emphasized as well the importance of long nights to latitudinal patterns (Fig. 12-15). We think Chang's work deserves much more attention.

Thus we have a number of plausible biophysical reasons as to why Costa Rican agriculture seems to be no longer responding very well to increased inputs and management. There may be others, too, including economic reasons such as (1) farmers are not interested in increasing yields if inputs are too expensive relative to the return, or (2) there may be a preference for low-

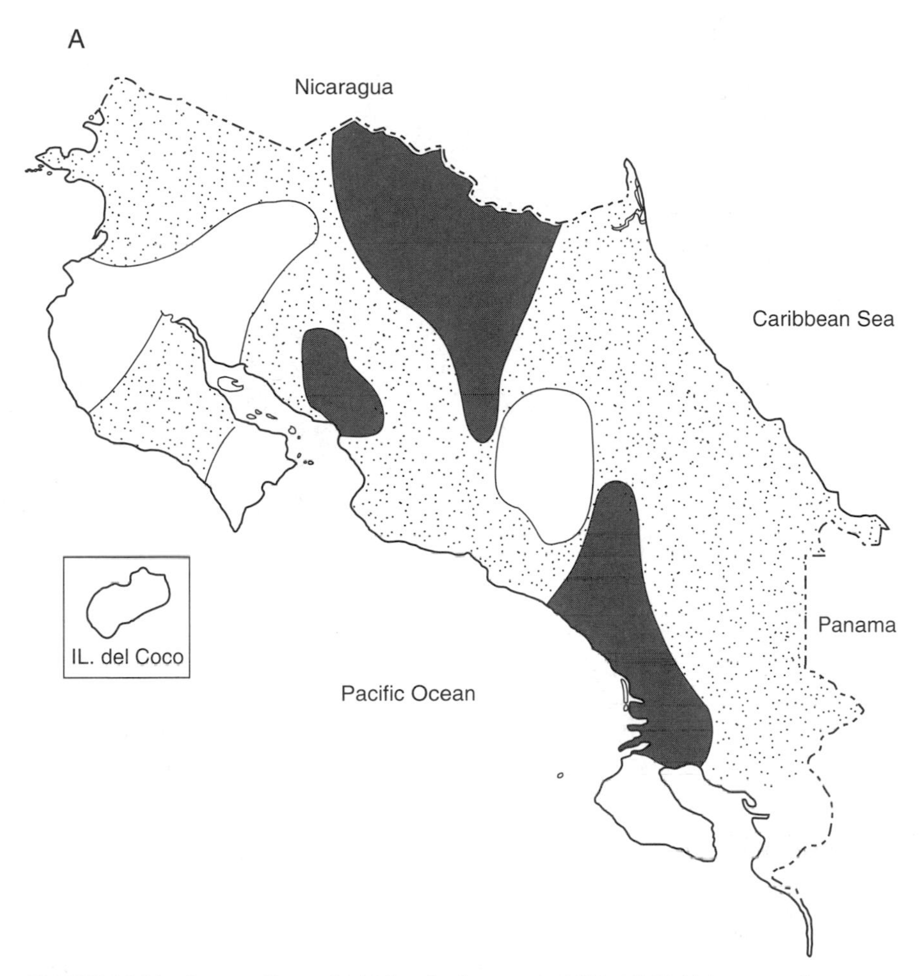


FIGURE 12-14 Same as Figure 12-10 but for bananas in 1973. (A) Yield in geographic space. (B) Yield in gradient space. Dark areas are where highest yields are possible; at least 10T/Ha/yr. Stippled areas yield at least 5 T/Ha/yr.

yielding, high-value crops; social reasons (if the farmers are getting by, perhaps they prefer leisure to higher yields); or still others. We tested the importance of biophysical constraints in the tropics vs other factors by plotting the yields of grains vs fertilizer inputs for both tropical and temperate countries. Our results (Figure 12-16; see also Ko *et al.*, 1998; and Hall, unpublished) show a clear asymptote in the response of grain yield to increasing fertilizer applications for various tropical countries where high inputs have been undertaken, most notably Nigeria and Venezuela (both rich oil-producing nations that have

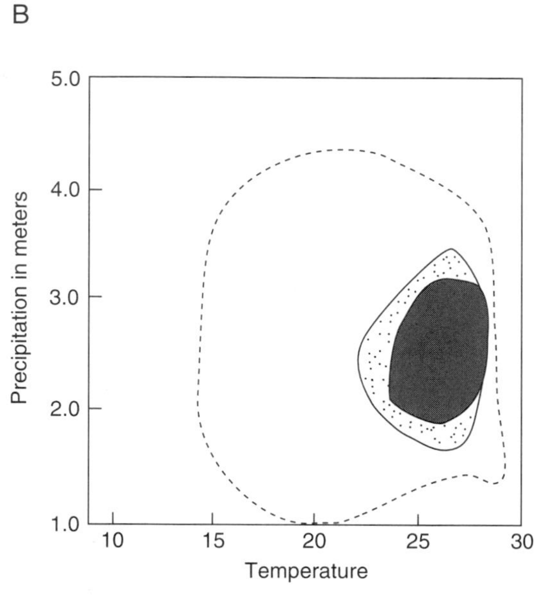


FIGURE 12-14 (*Continued*)

been able to afford lots of fertilizer). Thus we tend to favor the instrinsic climate limitations hypothesis.

## 2. Uncertainties in Estimates

An important additional factor is that our analysis so far is biased toward supporting too large a population because it is based on the present situation and does not include corrections for the observed decrease in yields per hectare due to progressive erosion and soil quality decline. This question is considered further in Chapter 15.

One might even want to question why society has undertaken all the agricultural research that has been going on, since yields per hectare no longer increase. The answer to that question may be that yields probably would have *decreased* without that research, and consequently that we are fortunate to have maintained agricultural productivity in recent years. Furthermore, new research on "softer" technologies (agroforestry, intercropping, sustainable agriculture, etc.) may generate a very different overall picture over the next few decades (e.g., Altieri and Masera, 1993). These questions are unresolved at this time, but it is clear to many that tropical agriculture production will be severely strained to keep up with the growing human population.

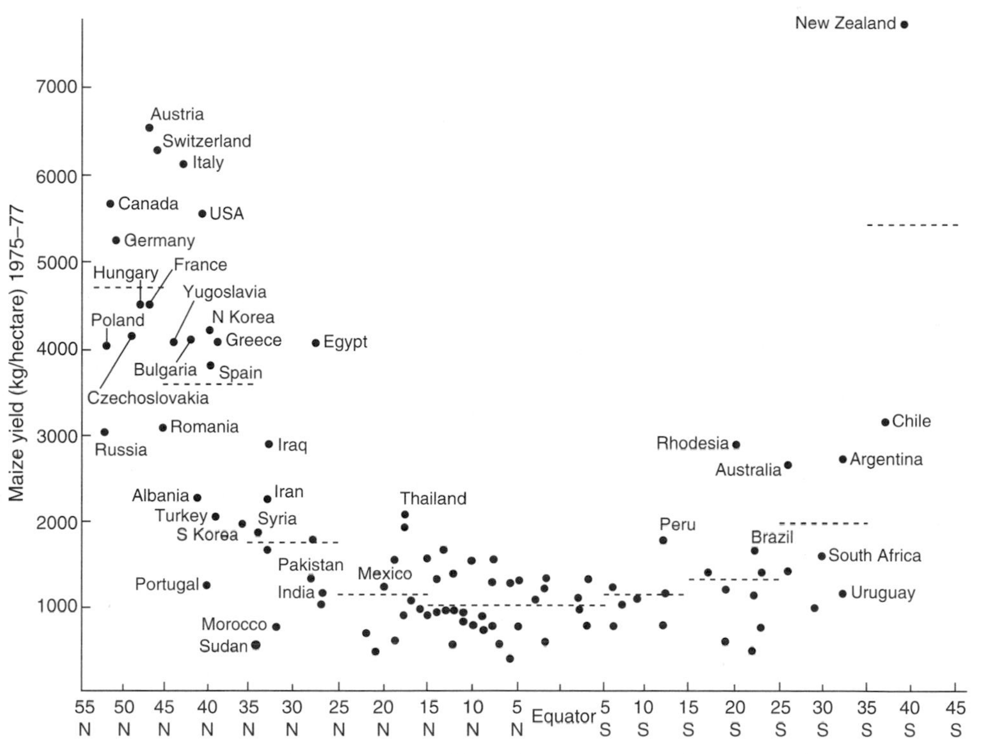


FIGURE 12-15 Corn yield as a function of latitude. ---, average yield for each 10° of latitude. (From Chang, 1981.)

## 3. Can We Increase Yields without Increasing the Use of Fossil-Fuel-Derived Inputs?

Is there any way to increase agricultural yield per hectare without an increase in the direct or indirect use of industrial energy? The answer for Costa Rica appears to be, at least so far, no. When fertilizers were decreased following price increases in the 1970s, yields per hectare flattened or decreased. What about new technologies? Before we attempt to answer that question it is useful to think of the growing crop plant in terms of its own energy relations. It must capture solar energy and invest this energy into sequestering nutrients, assimilating carbon, defending itself against insects, and taking up water. In a wild plant, these processes require substantial amounts of energy. With modern agricultural technology, we have increased the net yield of plants (i.e., the portion that we eat) but not generally the gross production (Gifford *et al.*, 1984). Fertilizers reduce the energy investments that plants would otherwise

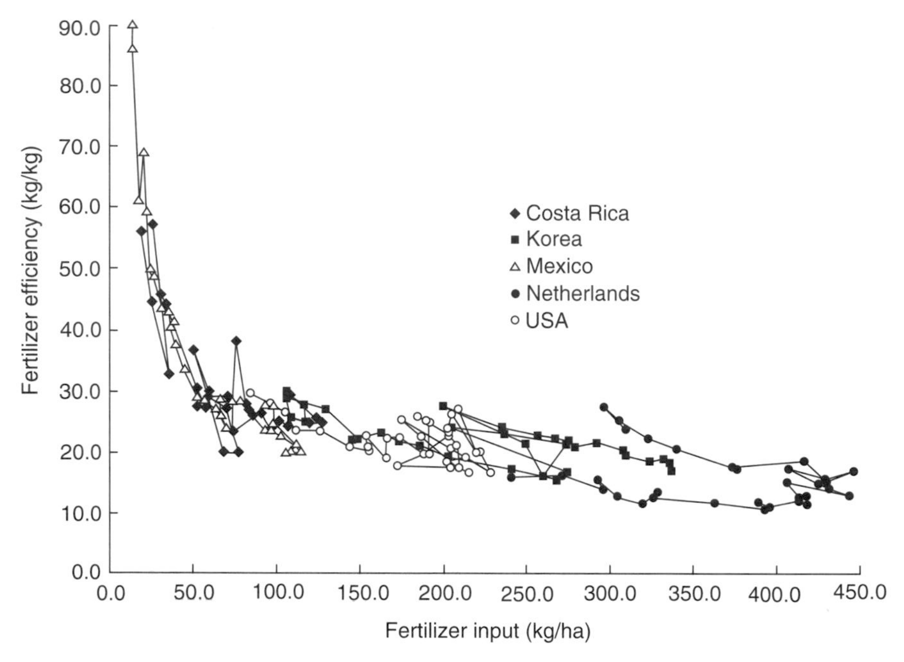


FIGURE 12-16 National fertilizer efficiency of yield per hectare for grains (estimated according to methods in Ko *et al.*, 1998) 1960–1996. Note that at least these tropical countries have an asymptotic yield value considerably lower than that for temperate countries.

use to grow longer roots to seek nutrients, irrigation allows breeders to breed plants with smaller root systems, and so on. Herbicides reduce the competition with other plants (weeds) for water, nutrients, and sunlight, and pesticides reduce the need for plants to invest energy into manufacturing their own alkaloids and other natural chemical defenses against insects. This is the essence of the "green revolution": the use of greatly increasing quantities of fertilizers, pesticides, irrigation, and herbicides along with the development of strains of crops that depend increasingly on these energy subsidies but that have more energy left over to go into the parts that we eat. Millennia of plant breeders have bred plants mostly for high yield, which has meant a diversion of the plant's energy budget away from investment in the means to get nutrients and water and to protect from pests. This process continues today. It works well, as long as the external supplies of the required inputs are available, and as long as plant breeders can keep ahead of evolving pests.

Unfortunately there is little evidence that we are producing higher yields without more inputs, at least at the national level. This is true for the data of China (Hall *et al.*, 1986), the United States (Hall *et al.*, 1986; Odum, 1989),

Japan, Israel (Odum, 1968), Mexico, and Korea (Ko *et al.*, 1998). Both Cleveland (1995) and Ko *et al.* (1998) found that per hectare yields actually had increased from about 1980 to 1993 in the United States and the Netherlands during a period of reduced fertilizer usage. However, this also was during a period of significant reduction in the area of land under cultivation, so that the reduced yield may have been due to the use of better average land. It is not clear whether new genetic engineering technologies or new (or old) types of "organic" farming will change this pattern if they are implemented on a large scale. Certainly it is a most important goal. But the absence of good yield data for large-scale applications makes these questions difficult to answer at this time. Of course human labor, another form of energy input, can be substituted for industrial energy, and this can increase the efficiency of yield relative to fossil energies. At the moment this is occurring on a large scale in Costa Rica already, with many farmworkers from other countries, especially Nicaragua. But more fertilizers are being used too, with relatively little effect on yield (Figure 12-7).

### 4. What Kind of Yields Could Costa Rica Attain for Each of Its Major Crops?

The total production, as well as the yields per hectare, for various crops appears to be saturating (Figures 12-4a–f). In a way this apparent saturation is unexpected because there is a large amount of potential agricultural land, some of it presumably good, left in Costa Rica, because agricultural technology is evolving rapidly, and because yields are considerably higher in at least some other parts of the world.

Given the strong dependency of agricultural production on climate, an interesting question to consider is the degree to which the land in Costa Rica is used optimally—that is, to what degree are crops raised in their optimal location, and how much climatically optimum land is available for further use? In order to try to answer these questions we have attempted to examine the total national potential of each crop if all appropriate land (based on a crop's needs for rainfall and temperature) were used for that crop alone (Table 12-5). The results indicate that roughly all of the land that is optimal for coffee or maize is in that use, but only 42% for sugar and 36% for rice. Thus there seems to be considerable land with appropriate climatic conditions available for the expansion of some crops. This analysis is incomplete, however, because although there is, for example, considerable room for expansion of bananas, much of that appropriate land not being used had been used previously to grow bananas, but was abandoned due to depleted soils or contamination with banana pathogens or chemicals. In addition, some of these lands are currently in other crops, reserves, or are forested, and may not be available for legal or

TABLE 12-5 Examples of the Limitations of Crops Based on Climatic and Soil Factors as Estimated by Ministerio of Agricultura and Ganaderia (in Thousands of Hectares)

| | Area | | |
|---|---|---|---|
| Crop by region | Ideal conditions | Possible conditions | Actually planted |
| Beans | | | |
| National | 10.0 | 57.6 | 50 |
| Central | 10.0 | 14.9 | |
| Chortega | 0.4 | 40.4 | |
| Central Pacific | 0.0 | 0.4 | |
| Brunca | 0.0 | 1.9 | |
| H. Atlantica | 0.0 | 0.0 | |
| H. Norte | 0.0 | 0.0 | |
| Coffee | | | |
| National | 54.3 | 184.6 | 110 |
| Central | 52.4 | 89.9 | |
| Chortega | 2.0 | 28.5 | |
| Central Pacific | 0.0 | 0.0 | |
| Brunca | 0.0 | 65.5 | |
| H. Atlantica | 0.0 | 0.3 | |
| H. Norte | 0.0 | 0.5 | |

*Note*. These areas, essentially climate zones, correspond essentially to the Central Valley (Central), Guanacaste (Brunca), Heuter Atlantica (Limon), northeastern and north central (H. Norte), and central Pacific regions. Source is Costa Rica (1993) Información Estadistica Vol. 7 1993 Cuadro 49 and 50.

other reasons. We have made some approximate corrections for these factors in Table 12-5.

But perhaps such analyses are spurious. Joseph Tosi, director emeritus of the Tropical Science Center, San Pedro, said upon hearing a seminar on the preliminary results of this analysis that although he thought the idea of agricultural production as a function of gradients was very interesting scientifically, he was not sure that it would tell us anything that an intelligent farmer did not know. Perhaps nearly all land is being used near its economic maximum, and we cannot expect too much additional yield. The exception seems to be the use of so much good land for pastures. It would seem that there is considerable room for additional crop production there. Even so it is clear that there is not enough good land to feed a doubled population *and* grow the export crops required to pay for the imported agricultural inputs that would be required at today's levels, no matter how the land was used.

# IV. AN ENERGY FLOW ANALYSIS OF AGRICULTURAL PRODUCTIVITY IN COSTA RICA

This section synthesizes much of the basic agricultural information for Costa Rica through the use of energy flow diagrams. The agriculture of a nation is complex, and we need tools here, as in all complex problems, to simplify the problems to a degree that allows our brain to understand the essence of the issue being considered. Probably the critical step in this process is the construction of systems flow diagrams, where the most important stocks and flows of the systems are formalized diagrammatically (Hall and Day, 1977; Odum, 1983). Normally energy stocks and flows are used, but the process may be done for other entities. We do this in order to help the investigator (and the reader) conceptualize the biophysical underpinnings of an economic system—in this case to help understand how the agricultural systems work from a biophysical as well as an economic basis, and in order to facilitate an understanding of how both a biophysical and an economic approach represent the same phenomena in different ways and with differing perspectives and degrees of completeness.

## A. Diagrammatic Overview of Entire Agricultural Economy

Figure 12-17 is an energy flow diagram representing general Costa Rican agricultural production units. In this figure the first critical input required for agriculture is solar energy and, of course, the land required to capture that energy. Solar energy is represented by the circle (source) on the left of the diagram. Crops are represented by bullet-shaped modules labeled "food" and "export" crop. Rural human consumers of the food crops, both on the farm and within the regional agricultural communities, are represented by hexagons. Since these consumers are also laborers, their own energy input to farming is represented by the solid feedback arrow going from the workers to the "arrow-head" symbols (formally called work gates) to the left of the crop. Modern agriculture also requires industrially derived inputs, and they are provided to farmers directly from regional cities (e.g., agricultural cooperatives). The suppliers in the cities get these inputs in turn from sales interactions with national and ultimately international suppliers. This flow (top right of Figure 12-17) represents a critical aspect of the economy of Costa Rica, for these inputs must be paid for by production within the system.

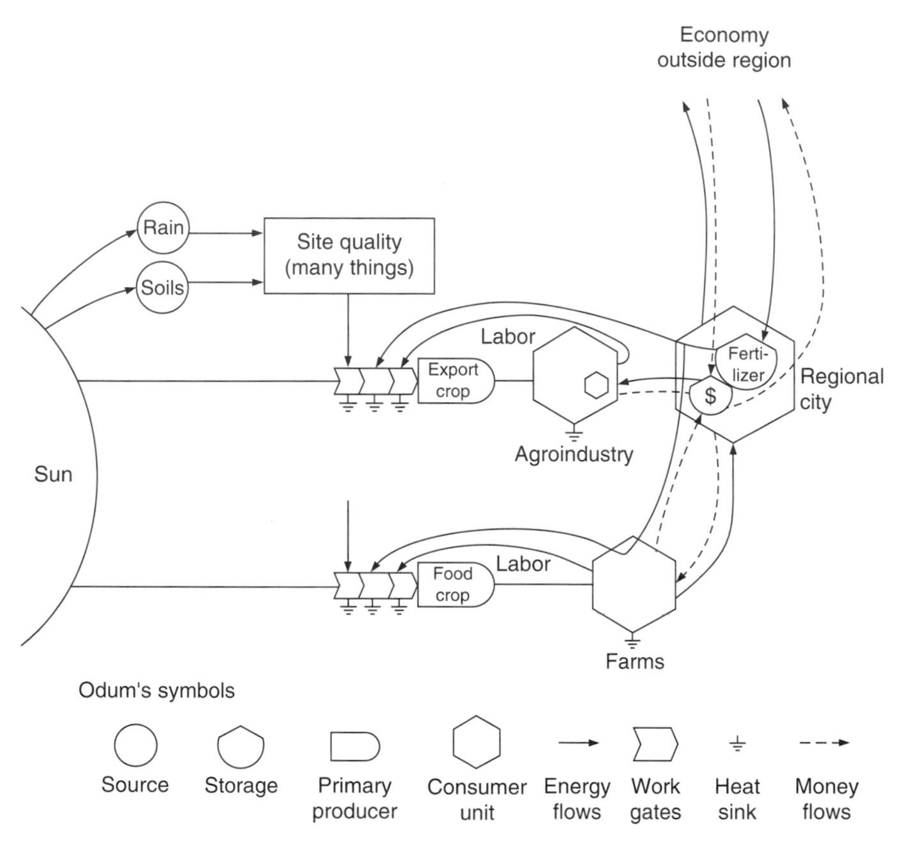


FIGURE 12-17 Energy flow diagram representing the general case of agricultural production units (one for local consumption and one for exports) and their relation to a consumer (farmhouse), local cities, and international trade. Circles represent sources of inputs from outside the system (solar energy, precipitation, fossil fuels), bullet-shaped modules represent plant community types (e.g., natural forest, maize), hexagons represent consumer units (e.g., people, cows, and cities), birdcage-type symbols represent storages (nutrients in soil, money in bank), arrows pointing to right or downward represent energy flows, arrows pointing to the left represent feedback or control flows where small amounts of high-quality energy are used to direct or control larger, lower quality flows, and dotted flows represent money flows, which are used to pay for energy or energy-derived goods and services (for a complete description see Odum, 1983).

Figure 12-18a, representing the national system in physical units, shows the area of each of the most important crops, and the most important inputs of fertilizer and outputs (crops) in terms of physical units (metric tons). The output of the crop, to the right of the bullet-shaped symbol, is shown being exported (the largest flow for coffee and bananas), used on the farm (a small component for all crops), and within the country (a larger component). This

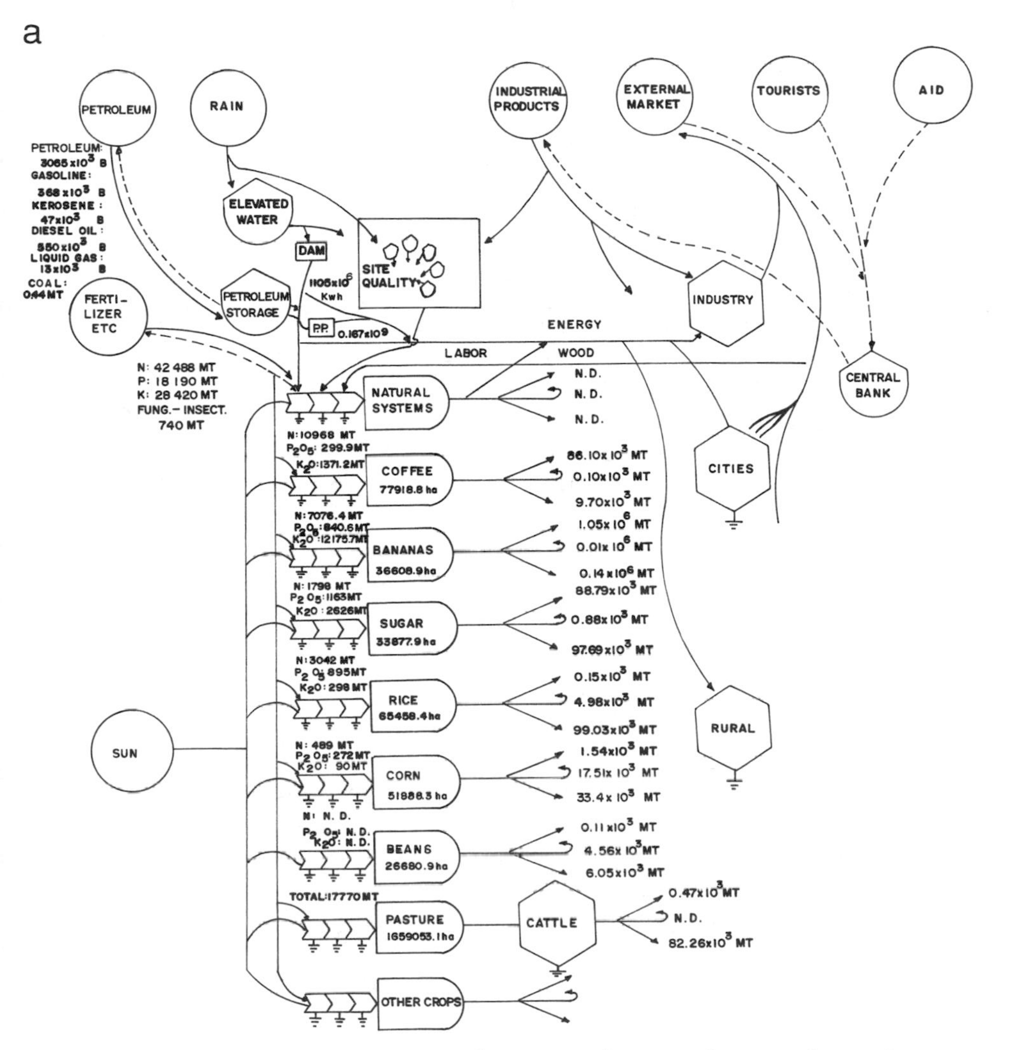


FIGURE 12-18 Flow diagrams representing the economy of Costa Rica for 1973 with an emphasis on agriculture. The three diagrams are very similar. Figure 12-18a represents the economy in terms of the flow of physical materials (i.e., tons of fertilizers or coffee). The units are B for barrels and MT for metric tons. Figure 12-18b represents the same flows in energy [all numbers are in TJ (1 TJ = $10^{12}$ J)] but energy embodied in materials is not shown. Figure 12-18c represents the same flows in money (in 1973 U.S. dollars). In general energy flows from left to right, and downward pointing arrows represent points at which energy is used (turned into heat). Bullet-shaped modules represent plant production units. Hexagons represent consumers (farmhouses, cities, and so on). Leftward pointing symbols represent "work gates," that is, places where one energy or resource flow influences or controls another. See text and Odum, 1983, for further description. The three arrows pointing to the right from each crop represent quantity exported, quantity used on farm, and quantity used inside Costa Rica from top down, respectively. (Artwork by Emilio Ortez.)

b

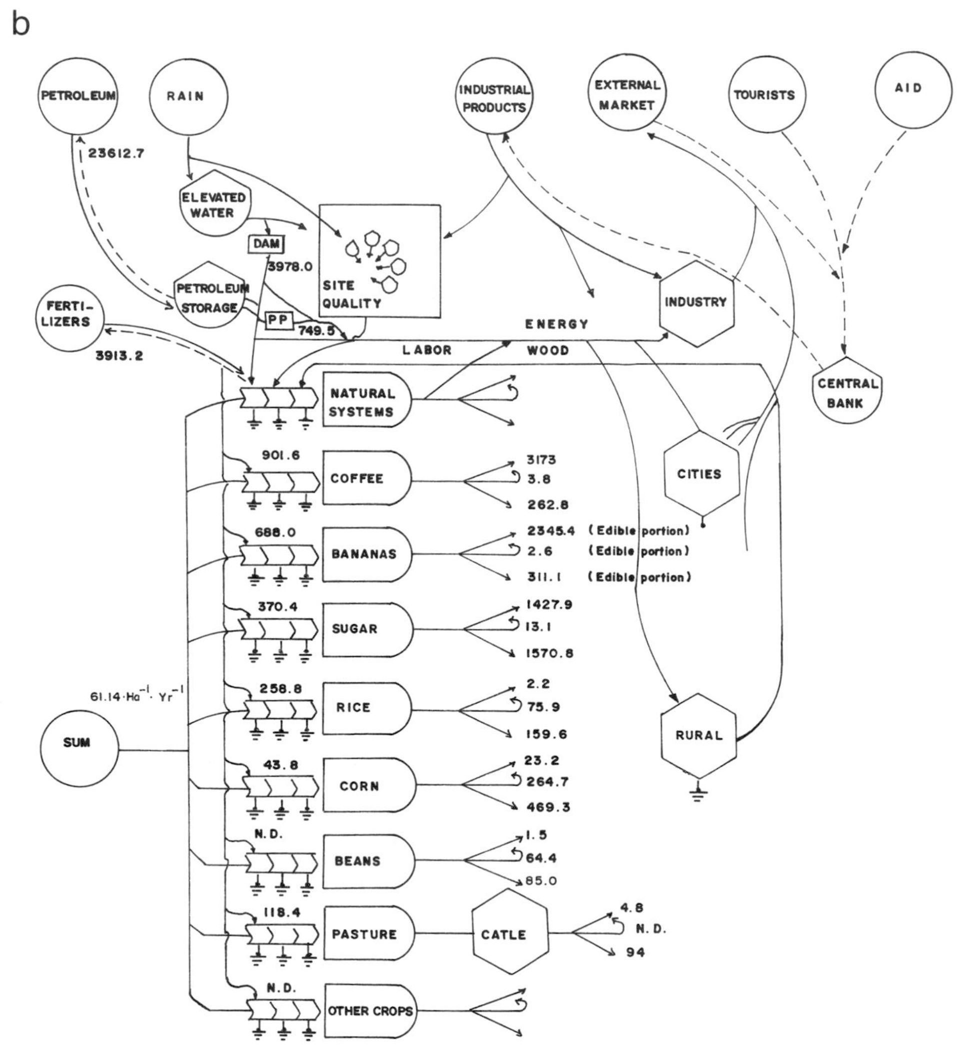


FIGURE 12-18 (*Continued*)

agricultural component of the economy is drawn within the broader context of the interaction of agriculture with cities, imports of fossil energy and internal production of hydropower, industries internally and abroad, and other major sources of foreign exchange.

Another component is the box near the top left labeled "site quality." This represents the importance of the variations in the local conditions, i.e., in

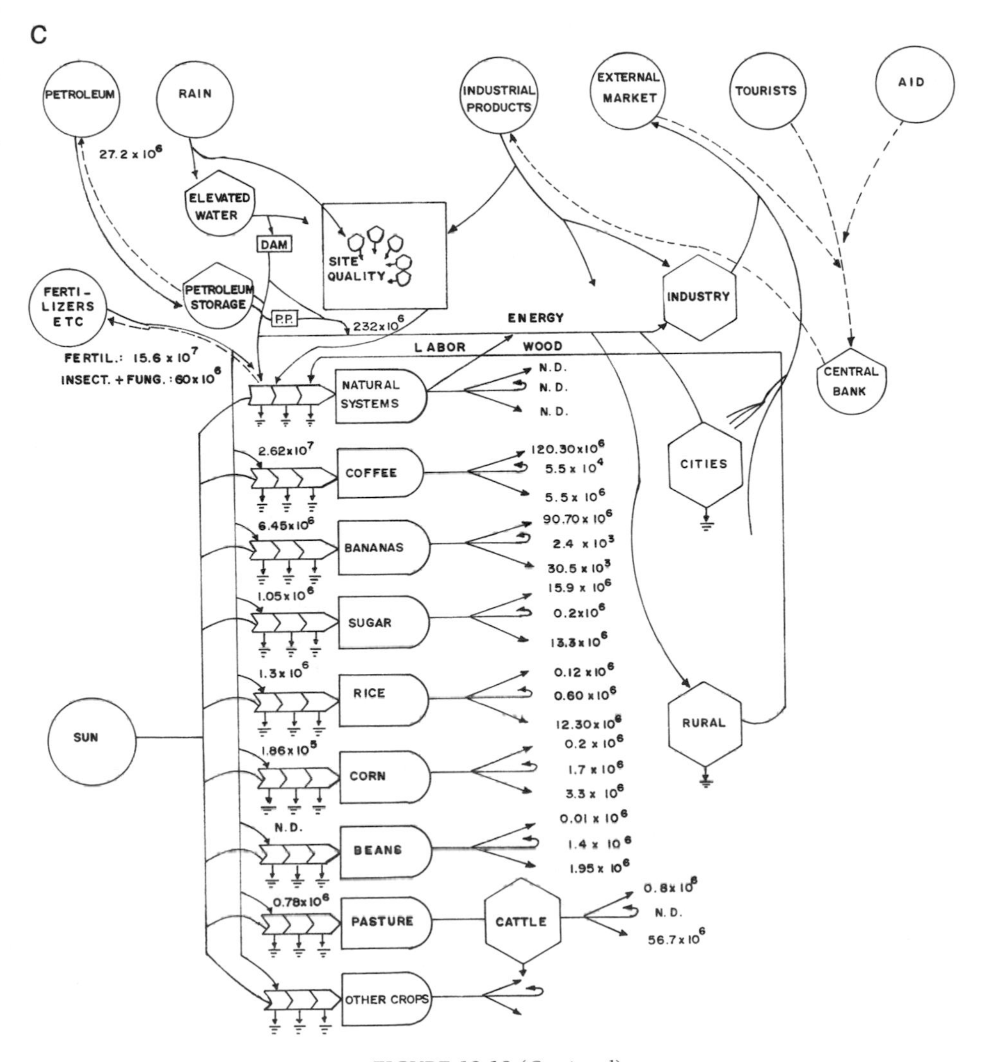


FIGURE 12-18 (*Continued*)

terms of meteorological, soil, and other factors as represented by location on gradient axes in Chapter 8 and in the previous section. These are presented by the leftmost "work gates" (interaction modules) in front of the crop symbols. A very important component of this analysis is determining which factors are responsible for regulating how well different crops grow in different regions, as discussed in the previous section.

Figure 12-18b shows the same relations, but this time represented as energy, where all crops, fossil energies, and so on have been converted into their approximate caloric equivalent. Finally the same relations are represented in Figure 12-18c in terms of monetary units. The flow of money goes opposite to energy (and associated materials). The important thing to recognize is that these three diagrams, based on three very different sets of things, are actually representing *exactly* the same economy, but from different perspectives. Thus the farmer sells his crop (a physical entity) to the city, receiving money (a medium of exchange, or a lien on goods and services) in return. The money is shown being held, at least temporarily, in regional banks. The farmer then uses some of that money to purchase fertilizer locally. A large part of that money is used by the retailer to purchase the fertilizer from distributors who must eventually, of course, purchase that fertilizer from sources in industrial countries. This is shown by the dotted lines leaving the national boundaries near the top of the figure. Thus overall we see a three-level hierarchy where goods (i.e., their material substance and their embodied or released energy) and economies interact over considerable spatial dimensions.

A similar analysis can be made for a whole host of other industrial products that are used on the farm. Figure 23-8 summarizes what these are for Costa Rica. In this list are a surprisingly large number of various products that are input to the agricultural systems. For example, a goal of Costa Rican agriculture was to produce all milk within the country. Given that 60% of the nation's area is in pastures (although only a much smaller proportion is in dairy pasture) that seems like a reasonable goal, and it has in fact been met. But although the milk itself is produced within Costa Rica, this production is hardly based entirely on internal resources, for a total of about 50,000 tons of feeds is imported each year, presumably mostly for dairy cows, in order to produce 500,000 tons of milk, more than 90% of which is water. Thus a clear conclusion from these analyses is that the energy flow analysis reinforces other aspects of our analysis in that the Costa Rican economy is inextricably linked with the industrial world, and is heavily dependent upon nonrenewable fossil fuels. In this case an intervening factor is the role of international banks in discouraging the use of local feeds and for encouraging the import of feeds (Chapter 23), so it is not clear at this time whether or not Costa Rica could in fact be able to generate its own dairy products from internal supplies.

## V. CONCLUSION

Is food production in Costa Rica sustainable? Can it be? Our analyses show the following:

1. Although Costa Rica is considered a rich agricultural land, self-sufficiency in food production, presumably one component of sustainability, is increasingly not occurring. Specifically the proportion of the Costa Rican diet that comes from food production within the country has decreased from nearly 100% in 1960 to about 80% in the 1980s, and to about 70% in the 1990s (some of this is due to forced "free trade"; see Chapter 24). Although it is possible to increase food production in Costa Rica, most of the good and even moderate quality available land is already being used, and much of the very best land is taken up through urbanization. Thus population growth itself makes it increasingly difficult for Costa Rica to sustain its own population through local agriculture. This problem is exacerbated as the demand for foreign products increases in Costa Rica as per capita affluence of at least part of the population increases.

2. Essentially all increases in food production per hectare occurred prior to 1985 and are correlated highly with increased fertilizer use. Food production per hectare has not increased since then.

3. Total national production of the most important food crops (except beans) has decreased since about 1980, and per capita production has decreased sharply, driven by increasing population and by declining agricultural production.

4. There has been a shift in land use to export crops, and these crops can feed many people indirectly. In a sense this is a perfectly reasonable sustainability concept as long as prices are favorable. (But see Chapter 24.)

5. The efficiency of production per unit fertilizer used for all crops collectively, as well as for individual crops, has declined substantially since the 1960s. Efficiency per hectare (at a given intensity of fertilizer) also has declined whenever there has been an increase in the area cropped. Hence we can say that generally efficiency of agricultural production is inverse to intensity of exploitation, a pattern we have observed for many resources (Hall *et al.*, 1986; Ko *et al.*, 1998). Thus increasing food production works against strategies of increasing efficiency.

6. Energy flow analysis shows the interactions of the biophysical and the monetary environment, and also the large (and increasing) dependence of agriculture on imports from the industrial world.

All of these trends argue against any increase in the potential for sustainable development in the agricultural sector of Costa Rica. On the other hand, if we are willing to accept the continued use (and associated depletion) of foreign-supplied agricultural inputs there is every indication that the present level of agricultural production can be sustained at a level capable of supporting a large proportion of the present Costa Rican population. There may be no viable argument as to why agricultural production should be based completely on a sustainable basis, at least until oil becomes much more expensive. It certainly makes no sense given current population levels. Rather, attention should be focused on protecting the most important agricultural resource, the soil itself. This is done in Chapters 10 and 15.

Finally this analysis adds a new dimension to the concept of "overpopulation." Although Costa Rica is not a crowded land, it is crowded relative to its need for, and ability to produce, agricultural products, including both food and export crops. For example, Costa Rica has about twice the population density of the United States. It also has a similar proportion of land in the best farmland category (although less in actual crops because of pastures on the best land), but only about one-third the agricultural production per hectare. Thus it can be considered as roughly six times more crowded relative to its agricultural potential, even not accounting for the dependence on imported agrochemicals. Costa Rica does not, and apparently more or less cannot, feed a growing population on its own farmland. By these analyses Costa Rica is a very crowded country.

## ACKNOWLEDGMENTS

We thank Alfredo Bolanos, Viviana Palmieri, and other staff at CATIE for research help, and Cutler Cleveland and Bernardo Aguilar for critical review.

## REFERENCES

Altieri, M., and O. Masera. 1993. Sustainable rural development in Latin America: Building from the bottom up. *Ecological Economics* 7:93–121.

Chang, J-H. 1981. Corn yield in relation to photoperiod, night temperature, and solar radiation. *Agric. Meteor.* 24:253–262.

Cleveland, C. J. 1995. Resource degradation, technical change, and the productivity of energy use in U.S. agriculture. *Ecological Economics* 13:185–201.

Cornell, J., and C. A. S. Hall. 1994. A systems approach to assessing the forces that generate tropical land use change. In J. O'Hara, M. Endara, T. Wong, C. Hopkins, and P. Maykish (Eds.), *Timber Certification: Implications for Tropical Forest Management*, pp. 35–43. School of Forestry and Environmental Studies, Yale University, New Haven, CT.

FAO. 1991. *Production Yearbook*. FAO, Rome.

FAO. 1992. *Fertilizer Use by Crop*. ESS/MISC/1992/3, FAO, Rome.

FERTICO. 1995. Areas de Cultivo y Consumo de Fertilizante en Toneladas metricas anos 1989 a 1995. Xeroxed document, FERTICO, San Jose, Costa Rica.

Gifford, R. M., J. H. Thorne, W. D. Hitz, and R. T. Giaquinta. 1984. Crop productivity and photoassimilate partitioning. *Science* 225:801–808.

Hall, C. A. S., and J. Day. 1977. *Ecosystem Modeling in Theory and Practice*. Wiley–Interscience, New York. Reprinted 1990, University Press of Colorado, Boulder, CO.

Hall, C. A. S., C. J. Cleveland, and R. K. Kaufmann. 1986. *Energy and Resource Quality: The Ecology of the Economic Process.* Wiley–Interscience, New York. Reprinted 1992, University Press of Colorado, Boulder, CO.

Hall, C. A. S., and M. Hall. 1992. The efficiency of land and energy use in tropical economies and agriculture. *Agriculture, Ecosystems and Environment* 46:1–30.

Hall, C. A. S., J-Y. Ko, C-L. Lee, and H. Q. Wang. 1998. Ricardo Lives: The inverse relation of resource exploitation intensity and efficiency in Costa Rican agriculture and its relation to sustainable development. Pp. 355–370 *in* S. Ulgiati (Ed.) Advances in Energy Studies. Energy flow and the Economy. MUSIS, Rome.

Ko, J-Y., C. A. S. Hall, and L. G. Lopez Lemus. 1998. Resource use rates and efficiency as indicators of regional sustainability: An examination of five countries. *Environmental Monitoring and Assessment.* 51:571–593.

Levitan, L. C. 1988. Land and energy constraints in the development of Costa Rican agriculture. Unpublished M.S. thesis, Cornell University, Ithaca, NY.

MacArthur, J. D. 1976. Some general characteristics of farming in a tropical environment. In H. Ruthenberg (Ed.), *Farming Systems in the Tropics,* pp. 19–27. Oxford University Press, Oxford.

MAG (Costa Rican Ministerio de Agricultura e Ganaderia). 1950, 1955, 1965, 1973, 1977, 1984. *Agricultural Census.* MAG, San Jose, Costa Rica.

Pimentel, D., C. Harvey, P. Resosudamo, K. Sinclair, D. Kurz, M. McNair, S. Crist, L. Shpritz, L. Fitton, R. Saffouri, and R. Blair. 1995. Environmental and economic costs of soil erosion and conservation benefits. *Science* 267:1117–1123.

Odum, H. T. 1967. *Energetics of World Food Production. Report of the Panel on the World Food Supply,* pp. 55–94. President's Science Advisory Committee, The White House, Washington, DC.

Odum, H. T. 1983. *Systems Ecology, an Introduction.* Wiley–Interscience, Wiley, New York. Reprinted 1994 by the University Press of Colorado, Niwot, CO.

Odum, H. T. 1989. Models for national, international and global systems policy. In L. C. Bratt and W. F. J. Van Lierop (Eds.), *Economic Ecological Modeling,* pp. 203–251. Elsevier Science, New York.

Pimentel, D. 1980. Handbook of Energy Utilization in Agriculture. CRC Press, Boca Raton, FL.

Ruthenberg, H. 1976. *Farming Systems in the Tropics,* 2nd ed. Oxford University Press, Oxford. 313 pp.

Statler, R. 1984. Crop specific fertilizer use. Personal communication, CATIE, Turrialba, Costa Rica.

Sundberg, U., and Silversides C. R. 1992. *Operational efficiency in Forestry,* Vol. I: *Analysis.* Kluwer Academic, Dordrecht/Norwell, MA.

Thrupp, L. A. 1991. Long term losses from accumulation of pesticide residues: A case of persistent copper toxicity in soils of Costa Rica. *Geoforum* 22:1–15.

Whiktaker, R. 1975. Communities and Ecosystems.

CHAPTER 13

# Biophysical Agricultural Assessment and Management Models for Developing Countries

GERRIT HOOGENBOOM, CARLOS LEÓN PÉREZ, DAVID ROSSITER, AND PATRICK VAN LAAKE



## I. INTRODUCTION

Scientists in the developed world, in an effort to develop tools that will support informed land use decision making, have created a number of very different land use analysis and agricultural production computer models. Such models should be increasingly useful in many parts of Latin America, where issues of sustainable land use are approaching critical dimensions. These include the need to maximize agricultural production while minimizing erosion, to use limited land areas in the most efficient manner, to reduce pollution pressure on the landscape, and to reduce deforestation—as well as optimize, in some sense, these and other goals. More generally such models can be used to analyze the possibilities and limitations for sustainability. We present here more or less the most important biophysically based (vs economically based) agricultural models currently available that hold considerable promise for addressing the problems of developing nations.

The models we review here are derived quite independently from one another and for very different purposes: ALES (Automated Land Evaluation

System) is a computerized framework for allowing expert judgment to estimate crop and consequent economic production, and DSSAT (Decision Support System for Agrotechnology Transfer) is an explicit suite of models of crop plant production. The general objectives are similar: how can we use agricultural land and other resources most efficiently? This often means most productivity, in terms of yield or monetary return per hectare or per unit input. If this can be accomplished then human welfare will be enhanced and there will be less need to develop remaining natural ecosystems.

ALES is designed to help assess optimum land use for a wide range of scales, from individual fields to regional groups of fields (see the El Laborador Settlement example in this chapter), nations [e.g., FAO-sponsored land evaluation of 82 crops for Papua New Guinea (Venema and Daink, 1992)], and continents [e.g., European Community study of general suitability for olives done by the Staring Centre (Van Lanen *et al.*, 1992b)]. It all depends on the level of detail of the planning objectives, the land utilization types, the land mapping units, and land data. For successful ALES application, all of these should be at the same level of abstraction as defined by Hoosbeek and Bryant (1992). The DSSAT models are routinely applied at the level of individual fields in order to help individual farmers make tactical decisions about what crop, if any, the land is best suited for. Since these models make relatively precise predictions, they require reasonably explicit information, such as the particular level of a soil nutrient, to achieve that precision. These models also can be scaled up to regional levels to look at agricultural production issues (Lal *et al.*, 1993).

## II. MODEL 1: ALES LAND USE POTENTIAL EVALUATION SYSTEM

### A. The ALES Concept and Framework

The Automated Land Evaluation System, or ALES, is a program for MS-DOS (PC-compatible) microcomputers (Rossiter, 1990; Rossiter and Van Wambeke, 1993). It allows land evaluators to build their own *expert systems*, in the sense of Waterman (1985), to evaluate the potential of land areas for different actual and proposed land uses, in both biophysical and economic terms. Each model is built by a different evaluator to satisfy local needs, using local data. ALES is very different from DSSAT in that relatively subjective, often local information is used to build the model response rather than more detailed experimentally derived information.

ALES is grounded in the concepts of *land evaluation*, which may be defined as:

1. The process of assessing land performance when used for specified purposes involving the execution and interpretation of surveys and studies of landforms, soils, vegetation, climate, and other aspects of land in order to identify and make a comparison of promising kinds of land use in terms applicable to the objectives of the evaluation (FAO, 1985, p. 212).

2. All methods for explaining or predicting the use potential of land. "[It provides] the information on the opportunities and constraints for the use of land as a basis for making decisions on its use and management" (Van Diepen *et al.*, 1991, p. 140).

Land evaluation includes any land classification method that attempts to predict land performance in physical or economic terms. Nowadays many methods, including ALES, are derived from the FAO's "Framework for Land Evaluation" (1976) and subsequent guidelines for major kinds of land use (FAO, 1983, 1984, 1985, 1991). Van Diepen (1991) provides a critical review of the Framework.

The ALES program has six major components:

1. A framework for a *knowledge base* describing proposed land uses, in both physical and economic terms (Figure 13-1). This is where the method for determining physical and economic suitability is entered by the evaluator. Land utilization types (LUTs) can be of any duration, with any mix of products within or among years. The products determine the "return" side of the financial balance sheet for the LUT, i.e., the expected return to the land user, should they implement the LUT on a given land area. The knowledge base also includes lists of inputs for each LUT which determine the "costs" side of the balance sheet. The key point is that levels of both inputs and products can vary among land types; the evaluator specifies how this variation should be determined in the knowledge base.

2. A framework for a *database* describing the land areas being evaluated; the total land area is divided into land mapping units.

3. An *inference mechanism* to relate these two, thereby computing the physical and economic suitability of a set of map units for a set of proposed land uses.

4. An *explanation facility* that enables model builders to understand and fine-tune their models by means of a "backward chain" of screens that explain each step of the decision process.

5. A *report generator*.

6. An *import/export* module that allows data to be exchanged with external databases, geographic information systems, and spreadsheets. This includes the *ALIDRIS* interface to the IDRISI geographic information system.

ALES program texts may be displayed in English, French, Spanish, or Indonesian.

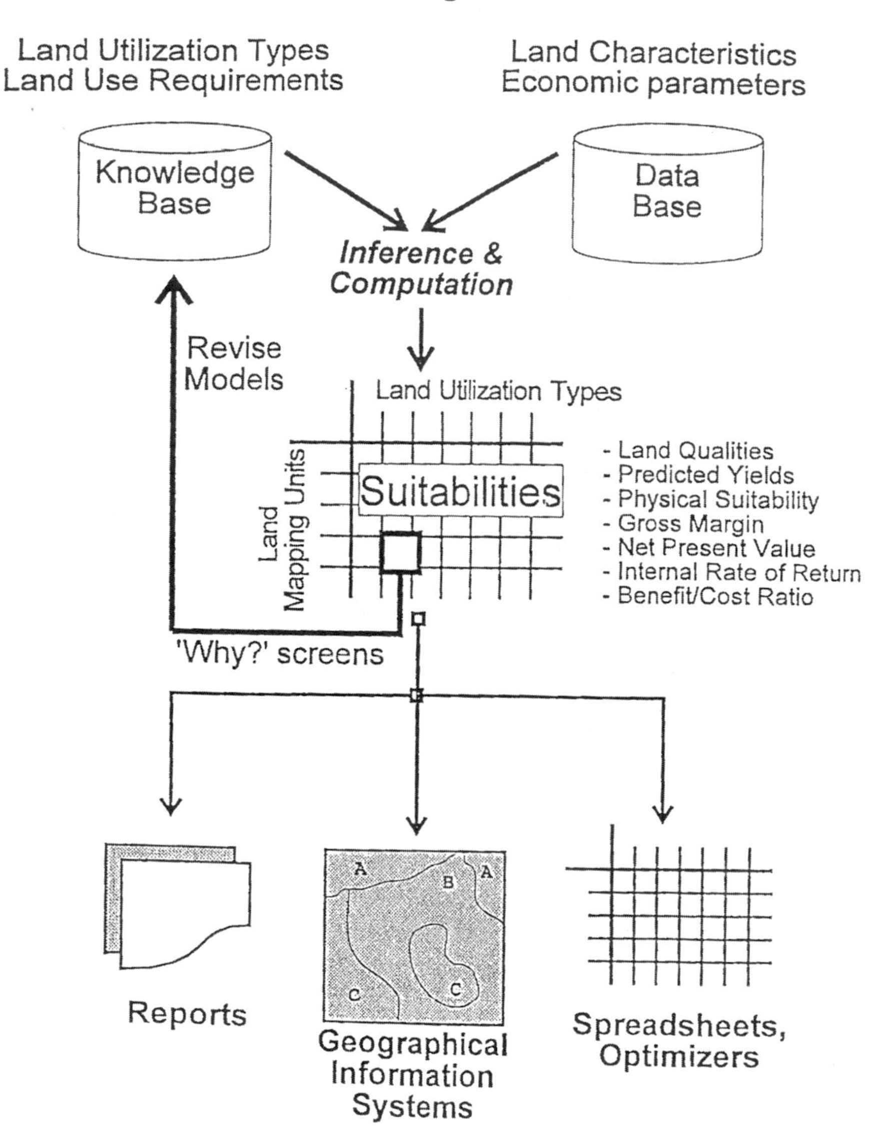


FIGURE 13-1 ALES program structure.

As explained in the previous section, ALES itself is not a model; instead, ALES provides a framework within which a certain class of models can be built based on the user's knowledge of the area considered. These are fundamentally *strategic* models of land performance, which should allow land managers to make sound decisions on land allocation. This is in contrast to a tactical model, which is used to assist day-to-day decision making, e.g., irrigation scheduling. The ALES model is a model of *expert judgment* about biophysical and economic reality, not a model of the reality itself (although the expert judgment is implicit even in a "mechanistic" model).

The strengths of the ALES approach to modeling are:

1. It can make use of *qualitative* and *semiquantitative* knowledge, as well as fully *quantitative* knowledge. For example, in the statement "steep slopes and silty surface soils greatly increase the risk of sheet erosion," we do not have to define "steep," "silty," or "greatly increase" except that we have to be able to identify these in the field. Many important land characteristics can in practice be defined only by nominal or ordinal classes.
2. It can make use of *indigenous* or other *local knowledge,* which very rarely is precisely quantified.
3. Models can be built *incrementally*, with the most important factors being described first, and less important or common factors being added only as necessary.
4. The expert system is *transparent*; i.e., it can explain how it arrived at a decision, step by step.

Some published studies using the program are Delsert (1993), DeRoller (1989), Johnson and Cramb (1991), León Pérez (1992), Maji (1992a,b), and Van Lanen *et al.* (1992a,b,c). The studies of León Pérez and Maji are examples from less-developed countries.

## B. Applications of ALES to Costa Rica: Growing Basic Grains and Watermelons under Various Soil Conservation Practices at El Laborador Settlement

The development of an agroecological modeling approach for land use planning in Costa Rica faced two major problems: (1) a dispersed and not very accurate database and (2) an official land use classification scheme that had little previous impact on decision making. We examined many different technologies and software options that we thought might be suitable for decision making and land use planning in developing countries. We ended up selecting the

ALES system because it met our needs without being restricted by these two basic constraints. ALES allows the construction of both a database and a knowledge base from information given in different formats and based on different levels of expertise, including that of the land use criteria currently being used. Although the modeling exercise requires expertise, ALES's consultation and evaluation mode is user friendly, making the program very useful for Costa Rican managers (for example, extension agents and regional planners). We also have found ALES useful for computer-literate farmers. Thus our conclusion is that ALES is a great performer relative to the other models we examined.

Our application of ALES in Costa Rica generated a debate among government officials over the utility of a "unique and official land use approach." But we concluded that the analysis framework built into ALES, which is based on the FAO land use approach, has the desired flexibility to run land use evaluations even for applications where the objectives and physical conditions are very different. We previously had found the manual FAO approach very useful, but much too time consuming. Its use via ALES, however, decreased the time to make an evaluation by perhaps a factor of 100.

### 1. El Laborador Settlement: An Example of the Application of ALES

El Laborador Settlement, one of the monitoring areas of the Costa Rican Ministry of Agriculture and Livestock (MAG) and FAO National Soil Conservation Program, is situated in the central Pacific region of Costa Rica in the vicinity of Jaco. In 1992 MAG, in conjunction with the FAO, started a project with the objective of promoting and demonstrating soil and water conservation practices that would improve the economic and biophysical sustainability of farming systems.

The Laborador project was established in 1992 as part of the Costa Rica Agrarian Development Institute's (IDA) program for land redistribution. The project includes 277 ha partitioned into 59 farms with an average size of 4.7 ha. Another way of looking at the same land area is that it encompasses eight land units, each composed of similar soils and slopes.

This region of Costa Rica has two marked seasons: (1) a summer or dry season from December to early May, during which agriculture is not possible and most of the farmers work on neighboring sugar cane plantations, extensive grazing operations, or irrigated farms, and (2) a wet season favorable to agriculture. Here peasant farmers initially grew beans and corn, still the predominant cropping system. By 1992 the residents had raised a Community Farmers Association and were receiving periodic visits from extension agents because they were encountering serious economic difficulties with their crops. The project was implemented to undertake analysis of watermelon and cattle ranching as possible new activities. It seemed possible that using ALES could aid farmers in finding a way to make farming more viable in this region.

## 2. The Criteria for Modeling

The objective of our evaluation was to determine the best crop management and soil conservation practices for the different land and land use types based on their biophysical and socioeconomic characteristics. Since our project started in 1992 some of the farmers were already maintaining production and economic records as well as notes on the most important problems they faced. The FAO–MAG technicians undertook a detailed soil survey, and helped the farmers to develop appropriate soil conservation practices, including the use of contour cropping, stone walls, and vetiver grass barriers. All of these data, as well as local expertise, were the basis of this land evaluation.

The prevalent agroecosystems were the primary basis for identifying which cropping (i.e., land use types) should be considered for modeling. These were: (1) basic grains, corn, and beans with vetiver grass barriers, or (2) watermelon with stone walls (20 cm high). Other crops were possible, but these two seemed to be the most important. Further consultations with national databases, farmers, and extension agents were integrated with the soil survey to build both the knowledge and the database for the evaluation. The role of local knowledge among farmers and technicians in developing this expert system for understanding the characteristics of the eight land use types modeled was of critical importance when defining the criteria to discriminate which factors should be used for decision making, the main objective of these models.

We started by listing the most significant biophysical and socioeconomic characteristics that determined productivity and economic success. We then reduced this list substantially by eliminating those which did not vary over the region examined, since they were not useful in reaching a decision. For example, even though rainfall is critical, it did not vary within this 277 ha; thus we did not consider it. The editing features of ALES allowed us to do this easily.

The land characteristics considered for this evaluation were grouped into three categories:

1. Climate and Management
    - 1.1. Consecutive dry months: 3 classes (<1, 1–6, >6 months)
    - 1.2. Present land use: 5 classes (annual crops, semipermanent crops, perennial crops, natural forest, grazing lands)
    - 1.3. Wind speed (km/hr): 3 classes (<15, 30–50, >50)
    - 1.4. Soil conservation practices: 5 classes (contour planting, live fence, individual terrace, stone barrier, agroforestry system)

2. Physical and Geographic
   2.1. Visual erosion (%): 5 classes (5–25, 25–50, 50–95, 95–100, >100)
   2.2. Fragments in soil (%): 7 classes (5–10, 10–15, 15–25, 25–50, 50–75, 75–90, >90)
   2.3. Slope (%): 5 classes (3–8, 8–15, 15–30, 30–75, >75)
   2.4. Soil depth (cm): 5 classes (30–60, 60–90, 90–120, 120–300, >300)
   2.5. Texture (% clay): 5 classes (0.1–0.2, 0.2–0.3, 0.3–0.4, 0.4–0.6, 0.6–1.0)
3. Soil Chemical Properties
   3.1. Calcium content (meq/100 ml): 3 classes (<5, 5–10, >10)
   3.2. Potassium content (meq/100 ml): 3 classes (<1, 1–5, >5)
   3.3. Phosphorus content (ppm): 3 classes (<10, 10–20, >20)
   3.4. Acidity saturation (%): 3 classes (<10, 10–50, >50)
   3.5. pH: 3 classes (<6, 6–7, >7)
   3.6. Sum of bases (%): 3 classes (<5, 5–10, >10)

Land use requirements were inferred from the land characteristics of areas where a particular crop grew well using decision trees. Table 13-1 shows the land characteristics that we used to determine each land use requirement. Every land use (e.g., crop) type has its own requirements; for example, watermelons require at least moderate moisture, at least moderate soil fertility (including high potassium), relatively low slopes (for obvious reasons), and so on. Basic grains should not get too wet while the grain is forming (or else they get a

**TABLE 13-1 Land Use Requirements and Land Characteristics for ALES Analysis**

| Land use requirement | Land characteristics |
|---|---|
| Available calcium | Saturated base, calcium content |
| Soil fertility | Saturated base, K concentration, acidity saturation |
| Susceptible to laminar erosion | Present land use, slope, visual erosion, soil conservation practices |
| Susceptible to plagues and diseases | Consecutive dry months |
| Susceptible to blowdown | Wind |
| Susceptible to conservation practices | Soil texture, rock fragments in soil, slope |
| Available potassium | Saturated base, K concentration |
| Available phosphorus | pH, phosphorus concentration |
| Drainage | Soil depth, soil texture |
| Drought risk | Consecutive dry months |

fungus) and need soils with higher nitrogen. This initial evaluation was the first step into a broader one, which is why some land use requirements are listed but not included in the present decision analysis.

Each of the eight cartographic, or land, units was defined as a biophysically and socioeconomically homogeneous mapping unit. The physical suitability of each cartographic land unit was derived for each land use type, in this case small grains and watermelons, using a decision tree within ALES. The physical decision tree assigns a suitability category (good, moderate, marginal, or none) to each land unit. For example, the basic grains model uses soil fertility and moisture, susceptibility to laminar erosion, and susceptibility to blowdown. A proportional yield decision tree estimates the output as, e.g., kg/ha of beans.

The watermelon model has a physical decision tree based on soil fertility, drainage, susceptibility to erosion, and susceptibility to conservation practices. Blowdown is not important for watermelons, again for obvious reasons, but drought risk effects the proportional yield tree.

If a land unit was classified as appropriate for a particular use according to physical criteria, its economic suitability was computed. The economic evaluation is a cost–benefit analysis built into ALES. The optimum yield is the best commercial yield found under similar agroecological and management conditions within Costa Rica. A proportional yield decision tree then reduces that yield according to the more precise physical and management conditions for that site to give explicit output per hectare. Once all the production inputs and outputs for every land use type on every land unit were entered into the program, an economic evaluation was carried out. Beside predicting crop yields, the benefit–cost analysis also derived financial indicators such as net present value, internal rate of return, benefit–cost ratio, and gross margin.

## 3. Results

Perhaps our most general result was that the El Laborador Settlement was hardly the best place in Costa Rica for agriculture, given the climatic and agroecological circumstances. Nevertheless farming there is a reality, and there remained a need to give answers for optimizing production and for giving the best possible advice to farmers.

Corn and beans were moderately suitable with respect to physical criteria in most of the land units. But on two land units susceptibility to blowdown from the existing high wind velocity was an important constraint for the growing of corn, which was otherwise the best crop. Although economic benefits exceeded costs, the economic analysis predicted a very low return on the investment, smaller than the opportunity cost of selling labor in neighboring farms on better land. The only rationale that we could give for continuing to grow these crops at El Laborador was the cultural factors associated with the farmers' desire to be in charge of their own subsistence.

We found with ALES, however, that watermelon was suitable on four land units, especially the ones with low slopes. A combination of soil texture, slope, and stones in the soil generated restrictions on the remaining land units. Three land units were classified as having only moderate potential for even watermelons because of their susceptibility to soil erosion, while one had low potential because of its poor response to soil conservation practices. The economic analysis showed a reasonably favorable prognosis for the physically suitable land units because of a good seasonal price for watermelons. Since the net income per hectare was almost five times that of the grain crops, our analysis supported an increasing shift to these export crops on most of the farms, and we recommended this to the appropriate farmers.

We found that the ALES software performed adequately for our objectives, even with the restrictions due to data-poor environment. Because of its success at El Laborador we have used it for determining land use recommendations in six other areas in Costa Rica. More generally we feel that expert-systems-based programs, decision support systems, computer simulations, and GIS's (geographic information systems) are valuable tools that should be seriously incorporated into training strategies at natural resources and agriculture-related institutions in Costa Rica and other developing countries. The Ministry of Agriculture of Costa Rica, in collaboration with FAO, already started and incorporated ALES into its land use planning strategies.

We have encountered more than a little resistance and criticism as we have tried to apply new tools (at least new in developing countries) like modeling. We agree with some of that criticism when it is directed toward models that are more academic than practical. But we do not believe that these criticisms apply to very practical and empirically based models such as ALES. ALES has demonstrated the ability to improve remarkably our decision making for land use planning at the regional level.

## C. Predicting National-Level Yields for Costa Rica

We analyzed the production potential for several important crops in Costa Rica, beans and maize, at the national level, using the procedures already described and the databases of Chapters 9 (meteorology), 10 (soils), and 12. Basically we solved the crop production equations for each of the 1-km$^2$ cells in selected areas of the country.

Through extrapolation our analyses indicated that if all of the nation was put into the production of maize, a maximum possible yield would be 4 million tons per year. If only that area (1.1 million ha) considered appropriate for annual crops (Land Class I) was used the production would be 1.5 million tons, enough food to feed 5 million people (Figure 13-2). If that area were planted to beans the yields would be 0.5 million tons per year, enough to feed less than 2 million people. If instead that land were used to generate foreign exchange (via coffee), and assuming a continuing global demand for Costa Rican coffee, then roughly $4.5 billion could be generated, enough to far more than pay for the imports required for the agricultural production if no other requirements were made for that foreign exchange. Thus this analysis indicates that the country of Costa Rica could feed its 1995 national population several times over based strictly on its agricultural productive capacity, but is very near the limits of its capacity to generate food while also paying for the imports required just to produce that food, let alone pay for other imports (see Figure 12-8, which comes to similar conclusions using a different model). These results are in approximate agreement with cruder estimates made in Chapters 3 and 11. Some very general predictions of maize yield for Costa Rica are given in Table 13-2.

## III. MODEL 2: DSSAT CROP SIMULATION MODELS

The International Benchmark Sites Network for Agrotechnology Transfer (IBSNAT) Project was funded in 1982 by the U.S. Agency for International Development (USAID) as a 10-year project. The main objective of the IBSNAT Project was to develop computerized simulation technology to generate information for addressing a wide range of agricultural, environmental, and economic problems (Uehara and Tsuji, 1998). In essence, crop yields are predicted as a function of crop physiological parameters, weather, and edaphic and management variables.

The technology was implemented on personal computers because of the ease of access to these hardware systems by people in the developing countries. The IBSNAT Project consisted of an international network of model developers located at the University of Hawaii, University of Florida, Michigan State University, International Fertilizer Development Center, University of Guelph, University of Edinburgh, University of Georgia, and several of the International Tropical Agricultural Research Centers, including Centro Internacional de Agricultura Tropical (CIAT) and the International Rice Research Institute (IRRI). In addition an elaborate network of model testers and model users developed during the project's operational period (Tsuji, 1998).

Under the auspices of the IBSNAT Project a special software system was developed, called Decision Support System for Agrotechnology Transfer Jones *et al.*, 1998). DSSAT includes a computer software shell which integrates crop simulation models, database entry and management utilities, and simulation application programs. The core of the system consists of crop simulation models for more than 15 of the most important international food crops (Table 13-3).

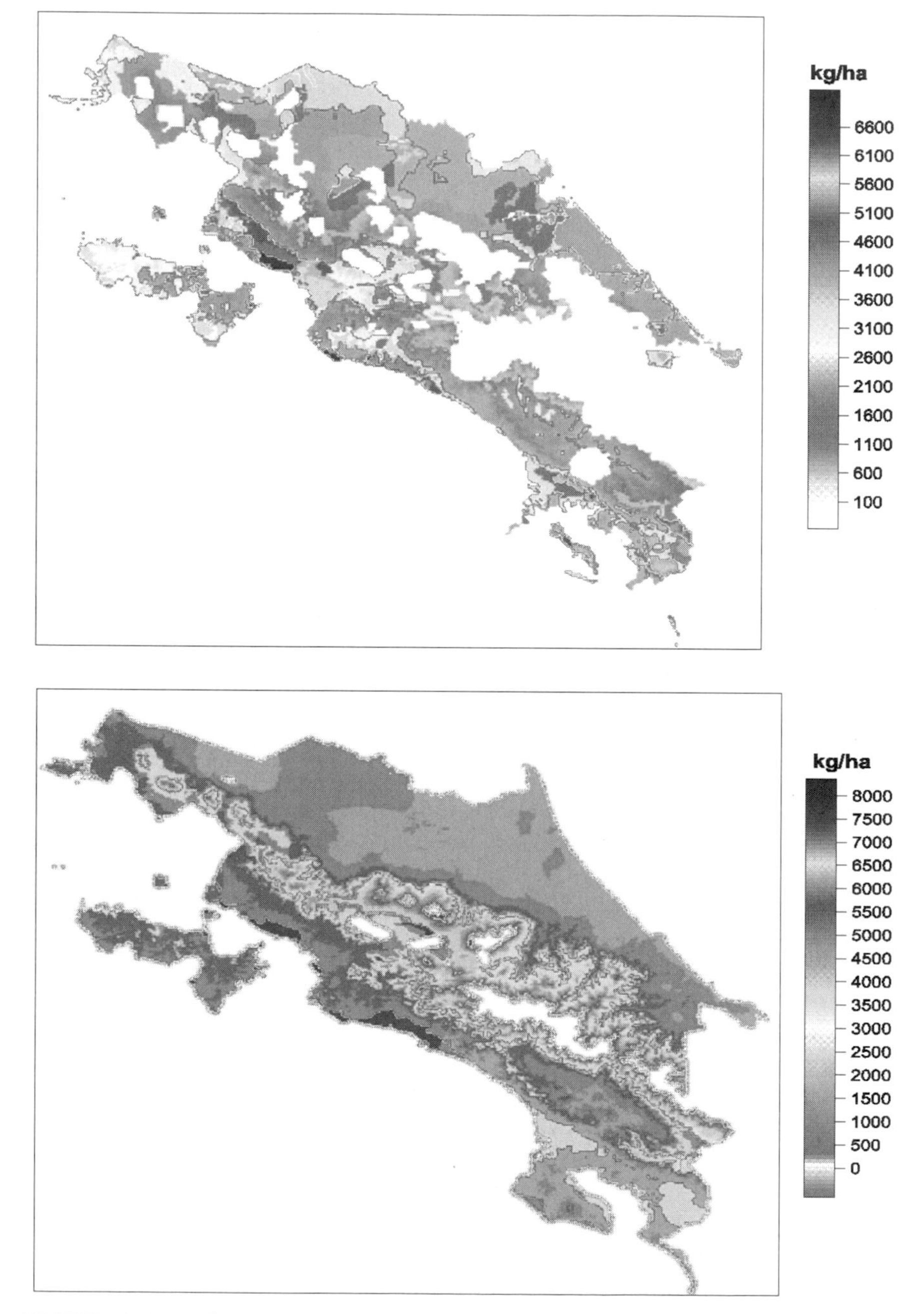


FIGURE 13-2 Predicted growth of (a) fertility-limited and (b) potential maize yield for the country of Costa Rica. (See CD for color version of figure.)

TABLE 13-2 Results of Various Agricultural Production Models for Average Conditions in Costa Rica Such as Might Be Found in the Western Central Valley on Soils of Moderate Quality

| Model | Crop | Fertilizer use[a] | Yield (T/ha/yr) |
|---|---|---|---|
| IBSNAT | Maize | None | 1.0 |
| | | Low | 5.0 |
| | | High | 10.0 |
| ALES | Maize | None | 0.5 |
| | | Low | 1.0–1.5 |
| | | High | 2.4–4.0 |
| ALES[b] | Maize | None | 0.25 |
| | | Low | 0.5–0.75 |
| | | High | 1.2–2.0 |
| FNCYPLT[c] | Maize | None | 0.4 |
| | | Low | 1.0 |
| | | High | 3.2 |

*Note.* Typical values for maize growth observed in this region are about 1–1.5 T/ha.
[a] Low, 20 kg/ha NPK; high, 200 kg/ha NPK.
[b] El Laborador Settlement.
[c] Chapter 5.

The models can be grouped into:

1. The grain cereal models, i.e., CERES-Barley (Otter-Nacke *et al.*, 1991), CERES-Maize (Ritchie *et al.*, 1989), CERES-Millet (Singh *et al.*, 1991), CERES-Rice (Singh *et al.*, 1993), CERES-Sorghum (Alagarswamy and Ritchie, 1991), and CERES-Wheat (Godwin *et al.*, 1989)
2. The grain legume models, i.e., BEANGRO (Hoogenboom *et al.*, 1994), PNUTGRO (Boote *et al.*, 1989), and SOYGRO (Jones *et al.*, 1989)
3. The root crop models, i.e., SUBSTOR-Aroids (Prasad *et al.*, 1991), GUMCAS (Matthews and Hunt, 1994), and SUBSTOR-Potato (Griffin *et al.*, 1993)

One of the unique features of DSSAT is that it uses a common shell to access a wide range of programs, and that this shell controls the access, execution, and linkages of these programs. The second unique feature is the common file structure and format used by all programs under the DSSAT shell. This will allow for easy exchange of information between the various program components or the use of one input or output file by more than one program. An example is the daily weather file. All individual crop models use the same weather file format. Therefore only 1 weather file for a particular site and year is needed, rather than many individual files. A program is also available to enter the weather data and

TABLE 13-3 The Main Crops Parameterized for Use in DSSAT

| | |
|---|---|
| Aroids | *Colocasia esculenta* L. [taro] and *Xanthosoma sagittifolium* L. [tannier] |
| Barley | *Hordeum vulgare* L. |
| Cassava | *Manihot esculenta* L. |
| Corn | *Zea mays* L. |
| Dry bean | *Phaseolus vulgaris* L. |
| Millet | *Pennisetum americanum* L. |
| Peanut | *Arachis hypogea* L. |
| Potato | *Solanum tuberosum* L. |
| Rice | *Oryza sativa* L. |
| Sorghum | *Sorghum bicolor* L. |
| Soybean | *Glycine max* [L.] Merr. |
| Wheat | *Triticum eastivum* L. |

generate the weather files into the required formats. Standard formats are available for both model input and model output files. Details of the input and output file formats have been described in Hunt *et al.* (1994).

The DSSAT crop simulation models require weather, soil, and crop management data and crop genetic parameters as input (Table 13-4). The latter are defined externally to each model and are provided for each crop. The models operate on daily time steps; i.e., plant growth and rates of change within the soil–plant–atmosphere system are calculated and integrated at daily time intervals. Simulation of the soil water balance can start before planting if field-measured data for soil water and nitrogen are available or if irrigation and/or fertilizers are applied. The actual plant component of the model is initiated at planting and the various phases of a plant's life cycle are simulated, including emergence, vegetative development, flower initiation and flowering, fruit set and seed development, and physiological and harvest maturity. The DSSAT crop models include detailed sections which simulate vegetative and reproductive development, the plant carbon balance, and the soil and plant water balance (Jones and Ritchie, 1990). The DSSAT models also include an option to simulate the nitrogen balance of the soil and plant system (Godwin and Singh, 1998).

Vegetative and reproductive development for all crops is predicted as a function of air temperature and in some cases as a function of photoperiod. Development rates also can be affected by environmental stresses such as drought or nitrogen shortages. For each crop, development is predicted through various phases. Each phase has a threshold value for degree days, thermal time, or photothermal time. Once the thermal or photothermal accumulator reaches this threshold, the model predicts that a new phase has occurred. The

TABLE 13-4 **Information Required to Run a Typical IBSNAT Model**

Weather information includes
- Daily total solar radiation
- Daily total precipitation
- Daily minimum and maximum air temperature

Soil characteristics include
- Albedo of bare soil
- Soil surface runoff as specified by the Soil Conservation Service (SCS) runoff curve number
- First-stage soil evaporation
- Variables which describe water permeability and drainage from the profile
- Each soil layer requires
  —a parameter value for layer thickness
  —saturated soil water content
  —drained upper limit of extractable plant water (field capacity)
  —lower limit of extractable plant (permanent wilting point)
  —initial soil water content at the start of the soil water balance simulation
  —a relative root weighting factor (Ritchie *et al.*, 1990)

Information required to describe crop management options include
- Cultivar selection
- Planting date
- Plant density, row spacing, and planting depth
- Soil profile identification

Depending on the type of model application
- Dates and amounts of irrigation
- Dates, amounts, and types of fertilizer

main phases for all crops include planting to emergence, emergence to start of flowering, flowering to beginning fruit set, and fruit set to physiological maturity. A detailed description of soybean development simulation can found in Jones *et al.* (1991); simulation of the development of the other two grain legume species is very similar. A description of development simulation for wheat, corn, and various other cereal crops can be found in Hodges (1991), Jones and Kiniry (1986), and Ritchie (1991).

All models include a detailed section that simulates plant tissue growth and partitioning. Processes which are simulated include photosynthesis, respiration, biomass growth and partitioning, and senescence. More details can be found in Hoogenboom *et al.* (1992), Jones and Kiniry (1986), and Jones and Ritchie (1990). The soil water balance simulation is identical for all DSSAT crop models. It uses the soil water model developed by Ritchie (1998) and includes the Priestley-Taylor (1972) equation to estimate potential evapotranspiration.

It is critical for any model application that field-measured experimental data are available for model calibration and testing. Detailed procedures for calibration and validation are presented in the models' user guides (Boote *et al.*, 1989; Godwin *et al.*, 1989; Hoogenboom *et al.*, 1991; Jones *et al.*, 1989; Otter-Nacke *et al.*, 1991; Ritchie *et al.*, 1989; Singh *et al.*, 1993). A special program has been developed to calibrate the models for a new cultivar which has not been previously used with any of the crop models (Hunt *et al.*, 1993). All DSSAT models have been extensively calibrated with various data sets collected in different environments. Because of the detailed information required, the models generally produce yield estimates that are very similar to what was actually observed (Hoogenboom *et al.*, 1992).

Once the models have been calibrated for a particular location and management regime, they can be used to address various management options, related to both tactical and strategic decision making. Tactical decisions include when to apply irrigation, and how much irrigation to apply. Strategic decisions include which crop and cultivar to plant, and when to plant the crop. In many cases this also includes how well one or more management alternatives, either before the growing season or during the growing season, respond to environmental conditions for a certain location. A special software program, called the strategy analysis program, has been included in DSSAT for these types of model applications (Thornton and Hoogenboom, 1994). The strategy analysis program allows the user to define one or more management options; once these have been defined the DSSAT system runs the crop models for these management conditions using long-term historical weather data. After the simulations are completed the program calculates means and standard deviations and generates frequency distributions for each predicted variable. The main variability of the predicted variables is caused by the variation of the weather conditions. With these outputs the user can compare selected management options, decide which option gives the desired optimum performance, and determine the risks associated with the selected management options due to the annual variation in weather conditions. Risk is defined as the probability of yield to be lower or higher than a present minimum or maximum acceptable level. For example, one can define the minimal acceptable risk for bean yield in Costa Rica to be 500 kg/ha. With this system one can predict the probability that yield is lower than 500 kg/ha for every combination of crop management scenarios.

Finally, we generated national-level potential yield maps for maize and beans from the national-level meteorological information in Chapter 10, the soil fertility data of Chapter 11, and the DSSAT model by solving for the environmental conditions on each square kilometer (Figure 13-2).

## IV. DISCUSSION

### A. Limitations in the Models

All models, even good ones, have inherent limitations. Some of them represent inevitable trade-offs. For example, ALES does not model photosynthesis with explicit equations. Instead, expert judgment about crop yield is modeled as the function of a local optimum, reduced by some fraction. The reduction fraction can be strictly multiplicative (Mitserlich) or limiting (Liebig), or can take any known interactions into account, depending on expert judgment backed up by experimental or survey results.

The DSSAT crop models, which are more mechanistic, instead use equations that are based on quantity of light absorbed and explicit equations for other environmental and crop parameters. Nevertheless the yield outputs of this model were broadly similar to those predicted from Hall's much cruder model (Chapter 5; Table 13-2). Since all the models are parameterized from real data they all are constrained to give results within the same general range.

A critical issue in model application is, of course, data requirements. ALES is an expert system, with the database to be defined by the evaluator, and has no fixed data requirements, in contrast to DSSAT. It is ideal in data-poor environments, and can even use strictly qualitative or linguistic variables, as long as these have some consistent meaning and can be related to land suitability. DSSAT crop models are much more precise and presumably accurate than ALES, but their data requirements are so large and so exacting that it cannot be used in most situations. On the other hand ALES, which is presumably less accurate, can be used just about anywhere by people without any particular training in modeling, or even science, as long as they have a good intuitive knowledge of the agricultural system under study.

ALES is intended as an integrating tool. More mechanistic models, such as those in the DSSAT, are ideal for quantifying both the local *optimum* yield and the *proportional* yield factors needed by ALES to predict yield. ALES can also be used to summarize land qualities that have been assessed with more sophisticated models of specific phenomena that affect land use, for example, chemical leaching to the groundwater by the LEACHM model (Hutson and Wagenet, 1991, 1992). On the other end, ALES results are intended for further analysis by optimization and multiple-criterion decision-making techniques. ALES makes predictions about map units or delineations, and a more general spatial model must be used to make statements about a geographical region *as a whole*.

## B. How We Can Improve These Models

The more detailed our data and the more mechanistic our models (*if* these can be calibrated locally), the more detailed our statements about land performance. Though requiring less detailed data and using less mechanistic descriptors of growth processes, it would make no sense to improve the mechanisms used in ALES because that would defeat its purpose. A good compromise is to use ALES for an initial semiquantitative evaluation, and then concentrate the mechanistic modeling effort where it will have the most leverage, i.e., where the

ALES model has indicated that a deeper analysis is warranted. An outstanding example of this approach is found in Van Lanen *et al.* (1992a).

## V. CONCLUSIONS

Although we have shown the efficacy and usefulness of the land evaluation and crop production models described here, we are not indifferent to the difficulty of convincing the public to follow their recommendations. There is an inherent social difficulty in applying even perfectly accurate models, especially where models show that a certain agricultural use might not be a good use for a particular land. For example, much of Costa Rica is far too steep for row crops, but it is very difficult to tell that to a campesino who must feed his family from that land regardless of consequences. Likewise a nation that loves children may be unwilling to accept the results of models that conclude that the nation may not be able to feed any more people. Thus the potential utility of biophysical analysis is constrained by subjective human factors, even though perhaps in the long run the biophysical results will have an explicit affect on those very people who choose to ignore model recommendations, predictions, and warnings.

## REFERENCES

Alagarswamy, G., and J. T. Ritchie. 1991. Phasic development in CERES-Sorghum model. In T. Hodges (Ed.), *Predicting Crop Phenology,* pp. 143–152. CRC Press, Boca Raton, FL.

Boote, K. J., J. W. Jones, G. Hoogenboom, G. G. Wilkerson, and S. S. S. Jagtap. 1989. *PNUTGRO V1.02: Peanut Crop Growth Simulation Model, User's Guide.* Florida Agricultural Experiment Station Journal no. 8420, University of Florida, Gainesville, FL. 76 pp.

Delsert, E. 1993. *Quelles possibilites pour l'utilisation du logiciel ALES dans le contexte de l'agriculture française?: Application a l'évaluation des potentialités du blé en Lorraine.* Ing. Agr., Institut Supérieur d'Agriculture, Université Catholique de Lille, Lille, France.

DeRoller, C. 1989. Farm scale land evaluation models for land use recommendations in the Guatemalan highlands. M.S. thesis, Cornell University, Ithaca, NY.

FAO. 1983. Guidelines for land evaluation for rainfed agriculture. *FAO Soils Bulletin,* no. 52. FAO, Rome.

FAO. 1984. *Land Evaluation for Forestry.* FAO, Rome.

FAO. 1985. *Guidelines: Land Evaluation for Irrigated Agriculture.* FAO, Rome.

FAO. 1991. *Guidelines: Land Evaluation for Extensive Grazing.* FAO, Rome.

Godwin, D. C., J. T. Ritchie, U. Singh, and L. Hunt. 1989. *A User's Guide to CERES Wheat: V2.10.* International Fertilizer Development Center, Muscle Shoals, AL.

Godwin, D. C., P. K. Thornton, J. W. Jones, U. Singh, S. S. Jagtap, and J. T. Ritchie. 1990. Using IBSNAT's DSSAT in strategy evaluation. In *Proceedings of IBSNAT,* pp. 59–71.

Godwin, D. C., and C. A. Jones. 1991. Nitrogen dynamics in the soil–plant systems. In J. Hanks and J. T. Ritchie (Eds.), *Modeling Soil and Plant Systems,* Chap. 13, pp. 289–321. *ASA Monograph,* Vol. 31. American Society of Agronomy, Madison, WI.

Godwin, D. C., and U. Singh. 1998. Nitrogen balance and crop response to nitrogen in upland and lowland cropping systems. In G. Y. Tsuji, G. Hoogenboom, and P. K. Thornton (Eds.),

*Understanding Options for Agricultural Production,* pp. 55–77. Kluwer Academic, Dordrecht, The Netherlands.

Griffin, T. S., B. S. Johnson, and J. T. Ritchie. 1993. *A Simulation Model for Potato Growth and Development: SUBSTOR-Potato V2.0.* Research Report Series 02. IBSNAT Project. Department of Agronomy and Soil Science, College of Tropical Agriculture and Human Resources, University of Hawaii, Honolulu, HI.

Hodges, T. (Ed.). 1991. *Predicting Crop Phenology.* CRC Press, Boca Raton, FL.

Hoogenboom, G., J. W. White, J. W. Jones, and K. J. Boote. 1991. *BEANGRO V1.01: Dry Bean Crop Growth Simulation Model. User's Guide.* Florida Agricultural Experiment Station Journal no. N-00379, University of Florida, Gainesville, FL. 122 pp.

Hoogenboom, G., J. W. Jones, and K. J. Boote. 1992. Modeling growth, development and yield of grain legumes using SOYGRO, PNUTGRO, and BEANGRO: A review. *Transactions of the ASAE* 35:2043–2056.

Hoogenboom, G., J. W. White, J. W. Jones, and K. J. Boote. 1994. BEANGRO, a process oriented dry bean model with a versatile user interface. *Agronomy Journal* 86:182–190.

Hoosbeek, M. R., and R. B. Bryant. 1992. Towards the quantitative modeling of pedogenesis—A review. *Geoderma* 55:183–210.

Hunt, L. A., S. Pararajasingham, J. W. Jones, G. Hoogenboom, D. T. Imamura, and R. M. Ogoshi. 1993. GENCALC: Software to facilitate the use of crop models for analyzing field experiments. *Agronomy Journal* 85:1090–1094.

Hunt, L. A., J. W. Jones, P. K. Thornton, G. Hoogenboom, D. T. Imamura, G. Y. Tsuji, and U. Singh. 1994. Accessing data, models and application programs. In: G. Y. Tsuji, G. Uehara, and S. Balas (Eds.), *DSSAT version 3, Volume 1.* pp. 21–110. University of Hawaii, Honolulu, HI.

Hutson, J. L., and R. J. Wagenet. 1991. Simulating nitrogen dynamics in soils using a deterministic model. *Soil Use and Management* 7:74–78.

Johnson, A. K. L., and R. A. Cramb. 1991. Development of a simulation based land evaluation system using crop modelling, expert systems and risk analysis. *Soil Use and Management* 7:239–245.

Jones, C. A., and J. R. Kiniry. 1986. *CERES-Maize: A Simulation Model of Maize Growth and Development.* Texas A&M University Press, College Station, TX.

Jones, J. W., K. J. Boote, G. Hoogenboom, S. S. Jagtap, and G. G. Wilkerson. 1989. *SOYGRO V5.42: Soybean Crop Growth Simulation Model. User's Guide.* Florida Agricultural Experiment Station Journal no. 8304, University of Florida, Gainesville, FL. 75 pp.

Jones, J. W., S. S. Jagtap, G. Hoogenboom, and G. Y. Tsuji. 1990. The structure and function of DSSAT. In *Proceedings of IBSNAT Symposium: Decision Support System for Agrotechnology Transfer, 17 October, 1989,* pp. 1–14. University of Hawaii, Honolulu, HI.

Jones, J. W., and J. T. Ritchie. 1990. Crop growth models. In G. J. Hoffman, T. A. Howell, and K. H. Solomon (Eds.), *Management of Farm Irrigation Systems,* Chap. 4, pp. 63–89. American Society of Agricultural Engineers, St. Joseph, MI.

Jones, J. W., K. J. Boote, S. S. Jagtap, and J. W. Mishoe. 1991. Soybean development. In J. Hanks and J. T. Ritchie (Eds.), *Modeling Soil and Plant Systems,* Chap. 5, pp. 71–90. *ASA Monograph,* Vol. 31. American Society of Agronomy, Madison, WI.

Jones, J. W., G. Y. Tsuji, G. Hoogenboom, L. A. Hunt, P. K. Thornton, P. W. Wilkens, D. T. Imamura, W. T. Bowen, and U. Singh. 1998. Decision support system for agrotechnology transfer; DSSAT v3. In G. Y. Tsuji, G. Hoogenboom, and P. K. Thornton (Eds.), *Understanding Options for Agricultural Production,* pp. 157–177. Kluwer Academic, Dordrecht, The Netherlands.

Lal, H., G. Hoogenboom, J-P., Calixte, J. W. Jones, and F. H. Beinoth. 1993. Using crop simulation models and GIS for regional productivity analysis. *Transactions of the ASAE* 36:175–184.

León Pérez, J. C. 1992. Aplicación del sistema automatizado para la evaluación de tierras-ALES, en un sector de la cuenca del río Sinú (Córdoba-Colombia). *Revista CIAF* 13:19–42.

Maji, A. K. 1992. A computerized approach for physical suitability evaluation of lands of Singhik sub-watershed, Sikkim. *Agropedology* 2:37–43.

Matthews, R. B., and L. A. Hunt. 1994. GUMCAS: A model describing the growth of cassava (Manihot esculenta L. Crantz). *Field Crops Research* 36:69–84.

Otter-Nacke, S., J. T. Ritchie, D. Godwin, and U. Singh. 1991. *A User's Guide to CERES Barley—V2.10.* International Fertilizer Development Center, Muscle Shoals, AL.

Prasad, H. K., U. Singh, and R. Goenaga. 1991. A simulation model for aroid growth and development. *Agronomy Abstracts* 7.

Ritchie, J. T., U. Singh, D. C. Godwin, and L. Hunt. 1989. *A User's Guide to CERES Maize—V2.10.* p. 77. International Fertilizer Development Center, Muscle Shoals, AL.

Ritchie, J. T., D. C. Godwin, and U. Singh. 1990. Soil and weather inputs for the IBSNAT crop models. In *Proceedings of IBSNAT Symposium: Decision Support System for Agrotechnology Transfer,* pp. 31–45. University of Hawaii, Honolulu, HI.

Ritchie, J. T. 1991. Wheat phasic development. In J. Hanks and J. T. Ritchie (Eds.), *Modeling Soil and Plant Systems,* Chap. 3, pp. 31–54. *ASA Monograph,* Vol. 31. American Society of Agronomy, Madison, WI.

Ritchie, J. T. 1998. Soil water balance and plant water stress. In G. Y. Tsuji, G. Hoogenboom, and P. K. Thornton (Eds.), *Understanding Options for Agricultural Production,* pp. 41–54. Kluwer Academic, Dordrecht, The Netherlands.

Rossiter, D. G. 1990. ALES: A framework for land evaluation using a microcomputer. *Soil Use and Management* 6:7–20.

Rossiter, D. G., and A. Van Wambeke. 1993. *Automated Land Evaluation System (ALES) Version 4 User's Manual,* July 1993 ed. Cornell University, Department of Soil, Crop and Atmospheric Sciences, Ithaca, NY.

Singh, U., J. T. Ritchie, and P. K. Thornton. 1991. CERES-CEREAL model for wheat, maize, sorghum, barley, and pearl millet. *Agronomy Abstracts* p. 78.

Singh, U., J. T. Ritchie, and D. C. Godwin. 1993. *A User's Guide to CERES Rice—V2.10.* International Fertilizer Development Center, Muscle Shoals, AL.

Thornton, P. K., and G. Hoogenboom. 1994. A computer program to analyze single-season crop model outputs. *Agron. J.* 86(5):860–868.

Tsuji, G. Y. 1998. Network management and information dissemination for agrotechnology transfer. In G. Y. Tsuji, G. Hoogenboom, and P. K. Thornton (Eds.), *Understanding Options for Agricultural Production,* pp. 367–381. Kluwer Academic, Dordrecht, The Netherlands.

Tsuji, G. Y., G. Uehara, and S. Balas. 1994. DSSAT v3 Vol 1, 2, 3. University of Hawaii, Honolulu, HI.

Uehara, G., and G. Y. Tsuji. 1993. The IBSNAT Project. In F. T. Penning de Vries, P. S. Teng, and K. Metselaar (Eds.), *Systems Approaches for Agricultural Development. Proceedings of the International Symposium (Bangkok, Thailand, December 1991),* pp. 505–513. Kluwer Academic, Dordrecht, The Netherlands.

Uehara, G., and G. Y. Tsuji. 1998. Overview of IBSNAT. In G. Y. Tsuji, G. Hoogenboom, and P. K. Thornton (Eds.), *Understanding Options for Agricultural Production,* pp. 1–7. Kluwer Academic, Dordrecht, The Netherlands.

Van Diepen, C. A., H. Van Keulen, J. Wolf, and J. A. A. Berkhout. 1991. Land evaluation: From intuition to quantification. In B. A. Steward (Ed.), *Advances in Soil Science,* pp. 139–204. Springer, New York.

Van Lanen, H. A. J., M. J. D. Hack-ten Broeke, J. Bouma, and W. J. de Groot. 1992a. A mixed qualitative/quantitative physical land evaluation methodology. *Geoderma* 55:37–54.

Van Lanen, H. A. J., C. A. Van Diepen, G. J. Reinds, and G. H. J. De Koning. 1992b. A comparison of qualitative and quantitative physical land evaluations, using an assessment of the potential for sugar beet growth in the European community. *Soil Use and Management* 8:80–89.

Venema, J. H., and F. Daink. 1992. *Papua New Guinea Land Evaluation Systems (PNGLES).* Papua New Guinea Department of Agriculture and Livestock, Port Moresby.

Waterman, D. A. 1985. *A Guide to Expert Systems.* Addison–Wesley, Reading, MA.

CHAPTER 14

# Geographical Synthesis of Data on Costa Rican Pastures and Their Potential for Improvement

MUHAMMAD IBRAHIM
*Research Professor, Area of Agroforestry, CATIE, Turrialba, Costa Rica*

SERGIO ABARCA
*Researcher, Ministry of Agriculture, Costa Rica*

OSCAR FLORES
*Research Assistant, CATIE, Turrialba, Costa Rica*



## I. INTRODUCTION

The intervention of humans in the ecosystems of Costa Rica was minimal before the Spanish conquest. Rather, as a natural component of the ecosystems, pre-Columbian humans extracted (through hunting and fishing) animal resources whose continued reproduction generally was assured. The new animal species introduced by colonization, however, were favored with privileged protection and encouragement by humans, including massive ecosystem restructuring. As a consequence, and because of their high reproduction rates and population growth, the introduced species eventually surpassed in number and replaced many of the native species, causing many additional changes to ecosystems (Crosby, 1986; Lujan, 1989; Janzen, 1991).

In Central America, including Costa Rica, human pressure on the use of land for animal agricultural activities has increased greatly during recent de-

*Quantifying Sustainable Development*

cades, raising many questions pertaining to sustainability. Domesticated animals, particularly cattle, are often regarded as a destabilizing factor in land use and a major contributor to environmental degradation. Their role in reducing the sustainability of land centers on issues of overgrazing, deforestation, soil compaction, and competition with wildlife. During the last century and especially in recent decades these issues have been raised increasingly as expansion to new agricultural frontiers continued, and indeed was encouraged by various monetary incentives.

Since 1950, Costa Rica has been one of the most rapidly deforested countries in Latin America (Sader and Joyce, 1988). Forest clearing can have detrimental effects ranging from loss of biodiversity (and with that possible cures for disease) to disturbance of regional weather patterns and the release of large quantities of $CO_2$ and other greenhouse gases into the atmosphere, potentially affecting global weather patterns (Veldkamp, 1993). Presently, approximately 45% of cleared land in Costa Rica is under permanent pasture (FAO, 1991). Much of this was cleared directly from the forests (Chapters 16 and 17). Now that much of the original forest is gone, and prior land use and lending practices have created enormous debt, we want to determine what land uses can best reverse previous trends and benefit the people of Costa Rica over the long term.

This chapter will focus on the long-term viability of raising livestock as a sustainable income source for the Costa Rican economy. Cattle were chosen for several reasons: there exists a large and expanding international market for beef products, and cattle production in Costa Rica requires relatively few external and labor inputs when compared to other cash crops such as coffee and bananas. The primary resources required for the production of cattle are pastureland and rain, both of which Costa Rica has in abundance (Ibrahim, 1994). Given these conditions most of this chapter is focused on answering two questions related to sustainability: does pasture productivity remain reasonably constant without the extensive use of resource and labor inputs, and can cattle production (beef and milk) create enough external revenue through export and tourist consumption to significantly impact Costa Rica's astronomical debt?

Animal production from pastures is the main land use in Costa Rica in terms of area used. The climate and soils in Costa Rica are such that there would not normally be any areas with natural treeless grassland. The originally prevailing vegetation types ranged from savannas or open woodlands in areas with a pronounced dry season, such as Guanacaste Province, to dense rain forests in areas of high rainfall without a pronounced dry season as found in the Atlantic zone. The savannas can be classified as a type of natural grasslands; over the decades of agricultural development trees have been removed from savannas in Guanacaste for cropping and to increase pasture production. Deforestation of rain forest was carried out in the northeastern parts of the country. As forest was cleared, including areas of low soil fertility, cattle ranching became the most important land use in Costa Rica. In Guanacaste beef production is the most important economic activity, and in the Atlantic zone the second most important after banana production. Dairying is practiced mostly on the plateau of San Carlos and on the slopes of the Poas, Irazu, and Turrialba volcanoes.

Any agricultural development requires a trade-off with preservation of undisturbed ecosystems, improved management of agriculturally productive areas, and rehabilitation of degraded lands. The objective of this chapter is to make an analysis of pasture evolution in Costa Rica, generate an estimate of what the yield potential is and in what ways yield is a function of various environmental and management characteristics, and examine quantitatively the degree to which activities associated with pastures do or do not contribute to the sustainability of the Costa Rican economy and environment.

## A. Historical Context

According to most of the relevant literature, the forestlands colonized by indigenous populations were not used in any important way for livestock production (Saenz, 1955). Apparently, these populations depended only on fishing and hunting wildlife to satisfy their daily protein requirements. Livestock was first introduced into Costa Rica in the mid-16th century (1560) by the Spanish colonizers, who were also responsible for the introduction of some grass species (*Paspalum conjugatum, Pas. notatum,* and *Axonopus compressus*) that are now naturalized in Costa Rica (Morales, 1992). These animals were not used for economic production but as a source of food for the population. For more than three centuries livestock was not an important economic land use activity in Costa Rica.

The development of the livestock sector as an important economic activity began in the second half of the 19th century with the introduction of dairy cows in the high-altitude zones of the Central Valley (Poas region) of Costa Rica. In the first half of the present century, beef breeds were introduced in the tropical lowlands together with more productive and improved grass species. Kikuyu grass (*Pennisetum clandestinum*) was introduced in the higher elevation zones (>700 m) where dairy production is most suited, and jaragua (*Hyparrhenia rufa*) in the seasonally dry lowland areas (Saenz, 1955; Lujan, 1989). Between 1880 and 1920, there were insignificant changes in the livestock industry, until the first livestock exhibition in 1930, which stimulated interest in beef and milk production (Saenz, 1955; Lujan, 1989). During this period the amount of beef produced satisfied only 40% of national demand.

### 1. Deforestation and Pasture Expansion

Between 1950 and 1984, Costa Rican forest cover shrunk by 74%, although deforestation rates appear to have decreased over the last two decades (Table

14-1; see Chapter 16) (Solorzano, 1995). Recent estimates are that annual deforestation fell from between 40,000 and 60,000 ha in the late 1970s and early 1980s to 18,000 ha between 1987 and 1992, and more recently to only 8500 ha (see Chapter 16) (Lutz *et al.*, 1993; Nunez, 1993; World Bank, 1993). A large proportion of this deforested land was transformed into pastures, either directly or after being used for crops (maize, beans, etc.), with most of the remainder used for continuous annual crop production by small farmers (Ledec, 1992; Walker *et al.*, 1993). In general deforested lands are cultivated with crops for the first five years and thereafter used as permanent pastures (Toledo, 1994). The expansion of pastures and beef production in Costa Rica and the rest of Central America was favored by an upsurge of beef prices, which increased by more than 30% between 1964 and 1973 (Trejos, 1992). This upsurge was coincident with a number of financial, credit, and subsidy incentives, but also a series of restrictive conditions. More than one-quarter of this rangeland is in Class I land, designated most suitable for annual crops (Hartshorn *et al.*, 1982). Grazing underutilizes the productive potential of this land, while at the same time compaction and erosion caused by livestock may bring about long-term degradation of its agricultural potential.

A geographical analysis of the evolution of land use systems in Costa Rica has shown that pasture expansion in the last 15 years occurred in different regions and ecosystems compared to the period between 1950 and 1979. During the earlier period, more than 60% of pasture expansion occurred in the Pacific and Central regions of Costa Rica, mainly in Guanacaste and Nicoya. These areas are classified as tropical dry areas with annual rainfalls of 1500 to 2000 mm and more than three dry months (Toledo, 1994). In these ecoregions, fire could be used easily to manage pasture to provide cattle feed of high nutritive value.

During the 1960s and 1970s cattle production expanded in the Alajuela, Guanacaste, and Perez Zeladon provinces. Since this time the expansion has

TABLE 14-1 Costa Rican Pasture Area and Cattle Population between 1949 and 1995

| Year | Pasture area (ha) | % national territory | Cattle population | Stocking rate (head/ha) |
|---|---|---|---|---|
| 1950 | 600,000 | 11.7 | 600,000 | 1.000 |
| 1973 | 1,558,053 | 30.5 | 1,693,912 | 1.087 |
| 1982 | 2,166,900 | 42.4 | 1,959,000 | 0.904 |
| 1984 | 2,229,100 | 43.6 | 2,079,000 | 0.933 |
| 1988 | 2,426,500 | 47.5 | 2,115,600 | 0.872 |
| 1994 | 2,000,000 | 39.1 | 1,593,600 | 0.797 |

*Sources:* For pasture area, Rodriguez and Vargas (1988), Van der Kamp (1990), Godoy (1997), and MAG (unpublished data). For cattle population, FAO (1980), Leonard (1987), Godoy (1997), and MAG (unpublished data).

moved eastward toward the Atlantic plains. Soils in the Atlantic zone range from very good to poor fertility, depending on the age of the volcanic deposition. The higher fertility soils are being used for agricultural production, thus leaving cattle production generally on fragile, infertile soils with too much rain for annual crops (Merelet *et al.*, 1992). The construction of a road in 1986 permitted rapid access from Limon to San Jose and promoted settlement and stimulated livestock production in the Atlantic zone. Rainfall and temperature conditions permit year-round production of forage in these areas, but soils are more susceptible to degradation because of nutrient leaching where rainfall is high.

## 2. Time Trends in Cattle Population and Pasture Area

The total cattle population and the area under pastures increased between 1950 and 1989, but a survey in 1994 showed that the area under pasture and cattle population recently has decreased significantly (Table 14-1), especially in the hillsides of the Pacific regions. The decline in cattle populations has led to a major increase in abandoned lands, which have become brushy, wooded areas and even secondary forest. In Costa Rica the area of secondary forest increased from 229,189 ha in 1984 to 388,341 in 1989, with most of this growth coming from abandoned pastures (TSC/WRI, 1991). These trends in the national statistics have been corroborated by two recent regional studies of land use in other areas of Costa Rica—Guacimo, Rio Jimenez, and Siquirres in Limon, and Arenal and Tempisque in Guanacaste—both of which showed a decline in pasture area, a greater presence of shrub vegetation within pastures, an increase in secondary forest, and an increase in croplands (Fallas and Morera, 1993; Huising, 1993; Kaimowitz, 1996). A high incidence of fire on abandoned pastures has been observed in the dry areas because of a heavy fuel load that accumulated over time. In addition there has been a reduction in the density of animals, especially in those areas that have been in pastures for the longest time.

In the Atlantic zone, recently grasslands have been increasingly used for banana plantations, palm heart cultivation, ornamental crops, and reforestation. Cultivated and heavily managed grasslands are becoming less abundant while there has been an increase in the area of grasslands classified as secondary grasslands (grasslands invaded by herb and scrub vegetation as a result of poor grazing and management). The expansion of palm heart cultivation in former pastures in the Atlantic zone of Costa Rica, even with a recent 50% reduction in the price of palm hearts, shows that this crop is still economically competitive with pastures (Figure 14-1).

A student project examining the use of pastures as a function of gradients of temperature and rainfall from 1950 through the 1984 using MAG (Ministerio de Agricultura y Ganaderia) surveys found that initial high densities of cattle (i.e., number of cows per hectare of pasture) were found in intermediate

FIGURE 14-1 Maps of land use in Neguev.

rainfall regions and, often, cooler temperatures (Kosmerl, 1996). As time progressed the density of cattle in these regions decreased, and pasture densities (hectares of pastures per hectare of province) moved toward wetter and warmer climates. This study is consistent with the idea that the best pastures, located in moderate climatic conditions, originally had high animal densities but that the ability of those pastures to support animals declined over time, forcing the pastures into suboptimal gradient space.

In another study Veldkamp and Fresco (1997) investigated Costa Rican land use and cover (in 1973 and 1984) using nested scale analysis. These authors noted that although much meat goes to the cities and is exported, pasture distribution is not very clearly driven by the urban population, but mostly by the rural population converting natural vegetation (forest) into pastures. The observation that the cattle density (correlation coefficient between pasture area and number of cattle equaled 0.98 at the canton level in 1973 and 1984) in Costa Rica is not related to biophysical factors suggests a rather extensive pasture management. Cattle density is far from maximum and could be more intensive and better optimized biophysically. This is also confirmed by the observation that cattle are also a status symbol and provide security for small holders. This indicates that deforestation for pasture expansion in Costa Rica over the last decades was not driven by land shortage caused by excessive cattle densities.

Meanwhile the area reforested in Costa Rica in forest plantations has also grown 76,465 ha since 1990, from 35,114 to 111,579 ha (MIRENEN, 1994).

Thousands of hectares formerly under pasture are now being used for peach palm production in the Atlantic zone of Costa Rica.

### 3. Trends over Time of Types of Cattle Production Systems

In Costa Rica there are three main cattle production systems: beef production, milk production, and dual-purpose, each with a different intensity of pasture utilization. Data on the distribution of animal numbers between these systems indicate the relative importance of each of the systems over time (Table 14-2). Over the past decades there has been a tendency for cattle farmers to shift from beef to dual-purpose production because of relatively higher prices paid for milk compared to beef.

Data presented by MAG (1989) showed that the beef cattle population decreased by 12% between 1982 and 1988 whereas the dual-purpose cattle population increased by 10.2% in the corresponding period. As profitability and the markets for beef have decreased in the past two decades, so has the level of investment in this activity. Nevertheless, the beef production system still predominates economically with respect to the capital investment in animals.

## B. Beef Production in Costa Rica

Beef production is carried out mostly on the tropical lowlands in the Pacific and Atlantic regions. It is characterized by low technology and is generally carried out on an extensive basis on farms greater than 75 ha in size. The use of inputs is low, labor use is scarce, and in general very little management is applied. The pastures are grazed continuously, and the stocking rate is on average 1 animal unit (AU)/ha (Jansen *et al.*, 1997), similar to that found in other Central American countries. The main source of feed for animals is pasture with only mineral supplementation. Forage legumes are not grown in

TABLE 14-2 Distribution of Female Cattle (Thousands) in the Main Production Systems in Costa Rica

| | Beef | | Milk | | Dual purpose | | |
|---|---|---|---|---|---|---|---|
| Year | No. | % | No. | % | No. | % | Total females |
| 1982 | 1052.3 | 68.7 | 255.2 | 16.8 | 222.3 | 14.5 | 1529.8 |
| 1988 | 849.9 | 56.7 | 277.0 | 18.6 | 369.3 | 24.7 | 1492.2 |

*Source:* MAG (1989).

the region, in spite of research showing that productivity can be improved with such high-nitrogen species.

## C. Milk Production Systems in Costa Rica

The environmental characteristics of the region are a major factor determining the ecology and characteristics of milk production systems in Costa Rica. In dairy production, most operations are small (less than 20 ha) and only 6.7% exceed 100 ha. Specialized milk production in Costa Rica is done in essentially two ecoregions: high altitudes and humid sites in the lowlands (for example, San Carlos and Rio Frio), and in the highlands around the volcanic central mountains of the country, generally over 1130 m above sea level, including the Poas, Coronado, and Santa Cruz (Turrialba) regions. In the highlands milk production is characterized by highly tecnified operations using breeds such as Holstein and Jersey that can produce between 3500 and 6000 kg per lactation. The animals are fed from fertilized improved pastures (mainly *Pe. clandestinum* and *Cynodon nlemfuensis*) supplemented with high amounts of concentrates, often imported. Stocking rates in these intensive systems are higher than those in other production systems (3 to 5 AU/ha) (Van der Grinten *et al.*, 1992). The producers are specialized in milk production and obtain between 90 and 95% of their income from this activity.

Milk production in the lowlands is carried out between 130 and 900 m above sea level. Rio Frio and San Carlos are the main milk-producing areas in the tropical lowlands. The technological level is less than that used in the highlands. The animals used are mostly genetically improved and have a production between 1900 and 2500 kg per lactation. Improved pastures are used with a lesser degree of supplementation (Holmann *et al.*, 1992).

The dual-purpose system is characterized by a level of technology midway between the dairy producers and the extensive and semiextensive beef producers. Production takes place in the lowlands below 900 m. Animals used are crossbreeds of tropical Zebu and specialized dairy breeds. Average milk production is low, between 420 and 1200 kg per lactation. Improved pastures are used as well as native pastures, but there is no supplementation. Although some income is received from milk, between 45 and 55% of income is derived from the sale of animals.

Milk production in Costa Rica has been increasing steadily during the last 23 years at an average annual rate of 4.5% (from 206 million kg in 1970 to 523 million in 1993). The high growth has allowed an annual increase of about 1.1% in the per capita milk consumption (from 120 kg in 1970 to 153 kg in 1993). In addition Costa Rica evolved from being a milk-deficient country (importing an average of 25 million kg in the 1970s, representing 10% of total production) to a net surplus country (net exports of 16 million kg in 1993, representing 3% of total milk production). Costa Rica has an estimated 19,422 dual-purpose farms and 15,047 specialized dairy farms (DGEC, 1987). Approximately 2500 farms are responsible for 50% of milk production in the country (French, 1994).

Like everything else, making a living for the dairy farmer is not easy. The cost of production per kilogram milk in the Atlantic region during 1993 was $0.25/kg, while the milk price at the farm gate was $0.29/kg, so that the net profit was 16%. Labor cost increased in real terms from U.S. $0.65/hr to $1.25/hr from 1970 to 1993 (Holmann *et al.*, 1992). Perhaps the real vulnerability is if fertilizers, animal feeds, or other technical inputs increase rapidly in price again as they did in the 1970s.

Nevertheless livestock production often remains part of the farm activity irrespective of the profitability or size of the farm. On small farms it is less a commercial activity and of less importance for income generation. The herd produces milk and calves for home consumption and for the market. The calves are sold when the need for money arises. Many Costa Ricans with a little land like to keep a few cows because they generate income without very much effort so that their owners can work at another job or have more leisure.

## II. PASTURES

### A. Pasture Characteristics

Forage is the cheapest feed source for almost all ruminant production in Costa Rica and elsewhere in Central America, and, in general, concentrates are used only to supplement high producing cows to sustain high milk yields in specialized dairy systems. In Costa Rica, cultivated pastures are found mainly in three ecoregions: (1) seasonally dry flat regions (e.g., Canas, Liberia), hillsides (e.g., Atenas, Esparza), and lowlands (e.g., the Puntarenas and Guanacaste regions); (2) Humid flat and hillside lowlands (e.g., San Carlos, Atlantic zone); and (3) highlands with volcanic soils (e.g., Poas, Curridabat, Santa Cruz). The main pasture species found in different regions in Costa Rica are shown in Table 14-3. In general pastures in Costa Rica are dominated by native and/or naturalized pasture grasses and to some extent by improved grasses such as *Brachiaria decumbens, Br. brizantha, Cy. nlemfuensis,* and *Pe. clandestinum.*

In the seasonally dry Pacific areas, pastures are managed extensively, with the dominant grass species being *H. rufa,* which is characterized by low productivity and quality (Sharma, 1992; Franco, 1997). Trees and shrubs are found dispersed in pastures, especially leguminous species that are well adapted to dry conditions. Examples of tree species that are most commonly found in pastures in the Pacific regions are *Enterolobium cyclocarpum* (Guanacaste), *Gliricidia sepium* (madero negro), *Pithecellobium saman* (genizaro), and *Gua-*

TABLE 14-3 Area of Pasture and Percentage Coverage of Principal Grass Species by Region in Costa Rica

| | Region[a] | | | | |
|---|---|---|---|---|---|
| | CH | CT | BR | HN | HA |
| Pasture area (ha) | 783,815 | 557,927 | 334,893 | 420,175 | 327,817 |
| Species | | | | | |
| *Hyparrenhia rufa* | 62.2 | 40.3 | 17.9 | 0.4 | 2.9 |
| *Cynodon nlemfuensis* | 11.6 | 7.9 | 12.5 | 23.5 | 13.6 |
| *Ischaemum ciliare* | 2.5 | 2.1 | 3.5 | 37.1 | 27.0 |
| *Brachiaria* spp. | 5.3 | 3.6 | 9.1 | 11.7 | 6.5 |

*Source:* SEPSA (1989).
[a] CH, Chorotega; CT, Central; BR, Brunca; HN, Nort Huetar; HA, Atlantic Huetar.

*zuma ulmifolia* (Guacimo). Trees provide shade to animals and feed (forage, litter, flower, pods) of a high nutritive value, especially during the dry season when there is a shortage of forage (CATIE/JICA, 1995).

In the Pacific regions rainfall is concentrated in only 6 months of the year (May–November) and this does not favor year-round production of forage. Annual yields are 7 to 10 tons dry matter (DM)/ha, of which more than 70% is produced in the 6-month rainy period. Severe feed shortage in the dry season results in heavy weight losses and, in extreme conditions, cow mortality (Sharma, 1992; CATIE/JICA, 1995). On farms which use improved technologies, excess forage produced in the wet season is conserved as hay to feed to animals in the dry season. Farmers on small and medium-sized farms usually sell a high percentage of their stock at the end of the dry season to reduce the stocking rates. In contrast, rainfall and temperature conditions in the humid tropics (e.g., Atlantic zone) favor high dry matter yields all year-round, provided that soil fertility is maintained. Under these conditions, improved grasses can produce more than 25 tons DM/ha/yr.

Nevertheless the largest percentage of Costa Rican pastures are characterized by native and naturalized grasses that are capable of producing 10 to 12 tons DM/ha/year (Table 14-3). Stocking rates on pastures vary mostly between 0.8 to 1.65 animals/ha (Table 14-4), with higher stocking rates observed in higher elevation milk production systems (Huising, 1993). The San Carlos region is a major milk-producing area and a high percentage of pastures are under improved grasses, including *Cy. nlemfuensis* and *Brachiaria* spp. (Morales, 1992).

It is not just the quantity of the forage that is important, however, but also its quality. A synthesis of data from naturalized and improved pastures grown under different temperatures and altitudes (i.e., in different locations in gradi-

**TABLE 14-4 Farm Types and Average Farm Size of Farms Dedicated to Livestock Production in the Atlantic Zone of Costa Rica**

| Production systems | Average farm size (ha) | Stocking rate (AU/ha) |
|---|---|---|
| Cow and calf | 256 | 1.13 |
| Fattening | 200 | 1.11 |
| Dual production | 47 | 0.85 |
| Milk production | 22 | 1.63 |

*Source:* Adapted from Huising (1993).

ent space) showed that the highest DM yields were obtained in the low altitudes, which are characterized by high rainfall and temperatures (Figure 14-2). In contrast the crude protein and digestibility of grasses tend to be lower in ecoregions with high temperatures and at low altitudes (Figure 14-3).

In the Atlantic zone of Costa Rica, with a mean annual rainfall of 4333 mm and temperatures of about 26°C, Huising (1993) defined two classes of grasslands with the aid of aerial photographs. One class corresponds to cultivated pasture with a homogeneous grass cover, characterized by little variation in height of the grass vegetation and only a small presence of shrubs and trees. The pasture types are ratana (*Ischaemum ciliare*), mixed pastures of ratana and natural grass (*Ax. compressus*), estrella (*Cy. nlemfuensis*) pastures, or pastures with *Brachiaria* species. In the humid tropics many farmers have attempted the cultivation of ratana to increase forage production, but this species has been found to be of low productivity and quality (Morales, 1992). Between 1982 and 1988, the area under ratana grass in Costa Rica expanded from 93,710 to 288,044 ha (SEPSA-CNP, 1990), but today it is not always looked upon so favorably and it is one of the most difficult species to control. According to Arosemena (1990) the dominance of ratana is associated with allelopathic effects on the surrounding vegetation.

The other class of grasslands in the Atlantic zone is considered noncultivated or neglected grasslands. It is characterized by mostly strong variation in the height of the grass vegetation, and the grasslands are often heavily invaded by shrubs. This class also includes spontaneous vegetation found in and around lagoons or in poorly drained depressions, or can be the consequence of negligence and poor maintenance of pastures. Species encountered are aleman (*Echinochloa polystacha*), guinea grass (*Panicum maximum*), *Pas. virgatum*, and *Andropogon bicornis*, among others.

Trees are also common in pastures in the humid tropics of Costa Rica, in particular, timber species (*Cordia alliodora* and *Cedrela odorata*) and fruit trees such as citrus (Leeuwen and Hofstede, 1995). A study conducted by Huising (1993) in the Atlantic zone found that smaller farm pastures were more likely

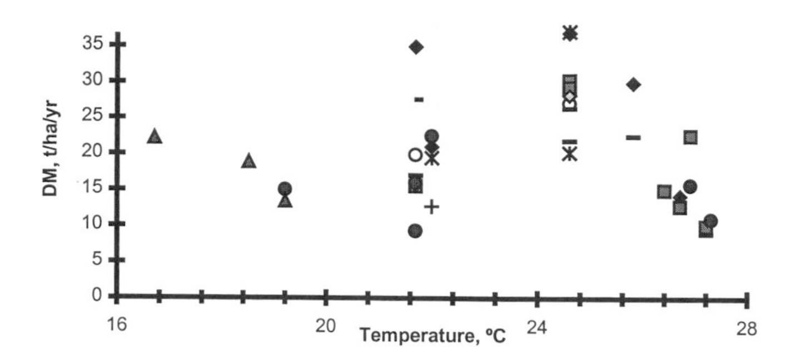


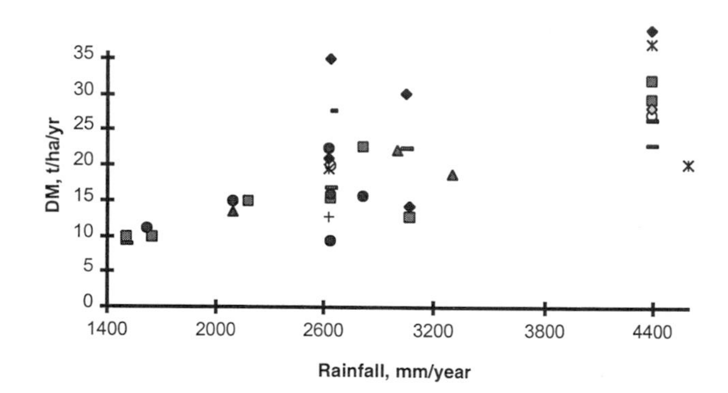


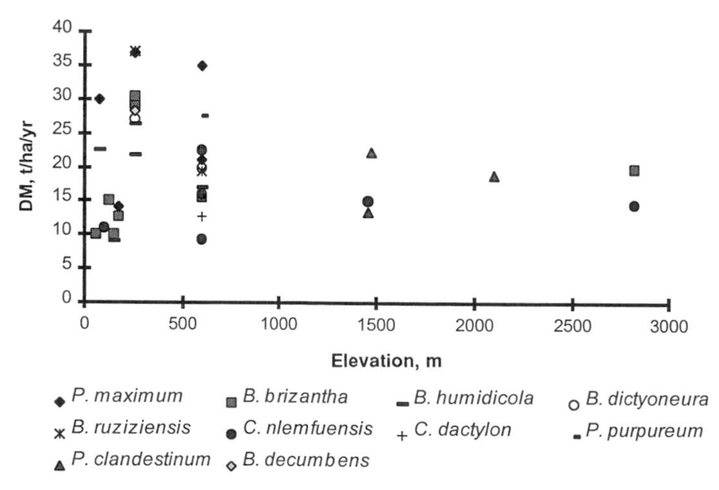


FIGURE 14-2 Variation in dry matter production (tons/ha/yr) of tropical grasses grown under different gradient conditions of temperature, rainfall, and elevation in Costa Rica (Costa Rican data).

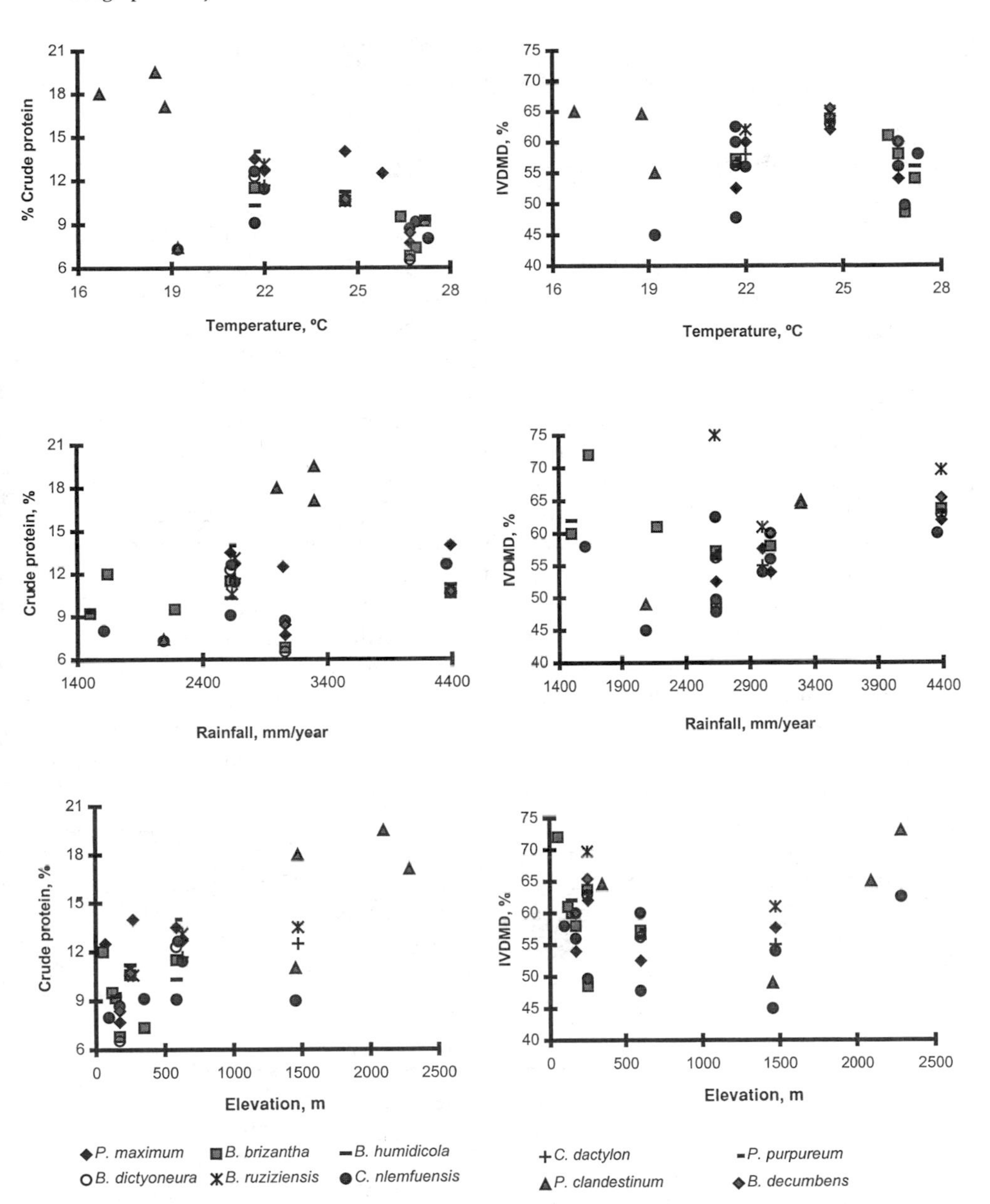


FIGURE 14-3 Variation in crude protein and *in vitro* dry matter digestibility (IVDMD) of tropical grasses grown under different gradient conditions of temperature, rainfall, and elevation in Costa Rica (Costa Rican data).

to be associated with fruit trees, while timber species tended to be more common with increasing farm size (Table 14-5).

## B. Pasture Degradation in Costa Rica

In Costa Rica and the rest of Latin America, cattle production has been one of the most criticized activities associated with deforestation and deterioration of natural resources. The cause for concern has been the observed high levels of soil erosion, deterioration of watersheds, and degradation of water sources. More than 70% of established pastures are in an advanced stage of degradation (Ibrahim, 1994).

Soil compaction is an especially important component of pasture degradation because it can reduce the productive lifetime of a pasture and decrease the potential of that land for other uses. Compaction of soil occurs in areas with high rainfall, and on soils with high peat, clay, or silt content (Davies *et al.*, 1989). Soil compaction can have the effect of raising the bulk density of the soil and decreasing the amount of macropores, which decreases the amount of soil aeration and inhibits satisfactory root development and desirable microbial activity (Brady, 1990). Plant development is impacted negatively by restricting root development. In one study conducted near Siquirres, total soil porosity in the topsoil decreased from an initial 50–55% in the forest to 20% after several years as pasture (Wielemaker and Lansu, 1991). In this study, however, pasture did maintain a higher porosity than did maize cropping (15%). The compaction was generally attributed to trampling by cows on water-saturated topsoils.

From another perspective the main causes for degradation are the lack of grass species adapted to heavy traffic and a lack of good grazing management. These problems lead to soil compaction and erosion in fragile areas. In the early 1970s a number of exotic grass species (*Pan. maximum, H. rufa, Cy.*

TABLE 14-5 Presence of Fruit Trees or Timber in Relation to Farm Size

| | Number of farms with | | |
|---|---|---|---|
| Farm size class (ha) | Pasture | Fruit trees | Timber |
| 1. 0–29 | 51 | 33 (65%) | 30 (59%) |
| 2. 10–60 | 51 | 33 (65%) | 41 (80%) |
| 3. 30–260 | 44 | 21 (48%) | 41 (87%) |
| 4. 85–1000 | 24 | 5 (21%) | 19 (79%) |
| 5. >400 | 7 | 0 (0%) | 7 (100%) |

*Source:* Huising (1993).

*nlemfuensis*, etc.) were imported from Africa. These species grow very well, but are extremely nutrient-demanding. Nutrient dynamic studies with these pastures showed degradation after five years of establishment. The degree of degradation was highly correlated with depletion of soil phosphorus and organic matter (Wielemaker and Lansu, 1991). The problems seemed to be exacerbated in the humid regions where rain is more than 2500 mm/yr. In these locations leaching of nutrients (N, K, etc.) from pastures can be very high if they are not properly managed.

In Costa Rica, a high cattle stocking density often is found on the hillsides of the Pacific region (e.g., Esparza, Hojancha, Atenas) and the problem of pasture degradation is more evident in these ecological zones. Overgrazing on these sites results in bare patches, and this, together with soil compaction, is associated with heavy soil erosion. Flores (1994) estimates soil losses of up to 50 tons/ha/yr in sloping areas. This would be roughly 250 tons of soil lost for every ton of beef produced! Seventeen percent of the land area of Costa Rica is considered severely eroded or degraded, more than 80% of this area is under pastures. Overgrazing also has led to the invasion and dominance by less palatable native species of lower nutritive value. The largest depreciation of land values occurs where annual crops and pastures are the main land use systems (Flores, 1994). Flores estimated that the loss due to overgrazing on pastures was 865 colónes/ha in 1989.

The problem of pasture deterioration is reflected in the decreasing stocking rate observed on Costa Rican pastures over time. Between 1973 and 1994, the mean stocking rate decreased from 1.087 to 0.797 animals/ha (Table 14-1), which implies a great pressure for additional land and natural resources in order to meet demand. If the stocking rate cannot be increased, the only way to increase production is to dedicate more land to pasture or to raise the productivity per animal, or both.

## C. Potential for Pasture Improvement in Costa Rica

Over the past years several international (CATIE and CIAT) and local (e.g., MAG, University of Costa Rica, and Technological Institute of San Carlos) institutions have been conducting research to develop low-input technologies for improving animal production in Costa Rica and in Latin America. The main objective is to increase animal production in a profitable manner without suffering unfavorable social and ecological consequences. Research was conducted in several sites that are representative of pasture production in the country. Examples include (1) high-rainfall areas with infertile soils (San Isidro); (2) high-rainfall areas with fertile soils (Guapiles); and (3) dry Pacific regions (Atenas).

Research activities focused on (1) selection of forage germ plasms that are well adapted to local edaphic and climatological conditions, (2) development of grass–legume mixtures, and (3) development of silvopastoral systems. In a sense the research is based on a gradients approach where the objective is to find the optimum sustainable system for each location within Costa Rican gradient space.

Improved high-yielding grasses have been selected for different ecological zones in Costa Rica. The grasses *Br. brizantha* and *Pan. maximum* CIAT 16061 are well adapted to fertile, well-drained soils, whereas *Br. humidicola* and *Br. dictyoneura* do well on acidic, infertile soils in the humid tropics. These species have the potential to produce more than 20 tons DM/ha/yr of high quality (Figures 14-2 and 14-3) and are capable of supporting higher stocking rates than existing species (Ibrahim, 1994). *Brachiaria dictyoneura* is known to maintain a good soil cover and is recommended for hillside areas with grazing to reduce soil erosion. Herbaceous legumes identified for pasture improvement were *Stylosanthes guianensis*, *Arachis pintoi*, and *Centrosema macrocarpum* (Ibrahim, 1994).

Additionally research efforts were made to select fodder trees and shrubs since these species generally are more adapted to dry conditions, and some are of higher nutritive value than most natural or native grasses dominating pastures in Costa Rica. Agronomic studies with *Gl. sepium* and *Erythrina berterona* showed that they are capable of producing more than 12 tons edible DM/ha/yr, with a crude protein content of more than 20% (CATIE, 1989, 1990; Pezo *et al.*, 1990). These species are commonly found on most cattle farms in the humid tropics of Costa Rica. *Gliricidia* is more adapted to dry conditions, and it is also found on pastures in the Guanacaste region. Other species with high fodder potential are *Morus alba* and *Cratylia argentea*.

In general Costa Rican pastures have substantially lower animal densities, per animal production, and reproduction rates than those in, for example, the United States (Hall, personal communication). This phenomenon may be caused by one or more factors, including the energy cost to the cows for moving on pastures on very steep slopes (which is generally the case except for the lowlands) and the low productivity and quality of the natural grasses. Good management can greatly increase the ability of a pasture to support additional production. Holmann *et al.* (1992) evaluated different scenarios to study dairy farms. These authors found that with improved pastures and application of 300 kg/ha of N the stocking rate could be increased from about 1 to 4 AU/ha. Therefore it is possible to increase herd size on farms under intensive grazing management as long as the fertilizer and grass management can be paid for.

### 1. Legume-Based Pastures

Legume-based pastures are important for sustainable development of pastures in Costa Rica. Legumes are of high nutritive value and they contribute to improvements in soil fertility and biology. Most of the studies on legume mixtures have been conducted in the humid tropics of Costa Rica. *Arachis*-based pastures represent the most promising forage technology for improving animal production in the humid ecosystems. Studies conducted by Hernandez *et al.* (1995) in the Atlantic zone of Costa Rica showed that live weight gains per hectare were fourfold higher with *Arachis* mixtures than with native pastures (Table 14-6) (Jansen *et al.*, 1997). The legume mixture sustained 3 AU/ha, which is threefold that reported for native pastures in the region (Jansen *et al.*, 1997). This means that cattle production can be practiced on reduced area, thereby liberating land for reforestation programs or agricultural activities.

The legumes also contribute to the improvement of soil fertility and biological activity (Torres, 1995). Ibrahim noted that *Arachis* fixed more than 100 kg/ha/yr of nitrogen in the humid tropics of Costa Rica, which is of significant importance for maintaining high pasture yields. Under similar conditions Torres (1995) found that legume-based pastures had higher earthworm populations than the grass monoculture (371 vs 195 $m^2$). These results demonstrate the potential of improved pastures for sustainable livestock development in Costa Rica.

## 2. Silvopastoral Systems

Most farming systems in Costa Rica include some agroforestry practices that contribute partially to their economic and biotic sustainability. Living fence posts (made from living trees) are found on more than 90% of cattle farms in Costa Rica, unlike most developed countries in which dead fence material is used. Living fence posts and other agrosilvopastoral systems provide forage,

TABLE 14-6 Mean Annual Live Weight Gain per Animal and per Hectare on *Brachiaria brizantha* Alone or with *Arachis pintoi* at Two Stocking Rates

| | Live weight gain (kg) | | | |
|---|---|---|---|---|
| | Per animal | | Per hectare | |
| | 1.5[a] | 3.0 | 1.5 | 3.0 |
| Pasture type | | | | |
| *B. brizantha* | 159 | 119 | 478 | 716 |
| *B. brizantha* + *A. pintoi* | 178 | 154 | 534 | 937 |
| LSD (*P* = 0.05) | NS[b] | 27.4 | NS | 145 |

*Source:* Hernandez *et al.* (1995).
[a] Stocking rate in units of AU/ha.
[b] NS, not significant.

shade for animals, and other benefits, including fuel wood, timber, or fruit production. Nitrogen-fixing trees improve soil fertility and more effectively recycle nutrients (Romero *et al.*, 1994). Use of forage species in cropping systems can provide additional sources of nutrition to the animals in those systems.

In general, cattle farmers in Costa Rica and elsewhere in Latin America inherited methods of tree clearing and management of pastures from early Spanish colonists. Most valuable timber and shade trees were cleared to increase forage productivity in pastures. However, today farmers are more conscious of the role trees play in providing shade for animals and more so of the value of precious timber trees (Viera and Barrios, 1997). Case studies in the Esparza and Hojancha regions showed that the density of valuable timber trees has increased in pastures over time, which was attributed to better management to promote regeneration of these species. A high percentage of timber extracted in Costa Rica comes from pastures (Table 14-7), and in the near future it is foreseen that pastures will become one of the main sources for timber production since the area under primary forest is eventually exhausted. Species commonly found in pastures that are of high value include *Ced. odorata, Co. allidora, Bombacopsis quinata,* and *Pi. saman.*

With increased timber production on cattle farms, conservationists will see livestock production in harmony with management of natural resources and not as a destabilizing factor in land use.

Current research in CATIE and the national agricultural research institutions is providing a better understanding of some of the more traditional agroforestry practices and their improvements. This is leading to the development of new alternatives involving woody perennials for more productive and sustainable animal agricultural systems. Moreover, increasing timber prices provide important economic incentives to adopt silvopastoral systems (Ibrahim, 1998; Jansen *et al.*, 1997). In addition to the forage, fuel wood, and timber qualities, other characteristics of the tree and shrub species have to be considered in developing silvopastoral systems.

**TABLE 14-7** Volume of Timber Extracted in Different Livestock Farms in Esparza during 1995

| Farm type | Volume extracted ($m^3$) | Area extracted (ha) | $m^3\ ha^{-1}$ |
|---|---|---|---|
| Large (>100 ha) | 1031.51 | 8057.00 | 0.13 |
| Medium (40–100 ha) | 956.10 | 1748.00 | 0.55 |
| Small (<40 ha) | 1027.39 | 762.41 | 1.35 |

*Source:* MINAE (unpublished); Viera and Barrios (1997).

## D. The Role of Improved Pastures in Restoring the Environment

Over several decades there has been increasing interest in studying the role of tropical pastures in emitting and sequestering $CO_2$. Unfortunately there are only a few studies that were undertaken in the humid tropics of Costa Rica. In the Atlantic zone of Costa Rica large areas of tropical lowland forest have been cleared in the last 40 years, releasing large quantities of carbon to the atmosphere from the tree biomass destroyed (Veldkamp *et al.*, 1992). In 1986 more than 50% of this cleared area was covered with pastures (Huising, 1993). The main species are *Ax. compressus, Pas. notatum,* and *I. ciliare* (Ibrahim, 1994). In these pastures, a study on the dynamics of the soil organic carbon using the $^{13}C$ method showed that deforestation, followed by 25 years of low productive pasture, caused an additional net loss of soil organic carbon between 1.5 and 21.8 Mg $ha^{-1}$, depending on soil type (Veldkamp, 1994). This is equivalent to a net soil organic carbon loss of 2 to 18%, and is roughly 10% of the losses directly from biomass (Chapter 17).

While degraded and unproductive native pastures have been important sources for $CO_2$ emissions, studies with improved grasses in the humid tropics of Costa Rica showed that conversely they can be important for carbon sequestration under favorable management. Most of these grasses have the $C_4$ Krantz pathway and under favorable climatic (i.e., temperature and rainfall) and soil conditions they are capable of producing up to 30 tons DM $ha^{-1}$ $yr^{-1}$ (Ibrahim, 1994), unlike unimproved grasses that yield 10 to 12 tons DM $ha^{-1}$ $yr^{-1}$ (Veldkamp, 1994). Some large part of this higher production is not being eaten but instead enters the soil where it acts as a carbon sink. Further studies are needed to assess the quantitative importance of this pathway.

Introducing improved pastures of *Brachiaria* to replace native pastures can reduce the net $CO_2$ emission by about 60% depending on soil types (Figure 14-4). Veldkamp noted that gross emissions from *Brachiaria* pastures were higher on Andisols than on Inceptisols, but the highest net emission was found for the unproductive *Axonopus* pasture on Andisols. In other studies, grass–legume and silvopastoral mixtures grown on volcanic soils in the Atlantic zone generated soil organic carbon levels similar to those measured under primary forest found in similar environments (Veldkamp, 1993; Ibrahim, 1994).

Pastures may also play an important role in emission of nitrous oxide ($N_2O$, another greenhouse gas; see Chapter 17). Working in the Atlantic lowlands of Costa Rica, Keller *et al.* (1993) found that $N_2O$ emissions first increased greatly, and then declined with time following the conversion of forest to pasture. Keller and Reiners (1994) discussed further the patterns of trace gas fluxes following deforestation, pasture use, and secondary succession of pasture to forest. They found that following forest clearance, soils of recently formed

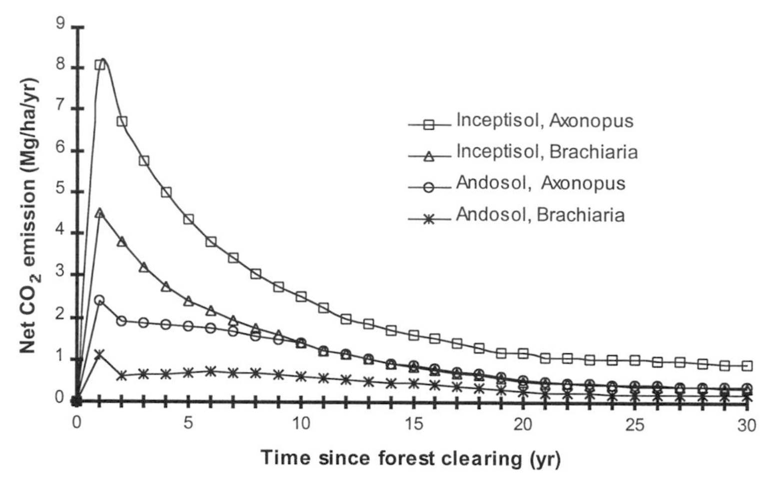


FIGURE 14-4 Simulated net $CO_2$ emissions for two soil types under a low productive (*Axonopus*) and high productive (*Brachiaria*) pasture. (Source: Veldkamp, 1993.)

(i.e., young) pastures have nitrous oxide ($N_2O$) and nitric oxide (NO) emissions equal to or greater than emissions from undisturbed forest soils. In pastures older than 10 years, however, $N_2O$ and NO soil emissions drop below levels observed from forest soils. These dynamic changes in $N_2O$ and NO emissions are caused by a combination of rapid decomposition of soil organic matter (SOM) following deforestation (Veldkamp, 1994) and high soil water content in pastures related to compaction of the topsoil (Keller *et al.*, 1993).

Because decomposition of SOM originally derived from the forest that once grew there provides the main supply of nitrogen in these pastures, productivity cannot be sustained in older pastures. Declines in pasture productivity lead some farmers to abandon old, unproductive pastures and other farmers to introduce experimental nitrogen management to sustain production. Native soil fertility plays a critical role in a farmer's decision. Farmers tend to abandon pastures on soils with low natural fertility, while on soils with the highest soil fertility pastures are replaced with more intensive agricultural uses (e.g., banana plantations).

In Costa Rica, any nitrogen management is rare in lowland pastures, but nitrogen fertilization is a common practice at higher altitudes on pastures used for dairy cattle. Nitrogen management affects soil—atmosphere nitrogen oxide fluxes depending on the source. Studies conducted on pastures in the humid tropics of Costa Rica showed that $N_2O$ flux from the fertilized pastures was significantly higher than that from legume and traditional pastures (Table 14-8). Although the mean annual $N_2O$ flux from legume pastures was almost

TABLE 14-8 Annual Mean and Standard Deviation of Soil Atmospheric $N_2O$ and NO Fluxes by Pastures Type

| Pasture type | $N_2O$ (ng N $cm^{-2}$ $h^{-1}$)[a] | NO (ng N $cm^{-2}$ $h^{-1}$) | WFPS (%) |
|---|---|---|---|
| Traditional pasture | 2.67 (2.88) | 0.94 (0.59) | 80.4 (5.4) |
| Legume pasture | 4.88 (4.12) | 1.29 (0.74) | 80.8 (7.6) |
| Fertilized pasture | 25.82 (10.60) | 5.30 (1.19) | 72.9 (7.2) |

*Source:* Veldkamp *et al.* (1996); values in parentheses are standard deviations.
[a] 1 ng $cm^{-2}$ $h^{-1}$ = 0.1 g $ha^{-1}$ $h^{-1}$.

twice as high as the flux from traditional pastures, this difference was not significant. On an annual basis, $N_2O$ nitrogen loss from the fertilized pastures was 22.6 kg $ha^{-1}$, which corresponds to 6.8% of the applied 300 kg $ha^{-1}$ of nitrogen. The annual NO nitrogen losses from fertilized pastures were 4.64 kg $ha^{-1}$, which corresponds to 1.3% of applied fertilizer.

The *Brachiaria* species, with their high productivity, tend to deplete soil mineral nitrogen, which may lower the nitrogen available to nitrifying and denitrifying microorganisms. The use of legumes in pasture may be an attractive alternative to lower nitrogen losses. However, successful implementation of grass–legume combinations also requires information on intensive management, for which the farmers presently lack the resources.

## III. ECONOMIC ISSUES

The exportation of beef has been an important source of hard currency for Costa Rica. Meat production increased from 1961 to 1980, but has since stabilized or perhaps declined some. Exportation also increased in the period 1961–1980 but has since decreased, lowering the importance of beef for generating foreign currency (French, 1994; Leonard, 1987). However, the growing number of tourists visiting Costa Rica has created an increasing demand for high-quality beef and increased internal beef consumption.

Meat, particularly beef, is an important component of the Costa Rican diet, for 70% of the meat consumed is beef. Costa Rican consumption of beef is about 26.6 kg per capita per year (Jarvis, 1986). Animal products (dairy and meat) provide 17% of the calories and 37% of the protein in the Costa Rican diet (Levitan, 1988). It would be difficult to find alternative domestic sources for the protein provided by these animal products since domestic production of beans is constrained by the lack of suitable land for their production. Costa Rica is perhaps the only Central American country that exports milk (Jarvis,

1986). As noted previously this milk production is partly dependent upon imported cattle feeds and fertilizers for pasture management.

Extensive beef production as a generator of foreign exchange is an inefficient use of land. Since the production of beef, with one animal per hectare, tends to be very low (roughly 500 kg wet weight or 100 kg dry weight per hectare per year), production per hectare for beef is much less than that from any other crop. The advantages of beef production, however, are that it takes very little human effort and almost no imported inputs, can take place on land too dry or wet for crops, and tends to generate less erosion than most crops. Another advantage of cattle is that domestic animals can constitute a store of value for individuals without access to banks or insurance (Waaijenberg, 1990). They are efficient and sustainable in the sense that animals convert crop residues not otherwise economically useful into food or income, while generating manure for fertilizer or fuel. They are not sustainable in the sense that the productivity of pastures tends to degrade over time unless supplemented with fertilizer.

With low prices of beef on the international market cattle farmers have been forced to adapt improved technologies and diversify production on their farms to generate additional income, especially small and medium-sized farmers. Studies conducted by Jansen *et al.* (1997) showed that improved legume-based pastures used for beef production are capable of producing internal rates of return of 117% (calculated for 5 years). Additionally the integration of valuable timber trees in pastures would make cattle production more profitable considering that timber prices have been increasing linearly in time (Figure 14-5).

## IV. CONCLUSION

In general we conclude that the present management of livestock using extensive, unimproved pasture systems is not a very efficient or sustainable use of the land because the productivity of the land degrades as the original nutrient inventory is used up. If we can intensify the production of the system with improved high-yielding grasses and silvopastoral technologies, each chosen for a particular region, we can meet the protein requirements of a growing Costa Rican population while allowing reforestation and conservation of natural resources. However, the adoption of these technologies by farmers will depend on prices of animal products and incentives paid to farmers for environmental services.

## ACKNOWLEDGMENTS

We thank Scott Bahr and Mark Kosmerl for contributions.

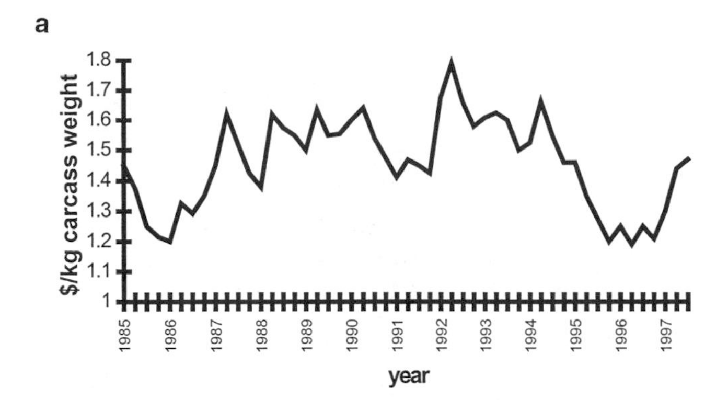


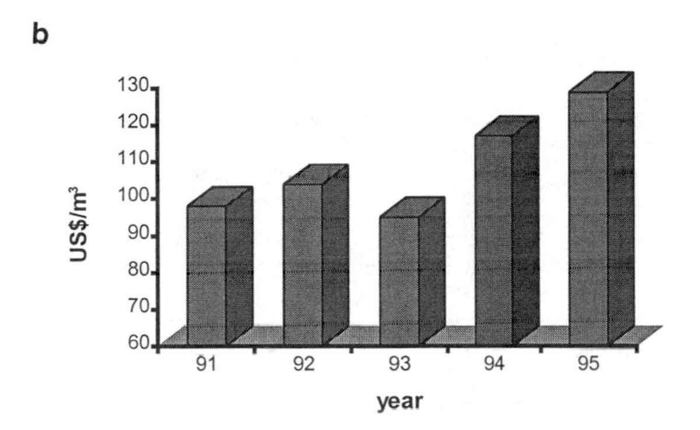


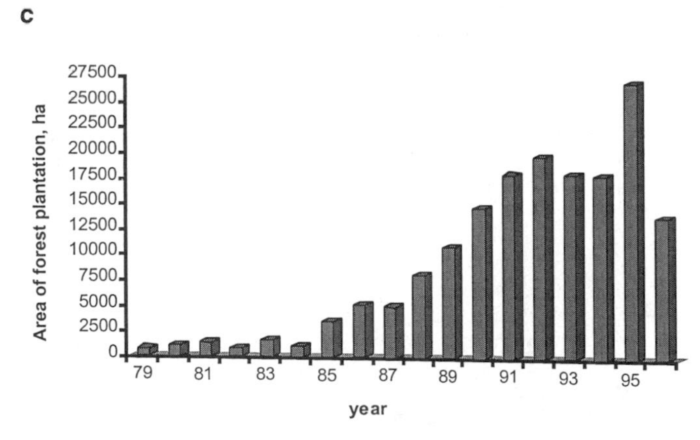


FIGURE 14-5 Variation in time in the price of (a) meat (Costa Rica), (b) fine quality timber (London market), and (c) area reforested in Costa Rica with incentives. [Source: for meat, Coopemontecillos (unpublished); for timber, Aguerre (1994); and for reforestation, Godoy (1997).]

## REFERENCES

Aguerre, M. 1996. La era de la madera barata toca a su fin. Una oportunidad para Argentina. ASORA ano 2(5): pp. 26–35.

Arosemena, E. 1990. Determinacion de mecanismos de interferencia por alelopatia y requerimientos externos e internos de fosforo en pasto ratana (Ischaemum indicum). Tesis Mag. Sci., CATIE, Turrialba, Costa Rica.

Brady, N. C. 1990. *The Nature and Properties of Soils*, pp. 109–110. Macmillan, New York.

CATIE (Centro Agronomico Tropical de Investigacion y Enseñanza). 1989. *Sistema regional de areas silvestres protegidas de America Central: Plan de accion 1989–2000*. CATIE, Turrialba, Costa Rica. 121 pp. with one map.

CATIE. 1990. *Projecto Sistemas Silvopastoriles para el Tropico humedo. Informe annual 1990*. CATIE, Turrialba, Costa Rica. 170 pp.

CATIE/JICA. 1995. *Identificacion y evaluacion de los Sistemas Agroforestales en el Canton de Hojancha, Gunacaste. Estudio de Caso*. CATIE, Turrialba, Costa Rica. 90 pp.

Crosby, A. 1986. *Ecological Imperialism. The Biological Expansion of Europe 900–1900*. Cambridge University Press, Cambridge, UK.

Davies, A., W. A. Adams, and D. Wilman. 1989. Soil compaction in permanent pasture and its amelioration by silting. *Journal of Agricultural Science* 113:189–197.

DGEC. 1987. *Agricultural Census 1984*. DGEC, San Jose, Costa Rica.

Fallas, J., and C. Morera. 1993. *Cambios en cobertura y evaluacion del uso de la tierra en la 1 etapa del proyecto de riego Arenal—Tempisque para los anos 1986 y 1992*. Universidad Nacional, Costa Rica.

FAO. 1980. *Production Yearbook*. FAO, Rome.

FAO. 1991. Production Yearbook 1991. Rome: FAO.

Flores, J. 1994. Current status and trends in the utilization of natural resources in Central America. In: Homan, J. (Ed.) Animal Agriculture and natural resources in Central America: strategies for sustainability proceedings, CATIE, Turrialba Costa Rica, pp. 23–40.

Franco, M. H. 1997. Evaluación de la calidad nutricional de *Cratylia argentea* como suplemento en el sistema de producción doble propósito en el trópico sub-humedo de Costa Rica. M.S. thesis, CATIE, Turrialba, Costa Rica. 74 pp.

French, J. 1994. Current status and trends in animal agriculture in Central America. In: Homan, J. (Ed.) Animal Agriculture and natural resources in Central America: strategies for sustainability proceedings, CATIE, Turrialba Costa Rica, pp. 9–22.

Godoy, J. C. 1997. Analisis economico y finaciero de los incentivos a la reforestacion otorgados por el gobierno de Costa Rica. M.S. thesis, CATIE, Turrialba, Costa Rica. 94 pp.

Hartshorn, G. S., L. Hartshorn, A. Atmella, L. D. Gomez, A. Mata, R. Morales, R. Ocampo, D. Pool, C. Quesada, C. Solera, R. Solarzano, G. Stiles, J. Tosi, A. Umaña, C. Villalobos, and R. Wells. 1982. *Costa Rica: Country Environmental Profile: A Field Study*. Tropical Science Center, San Jose, Costa Rica.

Hernandez, M., P. Argel, M. A. Ibrahim, and L. T. Mannetje. 1995. Pasture production, diet selection and liveweight gains of cattle grazing Brachiaria brizantha with or without Arachis pintoi at two different stocking rates in the Atlantic zone of Costa Rica. *Tropical Grasslands (Australia)* 29:134–141.

Holmann, F., Romero, F., Montenegro, J., Chana, C., Oviedo, E., and Banos, A. 1992. Rentabilidad de sistemas silvopastoriles con pequeños productores de leche en Costa Rica: primera aproximación. Turrialba 42(1):79–89.

Huising, J. 1993. Land use zones and land use patterns in the Atlantic Zone of Costa Rica. Ph.D. thesis, Wageningen Agricultural University, Wageningen, The Netherlands. 222 p.

Ibrahim, M. 1994. Compatibility, persistence and productivity of grass-legume mixtures for sustainable animal production in the Atlantic zone of Costa Rica. Ph.D. thesis, Wageningen Agricultural University, Wageningen, The Netherlands. 129 pp.

Ibrahim, M., and Schlönvoigt, A. 1998. Silvopastoral systems for degraded lands in the humid tropics: Environmentally friendly silvopastoral alternatives for optimising productivity of livestock farms: CATIE's experience. CATIE Actas de la IV semana cientifica, Turrialba Costa Rica, CATIE. pp. 77–282.

Jansen, H., Ibrahim, M., Nieuwenhuyse, A., Mannetje, L't., Joenje, M., and Abarca, S. 1997. The economics of improved pasture and silvopastoral technologies in the Atlantic Zone of Costa Rica. *Tropical Grasslands* 31:1–18.

Janzen, D. H. 1991. How to save tropical biodiversity. American Entomologist 37(3):159–171.

Jarvis, L. S. 1986. *Livestock Development in Latin America.* The World Bank, Washington, DC. 214 pp.

Kaimowitz, D. 1996. Livestock and deforestation in Central America in the 1980s: a policy perspective. Jakarata, Indonesia. Center for International Forestry Research. 88 p.

Keller, M., E. Veldkamp, A. M. Weitz, and W. A. Reiners. 1993. Effect of pasture age on soil trace-gas emissions from a deforested area of Costa Rica. *Nature* 365:244–246.

Keller, M., and W. A. Reiners. 1994. Soil-atmosphere exchange of nitrous oxide, nitric oxide and methane under secondary succession of pasture to forest in the Atlantic lowlands of Costa Rica. *Global Biogeochemical Cycles* 8(4):399–409.

Kosmerl, M. 1996. Gradient analysis of pastures of Costa Rica. Class project and independent study. SUNY ESF, Syracuse, NY.

Ledec, G. 1992. New directions for livestock policy: An environmental perspective. In T. Downing, S. Hecht, H. Pearson, and C. Garcia Dowining (Eds.), *Development or Destruction, the Conversion of Tropical Forest to Pasture in Latin America,* pp. 27–65. Westview Press, Boulder, CO.

Leeuwen, V. A. C., and A. M. Hofstede. 1995. *Forests, Trees and Farming in the Atlantic Zone of Costa Rica. An Evaluation of the Current and Future Integration of Trees and Forests in Farming Systems in the Atlantic Zone of Costa Rica.* Serie tecnica, Informe tecnico no. 257. CATIE/ WAU, Turrialba, Costa Rica. 48 pp.

Leonard, H. J. 1987. Natural Resources and Economic Development In Central America. International Institute For Environment And Development, Transaction Books.

Levitan, L. C. 1988. Land and energy constraints in the development of Costa Rican agriculture. Unpublished M.S. thesis, Cornell University, Ithaca, NY.

Lujan, R. E. 1989. *La Ganaderia de Costa Rica.* Cooperativa de Productores de Leche Dos Pinos, San Jose, Costa Rica.

Lutz, E. M., M. Vedova, L. Martinez, R. Ramon, A. Vasquez, L. Alvarado, R. Merino, R. Celis, and J. Huising. 1993. Interdisciplinary Fact-Finding on Current Deforestation in Costa Rica. Environment Working Paper no. 61, Environment Department, World Bank, Washington, DC.

MAG. 1989. *Encuesta Ganadera.* MAG, San Jose, Costa Rica.

Merelet, M., G. Farrell, J. M. Laurent, and C. Borge. 1992. *Identificacion de un programa regional de desarrollo sostenible en el tropico humedo, informe de consultoria.* Groupe de Recherche et d'Echanges Technologiques (GRET), Paris.

MIRENEM. 1994. *Opus Magna, MIRENEM 1990–1994.* MIRENEM, San Jose, Costa Rica.

Morales, J. L. 1992. Distribucion del pasto ratana (Ischaemum ciliare) en las tierras de pastoreo de Costa Rica. In *Seminario Taller, El pasto ratana (Ischaemum ciliare) en Costa Rica, Alternativa o problematica en nuestra ganaderia?* Cooperativa de Productores de Leche Dos Pinos, San Carlos, Costa Rica.

Nuñez, O. 1993. Deforestación en Costa Rica: la pesadilla y la esperanza. Esta semana (Abril 13–19):11–12.

Pezo, D., Kass, M., Benavides, J., Romero, F., and Chaves, C. 1990. Potential of legume tree fodders as animal feed in Central America. In: Devendra C, (Ed.) Shrubs and tree fodders for farm animals. Proceedings of a workshop in DENPASAR, Indonesia 24–29 July, 1989, IDRC, Canada, pp. 163–175.

Rodriguez, S., and E. Vargas. 1988. *El recurso forestal en Costa Rica, Politicas publicas y sociedad.* Editorial Universidad Nacional, Heredia, Costa Rica.

Romero, F., Benavides, J., Kass, M., and Pezo, D. 1994. Utilization of trees and bushes in ruminant production systems. In: Homan J (ed) Animal Agriculture and natural resources in Central America: strategies for sustainability proceedings, CATIE, Turrialba Costa Rica, pp. 205–218.

Sader, S. A., and A. T. Joyce. 1988. Deforestation rates and trends in Costa Rica 1940 to 1983. *Biotropica* 20:11–14.

Saenz M. A. 1955. *Los Forrajes de Costa Rica.* Editorial Universitaria de Universidad de Costa Rica, San Jose, Costa Rica.

SEPSA-CNP 1990. Encuesta Ganadería Nacional 1988. Secretaría Ejecutiva de Planifícación del Sector Agropecuario y de Recursos Naturales Renovables—Consejo Nacional de Producción, San José Costa Rica. 60 pp.

Sharma, P. 1992. *Conocimientos tradicionales agrosilvopastoriles y su adaptacion a la conservacion de los recursos naturales en la region Chorotega de Costa Rica.* Proyecto RENARM/manejo de Cuencas, CATIE, Turrialba, Costa Rica. 23 pp.

Solórzano, R. 1995. Breves diagnósticos y soluciones a corto plazo al sector forestal nacional. Notas técnicas y económicas No. 7. Centro Científico Tropical. San José, Costa Rica. 6 p.

Toledo, J. 1994. Livestock production on pasture: parameters of sustainability. In: Homan, J. (Ed.) Animal Agriculture and natural resources in Central America: strategies for sustainability proceedings, CATIE, Turrialba Costa Rica, pp. 137–159.

Torres, M. 1995. Caracteristicas fisicas, qumicas y biologicas en suelos bajo pasturas de *Brachiaria brizentha* sola y en asocio con *Arachis pintoi* despues de cuatro anos de pastoreo en el tropico humedo de Costa Rica. MSc thesis, CATIE, Turrialba Costa Rica, CATIE. 98 p.

Trejos, R. 1992. El comercio agropecuario extraregional. In C. Pomareda (Ed.), *La agricultura en el desarrollo economico de Centroamerica en los 90,* pp. 87–124. IICA, San Jose, Costa Rica.

TSC/WRI. 1991. *Accounts Overdue: Natural Resource Depreciation in Costa Rica.* World Resources Institute, Washington, DC.

Van der Grinten, P., Baayen, M., Villalobos, L., Dwinger, R., and Mannetje, L. 1992. Utilisation of kikuyu grass (*Pennisetum clandestinum*) pastures and dairy production in a high altitude region of Costa Rica. *Tropical Grasslands* 26(4):255–263.

Van der Kamp, E. J. 1990. Aspectos econmicos de la ganadera en pequea escala y de la ganadera de la carne en la zona atlêntica de Costa Rica. Field Report no. 51, CATIE/Wageningen/MAG. Costa Rica.

Veldkamp, A., and Fresco, L. 1997. Exploring land use scenarios, an alternative approach based on actual land use. *Agricultural systems* 55(1):1–17.

Veldkamp, E., A. M. Weitz, I. G. Staritsky, and E. J. Huising. 1992. Deforestation trends in the Atlantic zone of Costa Rica: A case study. *Land Degradation and Rehabilitation* 3:71–84. USA.

Veldkamp, Ed. 1993. Soil organic carbon dynamics in pastures established after deforestation in the humid tropics of Costa Rica. Ph.D thesis, Wageningen Agricultural University. The Netherlands. 122 p.

Veldkamp, E. 1994. Organic carbon turnover in three tropical soils under pasture after deforestation. *Soil Science of America Journal* 58:175–180.

Veldkamp, E., M. Keller, and M. Nunez. 1996. Effects of management on $N_2O$ and NO emissions from pasture soils in the humid tropics of Costa Rica. *Global Biogeochemical Cycles,* submitted.

Vera, C., and Barrios, C. 1997. Exploración sumaria de la producción de maderas en porteros de la zona ganadera de Esparza: especies, manejo y dinámica del componente maderables. Curso manejo forestal II, CATIE 1998, Turrialba Costa Rica.

Waaijenberg, H. (Ed.). 1990. *Rio Jimenez, ejemplo de la problemêtica agraria de la Zona Atlantica de Costa Rica. Un analisis con enfoque historico.* Programme paper no. 5, Atlantic Zone programme, CATIE/UAW/MAG, Turrialba, Costa Rica.

Walker, I., J. Suazo, A. Thomas, and H. Jean-Pois. 1993. *El impacto de las politicas de ajuste estructural sobre el medio ambiente en Honduras.* Posgrado Centroamericano en Economia, Universidad Nacional Autonoma de Honduras.

Wielemaker, W. G., and A. L. E. Lansu. 1991. Land-use changes affecting classification of a Costa Rican soil. *Soil Sci. Soc. Am. J.* 55:1621–1624.

World Bank. 1993. Costa Rica. Revisión al sector forestal, versión preliminar para discusión. Washington DC.

CHAPTER 15

# The Extent and Economic Impacts of Soil Erosion in Costa Rica

BENJAMIN D. RUBIN AND GLENN G. HYMAN



## I. INTRODUCTION

Ever-increasing population levels have called into question the adequacy of modern agricultural practices to provide enough food for future generations (Brown, 1981, 1994; Pimentel *et al.*, 1993, 1995). Two reasons why adequate yields might not be sustainable are that erosion and degradation of agricultural lands may cause severe decreases in agricultural yields (Pimentel *et al.*, 1995) and that potential increases in the cost of energy sources may make agricultural inputs and equipment that presently are being used to compensate for erosion unaffordable, at least to the poorer people of the world (Hall *et al.*, 1986).

Although soil erosion is a natural process, in some places human activities can increase rates of erosion by several hundred times over natural rates (Hartshorn *et al.*, 1982; Repetto and Cruz, 1991). In many areas, these anthropogenic activities have created a situation where the rate of erosion is much greater than the rate of soil formation. Soil erosion is caused by the forces of rain and wind, which act to dislodge soil particles and put them in motion. Rain erosion can include the effects of raindrops, which cause splash erosion, and runoff, which causes sheet erosion. Soil can be transported overland and in some areas gullies can form where erosive forces are concentrated (Christopherson, 1993). In most areas, including Costa Rica, rain is the most important cause of erosion; consequently most erosion models have focused on rain erosion.

Because of the importance of erosion as an environmental and economic problem, as well as its implications for the sustainability of agriculture, the purpose of this chapter is to (1) examine critically the methods used to model soil loss and its economic impacts, (2) apply such a soil loss model to the entire country of Costa Rica, and (3) use the model's results to examine the economic impacts of erosion in Costa Rica. Specifically, we examine the Universal Soil Loss Equation (USLE) (Wischmeier and Smith, 1965) and its various permutations and alternatives, and apply it to the entire country of Costa Rica using the spatial topographic and meteorological information given in Chapters 8 and 9. We also estimate the economic impacts of that erosion to the eroded site and to downstream ecosystems. Finally, we interpret our results from a biophysical, resource and energy-driven perspective, as well as from a more traditional (neoclassical) economic perspective.

Costa Rica is a good place to make an analysis of the effects of soil erosion. First, Costa Rican topography and climate facilitate erosion in many areas (Chapter 9) (EIU, 1995). Second, agriculture is a mainstay of the Costa Rican economy, accounting for 30% of the gross domestic product (GDP), 33% of national employment, and 55% of foreign exchange, and using 50% of the land as of the early 1990s (EIU, 1995; Repetto and Cruz, 1991). Third, the Costa Rican government has announced that it intends to become a model of sustainability for Central America (Chapter 2). Obviously for a country where agriculture is so important any such sustainability must include preserving soils.

## II. LITERATURE REVIEW

### A. Quantification of Soil Loss

The field of erosion modeling has been dominated by the use of the USLE since its inception 30 years ago. It was originally intended to predict erosion on a single field in the United States (Wischmeier and Smith, 1965) but has been used widely in other areas of the world and at other scales, as it is in this chapter (examples include Ewell and Stocking, 1975; Roose, 1980; Repetto and Cruz, 1991).

The USLE is empirical by nature; in other words, it does not attempt to represent explicitly the physical forces that cause soil erosion. Some researchers, such as Kirkby (1980) and Meyer and Wischmeier (1969), have worked on developing models that predict soil loss based on the physical mechanisms of erosion. These models are, of necessity, more complicated than the USLE and require more detailed data, but they have obvious conceptual advantages.

Over the last 20 to 30 years, erosion process models have become much more sophisticated. Kirkby (1980) believes that soil modeling will eventually be mostly process based. Fortunately, many of the data necessary for building empirical models will also be useful in building process models, so that the digital data bases constructed now will be useful in the future. Process-based erosion models, however, still lag behind empirical models in their versatility and applicability and require many data that are not necessary for operating empirical models. We believe that due to the relatively young stage of process-based modeling and the practical considerations involved in applying process models to large areas, improvements on already existing empirical models will be more effective in analyzing soil loss in Costa Rica.

Furthermore, although they do not cover all of the processes involved in erosion, it is useful to examine the six factors of the USLE here, because they represent the basic conditions that influence rates of soil loss.

$$A = R \times K \times L \times S \times C \times P,$$

where $A$ is average annual soil loss (T/yr), $R$ is the rainfall erosivity factor [(MJ $\times$ mm)/(ha $\times$ hr $\times$ yr)], $K$ is the soil erodibility factor [(T $\times$ ha $\times$ hr)/(MJ $\times$ mm)], $L$ is the slope length factor (dimensionless), $S$ is the slope gradient factor (dimensionless), $C$ is the crop management factor (dimensionless), and $P$ is the erosion control factor (dimensionless).

The USLE has been reviewed extensively in the literature (e.g., Hudson, 1971; Arnoldus, 1977; Kirkby, 1980; Mitchell and Bubenzer, 1980). One of the main difficulties in applying the USLE to different regions of the world has been the need to validate its results. In order to validate a model properly, it is necessary to compare the model's predictions to empirical measurements of the phenomenon being modeled. In some parts of the world, most notably the United States, the USLE and more advanced erosion models have been built and validated and are readily available (Renard *et al.*, 1994; Kautza *et al.*, 1995; Toy and Osterkamp, 1995; Yoder and Lown, 1995). The USLE has been revised to take into account specific slope configurations, as well as changes in vegetative cover and rainfall intensity throughout the year. Precise correction factors are available for specific regions of the United States. Many of these modifications have been combined into a computer software package named the Revised Universal Soil Loss Equation (RUSLE) (Renard *et al.*, 1991; Renard and Ferreira, 1993).

In many other places, however, this level of accuracy and precision is the exception rather than the rule (Moldenhauer, 1980; Hudson, 1980; Renard and Freidmund, 1994). The data necessary to calculate the USLE factors even in its original form have been a long time in the collecting for many developing regions (Low, 1967; Arnoldus, 1977; Renard and Freidmund, 1994), and the much needed modifications generally have not been made (Lal, 1987). This has resulted in a plethora of unvalidated and sometimes mutually contradictory estimates of soil erosion for the humid and dry tropics, which, ironically, are the regions where erosion is believed to be especially prevalent and costly. In response to these studies, several other authors have commented on the unreliability of estimates based on regressions that are highly extrapolated in time and space, or based on insufficient data (Hudson, 1980; Lal, 1987). They recommend that such estimates should be labeled clearly as such and used appropriately as "ballpark" figures or indicators of specific gaps in our measurements that require filling.

## B. Applications of the USLE in Costa Rica

The USLE has been applied to Costa Rica several times at various spatial scales (Hartshorn *et al.*, 1982; Repetto and Cruz, 1991; Jeffrey *et al.*, 1989). Jeffrey *et al.* (1989) made a countrywide map of predicted soil erosion based on the USLE. Their map shows that the developed areas found in the Central Valley, Guanacaste, the Nicoya Peninsula, and especially along the Pacific slope are where the most erosion occurs, and that the forested Talamancas and Caribbean slope experience comparatively less erosion. They conclude that 75% of the country experiences "tolerable" erosion rates (less than 10 T/ha/yr), and that 22% of the country has rates of 10 to 50 T/ha/yr. Three percent of the national territory experiences rates greater than 50 T/ha/yr.

Repetto and Cruz (1991) used soil classification, topography, and climate data from previously published maps of Costa Rica, but it is unclear from their description exactly at what spatial resolution they solved the USLE. Their *R* factor was derived from a 1 : 100,000 scale map of precipitation intensity. They derived the *K* and combined *LS* factors from 22 soil classes on a 1 : 200,000 scale map of Costa Rican soil types and land uses. Land use was derived from agricultural census data taken between 1963 and 1984. Repetto and Cruz (1991) assumed that the crop management factor, *C*, was 0.003 for forest land, 0.04 for pasture, 0.86 for perennial crops, and 0.34 for annual crops, but do not say how their values were derived. Repetto and Cruz (1991) estimate that the national soil loss from agricultural land has increased steadily for Costa Rica, from 121.8 million T/yr in 1970 to 188.6 million T/yr in 1984. These numbers represent approximately 92 T/ha, considerably higher than Jeffrey's numbers. Because Repetto and Cruz (1991) assume constant land use from 1984 to 1989, their estimates of erosion for these years are constant. Table 15-1 presents several measurements of erosion rates in Costa Rica.

## C. Economic Impacts of Soil Loss: On-Site Effects

### 1. Methods Used to Model On-Site Effects

The economic impacts of erosion are complex. Fundamentally, they can be divided into two categories: on-site and off-site costs. Among the on-site costs is potential reduction in soil fertility. This reduction can be caused by lowered water infiltration rates, water holding capacity, available nutrients, organic matter, soil biota, or soil depth (Pimentel *et al.*, 1995). Some of this loss in yield can be compensated for by increases in the application of fertilizer to replace nutrients, and, in fact, increasing fertilizer applications can increase yield, despite the effects of erosion, in some cases (Figure 12-4). If erosion is sufficiently severe, however, eventual declines in yield are inevitable, regardless of how much fertilizer is applied (Repetto and Cruz, 1991; Pimentel *et al.*, 1993; Alfsen *et al.*, 1996). The relationship between the amount of soil eroded from a given site and the subsequent changes in soil fertility, therefore, is complex.

Estimates of the effect of soil loss on crops are extremely variable. Several techniques have been proposed for making these estimates (see Lal, 1987, for a thorough review). One model, the erosion productivity impact calculator (EPIC) (Williams *et al.*, 1983, as cited in Lal, 1987; Williams and Renard, 1985), simulates the processes involved in erosion-related productivity declines using data similar to those required for the USLE. Pierce *et al.* (1983, 1984a,b) developed a different technique using a productivity index (PI) that is based on the physical and chemical soil requirements of a given crop and the ability of the soil at different depths to meet those requirements. Their model was developed in the United States; however, Rijsberman and Wolman (1985; Wolman, 1985) have recalibrated it for some other parts of the world. An alternate approach, used in the present analysis, is to avoid the problem of evaluating the effects of soil loss on crop yield and to simply measure the cost of erosion in terms of the cost of the nutrients that are removed from the soil (Repetto and Cruz, 1991).

### 2. Economic Analyses of Erosion in Costa Rica and Neighboring Countries

Several authors have estimated the economic impacts of erosion in Costa Rica and nearby areas (Wiggins, 1980; Repetto and Cruz, 1991; White *et al.*, 1994;

TABLE 15-1 Empirical Soil Erosion Plot Studies in Costa Rica

| Source | Location | Crop | Slope | Plot period (months) | Erosion rate (tons km yr) | Exceptional events* (tons km) | Bubnoff units** |
|---|---|---|---|---|---|---|---|
| Ives, 1951 | CATIE | various crops | 16% | 36 | 2.67 | 10097 | 1 |
| Ives, 1951 | CATIE | grass | 16% | 36 | 0.00 | 0 | 0 |
| Ives, 1951 | CATIE | bare soil | 16% | 36 | 8.00 | 12500 | 4 |
| Ives, 1951 | CATIE | various crops | 16% | 36 | 79.33 | 7810 | 40 |
| Ives, 1951 | CATIE | various crops | 45% | 36 | 274.33 | 8460 | 137 |
| Ives, 1951 | CATIE | grass | 45% | 36 | 0.00 | 0 | 0 |
| Ives, 1951 | CATIE | bare soil | 45% | 36 | 626.33 | 13100 | 313 |
| Ives, 1951 | CATIE | various crops | 45% | 36 | 413.67 | 11090 | 207 |
| Ives, 1951 | CATIE | sugar cane | 45% +/− | 36 | 0.00 | 0.00 | 0 |
| Ives, 1951 | CATIE | maize | 45% +/− | 36 | 0.00 | 0.00 | 0 |
| Ives, 1951 | CATIE | maize | 45% +/− | 36 | 0.00 | 0.00 | 0 |
| Ives, 1951 | CATIE | maize | 45% +/− | 36 | 0.00 | 0.00 | 0 |
| Ives, 1951 | CATIE | maize | 45% +/− | 36 | 0.00 | 0.00 | 0 |
| Rocha, 1977 | Bajo San Lucas | beans (weeded) | 40% | 12 | 172.00 | | 86 |
| Rocha, 1977 | Bajo San Lucas | beans (mulched) | 40% | 12 | 22.00 | | 11 |
| Rocha, 1977 | Bajo San Lucas | beans (mulched) | 40% | 12 | 27.00 | | 14 |

| | | | | | | | |
|---|---|---|---|---|---|---|---|
| Bermudez, 1980 | Turrialba "Florencia Sur" | coffee | 30% | 6 | 73.00 | | 37 |
| Bermudez, 1980 | Turrialba "Florencia Sur" | coffee with poro | 30% | 6 | 11.00 | | 6 |
| Bermudez, 1980 | Turrialba "Florencia Sur" | coffee with poro and laurel | 30% | 6 | 20.00 | | 10 |
| Berru, 1980 | La Suiza | pasture | 42–51% | 6 | 269.00 | | 135 |
| Romero, 1991 | La Suiza "La Selva" | beans | 69% | 6 | 5720.00 | 2860.00 | 2860 |
| Romero, 1991 | La Suiza "La Selva" | bare soil | 58% | 6 | 5800.00 | 2900.00 | 2900 |
| Garzon, 1991 | San Juan Sur | maize | 15–35% | 12 | 120.00 | | 60 |
| Faustino *et al.*, 1994 | San Juan Sur | beans/maize | 35% | 12 | 1.10 | | 1 |
| Faustino *et al.*, 1994 | San Juan Sur | beans/maize | 25% | 12 | 69.00 | | 35 |
| LeBeut, 1993 | San Juan Sur | bare soil | 25% | 12 | 0.08 | | 0 |
| Vahrson and Cervantes, 1991 | Puriscal | coffee | 56–59% | 12 | 18.00 | | 9 |
| Vahrson and Cervantes, 1991 | Puriscal | coffee w/shade | 56–59% | 12 | 150.00 | | 75 |
| Vahrson and Cervantes, 1991 | Puriscal | pasture | 56–59% | 12 | 37.00 | | 19 |
| Cortes *et al.*, 1987 | Cot/Tierra Blanca | potato, onion, carrot, pumpkin | 66% | 4 | 42600.00 | | 21300 |
| Cortes *et al.*, 1987 | Cot/Tierra Blanca | potato, onion, carrot, pumpkin | 51% | 4 | 4200.00 | | 2100 |
| Cortes *et al.*, 1987 | Cot/Tierra Blanca | potato, onion, carrot, pumpkin | 90% | 4 | 13200.00 | | 6600 |
| Cortes *et al.*, 1987 | Cot/Tierra Blanca | potato, onion, carrot, pumpkin | 40% | 4 | 1800.00 | | 900 |

* Values for the Ives (1951) study are reported for the 3 year plot period minus one exceptional storm under the "erosion rate" column. For the Ives study, the "exceptional events" column shows separate data for the Dec. 6, 1949, storm. Values for the Romero (1991) study have been listed in both columns to emphasize that the erosive events all occurred during 1 month.

** Bubnoff units were calculated from the "erosion rate" column where 1 B = 2 tons km yr (Saunders and Young, 1983).

Pimentel *et al.*, 1995; Alfsen *et al.*, 1996). The basis for the Repetto and Cruz analysis is that the cost of erosion will be reflected in increases in the amounts of fertilizers that need to be applied. They assume conservatively that all portions of the soil are removed at equal rates and, therefore, the amount of additional fertilizer required to restore the soil is equal to the soil's available nutrient concentration times the amount of soil removed. They intentionally neglect the cost of transporting fertilizer to the farm, because they believe it is small. In addition, they compare the results of their method to those of the EPIC for five cases in Costa Rica. They found that while neither method showed consistently greater or lesser productivity declines than the other, the EPIC estimate, when greater than the nutrient loss method estimate, was likely to be significantly greater.

Alfsen *et al.* (1996) use a different approach to estimate the costs of soil erosion for Nicaragua. Their approach considers the fact that even with unlimited amounts of fertilizer, erosion can still effect soil productivity. From their paper, however, it is not clear how much chemical fertilization is included in their estimates of decreased productivity.

## D. Economic Impacts of Soil Loss: Off-Site Effects

The off-site costs of erosion can be conveniently and usefully divided between (1) in-stream costs, including recreational damage, fisheries damage, damage to water storage facilities, and impacts on the costs of navigation, and (2) off-stream costs, that include sediment-related flood damage, damage to water conveyance facilities, and water treatment costs (Clark, 1985). One difficulty in estimating off-site effects is that most erosion models, such as the USLE, do not predict where the eroded soil eventually will be carried. Eroded soil can have substantial positive and negative economic effects (Chesters and Schierow, 1985; Clark, 1985; Libby, 1985; Whelan, 1989). For example, erosion is responsible for producing the alluvial floodplains of the world that contain some of our most productive agricultural soils; however, deposition of upslope soils on previously more fertile, downslope sites also can have negative impacts (Pimentel *et al.*, 1993).

Several authors have suggested that these costs are important (Hudson, 1971; Walker and Fleming, 1980; Clark, 1985; Southgate and Whitaker, 1994). Whelan (1989) reports that Tempisque Bay, in northwest Costa Rica, is seriously polluted with sediment, which decreases the dissolved oxygen content of the water, and with ammonium from the breakdown of fertilizers. Both of these problems are due in part to erosion and leaching from agriculture and pasture land.

Repetto and Cruz (1991) report on an analysis of the effects of sedimentation on the Cachi Dam and reservoir in Costa Rica. The total annual losses due to sedimentation-induced energy reductions, work stoppages, and maintenance costs amounted to $287,000 per year. Fortunately, erosion is less important in the Arenal Project, Costa Rica's major hydropower project, because of its steep sides and mostly vegetated watershed. However, the only complete quantitative estimate of off-site costs of erosion that we are aware of was conducted by Clark (1985). It covers erosion costs in the United States in 1980. Clark's estimates are conservative in the sense that there are several off-site effects that he does not include. For example, Crutzen and Ehhalt (1977) report that increased use of nitrogen fertilizer may contribute to the destruction of the ozone layer. If this is true, substituting eroded nutrients with fertilizer could have very high costs.

## III. METHODS

Our overall plan was to develop a spatial data set that allowed the quantification of soil loss for all of Costa Rica. Although our goal is ambitious given the serious difficulties of determining erosion over a large area, it is the only way to examine the impacts of soil loss on natural resources, the economy, and the agricultural sector. Therefore we present the following analysis of soil loss in Costa Rica as a best approximation given available information and existing models for estimating erosion. We have expanded on the work of Jeffrey *et al.* (1989) and Repetto and Cruz (1991) by using the USLE to estimate soil loss in Costa Rica. Our methodology differs from previous work in several important ways. We used slope information from a digital elevation model (Chapter 8) that gives a far more accurate spatial representation compared to the polygon maps of slope classes used in prior studies. Our rainfall information is based on a comprehensive study of 115 rainfall stations distributed throughout the national territory (Chapter 10) (Varhson and Derkson, 1990). Land cover information is based on a map interpreted from a nationwide aerial photograph survey (IMN, 1996). Therefore, our economic analysis is applied to a high-resolution spatially explicit map of erosion. The development of the data sets is described below.

### A. Quantifying Soil Loss

Rainfall intensity values (the *R* factor in the USLE) were compiled from a study of erosivity for 115 rainfall measuring stations distributed throughout Costa Rica (Vahrson, 1990) and are expressed in English units. Vahrson's

(1990) table of *R* factors includes the latitude and longitude of each pluviograph station along with the rainfall erosivity value. We used these locations to generate a point map of the stations. We georeferenced the point map to the Lambert projection to standardize the *R* factor map to other data held in the GIS. We used TIN (triangulated irregular network, Chapter 8) interpolation techniques to convert the point map to a continuous surface of rainfall intensity, expressed as the USLE *R* factor.

Soil erodibility (the *K* factor of the USLE) estimates were derived from a combination of two 1 : 200,000 scale soil maps. Most of the soil data come from Acon y Asociado's (1989) soil association map. But because this map did not cover protected areas in Costa Rica, data for protected areas had to be inserted from an earlier soil map (Perez *et al.*, 1978). To do this, we used a Boolean operation that determined if each grid cell was mapped in the 1989 soil map. If the grid cell was not mapped on the 1989 map, which meant it fell in a protected area such as a national park, we assigned the soil unit from the 1978 map location to that grid cell. *K* values were assigned to grid cells by inserting values for the representative soil type for each unit into Wischmeier and Smith's (1965) equation for calculating soil erodibility (*K*),

$$100\ K = (2.1\ M^{1.14}) \times 10^{4} \times (12 - a) + 3.25 \times (b - 2) + 2.5 \times (c - 2),$$

where M = (percent silt and fine sand) × (100% − percent clay), *a* = percent organic material, *b* is the structure class, and *c* is the permeability class. Structure and permeability classes were assigned according to Wischmeier and Smith (1978).

To calculate land cover (the *C* factor of the USLE), we obtained a digital land cover map interpreted from 1992 aerial photographs (MINAE, 1996). The factor values were assigned to each land cover according to values derived from a previous study (Jeffrey *et al.*, 1989; Derckson, 1990). The *S* factor was calculated using the standard equations from USLE documentation (Wischmeier and Smith, 1978).

We held *L* (slope length) and *P* (management practices) factors to 1 due to the difficulties of determining their values. The length of slopes in Costa Rica varies around the standard value of the USLE plot length (22.2 m). Our field observations indicate that on average slope lengths are less than the standard, suggesting that our results may overestimate erosion. Since *L* factor values are likely to be both above and below the standard value, however, the exclusion of the *L* factor may not be that important. It would be an extremely difficult task, if not impossible, to measure the farming practice factor, *P*, for an entire country. This suggests that our results again may overestimate erosion. We examine their possible importance through sensitivity analysis.

## B. Calculating the Economic Impacts of Soil Loss

Our economic analysis is based on the costs of replacing with commercial fertilizers the soil nutrients lost to erosion (Repetto and Cruz, 1991). We calculated the replacement cost of the nitrogen, potassium, and phosphorus that would be lost to erosion if each of these elements were removed in proportion to its abundance in the soil. As will be discussed, this is a very conservative measure of the costs of erosion.

In order to calculate the weight of fertilizer needed to replace lost nutrients, we first created maps of the available phosphorous and potassium. These maps are based on the average content of the nutrients in the soils of each Canton (Bertsch, 1986) and on a digital map (gridcell size = 1 $km^2$) of the districts of Costa Rica (Gregoire Leclerc, pers. com.) that we grouped into the larger political unit of Cantons. Next, we estimated available nitrogen content based on an organic matter content map (Gregoire Leclerc, pers. com.) such that available nitrogen equals 0.195% organic matter content (Reppetto and Cruz, 1991). We multiplied the erosion map by each of our nutrient coverages to estimate the weight of nutrients lost by erosion. We multiplied the weight of nutrients lost by the percentage of pure nutrient in commercial fertilizers and by commercial fertilizer efficiencies (Reppetto and Cruz, 1991) to obtain the weight of fertilizer needed to replace lost nutrients. Finally, we estimated the total cost of this fertilizer based on the cost per ton of fertilizer (U.S. Bureau of Census, 1993).

Both the quantity of eroded soil and the cost of that erosion are subject to changes over time. In order to estimate these changes, it is necessary to have estimates of erosion for at least two separate years. Several of the USLE factors are relatively constant from year to year, such as the slope factors ($S$ and $L$) and the soil erodibility factor ($K$). The other three factors, however, vary considerably over time. Since we have already assumed that the erosion control factor ($P$) is always 1 and since the rainfall intensity factor ($R$) is based on data that are combined over several years, we can estimate changes in total erosion from changes in crop factor ($C$).

Unfortunately, although there are some maps of land use for various years in Costa Rica, few of these data have been digitized, making it difficult to incorporate into spatially explicit, digital models. Therefore, we constructed $C$ factor maps for 1973 from data on the total area in each of five major land uses by district, within Costa Rica (MAG, 1973). The five land uses are maize, coffee, banana, pineapple, and pasture. From these data we constructed maximum and minimum estimates of erosion by arranging the land uses within each district so as to maximize or minimize erosion.

Our quantitative estimates of off-site erosion costs are obtained by using Clark's (1985) estimates of total impacts of erosion in the United States and dividing them by the total area of agricultural land in the United States in 1980 (U.S. Bureau of Census, 1993). We then applied the estimates on a per area basis to Costa Rica for 1973 and 1992.

# IV. RESULTS

## A. Quantity of Eroded Soil

We estimate that the total amount of soil erosion in 1992 was 22.2 million T/yr. Although this figure is considerably lower than the estimate of Repetto and Cruz (1991; 224.1 million T/yr.), it is closer to the estimates of Jeffrey *et al.* (1989) who conclude that 75% of Costa Rica experiences less than 10 T/ha/yr of erosion and that 97% experiences less than 50 T/ha/yr. The average erosion per area was 4.4 T/ha/yr; and erosion per hectare varied greatly for different land uses from 20.7 T/ha/yr on annual crops and 10.7 T/ha/yr under perennial crops to 2.6 T/ha/yr in forests (Table 15-2).

Using the minimum estimate of erosion in 1973, our results indicate that erosion has increased over time from 17.4 million T in 1973 to 22.2 million T in 1992 (Table 15-3). As discussed below, we believe that the minimum estimate of erosion in 1973 is more accurate than the maximum estimate of 90.5 million mt. The total amount of fertilizer applied in Costa Rica also increased by approximately 50% to 100% (Raphael Statler pers. com. and FERTICO, 1995) in the same time period (Table 15-3). As discussed below, a portion of this increase is necessitated by soil loss.

The overall spatial distribution of erosion for 1992 is shown in Figure 15-1. Erosion per hectare per year is greatest in the mountains and greater along the Pacific coast than along the Caribbean coast. Northern Costa Rica also experiences low levels of erosion.

**TABLE 15-2** Erosion Rates and Costs (1000 US $ 1990) by Land Use for 1992

| Land use | Area (km$^2$) | Erosion (million T) | Erosion rate (T/ha/yr) | On-site costs (1000 $US 1990) |
|---|---|---|---|---|
| Forest | 27,525 | 7.3 | 2.6 | 660 |
| Pasture | 16,575 | 6.9 | 4.2 | 796 |
| Perennial crops | 3695 | 4.0 | 10.8 | 493 |
| Annual crops | 1358 | 2.8 | 20.7 | 312 |
| Other | 1836 | 1.3 | 6.9 | 80 |
| Total | 50,989 | 22.2 | 4.4 | 2,342 |

TABLE 15-3 Erosion Rates and Costs in 1973 and 1992 (Minimum and Maximum Rates Depend on How Land Use Is Apportioned in Cantons. Minimum Is Probably the Better Estimate.)

| Year and estimate | Erosion (million T) | Fertilizer used (1000 T) | | | Fertilizer needed to replace eroded nutrients (1000 T) | | | Price of fertilizer ($/T) | | | Cost of erosion (1000 $US 1990) |
|---|---|---|---|---|---|---|---|---|---|---|---|
| | | N | P | K | N | P | K | N | P | K | |
| 1992 | 22.2 | 56.0 | 15.0 | 40.0 | 8.6 | 1.5 | 8.0 | $155.00 | $158.00 | $81.50 | 2222 |
| 1973 (min) | 17.4 | 34.0 | 10.0 | 20.0 | 6.7 | 1.3 | 6.4 | $48.30 | $42.50 | $31.50 | 580 |
| 1973 (max) | 90.5 | 34.0 | 10.0 | 20.0 | 33.9 | 6.4 | 32.7 | $48.30 | $42.50 | $31.50 | 2939 |

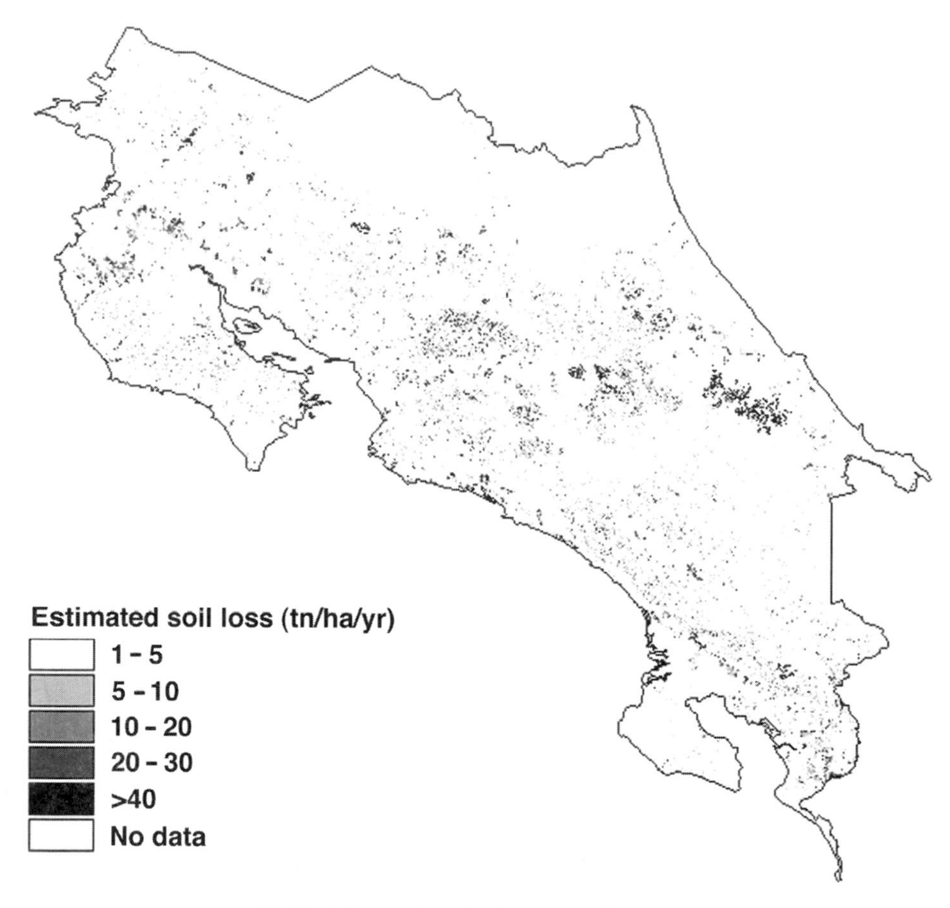


FIGURE 15-1 Map of soil erosion, 1992.

Figure 15-2 shows the spatial distribution of the *R*, *K*, and *S* USLE factors. It indicates that the rainfall is most intense along the Caribbean coast and is also very intense along the Pacific coast. The rainfall in northern Costa Rica and in the mountains is of lesser intensity. The erodibility of the soil is more varied on a small scale; however, in general the soil in the north and in the mountains is more easily eroded. The slopes are less steep in southern and central Costa Rica than in the north or along the Caribbean coast.

## B. Economic Impacts of Erosion

We estimate that the total cost of replacing nutrients lost to erosion in 1992 was \$2.3 million (Table 15-2). The nutrient content of pasture and agricultural

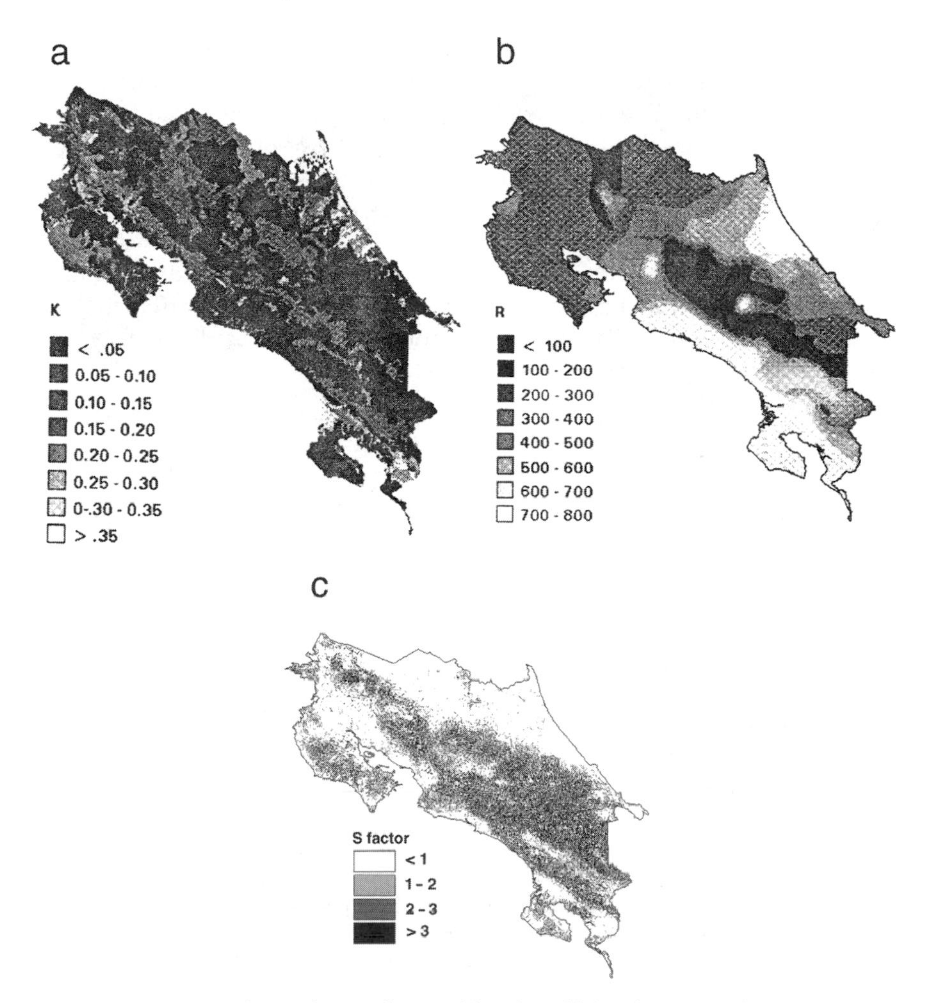


FIGURE 15-2 Maps of several USLE factors: (a) soil erodibility factor ($K$), (b) rain erosivity factor ($R$), and (c) slope factor (LS).

land is slightly higher than for forestland. This coupled with the increased erosion per area on developed land causes the cost of erosion per square kilometer to increase from $24 in the forest to $48 on pastures to $133 under perennial crops and $230 under annual crops.

According to the minimum estimate of erosion in 1973, the cost of erosion increased between 1973 and 1992 (Table 15-3). This is because the rate of erosion increased with intensification of agriculture and because the cost of a ton of eroded soil increased from approximately $0.03 in 1973 (regardless of which estimate is used) to $0.10 in 1992.

The off-site costs of erosion in 1992 are $1.2 million, or approximately 50% of the nutrient replacement costs. Of this, $840,000 accounts for in-stream damages and $370,000 for off-stream damages. When added to the on-site costs, this brings the total economic impact of erosion to $3.5 million per year. It is important to bear in mind, however, that the estimates of off-site effects are not nearly as precise as those for on-site effects. They are included to serve as a reminder that there are substantial costs to erosion that cannot be easily measured. Nevertheless we must conclude that the direct annual monetary cost of erosion as measured by our approach is not extremely large and hence is not inconsistent with sustainability. Whether this is true over the long run is another question considered next.

## V. DISCUSSION

### A. Biophysical vs Neoclassical Economic Approaches

At first glance it appears that technological advances since the industrial revolution have answered the challenge of our original question. Genetic engineering, agricultural mechanization, and the development of pesticides and fertilizers have increased yields per area. This success has been reflected in optimistic outlooks such as the one held by Seckler (1987, p. 89):

> In the long run, say two or more generations from now, it is not at all clear that the need for [agricultural land] will be nearly as acute as it is today. The food surplus problems of the developed nations are now also becoming the rule in some of the developing nations of Asia (especially in rice). Also, in most of the developing nations the percentage of the population that obtain their living from agriculture is decreasing, and in some countries the agricultural population is decreasing even in absolute numbers. The major exceptions to these generalizations are Nepal and Bangladesh in Asia and many of the African nations. They are exceptions because of the failure of economic growth, largely because of political and social problems, not resource problems. By the end of the next generation population growth will have slowed considerably, and, quite possibly, there will be major breakthroughs in biotechnology to produce more food from much less land.

Other researchers, however, have concluded that this extreme optimism is unwarranted (Ehrlich, 1968; Ehrlich and Ehrlich, 1970). For example, Hall and Hall (1993; see also Chapter 5), who worked in Costa Rica, concluded that "during the first quarter to half of the next century, it will be increasingly difficult, and eventually impossible, to feed a growing Costa Rican population no matter what (realistic) assumptions are made." p. 1.

While no one can predict future technology, we can say that thus far essentially all agrotechnology has been based on increases in energy intensity

(Hall *et al.*, 1986; Ko *et al.*, 1998). Some of the discrepancy between the two preceding points of view may be due to the approach that the different authors use in analyzing their respective issues. Seckler's paper describes methods of applying the principles of neoclassical economic analysis to problems of soil and water conservation. His work therefore employs the assumptions of his discipline, namely, that value is determined by the laws of supply and demand. Hall and Hall, however, apply a biophysical analysis of the efficacy of land and energy use in Costa Rica and other parts of the tropics. They believe that land, natural resource, and energy use are important because they are the basis for creating economic wealth and that limited resources limit potential economic growth. They might agree with Seckler that the economy could indeed respond to the challenges of needing to produce more food, and that indeed that might be possible. But they would argue that this would not occur through technology alone, but through the increasing industrialization of Costa Rica's agriculture. The rub is paying for the industrial inputs, which itself requires ever more agricultural land.

## B. Model Validity

The physical component of our model agrees closely with plot measurements of erosion in areas where such measurements have been made extensively (Table 15-1). For example, at CATIE, measurements range from 0 to 8 T/ha/yr on 16% slopes to as high as 400 to 600 T/ha/yr on 45% slopes. Our estimates average 5.5 T/ha/yr on 16% slopes and 244 T/ha/yr on the steepest slopes.

The economic component of our model is certainly conservative. First, we assume that there are no costs associated with fertilizer transport and application. Second, we do not calculate any cumulative effects of continued erosion over several years. Third, we assume that nitrogen, potassium, and phosphorus are removed only in proportion to their abundance in the soil. All accounts agree that erosion removes nutrients preferentially relative to other soil components (e.g., Lal, 1987; Repetto and Cruz, 1991; Pimentel *et al.*, 1993, 1995). Barrows and Kilmer (1963) report enrichment ratios (ER = concentration of material in runoff divided by concentration in soil) for several macronutrients in various locations, mostly in the United States. They report ER values ranging from 1.08 to 5.0 for nitrogen, 1.3 to 3.5 for soluble phosphorus, 5.4 for available potassium, 2.06 for calcium, and 1.39 for magnesium. One small caveat to this indicator is that ER can, of course, increase with soil conservation practices if such practices conserve soil particles more than soil nutrients. In some of the preceding cases, the authors specify whether soil conservation was being practiced. We have not included estimates where they specified that the soil was being conserved through active management. In no case do they report ER values < 1.

As already noted, our estimates of the off-site effects of erosion are very rough. Due to the lack of data on off-site effects, we are unable to formally validate these results. We interpret our results as an indication that the off-site costs of erosion may not be insignificant compared to the on-site costs.

## C. Interpretation of Soil Loss Map

Overall, our estimates of soil loss (Table 15-2) are in agreement with the estimates of Jeffrey *et al.* (1989), but are considerably lower than the estimates of Repetto and Cruz (1992). We believe that our estimates are consistent with general observations of the situation in Costa Rica. Tolerable rates of erosion (rates similar to the rate of soil formation) in Costa Rica are approximately 10 T/ha/yr (Jeffrey *et al.*, 1989; Repetto and Cruz, 1991). Personal observation suggests that while most of the nation of Costa Rica is experiencing tolerable rates of erosion, a significant portion is eroding at an unsustainable rate. Our data indicate that 28% of the national territory is experiencing greater than tolerable erosion (25% according to Jeffrey *et al.*, 1989). However, according to Repetto and Cruz (1991), the average site in Costa Rica is experiencing nine times the tolerable rate of erosion.

This is not to suggest that erosion does not pose a significant potential problem for the future. Most of the land experiencing intolerable rates of erosion is agricultural land. In fact, according to our analysis 60% of the land in perennial or annual crops was experiencing unsustainable rates of erosion in 1992. The cumulative effect of even small net soil loss is likely to be very important over long time periods (Chapter 5).

Our low estimate of erosion for 1973 (Table 15-3) indicates that rates of erosion are remaining constant or are slightly increasing with time. Even if we believe the maximum estimate erosion rates have decreased dramatically, however, we still believe that the low estimate is much more accurate (see below).

## D. Interpretation of Economic Analysis

Approximately 10–20% of the fertilizer applied in Costa Rica in 1992 (and also in 1973 according to our low erosion estimate) was used to ameliorate the effects of erosion (Table 15-3). According to our high estimate, almost all of the artificial fertilization was required to counteract erosion. The high erosion estimates of Repetto and Cruz (1991) lead to the same conclusion. If all the fertilizer applied was necessary to compensate for erosion, then crop yields would not be higher than "natural" yields without fertilization. High yields of most major crops (Chapter 12) contradict the maximum 1973 estimate and the estimates of Repetto and Cruz (1991).

Although at first glance our results may indicate that the economic effects of erosion are tolerable, there are at least three reasons for possible concern about the impacts of erosion in the future. First, as discussed above, economically important areas of agricultural land are undergoing rates of erosion greater than the rate of soil formation. The effects of this net loss are cumulative. Yields are maintained or increased on these lands for a period of time by adding 5 to 10 times the amount of fertilizer needed to replace the eroded macronutrients (Table 15-3). However, unless erosion is lessened, other components of the soil will eventually degrade to the point where they limit productivity. Yield decreases from erosion can have strong ramifications on many sectors of the economy (Alfsen *et al.*, 1996). Second, the cost of erosion in fertilizer is closely tied to the price of fossil fuels (note the fourfold increase in the cost of a similar amount of eroded soil between the minimum 1973 estimate and the 1992 estimate in Table 15-3). Despite the oil shortages of the 1970s, the market price of fossil fuels used to create commercial fertilizers is still well below the actual environmental and economic impact of their production (Cleveland, 1992). If the costs of fossil fuels ever rise toward their actual value, the economic impact of erosion will be felt much more keenly. Third, the off-site effects of erosion in Costa Rica are poorly understood. The estimates derived in this chapter are very crude and are the only estimates on the national scale of which we are aware. Because of this uncertainty about an important percentage of the cost of erosion, the total cost may be significantly greater than estimated.

## VI. CONCLUSIONS

This study and others have demonstrated that soil loss due to erosion imposes substantial costs on the Costa Rican economy. Erosion also poses serious questions about the meaning and feasibility of a sustainable Costa Rica. Therefore, additional steps need to be taken to develop a working understanding of the most important components of soil loss and to abate its negative effects.

Rates of soil erosion in Costa Rica are quite variable with topography, climate, soil type, and land use. We estimate that average losses per hectare of agricultural land are approximately 20.7 T/yr for annual crops, 10.8 T/yr for perennial crops, and 6.9 T/yr for pastures. Currently, the economic impact of erosion is significant, but not overwhelming. However, most (60%) of the agricultural land in Costa Rica is experiencing erosion rates that probably exceed the rate of soil formation. Therefore, present land use practices are not sustainable in many areas. In addition, the costs of erosion may increase substantially in the future if the prices of fossil fuel–based fertilizers increase.

## REFERENCES


Acon y Asociados. 1989. *Mapa de Suelos de Costa Rica. Escala 1 : 200,000*. Acon y Asociados, San Jose, Costa Rica.

Alfsen, K. H., M. A. DeFranco, S. Glomsrod, and T. Johnson. 1996. The cost of soil erosion in Nicaragua. *Ecological Economics* 16:129–146.

Arnoldus, H. M. J. 1977. Predicting soil losses due to rill and sheet erosion. In *FAO Guidelines for Watershed Management*. Forest Conservation and Wildlife Branch, Forest Resources Division, FAO, Rome.

Barrows, H. L., and V. J. Kilmer. 1963. Plant nutrient losses from soils by water erosion. *Advances in Agronomy* 15:303–316.

Berru, W. A. 1980. Evaluación de la Escorrentía Superficial y la Erosión en un Pastizal con Arboles Aislados en la Suiza, Turrialba, Costa Rica. Tesis. UCR/CATIE.

Brown, L. 1981. World population growth, soil erosion and food security. *Science* 214:955–1002.

Brown, L. R. 1994. *State of the World 1994*. Worldwatch, New York.

Chesters, G., and L. J. Schierow. 1985. A primer on nonpoint pollution. *Journal of Soil and Water Conservation* 40:9–13.

Christopherson, R. W. 1993. *Geosystems: An Introduction to Physical Geography*, 2nd ed. Macmillan, Engelwood Cliffs, NJ.

Clark, E. H., III. 1985. The off-site costs of soil erosion. *Journal of Soil and Water Conservation* 40:19–22.

Cleveland, C. J. 1992. Energy quality and energy surplus in the extraction of fossil fuels in the U. S. *Ecological Economics* 6:139–162.

Cortes, V. M., and G. Oconitrillo y L. Brenes. 1987. Calculo de tasas de erosíon hídrica en Cot y Tierra Blanca de Cartago. Informe Final Proyecto de Investigacíon 214-85-083. Universidad de Costa Rica. Departamento de Geografía.

Crutzen, P. J., and D. E. Ehhalt. 1977. Effects of nitrogen fertilizers and combustion on the stratospheric ozone layer. *Ambio* 6:112–117.

Derckson, P. M. 1991. *A Soil Erosion Mapping Exercise in Costa Rica: Purposes, Methodology, Results. En Taller de Erosion de Suelos*. Universidad Nacional, Heredia, Costa Rica.

Economist Intelligence Unit (EIU). 1995. Country Profile of Costa Rica and Panama. EIU. London.

Ehrlich, P. R. 1968. *The Population Bomb*. Ballantine Books, New York.

Ehrlich, P. R., and Erlich., A. H. 1970. Population, Resources and Environment. W. H. Freeman, San Francisco.

Ewell, H. A., and M. A. Stocking. 1975. Parameters for estimating annual runoff and soil loss from agricultural lands in Rhodesia. *Water Resources Research* 11:601–605.

Faustino, J., D. Kass, and A. Tineo. 1994. Erosíon hídrica y lixiviación en una rotación fríjol-maíz con prácticas de conservación de suelos, en tierras de laderas, Turrialba, Costa Rica. Publicaiciones del Proyecto Renarm/Manejo de Cuencas. Turrialba, CATIE.

FERTICO. 1995. Areas de Cultivo y Consumo de Fertilizante en Toneladas Metricas Anos 1989 a 1995. Xeroxed document. FERTICO, San José, Costa Rica.

Garzon, H. 1991. Evaluación de Erosión Hídrica y la Escorrentía Superficial Bajo Sistemas Agroforestales en Tierras de Ladera, Turrialba, Costa Rica. Tesis de Grado. Turrialba, CATIE.

Hall, C. A. S., C. J. Cleveland, and R. K. Kaufmann. 1986. *Energy and Resource Quality: The Ecology of the Economic Process.* Wiley–Interscience, New York. Reprinted 1992, University Press of Colorado, Boulder, CO.

Hall, C. A. S., and M. H. P. Hall. 1993. The efficiency of land and energy use in tropical economies and agriculture. *Agriculture, Ecosystems and Environment* 46:1–30.

Hartshorn, G. S., L. Hartshorn, A. Atmella, L. D. Gomez, A. Mata, R. Morales, R. Ocampo, D. Pool, C. Quesada, C. Solera, R. Solarzano, G. Stiles, J. Tosi, A. Umaña, C. Villalobos, and R. Wells. 1982. *Costa Rica: Country Environmental Profile: A Field Study.* Tropical Science Center, San Jose, Costa Rica.

Hudson, N. W. 1971. *Soil Conservation.* Cornell University, Ithaca, NY.

Hudson, N. W. 1980. Erosion prediction with insufficient data. In M. D. Boodt and D. Gabriels (Eds.), *Assessment of Erosion.* Wiley, New York.

Ives, N. 1951. Soil and water runoff studies in a tropical region. *Turrialba* 1(5):240–244.

Jeffrey, P. J., P. M. Derckson, A. Vasquez, and B. Sonneveld. 1989. Informe técnio no. 1-E. Apoyo al Servico Naciónal de Conservación de Suelos y Aguas de Costa Rica. FAO.

Kautza, T. J., Shertz, D. L., and Weesies, G. A. 1995. Lessons learned in RUSLE technology transfer and implementation. *J. Soil Water Cons.* 59(5), 490–493.

Kirkby, M. J. 1980. Modelling water erosion processes. In M. J. Kirkby and R. P. C. Morgan (Eds.), *Soil Erosion.* Wiley, New York.

Ko, J-Y., C. A. S. Hall, and L. G. Lopez Lemus. 1998. Resource use rates and efficiency as indicators of regional sustainability: An examination of five countries. *Environmental Monitoring and Assessment.* 51:571–593.

Lal, R. 1987. Effects of soil erosion on crop productivity. *Critical Reviews in Plant Science* 5:303–369.

Lebeuf, T. I. 1993. Sistemas Agroforestales con *Erythina fusca* Lour. y sus Effectos Sobre la Pérdida de suelo, la Escorrentía superficial y la Productión de los Cultivos Anuales en Tierras de Ladera, San Juan Sur, Turrialba, Costa Rica. Tesis. CATIE.

Libby, L. W. 1985. Paying the nonpoint pollution control bill. *Journal of Soil and Water Conservation* 40:33–36.

Low, F. K. 1967. Estimating potential erosion in developing countries. *Journal of Soil and Water Conservation* 22:147–148.

MAG (Ministerio de Agricultura y Ganderia de Costa Rica). 1973. *Agricultural Census.* MAG, San José, Costa Rica.

Meyer, L. D., and W. H. Wischmeier. 1969. Mathematical simulation of the process of soil erosion by water. *Am. Soc. Agric. Eng. Trans.* 12:754–758.

MINAE (Ministerio del Medio Ambiente y Energia). 1996. *Evaluacion Area de Cambio de Uso de la Tierra. Proyecto: Inventario Nacional de Emisiones de Gases con Efecto Invernadero.* MINAE, San Jose, Costa Rica.

Mitchell, J. K., and G. D. Bubenzer. 1980. Soil loss estimation. In M. J. Kirkby and R. P. C. Morgan (Eds.), *Soil Erosion.* Wiley, New York.

Moldenhauer, W. C. 1980. Developing erosion research programs in areas where limited data are available. In Boodt and Gabriels (Eds.), *Assessment of Erosion.* Wiley, New York.

Perez, S., A. Alvarado, and E. Ramirez. 1978. *Asociaciones de subgrupos de suelos de Costa Rica (mapa preliminar).* OPSA, San Jose, Costa Rica. [Scale 1 : 200.000, 9 h.]

Pierce, F. J., W. E. Larson, R. H. Dowdy, and W. A. P. Graham. 1983. Productivity of soils: Assessing long-term changes due to erosion. *Journal of Soil and Water Conservation* 38:39–44.

Pierce, F. J., R. H. Dowdy, W. E. Larson, and W. A. P. Graham. 1984a. Soil productivity in the corn belt: An assessment of erosion's long-term effects. *Journal of Soil and Water Conservation* 39:131–135.

Pierce, F. J., W. E. Larson, and R. H. Dowdy. 1984b. Soil loss tolerance: Maintenance of long-term soil productivity. *Journal of Soil and Water Conservation* 39:136–138.

Pimentel, D., J. Allen, A. Beers, L. Guinand, A. Hawkins, R. Linder, P. McLaughlin, B. Meer, D. Musonda, D. Perdue, S. Poisson, R. Salazar, S. Siebert, and K. Stoner. 1993. Soil erosion and agricultural productivity. pp. 277–292 in Pimentel, D. (ed.), *World Soil Erosion and Conservation*. Cambridge University Press, Cambridge.

Pimentel, D., C. Harvey, P. Resosudamo, K. Sinclair, D. Kurz, M. McNair, S. Crist, L. Shpritz, L. Fitton, R. Saffouri, and R. Blair. 1995. Environmental and economic costs of soil erosion and conservation benefits. *Science* 267:1117–1123.

Renard, K. G., G. R. Foster, G. A. Weesies, and J. P. Porter. 1991. RUSLE: Revised universal soil loss equation. *Journal of Soil and Water Conservation* 46:30–33.

Renard, K. G., and V. A. Ferreira. 1993. RUSLE model description and database sensitivity. *Journal of Environmental Quality* 22:458–466.

Renard, K. G., G. R. Foster, D. C. Yoder, and D. K. McCool. 1994. RUSLE revisited: Status, questions, answers and the future. *Journal of Soil and Water Conservation* 49:213–220.

Renard, K. G., and J. R. Freidmund. 1994. Using monthly precipitation data to estimate the R-factor in the revised USLE. *Journal of Hydrology* 157:287–306.

Repetto, R., and W. Cruz. 1991. *Accounts Overdue: Natural Resource Depletion in Costa Rica*. Tropical Science Center (TSC), San Jose, Costa Rica.

Rijsberman, F. R., and M. G. Wolman. 1985. Effect of erosion on soil productivity: An international comparison. *Journal of Soil and Water Conservation* 40:349–354.

Rocha, J. 1977. Erosión de Suelos de Pendientes Cultivadas con Maíz y Fríjol con Diferentes Grandos de Cobertuna Viva Dentro de una Plantación Forestal. Tesis. CATIE.

Romero, E. 1991. Evaluación de las Medidas Demonstrativas de Conservación de Suelos en la Finca "La Selva", Cuenca del Río Tuis, Costa Rica. Tesis. CATIE.

Roose, E. J. 1980. Approach to the definition of rain erosivity and soil erodibility in West Africa. In M. D. Boodt and D. Gabriels (Eds.), *Assessment of Erosion*. Wiley, New York.

Sauders, I., and A. Young. 1983. Rates of surface processes on slopes, slope retreat and denundation. *Earth Surface Processes and Landforms* 8:473–501.

Seckler, D. 1987. Economic costs and benefits of degradation and its repair. In P. Blaikie and H. Brookfield (Eds.), *Land Degradation and Society*. Methuen, London.

Southgate, D. S., and M. Whitaker. 1994. *Economic Progress and the Environment*. Oxford University Press, Oxford.

Toy, T. J., and W. R. Osterkamp. 1995. The applicability of RUSLE to geomorphic studies. *Journal of Soil and Water Conservation* 50:498–503.

U.S. Bureau of Census. 1993. Statistical Abstract of the United States (113th ed.). Washington, D.C.

Vahrson, W. G. 1990. El potencial erosivo de la lluvia en Costa Rica. *Agronomia Costarricense* 14:15–24.

Vahrson, W. G., and C. Cervantes. 1991. Tasa de escorrentía superficial y erosión laminar en Puriscal, Costa Rica. *Turrialba*. 41(3):396–402.

Vahrson, W. G., and P. Derckson. 1990. Intensidades criticas de lluvia para el diseño de obras de conservación de suelos en Costa Rica. *Agronomía Costarricense*. 14(2):141–150.

Walker, R. A., and G. Fleming. 1980. Sediment transport as part of total catchment modelling. In M. D. Boodt and D. Gabriels (Eds.), *Assessment of Erosion*. Wiley, New York.

Whelan, T. 1989. Environmental contamination in the Gulf of Nicoya, Costa Rica. *Ambio* 18(5):302–304.

White, D. C., J. B. Braden, and R. H. Hornbaker. 1994. Economics of sustainable agriculture. In J. L. Hatfield and D. L. Harlan (Eds.), *Sustainable Agricultural Systems*. CRC Press, Ames, IA.

Wiggins, S. L. 1980. The economics of soil conservation in the Acelhuate River Basin, El Salvador. In R. P. C. Morgan (Ed.), *Soil Conservation: Problems and Prospects*. Wiley, New York.

Williams, J. R., K. G. Renard, and P. T. Dyke. 1983. A new model for assessing erosion's effect on soil productivity. *Journal of Soil and Water Conservation* 38:381.

Williams, J. R., and K. G. Renard. 1985. Assessment of soil erosion and crop productivity with process models (EPIC). In R. F. Follett and B. A. Stewart (Eds.), *Soil Erosion and Crop Productivity*. American Society of Agronomy, Madison, WI.

Wischmeier, W. H., and D. D. Smith. 1965. *Predicting Rainfall-Erosion Losses from Cropland East of the Rocky Mountains*. Agricultural Handbook no. 282, U.S. Department of Agriculture, Washington, DC.

Wischmeier, W. H., and D. D. Smith. 1978. *Predicting Rainfall Erosion Losses*. Agricultural Handbook no. 537, Science and Education Administration, U.S. Department of Agriculture, Washington, DC. 58 pp.

Wolman, M. G. 1985. Soil erosion and crop productivity: A worldwide perspective. In R. F. Follett and B. A. Stewart (Eds.), *Soil Erosion and Crop Productivity*. American Society of Agronomy, Madison, WI.

Yoder, D., and J. Lown. 1995. The future of RUSLE: Inside the new revised universal soil loss equation. *Journal of Soil and Water Conservation* 50:484–489.

CHAPTER 16

# Land Use and Cover Change in Costa Rica

## A Geographic Perspective

G. Arturo Sanchez-Azofeifa



## I. INTRODUCTION

Developing countries are beginning to invest in systematic studies of land use and land cover change (LUCC) in order to guide development, assess environmental damage, and examine compliance or noncompliance with international treaties. Costa Rica has one of the most comprehensive resource inventory programs for doing so, and the success and failures of Costa Rica can be used to generate better surveys elsewhere.

Tropical deforestation has been reported as Costa Rica's most important LUCC issue (TSC, 1982; Sader and Joyce, 1988). Land use/cover trends in Costa Rica reflect the expansion of the agricultural and urban frontiers in a country associated with a high rate (2.5% per year) of population growth. The FAO (1990) reported that the deforestation rate of Costa Rica was 2.9% per year in the late 1980s, the fifth highest in the world. H. Leonard (1987) estimated that Costa Rica's deforestation rate was even higher, 3.9% per year for much of the period 1950–1984. Sader and Joyce (1988) found that deforestation between 1940 and 1983 resulted in the loss of 50% of the 1940 level of primary forest cover, defined as relatively undisturbed forest with an upper canopy covering more than 80% of the surface area.

Some authors suggest that the loss of forests will have a damaging impact on Costa Rica's economy because landscape modifications result in the loss of biodiversity and scenic values, increase erosion and reservoir siltation, and result in loss of agricultural topsoil (Quesada-Mateo, 1990; Solorzano *et al.*, 1991). Quesada-Mateo, in a comprehensive review of the environmental state of Costa Rica, concluded that deforestation is one of the most important causes of environmental degradation in Costa Rica. He identified the primary causes of deforestation as: (1) expansion of the agricultural frontier into critical and fragile forested areas, and (2) urban expansion. The Tropical Science Center (TSC) reported that 17% of the country was experiencing erosional processes caused by humans, that 24% of the country's surface was eroded due to LUCC processes, and that Costa Rica was losing an estimated 680 million tons of soil per year (TSC, 1982). They concluded that this erosion, which continues to occur in both the Pacific and the Atlantic watersheds, is jeopardizing Costa Rica's future agricultural productivity and its water resources infrastructure, and restricting future options for development. Therefore understanding the driving forces of LUCC and their impacts on natural and managed systems is critical for development, sustainable or not, in Costa Rica.

Costa Rica has taken important steps toward protecting representative areas of valuable natural and scenic habitats. Currently, 29% of Costa Rica's territory is under some degree of protection (Sanchez-Azofeifa *et al.*, 1998). Costa Rica's protected areas encompass more than 70 sites, including national parks, biological reserves, wildlife refuges, protected zones, and forest reserves (Umaña and Brandon, 1992). There is serious concern, however, about how land is managed outside national parks. Ramirez and Maldonado (1988) concluded that although a good national park system is in place, LUCC trends outside of the protected areas are unsustainable. Their 1988 report concluded that the continued expansion of the agricultural and cattle frontier, illegal deforestation, and squatter settlements are having serious detrimental impacts on the country's natural resource base. In addition, Sanchez-Azofeifa *et al.* (1998) suggested that although the relative value of natural reserves tended to curb deforestation, that deforestation rates were still high in the buffer zones around the parks and in areas without protection (see also Chapter 19).

The objective of this chapter is to review the past studies of LUCC in Costa Rica and compare them with estimates based on interpretation of Landsat thematic mapper (TM) satellite images. I will examine overall trends of deforestation as well as trends in important components of land use.

## II. OVERVIEW OF PAST AND PRESENT REMOTE SENSING AND GIS STUDIES IN COSTA RICA

Most of the current research in the area of remote sensing and geographic information systems (GIS's) in Central America has focused on Costa Rica. The country's political stability has permitted the development of several research projects in the area of tropical deforestation, habitat fragmentation, and rapid land use dynamics (Sader and Joyce, 1988; Luvall *et al.*, 1990; Sader *et al.*, 1991; E. Veldkamp *et al.*, 1992). The first attempt to implement a nationwide natural resources inventory in Costa Rica dates back to the late 1970s and early 1980s (R. Campbell *et al.*, 1979; Sader, 1980). This first attempt was coordinated by the Costa Rican government and the U.S. Agency for International Development (USAID). The first study was prompted by concerns of Costa Rica's government regarding accelerated rates of land use change and their associated impacts on the national environment. Campbell *et al.* (1979) implemented a three-phase project to design an operational natural resource inventory and information system for Costa Rica. This system used aerial photography and Landsat multispectral scanner (MSS) satellite scenes.

Sader's 1980 project was also part of the national natural resource inventory sponsored by USAID. The main goal of this project was to evaluate the applicability of Landsat MSS information for resource management in Costa Rica. Landsat MSS information was used as the primary data source to estimate the area of coffee lands in the Naranjo region. This study used a GIS and a stratified data base to improve coffee detection. Sader concluded that remote-sensing-derived information was able to estimate the area of coffee plantations in the area sampled with an error of 8%. In addition, this study reported that the Landsat MSS was able to identify mangrove and grassland classes reasonably well, but was considerably less accurate in the identification of forest and brush. Sader compared his results from remote sensing with aerial-photography-derived information and concluded that even though satellite studies provided important insights about the application of remote sensing technologies to Costa Rica, that at least at that time aerial photography was better suited for Costa Rica's interests.

Sader (1988a) conducted a second study related to application of remote sensing and GIS technology to Costa Rica. This study focused on the use of a multitemporal Landsat MSS to monitor and map forest change dynamics. He also compared satellite analysis to aerial photography interpretations. Finally he derived a normalized difference vegetation index (NDVI) for a portion of northeastern Costa Rica near the La Selva Research Station for selected dates in 1976, 1984, and 1986. Sader concluded that the use of multitemporal

NDVI, without cross reference to aerial photography, was not recommended for operational forest inventory programs where the objective was to update forest maps or forest area statistics. But he did find that a combination of two dates of imagery using NDVI was sufficient to detect deforestation.

Sader and Joyce conducted a far more ambitious third study, again using remote sensing. This study, a model for such analyses, resulted in the only comprehensive countrywide forest cover and deforestation analysis of Costa Rica to date (Sader and Joyce, 1988). They assessed "relatively undisturbed" natural forest, defined as forest with an upper canopy cover of 80% or more, for five dates from 1940 to 1983, using interpretation of aerial photography and digitizing of forest cover maps provided by several Costa Rican institutions. Maps and aerial photography interpretation were integrated into a GIS with other spatial data bases (roads, slope ranges, etc.). The minimum mapping unit was 55 ha (750-m grid). The study reported that total primary forest remaining in 1940, 1950, 1961, 1977, and 1983 was 67, 56, 45, 32, and 17% of Costa Rica's territory, respectively. They concluded that: (1) deforestation had initially occurred in tropical dry and moist life zones, but later tropical and premontane moist and wet zones were more affected, (2) by 1983, only the less accessible high-rainfall zones in rugged terrain retained relatively undisturbed forest, and (3) the reported forest area estimates pertained only to primary forest as portrayed by the source maps, so that total forest area in Costa Rica could be larger. The authors concluded that there was a need for future development of quantitative estimates of deforestation and afforestation for Costa Rica using remote sensing techniques.

Remote sensing and GIS technology has also been implemented in Costa Rica to monitor migratory bird habitats (Sader *et al.*, 1991). This project was implemented in northeastern Costa Rica, in the vicinity of the La Selva Research Station and the Braulio Carrillo National Park. An unsupervised Gaussian maximum likelihood classification was performed on a Landsat TM scene acquired on February 6, 1986. Results indicated an overall accuracy for derived forest and nonforest of 70% (kappa correction). They also found that the accuracy of forest classification was 93%. They concluded that newer satellite remote sensing was sufficiently accurate to be able to provide information about habitat availability and habitat conversion that could not be obtained by other means at that time. The study reported problems in classifying successional and secondary growth habitats when an unsupervised classification technique was used, although it could distinguish between major vegetation groups that are important for migratory birds. These authors concluded that they could improve the classification of habitat types through the use of a supervised classification, where sampling plots are selected using aerial photography in successional and secondary growth.

Several regional studies integrating remote sensing and other socioeconomic information using GIS's were conducted in Costa Rica during the 1990s (Mulders *et al.*, 1992; Veldkamp *et al.*, 1992; Alfaro *et al.*, 1994; A. Veldkamp and Fresco, 1994, 1995; Stoorvogel and Eppink, 1995; Schipper *et al.*, 1995). For example, Veldkamp (1992) presented a clear example of the potential of geographic information systems for studying and monitoring deforestation at the regional level in Costa Rica. This study links a quantitative inventory of deforestation to possible factors driving forest clearing, such as accessibility and soil quality. Aerial photography taken in 1952, 1960, 1971, and 1973 for a 395-km$^2$ area in the Atlantic region of Costa Rica was used. This study integrated several data layers, including soil fertility, river networks, and road distribution, making it possible to understand some of the dynamics of deforestation in the Atlantic region. The study concluded that aerial photography offers a good means for quantifying deforestation in Costa Rica.

More comprehensive mapping of deforestation on a regional scale was performed by the Timber and Forest Sector Cooperation Project. This project is part of a joint Costa Rican and German Government conservation project (COSEFORMA, 1994) funded by the German Agency for International Development (GTZ). The project focused on a 560,000-ha region in northern Costa Rica. The study's objective was to create a geographic data base to support a regional plan for forest development in the San Carlos region. The result of this project was a 1 : 50,000 scale map which indicated six different kinds of forest disturbance. This project, although useful, did not provide information regarding the state of forest fragmentation, nor the rate of tropical deforestation in the region.

During the 1990s, due to technology transfer in areas related to remote sensing and geographic information systems, several additional studies at the national and local level were implemented (IMN, 1994; CATIE, 1997). The National Meteorological Institute (IMN, 1994) produced two land cover maps using as reference years 1979 and 1992. These maps were generated by means of visual interpretation of 1 : 200,000 scale black and white prints of a Landsat multispectral scanner data set (1979) and a Landsat thematic mapper data set (1992). The data sets present important conceptual problems as well as classification mistakes that make the data not useful for detailed scientific studies. At the regional level, studies have been implemented by the Tropical Agronomic Center for Research and Teaching (CATIE, 1992). CATIE's work was implemented for the central volcanic region of the country. The goal was to monitor tropical deforestation and land use change at the area managed by the Foundation for Conservation of the Central Volcanic System (FUNDECOR). The study was implemented for data acquired between 1986 and 1992. The main problems with this data set are inconsistencies in the classification scheme that produce confused results, as well as a lack of a clear explanation of the methodology used in the study.

Finally, one of the most comprehensive studies has been completed by the Center on Sustainable Development (CIEDES, 1998). This study, using the methodology developed by the NASA Pathfinder Project, provided comprehensive wall-to-wall assessments of the country's forest cover (primary, secondary, plantations, and tropical dry forest). The study was implemented for Landsat thematic mapper information acquired between 1986 and 1997. This data set, currently in its final preparation stages, is considered a landmark for future conservation studies in Costa Rica. Its main results indicate a forest cover of ~40% of the national territory and a significant decline of the deforestation rate between 1986 and 1997 (~1% per year). The study reflects clearly the positive effects of institutional, legal, and financial incentives created by several Costa Rican governments on efforts to control tropical deforestation in the country.

In summary, deforestation and its consequences have been studied for at least two decades. This brief review shows that important progress has been made in applying remote sensing and GIS tools to address questions related to conservation biology, deforestation, and land use. The present study builds on this previous work, and provides the first comprehensive countrywide assessment of deforestation and forest fragmentation using new high-resolution (30 m) Landsat TM data.

## III. METHODS

I analyzed spatial patterns of land use change in Costa Rica using four independent, but comparable, techniques. First, I developed a geographic information system that included both computerized images of land use and a comprehensive natural resource and socioeconomic data set to study LUCC and its effect on the national and regional natural resource base (Sanchez-Azofeifa, 1996).

Second, I integrated agricultural census data for cattle and pasturelands into a common data set for the entire country. The main sources of this information were 1950, 1955, 1963, 1973, and 1984 agricultural censuses (Costa Rica, 1950, 1955, 1963, 1973, 1984). I derived time series analyses of heads of cattle, area in pastureland, area in sugarcane, and area in coffee at the province level from this information base (Figure 16-1). I next derived annual changes in each of these variables (Tables 16-1 to 16-4) and a statistical analysis to examine the changes in land use with various population and economic parameters. Information for bananas was not available during this study (but see Chapter 18) due to methodological approaches used by the Costa Rican Census Bureau to keep track of the extension of this cash crop.

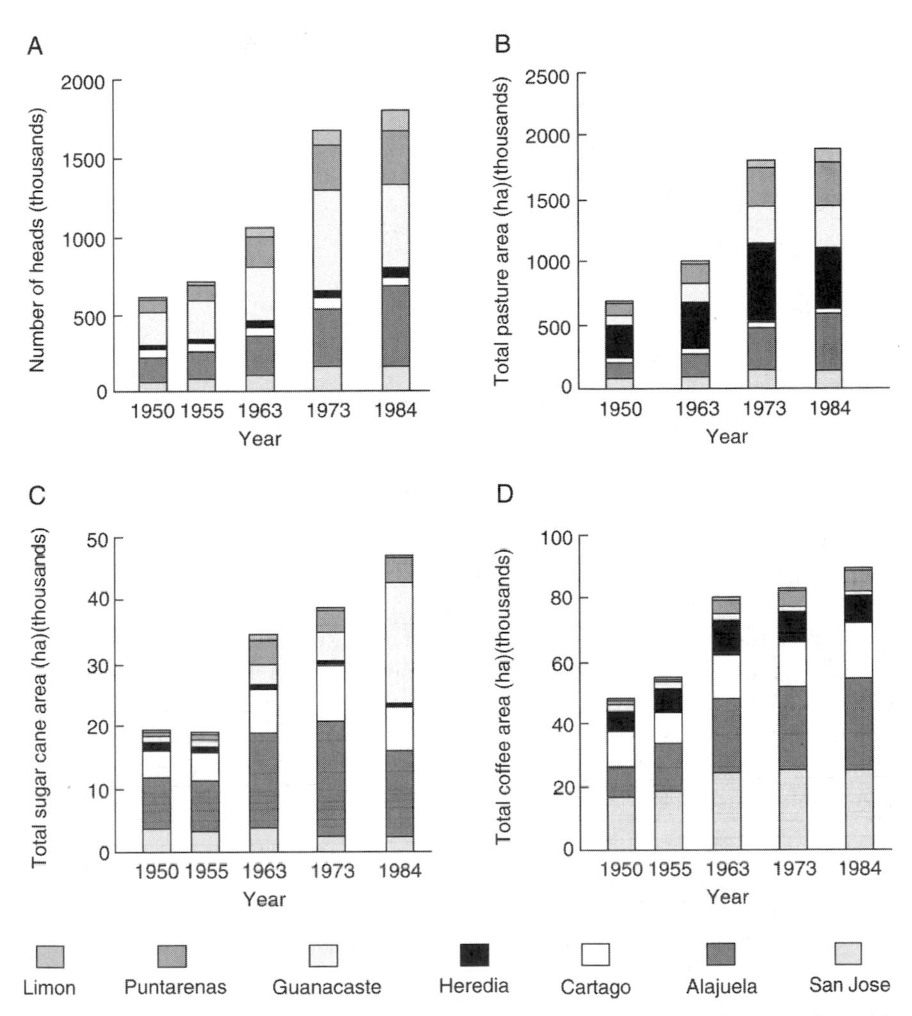


FIGURE 16-1 Time series of numbers of (a) head of cattle and areas in (b) pastureland, (c) sugarcane, and (d) coffee at the province level.

Official Costa Rican census data were last issued in 1984, and the failure to complete a census in 1994 has resulted in a serious gap in understanding of LUCC information.

Third, I integrated the 1984 official land use cover map produced by Costa Rica's National Geographic Institute (IGN, 1984) into my GIS as an independent estimate of the forest cover assessment presented in the agricultural census. Finally, I extracted forest cover information for 1991 using four Landsat

**TABLE 16-1 Costa Rica's Cattle Annual Growth Rates (%)**

| Province | Growth rate | | |
|---|---|---|---|
| | 1950–1963 | 1963–1973 | 1973–1984 |
| San Jose | 4.7 | 3.5 | 0.3 |
| Alajuela | 5.4 | 5.8 | 3.7 |
| Cartago | 0.7 | 3.3 | −0.7 |
| Heredia | 5.9 | 5.4 | 5.6 |
| Guanacaste | 4.9 | 6.5 | −1.8 |
| Puntarenas | 11.5 | 7.7 | 0.5 |
| Limon | 8.9 | 8.4 | 9.0 |
| Country | 5.6 | 6.1 | 0.8 |

*Source:* Costa Rica (1995).

thematic mapper satellite scenes acquired during that year. Images were processed using a supervised classification and a technique developed by the NASA Pathfinder Project for tropical deforestation known as in-pair processing (Chomentowski *et al.*, 1994). This technique permits the highly accurate quantification of forest. Images were first geocoded to a Lambert Conformal Conic projection, and later transformed from a raster to a vector format to allow for thematic and aerial quality control. A complete description of the methodology used can be found in Sanchez-Azofeifa (1996).

**TABLE 16-2 Costa Rica's Pastureland Annual Growth Rates (%)**

| Province | Growth rate | | |
|---|---|---|---|
| | 1950–1955 | 1955–1973 | 1973–1984 |
| San Jose | 5.91 | 3.82 | −0.31 |
| Alajuela | 6.26 | 5.23 | 3.12 |
| Cartago | 4.83 | 0.09 | −1.54 |
| Heredia | 1.99 | 10.18 | 5.13 |
| Guanacaste | 9.83 | 3.07 | −1.93 |
| Puntarenas | 13.36 | 6.81 | 1.19 |
| Limon | 17.30 | 4.01 | 6.40 |
| Country | 8.79 | 4.11 | 0.47 |

*Source:* Costa Rica (1995).

TABLE 16-3 Costa Rica's Sugarcane Annual Growth Rates (%)

| | Growth rate | | |
|---|---|---|---|
| Province | 1950–1963 | 1963–1973 | 1973–1984 |
| San Jose | 1.07 | −4.26 | 0.18 |
| Alajuela | 5.91 | 1.76 | −2.33 |
| Cartago | 5.30 | 2.10 | −2.09 |
| Heredia | −3.60 | 2.51 | −4.05 |
| Guanacaste | 12.31 | 2.73 | 27.93 |
| Puntarenas | 27.53 | −1.51 | 2.95 |
| Limon | 15.14 | −32.05 | 4.20 |
| Country | 5.86 | 1.09 | 2.00 |

*Source:* Costa Rica (1995).

# IV. RESULTS

## A. LUCC Dynamics in Costa Rica (1950–1984)

The FAO (1990a) has reported a Costa Rican deforestation rate of 2.9% per year for the late 1980s. For much of the period 1950–1984, Costa Rica's deforestation rate has been estimated to be 3.9% per year (H. Leonard, 1987). As estimated by Sader and Joyce (1988), deforestation between 1940 and 1983 resulted in the loss of 50% of the 1940 primary forest cover (Figure 16-2).

The FAO's (1990a) reported deforestation rate for Costa Rica of 2.9% per year for the late 1980s has ranked the country fifth in the world in terms of

TABLE 16-4 Costa Rica's Coffee Annual Growth Rates (%)

| | Growth rate | | |
|---|---|---|---|
| Province | 1950–1963 | 1963–1973 | 1973–1984 |
| San Jose | 3.19 | 0.69 | −0.15 |
| Alajuela | 11.99 | 0.87 | 1.31 |
| Cartago | 1.79 | 0.32 | 1.57 |
| Heredia | 2.35 | −1.48 | −0.66 |
| Guanacaste | 6.36 | −2.08 | −1.36 |
| Puntarenas | 54.40 | 2.67 | 2.12 |
| Limon | 42.79 | −0.86 | 8.31 |
| Country | 5.11 | 0.39 | 0.71 |

*Source:* Costa Rica (1995).

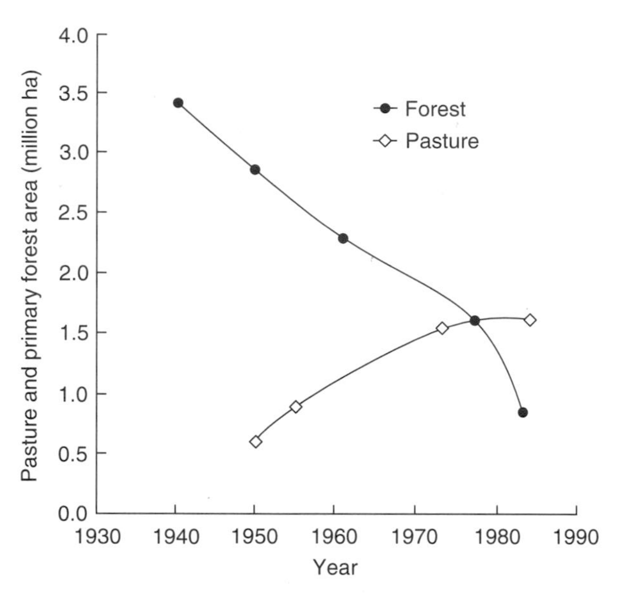


**FIGURE 16-2** Primary forest cover (from Sader and Joyce, 1988) and pasture area (from agricultural census) in Costa Rica between 1940 and 1984.

total forest loss. The dynamics of LUCC in Costa Rica for the last 30 years have been driven by expansion of the agricultural and cattle frontier and by urbanization (Centro Cientifico Tropical, 1982; Harrison, 1991; Quesada-Mateo, 1990; Ramirez and Maldonado, 1988). This LUCC process was encouraged by legislation which placed a low value on forest and encouraged agricultural development (Gaupp, 1992). Although there is clear agreement on the role played by policies oriented to support the cattle sector, disagreement exists regarding the role played by population growth. In a key study presented by Rosero (1997), it is indicated that population did not play an important role in the land cover change due to deforestation. In general, tropical deforestation, associated with expansion of the agricultural land base through national and international loans, was considered a necessity for economic growth and modernization. Deforestation was used as a tool to expand grazing land without regard for potential environmental degradation. The Tropical Science Center reported that by 1980 as much as 76% of all land with potential for growing annual crops was occupied by pastures (Centro Cientifico Tropical, 1982). The concept of using the forest as an open-access resource has produced high deforestation rates during the last 20 years.

## B. Economic Development Policies and LUCC

A principal theme of this book is that economic development policies and processes are linked closely with environmental conditions and changes. Data from the agricultural censuses allow for the analysis of the influence of economic development policy on LUCC dynamics. Five different economic phases can be identified during the 1950–1984 period (J. Aguilar, 1982; Quesada-Mateo, 1990): (1) a shift from an agrarian to an international export economy (1950–1963); (2) expansion of the internal market to regional agricultural trade markets (1963–1973); (3) import substitution and industrial development (1974–1978); (4) an important period of economic crisis (1978–1982); and (5) postcrisis economic recovery (1983–1989). The following paragraphs show how these economic phases were linked to land use change, and how all led to deforestation.

Figure 16-3 shows that the relative contribution of the agricultural sector to the national GNP decreased from 41% in 1950 to 20% in 1984. In 1950, only three crops, coffee, banana, and cocoa, represented 51% of the value of national agricultural production (Costa Rica, 1995). Exports of these three agricultural products accounted for 91% of the total export revenue (Costa Rica, 1995a). During the first phase, the country's economic growth was based on a policy of expanding the production of a few commercial crops such as coffee and sugarcane. These two crops experienced high growth rates during the 1950s (5.1 and 5.7%, respectively).

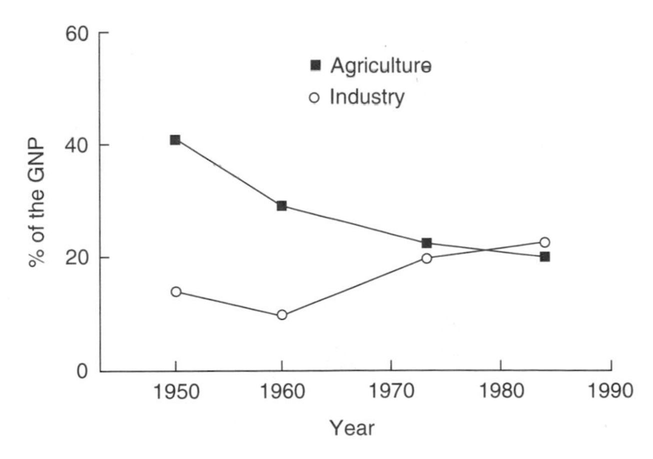


FIGURE 16-3 The relative contribution of the agricultural sector to the national GNP. (Source: Banco Central de Costa Rica, 1995.)

My analysis of Costa Rica's agricultural census during the study period indicates that the percentage of the national land base dedicated to both permanent crops and annual crops has been roughly constant (Figure 16-4). This is a result of a complex interaction of government policies, technologic advances, and, especially, inputs such as fertilizers which have increased yields. At the same time pastureland has expanded as a result of the existence of a cattle sector based on extensive exploitation of a resource rather than intensive use of the land. Deforestation was the main approach used to support the expansion of the cattle growth industry.

During the second phase, cattle production was a large part of a new approach to generating foreign exchange. The expansion of sugarcane and cattle production, for meat exports, became the main factors driving LUCC. In 1960, sugarcane and meat production had increased to 14.3% of the value of the country's GNP. Meanwhile, the contribution from coffee, banana, and cocoa declined to 44%. In 1963, Costa Rica became part of the Central America Common Market (MERCOCEN). At that time, new policies promoting industrialization were implemented. Thus the third phase began between 1974 and 1978. New policies aimed at transforming the national economy from an agrarian to an industrial one were implemented. These policies included investment credits, rapid depreciation rates, and low taxes on imported capital goods. During the period 1970–1979 the contribution of the agricultural sector to Costa Rica's GNP declined from 22.5 to 19% because the industrial sector's

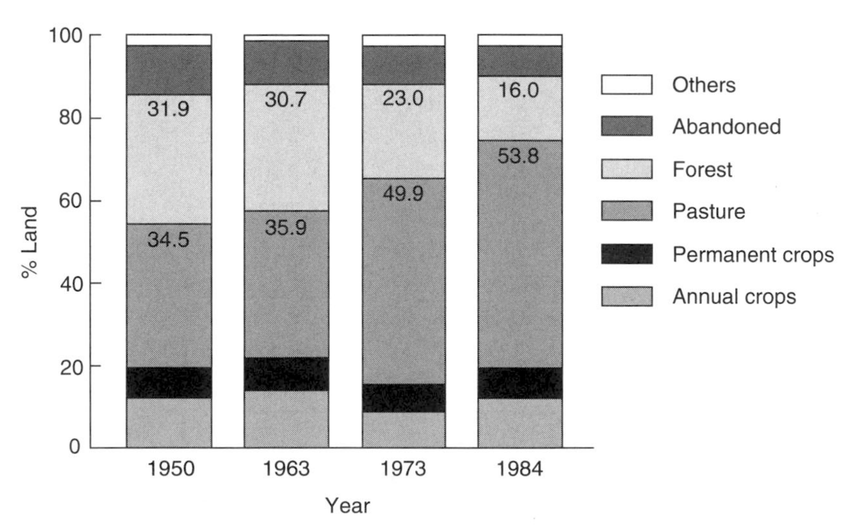


FIGURE 16-4 The proportion of the national land base dedicated to both permanent crops and annual crops.

annual growth rate was 9.6% while the agricultural sector grew at a much slower 5% per year.

The different economic development policies implemented at various times between 1950 and 1978 are reflected in different growth rates for the agricultural sector. This sector grew 3.2% per year between 1950 and 1962, 5.6% per year between 1962 and 1972, and 2.6% per year between 1972 and 1978 (Costa Rica, 1995). The exact reasons are complex, but Aguilar *et al.* (1982) concluded that the decrease observed after 1972 is related to four factors: (1) price increases in fertilizer and chemical additives derived from petroleum, (2) new taxes on banana production, (3) restrictions on expansion of coffee plantations, and (4) increased emphasis on industrial expansion.

Costa Rica's agricultural census (Costa Rica, 1995a) shows that farmlands grew 0.2% per year between 1910 and 1950. The change increased to 1.3% per year between 1950 and 1963, but dropped to 0.9% per year between 1963 and 1973. Most of the 1950–1963 high growth rate was due to land reclamation not directed at expanding agricultural land. During this period land reclamation policies by Costa Rica's Land Colonization Institute (ITCO) promoted most of the conversion from forest cover to other uses, especially subsistence farming systems.

## C. Pasture and Cattle Expansion

This section examines in more detail the specific forces behind land use and land cover changes. The expansion of cattle, and therefore pastureland, has been driven historically by external forces outside of Costa Rica. Schelhas (1991) indicated that "expansion of cattle pasture was driven by the increased U.S. demand for low grade beef and by the financing of cattle expansion by the international development banks in an effort to diversify Costa Rican exports beyond coffee and bananas." Similar conclusions have been reached by other authors (Quesada-Mateo, 1990). Most (1,558,000 ha) of Costa Rica's deforested land between 1950 and 1973 was turned into pasture (Costa Rica, 1950, 1973), so the percentage of the country in pasture area changed from 34 to 50% during this time period. By 1973, 80% of all agricultural land area was used for cattle ranching, and one-half of all farms were dedicated to this activity (Leon *et al.*, 1982). The dynamics of this LUCC process took place in all provinces (Figure 16-1). The period between 1950 and 1963 was characterized by an annual growth rate for cattle of 5.6%, and between 1963 and 1973 it was 6.1% (Table 16-1). These rates were driven by rapid increases in meat prices (2.7% per year) (Figure 16-5).

After Costa Rica's initial industrial transformation period (1973–1978), there was an important reduction in the cattle growth rate (0.8% between

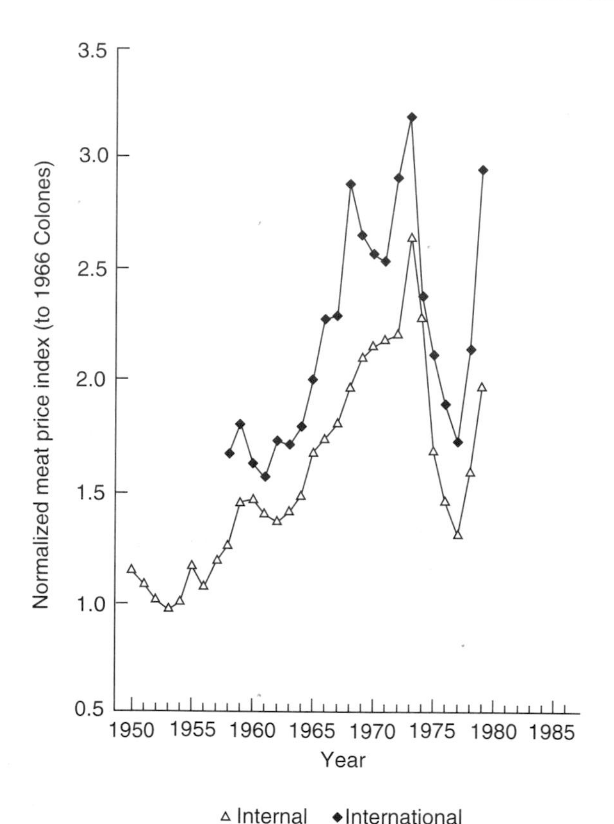


FIGURE 16-5 Domestic and international meat prices.

1973 and 1984; Table 16-1). Meat exports declined to represent only 8 and 10% of all exports by 1975 and 1979, respectively. This low growth rate can be related to several factors such as the elimination of government-supported loan programs during the 1970s and a decline in the international price of meat after 1974, when meat prices dropped 30 and 20% for internal and international consumption, respectively (Figure 16-5).

The 1973 oil embargo, which generated inflation in the United States, was identified as the main force behind the 20% reduction in international meat prices by Leon *et al.* (1982). Because Costa Rica sold most of its production to the U.S. market, which offered import incentives, Costa Rica's meat prices avoided a major drop. Nevertheless meat prices dropped to just over 60% relative to their 1973 prices during the 1974–1979 period; meanwhile other international markets dropped between 40 and 50% of the 1973 price (BIRF,

1971). This price drop did not affect total production, which continued to grow as a result of government incentives to the cattle sector in the form of loans (Lutz and Daly, 1991) (Figure 16-6a). Lutz and Daly note that of the total amount of money given by the government as agricultural loans, more than 50% was used by cattle ranchers. This group represented only 5% of all people who got benefits. The growth rate of pastureland has been estimated as 8.8% per year between 1950 and 1963, 4.1% for 1963–1973, and 0.5% between 1973 and 1984 (Table 16-2). During the first two periods, all provinces experienced high annual growth rates in pasture development (see Chapter 13).

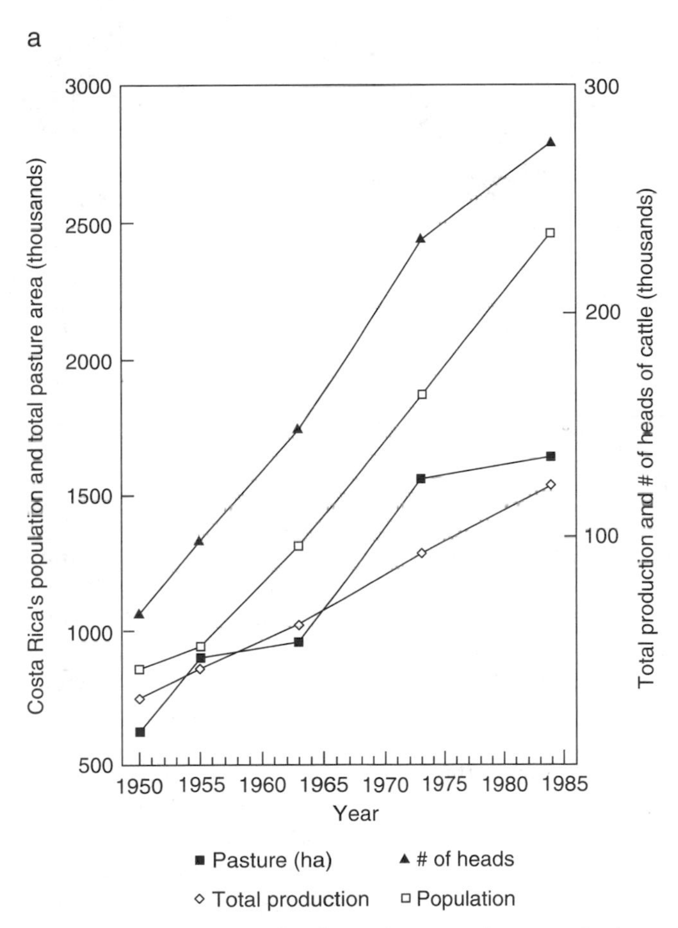


FIGURE 16-6 (a) Pastureland, heads of cattle, total meat production, and other population in Cost Rica. (Sources: Costa Rica, 1950, 1955, 1963, 1973, and 1984; Banco Central de Costa Rica, 1995.) (b) Percentage of the total meat production used for internal consumption and exports.

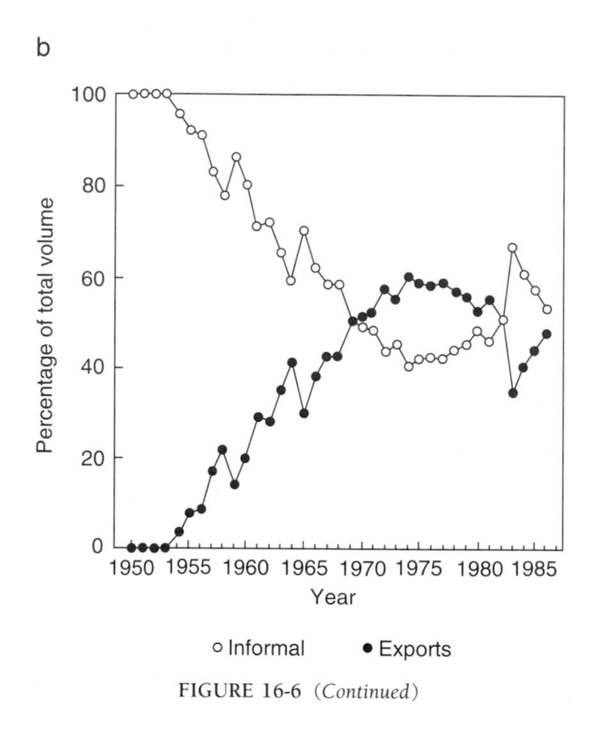


FIGURE 16-6 *(Continued)*

Expansion of pastureland and its direct relationship with the increase of international meat prices can be divided into the following development periods (Leon *et al.*, 1982):

a. 1950–1955. During this period the demand for pastureland exceeded supply. Farms dedicated to other less profitable agricultural uses were transformed into pastureland. During this period, pastureland increased at 1.92% per year. A total of 98,000 ha was transformed into pastureland, 30% from deforestation of primary forest and 70% from transformation of previously deforested land (Table 16-5). The meat price index for internal consumption (price per kilogram of standing cattle) hardly changed during this period (Figure 16-5).

b. 1955–1963. During this period, 16% of remaining primary forest was converted to crops and pasture. Twenty-eight percent of the total area deforested was dedicated to pasture. The meat price index changed for internal consumption from 1.18 in 1955 (constant 1966 colónes) to 1.43 in 1963 (Figure 16-5). International meat index prices changed from 1.09 to 1.72 during this period.

c. 1963–1973. During this period, a further 9% of remaining primary forest land was converted to agriculture. At the same time, land used for cattle growth increased 14%. Of this total, 25% (150,000 ha) was created from transformation

TABLE 16-5 Source of Land Use to Expand Pasture Area in Costa Rica

| Period | Type A (ha) | Type B (ha) | Type C (ha) | Type D (ha) |
|---|---|---|---|---|
| 1950–1955 | 68,000 | 30,000 | 98,000 | 39,000 |
| 1955–1963 | N.D. | 229,000 | 229,000 | 816,000 |
| 1963–1973 | 150,000 | 451,000 | 601,000 | 455,000 |

*Source:* Leon *et al.* (1982).
*Notes.* Type A, increment from lands already transformed into farms or already deforested. Type B, increment due to deforestation of primary forest. Type C, total increment on pastureland. Type D, total increment on farm area. N.D., not determined.

of previously deforested land. The international meat index price changed from 1.72 to 3.19, while the internal market price index increased from 1.43 to 2.66.

d. 1973–1984. During this period pastureland increased only 3.9%, even though total meat production, population, and the number of head of cattle continued to increase. This small growth rate can be related to the variations on the international price of meat during this period (Figure 16-5).

I examined the relation of pastureland to various possible causative agents for 1950, 1955, 1963, 1973, and 1984. Not surprisingly there was a high correlation with Costa Rica's population ($R^2 = 0.95$) and pasture area and total meat production ($R^2 = 0.97$) for both internal consumption and production for exports ($R^2 = 0.97$). A multiple regression found that total meat production was highly correlated ($R^2 = 0.96$) with Costa Rica's population and both international and internal meat prices.

$$\text{TMP} = -40{,}677 + 0.101\text{CRP} - 14{,}902\text{INTMP} - 4678\ \text{INTLMP},$$

where TMP is total meat production, CRP is Costa Rica's population, and INTMP and INTLMP are the normalized index of meat prices for internal consumption and exports, respectively [normalized to 1966 (real) colónes].

Figure 16-5 indicates that even though internal prices of meat increased during the study period, the rate of growth of the international price of meat increased even more, so that internal consumption changed from 100% of the total production in 1950 to 50% in 1969 (Figure 16-6b). Meat exports surpassed total meat production for internal consumption during 1970. The period 1970–1982 was characterized by a higher share of the meat market being used for exports rather than internal consumption. This situation changed after 1982 due to drops in the international price of meat and contractions of the local economy.

Pasture area is negatively highly correlated ($R^2 = 0.97$) with Costa Rica's population, total volume of meat exports (T), and total volume of internal consumption (T).

$$PL_i = 658{,}946 - 0.63CRP_i + 25.5VE + 17.2VI,$$

where $PL_i$ is the total pastureland (ha) for year $i$, CRP is Costa Rica's population in the $i$th year, and VE and VI are the total volume for exports and internal consumption (T), respectively. It is important to note that internal and international meat prices were not considered because of their correlation with VI and VE, respectively (Table 16-6).

The preceding correlation must be taken with caution, however, because meat prices for export were sustained artificially by government incentives. These incentives helped to avoid a major drop between 1972 and 1976. In addition, the rate of growth of Costa Rica's population and its per capita consumption did not change, creating a sustained growth in total meat production (Figure 16-6a).

## D. Annual and Permanent Crops

### 1. Sugarcane Production

Sugarcane is a highly productive crop that grows well in Costa Rica. It is used to make sugar, molasses, and rum. Sugarcane area grew 5.9% per year between 1950 and 1963 (Figure 16-7, Table 16-3); production changed from 38.2 MT/ha in 1955 to 46.3 MT/ha in 1963 due to the application of new technologies and the increasing use of inputs (Barboza, 1981). After this period of rapid growth, growth rates dropped to 1.1% per year between 1963 and 1973, and 2.0% between 1973 and 1984. However, sugarcane production had changed from 46.5 MT/ha in 1963 to 72 MT/ha by the 1980s. Sugarcane yield in Costa Rica is being estimated to be 66.3 metric tons (MT), which is above the world average of 53.5 MT/ha (Barboza, 1981). The number of farms used for sugarcane production decreased from 11,000 farms in 1950 to 9500 by 1973, although the total area of sugarcane production increased by 19,000 ha during the same period. By 1984, a total of 47,286 ha (7373 farms) were dedicated to sugarcane production (Costa Rica, 1984). Sugarcane production occurs primarily in the Alajuela and Guanacaste provinces. Alajuela was originally more important until 1973, when production shifted to Guanacaste due to the creation of CATSA (Central-Azucarera-del-Tempisque, S.A.), which started to exploit sugar production in the Tempisque River valley. By 1984, more than a third of the national sugarcane production was located in Guanacaste. The Guanacaste Irrigation Project has been partially responsible for this regional growth.

TABLE 16-6 Correlation between Pasture Area (1950, 1963, 1973, and 1984) and Other Socioeconomic Variables

| | Pasture | Costa Rican population | Volume of exports | Volume for internal use | Total volume | % volume internal | % volume exports |
|---|---|---|---|---|---|---|---|
| Population | 0.951 | | | | | | |
| Volume exports | 0.969 | 0.938 | | | | | |
| Volume internal | 0.785 | 0.900 | 0.705 | | | | |
| Volume total | 0.965 | 0.996 | 0.949 | 0.893 | | | |
| % volume, internal | −0.870 | −0.795 | −0.942 | −0.511 | −0.826 | | |
| % volume, exports | 0.870 | 0.795 | 0.942 | 0.511 | 0.826 | −1.000 | |
| Heads of cattle | 0.983 | 0.985 | 0.976 | 0.840 | 0.994 | −0.878 | 0.878 |

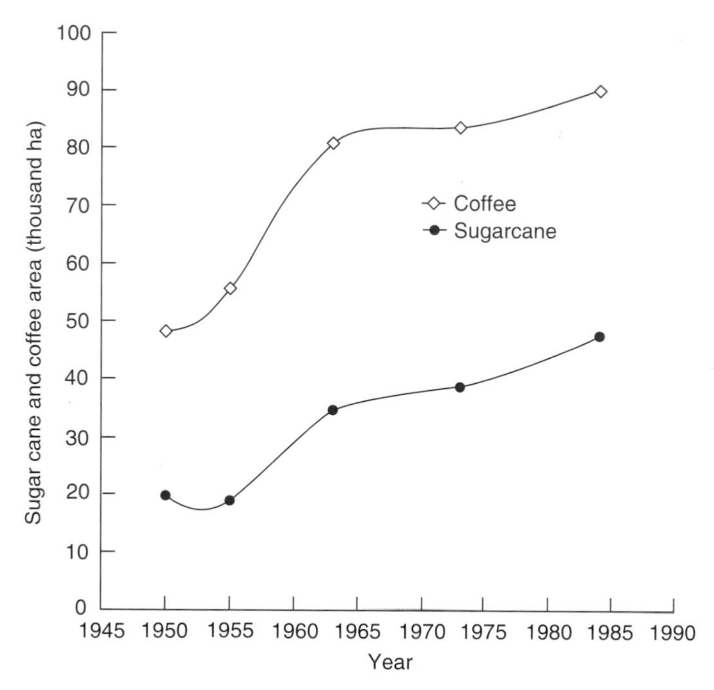


FIGURE 16-7 Total national coffee and sugarcane area.

## 2. Coffee Production

Currently, coffee plantations are located in Costa Rica's central region (68%), the South Pacific region (142%), and the central Pacific region (12.7%) (see Figure 11-10). The remaining 5.1% are distributed throughout the national territory. A more detailed analysis of spatial trends related to coffee plantations has been presented by Aguilar *et al.* (1982). As explained in Chapter 1 and in Schelhas (1991), expansion of coffee plantations was a driving force in Costa Rica's Central Valley during early colonizations. Seligson (1980) suggests that the number of coffee plantations started to multiply by 1820, mainly as a result of the aggregation of small farms into larger coffee plantations. During the 19th century, coffee production was the fundamental element for Costa Rica's internal economic growth.

Costa Rica's efficiency in coffee production has been increasing and is presently the highest in the world (6.3 MT/ha vs 2.8 MT/ha worldwide). Coffee production increased 5.1% per year between 1950 and 1973, 0.4% between 1963 and 1973, and 0.7% between 1973 and 1984 (Table 16-4). The small growth rates between 1963 and 1984 are explained by government policies aimed at controlling the extension and growth of coffee plantations (Aguilar *et al.*, 1982), but may also reflect the increases in fertilizer prices during that time.

Costa Rica's agricultural census data indicate that the total number of farms used for coffee production doubled between 1950 (15,222 farms) and 1963 (32,353). The average farm size was less than 10 ha. Total area grew 2.4% per year between 1950 and 1973, and changed from 48,885 ha in 1950 to 83,407 ha by 1973 (Figure 16-7). Between 1973 and 1984, few changes were observed in either the total number of farms (34,464 farms) or their total area (89,881 ha).

## E. Forest Cover Estimates from Remote Sensing

Four Landsat TM scenes with less than 20% cloud cover were used to map 1991 forest cover. These four scenes represented 93% of Costa Rica's territory. The analysis for 1991 showed that 29% (1,361,491 ha) of the country was covered with primary forest with a canopy density of ~80%, 54% (2,546,423 ha) was identified as nonforest, and 17% (800,687 ha) was covered with clouds. The highest cloud cover was in the northern zone and the Peninsula de Osa.

My results are significantly different from those of Sader and Joyce (1988), who reported that only 17% of the nation was covered by undisturbed forest in 1983, in contrast with the 29% reported in this study. There are several possible explanations for the difference. First, the Sader and Joyce (1988) study was based on digitized maps from different sources (which can contain important classification errors). Second, the minimum mapping unit selected for the 1983 study was 55 ha, in contrast with 3 ha in this study. Sader and Joyce were aware that their large mapping units could cause small areas of natural forest to be missed and suggested that their reported forest area could be underestimated. By overlaying the new 1991 forest cover image on maps of national parks, I calculate that of the total primary forest mapped, 29% of the mapped primary forest is protected by national parks, and 71% is outside of these protected areas. Existing forest cover is concentrated along the central cordillera, where high slopes make access difficult.

Four main forest islands, with areas of more than 103 ha, were found along the central cordillera. The presence of these forest islands is related to the existence of national parks and road construction along "pasos." Pasos is a Costa Rican term used to identify geographic depressions between the main mountain systems in the country. There is a strong correlation between location of roads and gaps in existing forest. This pattern has been documented by several authors (Sader and Joyce, 1988; Veldkamp *et al*,. 1992).

The distribution of primary forest at the province level follows. Provinces with the least primary forest cover were Guanacaste (3%), Alajuela (16%), Puntarenas, (19%), and San Jose (30%). The highest forest cover was found in Cartago (68%) and Limon (60%). The highest forest cover in Cartago is likely due to the presence of several conservation areas for water resource protection in the Reventazon and Pacuare river basins (Sanchez-Azofeifa and Harriss, 1994). Even though Limon Province has one of the highest remaining percentages of forest cover, the growth rates in pasture area have been estimated as 7% per year (DGEC, 1973, 1984). This trend suggests a potential conflict between conservation and land use change (deforestation) in the region.

### F. Rate of Forest Loss in Central Costa Rica (1986–1991)

My analysis also showed that deforestation produced an increase in island fragments during the study period. Between 1986 and 1991, the total number of forest islands between 3 and 50 ha and 50 and 100 ha increased by 524 and 45, respectively. In addition, 15 new islands with areas greater than 500 ha were created. Thus deforestation produced a net increase in the total area of fragmented forest islands for all categories with the exception of forest islands between 100–150 and 450–500 ha. Forest lost in islands with areas greater than 450 ha contributed to the increase in the number of forest islands in the smaller class ranges.

## V. DISCUSSION

Remote sensing and geographic information systems are important tools in communicating spatial and/or temporal trends to policymakers. Remote sensing information can provide a wide range of thematic information over a region. This information can be later integrated into a GIS and linked to other relevant *in situ* data sets.

Although the integrated data base presented in this paper is useful in identifying past LUCC trends, it has limited applications for understanding impacts and consequences of LUCC because of the nature of the original data sets and the process of sampling. The inconsistencies that were detected in the extension of forest cover puts into question the accuracy of the pastureland estimates presented in Figure 16-4. A comparison between the reported forest cover in the agricultural census and the official Costa Rica land use map provides additional information regarding the quality of the collected data

(Figure 16-8). Costa Rica's official land use map was prepared in 1984 by Costa Rica's National Geographic Institute (IGN, 1984). The IGN map is an interpretation of aerial photography and infrared aerial photography at several scales (Elizondo, 1985). The IGN map scale is 1:200,000.

The premise of this comparison was that if there is good agreement between the agricultural census and the reported forest cover on the official land use map, total reported forest at the county level will plot along a 1:1 line. Points under the 1:1 line will indicate underestimation, and points over it overestimation, of the agricultural census. Figure 16-8 indicates that, with few exceptions, the forest cover in Costa Rica is underestimated by the agricultural census. For example, for the Perez Zeledon, San Ramon, and San Carlos counties, forest cover is underestimated by 65,518, 49,180, and 90,486 ha, respectively. The correlation ($R^2$) between the agricultural census and the IGN map at the county level is only 0.42. This comparison explains the significant differences in total forest area between the agricultural census information and Sader and Joyce's (1988) satellite estimates (Figure 16-9 and Table 16-7). Moreover, this comparison questions the accuracy of all previous agricultural census data.

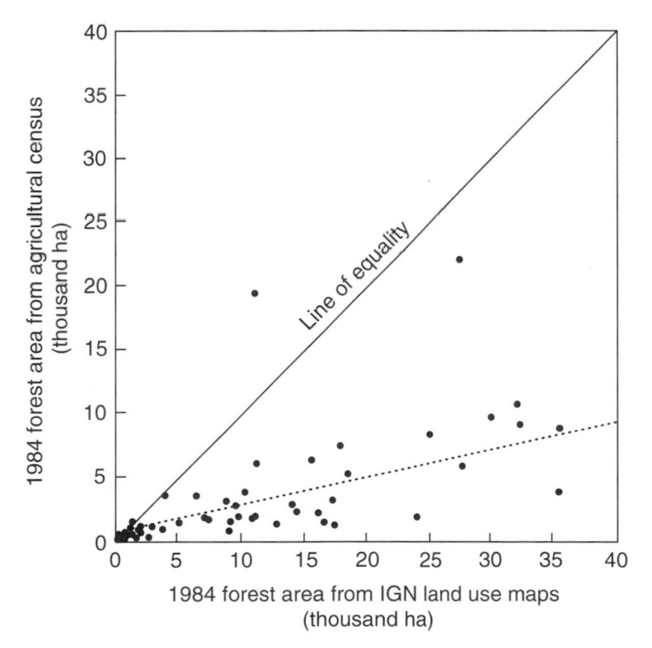


FIGURE 16-8 Comparison between the IGN agricultural census and the reported forest cover on the official IGN land use map. Points under the 1:1 line indicate underestimation and points over it overestimation by the agricultural census.

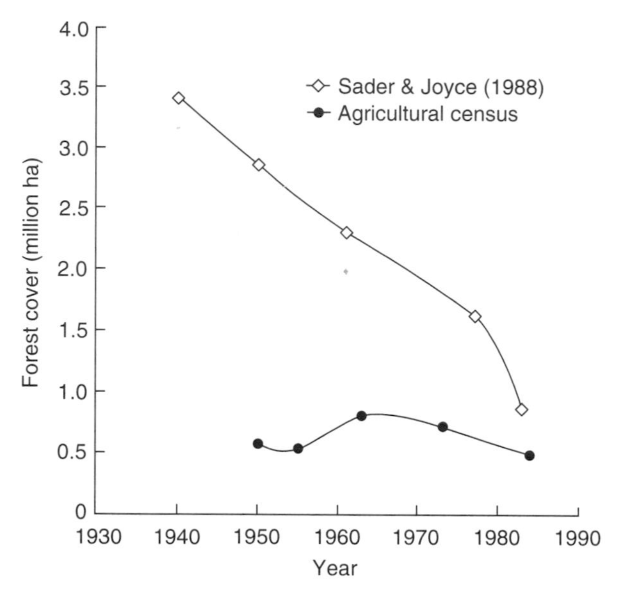


**FIGURE 16-9** Comparison between total forest area between the agricultural census information and Sader and Joyce's (1988) satellite estimates.

**TABLE 16-7 Comparison of Forest Cover Estimates from Different Sources and Methods**

| Source | Year | Method | Scale | Forest area (ha) |
|---|---|---|---|---|
| DGES[a] (1950) agricultural census | 1950 | Survey | N/A | 577,000 |
| DGES (1955) agricultural census | 1955 | Survey | N/A | 542,000 |
| DGES (1963) agricultural census | 1963 | Survey | N/A | 819,000 |
| DGES (1973) agricultural census | 1973 | Survey | N/A | 716,000 |
| DGES (1984) agricultural census | 1984 | Survey | N/A | 492,000 |
| Sader and Joyce (1988) | 1940 | Aerial photo inter. | 1 : 1,000,000 | 3,420,600 |
| Sader and Joyce (1988) | 1950 | Aerial photo inter. | 1 : 1,000,000 | 2,864,200 |
| Sader and Joyce (1988) | 1961 | Aerial photo inter. | 1 : 1,000,000 | 2,303,500 |
| Sader and Joyce (1988) | 1977 | MSS interpretation | 1 : 1,000,000 | 1,615,400 |
| Sader and Joyce (1988) | 1983 | MSS interpretation | 1 : 1,000,000 | 871,100 |
| Sanchez-Azofeifa (1996) | 1991 | TM interpretation | 1 : 250,000 | 1,359,000 |

[a] Direccion General de Estadisticas y Censos (Costa Rica, various years).

The causes for underestimation can be attributed to a lack of a statistical sampling scheme for census site selection.

I then checked the accuracy of IGN's forest cover maps by comparing them against the forest cover map generated for 1986 using a Landsat TM scene. Figure 16-10 indicates that there is a high correlation ($R^2$ = 0.87) between the IGN's land use maps and the information derived from remote sensing information. Significant differences were observed only for those counties which had high cloud cover.

Future remote sensing data sets would be valuable for enhancing understanding of both spatial and temporal trends in LUCC. Thematic remote sensing information should be analyzed for each year in which the agriculture census was and will be collected (e.g., 1984, 1996, 2006). The thematic information must have classes similar to those of the agricultural census (coffee, banana, pastureland, annual crops, forest, etc.). This will allow for refinement of collected census information at the spatial scale by means of remote sensing data such as Landsat thematic mapping. In addition, inter-census year information

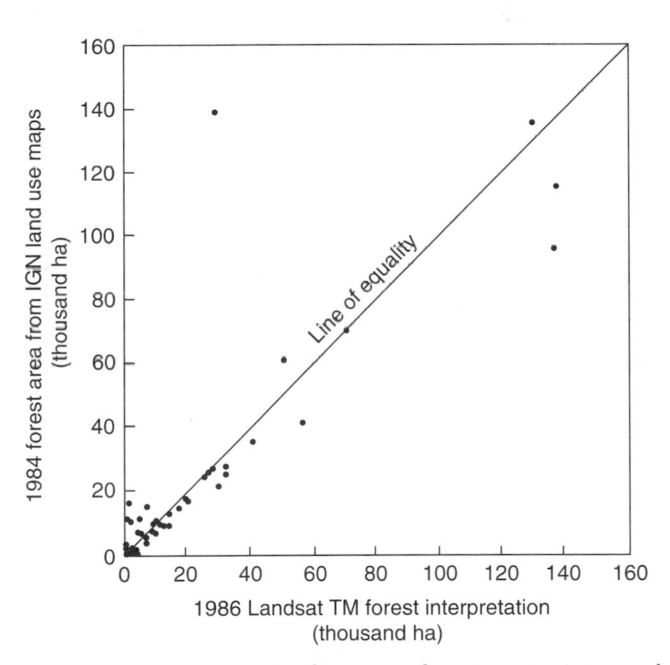


FIGURE 16-10 Comparison between 1984 forest cover from Costa Rica's National Geographic Institute (IGN, 1984) and forest cover derived from a 1986 Landsat TM scene (path 15/row 53) acquired on February 6, 1986. The high correlation ($R^2$ = 0.87) between the IGN information and the Landsat TM classification indicates that agricultural census data are underestimating current forest cover in Costa Rica.

can be generated from the same satellite sensor. This information will quantify trends in LUCC more accurately in both time and space. Selection of study sites can be accomplished by using partial sampling at the scene level.

An agricultural census provides aggregated information at the county or district level, without spatial variability of the different land use/cover types. A second-generation GIS data base will provide information regarding the spatial variability of crops and pastureland, granting more detailed information over the agriculture census.

Potential applications of the proposed second-generation data set will make it possible not only to identify potential regional conflicts between water resources development and LUCC, but also to have a more quantitative understanding of the spatial distribution of ongoing LUCC in a region (district, watershed, etc.).

This data base could also be combined with other topologic information such as slope distribution, precipitation fields, roads and river networks, and life zone classification. This will help in assessing the impact of several land use policies on the natural environment. Results such as those presented by Sanchez-Azofeifa and Harriss (1994) will be able to be reproduced in all drainage basins within the nation. The integration of this GIS/RS information within the objectives of national development programs will allow for more scientific decision making.

## VI. CONCLUSIONS

1. The analysis of the agricultural census data for a 34-year period documents how variable land cover/use change is over time. It also indicates that such change can be episodic, depending on social, economic, and technological factors. A better understanding of the forces, trends, and spatial distribution of different land use/cover changes can be achieved by integrating such surveys with remote-sensing-derived information and geographic information systems.
2. Improved decision making can be achieved if remote sensing information is integrated with agricultural census data. I recommend the creation of a second-generation data base as part of a more comprehensive operational plan for LUCC monitoring in Costa Rica. This data base must integrate current available information in GIS format with remote sensing thematic information. The remote sensing information must use the same classification scheme as the agricultural census. The remote sensing information must also be generated from a complete coverage of satellite scenes in the same year that the agricultural census is developed. The information

produced from the proposed second-generation GIS will provide a valuable independent data set for calibration and validation of land use change models such as USTED (Sustainable Land Use Development) (Stoorvogel and Eppink, 1995).

3. The second-generation GIS data base proposed in this paper will also help to make it possible to study several aspects related to watershed management. It will help in quantifying changes in land cover (e.g., deforestation) on a specific drainage basin, in addition to being a source of basic information for watershed characterization, prioritization, and land use planning by the Costa Rica's Electricity Institute.
4. The success or failure of future land use/cover change planning in Costa Rica will depend on an accurate resource inventory. This paper has demonstrated that forest information generated from agricultural census data is not accurate, and that it differs from independent estimates developed by other government agencies in Costa Rica. In addition, it has been demonstrated that forest cover information derived by means of remote sensing approaches estimates done by Costa Rica's National Geographic Institute.
5. I suggested that is time to generate a first-generation GIS system for other Central American countries. These systems can follow the same approaches that are recommended for building one for Costa Rica. The application of the first-generation GIS data base will have important regional considerations at the Central American level. Agricultural censuses are being collected in most of the countries, and national geographic institutes have their own political boundary information. The building of these first-generation data sets can be easily accomplished for each Central American country through national government agencies, national universities, or regional development agencies such as the Central American Commission for Environment and Development (CCAD), the Central American Committee for Water Resources (CRRH), or the Central American Project for Climate Change (PCCC).

## ACKNOWLEDGMENTS

I thank Dr. Robert C. Harriss, Dr. David Skole, Dr. Carlos Quesada, and Dr. Charlie Hall for their useful comments on this chapter. This work was funded by NASA's Interdisciplinary Research Program and by NASA's Mission to Planet Earth. Additional funding was provided by NASA's Pathfinder Tropical Deforestation Project, a Fulbright scholarship, the Vice-Presidency for Research of the University of Costa Rica, and a grant from the EOSAT Corporation. I thank Dr. Christine Orosz for her useful comments on this chapter.

## REFERENCES


Aguilar, J. 1982. *Cambios en el Uso de la Tierra en Costa Rica.* CONICIT, San Jose, Costa Rica.

Aguilar, F., C. B. Justo, and J. Leon. 1982. *El Desarrollo Tecnologico del Cafe en Costa Rica y La Politicas Cientifico Tecnologicas.* Consejo Nacional de Investigaciones Cientificas y Tecnologicas, Costa Rica.

Alfaro, R., J. Bouma, L. O. Fresco, D. M. Jansen, S. B. Kroonenberg, A. C. Leeuwen, R. A. Schipper, R. L. Sevenhuysen, J. J. Stoorvogel, and V. Watson. 1994. Sustainable land use planning in Costa Rica: A methodological case study on farm and regional level. In L. O. Fresco, L. Stroosnijder, J. Bouma, and H. V. Keulen (Eds.), *The Future of the Land: Mobilizing and Integrating Knowledge for Land Use Options,* pp. 183–202. Wiley, New York.

Barboza, C. V. 1981. Proyecto de instrumentos de politica y planification cientifica y tecnologica para centroamerica y Panama: Perfil No. 2: Desarrollo tecnologico en el cultivo de la cana de Azucar. CONITIC, San Jose, Costa Rica.

BIRF. 1971. Centro Cientifico Tropical. San Jose, Costa Rica.

Campbell, R. W. J., H. Rodriguez, and S. A. Sader. 1979. Design of a nationwide natural resource inventory and information system for Costa Rica.

CATIE. 1992. Personal communication.

CATIE. 1997. *Mapa de Uso de la Tierra de la Coordillera Volcanica Central (1992).* MINAE, San Jose, Costa Rica.

Centro Cientifico Tropical. 1982. Costa Rica: Perfil Ambiental, Estudio de Campo.

CIEDES. 1998. *Forest Cover Map.* FONAFIFO, San Jose.

Chomentowski, W., B. Salas, and D. Skole. 1994. Landsat Pathfinder project advances deforestation mapping. *GIS World* 7:34–38.

Costa Rica. 1950, 1955, 1963, 1973, 1984. *Censos Agricolas.* Direccion General de Estadisticas y Censos, San Jose, Costa Rica.

Costa Rica. 1973. *Censos Agricolas.* San Jose, Costa Rica: Direccion General de Estadisticas y Censos.

Costa Rica. 1995. Estadisticas de Desarrollo Economico. In Banco Central de Costa Rica, San Jose, Costa Rica.

COSEFORMA. 1994. *Inventario Forestal de la Region Huetar Norte: Resumen de Resultados.* Projecto de Cooperacion Tecnica entre la Republica de Alemania y La Republica de Costa Rica, San Jose, Costa Rica.

DGEC. 1973.

DGEC. 1987. *Agricultural Census 1984.* DGEC, San Jose, Costa Rica.

Elizondo, C. 1985. Programa de perception remota del Instituto Geografico Nacional de Costa Rica Informe General. In *Proceedings: 3rd Reunion Plenaria Selper, Santiago, Chile,* pp. 203–211.

FAO. 1990a. *Forest Resources Assessment.* FAO, Rome.

FAO. 1990b. *Centro America y los problemas del agro en el campo.* FAO, Santiago, Chile. 302 pp.

Gaupp, P. 1992. Ecology and development in the tropics. *Swiss Review of World Affairs,* Sept., 14–19.

Harrison, S. 1991. Population growth, land use and deforestation in Costa Rica, 1950–1984. *Interciencia* 16:83–93.

IGN. 1984. *Mapa de Uso de la Tierra.* IGN, San Jose, Costa Rica.

IMN. 1994. *Tablas Climatológicas para la Estación Fabio Baudrit.* IMN, San Jose, Costa Rica.

Leon, J., C. Barboza, and J. Aguilar. 1982. *Desarrollo tecnologico en la ganaderia de carne.* CONICIT, San Jose, Costa Rica.

Leonard, J. 1987. *Recursos naturales y desarrollo economico en America Central: Un perfil ambiental.* CATIE, Costa Rica, Turrialba.

Lutz, E., and H. Daly. 1991. Incentives, regulations, and sustainable land use in Costa Rica. *Environment and Resources Economics* 1:179–194.

Luvall, J. C., D. Lieberman, M. Lieberman, G. S. Hartshorn, and R. Peralta. 1990. Estimation of tropical forest canopy temperatures, response numbers, and evapotranspiration using an aircraft-based thermal sensor. *Photogrammetric Engineering & Remote Sensing* 56:1393–1401.

Mulders, M. A., S. De Bruin, and B. P. Schuiling. 1992. Structured approach to land cover mapping of the Atlantic zone of Costa Rica using single band TM data. *International Journal of Remote Sensing* 13:3017–3033.

Quesada-Mateo, C. 1990. *Estrategia de Conservacion para el Desarrollo Sostenible de Costa Rica (Costa Rica's Strategy for Sustainable Development)*. Ministerio de Recursos Naturales Energia y Minas, San Jose, Costa Rica.

Ramirez, A., and T. Maldonado (Eds.). 1988. *Desarrollo Socieconomico y el Ambiente Natural de Costa Rica: Situacion actual y perspectivas.* Garcia Hermanos, S.A., San Jose, Costa Rica.

Rosero-Bixby, L., and A. Palloni. 1998. Poblacion y deforestacion en Costa Rica, *Conservacion del Bosque en Costa Rica,* pp. 131–150. Academia Nacional de Ciencias, San Jose.

Sader, S. A. 1980. Methods of obtaining multiresource information from remote sensing and ancillary data sources for resources assessment in Costa Rica. In *Proceedings: Remote Sensing for Natural Resources Conference, Moscow, ID,* pp. 395–418.

Sader, S. A. 1988. Satellite digital image classification of forest change using three Landsat data sets, remote sensing for resource inventory, planning, and monitoring. In *Proceedings: 2nd Forest Service Remote Sensing Applications Conference, Slidell, LA, and Nstl, MS,* pp. 189–201.

Sader, S. A., and A. T. Joyce. 1988. Deforestation rates and trends in Costa Rica 1940 to 1983. *Biotropica* 20:11–14.

Sader, S. A., G. V. N. Powell, and J. H. Rappole. 1991. Migratory bird habitat through remote sensing. *International Journal of Remote Sensing* 12:363–372.

Sanchez-Azofeifa, G. A., and R. Harris. 1994. Remote sensing of watershed characteristics in Costa Rica. *International Journal of Water Resources Development* 10:117–130.

Sanchez-Azofeifa, G. A. 1996. *Assessing Land Use/Cover Change in Costa Rica.* Earth Sciences Department, University of New Hampshire, Durham, NH. 20 pp.

Schelhas, J. W. 1991. *Socio-economic and Biological Aspects of Land Use Adjacent to Braulio Carrillo National Park, Costa Rica.* School of Renewable Natural Resources, The University of Arizona, Tucson, AZ. 221 pp.

Schipper, R. A., D. M. Jansen, and J. J. Stoorvogel. 1995. Sub-regional lineal programming models in land use analysis: A case study of the Neguev settlement, Costa Rica. *Netherlands Journal of Agricultural Science* 43:83–109.

Seligson, M. A. 1980. *Peasants of Costa Rica and the Development of Agrarian Capitalism.* University of Wisconsin Press, Madison, WI.

Solorzano, R., R. de Camino, R. Woodward, J. Tosi, V. Watson, A. Vasquez, C. Villaobos, J. J. Jimenez, R. Repetto, and W. Cruz. 1991. *Accounts Overdue: Natural Resource Deletion in Costa Rica.* World Resources Institute, Washington, DC.

Stoorvogel, J. J., and G. P. Eppink. 1995. *Atlas de la Zona Atlantica Norte de Costa Rica.* Programa Zona Atlantica CATIE/UAV/MAG, Guapiles, Costa Rica.

TSC (Tropical Science Center). 1982. *Costa Rica: Perfil Ambiental, Estudio de Campo.*

Umaña, A. and K. Brandon. 1992. Inventing Institutions for Conservation: Lessons from Costa Rica. In S. Annis (Ed.), *Poverty, Natural Resources and Public Policy in Central America,* pp. 85–107. Transaction Publishers, Washington, DC.

Veldkamp, E., A. M. Weitz, I. G. Staritsky, and E. J. Huising. 1992. Deforestation trends in the Atlantic zone of Costa Rica: A case study. *Land Degradation and Rehabilitation* 3:71–84.

Veldkamp, A., and L. O. Fresco. 1994. *Modelling Land Use Changes and Their Temporal and Spatial Variability with CLUE.* Wageningen Agricultural University, Wageningen, The Netherlands.

CHAPTER 17

# Gradient Analysis of Biomass in Costa Rica and a First Estimate of Countrywide Emissions of Greenhouse Gases from Biomass Burning

E. H. HELMER AND SANDRA BROWN



## I. INTRODUCTION

The United Nations Framework Convention on Climate Change, signed at the UN Conference on Environment and Development in Rio de Janeiro during June 1992, came into effect on 21 March 1994, when more than 50 nations ratified the agreement. The objective of the agreement was to stabilize atmospheric concentrations of greenhouse gases (GHGs) and thereby prevent anthropogenic interference with the earth's climate. The carbon balances of the earth's forests are important for sustaining the earth's climate, among other things. Two of the agreement's major provisions are especially relevant to this analysis: "preparing national reports on how to reduce emissions and/or expand carbon sinks" and "providing financial and technical assistance to developing countries for inventories of greenhouse gas emissions" (Hecht and Tirpak, 1995).

An important component of sustainability issues is therefore the degree to which a region or nation is balanced with respect to atmospheric gases. The clearing and burning of tropical forests for expansion of pasture and agriculture or for other reasons, most of which are associated with economic expansion, releases GHGs such as $CO_2$, $CH_4$, and $N_2O$ to the atmosphere. Other gases such as CO and $NO_x$ also are released, which may cause respectively elevated ozone levels and acid precipitation. Both deforestation and associated emissions of GHGs have the potential to impact biogeochemical cycles (Crutzen and Andreae, 1990; Houghton, 1995), climate and hydrology at local and global scales (e.g., Salati and Vose, 1984; Fleming, 1986; Luvall and Uhl, 1990), and the availability and quality of land and water resources (Young, 1994). Many of these processes, such as global climate change, cross political boundaries.

Although each hectare of forest that is cleared and burned releases only small amounts of gas, the cumulative impact can be very large. The global total of C emissions from fossil fuel burning and cement manufacture in the decade of the 1980s was about 5500 Tg $yr^{-1}$ [1 Tg = $10^6$ Mg (or metric ton) = $10^{12}$ g] (Schimel *et al.*, 1995). Net C emissions from changes in land use during the same period have been estimated at 490 to 1600 Tg $yr^{-1}$, primarily from the tropics (Detwiler and Hall, 1988; Hall and Uhlig, 1991; Houghton *et al.*, 1995). Biomass burning, occurring mainly as fires for clearing of tropical forests and savannas and fuel wood burning, also releases trace GHGs. Such burning contributes about 40 Tg C as the trace gas methane annually, which is about 8% of global $CH_4$ emissions (Prather *et al.*, 1995). Methane molecules have about 20 times the warming potential of $CO_2$ molecules. Thus about 50% [(40 Tg $CH_4$-C $\times$ 20)/(1600 Tg $CO_2$-C)] of the warming potential from biomass burning derives from $CH_4$ emissions. While biomass burning contributes only about 3% of global emissions of the trace gas $N_2O$ (Prather *et al.*, 1995), the warming potential per molecule of $N_2O$ is about 200 times that of $CO_2$ (Lashof and Ahuja, 1990).

Estimates of net emissions of GHGs from biomass burning are uncertain for many reasons. Key uncertainties are (1) the quantity of biomass fuel originally on the areas burned (Hao *et al.*, 1991), (2) the portion of that fuel consumed and emitted as various gases in a burning event, and (3) the area impacted (Detwiler and Hall, 1988). Depending on forest type and human disturbances, biomass and thus fuel can vary from less than 40 Mg $ha^{-1}$ for a dry, open canopy forest, to greater than 500 Mg $ha^{-1}$ for a well-developed primary humid forest (S. Brown *et al.*, 1993; S. Brown, 1996). Thus GHG emission estimates from biomass burning are better when higher resolution data are used in the analysis, such as country-specific data subdivided along environmental gradients such as life zones (Holdridge, 1967). Burn combustion characteristics, which influence emissions from forest burning, also vary by forest type. Generally in more humid forests combustion is both less complete and less efficient. The result is that trace gas and particulate emissions are higher in humid as compared to dry forests (Kauffman *et al.*, 1992). Thus information specific to climatic forest formation is also important for estimating emissions of trace GHGs, such as $CH_4$ and $N_2O$.

However, past efforts to estimate GHG emissions from biomass burning generally have relied on global or regional data bases (e.g., Hao *et al.*, 1991). Although the Central American country of Costa Rica is small in area (50,060 $km^2$), it has several forest formations for which data on forest structure and the amount of area deforested are available. Between 1950 and 1984, the rate of deforestation in Costa Rica was one of the highest in the world at about 3.9% per year (see Chapter 16) (Harrison, 1991).

In addition, as ecologists and geographers (Tosi and Voertman, 1964; Sader and Joyce, 1988; Janzen, 1988) have noted, humans inhabit and clear drier forests preferentially to wet ones. Lower elevation forests in Costa Rica were also cleared preferentially. Therefore the country provides an excellent "microcosm" for an analysis of GHG emissions by forest type.

While both the Pacific and the Caribbean slopes have both a dry and a wet season, the Caribbean side tends to have a longer rainy season, and both slopes are influenced by various ocean winds (Chapter 9) (Stiles and Skutch, 1989). Under these diverse climatic and topographic conditions, many forest types have formed in areas of different temperature and rainfall regimes and elevation. With the steep environmental gradients and consequent range of forest formations in Costa Rica, our first objective in this chapter was to estimate the biomass of Costa Rican forest stands undisturbed by recent human activity (as far as we can tell) as a function of environmental gradients.

Our second objective was to estimate the release of GHGs to the atmosphere from forests developed under differing climatic conditions. We estimate emissions simply by assuming that 100% of aboveground biomass is burned eventually. This assumption simplifies the calculations for GHG emissions from burning. This method also is recommended in the Intergovernmental Panel on Climate Change and Organization for Economic Cooperation and Development (IPCC/OECD) methodology for inventory of GHG emissions (Houghton *et al.*, 1995). Estimating GHG emissions using this simple approach allows us to compare its results with those from more sophisticated models such as the one used in, for example, Detwiler and Hall (1988), which considers processes such as shifting cultivation.

Our final objective was to integrate the biophysical variability in emissions from biomass burning with the spatial and temporal pattern of how humans clear forests. Thus we evaluate the net result of the fact that environmental gradients affect both the rates and the pattern of deforestation, as well as the GHG emissions per hectare of forest cleared.

## II. METHODS

We derived biomass estimates from a series of previously sampled plots of tree diameters and densities, and analyzed the results as a function of environmental gradients and Holdridge life zones (Holdridge *et al.*, 1971). (The life zone approach is a particular gradient analysis using rain, evapotranspiration, and temperature.) From these estimates we derived national inventories of biomass using estimates of the areal extent of forest cover, by life zone, for 1940 and 1983 (Sader and Joyce, 1988). The change in biomass associated with forest removal between the two dates was used to estimate transfer to the atmosphere of trace gases, including carbon as $CO_2$, using emission factors based on biomass burned from previously published information (Ward *et al.*, 1992; Hao *et al.*, 1991).

### A. Aboveground Biomass Density

Costa Rica is a well-studied country from the perspective of its biota. Yet in order to determine whether its greenhouse gas emissions meet with the goals of the agreement reached during the UN Conference in Rio de Janiero, data are needed on the past, current, and potential biomass of its varied forest types. While the maximum biomass a forest can attain depends on its climate and soils (Chapter 7) (S. Brown and Lugo, 1982), the biomass present in a given stand is also dependent on its age and disturbance history. We wanted to estimate how biomass might have been distributed spatially across Costa Rica's climatically varied landscape when much of its forest was relatively undisturbed by human activities and therefore near a maximum potential biomass. Such data indicate the probable spatial distribution of forest biomass in the past and the biomass distribution possible under future scenarios.

We calculated aboveground forest biomass (AGB, in Mg $ha^{-1}$ of dry-weight biomass) from forest stand data collected in 1964–1966 (Holdridge *et al.*, 1971). The stands were described as "relatively undisturbed," and areas described in the text as having experienced obvious human disturbance are not included here in mean biomass values. Riparian and swamp sites were not included in the analysis because information on changes in their extent was not available. Detailed data also were not available on the structure or deforestation extent of wind-clipped, short-statured wet montane forest known as "elfin" forest (Lawton and Dryer, 1980) or other subalpine formations. Kappelle (1991) notes that these forests do not usually exceed 12 m in height. They have an extent of about 6,740 ha in Chirripó National Park (E. Helmer, 1999).

Holdridge *et al.* (1971) established 1 to 10 plots of 0.1 to 1.2 ha in size in each life zone. Nineteen of the 33 plots were 0.3–0.5 ha in size; 7 were 0.1–

0.2 ha; 5 were 0.6–0.8 ha; and 2 were 1 ha or larger. Within each of these plots, the diameter (dbh) and height for all trees with a dbh >2.5 cm were measured; species were also identified. The number of plots per life zone varied and are listed in Table 17-1.

The biomass values we calculated could over- or underestimate actual stand biomass because of the small plot sizes used. Plot sizes of 1 ha or more are generally recommended for estimation of biomass for an ecosystem in an ecological study, and for regional biomass estimates, more extensive forest inventories are desirable (S. Brown *et al.*, 1989). Only two plots were 1 ha or more in size. However, this data set was the only set available that surveyed in a consistent fashion forest stands representative of the various climatic forest formations in Costa Rica.

This series of previously sampled plots of tree diameters were distributed geographically along climatic gradients to represent the range of forest environments in Costa Rica (Holdridge *et al.*, 1971). We used techniques developed in S. Brown *et al.* (1989), that use regression equations, stratified by life zone, to calculate average mass per tree (in kg) from the tree diameters. The biomass regression equations for broadleaf forests were developed from a large data base that includes trees of many species harvested from forests from all three tropical regions of the world. It includes a total of 371 trees with a diameter range of 5 to 148 cm from 10 different literature sources, and it has been updated recently (S. Brown, 1997).

The data we used were in the form of stand tables listed in the vegetation data summary for each site (Holdridge *et al.*, 1971). These tables give the number of trees, as tree density in number/ha, by the following diameter classes: 2.5–7.5, 7.5–15, 15–30, 30–60, and >60 cm. Tree by tree data have since been destroyed, so we used the reported stand tables (number of trees per hectare by diameter class). Estimation of biomass from the stand tables and regression equations consisted of calculating the biomass of a tree with the dbh set at the midpoint of a given diameter class, multiplying the biomass by the number of stems in that class, repeating this calculation for each diameter class, and summing the biomass over all diameter classes. The quadratic mean diameter of each diameter class may be more representative of each class diameter distribution than the class midpoint (S. Brown, 1997). However, basal area was not reported by diameter class in the stand tables. As a result, we were not able to use the quadratic mean diameter for calculation of average biomass per tree.

Potential problems with the regression approach that we used to estimate biomass include: (1) the choice of the correct biomass regression equation to apply to a given site, (2) the small number of large-diameter trees used to develop the regression equations (e.g., for the moist tropical equation, the largest dbh was 148 cm, with only five trees >100 cm diameter; therefore

TABLE 17-1 Biomass Estimation by Life Zone

| Forest formation | Symbol | Number of sites | Total live aboveground biomass (nonweighted average)[a] (Mg $ha^{-1}$) | Belowground biomass[b] (Mg $ha^{-1}$) | Fine litter fraction[c] | Dead wood fraction[c] | Fine litter biomass (Mg $ha^{-1}$) | Dead wood biomass (Mg $ha^{-1}$) | Total biomass[d] (Mg $ha^{-1}$) |
|---|---|---|---|---|---|---|---|---|---|
| Dry, low | T-vdf | 2 | 14 | 5 | 0.074 | 0.034 | 1.1 | 0.5 | 21 |
| Dry, well-developed | T-df | 2 | 188 | 64 | 0.018 | 0.019 | 3.4 | 3.5 | 258 |
| Lowland moist | T-mf | 7 | 518 | 175 | 0.013 | 0.093 | 6.9 | 48.3 | 748 |
| Lowland wet | T-wf | 8 | 365 | 123 | 0.011 | 0.103 | 3.9 | 37.5 | 529 |
| Premontane moist | P-mf | 2 | 208 | 70 | 0.018 | 0.123 | 3.7 | 25.5 | 307 |
| Premontane wet | P-wf | 4 | 306 | 103 | 0.011 | 0.103 | 3.3 | 31.5 | 445 |
| Premontane rain | P-rf | 2 | 318 | 107 | 0.011 | 0.103 | 5.7 | 39.0 | 470 |
| Lower montane moist | LM-mf | 1 | 319 | 108 | 0.018 | 0.123 | 5.7 | 39.1 | 472 |
| Lower montane wet | LM-wf | 1 | 421 | 142 | 0.017 | 0.110 | 7.2 | 46.1 | 617 |
| Lower montane rain | LM-rf | 3 | 324 | 110 | 0.017 | 0.110 | 5.6 | 35.5 | 475 |
| Montane rain | M-rf | 1 | 309 | 104 | 0.017 | 0.110 | 5.3 | 33.9 | 453 |

[a] Calculated from stand tables in Holdridge *et al.* (1971) with methodology in Brown (1996).
[b] Calculated using equation in Cairns *et al.* (1997).
[c] Based on data in Delaney *et al.* (1997), Saldarriaga *et al.* (1988), and Kauffman *et al.* (1988).
[d] Aboveground plus belowground, plus fine and coarse litter.

diameters should be less than about 148 cm when using this equation), (3) the open-ended nature of the >60-cm diameter class in stand tables, (4) the wide and uneven width diameter classes in stand tables, (5) the difficulty with selecting an appropriate average diameter to represent a diameter class, and (6) the occasional missing diameter class (i.e., incomplete stand tables to a minimum diameter of 10 cm, which was not a problem with the Holdridge data set). We next describe how we addressed these shortcomings.

The biomass regression equations were stratified into four climate types or life zones: very dry, dry, moist, and wet. In lowland forests, rainfall can indicate the appropriate regression equation to use. In lowland dry, moist, and wet forests rainfall is, respectively, usually about <1500 mm, 1500 to 4000 mm, and >4000 mm annually (Holdridge, 1967), and these ranges can indicate which regression equation to use. In addition, S. Brown (1997) lists two possible dry forest equations: one for very dry sites with annual rainfall <900 mm (very dry forest), developed for a Mexican forest (original data source Martinez-Yrizar *et al.*, 1992), and one from a dry forest in India where rainfall exceeded 1200 mm.

In drier forest formations, forest stature is indicative of rainfall and dry season length, and can help determine which of the dry forest regression equations is most appropriate for estimating biomass. In addition, because the biomass regression equations do not specifically include height, we assumed that using regression equations derived from forests with heights similar to the stand of interest will yield the best biomass estimates. Therefore, we used forest stature, rather than rainfall, to indicate which of the two dry forest equations would be the most appropriate for biomass estimation in the dry zone Costa Rican stands.

The Costa Rican dry forest sites had very different structures; the closest rain gauge to them, at Cañas, recorded a relatively wet 1665 mm of precipitation annually, despite their classification as dry forests. In addition, the differences in these forests' stature, from an average of 5 m in height to an average of 22 m, indicated that the one rain gauge did not represent precipitation for these varied forest sites adequately. We assumed that the small-statured, savanna-like stands probably received much less rainfall than 1665 mm annually, while the well-developed stands probably received closer to that level.

Consequently, those short-statured forest sites, which averaged 5 m in height, were assumed to be very dry forests, and we used the corresponding equation to estimate their biomass. Their height was close to the average height of 7 m for forest which S. Brown (1997) used to develop the very dry forest regression equation. The well-developed dry sites averaged 13 and 22 m in height, closer to the average of 12 m for the forest from which the other dry forest regression equation was developed, where rainfall is >1200 mm. We also used the biomass regression equation for dry forest with rainfall

>1200 mm to estimate biomass of the premontane moist forest sites due to their short stature of 15 to 22 m.

We adjusted for the wide diameter class of 30–60 cm, as reported in the stand tables in Holdridge *et al.* (1971), by assuming the stands had a J-shaped diameter distribution. A J-shaped diameter distribution, in which tree number decreases asymptotically with diameter class, is characteristic of mature stands (Gillespie *et al.*, 1992). Gillespie *et al.* showed that this diameter distribution curve can be approximated by a straight line from one diameter class to the next. Using that approximation, we estimated tree density for two 15-cm subsets of the wide diameter class. In other words, we expected that using the midpoint of 45 cm for the 30- to 60-cm diameter class would overestimate biomass in that class if the majority of trees had diameters smaller than 45 cm, which is usually the case. Thus we developed a procedure, based on one presented in Gillespie *et al.* (1992), for dividing the number of trees in the 30- to 60-cm class into two smaller classes: 30–45 and 45–60 cm. The adjustment consisted of calculating the ratio of tree densities in two smaller diameter classes and assuming that this ratio was equal to the ratio between two larger, 15-cm-wide classes.

Another shortcoming is that Holdridge *et al.* (1971) did not specify the individual diameters of the largest trees in each plot. Rather, they reported number of trees in an open-ended diameter class of >60 cm. To use the biomass equations, we needed a midpoint diameter for that size class. We estimated that midpoint diameter by the following sequence of calculations: (1) Calculate the basal areas for all but the largest size class as the basal area for a tree with a diameter equal to the class midpoint, multiplied by the number of trees in that class. (2) Calculate the basal area for all trees in the largest size class as the difference between the sum of the basal areas calculated in the first step from the total stand basal area reported in Holdridge *et al.* (1971), which did not report basal area by diameter class. (3) Calculate the average diameter of the largest size class as the class basal area, calculated in step 2, divided by the number of trees in that size class.

This procedure did not work for 9 of the 33 plots because the sum of the basal areas of the smaller size classes, those below 60 cm dbh, equaled or exceeded the total stand basal area reported by Holdridge *et al.* (1971). The overestimation of basal area occurred even when trees were reported to be present in the >60-cm size class. Our estimation of basal area in the smaller diameter classes, from the midpoint of wide diameter classes, probably overestimated the basal area in each size class. In these cases the majority of trees in the various diameter classes probably had a smaller diameter than the class midpoints (Gillespie *et al.*, 1992). Therefore, because we were unable to calculate average diameter in the largest size class, we had to assume an arbitrary average diameter of 70 cm for the largest class in these stands. Then, we adjusted the final stand biomass estimate downward by the ratio of total reported basal area to the sum of the class basal areas calculated from diameter class midpoints, including basal area from the trees with the assumed 70-cm average diameter.

## B. Fine and Coarse Litter Biomass and Belowground Biomass

Belowground biomass (BGB) was estimated from aboveground live biomass using a regression equation developed from 160 data points from published studies of biomass of tropical, temperate, and boreal forests located worldwide (Cairns *et al.*, 1997). The equation for estimating root biomass, including stumps, is $BGB = \exp(-1.0850 + 0.9256 \times AGB)$, $r^2 = 0.83$. Fine and coarse litter was also estimated from AGB, by life zone, using data in Delaney *et al.* (1997), Saldarriaga *et al.* (1988), and Kauffman *et al.* (1988) for forests of Venezuela. From these studies, we calculated ratios of fine litter to AGB and dead wood to AGB by life zone (Table 17-1).

## C. Fuel Biomass and Consumption

Fuel biomass in the Costa Rican forests was assumed equal to AGB plus fine and coarse litter biomass (necromass). We did not correct fuel biomass with combustion factors because we assumed that with repeated burnings the total biomass would eventually be burned. This calculation simplifies the process by which burning consumes cleared forest biomass. The first time a forest is burned, a portion of coarse components such as tree boles remain on-site (Kauffman *et al.*, 1993), while most finer components, such as leaves, herbs, fine litter, and small branches, are largely consumed by fire (Kauffman *et al.*, 1992, 1988).

Different trace gas emission factors might apply to the remaining dead wood, which is consumed in later cycles of pasture burning. In addition, for any given forest type, the combustion factor will change with fuel moisture content and fire conditions (e.g., wind speed and topographic influences) (Kauffman *et al.*, 1992). Although repeated burning of regrowth would add to the emissions estimated here, shifting cultivation is now limited to relatively small regions in Costa Rica. Thus there is no need for complicated decay and regrowth dynamics such as in the carbon dynamics model used by Detwiler and Hall (1988).

## D. Trace Greenhouse Gas Emissions

We derived the greenhouse warming potential resulting from forest destruction by estimating the release of those GHGs and correcting for the relative greenhouse potential of each. Ward *et al.* (1992) measured the combustion characteristics for several compounds and two to three phases of burning intensity during experimental burns in Brazilian tropical forests and savannas. We used Ward's estimates for trace gas emissions to estimate emissions of GHGs.

Ward *et al.* (1992) used the mass ratio of $CO/CO_2$ in emissions from deforestation fires to calculate combustion efficiency values that varied from 0.84 to 0.95 in Amazonian forests (Figure 17-1). Combustion efficiencies varied from 0.92 to 0.95 during the flaming phases of drier savanna (cerrado) fires, which consumed about 97% of the carbon released from burning. Ward *et al.* (1992) also found that about 52% of all carbon released from fires in primary moist forest occurred during the flaming phase of the burn, in which combustion efficiency was 0.88, and that most fine fuels were consumed by fire. Combustion efficiency during the ensuing smoldering phases was lower, about 0.84–0.85.

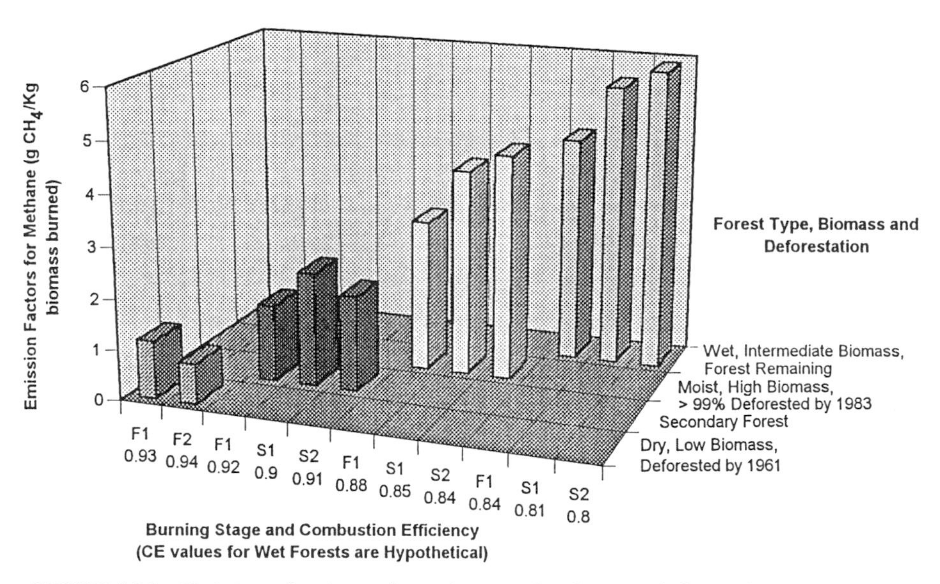


FIGURE 17-1 Variation of emission factor (see text for discussion) for methane by burning stage and moisture of forest formation. Mainly only wet forest formations remain in Costa Rica. F1 and F2 are the first and second flaming phases, respectively; S1 and S2 are the first and second smoldering phases. Combustion efficiency is the fraction of biomass carbon emitted as $CO_2$. (Data from Ward *et al.*, 1992.)

Thus Ward *et al.* (1992) found that the proportion of biomass converted into products of incomplete combustion varied by forest type. The emission factor (EF) for a combustion product is the mass of an emission product released per unit mass of fuel consumed (on a carbon or other basis). For particulate matter (PM) in smoke, and gases such as CO, $NO_x$, $CH_4$, and $H_2$, EFs increase with decreasing combustion efficiency (Ward *et al.*, 1992). In addition, combustion efficiency decreases and therefore EFs increase as a forest burn proceeds from the initial flaming to smoldering phases (Ward *et al.*, 1992).

We used the average of the EFs that Ward *et al.* (1992) published for the flaming phase and the two smoldering phases of moist primary forest for most of our estimates of emissions from burning (Table 17-2). For the very dry forests, we used the average of two EFs published for two flaming phases of Amazonian savanna (*Cerrado sensu stricto*) because the 4- to 8-m stand height for this formation is similar to the stand height of about 5 m for the dry, savanna-like forest site sampled by Holdridge *et al.* (1971). A source of error in using these EFs from Amazonian savanna is that they were measured from burning of a natural fuel bed rather than a slash fire (Ward *et al.*, 1992). However, comparable EF data from burning a slashed savanna were not available. We felt that these savanna EFs would be more representative than EFs from burning slash in a moist forest.

To estimate emissions of $N_2O$, we assumed that about 0.7% of the nitrogen in biomass is oxidized to $N_2O$, based on an average N/C ratio (mole/mole) of 1.4% in tropical forest vegetation (Hao *et al.*, 1991) and an average biomass C content of 50%. Molecules of $CH_4$ and $N_2O$ are assumed to have 20 and 200 times the atmospheric warming potential of each molecule of $CO_2$, respectively

**TABLE 17-2 Emission Factors (EF) of Greenhouse Gases (in kg/kg) for Different Forest Types and for Different Size Classes of Wood**[a]

| | Forest Type | |
|---|---|---|
| Compound | Moist[b] | Dry[c] |
| $PM\text{-}2.5_3$ | 0.0085 | 0.0045 |
| $CO_2$ | 1.582 | 1.713 |
| CO | 0.127 | 0.063 |
| NO-N | 0.0007 | 0.0005 |
| $CH_4$ | 0.00805 | 0.0016 |
| $H_2$ | 0.0036 | 0.0012 |

[a] From Ward *et al.* (1992).
[b] Moist primary forest
[c] Very dry, savanna-like (*Cerrado sensu stricto*)

(Lashof and Ahuja, 1990). To compare the total warming potential of these trace gases with $CO_2$ emissions, we multiplied the number of moles of each of these emitted compounds by their warming potential multipliers (20 and 200) and converted moles to grams of C.

## III. RESULTS AND DISCUSSION

### A. Biomass

The average estimated aboveground biomass ranged from 14.2 dry Mg (or metric tons) $ha^{-1}$ in the very dry forest to a high of 518 Mg $ha^{-1}$ in lowland moist forest (Table 17-1). The trends in biomass by life zone were consistent with our understanding of the influences that environmental gradients have on forest biomass accumulation (Figure 17-2). Biomass tends to be highest in moist tropical forests and lower where dry, very wet, or colder conditions limit potential accumulation (S. Brown and Lugo, 1982).

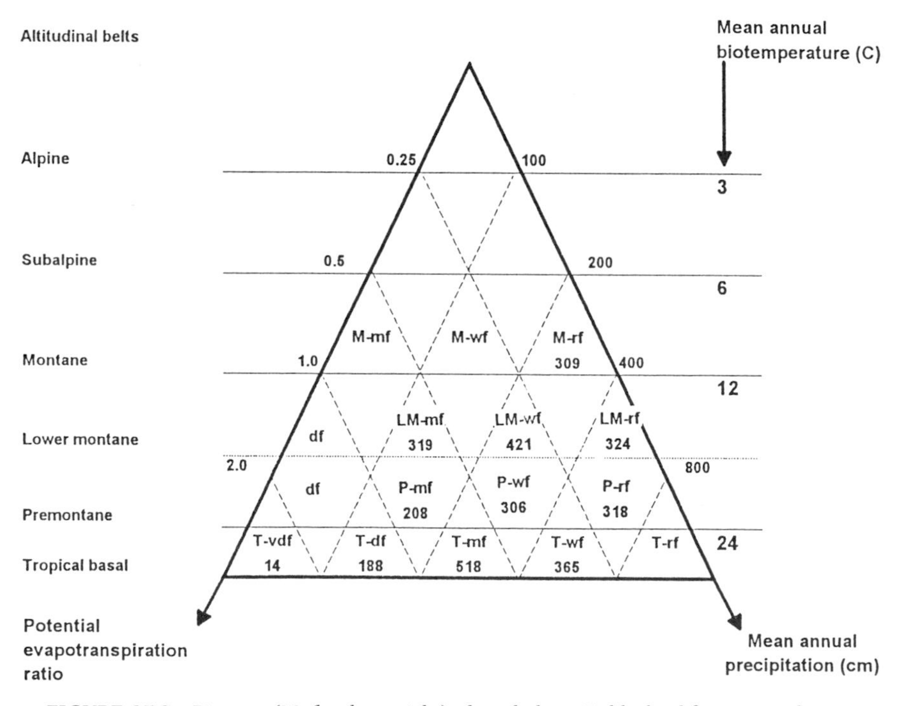


FIGURE 17-2 Biomass (Mg/ha dry weight) plotted along Holdridge life zone gradients.

The very dry, savanna-like plots had very low biomass values, averaging 16 Mg $ha^{-1}$. Productivity in these sites is limited by hot, dry conditions during dry months, and soil saturation during the rainy season. The dry forest sites, where moisture deficits are a major limitation to productivity, contained the next lowest biomass, averaging 188 Mg $ha^{-1}$ AGB. Published AGB values for other dry forest sites ranged between the very dry and dry Costa Rican forest plots. The Guanica Forest in Puerto Rico contains about 44.9 Mg $ha^{-1}$ AGB. In second-growth Brazilian dry forest, known as Caatinga, AGB was reported as 74 Mg $ha^{-1}$ (Kauffman *et al.*, 1993). Venezuelan very dry forest measured by Delaney *et al.* (1997) contained 140 Mg $ha^{-1}$.

Biomass, as a function of temperature and moisture, peaks in tropical moist forest. The Costa Rican lowland moist forest AGB, at an estimated 518 Mg $ha^{-1}$, is similar to the 513 Mg $ha^{-1}$ reported for the Ivory Coast by Huttel (1975), but is much higher than moist forest reported in other areas. Saldarriaga *et al.* (1988) reported 255 Mg $ha^{-1}$ for lowland moist forest in Venezuela. Fittkau and Klinge (1973) report 406 Mg $ha^{-1}$ for Brazilian moist forest. Golley *et al.* (1975) reported 264 and 378 Mg $ha^{-1}$ for two moist, transitional to dry, sites in Panama.

Lowland wet forest biomass in Costa Rica is intermediate between moist and dry forest biomass values, and averaged 365 Mg $ha^{-1}$, which is greater than the 322 Mg $ha^{-1}$ reported for Cambodian wet forest by Hozumi *et al.* (1969).

Moisture limitation was apparent in the two premontane moist sites, which averaged 208 Mg $ha^{-1}$ AGB. Their biomass is similar to generalized values reported in S. Brown and Lugo (1982), which averaged 241 Mg $ha^{-1}$. The Costa Rican lower montane moist site contained 319 Mg $ha^{-1}$ AGB, which is similar to the 346 Mg $ha^{-1}$ reported in a comparable Venezuelan forest (Delaney *et al.*, 1997).

All 10 higher elevation wet sites averaged 324 Mg $ha^{-1}$ AGB, which included all plots in premontane, lower montane, and montane wet and rain forest. These values were similar to the 310 Mg/ha reported for lower montane rain forest in New Guinea by Edwards and Grubb (1977). They were lower than what Brun (1976) reported for montane wet forest in Venezuela, which contained 347 Mg $ha^{-1}$. The average for the Costa Rican plots was somewhat higher than the 279 Mg $ha^{-1}$ reported for lower montane rain forest in Jamaica (Tanner, 1980) or the 198 to 223 Mg $ha^{-1}$ reported for subtropical Puerto Rican lower montane rain forest (Frangi and Lugo, 1985; Ovington and Olson, 1970). For the premontane wet forest at La Selva, Costa Rica, our estimate of 389 Mg $ha^{-1}$ was nearly identical to a previous estimate of 382 Mg $ha^{-1}$ (Jordan, 1985).

Also, we calculated biomass for a Costa Rican montane rain forest site based on a stand table presented in Jimenez *et al.* (1988). The calculation yielded an estimated AGB of 388 Mg $ha^{-1}$. That value is somewhat higher than our estimate of 309 Mg $ha^{-1}$ for AGB of a comparable forest analyzed in this study, and higher than the 314 Mg $ha^{-1}$ reported in Delaney *et al.* (1997) in a comparable Venezuelan forest.

Thus overall Costa Rica has biomass values that are either similar to, or in some cases considerably higher than, other tropical forests of their type.

## B. Greenhouse Gas Emissions

In the case of Costa Rica, our estimates of carbon emissions from deforestation over the period of study were higher than previous estimates. The discrepancy is due partly to the fact that our estimates of biomass density in extensively cleared forest types, when based on country-specific data and stratified by environmental gradients, were higher than the biomass estimates for all of the tropics used previously. This finding shows the importance of refining spatial analyses when undertaking analyses of greenhouse gas emissions.

The spatial pattern of GHG emissions reflects the general historical pattern of forest clearing and land development in Costa Rica, where the mesic and drier regions were developed before the wetter areas (Chapter 2). Thus trace gas emissions from deforestation are not proportional to the area of a given forest type. Sixty percent of trace gas emissions from deforestation and burning of the total fuel biomass between 1940 and 1983 (298 and 281 Tg, respectively) was from burning of lowland moist and wet forests (Table 17-3 and Figure 17-3). The burning of lowland dry and moist forests, and premontane moist zones, contributed respectively 0.5, 31, and 2% of emissions. Each of these three life zones, and especially the lowland moist forest, contributed greenhouse gases disproportionately to their area. Two were slightly, and the second very, disproportionate to their areas, which were already only 0.6, 15, and 2.5% of forest area in 1940 due to deforestation that had already started in those regions earlier (Chapter 16). Lower elevation, wet forests (T-wf, P-wf, P-rf) accounted for 66% of the total forest area in 1940, and together contributed 60% of emissions. In contrast, higher elevation wet forests (LM-wf, LM-rf, M-rf) composed 14% of the forest area in 1940, but contributed only 7% of emissions.

By 1983, no original forest biomass remained in dry and moist life zones. In addition, while higher elevation wet forest zones composed 12% of potential fuel biomass in 1940, they composed 29% of potential fuel biomass in 1983. Most of the remaining biomass was in lower elevation, wet forest zones (69%). Almost no biomass remains in those forest zones which were settled by humans first: the dry and moist zones. Productivity is low in both dry forests and in the higher elevation wet forests, but very high in moist forests. Thus, ironically, the moist forest regions, which have been settled for a long time, are also where

TABLE 17-3 Trace Gas Emissions by Biomass Burning in Costa Rica by Life Zone

| Forest formation[a] | Total aboveground biomass (live + dead) = total fuel biomass ($Mg\ ha^{-1}$) | Forest area cleared 1940–1983 ($\times 10\ km^2$) | Forest area remaining in 1983 ($\times 10\ km^2$) | Biomass consumption (by clearing/burning of forest 1940–1983) (Tg) | Biomass remaining (% of original) | PM-2.5 (Tg) | $CO_2$ (Tg) | CO (Tg) | $CH_4$ (Tg) | NO–N (Tg) | $H_2$ (Tg) |
|---|---|---|---|---|---|---|---|---|---|---|---|
| T-vdf | 16 | 21 | 0 | 0.34 | 0 | 0.002 | 0.6 | 0.02 | 0.001 | 0.0002 | 0.0004 |
| T-df | 195 | 21 | 0 | 4.2 | 0 | 0.033 | 6.6 | 0.53 | 0.033 | 0.003 | 0.015 |
| T-mf | 573 | 520 | 1 | 298 | 0 | 2.3 | 471 | 38 | 2.4 | 0.21 | 1.1 |
| T-wf | 406 | 691 | 282 | 281 | 29 | 2.2 | 444 | 36 | 2.3 | 0.20 | 1.010 |
| P-mf | 237 | 85 | 0 | 20 | 0 | 0.16 | 32 | 2.6 | 0.16 | 0.014 | 0.072 |
| P-wf | 341 | 656 | 224 | 224 | 26 | 1.8 | 354 | 28 | 1.8 | 0.16 | 0.805 |
| P-rf | 363 | 195 | 216 | 71 | 53 | 0.55 | 112 | 9 | 0.57 | 0.049 | 0.254 |
| LM-mf | 364 | 0.10 | 0 | 0.04 | 0 | 0.0003 | 0.06 | 0.005 | 0.0003 | 0.0000 | 0.0001 |
| LM-wf | 474 | 39 | 13 | 18 | 25 | 0.14 | 29 | 2.3 | 0.15 | 0.013 | 0.066 |
| LM-rf | 365 | 111 | 234 | 41 | 68 | 0.32 | 64 | 5.2 | 0.33 | 0.028 | 0.15 |
| M-rf | 348 | 35 | 79 | 12 | 70 | 0.10 | 19 | 1.5 | 0.10 | 0.008 | 0.044 |
| Total | | 2373 | 1048 | 968 | 29 | 7.6 | 1531 | 123 | 7.8 | 0.7 | 3.5 |
| Annual emissions[b] | | | | | | 0.18 | 36 | 2.9 | 0.18 | 0.02 | 0.08 |
| Annual C emissions | | | | | | | 8.9 | 1.2 | 0.14 | | |

[a] As in Table 17-1.
[b] Based on total emissions divided by number of years between 1940 and 1983.

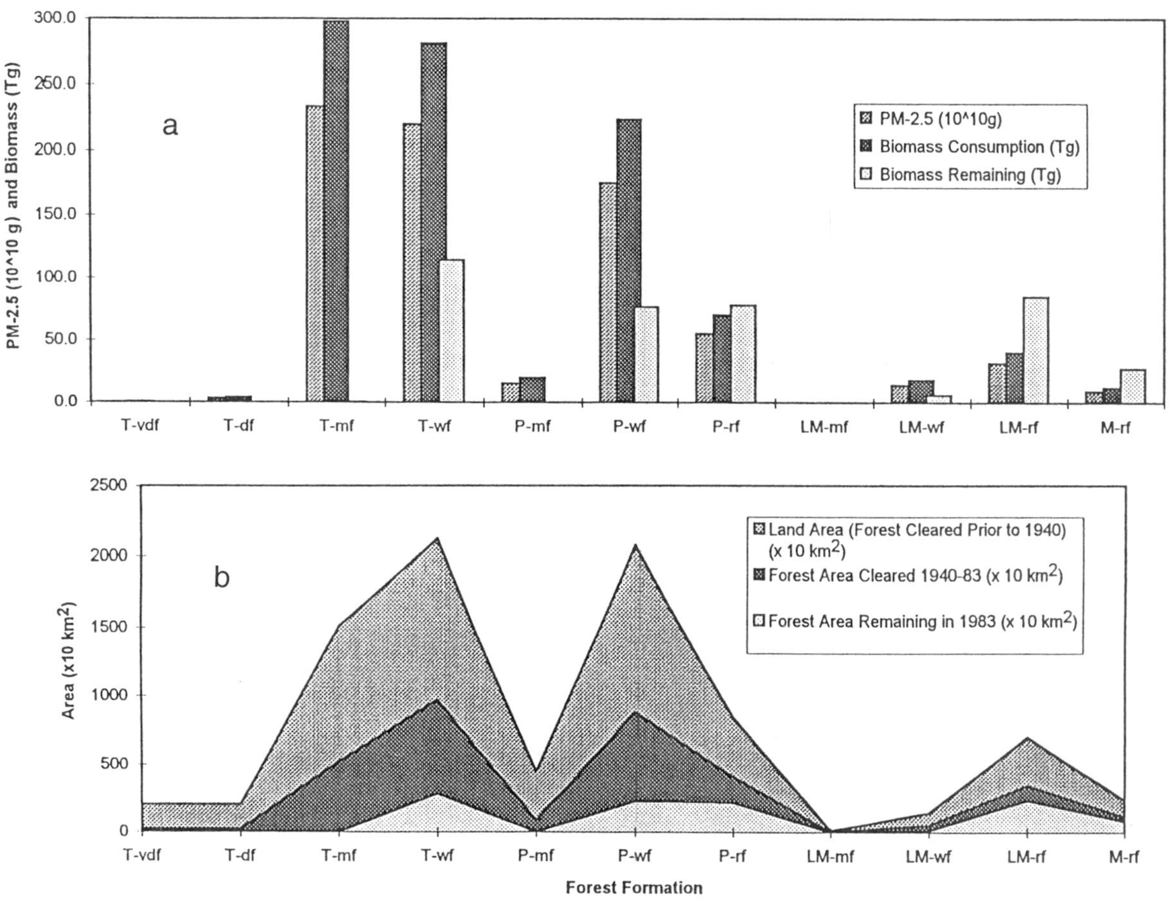


FIGURE 17-3 (a) Particulate emissions in smoke and biomass burned and remaining by forest formation. (b) Land area, forest area cleared, and forest area remaining by forest formation.

the most efficient carbon sequestration via secondary regrowth or plantation planting occurs or could occur.

We estimated annual release of carbon-containing compounds, over the 43-year period of 1940–1983, as 8.9 Tg $CO_2$-C, 1.2 Tg CO-C, and 0.14 Tg $CH_4$-C (Table 17-3). We also estimated an annual $N_2O$-N release of about $1.28 \times 10^{-3}$ Tg. We multiplied these molar emissions of $CH_4$ and $N_2O$ by their respective warming potential multipliers and converted all to mass of carbon equivalents to obtain the following annual "$CO_2$ greenhouse forcing equivalent" emissions: 2.8 Tg C from $CH_4$, and 0.11 Tg C from $N_2O$. These amounts sum to almost one-third of the 8.9 Tg C from $CO_2$. Therefore, in Costa Rica, emissions of $CH_4$ from incomplete combustion during biomass burning has about 30% of the warming potential of $CO_2$. Industrial emissions of $CO_2$ in Costa Rica in 1991 were roughly 8.87 Tg $CO_2$-C (WRI, 1994). Thus deforestation during this period appears to have been roughly the same magnitude as industrial activity as a source of greenhouse gases.

Tree plantations in tropical regions range in productivity from about 3 to 30 Mg $ha^{-1}$ $yr^{-1}$ of carbon (S. Brown *et al.*, 1986) in their first 30 years of growth. Costa Rica would need to plant between about 1000 and 10,000 $km^2$ of plantations, which amounts to between 2 and 20% of the country's total area, to offset annual deforestation emissions of the trace greenhouse gases $N_2O$ and $CH_4$ alone. One study of tree plantation area in Costa Rica estimated 400 $km^2$ in 1990, and about 37 $km^2$ is planted annually (FAO, 1993). Thus, in Costa Rica, plantations are only half as extensive as the minimum area needed to offset annual emissions of $N_2O$ and $CH_4$ from deforestation alone, and additional area would be needed to offset $CO_2$ emissions. Alternatively, Costa Rica could focus its efforts on ending deforestation. If that effort were successful, Costa Rica would "save" enough carbon to halve its total emissions, and recovering forests could compensate for much of the industrial emissions.

Hall *et al.* (1985) estimated $CO_2$ emissions from deforestation in Costa Rica from 1950 to 1980 using a computer program that corrected for the dynamics of decomposition and vegetation regrowth (where that took place). Their biomass estimates were derived from those published in S. Brown and Lugo (1982), which we mentioned previously, and were stratified into six life zones (five of which were present in Costa Rica). They did not estimate other trace gas emissions. They estimated a carbon release of $4.6 \pm 3$ Tg $yr^{-1}$ due to land use changes in Costa Rica between 1950 and 1980. Later, Hall and Uhlig (1991) revised estimates of biomass and estimated C release using the same dynamic C model. Biomass estimates were also stratified by life zone and were based on global, not national, estimates. Their revised estimate of C release in Costa Rica from land use change, mainly from deforestation, was 4.67 Tg $yr^{-1}$. They also narrowed the range of the previous estimate such that their low estimate was 4.2 Tg $yr^{-1}$ and their high estimate was 5.7 Tg $yr^{-1}$. The sum of all emission products containing carbon, from our study results, produced an estimate of 10 Tg $yr^{-1}$ of carbon. Our estimate of 8.9 Tg $yr^{-1}$ C from $CO_2$ alone is about twice as large as those earlier estimates.

According to Hall *et al.* (1985), roughly 20% of the difference between their high and low estimates is due to delays in carbon release from decomposition, plus the carbon absorbed in regrowth, for their more dynamic model. So our estimates, if corrected for this factor, are not very different from Hall *et al.*'s (1985) earlier estimates, but are somewhat higher than their later estimates. Much of the difference is due to the greater geographical specificity of our analysis, which generated higher carbon estimates for those areas deforested. Rather than using global estimates of biomass density as they did, we estimated biomass from surveys of Costa Rican forests. As discussed previously, our estimates of Costa Rican lowland wet and moist forest biomass density are about 25 and 60% higher, respectively, than the global averages for equivalent forest types that were reported in S. Brown and Lugo (1982). Those were the life zones in which the rate of deforestation and carbon release in Costa Rica was highest on average over the time periods investigated. All in all the two estimates are remarkably similar given the difficulty in such analyses and given the fact that the earlier estimates were based on global biomass averages.

## C. Errors in Estimating Greenhouse Gas Emissions

Because we used only one EF for each forest type, our analysis may either under- or overestimate trace gas emissions. A more accurate model might use EFs that are time-weighted according to the changing combustion efficiency of an entire burn (Figure 17-1). In addition, combustion efficiency increases as fuel becomes drier. Therefore, the EFs measured during a given tropical forest burn will tend to overestimate emissions from drier forest conditions and underestimate those occurring under wetter forest conditions.

Ideally, a model of emissions from biomass burning would incorporate several additional facets of the process under study, all of which vary spatially. For example, whereas the fuel loads shown in Table 17-3 are for total above-ground biomass, they are generally higher than the mass that would burn during one cut and burn event. In reality, not all biomass burns during one deforestation burning, and the portion that burns will vary by forest formation as well as by land use practices within a given region. For example, Kauffman *et al.* (1993) and Sampaio *et al.* (1993) estimated the biomass of slashed Brazilian tropical dry forest known as Caatinga to be about 74 Mg $ha^{-1}$. After the first slash fire, 16.4 Mg $ha^{-1}$ of wood debris remained on site. By contrast, for two primary evergreen forest sites in the Brazilian Amazon, total aboveground fuel biomass ranged from 292 to 435 Mg $ha^{-1}$ (Kauffman *et al.*, 1995). Fires resulted in the consumption of a much smaller portion, 42–57%, of aboveground biomass, including >99% of litter and root mat, but <50% of the coarse wood debris. This delayed oxidation of carbon is one reason that the analysis of Hall *et al.* (1985) gives a lower estimate than ours. The remaining biomass is, however, probably consumed by subsequent burns.

Socioeconomic processes also influence how burning proceeds within a region. The Rio Los Santos Forest Reserve, which is located in the southwestern portion of the Talamancan Mountain range, provides an example. In that region, partially burned tree boles which remain in pastures are gradually being converted to charcoal and subsequently burned for cooking (Kappelle and Juarez, 1995). Greater emission factors for products of incomplete combustion would be applicable to that biomass.

In addition, the combustion factor and combustion efficiency for a given fire within a particular ecosystem will vary according to wind and moisture conditions. Combustion factors of 78 to 95% were measured by Kauffman *et al.* (1993) for Brazilian Caatinga, and they depended on moisture conditions. Also, combustion efficiency decreases, and thus quantities of trace gas emissions other than $CO_2$ increase, as a burn event proceeds from flaming to smoldering phases. A model which incorporates that change in combustion efficiency throughout a burn would be somewhat more accurate.

Emission factors were not available to estimate trace gas emissions from tropical dry, wet, or rain forest life zones, therefore we had to use emission factors measured in a savanna and a moist forest. This aspect adds to uncertainty regarding incomplete combustion product emissions in the wet zones.

Given the various sources of uncertainty that we have described, a more accurate biomass burning and trace gas emissions model would incorporate the following: (1) the successively decreasing amount of original forest biomass which is burned during repeated slash and burn events, and the spatially variable burning fate of that biomass (such as in the Costa Rican example of gradual conversion of stumps and boles to charcoal); (2) pulses of biomass reaccumulation during fallow periods for those regions where slash and burn agriculture still occurs, such as indigenous reserves in the eastern portion of the Costa Rican Atlantic slope; (3) a gradual increase in combustion efficiency and fuel consumption as a disturbed site becomes less biomass-dense and drier; (4) knowledge of trace gas emission factors in very wet and dry forest formations; (5) knowledge of the portion of biomass first made into charcoal and then burned for fuel; (6) knowledge of the extent of field and pasture abandonment and subsequent secondary forest development; and (7) additional stratification of biomass and deforestation data along soil drainage and fertility gradients. Probably the greatest source of uncertainty in this study is simply the relatively poor database on biomass.

### D. Interaction between Patterns of Deforestation and Greenhouse Gas Emissions

By the end of the period of observation for this study, insignificant amounts of the dry forest types and lowland moist forests remained in Costa Rica. In contrast, montane and lower montane rain forests still have more original forest area remaining than has been cleared (Sader and Joyce, 1988). As lower elevation regions are saturated, deforestation in humid montane regions has become increasingly common in Costa Rica—as elsewhere in tropical America (e.g., Southgate and Basterrechea, 1992; Young, 1994).

Given the observation that most recent clearing and burning is in wetter tropical forests, the average emissions of incomplete combustion products per area of land deforested may increase globally. The cause for this increase would be the lower combustion efficiency with which wetter forests burn as compared to dry forests, and the consequent increase in products of incomplete combustion, such as $CH_4$. Thus, if tropical deforestation continues there probably will be an increase in trace GHGs released per square kilometer cleared.

## IV. CONCLUSIONS

A prominent feature of these results holds implications for future management of greenhouse gas emissions in Costa Rica and other tropical areas. If clearing of the dry and moist forest formations throughout the world has or will shift to clearing of wetter forests, much greater quantities of trace GHGs will be released (Figure 17-2). As a result, the average warming potential from tropical forest burning could increase.

Information gathering should include a thorough evaluation of carbon stores in dead plant materials (litter, organic soil horizons, dead wood). Such information could provide additional insight into an analysis such as this one. The high-elevation forests may, in some cases, have thick organic soil horizons and large amounts of downed wood. Therefore the biomass density remaining there might be even greater relative to other forest types than currently appears to be the case. These forests also recover their biomass more slowly than at lower (warmer) elevations (Kappelle *et al.*, 1996).

In the meantime, it appears that we have a pretty good handle on biomass and carbon release for Costa Rica, that we understand the reasons for the differences in different estimates, and that our estimates are good enough to make some conclusions. First, our estimate of 8.9 Tg carbon release per year, and another approximately 4 tons carbon equivalent as methane and nitrous oxide, is larger than the annual industrial $CO_2$ release of about 9 Tg. Second, if there is any hope for sustainability of resources in Costa Rica, the forests should be a good candidate, but in fact the forests are not only not a sink for carbon, they are a large source. Third, a reforestation program could make Costa Rica a carbon sink, but stopping deforestation appears to have a much larger impact, at least initially. Finally, the productivity of moist forests is sufficient to sequester all $CO_2$ released from fossil fuel burning if a very large proportion of the land area of the country was devoted to that.

## ACKNOWLEDGMENTS

We thank Charles Hall and J. Boone Kauffman for their technical reviews of a previous draft of this chapter. This document has been funded by the U.S. Environmental Protection Agency through National Network for Environmental Management Studies Fellowship U-914602-01-0 (E.H.H.). It has been subjected to the Agency's peer and administrative review and approved for publication as an EPA document. Mention of trade names or commercial products does not constitute endorsement or recommendation for use.

## REFERENCES

Brown, L. R. 1996. The acceleration of history. In *State of the World 1996*, pp. 3–20. W. W. Norton, New York.

Brown, L. R. 1997. Can we raise grain yields fast enough? *Worldwatch* 10:8–17.

Brown, S., and A. E. Lugo. 1982. The storage and production of organic matter in tropical forests and their role in the global carbon cycle. *Biotropica* 14:161–187.

Brown, S., A. E. Lugo, and J. Chapman. 1986. Biomass of tropical tree plantations and its implications for the global carbon budget. *Canadian Journal of Forest Research* 16:390–394.

Brown, S., A. J. R. Gillespie, and A. E. Lugo. 1989. Biomass estimation methods for tropical forests with applications to forest inventory data. *Forest Science* 35:881–902.

Brown, S., L. R. Iverson, A. Prasad, and D. Liu. 1993. Geographical distribution of carbon in biomass and soils of tropical Asian forests. *Geocarto International* 4:45–59.

Brun, R. 1976. Methodik und ergebnisse zur biomassenbestimmung eines nebelwald-okosystemsin den Venezolanischen Anden. In *Proc. Div. I, 16th IUFRO World Congress*, pp. 490–499. Oslo, Norway. International Union of Forest Research Organizations.

Cairns, M. S., S. Brown, E. H. Helmer, and G. A. Baumgardner. 1997. Root biomass allocation in the world's upland forests. *Oecologia* 111:1–11.

Crutzen, P. J., and M. O. Andreae. 1990. Biomass burning in the tropics: Impacts on atmospheric chemistry and biogeochemical cycles. *Science* 250:1669–1678.

Delaney, M., S. Brown, A. E. Lugo, A. Torres-Lezama, and N. Bello Quintero. 1997. The distribution of organic carbon in major components of forests located in six life zones of Venezuela. *Journal of Tropical Ecology*. 13:697–708.

Detwiler, R. P., and C. A. S. Hall. 1988. Tropical forests and the global carbon cycle. *Science* 239:42–47.

Edwards, P. J., and P. J. Grubb. 1977. Studies of mineral cycling in a montane rain forest in New Guinea. *J. Ecology* 65:943–969.

FAO. 1993. *Forest Resources Assessment 1990: Tropical Countries.* FAO Forestry Paper 112, FAO, Rome.

Fittkau, E. J., and N. H. Klinge. 1973. On biomass and trophic structure of the central Amazonian rain forest ecosystem. *Biotropica* 5:2014.

Fleming, T. H. 1986. Secular changes in Costa Rican rainfall: Correlation with elevation. *Journal of Tropical Ecology* 2:87–91.

Frangi, J. L., and A. E. Lugo. 1985. Ecosystem dynamics of a subtropical floodplain forest. *Ecological Monographs* 55:351–369.

Gillespie, A. J. R., S. Brown, and A. E. Lugo. 1992. Tropical forest biomass estimation from truncated stand tables. *Forest Ecology and Management* 48:69–87.

Golley, F. B., J. T. McGinnis, R. G. Clements, G. I. Child, and M. J. Duever. 1975. *Mineral Cycling in a Tropical Moist Forest Ecosystem.* University of Georgia Press, Athens, GA. 248 pp.

Hall, C. A. S., R. P. Detwiler, P. Bogdonoff, and S. Underhill. 1985. Land use change and carbon exchange in the tropics. 1. Detailed estimates for Costa Rica, Panama, Peru, and Bolivia. *Environmental Management* 9:313–334.

Hall, C. A. S., and J. Uhlig. 1991. Refining estimates of carbon released from tropical land-use change. *Canadian Journal of Forest Research* 21:118–131.

Hao, W. M., D. Scharffe, J. M. Lobert, and P. J. Crutzen. 1991. Emissions of $N_2O$ from the burning of biomass in an experimental system. *Geophysical Research Letters* 18:999–1002.

Harrison, S. 1991. Population growth, land use and deforestation in Costa Rica, 1950–1984. *Interciencia* 16:83–93.

Hecht, A. D., and D. Tirpak. 1995. Framework agreement on climate change: A scientific and policy history. *Climatic Change* 29:371–402.

Helmer, E. H. 1999. The landscape ecology of secondary tropical forest in Montane Costa Rica. Ecosystems (in press).

Holdridge, L. R. 1967. *Life Zone Ecology.* Tropical Science Center, San Jose, Costa Rica.

Houghton, J. T., L. G. Meira Filho, J. Bruce, Hoesung Lee, B. A. Callander, E. Haites, N. Harris, and K. Maskell. 1995. *Climate Change 1994, Radiative Forcing of Climate Change and an Evaluation of the IPCC IS92 Emission Scenarios.* Intergovernmental Panel on Climate Change, Cambridge University Press, Cambridge, UK. 339 pp.

Houghton, R. A. 1995. Land-use change and the carbon cycle. *Global Change Biology* 1:275–287.

Hozumi, K., Yoda, K., Kokawa, S., and Kira, T. 1969. Production ecology of tropical rain forests in souwestern Camodia. I. Plant Biomass. *Nature and Life in Southeast Asia* **6**, 1–54.

Huttel, C. 1975. Root distribution and biomass in three Ivory Coast rain forest plots. In F. B. Golley and E. Medina (Eds.), *Tropical Ecological Systems,* pp. 123–130. *Ecological Studies,* Vol. 11. Springer-Verlag, New York.

Janzen, D. H. 1988. Tropical dry forests: the most endangered major tropical ecosystem. In E. O. Wilson (Ed.), Biodiversity, pp. 130–137. Springer-Verlag, New York, NY.

Jimenez, W., A. Chaverri, R. Miranda, and I. Rojas. 1988. Aproximaciones silviculturales manejo un robledal (Quercus spp.). *Turrialba* 38:208–214.

Jordan, C. F. 1985. *Nutrient Cycling in Tropical Forest Ecosystems: Principles and Their Application in Management and Conservation.* John Wiley & Sons, Chichester, UK.

Kauffman, J. B., C. Uhl, and D. L. Cummings. 1988. Fire in the Venezuelan Amazon. 1. Fuel biomass and fire chemistry in the evergreen rainforest of Venezuela. *Oikos* 53:167–175.

Kauffman, J. B., K. M. Till, and R. W. Shea. 1992. Biogeochemistry of deforestation and biomass burning. In D. A. Dunnette and R. J. O'Brien (Eds.), *The Science of Global Change. The Impact of Human Activities on the Environment,* pp. 426–456. *ACS Symposium Series,* Vol. 483. American Chemical Society, Washington, DC.

Kauffman, J. B., R. L. Sanford, D. L. Cummings, I. H. Salcedo, and E. V. S. B. Sampaio. 1993. Biomass and nutrient dynamics associated with slash fires in neotropical dry forests. *Ecology* 74:140–151.

Kauffman, J. B., D. L. Cummings, D. E. Ward, and R. E. Babbit. 1995. Fire in the Brazilian Amazon: Biomass, nutrient pools and losses in slashed primary forests. *Oecologia* 104:397–408.

Kappelle, M. 1991. Distribucion altitudinal de la vegetacion del parque nacional Chirripi, Costa Rica. *Brenesia* 36:1–14.

Kappelle, M., and M. E. Juarez. 1995. Agro-ecological zonation along an altitudinal gradient in the montane belt of the Los Santos Forest Reserve in Costa Rica. *Mountain Research and Development* 15:19–37.

Kappelle, M., T. Geuze, M. E. Leal, and A. M. Cleef. 1996. Successional age and forest structure in a Costa Rican upper montane Quercus forest. *Journal of Tropical Ecology* 12(5):681–698.

Lashof, D. A., and D. R. Ahuja. 1990. Relative contributions of greenhouse gas emissions to global warning. *Nature* 344:529–531.

Lawton, R., and V. Dryer. 1980. The vegetation of the Monteverde cloud forest reserve. *Brenesia* 18:101–116.

Luvall, J. C., and C. Uhl. 1990. Transpiration rates for several woody successional species and for a pasture in the upper Amazon basin in Venezuela. *Acta Amazonica* 20:29–38.

Martinez-Yrizar, A., J. Sarukhan, A. Perez-Jimenex, E. Rincon, J. M. Maass, A. Solis-Magallanes, and L. Cervantes. 1992. Aboveground phytomass of a tropical deciduous forest on the coast of Jalisco, Mexico. *Journal of Tropical Ecology* 8:87–96.

Ovington, J. D., and J. S. Olson. 1970. Biomass and chemical content of El Verde lower Montane rain forest plants PPH53-H61. *In:* A Tropical Rainforest (H. I. Odum and R. F. Pigeion, Eds.). U.S. Atomic Energy Commission, Oak Ridge, Tennessee.

Prather, M., R. Derwent, D. Ehhalt, P. Fraser, E. Sanhueza, and X. Zhou. 1995. Other trace gases and atmospheric chemistry. In J. T. Houghton, L. G. Meira Filho, J. Bruce, Hoesung Lee, B. A. Callander, E. Haites, N. Harris, and K. Maskell (Eds.), *Climate Change 1994: Radiative Forcing of Climate Change and an Evaluation of the IPCC IS92 Emission Scenarios,* pp. 73–126. Intergovernmental Panel on Climate Change, Cambridge University Press, UK.

Sader, S. A., and A. T. Joyce. 1988. Deforestation rates and trends in Costa Rica 1940 to 1983. *Biotropica* 20:11–14.

Salati, E., and P. B. Vose. 1984. Amazon basin: A system in equilibrium. *Science* 225:129–138.

Saldarriaga, J. G., D. C. West, M. L. Tharp, and C. Uhl. 1988. Long-term chronosequence of forest succession in the upper Rio Negro of Colombia and Venezuela. *Journal of Ecology* 76:938–958.

Sampaio, E. V. S. B., I. H. Salcedo, and J. B. Kauffman. 1993. Effect of different fire severities on coppicing of caatinga vegetation in Serra Talhada, PE, Brazil. *Biotropica* 25:452–460.

Schime, D., I. G. Enting, M. Heimann, T. M. L. Wigley, D. Raynaud, D. Alves, and U. Diegenthaler. 1995. $CO_2$ and the carbon cycle. In J. T. Houghton, L. G. Meira Filho, J. Bruce, Hoesung Lee, B. A. Callander, E. Haites, N. Harris, and K. Maskell (Eds.), *Climate Change 1994: Radiative Forcing of Climate Change and an Evaluation of the IPCC IS92 Emission Scenarios,* pp. 35–72. Intergovernmental Panel on Climate Change, Cambridge University Press, Cambridge, UK.

Southgate, D. S., and Basterrechea. 1992. Population growth, public policy and resource degradation: The case of Guatemala. *AMBIO* 21:460–464.

Stiles, F. G., and A. F. Skutch. 1989. *A Guide to the Birds of Costa Rica.* Cornell University Press, Ithaca, NY. 511 pp.

Tanner, E. V. 1980. Studies on the biomass and productivity in a series of montane rain forests in Jamaica. *Journal of Ecology* 68:573–588.

Tosi, J., and R. F. Voertman. 1964. Some environmental factors in the economic development of the tropics. *Economic Geography* 40:189–205.

Ward, G. E., R. A. Susott, J. B. Kauffman, R. E. Babbitt, D. L. Cummings, B. Dias, B. N. Holben, Y. J. Kaufman, R. A. Rasmussen, and A. W. Setzer. 1992. Smoke and fire characteristics for cerrado and deforestation burns in Brazil: BASE-B experiment. *Journal of Geophysical Research* 97:14601–14619.

World Resources Institute. 1994. *World Resources 1994–95: A Guide to the Global Environment.* Oxford University Press, New York. 400 pp. with data diskette.

Young, K. R. 1994. Roads and the environmental degradation of tropical montane forests. *Conservation Biology* 8:972–976.

CHAPTER 18

# Forestry in Costa Rica and an Estimate of Energy Potentially Available from Forests

PATRICK G. MOTEL, PABLO MARTINEZ, AND MEEGAN CARROLL



## I. INTRODUCTION

Forestry is defined as the science, the art, and the practice of managing and using for human benefit the natural resources that occur on and in association with forestlands (i.e., tree-covered lands) (Sharpe, 1986). Forestry involves many aspects of natural resource management, including disciplines that deal with air, water, soil, fire, plants, animals, and humans in relationship to forested lands and the products and services they provide. The goal of most professional foresters is to create a truly sustainable system that allows continual production of valuable commodities for society while also maintaining more or less natural systems and their intrinsic values. Throughout most of the history of Costa Rica, the practice of forestry, as just defined, has not been practiced. The presence of well-managed, sustained-yield, commercial forests were and still are rare (Wille, 1990). This is not surprising, since forests overwhelmingly dominated most of the landscape until recently, and trees were there more or less for the taking. The reasons for this forest cutting are reviewed in Cornell and Hall (1994) and in Chapters 16 and 17.

Costa Rican tropical forests are among those having the largest biomass and highest diversity in the world. Nearly all of Costa Rica was forested in the early 1800s. But, as developed in detail in Chapters 14 and 15, the forests have been cut rapidly. According to Wille (1990), more than 90% of the forests cut in Costa Rica in the past 20 years have not regenerated and are now lost as forest. This is ironic, for in his opening address to the 50th anniversary of the Institute for Tropical Forestry in Puerto Rico, Leslie Holdridge, with Budowski, the grand old foresters of Costa Rica, said,

> My generation of foresters has failed. There should not be this crisis now about tropical deforestation. There is no need for cutting primary forests. All the things that we desire from tropical forests could be provided readily by well-managed secondary forests. We as foresters know well how to do that technically. Where we have failed is in the political implementation of what we know. (Holdridge, 1993)

## II. USE OF COSTA RICAN FORESTS AND THE FORESTRY INDUSTRY

The land and its vegetation have traditionally supplied a wide variety of natural goods and products for indigenous consumption and for industrial use (Tables 18-1 and 18-2). The main approach to forestry in Costa Rica has been simply

**TABLE 18-1 Traditional Use of Forest Products**[a]

| | |
|---|---|
| Agricultural uses: | shifting cultivation, forest grazing, nitrogen fixation, mulches, fruits and nuts, fuelwood and charcoal cooking, and other household uses |
| Building: | poles, pit sawing and sawmilling construction of farm buildings and housing, fencing, furniture, and joinery |
| Weaving materials: | ropes and strings, baskets, and furniture |
| Special woods and ashes: | carving, incense, chemicals, and glassmaking |
| Industrial uses: | sawlog lumber, joinery, furniture, packing, shipbuilding, mining, construction, and sleepers |
| Veneer: | logs, plywood, veneer, furniture, containers, and construction |
| Pulpwood: | newsprint, paperboard, printing and writing paper, containers, packaging, dissolving pulp, distillates, textiles, and clothing |
| Residues: | particle board, fiberboard, and wastepaper |
| Poles: | pit props and transmission poles |
| Charcoal: | chemicals, dry cells, and polyvinyl chloride (PVC) |
| Gums and resins: | naval stores, tannin, turpentine, distillates, resin, and essential oils |

[a] (U.S. Congress, 1988)

TABLE 18-2 Forest Products Consumption and Use[a]

| | |
|---|---|
| Roundwood and small roundwood: | wood in its natural state as felled or otherwise harvested, with or without bark, round, split, roughly squared, or in other forms (e.g., roots, stumps, and burls). |
| Fuelwood: | wood in the rough (from trunks and branches of trees) to be used as fuel for purposes such as cooking, heating, or power production. |
| Sawlogs: | logs, whether or not roughly squared, to be sawed lengthwise for the manufacture of sawed wood, railway sleepers (ties), or peeled or sliced for veneer (sawed wood includes sleepers; unplaned, planed, grooved, and tongued wood; boxboards; and lumber). |
| Veneer sheets: | thin sheets of wood of uniform thickness—rotary cut, sliced, or sawed—for use in plywood, laminated construction, veneer, etc. Plywood can be veneer plywood, core plywood, blockboard, laminate board, and batten board. Other plywood such as cellular board and composite plywood wood-based panels is an aggregate of the following commodities: veneer sheets, plywood, particleboard, and fiberboard. |

[a] (U.S. Congress, 1988)

the clearing of the primary vegetation and the converting of land to some other use. Often the trees felled from the primary forest are not utilized, but are simply cut and burned in place to make way for crop fields or pastures, or to assert legal ownership by "improvement of the land" (Hall *et al.*, 1985; Wille, 1990). The objectives of this type of "land management" rarely considered the various natural resources and services that the intact forest provided.

A second, less-common use of Costa Rica's primary forests was, and is, selective cutting (or "high grading") of valuable commercial trees. Throughout the tropics this usually involves entering an intact forest ecosystem and removing a small percentage of the vegetation, characteristically mahogany (*Swietenia macrophylla*) in the Americas (Figure 18-1) (Howard, 1995). This method often changes the composition, structure, and genetic diversity of the forest. Often far more trees are damaged than those removed; appropriate harvest techniques that minimize damage to the residual vegetation are seldom used in Costa Rica (Buschbacher, 1990). High grading should not be necessary in Costa Rica because 60–75% of the woody species found in the primary forests have commercial uses (Howard, 1995).

## A. Logging in Costa Rica

Historically the logging process in Costa Rica utilized axes and oxen, but these have been replaced by chain saws and bulldozers, particularly in the larger

**FIGURE 18-1** Mahogany trees on road near Siquerres. This was a common enough sight when the senior editor first visited Costa Rica in 1977 but was essentially never seen by the mid-1990s. (Photo by Charles Hall.)

operations (Sage, 1981). Transportation of logs is done primarily by trucks, but occasionally, in the province of Limon, by rail or rafting in the Tortuguero Canal and Toro Amarillo and Sarapiqui rivers (Sage, 1981).

There are three major methods of selling standing timber (Petriceks, 1975, in Sage, 1981): (1) The landowner does the logging and either sells the logs to truckers or keeps them as raw material for his own mill, (2) the logger purchases stumpage on a per unit basis or as a lump sum sale and takes the logs to the mills, or (3) the sawmill operator purchases timber from the landowner and carries out subsequent operations, including transport to the mill. When the landowner sells standing timber and does not perform any of the harvesting and milling operations, he receives only about 12% of the price paid by the consumer for the lumber due to the fact that most landowners do not have the resources to exploit and process the wood.

Logging operations have been notoriously inefficient. On average only 32% of the total volume of wood cut is processed by the forest industries, 19% is used as charcoal or fuel, and 49% is wasted (ACM, 1978, in Sage, 1981). Additional wood volume is lost in many of the mills because of old and inefficient equipment (Carlos Hernandez, personal communication).

## B. Trends in Timber Production

During the last three decades the total volume of timber production in Costa Rica has increased. Although some timber substitutes have appeared and remained in the market, such as concrete for building, metal for furniture, and petroleum derivatives for fuel, the rapid population growth and the increase in per capita income have increased the demand for forest products (Sage, 1981; Perez and Protti, 1978). In addition, the development and improvement of the rural road network has led to extractions of timber from previously inaccessible areas (Sage, 1981). For example, the U.S. government built a military road through previously undisturbed forests in northeastern Limon Province as part of the Reagan administration's efforts to counter Nicaraguan "Contra" activities. The almost immediate effect was rapid destruction of the primary forests of the region and their replacement by pastures during the early 1980s. Finally, the increasing use of bulldozers and chain saws has allowed more frequent and larger scale commercial logging operations.

The relentless harvest of the most valuable commercial tropical species has resulted in a shortage of forest products in Costa Rica (Howard, 1995). For some commercial species adequate supplies no longer exist (Lehmann, 1992, in Howard, 1995) and others are in danger of local extinction and are protected by law (Madrigal, 1993, in Howard, 1995). A survey by Sage (1981) found that even by the late 1970s some of the most valuable species, such as pochote (*Bombacopsis quinatum*), cristobal (*Platymisciun* sp.), cocobolo (*Dalbergia retusa*), caoba (*Swietenia humilis*), and cenizaro (*Pithecolobium saman*), had become more difficult to obtain and thus more expensive. As a result other species that in the past were considered less valuable have become widely used. For example, the Costa Rican plywood industry traditionally used Caobilla (*Guarea* sp.) as raw material until the 1970s or 1980s. Now it is too expensive, and ceiba (*Ceiba pentandra*) and other species are now used (Sage, 1981).

Another result is that Costa Rica has become a significant net importer of lumber (Figure 18-2), which is remarkable for a country that was nearly 100% forested a century ago. This change can be attributed to the wood needs for the expanding population and economy, traditional devaluation of forests in an agricultural-based culture, and a lack of forest management and planning. From 1964 to 1979 the value of exports of forest products was small, but by 1980 to 1992 the values increased substantially, from \$11.8 to \$24.1 million. During the same two time periods the value of imported forest products increased from \$7.8 to \$111.4 million. By some estimates Costa Rica's wood bill in the year 2000 will equal the costs of imported oil (Wille, 1990). The major imported forest product is, by far, paper products. Other imported

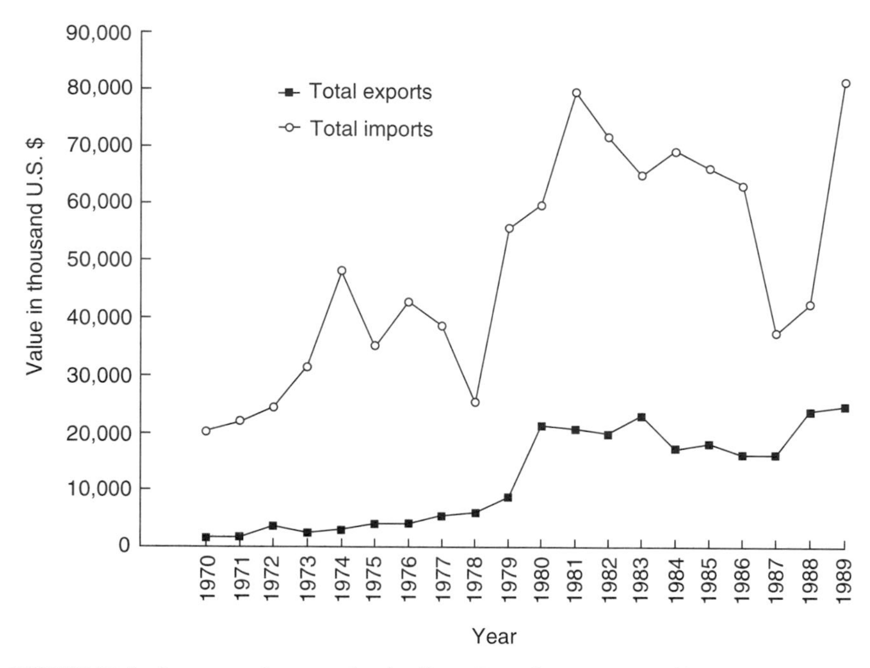


**FIGURE 18-2** Imports and exports for the Costa Rican forestry sector. (Source: FAO, 1993).

products include roundwood, wood-based panel, plywood, and paper products. Of course if paper products were made in Costa Rica many industrial components would still be required.

Nevertheless paper production in Costa Rica has increased steadily from 20 to 58 tons per year from 1970 to 1994. Important products are paper and paper board, newsprint, and tissue and other thin paper (FAO, 1993). A description and use of some major forest products from Costa Rica are given in Table 18-2. In 1993 the total quantity of forest products produced was about 9.6 million $m^3$ (FAO, 1993).

## C. Agroforestry

Another commonly used and relatively well-studied forestry activity in Costa Rica is the practice and development of agroforestry. One of the first definitions of agroforestry was "a sustainable management system which increases the yield of the land, combines the production of crops (including tree crops) and forest plants and/or animals simultaneously or sequentially on the same unit of land, and applies management practices that are compatible with the cultural practices of the local population" (Bene *et al.*, 1977; King and Chandler, 1978, in Steppler and Nair, 1987).

Agroforestry in Costa Rica, as in other parts of the world, is a very old practice. Various techniques of mixing trees with food crops were, and still are, well known by many Costa Rican landowners. Some examples include managed fallow in shifting cultivation, tropical mixed home gardens, and the mixing of trees and crops along ditches (chinampas) (Gliessman, 1981). Nitrogen-fixing alders (*Alnus acuminata*) have been planted in pastures in the highlands of Costa Rica for many decades. Budowski (1957) described successful windbreaks made up of cypress (*Cupressus lusitanica*) and laural (*Cordia alliodora*) in the high dairy regions and the wet lowlands, respectively, in Costa Rica. The journal *Tropical Agroforestry*, produced at CATIE in Turrialba, provides many interesting perspectives on agroforestry.

## D. The Potential for Plantations

Given the large amount of forest products Costa Rica imports and the relative scarcity of virgin forests, there is obviously a potential role for either secondary forests (Brown and Lugo, 1990) or plantations (Lugo *et al.*, 1988) for meeting the future fiber needs of the country. The potential for plantations, especially for meeting timber needs, depends a great deal on the ability of sawmills and carpenters to utilize the wood supplied to them. The very large diversity of the Costa Rican forest and the generally conservative nature of the users of wood have made supplying new timber sometimes difficult. One of the results has been that increasingly exotic trees, such as teak, a native of south Asia, are being grown. Haggar and Ewel (1997) studied the needs of the users of wood and the characteristics of a wide variety of Costa Rican species with the intention of finding native Costa Rican trees from the Atlantic zone that would be good candidates for plantations. They found four that grew especially well and had excellent characteristics for use (Figure 18-3). These are now being planted on a larger scale.

One of the most important aspects of plantations is matching the species of trees and the management techniques to the local climate. For example, teak (*Tectonis grandis*) is commonly planted in Guanacaste Province. Teak is a well-known, commercially valuable species with a lot of appeal to investors, as is obvious to many tourists visiting San José hotels. But teak does not grow especially well in most Costa Rican plantations (Carlos León, personal communication). Part of the reason may be that teak is originally from southeast Asia in regions that are very wet and very hot. Guanacaste has regions that

FIGURE 18-3 Very high growth rate of indigenous tree species. This tree, although not of dense wood, is less than five years old.

are wet and regions that are hot, but the wet areas tend to be in the mountains. Very preliminary data analyzed using a gradients approach suggest that teak grows well in Costa Rica only in regions that are both wet and hot (Figure 18-4). Unfortunately relatively little land area in Costa Rica has these characteristics. Thus any plantation approach to forest products must carefully match tree species with geographical characteristics.

## III. COULD THE FOREST PROVIDE THE ENERGY THAT COSTA RICA NEEDS FOR ITS ECONOMY?

Forestry has the potential to be a key component in the search for sustainable development in the tropics. Presently Costa Rica derives about 20% of its

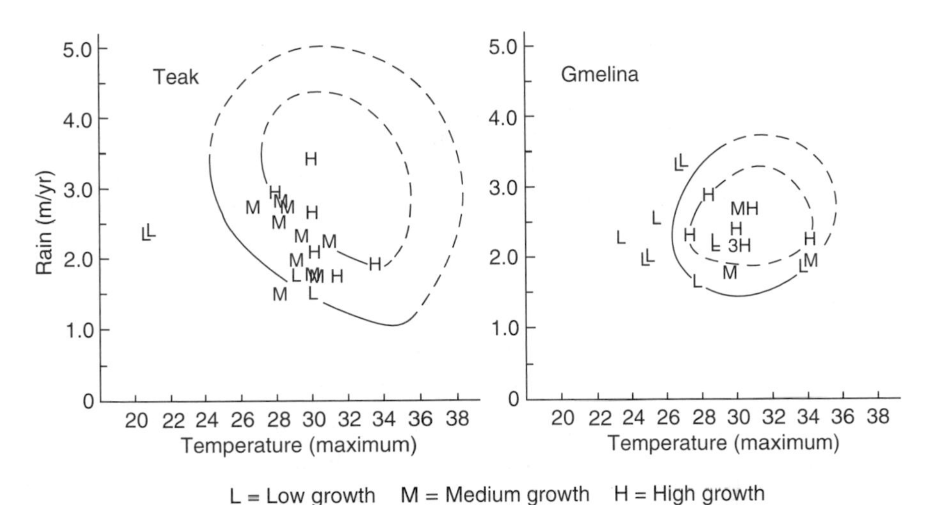


FIGURE 18-4 Teak and gmelina growth rates as a function of gradients of temperature and rain (data from CATIE). H, high growth; M, medium growth; L, low growth. Circles are preliminary estimates of growth rates in gradient space.

energy from biomass of one sort or another, a relatively high proportion compared to most nations (Figure 4-5). Most of this energy is derived from agricultural wastes, but a significant portion is from forest products. About 20% of Costa Rica still remains covered with dense forest, and much of the deforested land presently in pastures has the potential to support forests once again. Could the existing or potential forestlands supply the energy that Costa Rica needs to run its present economy, or even to support additional growth? This section will try to answer this question from a biophysical perspective by comparing the energy captured by Costa Rican forests with the consumption of energy in its economy.

A large reforestation program could increase energy self-sufficiency and also reduce the high rates of erosion and biodiversity loss. Some 2.5 million ha is used at this time for pastures in Costa Rica. If prices of meat continue to decrease, soil is depleted, or conservation initiatives become more powerful, these areas are likely to become secondary forest. They probably could provide a more sustainable and possibly profitable use in this role.

## A. Estimating Potential Wood Energy Available

Estimating the potential for sustainable energy from forest fuels requires (1) an estimate of energy produced (or fixed) by existing forests per year (or potentially produced by plantations), (2) an analysis of the suitability of forest fuels relative to the types of energy needed by society, and (3) an estimate of the energy cost of turning that wood into a useful energy product at the places it would be used. The energy return on investment (EROI) is a useful way to understand the potential of a fuel (Hall *et al.*, 1986). It is the ratio of the energy that a fuel supplies at its point of use to the energy that it took to derive that fuel.

The energy that Costa Rican forests sequester annually ($E_s$) can be estimated as

$$E_s = \text{forested hectares} \times \text{mean annual increment} \times \text{wood density} \times \text{energy density } \hat{E} \text{ of wood.}$$

The units are (kcal/yr) = hectares × ($m^3$/ha/yr) × (kg/$m^3$) × (kcal/kg).

We derived the number of hectares of both primary and secondary forests for "primary humid forests," along with their growth rates (Table 18-3), from Repetto (1992). Independent estimates of growth rates are given in Brown (1997), Brown and Lugo (1990), Lugo *et al.* (1988), and Caves (1991), who gave values for 11 species grown in the drier Guanacaste Province. The fastest growing species were *B. quinatum* and *Enterolobium ciclocaroum*. These values are only for trees with diameters larger than 10 cm because they were derived for timber potential. Edgar Ortez, a Costa Rican forester (personal communication), suggests that these timber volumes should be multiplied by 1.8 to give total biomass increments. We did not make this correction, since harvesting small trees is very energy intensive and expensive, but agree that the actual potential might be larger than the values we derive here. We used an increment value of 20 $m^3$/ha/yr for our calculations using the assumption that plantations would be planted with the fastest growing species.

The density of wood varies for each species. Brown (1997) compiled densities for many tropical woods. She gives the value of 0.35 g per cubic centimeter for *E. ciclocarpum* and three values for *B. quinatum*: 0.38, 0.45, and 0.51 g/

**TABLE 18-3** Estimates of Wood Volume Increments for Costa Rican Forests

| Forest type | Growth increment ($m^3$/ha/yr) | Source |
|---|---|---|
| Primary (humid) | 6.5 | Repetto (1992) |
| Secondary | 27.5–33.8 | Brown and Lugo (1984) |
| Secondary (dry) | 5.4–8.9 | Herrero |
| Plantations | | Chavez (1991) |
| *Bombacopsis quinatum* | 24.5 | |
| *Enterolobium ciclocaroum* | 15.5 | |

$cm^3$. Values for other species of secondary forest are close to 0.5 $g/cm^3$. Golley (1964) gives an energy density of 3.897 cal/g dry wood, and Chapman and Hall (1986) found a mean of about 3.7 for many North American fuelwoods. We used a value of 3.8. A much more detailed analysis by geography and/or life zone would be required to do this analysis more accurately.

## B. End Uses of Energy and the Need for Conversion

The major energy uses in Costa Rica are for transportation, industrial processing, and general domestic and commercial use. Transportation, responsible for roughly one-quarter to one-third of energy used, requires liquid fuels or, should there be an investment in light rail, electricity. Industrial process energy can be from many sources, including biomass. General domestic and commercial energy use is basically electricity. Hence a difficulty in using biomass energy is that there is relatively little demand for solid fuel directly, as that is already being met by existing biomass sources such as fuelwood, bagasse (sugarcane waste), and coffee endocarps (Chapter 21). Hence if biomass is to be very important, either society's use patterns would have to become very different or the biomass would have to be converted to either liquid fuels or electricity. Liquid fuel conversion would require the setting up of a massive and energy-consuming distillation process, and electric production would require the construction of giant thermal plants to burn wood and convert water to steam. These would have to be located near large water bodies to dissipate heat (about two-thirds of the energy generated at thermal power plant is converted to heat, not electricity, and has to be disposed of). Although there is a lot of rain in Costa Rica there are not large rivers near the population centers, which are located near the continental divide. Cogeneration, where waste heat is used for heating homes and so on, makes little sense in the tropics, although the heat might be used, for example, for drying crops.

Wood-produced electricity is unlikely to be competitive with hydroelectricity in Costa Rica, where hydroelectric sites are still plentiful. Either source is constrained more by a lack of capital than by the availability of the fuel itself. A more critical issue is the need for liquid fuels to displace expensive imported oil.

Thus an important issue becomes the EROI with which wood can be delivered to society in a liquid form. Wood can be used as firewood with a relatively high EROI, perhaps 30 to 1, but the gain is quickly lost if transportation distances are greater than 80 km (Smith and Corcoran, 1979). The conversion of wood into a liquid fuel is a much more complex process. Wood is composed of many complex and large molecules containing thousands of linked carbon atoms that have been selected for cohesiveness. Breaking these molecules apart into, say, eight (octane) or smaller fractions is a complex, difficult, energy-intensive and therefore expensive process. The wood can be cooked or treated with strong acids, as is done to make paper, or it can be subjected to bacterial digestion. In the latter case some of the energy in the wood is used as a fuel by the bacteria. In any case processed wood has a low hydrogen to carbon ratio, so that a liquid fuel, which requires a hydrogen–carbon ratio of about 2 : 1 by atoms, would require an extra source of hydrogen or the need to throw away a lot of carbon (Hall *et al.*, 1986). The main world source of hydrogen comes from stripping it out of natural gas. It also can be obtained by hydrolysis of water, which requires a great deal of electricity.

The energy return on investment of a fuel examines the energy delivered to society compared to the energy that society has to invest in obtaining it (Hall *et al.*, 1986). The EROI is equal to kilocalories of fuel extracted divided by the kilocalories of energy required to locate, extract, and refine that fuel.

Deriving the EROI for the use of wood or, especially, liquid fuels derived from wood for Costa Rica is difficult because there is little data and, to our knowledge, no national analysis. The EROI for a wood-based energy farm producing liquid fuels in the United States has been derived by Tillman (1978). Tillman's estimates assume hydrogenation and deriving liquid fuel for a national energy transportation system. Tillman found that the energy used in the conversion process was a large proportion of the energy gained as final product. His calculations for the process were that for every 1000 kcal originally cut from the forest, 800 kcal would be used to generate the product, including the energy used for forest management, harvest, transportation, and conversion. The EROI calculated from that data would be 0.2 for 1, which is very much lower that the 10 to 30 for 1 EROI characteristic of petroleum (Hall *et al.*, 1986). In this sense the biomass-derived fuels are not competitive economically with imported petroleum unless large subsidies or taxes are imposed to make sure the money stays in Costa Rica. Thus even though fuel derived this way is energetically inefficient, it could be done with local resources and hence not drain valuable foreign exchange. In addition, as shown in Table 18-4, the potential resource base is large enough, even if converted at 20% efficiency, to meet the

TABLE 18-4 Energy Potential of Forests from Costa Rica

| Forest type | Area ($10^6$ ha) | Increment ($m^3$/ha/yr) | Density T/$m^3$ | Energy density ($10^6$ kcal/T) | Potential supply | |
|---|---|---|---|---|---|---|
| | | | | | $10^{12}$ kcal | $10^{15}$ J |
| Primary | 1.7 | 6.5 | 0.45 | 3.8 | 18.9 | 79 |
| Secondary | 0.3 | 20.0 | 0.45 | 3.8 | 10.3 | 43 |

needs of Costa Rica for liquid fuels. The downside to the consumer, and to the economy, is that low-EROI fuels will be proportionately much more expensive monetarily, because each of the energy-consuming steps has to be paid for.

## C. Energy Potential of Costa Rican Forests

The absolute potential for energy production from Costa Rican forests is somewhat larger than the present energy use of the national economy (Table 18-5). Our estimate for wood production in primary forests is larger than the potential from secondary forests although that could be changed with massive abandonment of pastures.

Converting this energy to electricity, assuming a 25% use in transportation, a 10% management energy cost, and a conversion efficiency of 35%, could meet the entire electricity needs of Costa Rica (Figure 18-5).

The energy return on investment is a little more difficult to ascertain, but if we assume that Tillman's results are applicable (i.e., that 20% of the original resource can be turned into liquid fuel) then it would be barely possible biophysically to meet half the liquid energy requirements of Costa Rica (Table 18-5). This analysis does not include whatever energy costs might ensue from, e.g., the degradation of the soil resource that is not compensated for with fertilizer. In addition there would be a loss of whatever carbon dioxide absorption capacity would otherwise be represented by the forest growth so utilized.

# IV. DISCUSSION

Costa Rica, as most other tropical countries, has large areas that are nearly useless for sustainable agriculture but not for forestry. Tropical tree growth rates are often higher than those at higher latitudes. Therefore forestry could

TABLE 18-5 Biomass Energy Potentially Available from Costa Rican Forests Compared to Energy Needs of Costa Rica (in $10^{15}$ J)

| | | If converted to | | Energy use, early 1990s | |
|---|---|---|---|---|---|
| Forest type | Potential | Electricity | Liquid | Electricity | Liquid |
| Primary | 79 | 24 | 16 | — | — |
| Secondary | 43 | 13 | 9 | — | — |
| Total | 122 | 37 | 25 | 20 | 50 |

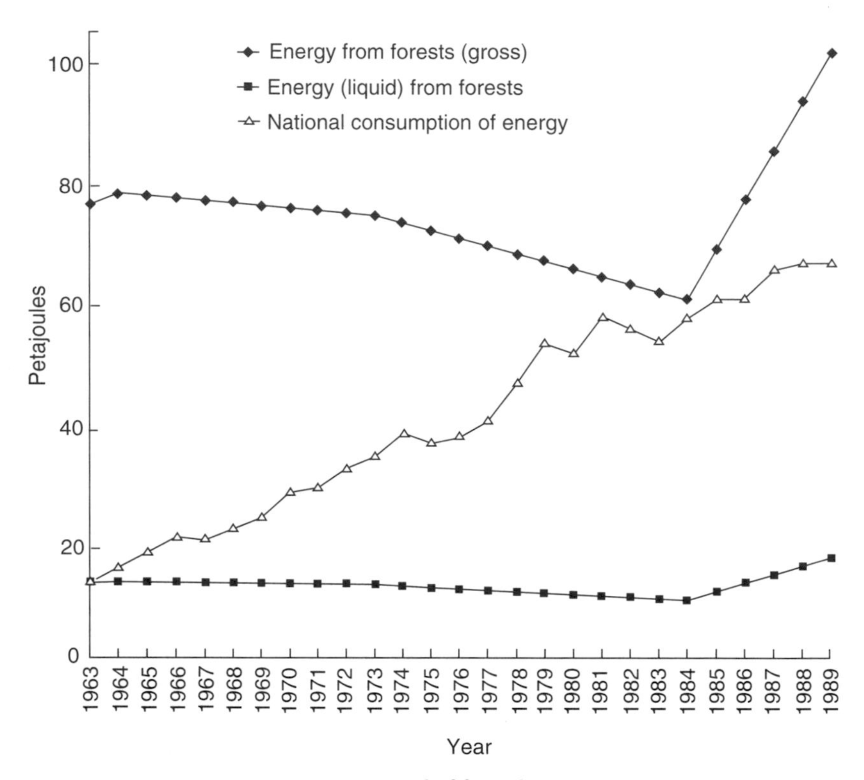


**FIGURE 18-5** Potential of forest biomass energy.

be the key needed for the sustainable development of Costa Rica, as in many other areas of the tropics. Since the energy produced this way, however, will almost certainly be many times as expensive as imported fossil fuels, because of the low energy return on investment, such fuels are unlikely to ever be economically competitive in an open international market. Therefore this potentially sustainable approach is unlikely to be implemented in today's global economy. Whether or not they should be subsidized is an issue open for debate.

A major problem is that many primary forests, perhaps the majority remaining, are in national parks and are protected (Chapter 19). Thus it is probably much better to look at secondary forests as the potential resource (Brown and Lugo, 1990; Lugo, 1990). But wood harvests are hardly environmentally benign (Chapman and Hall, 1986), although they might have much less impact than row agriculture or pastures. Continuous harvest drains a forest of its nutrient capital, and can cause the system to lose production over time. Leaving bark and leaves in the forests can reduce this problem (but also energy profits), as the boles are mostly carbon. A well-managed forest might even be a nutrient source if it were composed of species of the family Leguminosae, which fix nitrogen.

Thus there is considerable potential for forests to contribute to the sustainability of the energy situation in Costa Rica as elsewhere (Lonner and Tornquist, 1988). Realizing that potential, especially in today's free market climate, will be very difficult. And it is not even clear that energy forestry is desirable unless there are no other possibilities, since the land has other good uses, and energy forests, although not extremely destructive, are hardly environmentally benign. One possibility would be to let presently degraded land that has low utility grow back naturally, generating "energy in the bank" as a hedge against future increases in the price of energy, and in the meantime slightly reducing the increase in atmospheric $CO_2$. If the trees are not needed, at least the soil will be protected, biodiversity probably encouraged, and future timber harvests ensured.

There is a peculiar twist, however, to all of this. In the modern world we generally invest energy to get wood fiber (Chapman and Hall, 1986). At this time Costa Rica has a wood deficiency. Thus to use wood to get energy, instead of energy to get wood, implies that a forest's energy has become more valuable than the fiber. This is not yet the case.

## REFERENCES


ACM-OFIPLAN-INTERLOGOS. 1978. El sector forestal y Mderero en Costa Rica, San José, Costa Rica.

Bene, J. G., H. W. Beall and A. Cole. 1977. Trees, food and people: Land management in the tropics. IDRC-048e. International Development Research Center, Ottawa, Canada.

Brown, S., and A. E. Lugo. 1990. Tropical secondary forests. *Journal of Tropical Ecology* 6:1–32.

Brown, S. 1997. *Estimating Biomass and Biomass Change of Tropical Forests: A Primer*. FAO Forestry Paper 134, Rome, Italy. 55 pp.

Budowski. 1957. Quelques aspects de la situation forestière au Costa Rica. Boìs et Forêts les tropiques. 55: Sep.-Oct. 3–8.

Buschbacher, R. J. 1990. Natural Forest Management in the Humid Tropics: Ecological, Social, and Economic Considerations. *AMBIO* 19, 253–258.

Caves, S. E. 1991. Especies nativas aptas para la reforestacion. Guia agropecuaria de Costa Rica, no. 6, pp. 29–32. San Jose, Costa Rica.

Chapman, J., and C. A. S. Hall. 1986. Forest resources and energy use in the forest products industry. In C. A. S. Hall, C. Cleveland, and R. Kaufmann (Eds.), *Energy and Resource Quality: The Ecology of the Economic Process,* pp. 461–479. Wiley, New York. Reprinted 1992, University Press of Colorado, Niwot.

Cornell, J., and C. A. S. Hall. 1994. A systems approach to assessing the forces that generate tropical land use change. p. 35–43 in J. Ohara (ed). Timber certification: implications for tropical forest management. Yale School of Forestry.

Gliessman, S. R., R. Garcia, and M. Amadon. 1981. The ecological basis for the application of traditional agricultural technology in the management of tropical agroecosystems. *Agroecosystems* 7:173–185.

Golley, F. B. 1964. Caloric value of wet tropical vegetation. *Ecol.* 50:517–519.

Haggar, J. P., and J. Ewel. 1997. Primary productivity and resource partitioning in model tropical ecosystem. *Ecology* 78:1211–1221.

Hall, C. A. S., and D. L. DeAngelis. 1985. Models in ecology: Paradigms found or paradigms lost? *Bulletin of the Ecological Society of America* 66:339–346.

Hall, C. A. S., R. P. Detwiler, P. Bogdonoff, and S. Underhill. 1985. Land use change and carbon exchange in the tropics. 1. Detailed estimates for Costa Rica, Panama, Peru, and Bolivia. *Environmental Management* 9:313–334.

Hall, C. A. S., C. J. Cleveland, and R. K. Kaufmann. 1986. *Energy and Resource Quality: The Ecology of the Economic Process.* Wiley–Interscience, New York. Reprinted 1992, University Press of Colorado, Boulder, CO.

Holdridge, L. 1993. Opening address to the 50th anniversary of the Institute for Tropical Forestry in Puerto Rico, June 4, 1993.

Howard, A. F. 1995. Price trends for stumpage and selected agricultural products in Costa Rica. *Forest Ecology and Management* 75:101–110.

FAO. 1993. *Forest Resources Assessment 1990: Tropical Countries.* FAO Forestry Paper 112, FAO, Rome.

King and Chandler. 1978. In Steppler and Nair (eds), 1987.

Lehmann, M. P. (1992).Deforestation and Changing Land-Use Patterns in Costa Rica. *In:* Changing Tropical Forests (H. K. Steen and R. P. Tucher, Eds.), pp. 58–76. Forest History Society, Durham, North Carolina.

Lonner, G., and A. Tornquist (Eds.). 1988. *Economic Evaluations of Biomass-Oriented Systems for Fuel.* International Energy Agency, Task III, Applications of Systems Analysis. SIMS, Swedish University of Agricultural Studies, Uppsala

Lugo, A. E., S. Brown, and J. Chapman. 1988. An analytical review of production rates and stemwood biomass of tropical forest plantations. *Forest Ecology and Management* 23:179–200.

Madrigal, Q. J. 1993. Arboles Maderables en Peligro Extinction en Costa Rica. INCAFO, San Jose, Costa Rica.

Pérez, S., and F. Protti. 1978. Comportiemento del sector forestal durante et período 1950–1977. OPSA, San José, Costa Rica.

Petriceks, J. 1975. Situación de la economío forestal en el pais. PNUD/FAO/COS/721013 Documento de Trabajo No. 3 San José Costa Rica.

Repetto, R. 1992. Accounting for environmental assets. *Scientific American,* June, 94–100.

Sage, L. F. 1981. An economic analysis of the forestry sector in Costa Rica. MS Thesis, SUNY Environmental Science and Forestry, Syracuse, N.Y.

Sharpe, G. W. 1986. Introduction to Forestry, 5th Edition. Mcgraw Hill, New York.

Smith, N., and T. Corcoran. 1979. *American Chemical Society Proceedings,* Vol. 21, no. 2: *The Energy Analysis of Wood for Fuel Applications.* Fuel Division, Washington, DC.

Steppler, H. A., and Nair (eds.). 1987. Agroforestry: A decade of development. Nairobi-ICRAF (International Council for Researching Agroforestry).

Tillman, D. A. 1978. *Wood as an Energy Resource.* Academic Press, New York.

United Nation, 1993. Statistical yearbook for Latin America and the Carribean. Economic Commission for Latin America and the Carribean, UN, N.Y. 778 pp.

Wille C. 1990. Trees on Trial in Central America. American Forests. Sept./Oct. 21–24.

CHAPTER 19

# Assessing the Role of Parks for Protecting Forest Resources Using GIS and Spatial Modeling

JOSEPH D. CORNELL



## I. INTRODUCTION

Two conflicting views of conservation in Costa Rica exist. One view is that Costa Rica is a "model for the preservation of biodiversity in the tropics" (Boza p. 239, 1993). The other view is that—despite its conservation policies—habitat destruction and the depletion of natural resources continues in Costa Rica at a rapid rate (Hunter, 1994), calling into question Costa Rica's commitment to conservation. Which view is correct? The answer is that both contain a great deal of truth (Boza *et al.*, 1995; Quesada-Mateo and Solis-Rivera, 1990).

Costa Rica has set aside about 12% of its lands in a national park system (Boza, 1993). At the same time deforestation and habitat conversion continue in Costa Rica (Chapter 16), and areas around and even within national parks have not been immune. As another example of this paradox, Costa Rica has a program for assessing biodiversity (INBIO) that is the envy of other nations (Tangley, 1990). At the same time, significant areas of Costa Rican forests are being destroyed before any such biological inventories can be made (Hunter, 1994). Finally, even though 12% of Costa Rica's national area has legal protection within a national park, this tells us little of what, if anything, there is within that park to protect. A significant portion of the areas within national parks in Costa Rica were deforested well before they were designated as parks, while other parks are located in areas that are so inaccessible or otherwise unsuitable for other land uses that they have never been in great danger of being deforested.

What we are left with is a paradoxical view that is all too common in the developing world. Conservation and development are occurring literally side by side. Our difficulties in resolving this paradox and revealing just how committed to conservation a country such as Costa Rica is lie in assessing how well conservation goals are being met in the midst of economic development and land use change.

In this chapter I examine the role of national parks in protecting forest resources in Costa Rica. The national parks system is the flagship of conservation efforts in Costa Rica, and is the underlying policy instrument for much of the rest of Costa Rica's conservation plan, including the preservation of biodiversity, the conservation of natural resources, and the promotion of ecotourism.

In my own opinion, the primary way in which the national park system supports other conservation efforts is through the protection of forests. Costa Rica's forest resources are the base on which other conservation efforts and ecotourism are built. The idea here is that without the forest, you do not have the biodiversity and the potential for ecotourism (Holl *et al.*, 1995). And without the national park system, you do not have the forests. Again, the paradox is that development continues around, and sometimes within, protected areas. This chapter presents an attempt to quantify the relationship between conservation and national park protection, and so resolve this paradox. To accomplish this I used GIS techniques to assess what these parks protect in terms of where they are located and the degree to which they were or have been deforested. I also assessed the desirability of park areas for other land uses using a spatial model of land use change in Costa Rica with and without park protection.

## A. Deforestation in Costa Rica

According to Zon and Sparhawk (1923), in 1890 about 95% of Costa Rica was forested. They estimated that between 1890 and 1920, the population approximately doubled (to about 450,000 persons) and the area in forest cover decreased to around 75% (36,423 km$^2$), or roughly 8.1 ha per capita. The reliability of the data from Zon and Sparhawk, which were compiled from forestry reports and other secondary sources, and not from measurements such as photogrammetry, is questionable however.

By 1940, aerial photography began to become more accessible and has since been used to track the progressive loss of forest cover in Costa Rica (S. Pierce, 1992). Between 1940 and 1983, Costa Rica lost about 50% of its closed cover (defined as >80% canopy cover) forest resources (Sader and Joyce, 1988). These forests represent the most intact and least disturbed forest areas.

Degradation of these forests, however, is often confused with absolute deforestation. When I use the term deforestation, I am talking about the loss of virtually 100% canopy cover. So, Sader and Joyce's estimate that Costa Rica had only 17% of its original closed canopy forest in 1983 does not translate into an absolute deforestation of 83%.

Still, between 1970 and 1989 the rate of deforestation in Costa Rica was about 31,800 ha per year, roughly two to three times higher than the international average [TSC/WRI (1991); but see other estimates in Chapters 16 and 17]. The World Resources Institute (TSC/WRI, 1991) estimated that in 1966 forests (both closed and open) covered 58.5% of Costa Rica, but that by 1989, only 42.9% of all forest cover remained. According to the FAO (the Food and Agricultural Organization of the United Nations), however, only 28% of the country was forested in 1990, based upon their analysis of remote sensing imagery from 1981 and 1990 (FAO, 1993).

The differences between these forest inventories are probably a result of differences in methodology and interpretation, something that has been an ongoing problem in this field for some time (Molofsky *et al.*, 1986). From my own analysis, however, I concluded that by the mid-1980s Costa Rica had lost approximately 50% of all forest cover and close to 70% of its closed cover forests. The primary reason for this progressive loss of forest cover in Costa Rica can be attributed directly to one factor: the expansion of cattle production.

As developed in Chapters 2, 14, and 24 of this volume, cattle production was subsidized heavily in the 1970s and the 1980s. With the collapse of credit for large-scale cattle production in the 1990s, pasturelands are returning to scrubby forest, but the process is limited by the recalcitrance of introduced grass species and the lack of suitable local seed sources for native tree species. Some areas are being replanted with either plantations or even native tree species. Forest loss and conversion continue, however, particularly due to increased banana production in the lowlands, but also due to increased development, road development, and other farm and nonfarm uses (Hunter, 1994).

In contrast to cattle production, extraction of forest products has not been a major cause of deforestation (TSC/WRI, 1991). Forest exports account for only a small fraction of Costa Rica's foreign exchange, with the bulk coming from cash crops such as coffee and bananas, and to a lesser extent, sugar cane, fruits, and vegetables. This has been true for several reasons, including the fact that many of the commercial stocks of tropical hardwoods were depleted before this century throughout most of Central America, and because of the wasteful nature of land conversion in Central America.[1]

## B. The Parks System in Costa Rica

The parks system in Costa Rica has developed over time from an assortment of special protection areas into a relatively unified network of national parks, forests, and bioreserves. The first parks were created in 1970–1971 and included Poás Volcano National Park, Cahuita National Park, Santa Rosa National Park, and Tortuguero National Park. These four areas were chosen on the basis of a variety of criteria such as scenic beauty, historical significance, and importance for conservation (Boza, 1993). Funding for the parks was obtained from the Costa Rican Legislative Assembly and from international donors such as the World Wildlife Fund and international aid agencies.

Also, a National Park Fund was created within Costa Rica's National Park Service to help promote the idea of a national park system, and to administer funds received and separate them from general revenues under the control of the Costa Rican legislature. This has provided a measure of autonomy for the national parks and protected the moneys they receive in user fees and in donations from being diverted to other uses by the government. To date, over U.S. $70 million has been raised to purchase and administer parklands within Costa Rica. As a result, Costa Rica now boasts more than 30 national parks covering about 12% of the entire country.

Costa Rica also has drafted laws to ensure that, once incorporated into a national park, land cannot be taken out of the parks system except by legislative decree. Again, on the one hand, this appears to be serious support for conservation on the part of Costa Rica. On the other hand, however, we need to know more about what resources those parks contain, and where they are located, to fully assess Costa Rica's commitment to conservation.

[1] As a rule, about 90% of the trees that are cut down in Central America are reduced to ashes or allowed to rot on site (TSC/WRI, 1991). Only a small fraction of commercially valuable tree species are harvested and exported or used internally. As a result, Costa Rica, like most of the other countries in Central America, is a net importer of wood products. Costa Rica imports most of the paper pulp, plywood, and construction timber it uses. Ironically, the wasteful nature of forestry production outside of the national park system has led to increased pressure on stands of commercially valuable timber within the parks. Often, when passing a truck on the highways laden with large specimens of tropical hardwoods, my Costa Rican companions would point and tell me that those trees probably came from a national park. Anecdotes such as this are hard to substantiate without a thorough knowledge of Costa Rica's parks and the forest resources within them, which is of course the whole point of this study.

## C. How Does an Area Become a National Park?

A cynical view of conservation might suggest that protected forests occur where forestlands are easiest to protect. This viewpoint includes the idea that once a country begins to realize that it is losing its forests at an alarming rate, the only forests left to protect are those that no one wanted in the first place. Hence, forests on rugged slopes and at higher elevations which are typically not desirable for other uses such as pasture and agriculture become the first candidates for protection.[2] In other words, these areas are chosen almost by default without much recourse to other criteria. In turn, these areas remain relatively intact due precisely to their continued undesirability from an agricultural perspective.

An alternate method of choosing areas for protection is to pick those areas that exhibit extraordinary potential for preserving particular species or ecosystems. This method might be termed the "conservation priority" method and is based on assessments that identify particular areas important for achieving specific conservation goals. The Corcovado biosphere reserve in the Osa Peninsula region of Costa Rica can be considered such a priority conservation area.

We now have two visions of how forest areas are chosen for protection: by default and/or by design. Certainly, some of the most important areas for conservation are those that actually have intact forest, so we cannot rule out the importance of protecting forestlands such as those in mountainous or otherwise inaccessible regions which might be considered undesirable for other land uses. Likewise, we cannot overemphasize the importance of protecting specific areas that would otherwise be converted too readily to other land uses such as agriculture. In the next section I explore the factors that might explain the placement of park areas within Costa Rica and the role that park status plays in protecting forest resources.

[2] This is certainly true for the United States where "protected areas comprise only a tiny fraction of the United States' land mass, and most of the acreage is in the western mountains and deserts" (Gavin and Sherman, 1995, p. 1343). In the United States, these western protected areas were precisely those that remained sparsely settled well into the 20th century, due in great part to their inaccessibility and/or their undesirability. Mountain ranges and deserts of course have great conservation potential, but their potential for agriculture is constrained. Instead, coastal areas that could support rain-fed agriculture were the first areas in the United States to become heavily populated and converted to agriculture, leaving the mountains and the deserts of the west relatively untouched.

## II. USING A GIS TO ASSESS THE QUALITY OF NATIONAL PARKLANDS

In order to assess the quality and extent of forest resources in Costa Rica, I used a GIS data base for Costa Rica developed as part of an analysis of historical land use change and carbon exchange in Central America from 1880 to the present (Cornell, 2000). The database includes maps derived from Sader and Joyce (1988) showing the distribution of closed canopy forests in Costa Rica for the periods 1961, 1977, and 1983. Again, I cannot overemphasize that the maps from Sader and Joyce do not represent deforestation in Costa Rica per se, but rather the loss of closed canopy forest (i.e., those forests with 80% or more crown cover). This point is important as these maps have been used inaccurately by many authors (for example, Terborgh, 1992, p. 195) to represent the distribution of "deforestation" in Costa Rica. Deforestation is usually considered as the loss of virtually all forest cover.

This means that the areas shown as "forest" in most publications using this series of maps are not necessarily untouched; they may be missing up to 20% of their original crown cover. Indeed because people tend to take the large trees first, these areas may also be missing an even larger percentage of their original biomass ( Brown, 1996). An even more important point to note is that the areas marked as "deforested" on most secondhand representations of this series of maps are not necessarily deforested: they may just have lost more than 20% of crown cover.

This is an important distinction, particularly to Costa Ricans who, upon looking at the maps I have produced, have pointed out that many of the areas marked as "deforested" by Sader and Joyce in fact have significant forest cover (R. Tiffer and C. A. S. Hall, personal communication, as well as personal observation). The maps by Sader and Joyce, however, are extremely useful in the context of conservation. If we would like to conserve or preserve anything at all, relatively undisturbed, close canopy forests should be a priority.

Loss of even 20% of the forest certainly means changes in plant and animal diversities and abundances as organisms adjust to disturbance. "Disturbance" implies changes that disrupt currently prevailing, natural conditions, and can run the gamut from full-scale deforestation to seemingly harmless intrusions. Even so slight a disturbance as increasing the number of visitor days in an area, without significantly disturbing the forest itself, has been reported to alter bird diversity markedly (Klein *et al.*, 1995). As such, the distribution of relatively undisturbed, closed canopy forest can be used as one measure of successful conservation.

Additional maps on the distribution of spatial features such as elevation, soil types, and road networks were obtained by digitizing maps from the

Alliance for Progress series (1964). This series of maps was produced by Costa Rica in cooperation with the U.S. Army Corps of Engineers in the 1960s and represents a valuable source of spatial information. Each map is based upon an identical base map which allowed easy comparison between maps and greatly aided in their incorporation into the digital data base. Lastly, the map of national park areas was created by digitizing the map from Boza (1993). All maps were manually digitized in ERDAS. Raster grid-cell images (ERDAS ".gis" files) were produced and converted into IDRISI (".img") image files. All map comparisons were done in IDRISI. The resolution of each grid cell is 1 km$^2$ and each image consists of 155,829 grid cells (409 rows by 381 columns).

## A. Forest Resources in National Parks, 1961–1983

The parks system as it is today in Costa Rica did not exist until the 1970s or later, so it is not useful to compare the distribution of parks with forest cover in 1940 or 1950 (which otherwise would be possible using the data from Sader and Joyce). However, it is clear that closed cover forests were disappearing rapidly during that same time period. Furthermore, some Costa Ricans such as Mario Boza were aware of this rapid loss and took steps to slow or stop its spread. The areas they chose for protection were often areas that still had significant forest resources or otherwise had high potential for conservation (Boza, 1993).

It is appropriate, therefore, to assess the distribution of forest resources immediately before the creation of the first national parks in Costa Rica (ca. the 1960s). I did this by comparing the map showing the distribution of national parks with a map showing the distribution of land cover in 1965 derived from the Alliance for Progress series and with the map from Sader and Joyce (1988) showing the distribution of closed cover forests in 1961.

The map of land cover includes information on forests, agriculture, and pasture in 1965 and shows that approximately 19,665 km$^2$, or 37.8%, of Costa Rica had been converted into some form of agricultural land use, including areas with mixed forests and agriculture. The map also shows that approximately 29,411 km$^2$, or 56.6%, of land cover consisted of either dense broadleaved evergreen forest or mixed dense to open broadleaf evergreen and deciduous oak forest. In addition, 2905 km$^2$, or 5.6%, of land cover consisted of mangrove and palm forest (Table 19-1).

To assess the quality of forest resources in those areas that would eventually become parks, I merged the map showing the distribution of closed cover forests in 1961 with the map of land cover for 1965. This gave me a map

TABLE 19-1 Comparison of National Parks with Land Cover in 1965

| Land cover type | Total area ($km^2$) | Area in parks ($km^2$) | Percentage of total park area |
|---|---|---|---|
| Dense, broadleaf evergreen forest | 25,855 | 3326 | 41 |
| Open deciduous forest, savanna, and crops | 9,469 | 899 | 11 |
| Dense to open broadleaf evergreen and oak forest | 3,556 | 1939 | 24 |
| Mangrove and palm flooded forest | 2,905 | 1046 | 13 |
| Scrub and grassland | 220 | 119 | 1 |
| Crops and upland pasture | 7,950 | 663 | 8 |
| Coffee plantations | 1,041 | 25 | <1 |
| Banana plantations | 987 | 93 | 1 |
| Total | 51,983 | 8110 | 100 |

showing the probable distribution of both closed cover and more open forest resources circa 1961. Comparison of the map of national parks with this new map (Figure 19-1) showed that 5388 $km^2$, or 66.4%, of the total area that would eventually become national parks contained closed cover forest in 1961 (see also Table 19-1). Furthermore, another 1446 $km^2$ (an additional 18%) of the eventual parks area contained dense or open to dense forest, mangroves, or palm forests. This means that as much as 10 years before being protected, 15.6% of the area that would eventually become a national park had already been deforested completely and an additional 18% of those forests had lost more than 20% of canopy cover.

The loss of closed cover forests both inside and outside of national parks continued between 1977 and 1983 (see also Figure 19-2). Between 1961 and 1977, in Costa Rica as a whole, 5676 $km^2$ of closed cover forest was degraded or lost (an area 500 $km^2$ larger than all the closed cover forest contained in national parks in 1961). This represents about a 1.5% loss per year for the entire country. Within national parks during the same period, however, only 212 $km^2$ of closed cover forest parks was degraded or lost.

Between 1977 and 1983, another 8288 $km^2$, or 54%, of the remaining closed cover forests in the entire country was degraded or lost. Clearly, the loss of closed cover forests accelerated during this period (about 9% per year), as compared with the period 1961 to 1977 (1.5% per year). Within national parks approximately 762 $km^2$ of the remaining closed cover forests was degraded or lost (a loss of about 2.5% per year). Because the rate of loss within parks was much lower than that outside of parks, the percentage of remaining closed

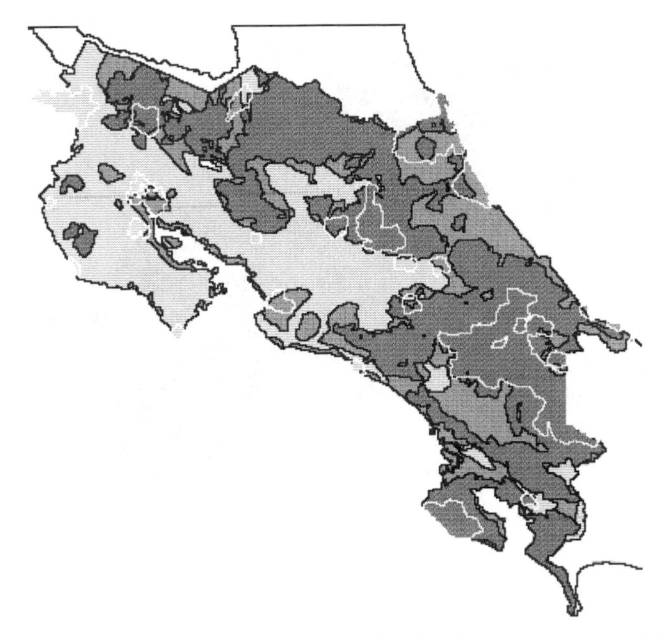

FIGURE 19-1 Comparison of national parks with closed and open forest, circa 1961. Dark gray represents closed cover forests and medium gray represents open forest. Park areas are outlined in white.

cover forest resources within Costa Rica's national parks nearly doubled between 1961 and 1983, despite some losses (see Table 19-2).

In 1961, 22.7% of Costa Rica's closed canopy forests were located within parks, but by 1983 the figure rose to 45.1%. In Costa Rica as a whole, degradation or loss of closed cover forests between 1961 and 1983 occurred at a rate of 2.67% per year (634.7 $km^2\ yr^{-1}$). But within national parks, the rate of loss was much lower at 0.82% per year (about 44.2 $km^2\ yr^{-1}$).

## B. Comparison of Elevation and the Distribution of Parks

In the tropics a gradient of land use exists where lower elevation lands are often more desirable (Hall *et al.*, 1995). An analysis of changes in forest cover (i.e., changes in land use) versus elevation therefore can be very revealing. For example, when I compared the distribution of Costa Rica's national parks

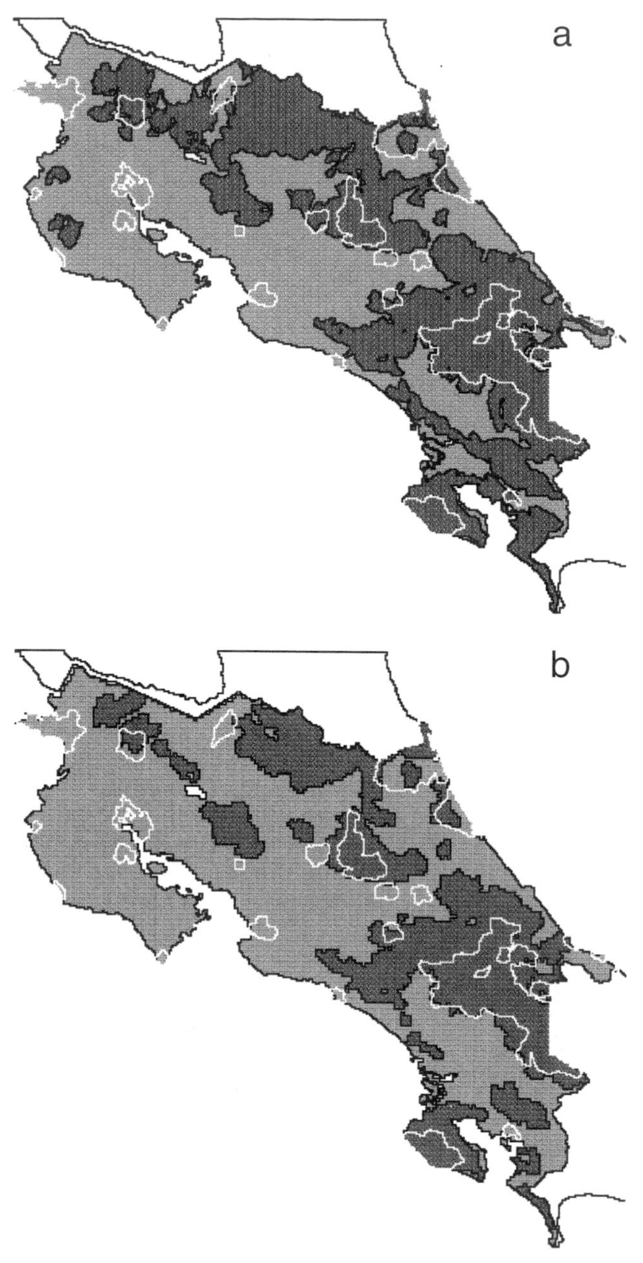


FIGURE 19-2 Comparisons of national parks with closed forest in (a) 1961, (b) 1977, and (c) 1983. Dark gray represents closed cover (>80%) forests. Park areas are outlined in white.

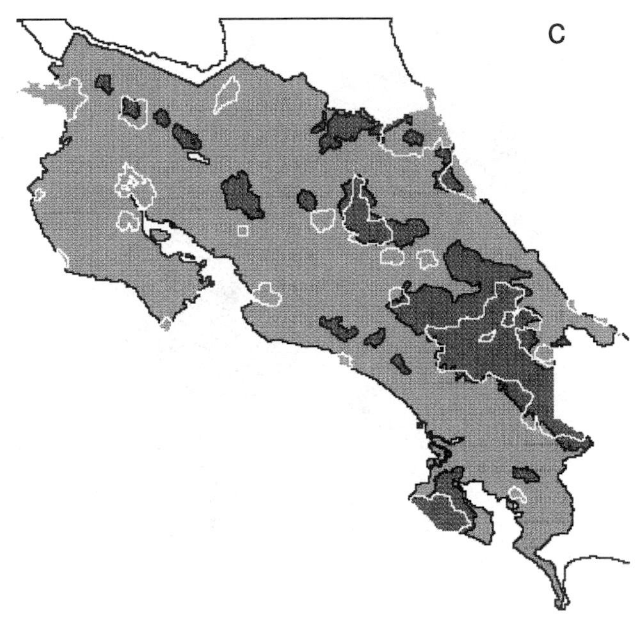


FIGURE 19-2 (*Continued*)

with a map showing elevational classes (Figure 19-3), I found that slightly more than half (53.4%) of the total area within national parks was located at or below 1000 m (Table 19-3). This area represents about 31% of the total lowland area (below 1000 m) of Costa Rica. The remaining 46.6% of the total area of national parks is distributed from 1000 to over 3500 m in elevation above sea level. What is interesting to note is the fact that at each successive elevation class above 1000 m, national parks almost always account for an increasingly larger percentage of the total area at that elevation. For example, although parks located above 2500 m account for less than 15% of the total area within the park system, they account for nearly 71% of the total area at

TABLE 19-2 Loss of Closed Cover Forests in Parks, 1961–1983

| Year | Total area in closed forest ($km^2$) | Percent total area | Forest area in parks ($km^2$) | Percent of park area | Percent of remaining forest in parks |
|---|---|---|---|---|---|
| 1961 | 23,747 | 45.7 | 5388 | 66.4 | 22.7 |
| 1977 | 18,071 | 34.8 | 5176 | 63.8 | 28.6 |
| 1983 | 9,783 | 18.8 | 4414 | 54.4 | 45.1 |

FIGURE 19-3 Comparison of park distribution with elevation. Darker areas indicate lower elevations. Park areas are outlined in white. About half of the total area within national parks occurs higher than 1000 m above sea level.

TABLE 19-3 Distribution of National Parks by Elevation Class

| Elevation class[a] | Area in parks ($km^2$) | Total area in Costa Rica ($km^2$) | Percent total park area | Percent of elevation class within parks |
|---|---|---|---|---|
| 0–100 | 2631 | 18,082 | 32.4 | 14.6 |
| 100–500 | 900 | 13,874 | 11.1 | 6.5 |
| 500–1000 | 806 | 8,425 | 9.9 | 9.8 |
| 1000–1500 | 948 | 5,701 | 11.7 | 16.6 |
| 1500–2000 | 906 | 2,551 | 11.2 | 35.5 |
| 2000–2500 | 835 | 1,857 | 10.3 | 44.8 |
| 2500–3000 | 722 | 1,020 | 8.9 | 70.8 |
| 3000–3500 | 290 | 443 | 3.6 | 65.5 |
| 3500–4000 | 72 | 72 | 0.9 | 100.0 |

[a] Meters above sea level.

or above that elevation. And at the highest elevations at or above 3500 m above sea level, 100% of the area, about 72,000 km$^2$, is located within a national park. Obviously, the higher up you go in elevation in Costa Rica, the greater the likelihood that you are inside a national park.

This could be interpreted to support the proposition that people "protect" what they are not going to use anyway. Most agriculture in Costa Rica, for cash crops and for subsistence, occurs at or below 2000 m. The obvious counter to this argument, however, is that high-elevation areas may have high conservation value, even if they are not particularly attractive for agriculture or pasture (Huston, 1993).

Still, when I compared other maps of land use with the map of topography, it was clear that most agricultural land use, with the exception of upland, seasonal pasture, occurs at lower rather than higher elevations. For example, if we exclude pasture and coffee production, two types of land use which can take advantage of cooler areas, 97.8% of all other land use in 1965 was located below 1000 m in elevation, and the bulk of that (92.3% of the total) occurred below 500 m.

Even if we include higher elevation areas used for coffee and upland pasture, 91.5% of all land use was located below 1500 m, with the bulk of that (81% of the total) occurring below 1000 m. Clearly, most agricultural land use in Costa Rica occurs in the lowlands and not at higher elevations, leading me to conclude that there is a gradient of land use pressure in Costa Rica as a function of elevation, with lower elevations being under greater pressure for conversion. I also conclude that the lack of pressure on higher elevation sites makes them easier to protect, although not necessarily less valuable for conservation (Huston, 1993).

Due to the uncertainty regarding the land use change pressure on high-elevation areas, the analysis of parks versus elevation may not be a good indicator of how well park protection translates into forest protection. To evaluate the effectiveness of park protection in the face of real pressure from land use change, it is necessary to evaluate the effect of park protection on lowland forest resources.

## III. USING SPATIAL MODELING TO ASSESS RESISTANCE TO LAND USE CHANGE

To assess the efficacy of parks for protecting closed cover forest resources, including lowland forests, I used GEOMOD2, a spatial model of land use change (Pontius, 1994; Hall *et al.*, 1995). GEOMOD2 is a simulation model that compares various spatial drivers such as maps of soil type or rainfall with maps showing the distribution of land use change over time. Each spatial driver map represents some factor or relationship associated with human land use, such as the perceived relationship between population density and land use change, or the relationship between soil type and productivity. GEOMOD2 first compares each driver map to the maps of land use change over time and then adjusts preassigned "weight values" for each class value within a map. Some classes are more highly correlated with land use change, indicating that areas associated with those map classes are used preferentially.

After adjusting the weighting factors for each map, GEOMOD2 simulates land use change for a particular time period. The model also evaluates which spatial drivers are most effective for simulating the distribution of land use change over time and space by comparing the results of its subroutines driven with selected combinations of spatial drivers with an estimated success rate of a random model of land use change.

This comparison is summarized using the "kappa" coefficient, which was developed to summarize the results of accuracy assessments used to evaluate land use or land cover classifications obtained by remote sensing (Stehman, 1996). In the case of remote sensing data, the kappa coefficient is a measure of the agreement between the class assigned to each pixel derived from remote sensing and the actual class value based upon ground truthing.

In GEOMOD2, the kappa coefficient is used as a measure of the model's actual success rate compared to the expected success rate due to chance. Within a large spatial data set where a great deal of change is occurring, even a random model will choose some of the areas to be converted correctly. The kappa coefficient therefore is a powerful and essential tool for evaluating GEOMOD2's performance, since percent success due to chance can be substantial (Pontius, 1994).

## A. Using GEOMOD2 to Predict Historical Patterns of Land Use Change

Pontius (1994) developed GEOMOD2 from previous models of tropical land use change developed by Charles Hall and others (Hall *et al.*, 1995). Pontius improved upon these models by adding subroutines to evaluate statistically the usefulness of different spatial drivers for simulating land use change. GEOMOD2 does this by comparing the distribution of classes with a spatial driver map with the distribution of land use change over two or more points in time and then assigns a "friction" value to each class within each driver map based upon its correlation with land use change. Pontius also used the same data set created by Cornell (2000) to evaluate the performance of GEOMOD2 itself. The data base for Costa Rica is unusual in that it contains many different maps for use as spatial drivers, as well as maps showing the distribution of land use for five points in time. With this data base, GEOMOD2 was able to predict the loss of closed cover forests in Costa Rica with an accuracy of about 86% and a 0.36 kappa coefficient.

### B. The Effect of National Parks on Model Accuracy

For this study, I added the map of national parks as a new spatial driver for use in GEOMOD2. With national parks as a spatial driver, GEOMOD2's performance increased to almost 90% with a 0.49 kappa coefficient. Most notable of all was the improvement in the distribution of land use change in the lowlands around the Osa Peninsula. Previously, without using the map of national parks as a spatial driver, GEOMOD2 would consistently deforest all lands in the Osa Peninsula. This shows that GEOMOD2 considers this area to be highly desirable for agriculture, based upon spatial drivers such as soil type, soil moisture, and Holdridge life zone types. GEOMOD2 arrives at this conclusion by weighting each class within each spatial driver based upon its comparison of each driver with the distribution of previous land use change.

When I added the map of national parks as a new spatial driver, however, GEOMOD2 recognized that deforesting park areas were no longer so desirable. Therefore, GEOMOD2 deforests or degrades only a small portion of the closed cover forests on the Osa Peninsula (Figure 19-4). This is an important result as it shows that the distribution of parks is not only useful as a spatial driver, but that parks do protect lowland forests in the face of otherwise tremendous land use pressure.

## IV. CONCLUSIONS

The loss of closed canopy forest in Costa Rica between 1961 and 1983 occurred at a rate of 2.67% per year (634.7 $km^2$ $yr^{-1}$). Within national parks, however, the rate of loss was much lower at 0.82% per year (about 44.2 $km^2$ $yr^{-1}$). In 1961, 22.7% of Costa Rica's closed canopy forests were located within parks, but by 1983 this was about 45.1%. Elevation appears to have played a role in both the establishment and the efficacy of national parks. The results of the GIS analysis show that national parks are located disproportionately in less-accessible regions, such as those with rugged mountainous terrain. Parks may exist principally in these areas because no one wants to develop them anyway. Spatial simulations of land use change for this period, however, show that national parks were successful in protecting several areas that the simulation

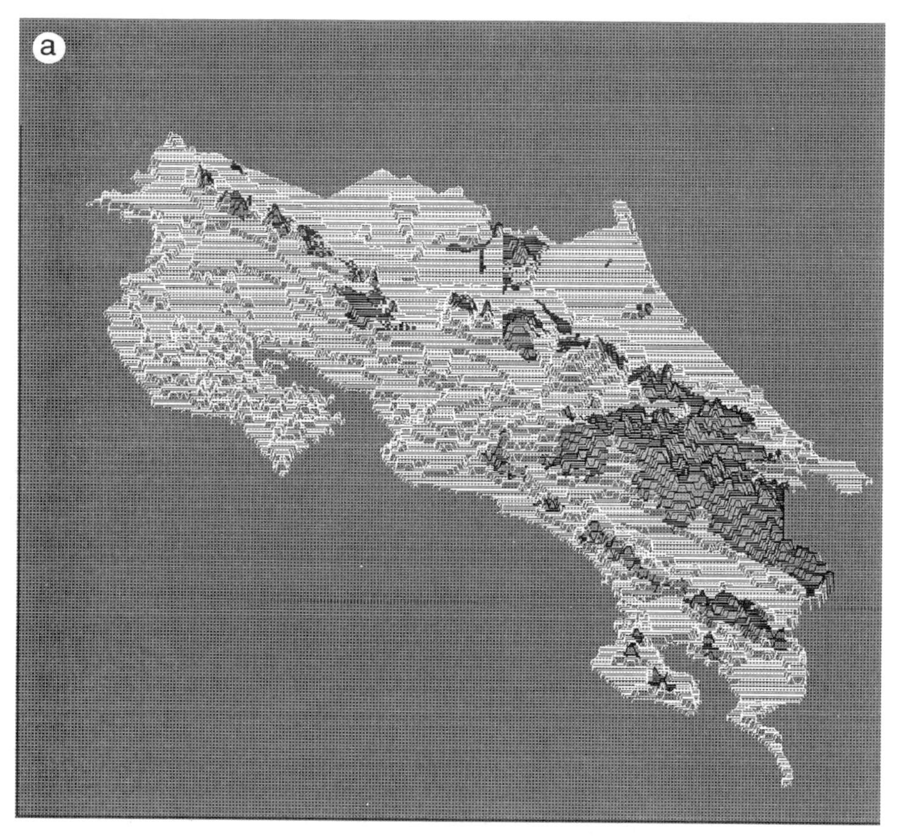


FIGURE 19-4 (a) Results of GEOMOD2 without using national parks as a spatial driver. GEOMOD2 predicts that most lowland areas would be good candidates for conversion. (b) Results of GEOMOD2 using national parks as a spatial driver. Addition of national parks as a spatial driver improves GEOMOD2's accuracy by preventing the conversion of some lowland areas such as Corcovado Park on the Osa Peninsula.

model would develop relatively soon, specifically, Corcovado Park on the Osa Peninsula.

From this analysis I conclude that national parks in Costa Rica were an effective, if not perfect, source of protection for closed canopy forest resources within national parks well into the mid-1980s. Some forest loss within national parks did occur, but at a much slower rate than in the rest of Costa Rica. Despite some loss within national parks, the percentage of Costa Rica's remaining closed canopy forests within national parks nearly doubled between 1961 and 1983. This amply illustrates the paradox of Costa Rica as a protector and an exploiter of its forest resources. Forest protection is occurring despite enormous pressures to deforest. At the same time, seemingly nothing can stop the continued

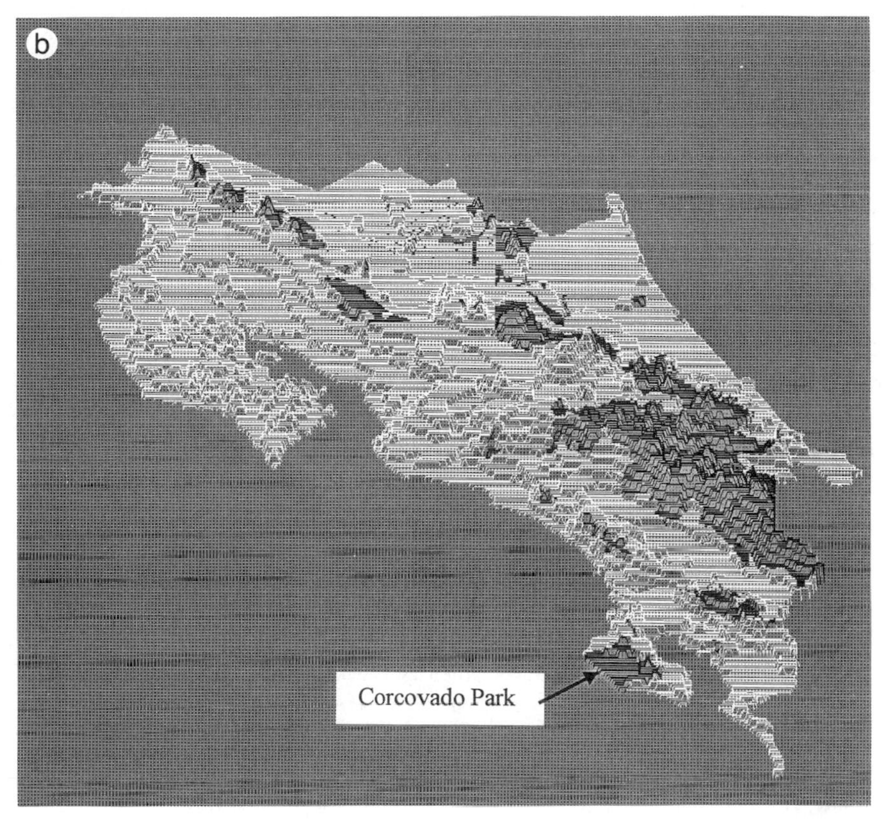


FIGURE 19-4 *(Continued)*

loss of closed canopy forest in Costa Rica, sometimes even within protected areas.

National parks were effective in protecting some particularly important lowland forest resources in Costa Rica, but a large proportion of the areas under protection were either not forested, as in the Guanacaste Reserve, or are not very desirable or accessible. As road building and development continues, however, even the more remote areas of Costa Rica, such as the forests on the Atlantic coast, have been opened up to development. Areas outside the national park system continue to be converted wholesale, primarily for banana production, as in the Sarapiqui region.

How do you measure a country's commitment to conservation? Do you look at the amount of land set aside in parks, or do you look at the amount of land that continues to be lost each year? The answer is that you must look at both and weigh the consequences for conservation. In my opinion, Costa Rica is committed to conservation, but that commitment has been sorely tested and will continue to be tested. The immediate future of Costa Rica's forest resources comes down to this: continued losses outside the national park system mean that the importance of national parks will increase even as more pressures are put upon the resources they protect.

## ACKNOWLEDGMENTS

I thank Ruth Tiffer de Sotomayor for her insight during the preparation of this manuscript. It is always good to have someone from the country you are studying to tell you when you are dead wrong. I thank Jeff Mangle for his help on an earlier version of this analysis. Lastly, I thank Dr. Therese Donovan for reviewing this chapter and providing a great deal of help in making certain of my ideas clearer.

## REFERENCES

Boza, M. A. 1993. Conservation in action: Past, present, and future of the national park system of Costa Rica. *Conservation Biology* 7:239–247.

Boza, M. A., D. Jukofsky, and C. Wille. 1995. Costa Rica is a laboratory, not Ecotopia. *Conservation Biology* 9:684–685.

Brown, L. R. 1996. The acceleration of history. In *State of the World 1996,* pp. 3–20. W. W. Norton, New York.

Cornell, J. D. 2000. Land-use change and carbon exchange in Central America 1880–2000: A new analysis using GIS and spatial modeling. Ph.D. thesis, SUNY ESF, New York.

FAO. 1993. *Forest Resources Assessment 1990: Tropical Countries.* FAO Forestry Paper 112, FAO, Rome.

Hunter, J. R. 1994. Is Costa Rica truly conservation-minded? *Conservation Biology* 8:592–595.

Huston, M. 1993. Biological diversity, soils, and economics. *Science* 262:1676–1680.

Klein, M. L., S. R. Humphrey, and H. F. Percival. 1995. Effects of ecotourism on distribution of waterbirds in a wildlife refuge. *Conservation Biology* 9:1454–1465.

Molofsky, J., C. A. S. Hall, and N. Myers. 1986. *A Comparison of Tropical Forest Surveys.* Office of Energy Research, Carbon Dioxide Research Division, U.S. DOE, Washington, DC.

Pierce, S. M. 1992. Environmental history of La Selva Biological Station. In H. K. Steen and R. P. Tucker (Eds.), *Changing Tropical Forests,* pp. 40–57. Forest History Society, Durham, NC.

Pontius, R. G. 1994. Modeling tropical land-use change and assessing policies to reduce carbon dioxide release from Africa. Ph.D. thesis, State University of New York, College of Environmental Science and Forestry, New York.

Quesada-Mateo, C., and V. Solis-Rivera. 1990. Costa Rica's national strategy for sustainable development: A summary. *Futures* 22:396–416.

Sader, S. A., and A. T. Joyce. 1988. Deforestation rates and trends in Costa Rica 1940 to 1983. *Biotropica* 20:11–14.

Stehman, S. V. 1996. Estimating the kappa coefficient and its variance under stratified random sampling. *Photogrammetric Engineer. & Remote Sensing* 62:401–402.

Tangley, L. 1990. Cataloging Costa Rica's diversity. *Bioscience* 40(9):633–639.

Terborgh, J. 1992. *Diversity and the Tropical Rainforest.* Scientific American Library Series no. 38, Scientific American, New York. 242 pp.

TSC/WRI. 1991. *Accounts Overdue: Natural Resource Depreciation in Costa Rica.* World Resources Institute, Washington, DC.

Zon, R., and W. Sparhawk. 1923. *Forest Resources of the World.* McGraw–Hill, New York.

# SECTION VI

# *Biophysical Analysis of Major Components of the Economy*

This final section returns to the major issues of Costa Rica as introduced in Chapters 1, 2, and 5, and especially the dominating effect of foreign trade, by applying the biophysical approach introduced in Chapter 3 and the geographical data bases developed in Sections III and IV to the major industries of Costa Rica that relate to issues of foreign trade. These are the banana, coffee, and tourism industries. We then include our only explicitly "political" chapter since it became obvious to us during our analyses that the biophysical reality of Costa Rica is not simply part of the internal operations of that society but was to a large degree imposed by political and economic actors outside the nation. Thus our attempts to undertake a strictly biophysical analysis were undermined by the reality that we should have recognized all along. The final chapter of this book summarizes the conclusions of all other chapters from a biophysical perspective and gives our perspective on what can and what cannot work as a basis for sustainable development, or indeed a tropical economy in general. In this final chapter we break our commitment to a strictly analytical approach by making our one and only policy recommendation.

CHAPTER 20

# The Costa Rican Banana Industry

## Can It Be Sustainable?

CARLOS HERNÁNDEZ, SCOTT G. WITTER, CHARLES A. S. HALL, AND CYNTHIA FRIDGEN



## I. INTRODUCTION

Costa Rica's 20th century development paradigm has been based in large part on growth in agricultural production. Historically, this development has tended to emphasize economic gains over ecosystem integrity. Human wants and needs have been met with little regard to what has been considered previously as an inexhaustible natural resource base. As developed elsewhere in this book, the modernization and industrialization of Costa Rica has required increasing amounts of foreign exchange, and this has been based principally on agricultural products.

Costa Rica's export economy historically was based on coffee. At the beginning of the 1980s, however, increased competition from other countries, increased mechanization of the production system, widespread overproduction, and a drop in market prices reduced greatly the number of coffee plantations, and hence the importance of coffee to Costa Rica's exports. Sugar and meat also have been important, but international quotas and boycotts have kept them from being major income generators.

As of this writing bananas are the most important export crop. In 1993 they generated U.S. $531.1 million, followed by coffee at $203.5 million, meat at $60 million, and sugar at $25.3 million (Figure 20-1). The economic feasibility of producing these commodities is in danger because of worldwide overproduction and competition from countries that have fewer internal input price distortions, lower standards of living, and fewer environmental protection regulations in place. But the problem is also that Costa Rica is running out of alternative cash crops that are competitive in the export market.

Costa Rica is the second largest producer of bananas for the export market in the world [CORBANA (Corporacion Bananera Nacional), 1993] (Figure 20-2a). Bananas provide an important source of benefits to the citizens of Costa Rica both directly and indirectly. Banana producers provide many on- and off-site jobs, plus banana worker communities receive health care, recreation facilities, and schools from the producer companies. Additionally, a complex and substantial infrastructure of general use roads, bridges, ports, etc., has been built to facilitate the marketing of bananas for the export market. As elsewhere in the world there has been a great degree of disparity between those who have benefited from development and those who have not. This is especially true of the banana industry where historically a great deal of the benefit has accrued to transnational companies (producers, transporters, marketers) and a small privileged class of nationals who have regarded Costa Rica as little more than a source of inexpensive resources to be used for their own

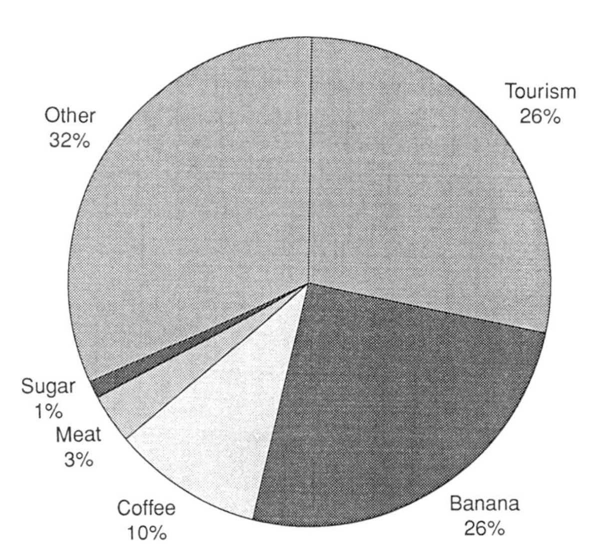


FIGURE 20-1 The major contributors to the generation of Costa Rican foreign exchange. (Source: Banco Central de Costa Rica, 1993; Instituto Costarricense de Turismo, 1995.)

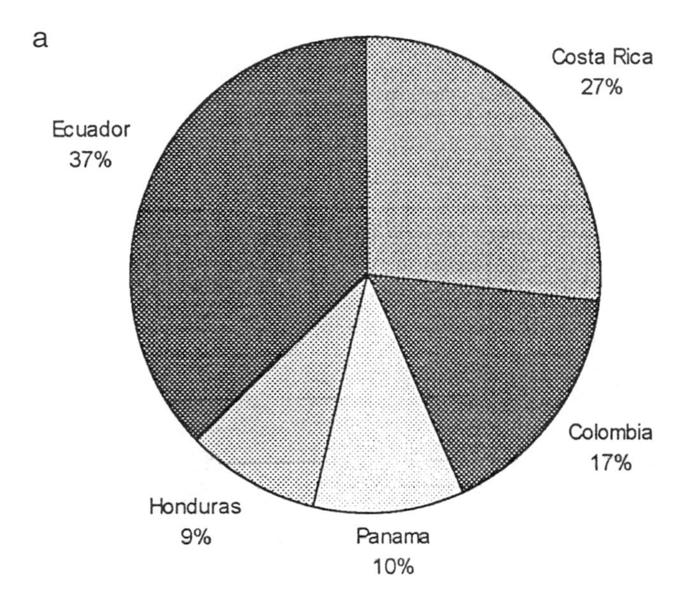


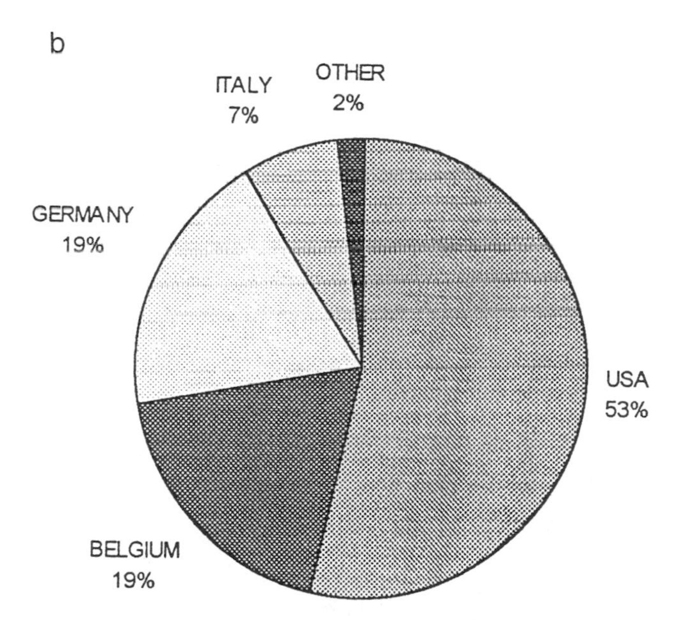


FIGURE 20-2 (a) Banana production by country. (b) Destination of Costa Rican banana exports. (Source: CORBANA, 1993.)

economic gain. The wealthy have been able to muster the majority of the power, and thus are very influential with a government eager to maintain economic growth. Change has been gradual in coming and often accompanied by conflict, but current policies stressing enforcement of new labor and environmental laws are making positive differences.

Nevertheless the overall benefits from bananas are substantial. But so are the environmental problems and health costs associated with their production. The primary reason that the costs are high is because the current production system is not sustainable. Although this statement is controversial and not universally accepted in Costa Rica, we will show how the present system fails to meet the following three fundamental conditions of sustainable development: (1) economic feasibility, (2) environmental soundness, and (3) social appropriateness and equity.

A simplistic recommendation to address these problems is to enforce and reinforce existing legislation to protect the environment, labor, and health; introduce production quotas; and interject land use regulations. An important controversy in this solution lies in various preconceptions of what environmental protection means. Preservationists, especially non-Costa Rican ones, typically want to adopt the American John Muir's approach to Costa Rica's natural resources, conserving and preserving the abundant biodiversity of flora and fauna for future foreign generations (and to a lesser extent Costa Ricans) with little regard for the needs of the citizens. Many Costa Ricans regard this attitude as a situation not unlike the "British Old Boys Club" that existed during the late 1800s and early 1900s when Africa's big game animal populations were managed as European private hunting preserves in the name of sportsmanship and conservation (Adams, 1992). In many cases it appears to Costa Ricans that preservationists are simply substituting ecotourism for sportsmanship.

An important problem of enforcing environmental regulations facing Costa Rica is that if the transnational companies that produce bananas are too restricted by legislation, and the cost of production increases beyond the international market price, they may, as they have in the past, move to another country with fewer and/or less restrictive environmental laws. In Costa Rica this would result in a drop of approximately 20% in hard currency generation. This fear has allowed the "five capital sins" of banana production to evolve and prosper (Hernández and Witter, 1996):

1. Disruption of the social and cultural base of those people living in the production region
2. Generation of large volumes of degradable, nondegradable, and hazardous waste at the production site
3. Excessive and often irrational use of agrochemicals

4. Increased deforestation (directly and indirectly caused by banana plantation expansion)
5. Increased erosion of the most productive soils

We examine these "sins," focusing especially on the ones related to the national economy and the environment. Our analysis begins with a brief history of banana production in Costa Rica, considers the banana ecosystem, and then looks at the economic and environmental aspects of banana production in depth.

## A. A Brief History of Banana Production

The U.S. citizen Minor Keith started the first commercial banana plantation in Costa Rica in 1872, in the Zent Valley. Seven years later Costa Rica was exporting bananas regularly to the United States (Soto, 1992). To increase production the Costa Rican government began to offer generous land concessions, tax exemptions, and other incentives to attract transnational companies.

Minor Keith and the Boston Fruit Company joined forces in 1899 to form the United Fruit Company. This partnership made it possible to develop a worldwide market for bananas, and initiated United Fruit's dominance over the economies and politics of large sections of Central America. By 1920, United Fruit had producing farms in most of Central America, and by 1930, United Fruit owned about 4% of the total territory of Honduras, Guatemala, Costa Rica, and Panama. It was this dominance of the region that brought about the humiliating characterization of these countries as the "Banana Republics," a perception still held by much of the world.

There is only a finite amount of land area within Costa Rica suitable for conventional banana production because bananas for the international markets can be commercially produced only on the best soils and in the optimum climate (Figure 2-1). Banana plantations require flat terrain and deep (no less than 1.20 m), well-structured, and well-drained soils (humid but not saturated) with a high balance of nutrients (especially potassium) and a pH between 6 and 7.5 (Soto, 1992). The best soils are those found in the alluvial plains and on volcanic ash deposits (Soto, 1992) (Figure 20-3). In addition high yields require quite circumscribed climatic conditions (Fig. 12-13). To maintain commercial production rates (2000 or more boxes per hectare), it is necessary to add significant amounts of fertilizer to the soil throughout the entire growth cycle.

Unregulated exploitation strategies of maximum production with minimum protection typified early transnational operations. Eventually, however, production efficiency dropped as the original soil fertility was exhausted. When

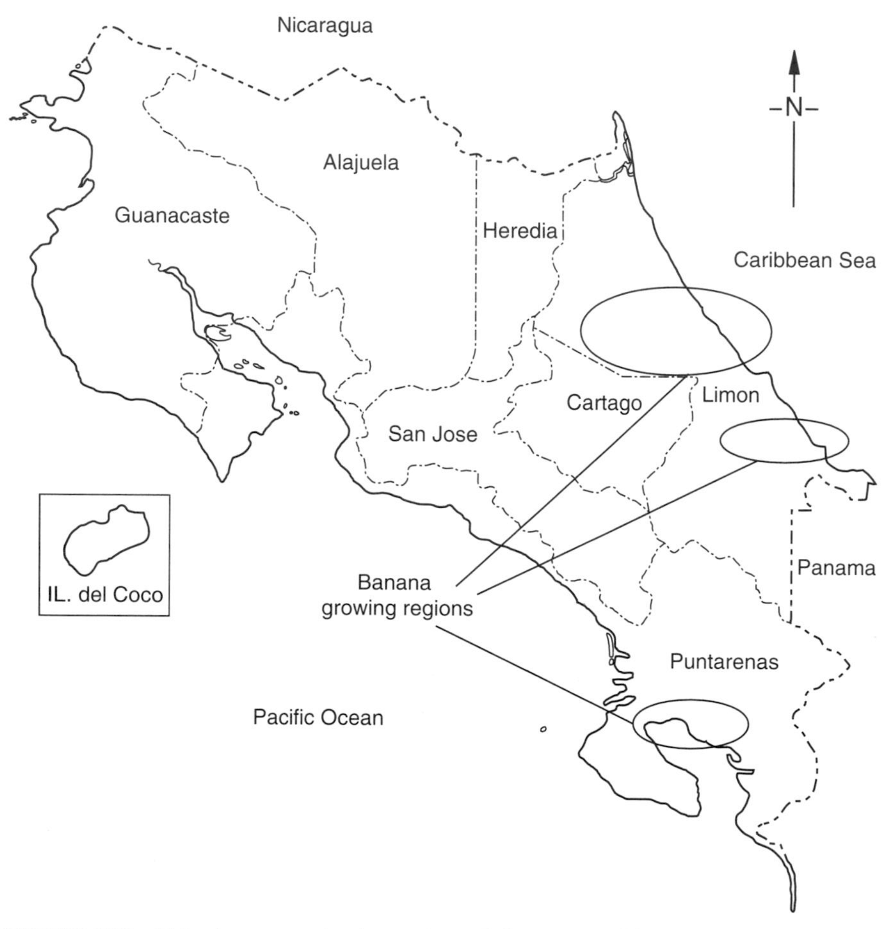


**FIGURE 20-3** Major banana production regions of Costa Rica. Although bananas of various types will grow over all of Costa Rica these are the only regions where major commercial production of export-quality bananas are grown.

the "Panama" banana disease (a fungus that attacks the fruit of the banana plant) forced production down further, United Fruit began abandoning plantations along the Atlantic coast of Costa Rica and transferring operations to the Pacific coast (Kepner, 1935; McCann, 1976; Casey, 1979).

By 1956, the government of Costa Rica had become concerned about the growing number of banana plantations abandoned due to the "Panama disease," inefficient production technologies, and, in general, the careless use of the country's natural resources. The need for a shift in the banana production process was apparent, and so began the second era of banana production in

Costa Rica. In an attempt to offset the downward trends, the government recruited Standard Fruit Company to not only establish its own in-country plantations but to also begin purchasing fruit from Costa Rican producers (McCann, 1976; Bourgois, 1989).

Standard Fruit introduced the Valery banana clone, which was resistant to the Panama disease, had higher yields, had favorable packing and shipping characteristics, and generated a fruit that the international consumer would buy. This transformation was not without side effects, however. The Valery banana required considerably more chemical inputs, as well as intensive field and processing management (i.e., plastic bags to cover the fruit, special packaging, and atmospheric controls during shipping). Therefore, while the addition of Standard Fruit introduced much needed competition to the banana economy and benefits to the national producers, it changed banana production from a low-chemical-input system to one dependent on large volumes of pesticides, herbicides, and fertilizers. This created a production system that has used chemicals in volumes and frequencies designed to eliminate the risk of reduced yields.

The nationalization of the banana industry of Ecuador in 1965 resulted in the migration of other transnational companies to Costa Rica, further intensifying regional competition and the need to produce more per hectare (Lauer, 1989). A decade later, United Fruit was reorganized, becoming the United Brands Company, while introducing the Chiquita label to Costa Rican products. In the 1980s, United Brands closed its operations on the Pacific coast of Costa Rica and returned to the Atlantic under the name of Compañia Bananera del Atlantic, Ltd. (COBAL). COBAL aggressively expanded banana production along the Atlantic coast and is now the second largest producer in the country.

In 1992, 60% of the total harvest came from transnational plantations. In 1995, they produced 40% of the total output, and maintain control over 98% of the exports through shipping and marketing agreements (Figures 20-4 and 20-5). National producers continue to be required to use prescribed production practices and meet strict quality standards which require the use of large volumes of chemical inputs by the transnational companies who purchase and market the fruit.

## B. The Present Banana Ecosystem

Costa Rica is located between 8° and 9° north latitude and 83° to 86° west longitude. It encompasses 51,032 km$^2$. The interception of the trade winds by the central mountain range produces large volumes of precipitation, especially on the north slopes. A number of rapidly flowing rivers originate in the mountains and have for centuries chiseled the volcanic-rich sediments of

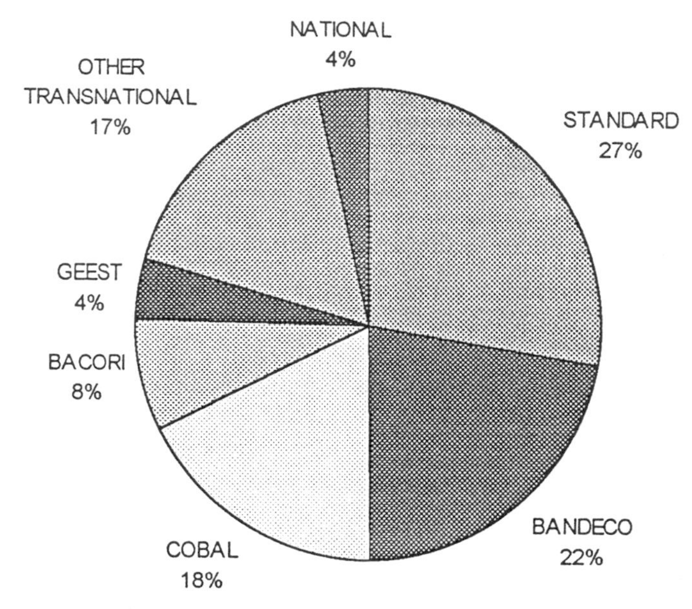


FIGURE 20-4 Banana exports arranged by marketing company. Standard Fruit is an international company, and the rest are Costa Rican. (Source: CORBANA, 1993.)

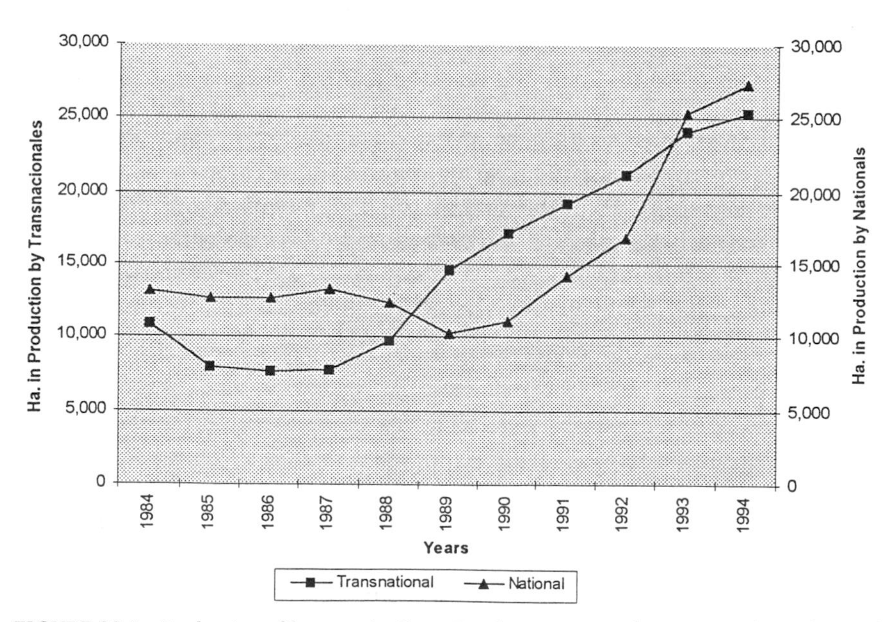


FIGURE 20-5 Production of bananas in Costa Rica by transnational companies (triangles) and by national companies (squares). (Source: CORBANA, 1993.)

the mountains and deposited them along their lowland floodplains. These sediments have formed patches of deep rich soils which are ideally suited for banana production. It is estimated that 98.5% of the total area planted in bananas is located along the Atlantic coastal plains (CORBANA, 1993) (Figure 20-3).

This ecosystem, or more precisely group of ecosystems, provides humans with the habitat and resources necessary to produce large quantities of bananas when using advanced packages of industrial technology. This development has brought about both positive and negative impacts. Figure 20-6 is a diagrammatic model of a banana production system in Costa Rica. Three major components within the ecosystem are identified: the plantation, the packing plant, and the housing facilities for the workers. It is important to emphasize that conventional banana production for the export market is a perennial monoculture system. It is an open system where the resource inputs have a large

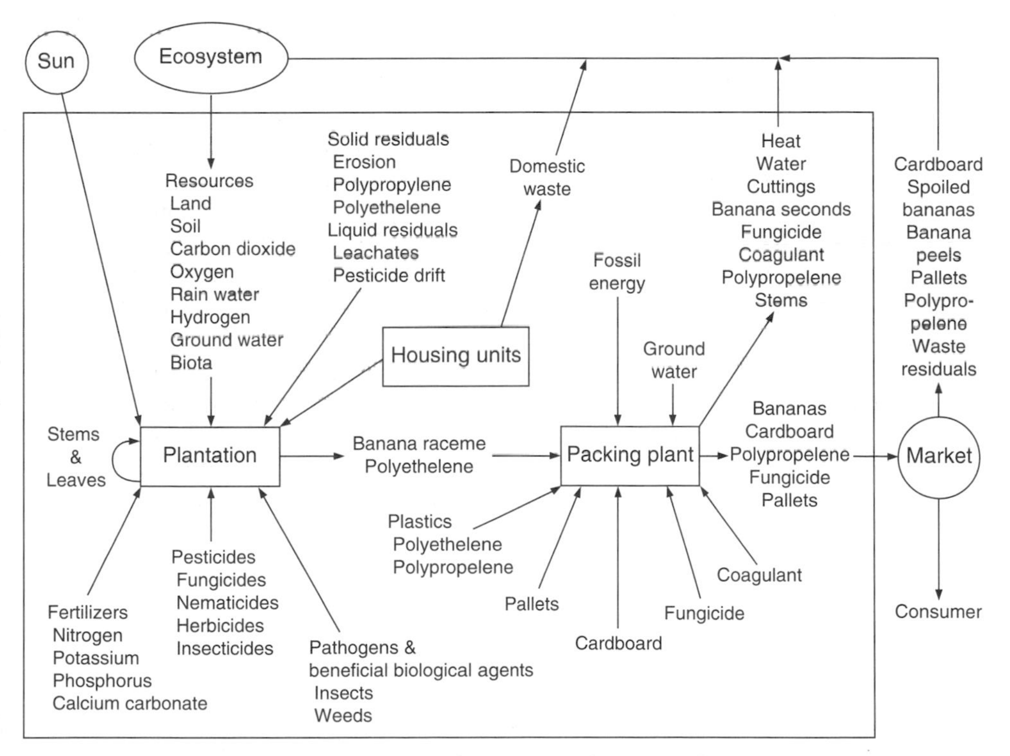


FIGURE 20-6 Diagrammatic overview of Costa Rican banana production ecosystem. Boxes represent major components of the ecosystem, and flows (arrows) represent the inputs required to maintain those flows and the products, intended or otherwise, of the banana production process.

direct influence on the outputs. As natural resources are exploited and the monoculture system persists, the output of bananas requires an increasing amount of inputs.

It is necessary to remember that besides the desired outputs, the fruit, the ecosystem generates output wastes that must be managed and in many cases recycled if the banana production system is to be sustainable. In terms of resource and money flows, the banana ecosystem can be thought of as a sieve. A large quantity of natural resources (sun, soil, rain, added chemicals, labor) pass through. The sieve retains a small percentage of the total flow as bananas, and even less as benefits for Costa Rica. The resulting pollution and degradation are not factored into any benefit–cost analysis of banana production, and if they were it is doubtful that the production of bananas would result in a positive resource flow to the nation. Even if the disbenefits were so monetized and factored in, and the results showed a negative money flow, bananas would still be grown since Costa Rica has few significant alternatives for producing money and especially foreign exchange. Costa Rica, like most nations, cannot support its present population and economy on interest but must use up capital such as soils and fossil fuels. That is just the way it is.

Monoculture ecosystems, like banana plantations, increase the concentration of food sources for other organisms (i.e., insects, bacteria, and fungi). As a result of an abundant food supply, such organisms multiply readily and begin to compete with humans for the harvest. Consequently, high-production systems require high levels of inputs not only to produce a product, but also to eliminate competing organisms (Cunningham and Saigo, 1990; Peakall, 1992).

Fertilizers, herbicides, fungicides, and insecticides are not used entirely by the plants, nor are they necessarily retained within the plantation ecosystem. Some portion of each leaves the production system as liquid leachates, surface runoff, erosion, or gases. These by-products often have a dramatic impact on the environment, especially as they pass through surface water systems (Table 20-1). In addition, pests and soil organisms appear to adjust to biocides by developing a resistance to the chemicals (McKenry, 1991; Peakall, 1992). This generates a negative feedback by creating the need for higher quantities of inputs and new chemicals to maintain production levels.

## II. THE PRESENT ROLE OF THE BANANA INDUSTRY IN THE COSTA RICAN ECONOMY

The banana industry plays a key role in Costa Rica's economy. It is responsible for 35% of the total value generated by the entire agricultural sector of the country (ANAPROBAN, 1995). It is estimated that 40,000 workers (approximately 3.7% of the workforce) are employed directly by the industry. The

TABLE 20-1 A Banana Plantation's Waste Stream for 1993

| Description | Tons of waste generated per year | Tons exported 1993 | Tons of waste per 100 tons exported = *TW* (Te × .01)* |
|---|---|---|---|
| Banana production for 1993 in tons | | 952,776 | |
| Twine | 2100 | | 0.22 |
| Plastic bags | 2801 | | 0.29 |
| Packing material | 3211 | | 0.34 |
| Total nondegradable | 8112 | | 0.85 |
| Crown and flowers | 24,505 | | 2.57 |
| Raceme's stems | 143,528 | | 15.06 |
| Fruit rejected | 317,592 | | 33.33 |
| Pesticides (estimated 25% loss)** | 2369 | | .25 |
| Nitrogen | 771 | | .08 |
| Agrochemical containers | Information not available | | |
| Subtotal | 502,620 | | |
| Total waste requiring treatment** | 502,620 | | 52.75 |

* TW/(TE/0.01) = Waste in tons × .01 tons of banana exported in 1993.
** The estimate represents a best guess by the authors.

mean income, including federally required health insurance, is U.S. $433 per worker per year, for a total of $17,320,000 annually. Another 100,000 workers (approximately 9.2% of the national workforce) are employed in related service industries, increasing the mean income generated annually to U.S. $60,620,000 (ANAPROBAN, 1995).

In 1993, the Central Bank of Costa Rica reported that banana exports generated U.S. $536 million, 40% of the total value of all exports.

## A. The Squeeze on Banana Producers

Higher costs of operation and increased area in production per hectare have forced a reduction in farming inputs, which in turn has resulted in a drop in production per hectare (Fig. 12-4a). For a period of time, producers have, as a result, brought more and more marginal lands into production to maintain overall yield to meet production quotas. Farming more land has further eroded the sustainability of the banana production system. Energy costs and interest rates continue to increase and are for the most part driven by exterior forces. Costa Rica has no known oil reserves and must depend on oil imports and hydroelectricity for power. Interest rates in 1995 reached 38% per year, making it harder for newer and marginal producers to survive in current market conditions. To lessen the squeeze, the Costa Rican government's banks have changed banana loans to U.S. dollars (1997), allowing for a reduction in the interest rates to a few points over prime. Meanwhile the impact of black Sigatoka disease in Costa Rica is between $1000 to $1200 per hectare. This has triggered a series of research trials in an attempt to control this fungus in a more cost-effective manner. In 1997, the expanded use of integrated pest management (IPM) has made it possible to hold costs at approximately the same level.

Salaries and benefits for plantation workers are negotiated on a biannual bases. They are based on negotiated rates between the private and public labor sectors, and adjusted for the current rate of inflation. The increased costs are paid by the producer, who in turn has historically had a fixed price contract with the transnational shipping companies. The rates of inflation and devaluation of the Costa Rican currency have continued to increase. As long as the devaluation rate is higher than the inflation rate the producer is somewhat compensated because his income is paid in U.S. dollars, which have historically been more stable (Figure 20-7). Thus the producer relies on an increased domestic devaluation rate that is higher than the inflation rate to make a profit (i.e., he can

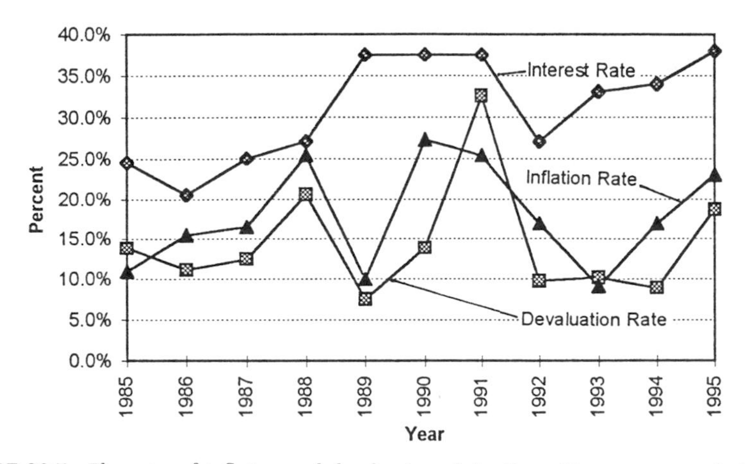


FIGURE 20-7 The rates of inflation and devaluation of the Costa Rican currency. As long as the devaluation rate is higher than the inflation rate banana producers are somewhat compensated because their income is paid in U.S. dollars, which historically have been more stable. (Source: CORBANA, 1995.)

buy more Costa Rican currency for fewer U.S. dollars). Unfortunately, this system contributes to national inflation and a continued devaluation spiral that is difficult to control. The impact is greatest on the new plantation owners, who are paying higher land rent costs. The government is seeking alternative methods to check inflation and manage its impact.

At the same time retailers and consumers in the United States and Europe are willing to substitute apples or another fruit for bananas if the market price becomes too high. Limited interviews of retail produce managers in major Michigan grocery chains in 1994 indicated that when bananas reach $0.45 per pound (versus the more generally encountered price of $0.29 per pound) consumers will purchase domestic fruit, such as apples or oranges. Of course many consumers would continue to eat bananas because they are a tasty, low-fat, high-quality food and an excellent source of dietary potassium.

Governmental incentive programs for banana producers worldwide have resulted in overproduction, lower prices, and greater rejection rates by transnational shippers and marketers. This, combined with export and import restrictions based on world politics, loss of soil organisms needed to decompose detrital material, and climate changes, both forecasted and actual, makes banana production a risky business for the producer. For example, severe floods in 1996 eliminated the road between the Central Valley and the Pacific producing areas, requiring the already hard-pressed government to seek expensive new loans to repair the road.

## B. Plight of the Government

While the future of Costa Rica's banana producers is not bright, the government's plight begins with the need to increase production to service a growing international and domestic debt (Fig. 4-10a). As of March 1995, each of Costa Rica's 3.4 million citizens had a national international debt of U.S. $1000 and a domestic debt of U.S. $412 (Monge, 1995). This does not include personal debts on credit cards and so forth, which are also very large. While such a national debt may not be large in comparison to other developing nations [e.g., Honduras at $3100 per capita, Panama at $6400, and Colombia at $14,100 (all U.S. dollars)], it has not been possible to service both the interest and the principle payments of the debt, and the total continues to rise (Fig. 4-10; (Monge, 1995).

Costa Rican foreign debt problems began in the early 1970s, when the large international banks had a huge supply of petrodollars. The banks were more than willing to make large loans at low interest rates and with long-term repayment schedules. For Costa Rica, this coincided with a period of high coffee production and high international prices. The increased prosperity gave the government the opportunity to invest in large development projects that were characterized by high risk, huge infrastructures, high salaries, and numerous expensive consultants. These undertakings often failed, most often because of mismanagement. All of this led to the development of a new rich class of Costa Ricans who were able to take advantage of the system because of their positions. They consumed from Costa Rica and banked abroad, draining capital from the country.

## C. Productivity, Production Costs, and Income

Since 1977, the amount of land devoted to banana production has steadily grown. The largest increase in area took place between 1989 (24,772 ha) and 1994 (52,737 ha), an increase of 213%. During the same six-year period, yield per hectare dropped by 28% (Figure 20-8). After peaking in 1994, the amount of land in banana production started to decline, and by late 1996 there was 52,166 ha in production (CORBANA, 1996).

## D. The Net Contribution of Bananas to the Generation of Foreign Exchange for Costa Rica

Chapters 2 and 5 have documented the critical importance of foreign exchange for the present and future economic conditions of Costa Rica, including the

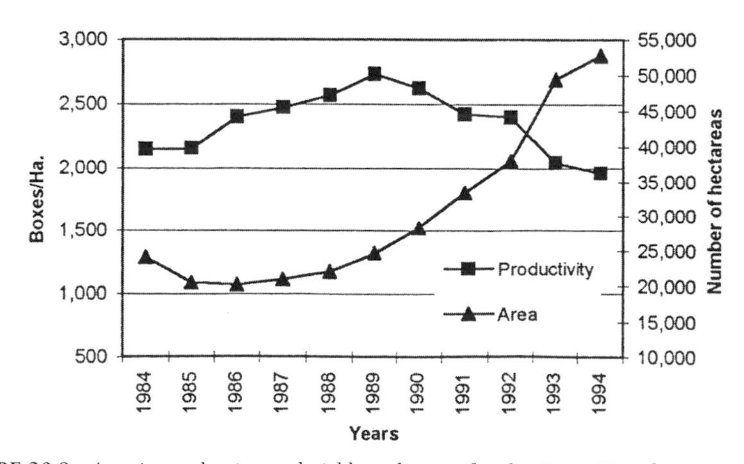


FIGURE 20-8 Area in production and yield per hectare for the Costa Rican banana industry.

importance of inputs to agriculture itself. Chapter 23, as well as this chapter, develops the importance of bananas for generating this foreign exchange. Specifically, in all years from 1983 to 1994 bananas were the most important generator of foreign exchange, and in the years since 1994 bananas have been a close second.

In earlier work on the energetics of Costa Rican agriculture, undertaken at CATIE by Hall and Schlichter (e.g. see Chapter 5), it became obvious that, in general, agriculture in Costa Rica was tremendously dependent upon energy-intensive inputs from the industrial world. This perspective was developed more explicitly, especially for coffee, in Levitan (1988). The following takes this general methodology and applies it to bananas.

The first step in attempting to understand a given entity or process from a systems perspective is to develop a systems flow diagram. While it might be possible to undertake a comprehensive systems analysis by some other approach, the energy flow approach is especially useful because, unlike monetary flows, all important pathways must be included and quantified for an accurate analysis, and no unaccounted-for subsidies are possible. Thus we generated an energy flow diagram for a typical banana plantation, first using boxes and arrows (Figure 20-6) and then using the symbols of Odum (1983). Since one of the rules for drawing a systems flow diagram is to define boundaries, it was necessary to focus on the transnational boundary flows of energy and materials. This also confronted us with a series of questions about the degree to which bananas might be dependent upon fossil fuel inputs. The result was the non-quantified skeleton, that is, the symbols and their connections, of Figure 20-9. We were rather surprised to see the many, many ways that banana production in Costa Rica was related to processes originating in the industrial world (i.e., outside of Costa Rica). While this may not be surprising to any grower of bananas, the total magnitude and complexity of the process was revealed only after we had made such a diagram.

When quantifying the energy flows represented in that figure, we immediately came upon a problem: almost all data at our disposal were in economic units. However, Hall *et al.* (1986) and others (Hannon, 1982) had previously quantified many energy flows from economic data for which conversions are available. Thus we looked at cost data from surrounding plantations, and derived economic data for the flows represented in Figure 20-9. The results are the numbers now in that figure. Thus came our second shock, although again it is probably obvious to any banana manager: about 50 cents of each dollar derived from banana sales *leaves* Costa Rica even *before* the banana is produced as money to purchase the inputs required to grow bananas.

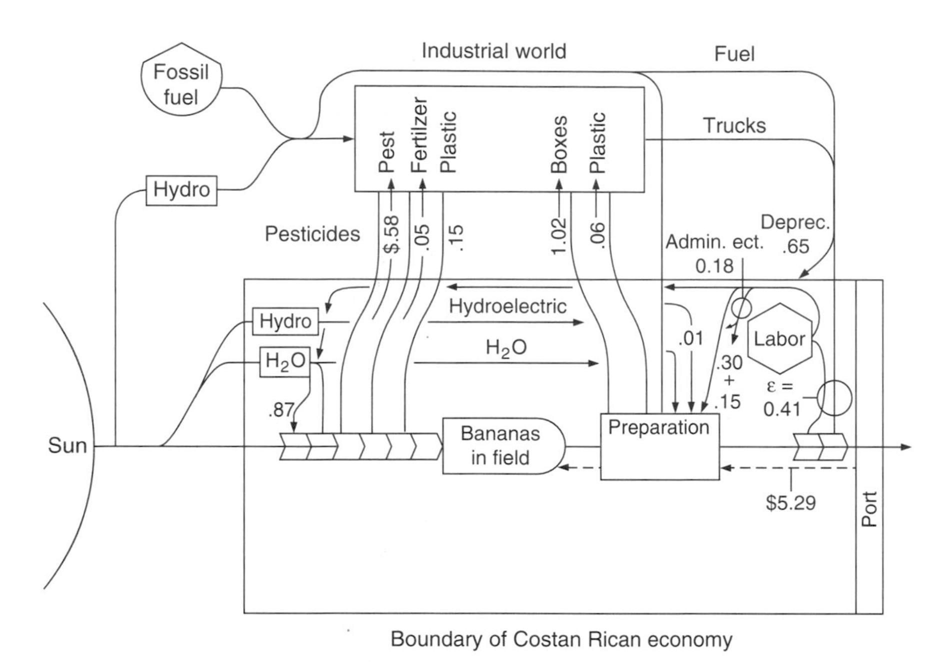


FIGURE 20-9 Diagrammatic model of energy in the Costa Rican banana production system. Symbols are from Odum (e.g., 1983). This figure shows the flow of energy from the sun (left) to the plantation, to the packing plant, and to the port. Of particular importance here is the boundary between the nation and the rest of the world, represented by the large box, and the "work gates" (arrowhead-shaped modules) representing points where energy inputs control the main energy flow from sun to bananas at the port. The economic data represented on the diagram represent values derived from Earth College records, and are thought to be representative of the Costa Rican banana industry.

That relation can be expanded to the entire national banana industry, and what it means is that for the roughly \$600 million earned from the exports of bananas from Costa Rica only about \$300 million is *net*, or genuine, foreign exchange. The rest is required to grow the bananas! Or at least to grow bananas that are competitive in the world market, which means effectively those northern industrial countries represented in Figure 20-2b. We believe that this concept of net foreign exchange may be one of the most important findings of this book.

## E. Environmental Problems Associated with Banana Production

Most of our analysis to date has dealt with only the directly monetized costs and benefits of banana production. There is also another complete set of relations involved with banana production systems have to do with the generally nonmonetized (or at best imperfectly monetized) aspects related to waste production.

## 1. Non-Point and Point Source Impacts

First some definitions are needed. *Degradable waste* is any material that is stabilized and assimilated by the ecosystem in a period of less than two years. *Hazardous waste* is any solid, gas, or liquid that presents a risk of degrading the health of humans or wildlife. Hernández and Witter (1996) grouped waste generated from banana plantations into five primary categories:

1. Non-point source waste (solid, liquids, and gases)—degradable and nondegradable, and hazardous and nonhazardous
2. Solid point source degradable waste
3. Liquid point source degradable waste—hazardous and nonhazardous
4. Solid point source nondegradable waste
5. Point source hazardous waste (solids, liquids, and gases)

Environmental degradation resulting from *non-point source waste* (solid, liquids, and gases, both degradable and nondegradable, and hazardous and nonhazardous) is the most difficult to track and quantify. Non-point source wastes are generated through the use of commercial and natural fertilizers. According to Soto (1992) the mean use of fertilizers on bananas in Costa Rica (in kilograms per hectare per year) is as follows: N = 56.3, $P_2O_5$ = 24.3, $K_2O$ = 220.4, CaO = 8.8, and MgO = 20.7 kg/ha/yr.

In addition the use of nematocides such as Counter, Furadan, and Tilt is about 55 kg per hectare each, which is the EARTH College rate (1996). Herbicides are used at about 6.34 liters per hectare based on the EARTH College rate (1996). Generally only one insecticide (chlorpyrifos) is used widely and it is impregnated in the blue plastic bags used to cover the raceme at 1% of the bag weight.

It is important to note that pesticides are used and abused widely in Costa Rica generally. Studies conducted by the National University (Chaverri and Blanco, 1995) show that the application rate of formulated pesticides for the country in 1993 was 27.48 kg per cultivated hectare (12.10 kg of active ingredient per hectare).

Historically, it appears that the chemical use levels recommended have been ignored by producers under the philosophy of minimizing risk. While the guidelines covering the type of chemicals permitted for use in Costa Rica are regulated by the government and are essentially identical to the standards in the United States, the transnational companies have traditionally determined the volume, kinds, and frequency of the chemicals used in their own plantations as well as those of their independent producers. Failure by an independent producer to use these prescribed chemical cocktails has resulted in higher rejection rates of that plantation's crop, principally for cosmetic reasons.

Fungicides to control roya (coffee) and Sigatoka (bananas) account for about 49% of the chemicals used nationally. Herbicides account for 26% and insecticides for 20%. Approximately 10% of the chemicals used are classified as extremely toxic by the World Heath Organization, 8% as highly toxic, 24% as moderately toxic, 8% as lightly toxic, and 50% as low risk; a few have not been classified. It is important to emphasize that the classification is based on appropriate use, and that all pesticides are hazardous if not used in accordance with recommended procedures. Most of these pesticides are imported and cost the country an estimated $80 million per year. Of this amount, approximately $40 million is spent on fungicides for use on coffee and bananas.

The quantity of chemical deposition depends on the amount and frequency of application, type of chemical, soil type, precipitation, temperature, wind velocity, location, and method of application. Some of the agrochemicals metabolize quickly into the ecosystem and are very difficult to trace unless sampling is done on a daily basis, starting from the time of application. Wastes are transported outside the plantation by wind, runoff, and infiltration. Quantification of the amounts and concentrations is very difficult. Currently the best available methods for determining the aerial extent of the pollution is to sample and test surface water, plant tissue, and surface soils located both within and outside of the plantation. According to Stover and Simmonds (1987) fungicide applications occur some 45 times per growth cycle. Because fungicides are applied from the air a portion is lost to wind drift and ends up in the surface waters of Costa Rica. The exact amounts lost are not known, but are highly debated.

These pesticides can have a large impact on human health. Two recent studies quantified the number of cases treated for chemical poisonings by crop type. Quiros *et al.* (1994) found that 21.1% of the cases could be attributed to coffee, 15.6% to rice, 13.3% to bananas, and the balance to other crops. The Costa Rican Ministry of Health (Ministerio de Salud, 1994), however, reported bananas as being responsible for the largest number of cases reported at 36.4%. Insecticides accounted for 47.9% of the cases reported and herbicides for 34.3%. Parquet, responsible for 26.6% of the poisonings, was used frequently in banana production until 1999. As a result, most producers have sought an alternative herbicide.

The pseudostem of the banana plant (leaves and leaf veins) represents the major *non-point source degradable residual.* These residuals are an important source of nutrients for the plantation in the medium and long term. Obando *et al.* (1996) quotes Godefroy (1974) and Twyford and Wamsley (1974) as reporting nutrient losses from residuals in intensive banana production systems as 199, 23, 660, 126, 76, and 50 kg/ha/yr of O, N, P, K, Ca, Mg, and S, respectively, and at least 50% of these amounts are released back to the soil. Since there is no reason to extract these residuals from the plantation they are left *in situ* to decompose and be incorporated into the natural nutrient cycle of the plantation. For this reason, the 4 million plus tons of these residuals (countrywide, annually) are treated separately (Table 20-1) and are not accounted for as waste.

The fruit stalks constitute a major *point source degradable waste*. Disposal methods for the stalk depend on where the waste is generated. Some plantations detach the fruit from the stalk in the field, but the majority remove it at the packing plant. Because most plantations remove stalks at the packing plant, they were often transported to open-air dumps rather than back to the fields. These dumps are usually located along a river. The stalks deteriorate rapidly, and the waste is eventually carried away by floodwaters.

Since 1993, a significant change in disposal procedures was implemented on the majority of both transnationally and nationally owned plantations. For the first time on a wide scale stalks were used as mulch to increase organic materials within plantation soils. New and exciting technologies are being researched and developed to compost stems and discarded fruit (Obando *et al.*, 1996; Ruiz *et al.*, 1996).

From 16 to 25% of the bananas produced do not meet export standards because of handling bruises, scars, stains, crown infections, mutilated fingers, and unacceptable dimensions. Before the market became saturated, 5 to 16% were exported as second-class fruit at a 25% price reduction. This is no longer the case and bananas that do not qualify as premium and first-class fruit are discarded. In the past most plantations deposited the remaining bananas in open-air dumps. Today most of the transnational and larger national companies deposit the fruit (that which cannot be sold in local markets or processed, such as with Gerber) in trenches designed to minimize nonpoint environmental problems.

*Point source liquid waste* originates at the packing plant. Large quantities of water are used to cushion the fruit during handling and transportation from one section of the processing to the next, and to clean the fruit and processing facilities. Bananas have a female flower attached at the tip of the fruit at the time of harvest. When the flower is removed at the packing plant, the fruit oozes a white milky latex from the wound. If not handled properly, the latex will stain the fruit and make it unacceptable for the export market. The same is true from cuts made during the detachment of the fruit from the stalk. It is necessary to wash the fruit as soon as possible after removing the flower and detaching it from the stalk. The fruit will continue to secrete latex unless alum (aluminum sulfate) is applied to the wounds. As of 1997, Article 132 of the Wildlife Conservation Law requires treatment of all processing water.

The wounds are also susceptible to fungus, making it is necessary to treat the crown of the banana with a fungicide (Mertak). Alum and Mertak both contaminate the surface water downstream from the packing plants unless the wash water is filtered and recycled. Water treatment at EARTH College has been successful in removing essentially all of the biocides used during processing.

EARTH College's packing plant uses approximately 12 liters of water per second to process 4000 boxes of banana during a 10-hr day (108 liters per box of bananas shipped). When extrapolated to the yearly production of Costa Rica, the total water required for banana production is 5.7 million liters of water, most of which was historically returned to the environment untreated. Recently many of the larger packing plants have installed filter systems similar to the one developed at EARTH College by Carlos Hernández and Edgar Matamoros. These systems remove solids and most of the latex at a very low cost for construction and operation, and are now widely used.

Most plantations use deep well water for their washes. Some plantations have well water quality problems. Excess iron in the water stains the fruit. Chlorine is added to precipitate the iron, and this in turn becomes part of the waste stream. Because of the costs to run the pumps and to buy the chlorine, some of the plantations have started to recycle the water by passing it through a water purification system.

Blue plastic bags and twine are the primary *point source, nondegradable wastes* from banana production. Plastic bags have been used in banana production for 30 years (see the CD). Workers attach bags over the banana fingers as soon as they start pointing upward. The bags perform four basic functions: (1) they are a physical barrier to insects and contaminants (e.g., fungicides), (2) they serve as a chemical barrier to insects, (3) the plastic creates a microenvironment that stimulates fruit development, and (4) they protect fruit from aerial spraying. Bananas produced without plastic bags have a higher number of skin blemishes, and as a result have a 10% higher market rejection rate than do those produced in a bag. This 10% is worth over $50 million per year to Costa Rican producers, and as a result the bags will continue to be used.

An exception is Platanera Río Sixaola in Germany, which is marketing a banana that uses chemical-free plastic bags. Its seal reads "die faire banane" ("the honest banana"). Consumers are told that the banana skin has blemishes because no insecticides are used. Over time it has developed a market niche, but has had problems obtaining import licenses from the European Common Market. This is proof that there are market niches for more environmentally sound bananas, but that government interference and protectionism are hampering the creation of an environmentally friendly banana.

Plastic twine is used to anchor the bananas, so that the weight of the fruit and the wind will not collapse the plant. Traditional harvesting methods include discarding the twine directly onto the plantation floor. The twine is not biodegradable and accumulates over time. Since 1993 producers have been collecting the twine and removing it from the plantation.

Storage of agrochemical containers represents a major *hazardous waste* to Costa Rica's environment. It is difficult to project the magnitude of the container problem on a national scale because of the large variation in the size and type of containers available. The primary environmental problems associated with agrochemical containers are spillage during handling and leakage from improperly stored containers.

The government of Costa Rica will not issue permits to plantations for the incineration of hazardous materials, including plastic containers. Therefore, most empty insecticide and herbicide containers are stored on site. Improper handling and storage of these chemicals represents a significant environmental and health problem. The amount of data on the actual and potential health impacts for the workers and people living within the production region from these chemicals is growing (Peakall, 1992; Ragsdale and Sisler, 1991; Hilje *et al.*, 1992). In time the costs associated with increased medical problems will have to be factored into the benefit–cost equation of banana production.

Currently only two companies produce reusable containers in Costa Rica, Tilt's Farm Pack fungicide and Rhone-Poulenc's Surefill returnable nematocide containers. Banana producers are beginning to apply pressure on the agrochemical companies to either produce reusable containers or to take them back, as storage of used containers is becoming a major problem. While there are alternatives to these problems, they represent a change in management and additional costs to the producer which the transnational marketers, foreign retailers, and consumers historically have not been willing to share (Hernández and Witter, 1996). In the past there has been little incentive for producers to change production practices that cut into their profit margins. Enforcement of existing laws regarding chemical use and storage in Costa Rica, as elsewhere in the world, suffers from being understaffed and underfunded. Economic stress on the market can lead to fewer environmental safeguards—perhaps the ultimate free market slogan is "Live for Today, Tomorrow's Problems Will Be Paid for by Someone Else."

## 2. Producing an Environmentally Friendly Banana

There are technological packages that can minimize these environmental side effects. These technologies, while more sustainable, impact yields and erode the producer's profit margin. Furthermore, these technologies tend to compromise the aesthetic quality of the banana skin. If all banana producers worldwide were to adopt these technologies the costs could be transferred onto the consumers. This would induce some market fluctuations, but the impact would be borne across all producers and consumers, lessening the overall impact on individuals and countries. Many producers' commercial banana production and processing facilities have obtained the "ECO-OK" seal from the Rainforest Alliance. Researchers there estimate that an increase of $0.02 per pound to the consumer would cover the costs and incentives for the producers to use more environmentally friendly production methods.

The problem is that most banana marketing companies do not recognize an immediate economic benefit for marketing an environmentally and people-friendly banana. Individual producers lack incentives to produce ecofriendly bananas because the additional efforts are not compensated for in the quantity demanded or the market price paid by the wholesalers and consumers. The existing market structure is a well-integrated vertical system controlled by transnational corporations, who have historically controlled a major segment of production and most of the exports. History and economic theory have shown that commercial retailing operations will purchase the highest quality at the lowest price to remain competitive in their local markets. The retailer purchases bananas using a consumer-based quality, service, and price model. For example, retailers, especially in the United States, want to buy a 20- to 25-cm banana without skin blemishes that is priced below apples and oranges. Unlike apples and oranges, there is only one variety of bananas offered in the stores, thus the only choice is brand name, even though many other excellent varieties are available in Costa Rica. Because there is no direct contact between consumers and producer, or any reason for the retailer to educate their consumers about the benefits of a more environmental/people-friendly banana production system, the production system remains controlled by retail market price and the transnational companies. This is the ultimate triumph of the system of economic thought known as neoclassical economics—that production is not of interest, and consumption and consumers reign supreme [see critique of neoclassical economics in Hall (1992) and Chapter 3].

There may be some hope. Surveys conducted by the Roper Organization concluded that young affluent married women living in the northeastern or western portions of the United States are willing to pay more for environmentally safe products (Schwartz and Miller, 1991). But of the 750 persons surveyed as a sample of the general U.S. population, 70% were more interested in shopping convenience than environmental safety, and 53% were unwilling to pay more for environmentally safe products (Schwartz and Miller, 1991). Americans find it easier to express environmental opinions than to make lifestyle changes (Market Research Services, 1990). In general, European consumers are more sensitive to environmental- and people-friendly products, but nevertheless do not recognize an environmentally friendly label, such as ECO-OK, and are only willing to pay more for organic bananas. European consumers together with the Americans make up over 90% of Costa Rica's international banana market (Figure 20-2b). Neither market is yet ready for differentiation based on production methods other than "organic" (ECO-OK/ Project Werkstatt, 1996). The key is to establish a link between the consumer and the producer in order to influence the consumer's preference and the producer's awareness through communication and education.

In the early 1990s, Costa Rica's banana industry recognized that it had an environmental problem and that it needed to take steps to rectify the actions of the past. The Banana Environmental Commission (CAB) was formed as a scientific advisory committee to the banana industry. It is made up of representatives from the banana producers, marketing firms, universities (e.g., Dr. Hernández represents EARTH College on CAB), and the government. Money to operate CAB has come from CORBANA, which has a governmental mandate to devise environmental policies for the banana industry.

The commission initially was responsible for assessing environmental problems associated with banana production and distributing the results to Costa Rican banana producers. It prepared a document entitled "Principles and Goals of Environmental Management of the Banana Industry" that was later adopted by CORBANA and the government of Costa Rica. As part of the principles, CAB was empowered to conduct periodic inspections of plantations to ensure compliance. With few exceptions, Costa Rican producers have been lagging behind their transnational counterparts. For example, out of 97 plantations owned by transnational companies surveyed, 98.5% followed the recommended solid waste management practices prescribed. In contrast, only 28% of 25 independent producers complied with recommended solid waste management practices (Ruiz, 1998).

Funding CAB has been a major problem. Producers and marketing firms are already paying an environmental tax to the government of Costa Rica for each box of bananas produced. Originally this tax was supposed to pay for environmental research and development. Congress has since modified the laws related to managing these funds and the revenues have been diverted to other non-banana-related purposes. As a result, producers and exporters have refused to pay additional fees to protect the environment. In 1995 CORBANA was unable to provide an operating budget for CAB and the organization was dormant. Funding for research and development was allocated for 1996 and CAB reactivated. There is no certainty that funds will continue to flow for the future operation of CAB, which precludes any long-term planning.

For the foreseeable future Costa Rica must search for environmentally and cost-effective technologies that will help to make banana production feasible and thus sustainable. The questions regarding the need to service its foreign debt, and the hope that foreign consumers will rapidly evolve an environmental conscience that will induce them to seek more expensive Costa Rican bananas to protect Costa Rica's environment and people, are, for all practical purposes, "moot points" in the foreseeable future.

### F. Sustainable Banana Production

Within Costa Rica and throughout the humid tropics the concept of sustainable development is gaining momentum as countries face the reality that most renewable natural resources are quite limited (Harwood *et al.*, 1992; Hernández and Witter, 1996). The old economic theory that as one resource is exhausted the market will allow another to take its place is being reexamined. The reality that the fruits of economic development have been infected with the rot of environmental deterioration started in the United States during the 1960s (Knesse, 1986; Hall, 1992) and is now understood clearly among many of the environmental agencies and universities found throughout the humid tropics. The more difficult topic is whether "ecological economics" and "sustainable development" offer a different, less environmentally damaging way for growing populations to find reasonable economic security, or whether these terms provide only a feel-good smoke screen (Hall, 1992) (Chapter 24).

From an ecodevelopment perspective, if bananas are grown on the best soils where they have all of the solar radiation, nutrients, and moisture needed, their growth should be sustainable. If the producers use the best available management practices and the government removes obstacles, the negative impact of banana production on its laborers, people living in these regions, and the environment should be minimized. Costa Rica must take immediate steps to improve its own environment and at the same time develop long-term strategies for managing its debt load and diversifying its export portfolio.

## III. TOWARD A SUSTAINABLE BANANA PRODUCTION FUTURE

The purpose of this chapter is to document the existing environmental, economic, and human problems associated with commercial banana production and to propose realistic and possible methods and strategies that hopefully will make banana production more sustainable using an ecodevelopment approach. Any technique based upon ecological considerations and which optimizes the integrity of both an ecosystem and human-made systems can be classified an ecotechnology. If technologies involve a unit process (such as scrubbers, sand filters, digesters, sediment traps, and drainage canals) that takes advantage of natural ecosystem functions, they can be classified as environmental technologies (Mitsch and Jorgensen, 1989).

### A. The Need for Cleaner Technologies

Cleaner technology is defined as production procedures that increase the contribution of an industry to the economic and social well-being of present and future generations without endangering fundamental economic and ecological processes (ONUDI, 1992). Cleaner technologies have three basic objectives: (1) to generate less environmental pollution (water, air, and soil) per unit of production, (2) to generate less waste per unit of production, and (3) to use fewer natural resources (water, energy, and raw materials) per unit of production.

Cleaner technologies do not mean zero pollution. Cleaner technologies shift the emphasis from the reactive "end-of-pipe" solutions to the proactive "process-integrated" total quality management solutions that are constantly changing in the quest for excellence.

The Ministry of Science and Technology of Costa Rica has recognized the short-term benefits of cleaner technologies, which can reduce pollution and waste generation by 30% with the following overall benefits:

1. Savings in the costs required to eliminate dangerous waste and residual waters
2. Savings in end-of-pipe infrastructure necessary to abate present pollution
3. Savings in future costs of cleaning groundwaters and soils
4. Fewer required inputs through rational use of resources and efficient processes
5. New and better opportunities to compete in ecofriendly markets

## B. A More Sustainable Future Through Integrated Farm Management

Tropical landscapes are a diverse and complex mosaic of communities. Extensive monocultures change the landscape considerably by simplifying it and reducing biodiversity. Banana plantations in Costa Rica are very large and use the same soils over and over again, year after year. This typical, widespread practice has caused significant soil erosion, soil structure degradation, and increased numbers of pests throughout Costa Rica's banana-producing regions. New technologies and management strategies are needed to reverse this problem.

Integrated farm management through such practices as crop diversification, crop rotation, intercropping, and biological filters designed to separate intensive agricultural areas from natural ones helps minimize the effects of ecosystem intervention. To achieve both high yields and ecologically sound production one must begin with a clear understanding of the ecology of bananas. An analysis of the plant's water, soil, temperature, and sunlight requirements will indicate those geographic areas best suited for maximum banana production. Secondly, by attempting to simulate the natural production cycle, we can minimize disturbance to the ecosystem caused by farming practices. Integrated farm management, and its subcomponents of integrated waste management (IWM) and integrated pest management (IPM), incorporates these goals. Specifically we recommend the following strategies in order to ameliorate some of the environmental problems already witnessed in Costa Rica's banana growing regions and to help ensure a sustainable future.

1. Fields that produce a low economic yield (<2000 boxes/ha) should have the banana plants destroyed and the fields left in fallow for at least two years. This should be followed with a compatible crop rotation with bananas being left in the field no longer than five years.

2. Biological barriers or environmental filters should be incorporated into the production system in ~15-m strips. These filters should be other cash crops to keep the cash flow at a sustainable level to maintain the plantation. The filters should be designed to reduce soil erosion, to capture nutrients flows, to contain commercial agrochemicals mobilized via runoff, and include fast-growing trees species to help capture agrochemicals caught in wind drift. These filters will also help act as wind barriers, which will in turn capture wind-borne Sigatoka spores. The trees can be sold in the wood futures market and add to the economic sustainability of the plantation. Types of filters, including their composition, and their economic benefits will vary with each plantation and will require continued study.

3. Drainage controls are very important to maintain plant vigor and reduce the number of fungal and insect pests. To limit erosion, drainage canals should be lined with a dense grass cover. Runoff from the plantation should be filtered through a wetland (natural or constructed) to reduce the sediment and nutrient content of the water. Care should be taken to make sure to include 15-m-wide vegetation barriers along the main drainage system, so as to limit downstream erosion.

4. Fertilization will continue to play an important role in high-yield banana production. The frequency of its application, quantity applied, and method of application must be managed closely. Bananas have a limited nutrient intake rate and produce better with a balanced level of nutrients in the soils throughout the growth cycle. Applications should be frequent and in appropriate dosages based on frequent soil tests.

5. The introduction of organic matter into the plantation plays a significant role in maintaining the humus content of the soil and limiting erosion. Increased detritus increases biological activity, as well as biodiversity.

6. Organic waste collected at the packing plant should be returned to the plantation but should not be distributed until it has been composted to reduce fungi, pests, etc. Composting increases the temperature of the organic matter sufficiently to kill most pathogens found in this waste and provides a

concentrated nutrient package to be used with the bananas, crops, and trees in the filter system. Composting in high-rainfall areas is difficult because excessive humidity saturates the mix and displaces oxygen. Research is needed to find economic alternatives to open-air composting.

7. Another alternative is to feed the organic waste to worms (e.g., California Reds), which consume the nutrients and defecate humus material. Worms are very efficient processors of waste. Each worm can consume half its weight per day in waste. The average weight of a worm is 800 mg.

8. A very important part of integrated farm management is managing insects and weeds with low levels of commercial agrochemicals. Chemical pest control should be checked by frequent (every 10–14 days) field evaluations. Crop tolerance thresholds should be monitored closely to determine the types and amounts of agrochemicals required.

9. Careful applications of agrochemicals are critical if large amounts are not to be lost via wind drift and surface runoff. Precise applications have both an economic and an environmental implication. Agrochemicals which do not reach the affected banana plant become waste residuals in the system. Because it is nonpoint source waste, remedial actions are limited and expensive to implement. Spraying fungicides (black Sigatoka control) from helicopters reduces the amount lost to wind drift.

10. It is also very important to introduce biological insect control. The University of Costa Rica has an Integrated Fruit Fly Management Project which has had excellent results with oranges and mangoes. Two methods showing promising results are: (1) to produce sterile male insects by exposing them to radiation, and (2) to produce a parasitoid that lays eggs on the fly pupae and destroys them.

11. Similar procedures can be used to treat nematodes. Certain species of nematodes lay their eggs on other nematodes and the larvae feed on the host until it dies. Additional research is needed to identify more parasitoid species and safe application procedures. Biological control can also be applied by using pest-control-specific fungi and bacteria. In the case of nematodes, the fungus *Paecilomydes lilacinus* has been used very successfully (Tabora, 1996).

Unfortunately, biological control is difficult on plantations that also use agrochemicals. These same chemicals are toxic not only to the pest, but also to the biological agent. Staggered applications are needed to gain the optimum benefit and to prevent the reintroduction of pests from untreated plantations (Tabora, 1996). Biotechnology involves alterations of genetic structures of plants, insects, etc., to develop immunity to diseases, fungus, and insects. It holds great potential for all agricultural enterprises, but is not without controversy and ethical considerations.

## IV. VISIONS AND GOALS

Banana production in Costa Rica is currently not sustainable. New and more environmentally friendly, cost-effective, and energy efficient technologies are needed before wide-scale sustainability can be a reality for the producer. Still, the large-scale production of "ecosafe" bananas is difficult, if not impossible, under the ecological conditions found in the humid tropics and the economic philosophy dominating the world markets. Future efforts aimed at sustainability should focus on the application of clean technologies which make use of natural resources in a rational manner and decrease producers' dependency on agrochemicals. Ecotechnology, environmental technologies, and biotechnology can play important roles in this new sustainable banana production paradigm.

At this time, however, the banana industry is suffering economic and environmental problems that are keeping new plantations from being sustainable. New plantations face extremely high interest rates, increasing labor costs, and transnational purchasers who control the quality and quantity of the product that is bought. Currently, to maintain market quality standards, all plantations are forced to use excessive amounts of agrochemicals, which in turn pollute the plantation's and the nation's soils, rivers, groundwater, and laborers.

There are not enough research dollars available to develop the base-line information needed to help producers overcome the economic and environmental problems they are facing. The only sector of the banana industry that is profiting from this cycle is the agrochemical companies. There is little incentive for these companies to develop or support the development of less chemically intensive banana production.

The research and product development that is needed will have to come from within the industry and through partnerships between research institutions in Costa Rica, the United States, and Europe. Together these partnerships are going to have to solicit support form the multilateral developing agencies to initiate a series of interconnected pilot projects focused on specific producer-related problems. Such partnerships will make it much easier to share needed information without worrying about trade secrets or market shares. Research results can be published and shared by all. Universities in Costa Rica has been working with the European Union to solicit just this type of support. The success of this effort will depend upon the leadership within Costa Rica's research community and the support of the donors to achieve a more sustainable banana production system.

Not all transnationals operate without regard for Costa Rica's environment and citizens. Transnational companies play a very important role in the banana industry and should be kept as part of the banana industry in Costa Rica.

Costa Rica, however, must find a way to redefine this historic partnership. Costa Rica needs to be able to manage her resources in a sustainable manner and maintain a profitable partnership with the banana producers and the world's consumers. To do this she will have to become the center for international conventions on sustainable banana production and establish bilateral and multilateral trade agreements to ensure Costa Rica's future as a major banana producer.

Costa Rica must promote the development of domestically owned value-added processing of bananas. There is a tremendous potential for the development of baby food industries and other banana products such as starch, flower, alcohol, vinegar, syrup, juice, and snacks. Costa Rica has a well-educated workforce that is well suited to adopting new high-tech processes necessary to make a valued-added industry grow.

The government of Costa Rica also needs to identify and promote crop diversification and the elimination of monocropping of bananas. This is especially true of marginal lands. The development of the new futures market in the forestry sector is an excellent example of how to creatively manage these lands.

Costa Rica cannot wait for the world community to modify its idea of what a banana should look and/or taste like. The government can no longer be a passive partner in the marketing of bananas or the establishment of research and education programs regarding banana production. Bananas are and should be an important part of the Costa Rican economy. Costa Rica must provide the leadership necessary to make banana production sustainable for the long term.

## REFERENCES

Adams, W. M. 1992. *Green Development: Environment and Sustainability in the Third World*. Routledge, London/New York.

ANAPROBAN. 1995. Proposal to the Executive Branch by the Banana Industry to Jose Maria Figueres, President of Costa Rica, April 1995.

Banco Central de Costa Rica. 1993. Productos de Exportacion. Institute Costarricense de Turismo. San Jose, Costa Rica.

Bourgois, P. I. 1989. *Ethnicity at Work: Divided Labor on a Central American Banana Plantation*. Johns Hopkins University Press, Baltimore, MD.

Casey, G. J. 1979. *Limon, 1880–1940: Un Estudio de la Industria Bananera en Costa Rica*. Editorial Costa Rica, San Jose, Costa Rica.

Chaverri, F., and J. Blanco. 1995. *Inportacion, Formulacion y Uso de Plaguicidas en Costa Rica. Periodo 1992–1993*. Ministerio de Salud and de Agricultura y Ganaderia, San Jose, Costa Rica.

CORBANA (Corporacion Bananera Nacional). 1995. Sigatoka Negra, un grave problema en los bananales de Costa Rica. *Carta Informativa*, Ao 2, N9.

Cunnigham, W. P., and B. Saigo. 1990. *Environmental Science*. Brown Publishers, Dubuque, IA.

ECO-OK/Project Werkstatt. 1996. Promoting ecological landuse in the world. Paper presented at Conference at Bahaus Dessau, Germany, March 1996.

Godefroy, J. 1974. Evolution de la Matiere Organique du sol sous Culture du bananier et de L'ananas. Relations avec la structure et la capacite d'echange cationique. Francia, Nancy. Université de Nancy. pp. 166.

Hall, C. A. S., C. J. Cleveland, and R. K. Kaufmann. 1986. *Energy and Resource Quality: The Ecology of the Economic Process.* Wiley–Interscience, New York. Reprinted 1992, University Press of Colorado, Boulder, CO.

Hall, C. A. S. 1992. Economic development or developing economics: What are our priorities? In M. Wali (Ed.), *Ecosystem Rehabilitation,* Vol. I. *Policy Issues,* pp. 101–126. SPB Publishing, The Hague, The Netherlands.

Hannon, B. 1982. Analysis of the energy costs of economic activities: 1963–2000. *Energy Systems Policy Journal* 6:249–278.

Harwood, R. *et al.* 1992. Sustainable Agriculture and the Environment in the Humid Tropics. Committee on Sustainable Agriculture and the Environment in the Humid Tropics. National Academic Press, Washington, DC.

Hernández, C. E., and S. G. Witter. 1996. Evaluating and managing the environmental impact of banana production in Costa Rica: A systems approach. *Ambio* 25:171–178.

Hilje, I., J. Castillo, L. Thrupp, and I. Wesseling. 1992. *El uso de los Plaguicidas en Costa Rica.* EUNEO, Heliconia, San Jose, Costa Rica.

Instituto Costarricense de Turismo (ICT). 1995. *Annuario Estadistico 1994* [and other volumes]. ICT, San Jose, Costa Rica.

Kepner, C. D. 1935. *The Banana Empire: A Case Study of Economic Imperialism.* Vanguard Press, New York.

Knesse, A. V. 1986. *Measuring the Benefits of Clean Air and Water.* Resources for the Future, Inc., Washington, DC.

Lauer, W. 1989. Climate and weather. In H. Lieth and M. J. A. Werger (Eds.), *Ecosystems of the World 14B: Tropical Rain Forest Ecosystems,* pp. 7–49. Elsevier Science, Amsterdam.

Levitan, L. C. 1988. Land and energy constraints in the development of Costa Rican agriculture. Unpublished M.S. thesis, Cornell University, Ithaca, NY.

Market Research Services, Inc. 1990. The Green Shopping Revolution: How Solid Waste Issues Are Affecting Consumer Behavior. Food Marketing Institute and Better Homes and Gardens Magazine.

McCann, T. P. 1976. *An American Company: The Tragedy of United Fruit.* Crown Publishers, New York.

McKenry, M. W. 1991. The nature, mode of action, and biological activity of nematicides. In D. Pimentel (Ed.), *CRC Handbook of Pest Management,* pp. 461–496. CRC Press, Boca Raton, FL.

Ministerio de Salud. 1994. *Reporte Oficial Intoxicaciones con Plaguicidas, 1994.* Ministerio de Salud, San Jose, Costa Rica.

Mitsch, W., and S. Jorgensen. 1989. *Ecological Engineering: An Introduction to Ecotechnology.* Wiley, New York.

Monge, C. 1995. Costa Rica se Ahoga en Deudas. *La Pensa Libre,* July 11, 1996 p. 3.

Obando, R., and R. Vargas. 1996. Biotransformacion de los Desechos Organicos. In Direccion de Investigaciones y Servicios Agricolas. CORBANA. San Jose, Costa Rica. pp. 29–37.

Odum, H. T. 1983. *Systems Ecology, an Introduction.* Wiley–Interscience, Wiley, New York. Reprinted 1996 by the University Press of Colorado, Niwot, CO.

ONUDI. 1992. Memoria de la Conferencia sobre el Desarrollo Industrial Ecologicamente Sostenible. Organization de las Naciones Unidas para el Desarrollo Industrial, Document: PI/112.

Peakall, D. 1992. *Animal Biomarkers as Pollution Indicators.* Chapman & Hall, London.

Quiros, D., A. Salas, and Y. Leveridge. 1994. In toxicaciones con Plaguicidas en Costa Rica. Centro Nacional de Control de Intoxicaciones, Hospital de Niños. San Jose, Costa Rica.

Ragsdale, N. N., and H. D. Sisler. 1991. The nature, modes of action, and toxicity of fungicides. In D. Pimentel (Ed.), *CRC Handbook of Pest Management,* pp. 461–496. CRC Press, Boca Raton, FL.

Ruiz, R. 1998. Personal Interview at CAB. San Jose, Costa Rica.

Ruiz, R., R. Obando, V. H. Cerdas, and R. Vargas. 1996. Efecto del Efluente de Elaboracion de Compost con Base en Banano de Desecho y Pinzote Sobre la Germinacion y Desarrolo de Plantulas de Maiz y Frijol. In *Direccion de Investigaciones y Servicios Agricolas,* pp. 33–37. CORBANA, San Jose, Costa Rica.

Schwartz, J., and T. Miller. 1991. The Earth's best friends. *American Demographics,* Feb., 13–26.

Soto, M. 1992. *Bananos: Cultivo y Comercializacion.* Litografia e Imprenta LIL, S. A., San Jose, Costa Rica.

Stover, R. H., and N. W. Simmonds. 1987. *Bananas.* Longman Group, UK.

Tabora, P. 1996. Personal Interview. EARTH College, San Jose, Costa Rica.

Twyford, I. T., and D. Walmsley. 1974. The Mineral Composition of the Robusta Banana Plant. Volumes II, III, IV. *Paises Bajos* **41**(3), 493–508.

Vargas, R., and Flores, C. L. 1996. Retribution Nutricional de los Residues Hoyas, Venasdehoyas, Pseudotallo y Pinzote de Banana. (Musa AAA) en fincas de dlfferentesedades de cultivo. *In:* Revista Corbana, pp. 33–48. CORBANA, San Jose, Costa Rica.

CHAPTER 21

# The Costa Rican Coffee Industry

BERNARDO AGUILAR AND JULIE KLOCKER



## I. INTRODUCTION

As stated pervasively in this book, the Costa Rican economy has been and remains largely dependent on exporting agricultural commodities for the generation of foreign exchange. The principal export commodities have been traditionally coffee, bananas, sugar, and meat. Historically, coffee has been one of Costa Rica's chief sources of foreign exchange, generating between 60 and 90% of all foreign exchange earnings in the early part of the 20th century (Seligson, 1980). While coffee's economic importance has diminished with the expansion of other economic activities, it is still critical for the economy in Costa Rica as well as many other developing tropical nations. Between 1972 and 1996, coffee generated 10–36% of total Costa Rican export revenue (Figure 21-1). The importance of coffee to Costa Rica as well as other producing nations cannot be overstated. Many nations that produce coffee lack other cash crops or industry and therefore rely heavily on coffee for foreign exchange and economic stability. In worldwide trade, coffee is the most important agricultural commodity and is second only to petroleum among all commodities traded (Geer, 1971; Smith *et al.*, 1992; Stocke, 1995). Because of its enormous

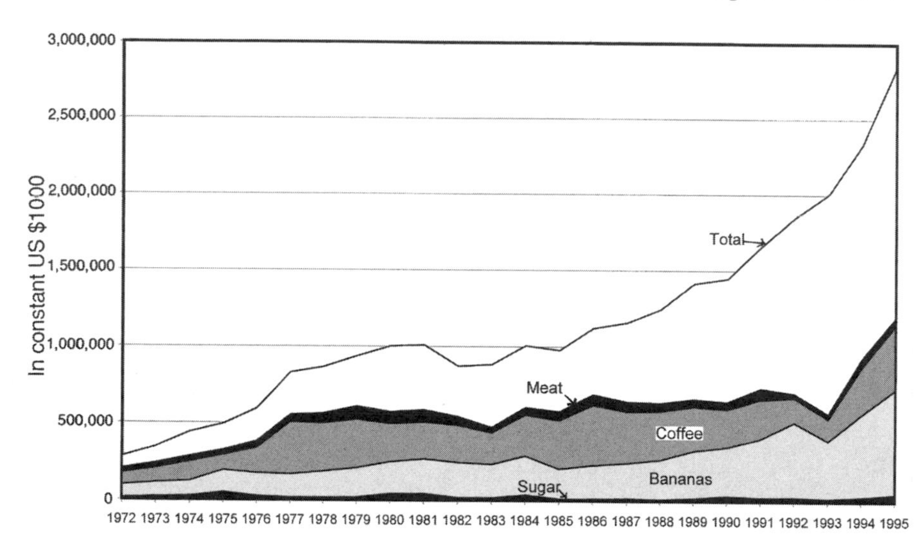


**FIGURE 21-1** Foreign exchange earnings from traditional agricultural products compared to total export revenue (1972–1995).

role in the economy of producing nations and, in fact, the global economy, the sustainability of coffee production is of great concern.

New aspects of the sustainability of coffee production have received considerable attention in recent years. Forest cover is critical habitat for migratory birds. Therefore, many environmental organizations promote "bird-friendly" coffee, otherwise known as "shade coffee," as a sustainable product. In theory, coffee plantations that provide a canopy cover mimic a natural forest by incorporating shade trees. "Ecofriendly" groups also strive for "environmentally friendly" plantations that limit the use of chemicals in coffee production (e.g., Boyce *et al.*, 1993). Finally, there are other groups that address the sustainability of the social welfare of producers by demanding fair pricing for growers (e.g., Appropriate Technology International, 1993).

This chapter adds to the existing literature by examining the long-term sustainability of coffee production both economically and environmentally. We examine various growing methods in order to understand the degree to which each does or does not contribute environmental impacts. In order to achieve this goal we survey the main social, economic, and ecological trends in coffee production in Costa Rica. By using an historical review we examine the role of coffee for the development of political, social, and economic structures, as well as identify the importance of the global economy on a country dependent on agricultural commodities for economic stability. We also use a biophysical approach to determine production as a function of inputs and resource quality. We examine the efficiency of various growing methods to compare the advantages of these methods.

Finally, we examine the possibilities and limits of alternatives to traditional coffee production. This is a particularly important aspect in the discussion of sustainability as the extremely competitive market can generate consequences beyond the limits of traditional economic analysis.

## A. Historical Survey

After independence from the Spanish Crown in 1821, the Costa Rican government found itself very poor with little access to revenue-generating commodities (Seligson, 1980). Other Central American countries had established trade opportunities with Europe with indigo and cochineal dye or precious metals, but the Spanish Crown had banned Costa Rica from producing the dyes and there were no gold or silver deposits there such as in Honduras. So when it was demonstrated that coffee could be grown and marketed successfully, cultivation in the Central Valley expanded rapidly.

Coffee production also expanded throughout the rest of Central America on large haciendas. The hacienda plantation style was the most profitable for the large landowners who benefited from harnessing the large population of indigenous people, initially as slaves and later as wage laborers. Costa Rica was sparsely populated, however, and had few natives or immigrants. Further, as there had not been any major trade or industry, most of the population maintained small parcels of land for subsistence and local market trade (Samper, 1990). As coffee proved profitable, more and more landowners devoted more and more of their own land to coffee production. By 1853, Costa Rica was exporting more than 3 million kg a year (Williams, 1994). As the economy began to grow the government took an active role in the promotion of coffee through generous land reform policies that favored the continuation of the smallholder settlement style.

These land reforms were meant to favor democratic land policies. Over time, however, they often resulted in concentrated landownership. This concentration was due to several factors. There was a labor shortage in Costa Rica so many larger, wealthier plantations offered top wages to laborers. In addition, land prices in the Central Valley were high due to the coffee boom. Many smaller landowners chose to sell their land and work on plantations where they received high wages. To a lesser extent, mortgage foreclosures also encouraged land concentration (Samper, 1990). By the mid-1880s, 71% of the agricultural population were landless laborers (Seligson, 1980). The social structure which emerged by the late 1800s was one of an elite class that monopolized coffee processing, access to credit, and product marketing, and a peasant class of wage laborers.

Coffee became "brown gold" (Seligson, 1980) for Costa Rica, financing infrastructure development as well as acting as a catalyst for trade of other world goods. As smallholders increased the acreage devoted to coffee production, less and less land was used for subsistence crops. By 1854 most of the flour consumed in Costa Rica was imported from Chile (Seligson, 1980). The market became more important and with that emerged a powerful mercantile class. Coffee growers became dependent on coffee as a cash crop in order to purchase required supplies, including food.

While in many areas of Central America the maintenance of the social structure took place through dominance, violence, and repression, the Costa Rican elite maintained their position through a decidedly less violent approach. The coffee elite dominated the political situation by holding important political offices and appeased the campesinos and peasants by enacting minimum wage laws and land reform policies that encouraged settlement in outlying areas of the Central Valley (Samper, 1990), and later, outside the Central Valley (Williams, 1994).

By the 1900s Costa Rica had evolved from an unorganized, financially constrained country to one with a coherent national state, modest wealth, and solid infrastructure. Coffee had paved the way for the sufficiency of the government and it was also because of coffee that Costa Rica become firmly integrated into the world market. This integration, while facilitating development and economic growth, also had a price that became increasingly evident in the early 20th century. Dependency on the global market created vulnerability to global depressions and disruptions. This was particularly the case for the demand for coffee which, because it has no nutritional value, was reduced in times of restrained global economic activity. Coffee was influenced further by global activities because market pricing and distribution controls occur principally in the "north." The vulnerability of the Costa Rican economy to factors outside its control was first experienced during the Great Depression and furthered greatly by World War II when European markets were virtually closed to Costa Rican goods. The global crisis hit the Costa Rican economy through a decline in coffee prices and a subsequent reduction in capital flow (Samper, 1995). Even before these occurrences, however, the instability of the coffee market became evident due to global overproduction and decreased demand that caused fluctuations in price. The Inter-American Coffee Agreement in 1940 attempted to control the irregularity of supply, demand, and pricing. This agreement guaranteed that the United States would purchase the majority of Central American coffee at a low cost in order to stabilize the economies of the region.

Stunned by the vulnerability of its economy, the government of Costa Rica attempted to reduce its dependence on world market systems and a single export crop through the import-substitution measures discussed in Chapters 2 and 23. Along with import substitution, the government had access to United States-supported loans to build internal infrastructure. The Costa Rican government established the Costa Rica Coffee Institute (ICAFE) in 1961 with this increased capital flow. The institute was to regulate internal pricing structures, ensuring farmers were paid fair prices, and conduct research in production and processing methods through a parallel research center (CICAFE). The institute was also to run extension programs to disseminate new technologies and discuss relevant issues with both growers and processors. The research institute appeared very successful as Costa Rica achieved some of the highest coffee yields per hectare in the world.

Meanwhile there were deliberate efforts by smallholders to dilute the financial power of the coffee elite. The most important movement was the establishment of cooperatives in the 1960s after "tense" relations between producers and processors. The more marginal farmers outside the Central Valley had been hardest hit by these relations and provided the initiative for the establishment of coffee cooperatives (Winson, 1989). These cooperatives broke the hold of the coffee elite by establishing their own processing plants and provided credit assistance to members. The cooperative effort was aided by the establishment of the Federation of Cooperatives of Coffee Growers (FEDECOOP) in 1962 which provided necessary marketing and technical assistance. The cooperative movement received further assistance by laws that granted concessions to cooperatives by way of relief from land and customs taxes as well as special electricity rates. By the 1970s about one-third of the coffee processed was done at the cooperative plants (Winson, 1989).

The International Coffee Agreement (ICA) in 1962 (amended in 1968, 1976, and 1983) provided some stability to the market (Marshall, 1985). It was developed after dramatic price increases of coffee in the 1950s prompted increased production that the market could not absorb. The agreements were to regulate the price of coffee by establishing production quotas for coffee-growing countries (Paige, 1997). Costa Rica experienced rapidly increased production during this time (Winson, 1989). By the time of the 1989 negotiations Costa Rica felt that its quota was based on historical production, not current production. Approximately 40% of Costa Rican coffee was being sold to non-ICA markets at significantly reduced prices (Paige, 1997). Several producer countries were following similar practices. Debates for establishing new quota systems ultimately resulted in the collapse of the agreement in 1989. The "stability" provided by the agreement was evident by the fall in coffee prices to depression prices, in real terms (Paige, 1997). However, Costa Rica grows the most desired of all coffee beans, the mild arabica, and has perfect climatic growing conditions and rich volcanic soil. Good growing conditions result in large beans that contribute to excellence in flavor. The result is that Tico coffee is among the best in the world. Therefore, although international markets may be turbulent, Costa Rican coffee remains a much sought after commodity.

But the oil crises in 1973 and 1979 and the ensuing global recession generated the most protracted recession that Costa Rica had experienced since independence. Its direct effects lasted into the late 1980s and the indirect effects, including debt, continue today. The International Coffee Agreement fell apart in the late 1980s, causing further hardship on both growers and the country as a whole. While technically Costa Rica has recovered from the recession, the coffee market continues to be plagued with problems of global overproduction and wavering demand. These factors result in unstable price levels for coffee, while production in Costa Rica has varied only slightly (Table 21-1). Such problems continued even after the creation of a new producers association in early 1995 (Association of Coffee Producing Countries). Costa Rica has decreased its dependence on coffee as the major source of foreign exchange by increasing banana production, nontraditional agricultural production, industry, manufacturing, and tourism. Nevertheless, coffee remains an important part of the socioeconomic fabric of the country and generated 12.8% of foreign exchange in 1996 (Figure 21-1).

## B. Producing a Cup of Coffee in Costa Rica

Coffee has its origins in Africa where there are over 20 species of the genus *Coffea* (Anthony *et al.*, 1993). Today, only two of these species, *Coffea canephora,* also known as *robusta,* and *Coffea arabica,* are used for marketable

**TABLE 21-1 Coffee Production and Pricing**

| Growing season | Production (millions 60-lb bags) | Price/60-lb bag (U.S. dollars) |
|---|---|---|
| 1992/1993 | 2.662 | $ 81.10 |
| 1993/1994 | 2.475 | $103.22 |
| 1994/1995 | 2.472 | $230.00 |
| 1995/1996 | 2.451 | $159.00[a] |
| 1996/1997 | 2.376 | |
| 1997/1998 | 2.55[a] | |

Data from USDA Annual Coffee Report: Costa Rica, various years.
[a] Estimate.

coffee production (Cros *et al.*, 1993; Charrier and Berthaud, 1985). About 75% of the world coffee traded is arabica (van der Vossen, 1985). The robusta species is more resistant to diseases such as leaf rust, *Hemileia vastatrix,* but the beans are more bitter and have lower acidity than arabica species (Clifford, 1985). Robusta is used mainly for instant and bulk coffee, and as a base for blends (Clarke, 1985).

Coffee is, in its natural state, a subcanopy plant, producing few flowers and therefore few beans (Cannell, 1985). Through agronomic manipulation, coffee was found to be more productive using interspersed shade cover. This system provided a microclimate that protected the plant from exposure to wide ranges of temperature variations while encouraging more flowering than complete shade. Before the 1950s the arabica variety was widely used (Rice and Ward, 1996). Coffee plantations included timber or fruit trees planted for cover, which also buffered farmers from the loss of harvest due to the cyclical nature of coffee plants.

A new variety of arabica coffee called caturra was introduced from Brazil in the 1950s. This variety responded very well to fertilizer applications, and yields increased up to 30%. Shade trees could be eliminated and plant densities increased from approximately 1200 to between 4000 and 7000 per hectare to increase yields even further (Rice and Ward, 1996). Using this variety and aided by Costa Rica Coffee Institute research and dissemination efforts, Costa Rica achieved among the highest yields per hectare in the entire world (Winson, 1989) (Figure 12-4c). From 1970 to 1980 the United States invested more than $80 million in USAID projects for coffee producers in Central America in an effort to increase coffee production (Rice and Ward, 1996).

While coffee is grown in many areas of Costa Rica, we focus on the Central Valley as it was the first, and traditionally is seen as the prime, area for coffee production. The Central Valley provides excellent growing conditions for the preferred arabica varieties of coffee, such as Caturra, Catimor, and Catuai. Earlier volcanic activity resulted in soils rich in nutrients. The climate is optimal in essentially all respects, so that the coffee is at the "center of its gradient space" (Figure 12-10). Temperatures are around 19°C with little risk of frost. The rainfall of between 1000 and 3000 mm/yr is also optimal. Although coffee will grow nearly anywhere in Costa Rica, the elevations below 500 and higher than 1700 m above sea level are decidedly nonoptimal (ICAFE, 1989). In elevations over 1000 m, the typical afternoon cloud cover shields the plant from otherwise excessive heat and light. Since permanent shade decreases yields due to reduction of solar radiation on flowering nodes (Cannell, 1985), most plantations in the higher areas are unshaded. The rows of coffee are planted very close together so that the lower branches are shaded by the upper branches, forming a self-shading system. This system relieves stress on the plant while allowing flowering nodes to receive sunlight. ICAFE recommends plantings of up to 7000 plants/ha in the Central Valley and fertilizer applications in the range of 500 to 1000 kg/ha/yr of complete fertilizer, or 300 kg/ha/yr of nitrogen, when grown without shade (ICAFE, 1989). Harvest rates with this application range between 1.0 and 1.5 tons/ha. Essentially, 1 ton of fertilizer worth about $350 is traded for 1 ton of coffee worth about $2600 in trade.

As one moves away from the optimal conditions into suboptimal areas, more management options may be necessary to maintain production. ICAFE recommends that shade trees be incorporated into the coffee plantation in lower elevation areas, such as Turrialba. The trees function in three ways: first by creating a microclimate that regulates temperature, wind, and moisture extremes; second, by accumulating biomass that serves as a source of nutrients when used as litter; and third, often times, by fixing nitrogen with some species. These lower elevations are warmer and receive more direct sunlight than the highlands. The shade trees lower the respiration rates for the coffee plants, thus allowing them to use less energy for their maintenance (Beer, 1987). In areas of steep terrain, tree roots decrease soil erosion by providing stability to the soil. In the Turrialba region the recommended density of poro (*Erythrina* spp.) trees, one of the most common nitrogen-fixing trees, is 220 trees/ha. Recommended rates of fertilizer application are lower, 700 kg/ha. The poro trees provide approximately 30 metric tons of biomass a year and are commonly cut twice a year, the biomass allowed to decompose in place.

The system of using trees is also recommended to buffer small farmers against market price fluctuations. When the market prices are low, the shade trees are not trimmed, and therefore the rate of fertilizer application can be decreased. Another consideration is that small farmers can receive other financial benefits from trees if the species used is one of economic importance, such as for a timber or fruit-bearing tree. A negative aspect of incorporating trees is that moisture retention is increased, which increases the incidence of fungus problems. In unshaded areas of the west side of the Central Valley fungicides may be used only three times a year whereas rates increase to six or seven times a year in the more shaded plots. The same type of relation may exist with pests, requiring more inputs of pesticides and nematocides.

Recently there has been much discussion regarding the environmental health of full-sun versus shade coffee plantations. The development of these concerns is coupled with a myriad of seemingly contradictory studies on yields, erosion, leaching, and other aspects of sun and shade plantations. In fact, most of these studies do not contradict one another. The primary explanation of the seemingly contradictory results are the site-specific considerations of climate and soil (Figure 21-2). In other words, there is no one method or condition that dictates the growing method. In addition to site-specific considerations there are a number of socioeconomic concerns for the coffee grower that also influence plantation style (Figure 21-3). Another factor is that there

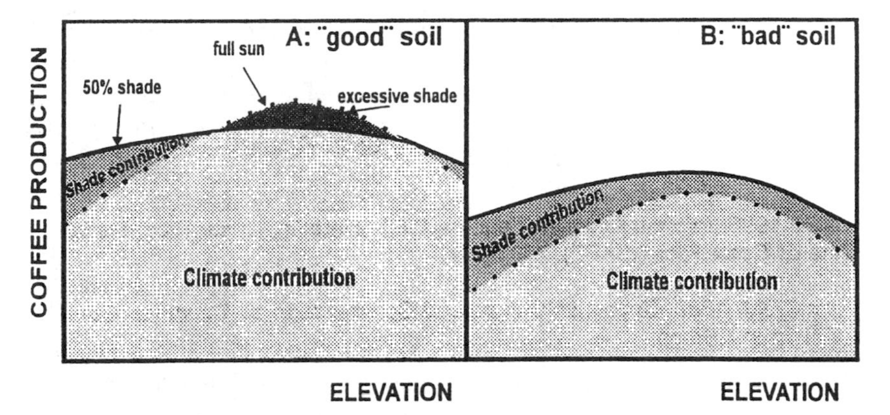


FIGURE 21-2 Idealized hypothetical coffee production in sun and under trees giving 50% shade as a function of elevation without (A) and with (B) limitations of rooting depth, nutrients, or moisture.

are trade-offs in farmers' objectives. According to Muschler (1997) most literature on coffee is in agreement that the production in diverse, shaded systems is both lower and typically more stable than that in sun. Small farmers are interested in both stability and economic return.

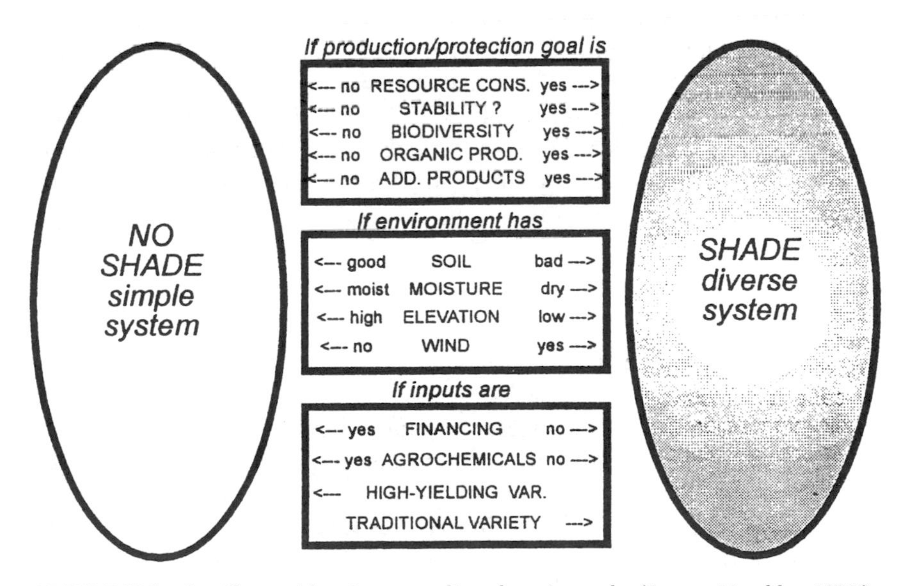


FIGURE 21-3 Specific considerations regarding plantation style. (Source: Muschler, 1997.)

It is normally 5–6 years after the initial planting before a profitable harvest is achieved. Seedlings require 1 year in a nursery and then are transplanted and maintained for 4 to 5 years. Berries will develop after several years, but not enough for a profitable harvest. From the time of transplantation, the crop requires intensive weeding, pruning, pest control, and fertilizer applications, as well as eventual harvest activity. The average life span for the typical caturra coffee plant is 30 years (Haarer, 1963).

The labor requirements for the production of coffee are large, larger than essentially all subsistence crops and most other export crops. The labor required to maintain a hectare of coffee is about five times that of beans, three times that of corn, and two times that of sugar or cotton (Williams, 1995). For every hectare brought into production, 706 man-days are required before initial harvests are experienced. Once land is dedicated to the production of coffee, it precludes that land from rapid turnover to other crops even if coffee prices fall. Coffee plants are productive for approximately 15–30 years and are expensive to replace. If there is market turmoil, farmers may opt to maintain plantations at a minimal level, which still requires intensive labor, by cutting back the trees. When prices rise the trees recover rapidly by coppicing.

The processing element in coffee production is also very intensive. Harvesting occurs from July through March, depending on elevation and weather. Once the cherries (the bean with the husks) are picked they must be transported to the processing mills within 24 hours or the beans will sour. The first step in processing separates the healthy from unhealthy cherries. Traditionally, pulpers stripped the outer layer and the mucilage, which is high in sugar content, through fermentation. This process uses enormous amounts of water and the waste products were traditionally channeled into nearby rivers. Now the cherries are stripped mechanically and the mucilage is removed immediately, resulting in a better tasting bean. The beans are then dried slowly using the outer layer, or endocarp, as a fuel source. After this process the unroasted "green bean" is ready for export.

## C. The Environmental Impact of Coffee Production and Processing

While it is true that Costa Rica achieved some of the highest yields per hectare in the world through intense management and industrialization, the process was accompanied by changes in the landscape and environment. This modernization of coffee production began in the 1950s when new varieties of coffee were engineered which resulted in increased yields but only when supplemented with large inputs of nitrogen fertilizers and sometimes pesticides. One of the consequences of that process was the removal of shade trees in the Central Valley when the caturra species was introduced to Costa Rica.

The introduction of the use of chemicals introduces environmental and health concerns. Overapplication of nitrogen fertilizers results in runoff of excess nutrients into local water supplies. Application without proper clothing or equipment jeopardizes human health. Although coffee, as a perennial crop, stabilizes the soil much more than annual crops, erosion appears to increase when plantations are converted to full sun (366 kg erosion per hectare per six months), shade *Erythrina* (59 kg/6 months), or shade *Cordia* (104 kg/6 months) (Beer, 1985). The rate of erosion may be decreased, however, with proper management practices such as the use of mulch and ground cover.

Fungicides are required to protect the plants against infestations. The FAO estimates that in 1984 the recommended application of 3.7 kg/ha/yr resulted in the application of 213 tons/yr in the Tarcoles River Basin, where most of the coffee is grown. This figure assumes that applications correspond with recommendations, but it is believed that overapplications are fairly common (Coto, 1993).

One of the most environmentally destructive aspects of production is the processing of coffee. Almost all coffee in Costa Rica undergoes the wet method of processing which uses very large amounts of water in order to break down the mucilage (Figure 21-4). When the water is returned to the creek it carries with it contaminants. In earlier decades the waste was 276,000 kg/day, which was more than half of the organic contamination load of the main watersheds and more than double that of all other industrial and household wastes (TSC, 1982). Furthermore, most of the contamination to the rivers occurs during the dry season when rivers are at the lowest level, resulting in the formation of anoxic areas. The reduction in dissolved oxygen almost certainly "affects the density and diversity of aquatic fauna" (Coto, 1993).

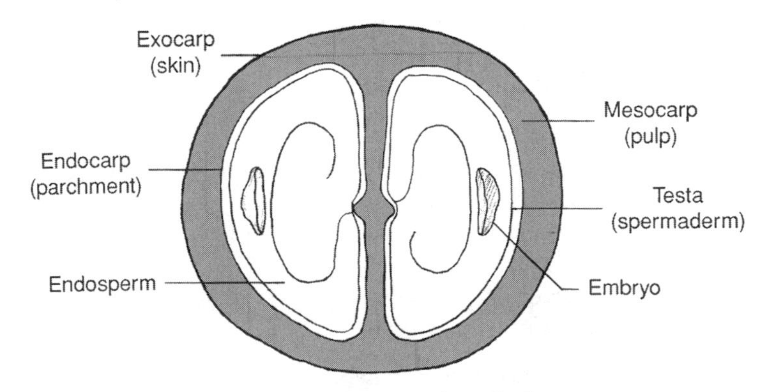


FIGURE 21-4 Diagram of a coffee berry.

There have been many investigations into alternative uses for the organic wastes. Endocarps are now dried and used as fuel for processing. The dried pulp is used as a mulch on coffee plantations. There is further investigation as to whether the by-products of anaerobic digestion can be harvested successfully as an alternative source of fuel. The ICAFE, in conjunction with several other national branches of the government, is mandating the inclusion of water treatment and sludge handling at all processing plants (Agricultural Engineer Rolando Vasquez Morera, personal communication).

## II. HOW SUSTAINABLE IS SUSTAINABLE? A BIOPHYSICAL PERSPECTIVE

An aspect that has not received a lot of attention in previous sustainability analyses of this crop is energy efficiency. Little is known on the intensity of energy throughput across the different stages of the whole cycle of production. Previous analyses of coffee sustainability have been limited to the examination of traditional social–environmental impacts, including economic profitability, pesticides and their effects on the environment, the health of humans and other species, water pollution, soil erosion, and the environmental reputation of the country (Boyce *et al.*, 1993). A perspective of energy and materials throughput and its relation to environmental degradation is necessary in addition to these traditional approaches. For example, land degradation may increase as production is expanded into suboptimal areas due to market incentives or land use changes. It may also increase as more intensive technology packages are adopted to increase productivity in the areas already planted.

Newer production technologies may become increasingly more dependent on energy inputs. If the same holds true for current and future production, this would increase the dependence level of Costa Rica significantly since it imports all of its crude oil and petroleum-derived products.

The following questions seem to be especially important in relation to Costa Rica's dependence on coffee production:

1. How efficient is coffee production in Costa Rica in terms of the tons of coffee produced or dollars of revenue generated per unit of energy input and how has this changed over time?
2. Does production in suboptimal areas require higher levels of inputs to compensate for suboptimal land qualities?
3. Do alternative cropping systems offer more energy-efficient options for the management of this crop?

There has been a lack of this type of analysis in Costa Rica except for a few unpublished studies. Therefore, we attempt to answer these questions based on an assessment of the energy efficiency for the different production and processing systems mentioned.

## A. Methods

As previously suggested in Chapter 3 and others, biophysical benefit–cost accounting can be done through quantifying the relation between outputs and inputs, in a fashion similar to an economic benefit–cost ratio but in terms of energy. Hall and Hall (1993, p. 4) state, "a valid index of productivity is energy efficiency, or productivity per unit of energy used, since more than 99 percent of the energy used in the process of economic production is not human labor but industrial use (i.e. from fossil fuels, nuclear and hydropower)." Daly (1990) also suggests an ecological economic efficiency measurement using the ratio between man-made capital services gained and natural capital services sacrificed. This relationship of outputs and inputs can be formalized as

$$EE_t = QE_t/EU_t$$

where $EE_t$ is energy Efficiency at time t, $QE_t$ is the output from time t in energy units, and $EU_t$ is Energy used during time *t*.

All inputs and outputs, in physical units, are converted into energy units while correcting for the quality differences between the different energy sources used. Cleveland (1991, 1995) defines the energy use component as

$$EU_t = DE_t + IE_t$$

where $DE_t$ is quantity of energy used at that site from sources such as fossil fuels and electricity in the economic activity during time *t*, and $IE_t$ is indirect energy used in the process during time *t*.

Indirect energy use is the energy used elsewhere in the global economy to produce inputs. The advantage of including indirect energy is to account for the complete use of energy in the system. This embodied energy approach, as suggested previously, also accounts for environmental impact in a more comprehensive and hence accurate way.

Our energy efficiency analysis examines the entire industry, including both the production and the processing of the beans to the "green bean" stage at the port. The driving motivation of the entire process is to obtain a green bean of exportable quality. Therefore, all decisions that effect energy efficiency are made with that objective in mind. For example, a farmer may decide to increase the use of fertilizers to achieve the bean size that is required by the processor, or "beneficio," standards because the beneficio wants only those beans of high quality that sell well in the international market.

As our previous explanation suggests, the coffee cycle can be divided into two main subsystems or stages: production and processing (Figure 21-5). Energy and materials flow into the subsystems and yield green coffee beans as a final product, as well as waste and heat emission by-products. The entire cycle can be divided into several substages (Figure 21-6). The forms of energy used principally during the production stages are indirect. Given the traditional technological package (ICAFE, 1989; Rojas, 1996), actvities such as fertilization, weed and disease control, and fungus and pest control use energy mostly indirectly in the form of agrochemicals. The only energy used directly in production is diesel or electric pumps for irrigation during the dry season. After the harvest, more direct energy is used, supplied by fossil fuels, electricity, and biomass for drying and processing. There is also some embodied energy in machinery and some energy used indirectly for lubrication. Figure 21-6 does not include the human energy as work used in all stages, nor the energy required to support or transport the workers.

The first constraint that affected our analysis was the lack of consistent agroindustrial input and energy data bases. This has affected several studies done for Costa Rica to date (Aguilar *et al.*, 1994, 1996; Hartman *et al.*, 1996; Cague *et al.*, 1997). None of these studies could account for energy use before the 1979–1980 harvest year. The necessity of using proxy variables due to lack of explicit data available for specific input variables has yielded efficiency estimates that are not necessarily reliable. For example, Aguilar *et al.* (1994) analyzed the energy efficiency for specific coffee-producing regions in Costa

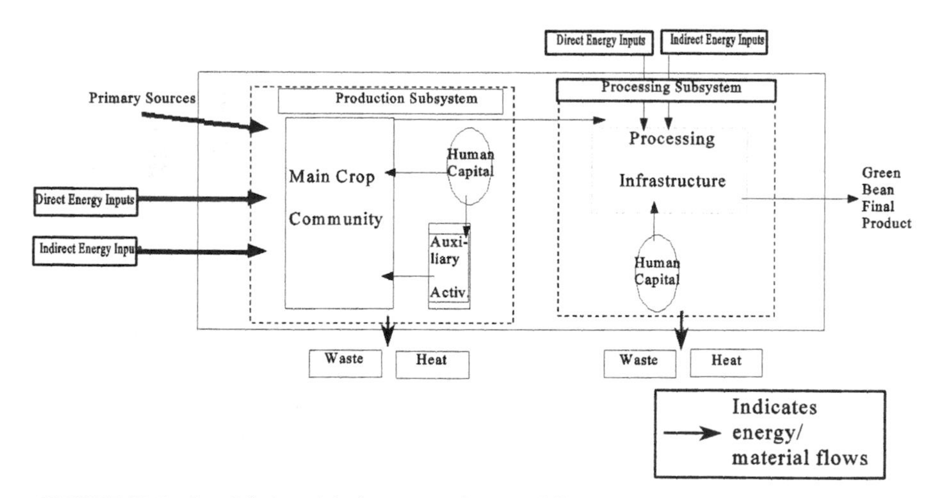


FIGURE 21-5 Simplified model of energy and material flows in an agroindustrial coffee system in Costa Rica assuming a competitive process. Primary sources are those that do not require human intervention to render energy services to the system: sun, rain, and so on.

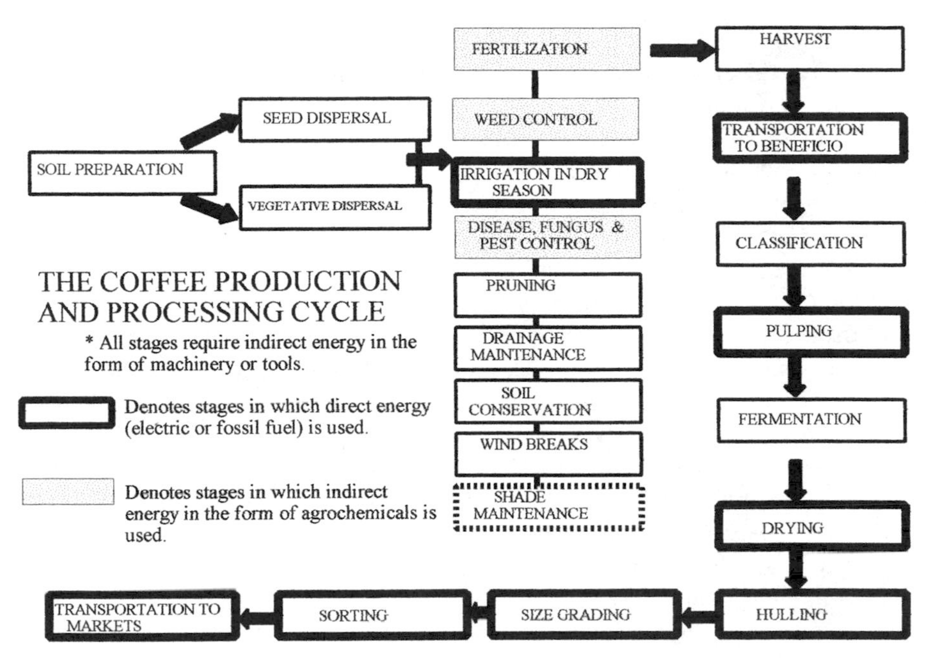


FIGURE 21-6 Detailed description of the coffee production and processing cycle, including the types of energy most frequently used at each stage.

Rica. The study was able to account only for the direct electricity and fossil fuel used. Since the largest component of energy used directly in coffee processing is firewood and coffee endocarps, there may be substantial error in the analysis. An additional problem is that the estimates of direct electric energy use are based on aggregate figures for the entire agricultural sector in the regions analyzed, not on numbers specific to coffee. Also, adjusted total county fuel use was a proxy for specific use in coffee production in the absence of better data.

Later studies have tried to solve these limitations through substantial field data collection and careful use of secondary sources of data. This was our approach.

We estimated energy efficiency in two ways. First, we estimated it in terms of metric tons produced per terajoules of inputs (Ton/TJt). In order to account separately for the combined effect of international price fluctuations and monetary policies in the country, we also estimated efficiency in terms of dollars of revenue per terajoules of inputs ($/TJt). In order to fully understand the different influences that different inputs have on each stage of the cycle, we also calculated these two versions of energy efficiency for the production and processing stages. Therefore, we report estimates for efficiency in dollars per terajoule in production ($/TJp) and processing ($/TJb). The same division is done for efficiency in tons per terajoule in production (Ton/TJp) and efficiency in tons per terajoule in processing (Ton/TJb).

In the estimates of dollars per terajoules of inputs ($/TJt, $/TJp, and $/TJb) we kept our numerator in nominal or current terms (not deflated). We do this to capture the effects of exchange rate fluctuations on production and processing decisions.

We analyzed production for the 1980/1981 through 1995/1996 harvest years. Our accounts include energy consumed directly in the transportation of the beans to the beneficio and materials to the farms. We also estimated the indirect energy included in fertilizers (soil and foliar) and other agrochemicals (fungicides, nematocides, insecticides, herbicides, and moisturizers). We use Hannon *et al.*'s (1977) data for the energy intensity of fertilizers and other agrochemical products, requiring that we deflate all our dollar values to 1977 dollars. To do this we derived an index of price changes for these products based on price reports from the Central Bank of Costa Rica. Intensities per volume for direct energy inputs were taken from Costa Rica's Ministry of the Environment reports.

Original data for production in Costa Rica are reported by ICAFE in fanegas. A fanega is the unit of volume used by processing plants or beneficios when receiving coffee from producers. It is equivalent to 2 double hectoliters and weighs 558 lb. When processed according to present standards it yields approximately 100 lb of dried green beans. We converted these figures to their equivalents in metric tons using 1 T = 19.68 fanegas.

Data for coffee processing efficiency are taken from Cague *et al.* (1997) and include the 1984/1985 through 1995/1996 periods. They gathered data for all of the processing plants in Costa Rica from ICAFE. We converted their data to dollars and tons per terajoule of input for consistency. These figures account only for direct energy inputs.

Estimates of firewood and endocarp use were derived by qualified personnel from ICAFE and from some beneficios included in their study. This yields rough estimates of the quantity used for each year. We assume that the variability in energy content of these fuels was small. We made a small correction in Cague *et al.*'s (1997) use of producer prices when calculating $/TJb.

Boyce *et al.* (1993) examined the accuracy of official Costa Rican cost statistics in representing reality, since their different time series are not consistent, and for some periods the only data available are that of cost models. We use cost-based estimates of inputs for the periods after 1989/1990 for lack of better data. ICAFE officials claim that the published data reflect a combination of technological recommendations and the real costs of production that they collect annually. Further, they claim their data reflect the effects of input price fluctuations and general inflation over the production cycle (Rojas, 1996). Therefore, we consider this information to have some utility.

## B. Results for the Production and Processing Stages

The efficiency of coffee production in terms of dollars per terajoule of energy used ($/TJp)decreased by 57% between 1980/1981 and 1995/1996 (Figure 21-7 and Table 21-2). There is a similar decline of 60% (Table 21-3) for tons produced per terajoule of energy used (Ton/TJp). This decrease in efficiency was accompanied by a 29% increase in production and a 50% decline in real producer prices.

To understand this trend better we disaggregated Ton/TJp according to the different inputs used. Fertilizer inputs are the largest component of Ton/TJp, on average, 72% of the total energy use throughout the period (Figure 21-8). Their use tended to increase over time from 34 to 73 GJ per ton produced (a 114% increase).

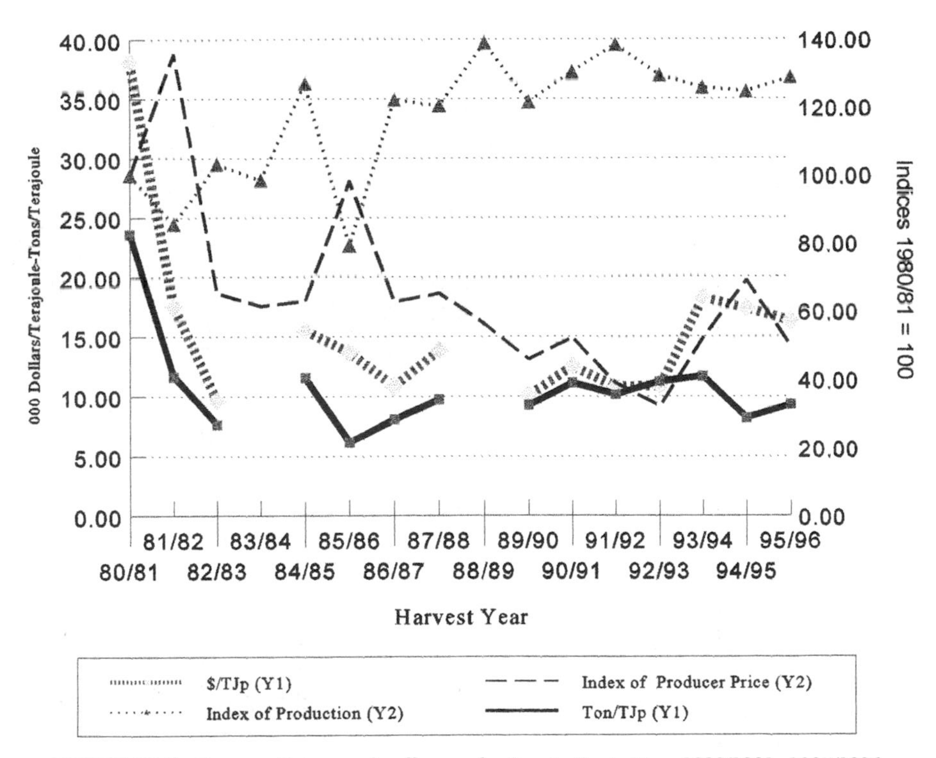


FIGURE 21-7 Energy efficiency of coffee production in Costa Rica, 1980/1981–1995/1996.

TABLE 21-2 Comparative Figures for Energy Efficiency in Dollars per Energy Used in Coffee Production and Processing and Costa Rican Agriculture[a]

| Harvest year | Coffee production ($/TJ) | Coffee production ($/TJ, direct) | Coffee process. ($/TJ) | Coffee process. ($/TJ, direct) | Total coffee ($/TJ) | Total coffee ($/TJ, direct) | Agriculture ($/TJ, direct) |
|---|---|---|---|---|---|---|---|
| 1980/1981 | 38,104.14 | 838,128.90 | N/A | N/A | N/A | N/A | 101,700.00 |
| 1981/1982 | 17,526.98 | 740,421.37 | N/A | N/A | N/A | N/A | 100,840.00 |
| 1982/1983 | 9,603.83 | 535,828.19 | N/A | N/A | N/A | N/A | 126,460.00 |
| 1983/1984 | N/A | N/A | N/A | N/A | N/A | N/A | 138,950.00 |
| 1984/1985 | 15,490.43 | 477,236.86 | 156,506.77 | 156,506.77 | 14,095.33 | 117,920.42 | 140,590.00 |
| 1985/1986 | 13,669.17 | 497,773.28 | 178,922.88 | 178,922.88 | 12,699.00 | 131,692.50 | 144,850.00 |
| 1986/1987 | 10,792.02 | 436,918.62 | 108,277.84 | 108,277.84 | 9,813.87 | 86,901.69 | 137,360.00 |
| 1987/1988 | 13,839.03 | 533,111.20 | 116,238.52 | 116,238.52 | 12,366.69 | 95,512.98 | 135,030.00 |
| 1988/1989 | N/A | N/A | 113,681.58 | 113,681.58 | N/A | N/A | 141,420.00 |
| 1989/1990 | 10,060.53 | 215,451.36 | 87,210.57 | 87,210.57 | 9,020.00 | 62,100.54 | 145,730.00 |
| 1990/1991 | 12,468.54 | 248,620.54 | 91,667.43 | 91,667.43 | 10,975.64 | 66,996.32 | 139,770.00 |
| 1991/1992 | 10,789.56 | 232,853.11 | 86,281.28 | 86,281.28 | 9,590.29 | 62,970.10 | 151,270.00 |
| 1992/1993 | 10,834.91 | 171,806.36 | 78,901.69 | 78,901.69 | 9,526.69 | 54,089.12 | 153,210.00 |
| 1993/1994 | 18,263.26 | 242,446.07 | 126,190.93 | 126,190.93 | 15,954.25 | 83,024.56 | N/A |
| 1994/1995 | 17,296.28 | 201,665.77 | 174,064.40 | 174,064.40 | 15,732.94 | 93,447.47 | N/A |
| 1995/1996 | 16,266.31 | 207,045.91 | N/A | N/A | N/A | N/A | N/A |

[a] The coffee year begins on October 1. Agriculture is presented in calendar years, therefore it is matched with the harvest year, which includes the majority of the corresponding calendar year.

TABLE 21-3 Energy Efficiency in Tons per Megajoule of Energy Used in the Coffee Cycle[a]

| Harvest year | Total (ton/TJ) production | Total (ton/TJ) processing | Total (ton/TJ) | Ton/TJ fertilizers in production | Ton/TJ other agrochemicals in production | Ton/TJ fossil fuels in production | Ton/TJ firewood in process. | Ton/TJ endocarps in process. | Ton/TJ electricity in process. | Ton/TJ fossil fuels in process. |
|---|---|---|---|---|---|---|---|---|---|---|
| 1980/1981 | 23.60 | N/A | N/A | 29.36 | 156.78 | 519.94 | N/A | N/A | N/A | N/A |
| 1981/1982 | 11.63 | N/A | N/A | 13.44 | 104.89 | 491.21 | N/A | N/A | N/A | N/A |
| 1982/1983 | 7.63 | N/A | N/A | 10.09 | 33.76 | 425.68 | N/A | N/A | N/A | N/A |
| 1983/1984 | N/A | N/A | N/A | N/A | N/A | N/A | N/A | N/A | N/A | N/A |
| 1984/1985 | 11.54 | 82.67 | 10.13 | 13.97 | 81.64 | 355.54 | 167.99 | 233.93 | 1184.31 | 938.95 |
| 1985/1986 | 6.15 | 80.27 | 5.71 | 7.74 | 34.60 | 224.00 | 167.99 | 233.93 | 1026.23 | 801.40 |
| 1986/1987 | 8.06 | 80.90 | 7.33 | 12.63 | 23.94 | 326.44 | 167.99 | 233.93 | 1103.64 | 829.24 |
| 1987/1988 | 9.74 | 81.79 | 8.70 | 12.44 | 50.85 | 375.10 | 167.99 | 233.93 | 1072.33 | 946.96 |
| 1988/1989 | N/A | 81.65 | N/A | N/A | N/A | N/A | 167.99 | 233.93 | 1131.15 | 887.66 |
| 1989/1990 | 9.28 | 80.42 | 8.32 | 13.85 | 32.72 | 198.67 | 167.99 | 233.93 | 1067.28 | 729.99 |
| 1990/1991 | 11.12 | 81.76 | 9.79 | 16.33 | 41.38 | 221.76 | 167.99 | 233.93 | 1106.32 | 912.66 |
| 1991/1992 | 10.15 | 81.19 | 9.02 | 15.22 | 35.42 | 219.12 | 167.99 | 233.93 | 1090.37 | 855.83 |
| 1992/1993 | 11.22 | 81.70 | 9.86 | 17.33 | 38.76 | 177.90 | 167.99 | 233.93 | 1023.91 | 969.21 |
| 1993/1994 | 11.65 | 80.49 | 10.18 | 18.98 | 37.48 | 154.64 | 167.99 | 233.93 | 1037.97 | 814.35 |
| 1994/1995 | 8.19 | 82.43 | 7.45 | 12.24 | 33.43 | 95.90 | 167.99 | 233.93 | 1225.92 | 921.94 |
| 1995/1996 | 9.32 | N/A | N/A | 13.64 | 39.09 | 118.59 | N/A | N/A | N/A | N/A |

[a] Values of MJ/Fanega in processing are from Cague *et al*. (1997) and converted to metric tons/TJ of energy used.

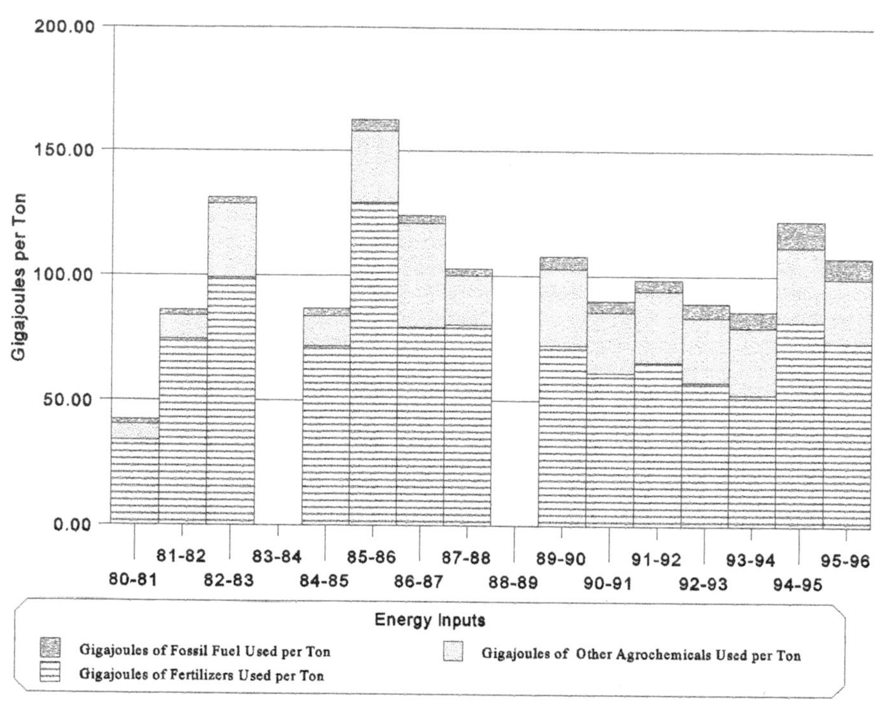


FIGURE 21-8 Energy inputs, production levels, and producer prices in coffee production, 1980/1981–1995/1996.

Other agrochemical inputs account for an average 23% of total energy use. They increased by 300% from 6.4 to 25.6 GJ used per ton of production. Fossil fuel intensity also grew through the period, going from 1.9 to 10.5 GJ per metric ton produced (a 443% increase). It accounted for 4.6% of the energy costs, on average, for the period.

Coffee farmers say that fluctuations in fertilizer and agrochemical applications are driven partly by the price received for coffee. Specifically, ICAFE reports that for the 1992/1993 and 1993/1994 crop years producers applied fertilizers at levels that were between 50 and 75% of the recommended quantities of the coffee bureau due to a decrease in profitability following the downfall of the International Coffee Agreement in 1989 (Alvarado, 1994; ICAFE, 1995). Nevertheless, as Figure 21-7 and Table 21-2 show, there were no substantial increases in Ton/TJp or $/TJp during those periods in which real producer price was at its lowest levels.

It is clear though that after the recuperation in prices in the middle nineties, $/TJp increased up to $16,266/TJ. Each ton of the product was worth more. Yet, the level of production was maintained through a more than proportional increase in energy use that actually lowered the biophysical efficiency to 9.32 tons/TJ of energy used.

Results of the processing stage do not show much variation in energy efficiency in tons processed per terajoules used (Ton/TJp) (Table 21-3). The largest components of the energy expense accounted for in the processing stage are firewood and endocarps [an average 84% of the energy cost per metric ton (Figure 21-9) per unit]. Efficiency, in U.S. dollars gained per terajoule of energy used in processing ($/TJp), decreases consistently from 1985/1986 to 1992/1993 (from $178,923 to $78,902/TJ). This 126% decrease is correlated with price fluctuations, which show a significant influence on $/TJp ($p = 0.001$) during the period analyzed (Figure 21-10).

## C. Results of the Entire Cycle

We derived crude estimates for the efficiency of the complete coffee cycle in Costa Rica from combining the preceding data for the 1984/1985 to 1994/1995 period (Figure 21-11). Total efficiency, in tons per terajoule (Ton/TJt), shows a significant inverse dependence on the level of coffee production ($\beta = 2.66 \times 10^{-6}$, $p = 0.0102$). Because fluctuations in Ton/TJt show a parallel trend to variations in production, it follows that Ton/TJt will vary inversely

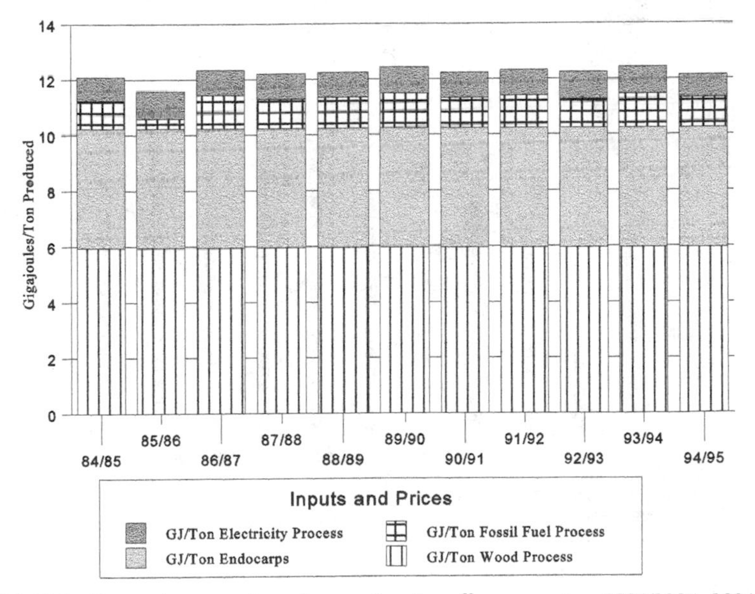


FIGURE 21-9 Energy inputs and producer prices in coffee processing, 1984/1985–1994/1995.

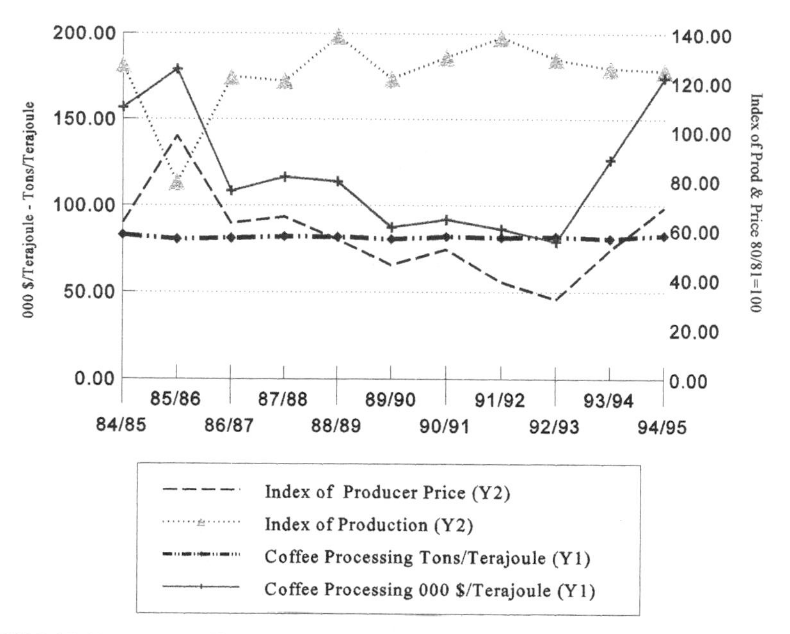


FIGURE 21-10 Energy efficiency of coffee processing in Costa Rica, 1984/1985–1994/1995.

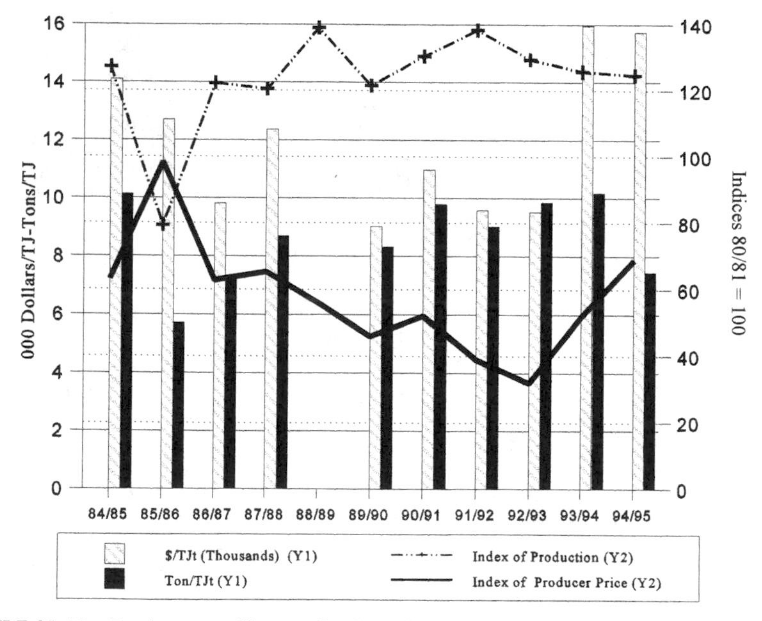


FIGURE 21-11 Total energy efficiency for the coffee production and processing cycle in Costa Rica, 1984/1985–1994/1995.

to price changes, a trend that also proved significant ($\beta = -5.82 \times 10^{-2}$, $p = 0.015$). Total efficiency in dollars per terajoule of energy use ($/TJt) did not show a clear pattern or significant correlation to production or real producer prices. This was probably an effect of inflation distortions.

The main energy costs for coffee are imported inputs. The inclusion of machinery costs would increase this. On average, fertilizers represent 62.5% of the total energy costs accounted for per each metric ton produced and processed. Other agrochemicals used are second (22.61%). Direct fossil fuel use is 4.72% and firewood 4.68%. All agrochemicals (fertilizers and others) are 86% of the energy costs. Biomass sources and electricity, which might be seen as more sustainable sources of energy in Costa Rica, are among the lowest components of total energy intensity (3.36 and 0.72%, respectively).

Given the importance of fossil fuels and electricity for Costa Rica, it seemed relevant to explore specifically the effects of direct energy prices on efficiency (Figure 21-12). We found no significant correlations. Other factors seem to have an influence on the behavior of direct energy efficiency. For example, during the period of real price increases after 1992/1993, efficiency increased in terms of dollars per terajoule of direct energy consumed by almost 40%. In this same period, electricity prices increased in real terms by almost 10%.

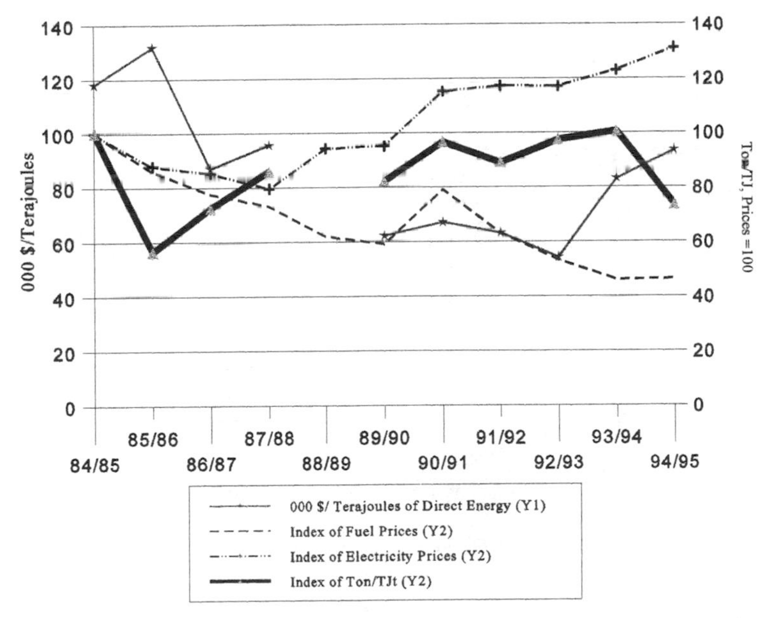


FIGURE 21-12 Direct energy efficiency in the coffee production and processing cycle and its relation to energy prices.

A parallel decrease in real fossil fuel prices also occurred. But Ton/TJt did not increase. In fact, from 1993/1994 to 1994/1995 it decreased by almost 25%.

Chapter 12 presented estimates of the relation between the efficiency of coffee production and that of Costa Rican agriculture. Aguilar *et al.* (1996) derived estimates for the direct energy efficiency of Costa Rican agriculture that allow for a comparison between agriculture as a whole and coffee according to energy types. Total efficiency (as dollars of income per terajoule of direct energy used) is much lower in coffee than in general agriculture (Figure 21-13). On average, coffee is 1.85 times less efficient in using direct energy than the agricultural sector as a whole. An indirect energy use comparison can be made by looking at fertilizer applications (kilograms per hectare). Coffee production uses much more fertilizer compared to general Costa Rican agriculture (Figure 21-14). Thus we can say that export crops are less sustainable in terms of energy than other crops.

## D. Discussion

It appears that, perversely, farmers are becoming less efficient with declining real producer prices by trying to increase their total revenue by increasing

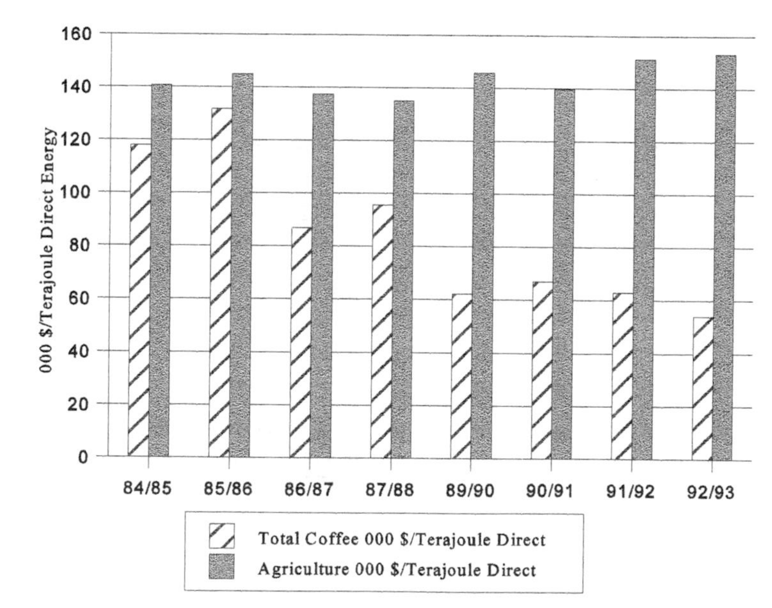


FIGURE 21-13 Comparison of direct energy efficiency between coffee production and processing and agriculture in Costa Rica.

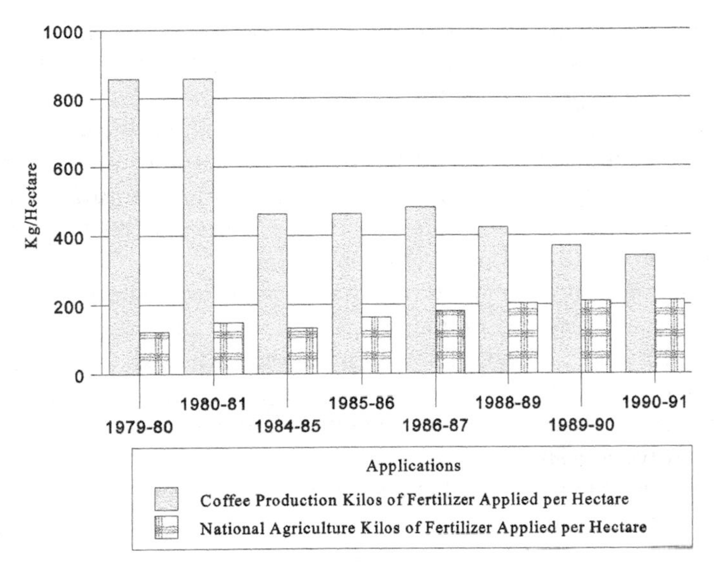


FIGURE 21-14 Comparison of fertilizer applications in kg/ha between the coffee production cycle and agriculture in Costa Rica.

production even if that means lower energy efficiency methods. The individual linear relation of Ton/TJt with real producer prices and with quantity produced supports this trend statistically. Efficiency in dollars of revenue per energy used fluctuates parallel to producer prices. This shows that farmers are concerned with generating revenue, not with decreasing the amount of energy inputs needed for each additional unit produced.

The reappearance of agroforestry alternatives for coffee production (use of shaded instead of nonshaded systems) may be having an influence on the energy efficiency results reported for production. As said before, unshaded coffee systems may require larger quantities of agrochemical applications per ton of coffee produced. Even if the system requires less fertilizer, farmers may be following price fluctuations by trying to increase yields to their maximum possible. The nature of the shaded system appears to require higher rates of application of other agrochemicals to prevent disease and pest infestations.

Cague *et al.* (1997) give a possible explanation for the differences in efficiency between production and processing observed in this study. The largest components of the total energy costs for processing were firewood and endocarps. One would be tempted to believe that this means processing is much less energy intensive than production because the first stage depends highly on fertilizers and the second depends mostly on firewood and endocarps. Coffee bean endocarps are a by-product of the processing itself, and firewood used in this process is mostly from shade trees used in the coffee farms (Cague *et al.*, 1997). Yet, even if we assumed that all firewood came from shade trees, which would not be likely in areas where shade is not used, the accounting of embodied energy is incomplete. We did not account for the indirect energy embodied in the machinery used in processing. Not only would this factor increase the energy intensity, but it would add to the imported components of such costs. Unfortunately, information to account for this factor was not available at the time the study was done.

Specific machinery use and transportation needs also affect the results. A higher crop year may require the use of additional machinery for processes that usually rely just on natural forces. Additionally, a small crop might mean lower transportation costs. For example, during the last two years, the cooperative beneficio of Atenas County has invested in an impressive water treatment system that reduces significantly its river pollution impact. Paradoxically, this might show up as an increased energy cost for those years.

The literature states that an increase of the proportion of biomass as a source of energy for coffee processing is a technological improvement (Koss, 1989). Before the 1980s, fossil fuel derivatives were used to dry almost all coffee in Costa Rica. Fuel price increases prompted a shift to the use of firewood and endocarps.

Endocarps could supply all the energy needs of coffee processing if the levels of energy efficiency (heat/drying) from the existing drying technology could be increased. It is an optimal fuel not only because it is a by-product, but because its burning characteristics can be regulated through automatic methods by keeping temperature and burning conditions optimal. Firewood does not offer this possibility since it has to be piled and its properties (i.e., humidity) vary constantly (Koss, 1989). Many beneficios store their firewood in covered areas for one or more harvest years. The slow drying allows better combustion characteristics and increase its caloric power (Koss, 1989).

It is interesting to compare the figures obtained versus the value added in monetary terms for each stage of coffee production and processing. From the total value of green beans, 25% is added by production when done by small producers (3% of the total value chain) (Appropriate Technology International, 1993). Given the trends found here in terms of energy costs and efficiency, this situation would be unsustainable in Costa Rica, especially if the price of oil increases again. The strict regulation of producer prices that exists in Costa Rica means that the profit margin that the processor receives is no more than 9%, after taxes and other deductions, of the price received from international sales. So, Costa Rican producers get about 23% of the total value chain (relative to retail prices). As our results show, even if Costa Rican farmers are better off, they are still fairly sensitive to price changes and so is the efficiency of their production.

A question for study seems to arise as to the effects of this pricing system on the environmental impact of production if energy prices increase.

As stated, local energy prices did not affect the trends observed of direct energy use. This probably reflects direct energy's small percentage of all energy consumed. It could be a result of the strong role that the Costa Rican government has in determining energy prices in Costa Rica. The subsidies for larger consumers could be providing an incentive not to worry about price trends.

In summary, accounting for energy costs questions the supposed comparative advantages of producing this crop in Costa Rica. Detailed regional analysis provides more insight into this statement.

## III. WHY AREAS WITH SUBOPTIMAL GROWING CONDITIONS GET HIGHER SELLING PRICES: A CASE STUDY IN THE WEST CENTRAL VALLEY

As previously suggested, a consequence of the importance of coffee to Costa Rica, and its increasing production, has been its expansion to areas that are not climatically optimal. Some of these suboptimal areas are receiving some of the highest producer prices in the country (Aguilar *et al.*, 1994, 1996; Cague *et al.*, 1997). It seems that, given the pricing system, these areas would have to invest higher amounts of energy to produce a competitive green bean. This would suggest that this pricing system could be promoting perverse environmental incentives.

We examined this problem in the Atenas region in the western Central Valley. Atenas is located around 800 m above sea level. Its precipitation ranges between 1400 and 2200 mm per year. The dry season extends from late November to early May, with January to March being the drier months. Average humidity is 79% annually, with June, September, and October averaging 87–88% humidity (IMN, 1994). Thus the zone is in the lower range of the optimum levels. Nevertheless, two beneficios located in this county receive 5–10% higher producer prices than the national average (Table 21-4).

Cague *et al.* (1997) examined the trends of coffee beneficios in Atenas compared to the national average. They gathered cost data directly from the beneficios (one cooperative and one private individual). This added to the accuracy of the estimates (Figure 21-15). They found that the county of Atenas shows a positive correlation between energy efficiency (in dollars per terajoule of energy used) and producer prices.

Consistently, between 1984/1984 and 1994/1995 the efficiency levels of the county were higher than the national average (almost 25% more efficient on

TABLE 21-4 Comparison of Producer Prices (Nominal Colónes per Fanega) of Atenas County and Costa Rica[a]

| Harvest year | Producer price, cooperative beneficio (Atenas) | Producer price, private individual beneficio (Atenas) | Average producer price, Costa Rica |
|---|---|---|---|
| 1984/1985 | 3,615.80 | 3,192.62 | 4,612.88 |
| 1985/1986 | 6,246.24 | 5,165.14 | 5,953.58 |
| 1986/1987 | 4,435.38 | 4,314.62 | 4,205.40 |
| 1987/1988 | 5,763.58 | 4,929.42 | 5,126.58 |
| 1988/1989 | 6,028.36 | 5,160.78 | 5,328.16 |
| 1989/1990 | 5,874.04 | 4,853.74 | 5,095.04 |
| 1990/1991 | 8,166.12 | 7,159.48 | 6,889.10 |
| 1991/1992 | 7,784.76 | 8,117.46 | 6,626.26 |
| 1992/1993 | 7,515.00 | 6,483.98 | 6,635.70 |
| 1993/1994 | 12,128.00 | 11,703.60 | 11,742.52 |
| 1994/1995 | 21,173.20 | 21,587.20 | 18,673.06 |
| 1995/1996 | 19,148.44 | 19,384.90 | 16,445.41 |
| Average | 8,989.91 | 8,504.41 | 8,111.14 |

[a] Data from Cague *et al.* (1997).

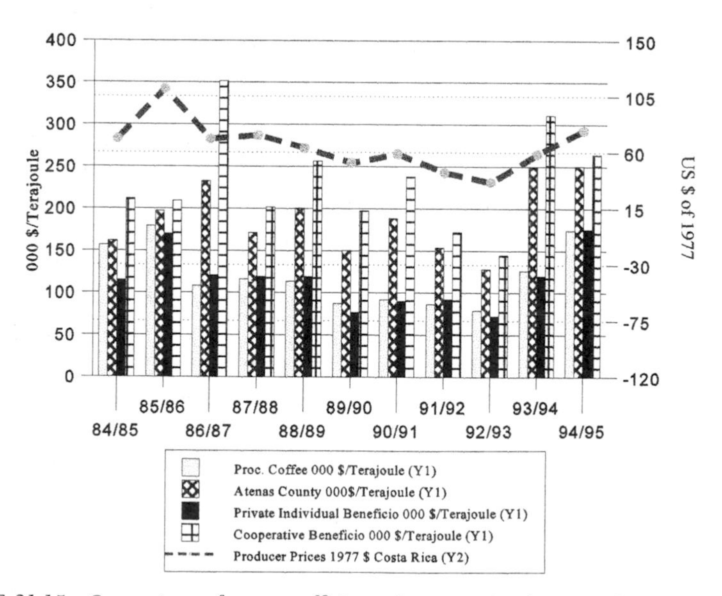


FIGURE 21-15 Comparison of energy efficiency in processing between Costa Rica's overall production and Atenas County beneficios.

average). Both beneficios do not represent this trend. The efficiency of the cooperative beneficio is almost 100% higher than the national average for this period.

Yet, producer prices do not reflect these differences. For example, in the last two years of the analysis, the private individual beneficio was receiving higher producer prices even with lower energy efficiencies than the total country average (Table 21-4, Figure 21-15).

New drying ovens introduced by the cooperative beneficio two years ago might also affect these trends (Cague *et al.*, 1997). The ovens of the private beneficio are of the direct fire type where the furnace is next to the drying chamber (Rojas, 1997). This type of oven lowers the quality of the bean more than the more modern indirect fire ovens that keep the flow of combustion gases separate from the drying air (Koss, 1989).

On the production side, the energy efficiency (U.S. dollars/TJ of energy used) in Atenas can be compared to the national figures of Hartman *et al.* (1996), who calculated direct and indirect energy use (in kilocalories per dollar) for coffee producers. They surveyed 15 producers in the county. After converting their numbers, our efficiency figures for the 1995/1996 national crop are lower (Figure 21-16).

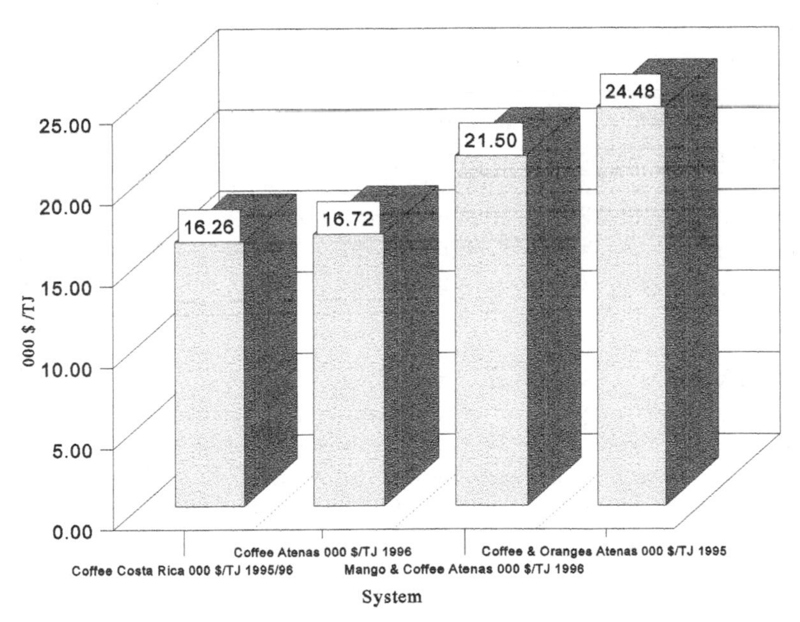


FIGURE 21-16 Comparison of energy efficiency in coffee production in different systems, 1984/1985–1995/1996.

Both Hartman *et al.* (1996) and Aguilar *et al.* (1996) present two other interesting technological possibilities facilitated by the climatic conditions of this region. Oranges and mangoes are also produced there and many farmers have chosen to hedge their agricultural risk by polycropping their coffee farms with these fruit trees.

Hartman *et al.* (1996) found no statistical difference between the efficiency of the coffee-only system and the polycrop with mango system in Atenas. The efficiency of the system in 1996 was \$21,500 per terajoule of energy used. Aguilar *et al.* (1996) did a similar exercise for the coffee–orange polycrop and found out that the system had an efficiency of \$24,480 per terajoule of energy used.

## IV. LAND USE CHANGES

An important factor in future coffee production in Costa Rica is the long-term land use changes of the Central Valley. This region has the most productive agricultural soils in Costa Rica, yet it also contains the largest urban areas. A study done by the Tropical Science Center (1982) found that of the 67,700 ha of the Greater Metropolitan Area, 15,500 ha were already residential, 14,000 ha were occupied by industry, parks, and airports, and 11,000 ha have been designated as priority areas for urbanization. This area still contains 23,000 ha of the highest quality land suitable for agriculture. As the population of the Central Valley increases there will be parallel growths in industry and infrastructure that will continue to consume more of the prime area, forcing coffee production to move into increasingly suboptimal areas.

In Chapter 12, as production moves out of optimal ranges more inputs will be required to maintain the same level of productivity. Both the economic and the environmental sustainability of coffee production need to be examined in the context of land with decreasing intrinsic value as occurs with urbanization.

## V. SUMMARY AND CONCLUSIONS

Coffee, as a primary commodity with a very unstable international market, has contributed to increases in the economic dependence of Costa Rica. Even as coffee generates a substantial amount of Costa Rica's export revenue, it also increases Costa Rica's trade deficit since its cultivation is highly dependent on imported technology.

The crop is known for its agrochemical impact and water pollution problems. Many of these problems are being reduced as endocarps are used more and more for fuel, and as waste treatment for by-products increases. This has decreased dependence on oil. This study has shown that the crop appears significantly less energy efficient than the agricultural sector as a whole. Such costs are concentrated in the production stage, where producers are decreasing the efficiency in their production in order to respond to price fluctuations. Through the years some technological advance is evident, and some increases in efficiency are occurring with processing. Nevertheless, reliance on fertilizer inputs especially remains high and is increasing and total efficiency is decreasing. Thus the coffee industry is as vulnerable to future energy price increases as it was in the 1970s.

Biophysical exploration of alternative production technologies provides further insight into ways of improving the energy efficiency of the crop. One approach is polycropping with fruit trees. Other studies stress the sustainability of shaded (Vandermeer and Perfecto, 1995) and organic (Boyce *et al.*, 1993) systems. These approaches have not yet been studied through an embodied energy approach (but see Klocker, 1999).

More remains to be studied on the efficiency levels of production and processing in suboptimal areas. Nevertheless, our results suggest that technological improvements offer some potential benefits in those areas that can compensate for adverse climatic conditions. As area planted shifts more to these regions, the pressure for higher yields might decrease efficiency levels. Technological improvements may be possible, but we were not able to identify for certain what might work.

Another element that needs to be studied is the cultural value of this crop. It has been part of the Costa Rican culture for over 100 years. Any proposal for sustainability regarding this crop should take into account the potential consequences that technology changes or even the complete substitution of the crop might have on cultural continuity. But it is clear that social and cultural sustainability have been purchased, at least to some degree, at the expense of energy sustainability.

## ACKNOWLEDGMENTS

We thank Eric McClanahan, Lois Leviton, Shauna Swantz, and Stephanie Katz for insights and preliminary analysis. We also thank Ligia Umaña for her continuous support.

## REFERENCES

Aguilar, B. J., T. Gillespie, C. Waddick, C. Williams, E. Rodman, E. Jones, and D. Fuchman. 1996. A biophysical assessment of tropical crops according to trade levels: Potential implications of multi-cropping nationally consumed and exported crops in developing countries.

In *Proceedings, Fourth Biennial Meeting of the International Society for Ecological Economics—Designing Sustainability: Building Partnerships among Society, Business and the Environment, Aug. 4–7, 1996, Boston University, Boston, MA*, pp. 4–7. International Society for Ecological Economics/Center for Energy and Environmental Studies/Boston University, Boston.

Appropriate Technology International. 1993. *Central American Coffee Initiative Regional Program to Increase Value Added Chain by Small Producers and Reduce Environmental Impact of Processing.* API, Washington, DC.

Beer, J., R. Muschler, D. Kass, and E. Somarriba. 1998. Shade management in coffee and cacao plantations. *Agroforestry Systems* 38:139–164.

Boyce, J., A. Fernández, E. Furst, and O. Segura. 1993. *Sustentabilidad de la Producción Cafetalera Costarricense y Conveniencia del Café Orgánico como Alternativa.* Universidad Nacional de Costa Rica, Heredia, Costa Rica.

Cague, R., S. Langley, M. McCormack, and V. Vitiello. 1997. Coffee Processing Efficiency in Costa Rica: Do Liquidation Prices Truly Reflect the Real Cost? Unpublished report, School for Field Studies, CSDS, Atenas, Costa Rica.

Cannell, M. G. R. 1985. Physiology of the coffee crop. In M. N. Clifford and K. C. Willson (Eds.), *Coffee: Botany, Biochemistry and Production of Beans and Beverage.* AVI Publishing, Westport, CT. 457 pp.

Charrier, A., and J. Berthaud. 1985. Botanical classification of coffee. In M. N. Clifford and K. C. Willson (Eds.), *Coffee: Botany, Biochemistry and Production of Beans and Beverage.* AVI Publishing, Westport, CT. 457 pp.

Clark, E. H., III. 1985. The off-site costs of soil erosion. *Journal of Soil and Water Conservation* 40:19–22.

Clarke, R. J. 1985. The technology of converting green coffee into the beverage. *In:* Coffee: Botany, biochemistry and production of beans and beverage (M. N. Clifford and K. C. Willson, Eds.), 457 pp. AVI Publishing Co., Westport, Connecticut.

Cleveland, C. J. 1991. Natural resource scarcity and economic growth revisited: Economic and biophysical perspectives. In R. Costanza (Ed.), *Ecological Economics: The Science and Management of Sustainability,* pp. 309–317. Columbia University Press, New York.

Cleveland, C. J. 1995. Resource degradation, technical change, and the productivity of energy use in U.S. agriculture. *Ecological Economics* 13:185–201.

Clifford, M. N. 1985. Chemical and physical aspects of green coffee and coffee products. In M. N. Clifford and K. C. Willson (Eds.), *Coffee: Botany, Biochemistry and Production of Beans and Beverage.* AVI Publishing, Westport, CT, pp. 305–374.

Coto, J. M. C. 1993. Water pollution in Costa Rica by residues from coffee processing and hog production. *In:* Proceedings of the FAO expert consultation. Santiago, Chile. October 20-23, 1992. Water Reports, #1. FAO 1993.

Cros, J., Lashermes, Ph., Marmey, Ph., Anthony, F., Hamon, S., and Charrier., A. 1993. Molecular analysis of genetic diversity and phylogenetic relationships in *Coffea.* OSTROM, Laboratoire de Resources Genetiques et Amelioration des Plantes Tropicales, Montpelier, France.

Daly, H. 1990. Toward some operational principles of sustainable development. *Ecological Economics* 2:1–6.

Geer, T. 1971. *An Oligopoly: The World Coffee Economy and Stabilization Schemes.* Dunellen, New York. 323 pp.

Haarer, A. E. 1963. *Coffee Growing.* Oxford University Press, New York.

Hall, C. A. S., and M. H. Hall. 1993. The efficiency of land and energy use in tropical economies and agriculture. *Agriculture, Ecosystems and Environment* 46:1–30.

Hannon, B. M., C. Harrington, R. W. Howell, and K. Kirkpatrick. 1976. *The Dollar, Energy, and Employment Costs of Protein Consumption.* Center for Advanced Computation, University of Illinois at Urbana-Champaign, Urbana, IL. 81 pp.

Hartman, J., T. Keil, J. Kind, J. Lemly, and B. Nelson. 1996. A Biophysical Analysis of Coffee and Mango Polycropping in the Atenas Region. Unpublished report, School for Field Studies, CSDS, Atenas, Costa Rica.

ICAFE (Instituto de Café de Costa Rica). 1989. *Manual de recomendaciones para el cultivo del cafe.* Programa Cooperativo ICAFE–MAG, San Jose, Costa Rica. 122 pp.

ICAFE Institudo de Café de Costa Rica. 1995. Informe sabre la Actividad Cafetatera de Costa Rica. TCAFE, San Jose, Costa Rica

ICAFE. 1995. *Informe sobre la Actividad Cafetalera de Costa Rica.* ICAFE, San Jose, Costa Rica.

IMN. 1994. *Tablas Climatológicas para la Estación Fabio Baudrit.* IMN, San Jose, Costa Rica.

Instituto de Café de Costa Rica (ICAFE). 1995. Informe sobre la Actividad Cafetalera de Costa Rica, ICAFE, San José, Costa Rica.

Koss, L. 1989. *Evaluación de los Sistemas de Generación de Calor para Secado de Café.* Centro de Investigaciones en Café, Heredia, Costa Rica.

Marshall, C. F. 1985. World coffee trade. In M. N. Clifford and K. C. Willson (Eds.), *Coffee: Botany, Biochemistry and Production of Beans and Beverage.* AVI Publishing, Westport, CT. 457 pp.

Muschler, R. G. 1997. Shade or sun for ecologically sustainable coffee production: A summary of environmental key factors. (J. Beer, Ed.) Proceedings from a seminar: "Tropical agroforestry: Two decades of development". Indianapolis, IN. November 1997 (in press).

Paige, J. M. 1997. *Coffee and Power: Revolution and the Rise of Democracy in Central America.* Harvard University Press, Cambridge, MA. 432 pp.

Rice, R. A., and J. R. Ward. 1996. *Coffee, Conservation, and Commerce in the Western Hemisphere.* Natural Resources Defense Council and Smithsonian Migratory Bird Center and Weadon Progressive Communication Services. 47 pp.

Rojas, G. 1996. *Modelo de Costos de Producción de Café.* Instituto del Café de Costa Rica, San Jose, Costa Rica.

Rojas, R. 1997. Owner of the Rolando Rojas Beneficio in Atenas. Personal communication, Aug. 21.

Samper, M. 1990. *Generations of Settlers: Rural Households and Markets on the Costa Rican Frontier, 1850–1935.* Westview Press, San Francisco.

Samper, M. (Mario Samper Kutschbach). 1995. In difficult times: Columbian and Costa Rican coffee growers from prosperity to crisis, 1920–1936. *In:* Coffee, society, and power in Latin America. (W. Roseberry, L. Gudmundson, and M. Samper Kutschbach, Eds.) 304 pp. The Johns Hopkins University Press. Baltimore, Maryland.

Seligson, M. A. 1980. *Peasants of Costa Rica and the Development of Agrarian Capitalism.* University of Wisconsin Press, Madison, WI.

Stocke, V. 1995. The labors of coffee in Latin America: The hidden charm of family labor and self-provisioning. *In:* Coffee, society, and power in Latin America. (W. Roseberry, L. Gudmonson, and M. Samper Kutschbach, Eds.) 304 pp. The Johns Hopkins University Press, Baltimore, Maryland.

TSC (Tropical Science Center). 1982. *Costa Rica: Perfil Ambiental, Estudio de Campo.*

Van der Vossen and A. M. Herbert. 1985. Coffee selection and breeding. In M. N. Clifford and K. C. Willson (Eds.), *Coffee: Botany, Biochemistry and Production of Beans and Beverage.* AVI Publishing, Westport, CT. 457 pp.

Vandermeer, J., and I. Perfecto. 1995. *Breakfast of Biodiversity. The Truth about Rainforest Destruction.* Institute for Food & Development Policy, Oakland, CA. 185 pp.

Williams, R. G. 1994. *States and Social Evolution: Coffee and the Rise of National Governments in Central America.* University of North Carolina Press, Chapel Hill, NC. 357 pp.

Winson, A. 1989. *Coffee and Democracy in Modern Costa Rica.* St. Martin's Press, New York.

CHAPTER 22

# Costa Rican Industry

## Characteristics, History, and Potential for Sustainability

TIMM KROEGER



## I. INTRODUCTION

This chapter examines the industrial component of the Costa Rican economy in three sections. The first quantitatively and qualitatively reviews historical aspects and trends, including the impact of internally and externally derived policies. The second examines several emerging industries, including tourism, textiles, and computers. The third examines Costa Rican industry from the perspective of economic and biophysical sustainability. Although we are aware that the term "industry" is frequently used to include the agricultural sector of an economy, in the analysis in this chapter we refer to industry in a narrow definition of the term. In doing so, we use it to denote fossil-fuel-, biomass-, or hydroelectric-energy-powered factory-based production processes. Additionally, we do include tourism as an industry. Agriculture is covered separately in Chapters 4, 5, and 12.

## II. THE HISTORY OF THE COSTA RICAN INDUSTRIAL SECTOR

The Costa Rican economy, or at least that part associated with European immigrants and those indigenous people associated with them, has always had to face the principal problem of operating from a small country with limited industrial resources. A continuing problem has been the generation of foreign exchange to pay for expensive industrial inputs from the industrialized world. These imports originally included iron tools, and in this century nearly all transportation items, including automobiles, trucks, buses, and, of course, the fossil fuels their operation is based on. It also includes most machinery needed for the domestic production of heavy goods. Prior to the 1960s, the Costa Rican export economy was based primarily on coffee and bananas, and secondarily on sugar and beef. These exports did not require at that time much in the way of imported inputs, and generated a limited amount of foreign exchange that was in large part used up by the purchases and maintenance of imported machinery and agricultural inputs. The establishment of the Central American Common Market (CACM) in 1960 was seen as a strategy to reduce the Central American countries' dependence on their primary commodity exports, and to expand the intraregional market for products manufactured in CACM member countries in order to promote regional import-substitution industrialization (ISI) (Hamilton and Thompson, 1994). The principal components of the national industry over time are given in Figure 22-1.

The rationale for ISI was to encourage the domestic and regional manufacture, rather than the importation, of basic goods in order to build up a modern domestic industrial sector and diminish Costa Rica's dependency on extra-regional markets. The necessary next step, as was undertaken successfully by, for example, the newly industrializing countries in East Asia, was to reorient the matured industries toward expanded exports to generate, as well as save, foreign exchange. Initially Costa Rica was able to achieve these goals to some degree: its manufacturing capacities and export structure and markets became more diversified, particularly with the production and export of processed foods, textiles, furniture and wood products, and chemicals to other Central American countries (Wilkie *et al.*, 1995; Hamilton and Thompson, 1994). The country's intraregional exports increased from about 5% of its total exports in 1960 to a high of more than 34% in 1983 (Wilkie *et al.*, 1995). Since then, however, this percentage has been falling gradually (to about 17% in 1991), primarily because Costa Rica has been focusing its exports on markets in industrialized countries. The reasons for this switch of the markets targeted for the country's exports are outlined in Chapter 24.

According to Hamilton and Thompson (1994), however, export dependence remained high and in fact *increased* as a result of ISI and the influx of foreign

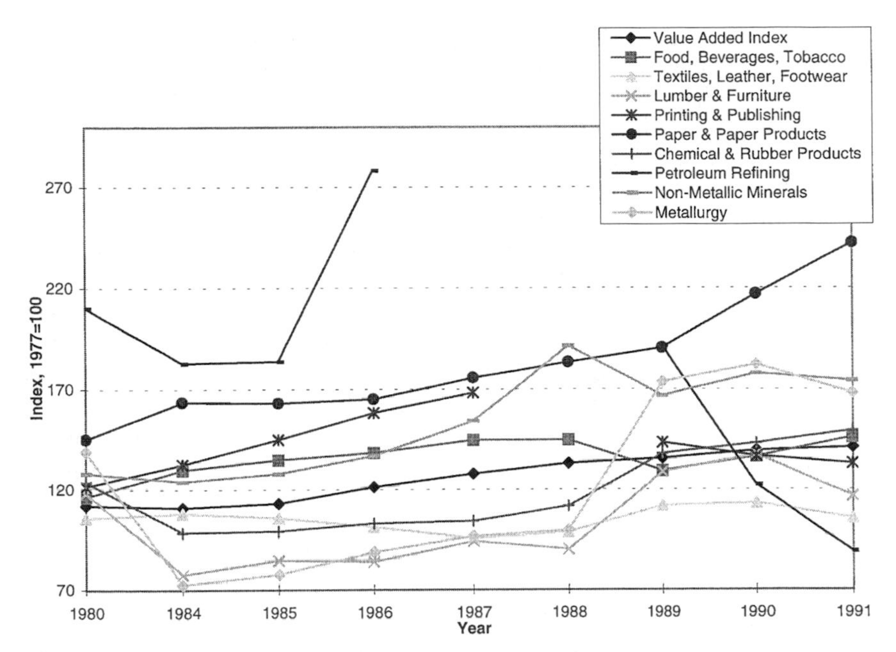


FIGURE 22-1 The principal components of the Costa Rican manufacturing industry over time. (Source: *Statistical Abstract of Latin America,* 1989, 1995. For inflation correction, the following source was used: *Economic Indicators. August 1997.* Prepared for the Joint Economic Committee by the Council of Economic Advisors. 1997. U.S. Government Printing Office, Washington, DC.)

corporations and loans. Hamilton and Thompson (1994) and Solis (1996) argue that this was a consequence of the absence of state autonomy vis-à-vis powerful interest groups, including the domestic agroexport oligarchy and the new industrialists. New redistributive measures, including a reduction in protectionism in domestic markets, and tax reforms, that were necessary if ISI were to work, were precluded by these powerful interests. Furthermore, the continued protection of consumer industries dependent upon imported goods and the failure to protect intermediate and capital goods industries played an important role in the inefficiency and low competitiveness of some domestic industries.

Fundamental to all of these problems was that while Costa Rican industry could be ordered or encouraged by fiat to produce whatever industrial products were desired, one could not change by fiat the fact that Costa Rica had almost none of the essential industrial raw or intermediate goods required to generate the industrial products. Thus the importing of finished products was replaced with the importing of intermediate goods from which the final goods could be manufactured, and the import dependence continued (Figure 22-2). In some cases, where Costa Rican industry was especially inefficient, imports increased. More important, however, were the "oil crises" of 1973 and 1979. In their wake all of the imported goods became more expensive as a result of increases in the price of petroleum to the manufacturers, and Costa Rican borrowing on international financial markets increased. Concurrently, and as a direct function of the price increases of petroleum, international commercial banks had a large supply of petrodollars available at low interest rates and long payment terms. Costa Rica, like many other poorer countries, took advantage of the funds available for infrastructure development and service improvement. This period coincided with high prices for coffee on international markets that gave policymakers the impression that needed finance capital could be obtained cheaply by borrowing on international capital markets. Such a strategy can be successful if the borrowed capital is invested in such ways that the resulting economic growth in the borrowing country results in higher government revenues which in turn allow the repayment of the loans. In Costa Rica, however, the money was used to a large degree to maintain consumption, including the highly valued health and education programs. Then, unfortunately, interest rates on international capital markets increased markedly in the early 1980s [Economist Intelligence Unit (EIU), 1995; Carley and Christie, 1992]. Since the borrowed capital from the cheap petrodollar loan period had not been paid back, the cost of servicing that debt increased sharply.

All of these factors led to a failure of the ISI strategy with devastating effects on Costa Rica. The country incurred rapidly rising international debt and, by the early 1980s, had the highest per capita debt in the world (Hamilton and Thompson, 1994). In 1981, Costa Rica had to default on the payment of its foreign debt. The consequences of the country's financial crisis are developed in Chapter 24.

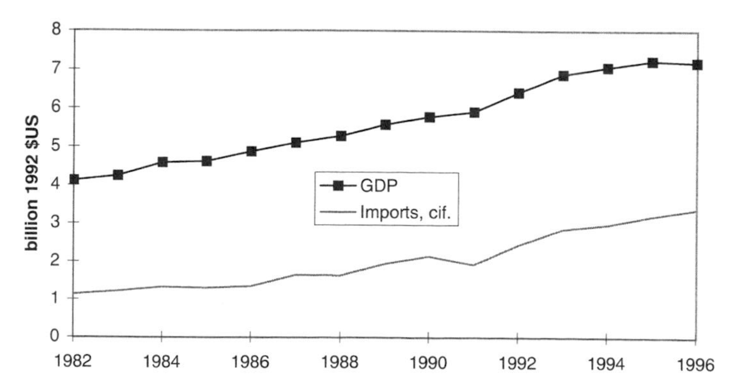


FIGURE 22-2 The ratio of imports to GDP. (Source: IMF *International Financial Statistics*, 1998). cif., "cost, insurance, and freight." The imports/GDP ratio increased from 28 percent in 1982 to 47 percent in 1996.

## III. WILL EMERGING INDUSTRIES BE ANY DIFFERENT?

It seems that the traditional industries, including the now traditional "nontraditional" ones, such as ornamental plants and cut flowers, pineapples, melons, shellfish, fruit preparations, pharmaceuticals, rubber seals, electronics, and clothing, have hardly solved Costa Rica's problems with international debt (see Figure 23-1).

While as of 1997 official government policy had turned back to the encouragement of "traditional" export crops, there are new industries that might offer considerable hope. We consider here three: tourism, textiles, and computers and telecommunication.

### A. Tourism

In 1995 tourism overtook bananas as the single most important generator of foreign exchange for Costa Rica (EIU, 1995). The tourism industry attracted over $1 billion in investment in the first half of the 1990s (EIU, 1995). Clearly tourism is booming in Costa Rica, as evidenced by the number of international arrivals in the country, which in 1994 topped 760,000 (Figure 22-3). A new twist on this old industry is the emergence of ecotourism.

The continuing sharp increase in international tourist arrivals since 1986, when they totaled 260,000, is due primarily to an increase in North American arrivals, but also from increasing numbers of Central American and European visitors [Instituto Costarricense de Turismo (ICT), 1995]. Preliminary figures for 1995, however, show a remarkable slowdown of this growth trend to an increase of less than 5% over the previous year's numbers (ICT, 1995). Aylward

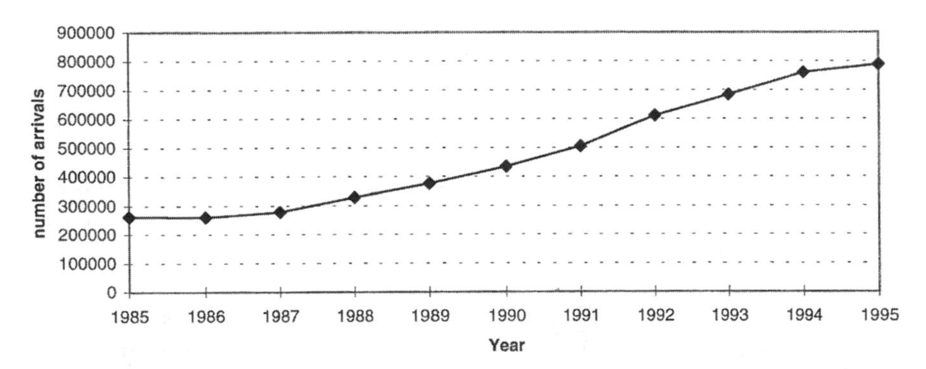


**FIGURE 22-3** Number of international arrivals by plane in Costa Rica. (Source: ICT.)

*et al.* (1996) attribute this leveling off to a number of factors: the negative coverage in the media of increasing crime in the country, particularly violence toward tourists, and possibly the rising entrance fees to Costa Rican national parks and preserves. Also, as wages, prices, and crowding have increased in Costa Rica, and as other countries are focusing on the lucrative market, tourists are finding other, often less expensive locations, especially in other Central American countries and the Caribbean (Aylward *et al.*, 1996).

Any attempt to identify the "ecotourists" among tourists is necessarily dependent upon a particular definition of ecotourism. If we follow Aylward *et al.* (1996) in deciding that visits to national parks, private reserves, and other "protected" areas are indicative of ecotourism in Costa Rica, as opposed to visits to the more popular beaches and volcano parks, we arrive at about 100,000 "ecotourists" in 1992 out of 636,000 total international arrivals.

Aylward *et al.* (1996) studied ecotourism at the Monteverde Cloud Forest Preserve, and arrived at the conclusion that the private preserve appears to be an example of sustainable ecotourism. This judgment is based on their observation that the preserve has managed to maintain a very low level of impact on the local ecosystems while managing to attract a stream of visitors that reached 50,000 in 1992 and has since remained about constant (Aylward *et al.*, 1996). From the entrance fees, the private nonprofit preserve is able to earn an annual financial surplus which is reinvested in improving the quality and quantity of the services provided.

## B. Textiles

Textiles and apparel are the main products of the Costa Rican maquila (light export) industries. Costa Rican textile and apparel exports benefit from the Caribbean Basin Initiative (CBI) legislation of the United States, which grants privileges to certain products from CBI countries with respect to their import into the United States. The CBI countries are at a disadvantage when compared to Mexico, however, which, as a member of the North American Free Trade Agreement (NAFTA), is exempt from any import quotas to the United States. The resulting disadvantage for CBI countries is evidenced by the fact that for January–September 1996, Mexican apparel exports to the United States rose by over 35% compared to the same period of the previous year (EIU, 1997). For textiles and apparel combined, the figures are a 26% increase for Mexico and a 9% increase for the CBI countries. Costa Rica actually experienced a loss of 7% in the same period (EIU, 1997).

The increase in value added in the Costa Rican maquila industry (Figure 22-4) is mirrored only partly in the development of its textile component. The latter has experienced a 39.4% decline in value of production from 1993 to

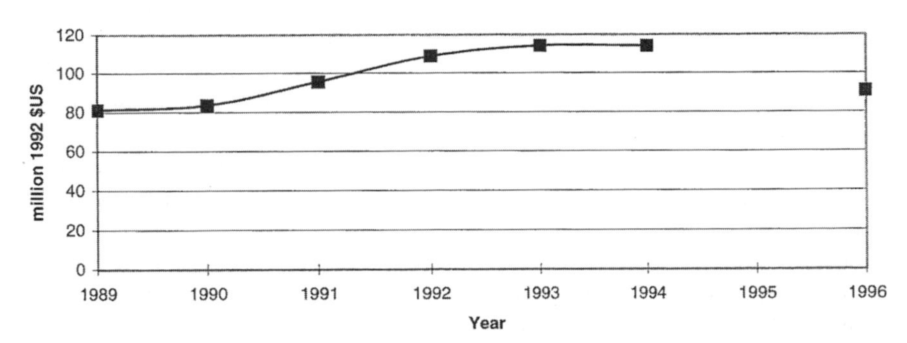


FIGURE 22-4 Value added by the Costa Rican maquila industry. (Source: EIU, 1995, 1997.)

1994 (EIU, 1995), followed by another decline of 7% from 1994 to 1995. Meanwhile, the textile industry is undergoing a process of substituting "easy" assembly of T-shirts for "difficult" cutting and assembling processes, such as those needed for fine woolen winter wear (EIU, 1995). "Easy" assembly in Costa Rica is becoming increasingly less competitive with other Central American locations, as evidenced by the massive relocations of textile firms from Costa Rica to the latter (EIU, 1996b).

## C. Computers and Telecommunication

Costa Rica's stable government, its well-educated population by comparison with other low/middle-income countries, good communications and air transport infrastructure, low wages, and amenities make it a potentially attractive place for computer and telecommunications companies to locate. As of the writing of this chapter several important companies have recently done so. Acer America Corporation, a Taiwanese computer company, established a customer service center near Heredia. Costa Rican personnel trained by ACER will answer telephone questions from Acer's North American clients (EIU, 1996a). Also, the U.S. telecommunications equipment producer Motorola is expanding its local operation in Costa Rica with a $3 million investment. This expansion is estimated to require the employment of an additional 750 people, ranging from workers to electronic engineers (EIU, 1997).

A potentially very important new move is that Intel, a leading manufacturer of ultrasophisticated computer chips, is constructing a $500 million manufacturing facility near the San Jose airport. The total investment is the largest ever made by a single company in Costa Rican history. The Intel plant is an assembly and test facility for the final stages of chip production. The facility will employ 2000 Costa Ricans, 400 of whom will be engineers. Within five years the workforce is planned to increase to 3500 people (EIU, 1997). If we assume that the 2000 Costa Ricans make a locally high average salary of $10,000 per year then this company would bring in $20 million of foreign exchange, about 3% of that from bananas.

Following global patterns, the location of the Intel plant in Costa Rica does not come for free. The company receives the "zona franca" (free-trade zone) status for its plant. This entails a 100% exemption from export taxes for the first eight years, and a 50% reduction of export taxes for the following four years. Furthermore, the exemption granted from land and permit taxes represents a subsidy that costs the municipal government roughly $600,000. Also, the Ministry of Transportation will construct access roads that connect the plant with the main highway of the country, the Pan American Highway, and the government is building a special electric substation at an estimated cost of $2 million to satisfy Intel's high electricity demand. Moreover, for Intel and other high-technology industries, electricity fees will be reduced by 27% from $0.68 to $0.50 per kilowatt-hour. Finally, the government will invest in developing education programs that match Intel's technological needs.

## IV. SUSTAINABILITY OF THE COSTA RICAN INDUSTRIAL SECTOR

There are two important components of sustainability related to Costa Rican industry. The first, "economic sustainability," is the degree to which the industry is financially sustainable. The second criterion for sustainability, "biophysical sustainability," is more stringent and relates to whether there will be available indefinite supplies of necessary raw materials (current and substitutes) and unlimited or at least sufficient environmental absorption capacity to allow the continuation of current trends in the throughput of material and the output of waste. Because Costa Rican industry is to a large degree dependent upon the imported resources it requires as inputs, the ultimate assessment of its sustainability cannot be carried out without an analysis of the biophysical underpinnings of the country's economy, and the relation between biophysical sustainability and the relative prices of necessary raw materials and products generated.

### A. Sustainability from an Economic Perspective

The structure of Costa Rica's industrial sector was and is influenced by all of the past policies, opportunities, and problems discussed here and in Chapters

23 and 24. A net result is that Costa Rica now is the most industrialized country in Central America, with more people employed in the industrial sector (manufacturing, construction, commerce, and services) than in the agricultural sector (EIU, 1995; Tardanico and Lungo, 1995). Also, the industrial sector contributes a larger share to the country's gross domestic product than does agriculture, and industrial exports, excluding maquila (offshore assembly) exports, are increasing sharply (EIU, 1995). Costa Rica's largest single industry as of the mid-1990s was food processing. Other important ones are chemical products, textiles, and metal processing (EIU, 1995).

Based on the history of the industry to date, the economic sustainability of Costa Rican industry appears questionable. The industrial sector remains heavily dependent on imported materials (Table 22-1) (EIU, 1997), and it continues to be completely at the mercy of economic factors external to the influence of the country. A good example is the textile industry, on which Costa Rica has unfortunately built a good part of its export promotion program. It is totally dependent on foreign markets for these exports, especially the United States, which remains Costa Rica's main export (and import) partner (Wilkie *et al.*, 1995). But with the establishment of the North American Free Trade Agreement, Costa Rican garment exports now are threatened by a Mexican garment industry that is not subject to import quotas and tariffs, and hence has gained in relative competitiveness compared to the garment industries in non-NAFTA countries such as Costa Rica. Also, textile firms are rather flexible with regard to their production location. Therefore, neighboring Central American countries with unemployment rates of up to 60% and correspondingly lower wages are becoming increasingly attractive to these industries. As a result, the number of enterprises in the Costa Rican textile industry has declined rapidly over the past five years, from 1011 companies in 1990 to only 679 in 1995 (EIU, 1996b). Moreover, Costa Rican textile exports are constantly threatened by arbitrarily imposed import restrictions in target markets. One example is the quotas imposed by the United States on textile imports from Costa Rica in 1997. Although the World Trade Organization (WTO) demands that the United States lift the restrictions on grounds that these violate the General Agreement on Tariffs and Trade (GATT), the United States, as of this writing, shows no inclination to follow this demand.

The computer and telecommunications sector is characterized by a somewhat lesser flexibility of relocation. However, the global pattern clearly indicates the gradual shift of high-technology industrial production to countries with progressively lower wage levels (Greider, 1997). Hence, Costa Rica is subject to the relocation threats that rising wage levels entail.

Tourism has become very important in Costa Rica, but the slowdown in the increase in international arrivals in Costa Rica has led the information ministry to publish a series of commitments to support the tourism sector.

These include a $15 million investment in the promotion of Costa Rica as a tourist destination; increased security measures in areas of tourism; and improvements in tourism-related infrastructure such as airports, ports, roads, and street signs (EIU, 1996a). However, as the Economist Intelligence Unit points out, Costa Rica has traditionally targeted the "well-to-do tourist" (EIU, 1995). In this bracket, as well as in the ecotourism category, the country competes with a number of other locations for a limited pool of potential international customers and their foreign exchange. This competition can already be seen between different locations within the country itself. In a case study of the privately owned Monteverde Cloud Forest Preserve by Aylward *et al.* (1996), the authors presume that the revised fee structure of the national parks may have led to a loss in revenue for the private preserve, by redirecting visitors from the preserve toward the national parks system. The authors find that the Preserve and the Park System may be complementary goods for the high value, low volume ecotourists that Costa Rica is trying to attract.

Based on the previous sections, it seems clear that Costa Rica is not less dependent on its exports than it was at the beginning of the Structural Adjustment Policies and their redirection of the country's strategy for industrialization. The dependence has only shifted to different products, since as of 1994 nontraditionals made up 54% of the country's export value (EIU, 1995) (Table 22-1).

**TABLE 22-1 The Dependence of Traditional and Nontraditional Export Industries in Costa Rica on Imported Materials**

| | Values in million 1992 $U.S. | | |
|---|---|---|---|
| | 1991 | 1992 | 1993 |
| Main exports fob. | | | |
| Traditionals | 782.1 | 802.1 | 836.6 |
| Nontraditionals | 744.9 | 905.6 | 1119.2 |
| Maquila value added | 96.5 | 108.9 | 114.2 |
| Total | 1623.5 | 1816.6 | 2070.0 |
| Main imports cif. | | | |
| Raw materials | 825.3 | 982.7 | 1046.2 |
| Fuels and lubricants | 156.3 | 158.9 | 172.0 |
| Capital goods | 402.7 | 558.8 | 712.8 |
| Construction materials | 73.8 | 86.2 | 91.9 |
| Total | 1458.1 | 1786.6 | 2023.0 |
| Balance (exports − imports) | 165.4 | 30.0 | 46.9 |
| Consumer goods imports (not included) | 448.5 | 644.0 | 812.5 |

*Source:* EIU (1995).
fob, "free on board" (i.e., dockside price); cif, cost, insurance, freight.

Clark (1995) assesses the potential economic sustainability of Costa Rican export industries in terms of, first, their maintaining at least the proportion of GDP reached after the take-off period, and second, the existence of any political or economic factors that might threaten the industries. The first condition seems to be fulfilled: manufacturing industries have held their share of the GDP since 1988, and nontraditional exports have increased their share of the GDP slightly (EIU, 1995; Clark, 1995) (Figure 22-5). The second condition, however, reveals that the empirical indicators for the future of the export industry, i.e., export market volatility and import dependence, presage an insecure future: the heavy dependence on export markets (and the problems that come with it, as already outlined) as well as on imports for the industrial sector clearly poses a threat to the successful further development and even maintenance of these industries.

It is also important to analyze the economic sustainability of the structure of Costa Rican industry in terms of its effects on the country's international debt and its trade balance, since it certainly has a major influence on both. Here, too, the record is not unequivocal. Costa Rica's total debt grew by 41% from 1980 to 1994, when it was $3.8 billion, and is expected to have declined slightly to $3.5 billion by 1996 (EIU, 1997). The trade balance has been negative every year since 1987 (see Figure 24-1). Certainly, the country's financial problems have not been solved. On the other hand, one might say that for the time being they seem to be no longer increasing.

In summary, the present situation of the industrial sector of Costa Rica is characterized by:

1. A heavy dependence on imports of raw materials and manufactures for industrial production

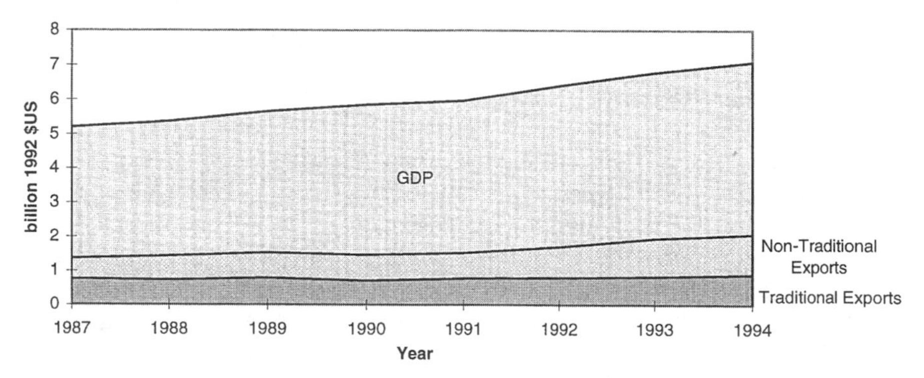


FIGURE 22-5 Costa Rican exports as share of GDP. (Source: EIU, 1995; Inter-American Development Bank, 1995.)

2. Increasing trade deficits resulting from (1)
3. A heavy dependence on volatile export markets for sale of primary commodities and tariff/quota-regulated manufactures
4. Increased volume of fossil-fuel-based trade to and from Costa Rica with vulnerability to potential changes in fossil fuel prices (C. Campbell, 1997a; Campbell and Laherrere, 1998)

None of these indicators are consistent with a sustainable economy.

## B. Sustainability from a Biophysical Perspective

Finally we should consider not only economic sustainability, but also environmental and resource sustainability as these ultimately must underlie any economic sustainability. Hence, in this section we focus on the biophysical foundations of Costa Rican industry, analyzing the flow of energy and matter through the Costa Rican ecosystem.

Costa Rica's industrial structure is a subsystem of the Costa Rican ecosystem that is, in turn, a subsystem of the global ecosystem. Examples of such a model can be seen in Figures 21-5 and 21-6, flowcharts of the country's coffee industry (see also Figure 12-18). We can see from this figure that the structure of Costa Rican industry is linked tightly to the industrialized countries, and that it is inevitably limited by the availability and relative price of resources in the whole system, i.e., the global ecosystem. Furthermore, it is clear that the development of Costa Rican industry is as dependent on activities and policies in other systems as it is on activities and policies in the Costa Rican industrial sector itself. This includes both other nations and other sectors of the Costa Rican economy, all of which compete for scarce resources such as matter and energy in the form of capital, workers, water, land area, and fossil and nonfossil fuels.

As is evident from the analyses in most chapters of this book, all of Costa Rican industry (and agriculture) is heavily dependent upon the use of fossil fuels, and on fossil-fuel-derived and fossil-fuel-intense products manufactured elsewhere. Costa Rican industry uses, and contributes to the using up of, fossil fuels and their derivatives, fossil-fuel-intense industrial products, groundwater, metals and minerals, forests and forest products, soil, etc. This occurs both directly, e.g., in manufacturing both in Costa Rica and in the industrial countries, and indirectly, e.g., in growing export crops to pay for the imported materials and fuels. Thus at this ultimate level, Costa Rican industry cannot possibly be characterized as sustainable, and as imports and consumption continue to increase every year it is becoming ever less so. The same argument, of course, holds true for every industrializing and industrialized country in the world.

There are, of course, many who would not agree with the biophysical definition of sustainability that is employed here. Instead, they argue for using other adjectives to denote their preferred concept of sustainability, such as "economic" or "institutional." One needs to be aware, however, that the usage of the term "sustainability," if not accompanied by any further temporal specification, implies that the biophysical resources of the global system are sustained. If they are not, unsustainability is the result. This unsustainability affects all systems of the earth, and particularly human-dominated systems characterized by their complete dependence on the surrounding ecological systems. Hence, biophysical unsustainability deprives any notion of "institutional sustainability," "economic sustainability," and other popular constructs of any potential empirical relevance.

This argument also, or rather particularly, holds true for the tourism sector. I examine as one component of the sustainability of tourism the amounts of fossil fuels that are consumed by transporting the tourists into and out of Costa Rica by plane. In 1994, over 40% of Costa Rica's international visitors originated in North America. The respective numbers for Central America and Europe were 30 and 18%, respectively. The remaining 10% of visitors were from other countries (ICT, 1995). Let us assume an average 4-hr flight to and from San Jose, e.g., Philadelphia to San Jose. This seems to be a conservative estimate of the average tourist flight distance to/from Costa Rica, considering that all Canadian and European visitors have to travel far longer distances to reach the country. Costa Rica's national airline (LACSA) operates many older Boeing 727 airplanes that use about 10,000 lb of fuel (5 metric tons) per hour. This translates into roughly 3.9 metric tons of carbon released per hour by the plane. A 727 carries 170 passengers maximum. At a normal load factor of 0.7, this is about 120 passengers. For a 4-hr flight, this translates into about 260 kg carbon per person for a round trip, counting only direct fuel consumption and not the energy to build and service the airplane, airports, and associated ground transportation, or lodging or tourist travel within Costa Rica. This results in roughly 202,000 metric tons of carbon released by 770,000 passengers per year. If a more modern Boeing 757 airplane is used the fuel use and associated carbon release would be about two-thirds of the given amounts. Most North American and European visitors to Costa Rica travel on planes more efficient than a 727. However, as already mentioned, their travel distance is considerably longer. Therefore, the numbers just calculated can be used as rough averages for direct fuel consumption and the correlated carbon emissions.

Thus the tourism industry is hardly sustainable if we consider the fossil fuel resources used in getting the tourist into and out of Costa Rica, let alone to various points within the country. This is so for two reasons. First, petroleum is a finite and not readily substitutable resource, so any activity dependent on its consumption is by definition unsustainable (tourist airplanes powered by renewable energy are not in sight). Second, the combustion of fossil fuels contributes to the increase of carbon in the earth's atmosphere, and is seen as one causal factor in the atmosphere's warming (National Science and Technology Council, 1996).

Considering only the second factor, we can assess the possible sustainability of the carbon released by the tourist visits by calculating the carbon that might be taken up by either a mature or a growing tropical forest. A mature tropical rain forest in equilibrium has been estimated to pump perhaps 100 kg carbon per hectare per year into deep storage, such as the ocean (Hall *et al.*, 1992b). About 3 ha of such a forest per round trip per person traveling on a plane would be needed to compensate for carbon emissions. The area needed expands to more than 2.3 million ha of mature tropical rain forest if it is to compensate for the 770,000 international arrivals per year in Costa Rica. This, however, is almost half the total land area of Costa Rica, and considerably more forest than exists in all of the country.

Perhaps it would be more realistic to consider the carbon absorptive capacity of the recovering dry forests of Guanacaste. Reforestation of existing pasturelands could sequester much more carbon per hectare than a mature forest. A reasonably growing tropical dry forest (the ones currently recovering) might sequester 5 metric tons of carbon per hectare per year. The 770,000 international arrivals per year then would require 46,000 ha of recovering dry forest for absorption of the carbon released by their visits. As the respective reforested area approaches maturity and its carbon-sequestering ability diminishes, another area would need to be reforested to offset this effect and maintain the carbon absorption. Obviously, this process cannot continue for long, since reforestation competes with agriculture, urban areas, and other land uses. Considering these numbers, the claim of the sustainability of tourism (or ecotourism, for that matter) seems to be illusory.

A word seems to be in order to avoid misinterpretations regarding the author's position on ecotourism, in Costa Rica and elsewhere. Ecotourism, if it helps to preserve the integrity of biotic systems, and natural resources in general, is clearly preferable to conventional or mainstream tourism. The author unequivocally agrees with this and sincerely hopes that the percentage of ecotourists will substantially rise in the future. One should, however, be careful not to confuse preferability with sustainability.

## C. Could Industrial Sustainability Be Achieved in Costa Rica?

One might conclude that with a different set of policies, based on biophysical analyses of the country's resource endowment, a more sustainable industrial structure could have been established in Costa Rica. On the other hand one needs to consider and critically examine the possibility that an industrial expansion sufficient to eliminate the debt problem might simply be impossible within the biophysical limits of Costa Rica: that it simply is not possible for the nation to generate a comfortable amount of economic wealth for its continually growing population without a considerable increase in both economic and environmental nonsustainability.

The fundamental problem, besides a steadily growing population, is the requirement for high-quality land and other resource inputs to generate the foreign exchange needed to pay for imported industrial inputs, agricultural inputs, and consumer goods. As pointed out in the previous section, these foreign exchange uses compete with each other for the scarce natural resources, e.g., high-quality land. Costa Rican industry needs the financial streams flowing from the export commodities grown on this land in order to pay for imported inputs. At the same time, the land would be needed to grow the forests necessary to make the Costa Rican tourism industry sustainable, if only from the perspective of sequestering the carbon released by tourist planes. Alternatively, although not explicitly the concern of this chapter, biophysically sustainable agriculture would impose significant, and probably unfulfillable, additional requirements for land, thereby interfering with the direct and indirect demands for land exerted by industry, consumer goods, and tourism.

From a systems perspective, the possibility of future sustainability of Costa Rica's present style of industrial development could be hoped for only by opening up access to a nondiminishing energy supply that does not degrade the country's and the earth's ecosystems. The possibility of such a supply might exist in Costa Rica in the form of fuelwood (Chapter 18) and hydroelectric power. However, the question of its biophysical sustainability is not answered yet (Chapter 26), and it certainly could not support for long the current path of development with its rising demand for high-quality fuels, especially electricity. For the world as a whole, on its current trajectory of "development," such a nondiminishing, biophysically sustainable energy supply is, as of yet, beyond imagination (Hall *et al.*, 1986; Peet, 1992). Therefore, an analysis of the present situation cannot be based on the presumption of the future availability of such an energy source, and neither can any policy built on such presumption be considered wise.

The preceding analysis tells us something that was not obvious from our previous analysis of the industrial sector from a monetary perspective. This comes as no surprise, since monetary analyses are often disconnected from the underlying biophysical realities of economic systems. Its principal value is in helping to understand the biophysical causes of the economic difficulties that Costa Rica has encountered. An interesting question is whether the generation of biophysical assessments in the past would have guided these development policies in a more effective, or even sustainable, way, and if they could do so in the future. I like to think so.

## ACKNOWLEDGMENTS

The author thanks Doug Havnaer, commercial airline pilot and brother-in-law of the editor for the information regarding airplane fuel use and load data, and Bernardo Aguilar, for the provision of specific data on the new INTEL plant.

## REFERENCES

Aylward, B., K. Allen, J. Echeverria, and J. Tosi. 1996. Sustainable ecotourism in Costa Rica: The Monteverde Cloud Forest Preserve. *Biodiversity and Conservation* 5:315–343.

Campbell, C. J. 1997. Depletion patterns show change due for production of conventional oil. *Oil and Gas Journal* 29:33–37.

Carley, M., and I. Christie. 1992. *Managing Sustainable Development.* Earthscan Publications, London.

Clark, M. A. 1995. Nontraditional export promotion in Costa Rica: Sustaining export-led growth. *Journal of Interamerican Studies & World Affairs* 37:181–223.

Economist Intelligence Unit (EIU). 1995. *EIU Country Profile 1995–1996: Costa Rica.* EIU, London, UK.

Economist Intelligence Unit (EIU). 1996. *EIU Country Profile 1996–1997: Costa Rica.* EIU, London, UK.

Economist Intelligence Unit (EIU). 1997. *EIU Country Report 1st Quarter 1997: Costa Rica.* EIU, London, UK.

Greider, W. 1997. *One World, Ready or Not. The Manic Logic of Global Capitalism.* Simon & Schuster, New York.

Hall, C. A. S., C. J. Cleveland, and R. K. Kaufmann. 1986. *Energy and Resource Quality: The Ecology of the Economic Process.* Wiley–Interscience, New York. Reprinted 1992, University Press of Colorado, Boulder, CO.

Hall, C. A. S. 1992. Economic development or developing economics: What are our priorities? In M. Wali (Ed.), *Ecosystem Rehabilitation,* Vol. I. *Policy Issues,* pp. 101–126. SPB Publishing, The Hague, The Netherlands.

Hamilton, N., and C. Thompson. 1994. Export promotion in a regional context: America and Southern Africa. *World Development* 22(9):1379–1392.

Instituto Costarricense de Turismo (ICT). 1995. *Anuario Estadistico 1994.* ICT, San Jose, Costa Rica.

Inter-American Development Bank. 1994. *Economic and Social Progress in Latin America. 1994 Report.* Inter American Development Bank, Washington, DC.

Inter-American Development Bank. 1995. *Economic and Social Progress in Latin America. 1995 Report.* Inter American Development Bank, Washington, DC.

International Monetary Fund. 1998. International Financial Statistics Yearbook 1997. International Monetary Fund, Washington, DC.

National Science and Technology Council, Subcommittee on Global Change Research, Committee on Environment and Natural Resources Research. 1996. *Our Changing Planet. The FY 1996 US Global Change Research Program*. Office of Science and Technology Policy, Washington, DC.

Peet, J. 1992. *Energy and the Ecological Economics of Sustainability*. Island Press, Washington, DC.

Tardanico, R., and M. Lungo. 1995. Local dimensions of global restructuring: Changing labour-market contours in urban Costa Rica. *International Journal of Urban and Regional Research* 19:223–249.

Wilkie, J. W. , and Ochoa, E. (Eds.) 1989. Statistical Abstract of Latin America, Vol 27. University of California, Los Angeles.

Wilkie, J. W., C. A. Contreras, and C. Komisaruk (Eds.). 1995. *Statistical Abstract of Latin America,* Vol. 31. University of California, Los Angeles.

CHAPTER 23

# The Internationalization of the Costa Rican Economy: A Two-Edged Sword

DAWN R. MONTANYE, JUAN-RAPHAEL VARGAS, AND CHARLES A. S. HALL



## I. INTRODUCTION

International trade has been growing throughout the world. For developing countries, increased trade through liberalization policies has been prescribed by lending organizations and many economists to facilitate economic growth and alleviate debt crises. Ironically, increased trade often has been associated with increased foreign debt and trade imbalance (Ritchie, 1992; Morris, 1990). Nevertheless, many pressures continue to boost those sectors that provide tradable commodities. This is at least as true for Costa Rica as it is for most other parts of the globe. This chapter examines trade in Costa Rica and its relation to the country's foreign debt situation.

### A. DEVELOPMENT AND THE WORLD MARKET

Development through economic growth, at least as it has occurred over the last two centuries, is nearly synonymous with industrialization (Cottrell, 1955;

Hall, 1992a). The way a country industrializes traditionally has been determined by a country's approach to trade. Two different trade strategies have been typically employed by governments to move toward development and change the pattern of industrialization: import substitution and export-led growth (Gillis *et al.*, 1996).

Import substitution replaces imports with domestically produced goods. The process involves identifying domestic markets for development as determined by those dominated by imports. The mastery of production technology is encouraged by instituting policies that are designed to ensure domestic profitability. Competition from imports is alleviated by erecting tariffs and/or quotas on imports. The anticipated result is the emergence and development of the domestic industrial sector.

Export-led growth (ELG) is seen as a way to sustain industrialization and support higher levels of production than import substitution. The neoclassical economics concept of regional comparative advantage through specialization is the conceptual framework behind the ELG strategies (Chapters 3 and 24). These include identifying or developing areas of a country's economic sector that enjoy global comparative advantage and offering export incentives such as tax exemptions, tariff reductions on the imports required for production, and incentives to attract export-oriented foreign investment.

In the developing world, Latin America was the first to explore the potential of an import-substitution strategy following the disruption of the profitability of their primary exports during the Great Depression and World War II. For many emerging industrial countries in the 1940s and 1950s, industrialization through import substitution was considered the best way to develop. It was believed that this type of trade structure allowed for greater growth potential and was safer and more predictable than participating in the international market (Rottenberg, 1993). A policy of import substitution was practiced for many decades until the 1970s when the governments of many developing countries loosened protectionist policies, encouraging expanded trade. Often, however, these changes were limited in scope.

In the early 1980s, when the foreign debt burden increased, many developing countries began to pursue export-led growth and trade liberalization strategies for development and income generation. The major reforms toward export-led growth strategies shifted the focus of economic production from import substitution for the domestic market to manufacturing for export to foreign markets.

Trade liberalization facilitated the production and trade of exports by weakening trade barriers. Policies included establishing free trade zones and eliminating price "distortions" to reflect the "free market value." According to this approach, the resultant increase in trade through greater participation in the world market would ensure the generation of foreign exchange, enhance economic stability, and improve access to foreign-derived industrial technology.

These reforms were reinforced strongly through the 1980s by major pressures from creditor nations to generate foreign exchange for debt service (Inter-American Development Bank, 1996). In addition, agreements with multilateral lending organizations had as one of their principle objectives of adopting outward-oriented strategies, the improvements in the balance of payments (U.S. Congress, 1994). The World Bank noted in 1982 that free trade was adopted by several countries in order to restore equilibrium between aggregate supply and demand, narrow the imbalance in foreign trade, and reduce recourse to foreign borrowing.

## B. Trade in Costa Rica

In the late 1950s there was an intense debate in Costa Rica about how to develop economically. The country's economy was small, and as a consequence vulnerable if opened to increased trade in the global market. The export agricultural sector was strong, but narrow in scope with coffee and bananas making up 97% of exports. In 1963 the country decided to join the Central American Common Market (CACM), creating trade avenues for diversifying agricultural and manufacturing production and opportunities for increased industrialization. Between 1963 and 1973 exports of manufactured goods rose from 13.9 to 30.6% of total exports with about four-fifths of production going to CACM partners. Exports of beef and sugar together grew from 11.4 to 15.8% of total exports. Although these increases represented a boon for particular commodities, traditional agricultural products of coffee and bananas still held the lion's share of exports, representing 53.5% of total exports in 1973 (Rottenberg, 1993).

Because of Costa Rica's small economy, including a small domestic market, a specialized export market, and a dependence on imports, many in the country understood both their vulnerability to the world market, but also that "trade must serve as the economy's engine for growth" (Rottenberg, 1993, p. 96). The government of Costa Rica chose to encourage highly protected import-substitution industries, while exporting globally a few agricultural products (coffee, bananas, beef, and sugar) and engaging in trade with other Central American countries through the CACM.

Although there was free trade among the customs union partners, a common, highly protective external tariff barrier existed for imports from all other countries. This barrier, however, existed mainly for final consumer goods, with much weaker constraints imposed on imports of raw and intermediate goods. As a result the manufacturing sector became very reliant on these imported raw and intermediate inputs to production (Rottenberg, 1993).

## C. Internal Affairs

Since the 1940s, the Costa Rican government had been investing in social programs, energy, and transport, and in addition did not have to contend with internal political unrest. The early 1960s and late 1970s marked a great expansion of Costa Rica's public sector and large investments in the development of national infrastructure and state production activities. The development of national government programs of universal social security, health and maternity services, and education were made available to greater numbers of Costa Ricans. Meanwhile, increased government spending, diversification of the agricultural sector, the coffee boom of 1976–1977, and growth of manufacturing helped to stimulate the economy and facilitate growth for the country as a whole as well as on a per capita basis (Chapters 2 and 4).

## D. Crisis and Structural Changes

The oil price increases of the late 1970s and early 1980s and a decline in coffee prices in the early 1980s caused an economic crisis for Costa Rica (Chapter 2) and brought about heavy government borrowing to maintain stability. Costa Rica's small economy was extremely vulnerable to fluctuations in world market prices for both imports and exports. Support of the industrial sector and the costs of supporting an ever burgeoning bureaucracy meant large expenditures for the Costa Rican government. In addition to high oil prices and less income, the country was left with a large balance of payment deficit. The government chose to increase borrowing from foreign commercial banks instead of weakening public institutions or public services. When foreign exchange was exhausted after 1980, the government expanded credit. Overvaluation of the colón meant that importers of Costa Rican goods were now paying comparatively more, and as a result exports decreased.

In 1982, the external debt was approaching U.S. $3 billion, nearly equal to the real GDP, which was falling by 7% per year (Figure 23-1), and average annual inflation was up to 109%. Costa Rica turned to the World Bank, the International Monetary Fund (IMF), and the U.S. Agency for International Development (USAID) for assistance. As part of the loan agreements Costa Rica was forced to implement "structural adjustment" policies (Chapter 24) and import substitution was eliminated in favor of export-led growth strategies.

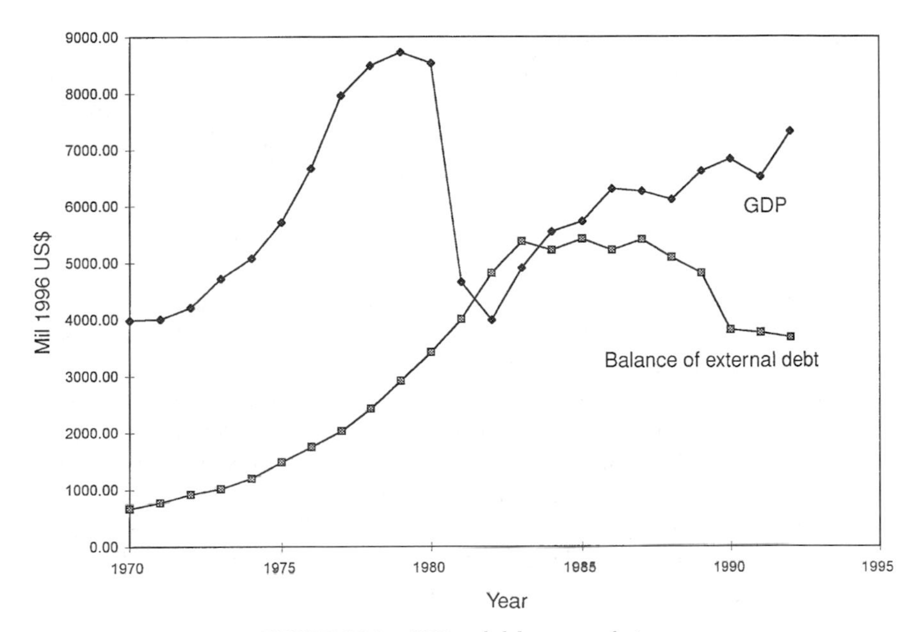


FIGURE 23-1 GDP and debt accumulation.

To encourage exports, the Costa Rican government put in place fiscal benefits such as subsidies and tax exemptions for exporters, and eliminated tariff and nontariff barriers to imports, allowing the freer flow of the imports required for export production. In addition, trade zone liberalization was instituted and tax break incentives were offered to producers of exports. For example, assembly plants operating in free trade zones that exported over half of their production were exempt from taxation for 10 years, and benefited from the priority given by the government to the construction of infrastructure required for transport (Hansen-Kuhn, 1993). These plants, known as maquiladoras, included food processing, chemical products, textiles, and metal processing industries [Economist Intelligence Unit (EIU), 1995]. Export contracts, know as CATs (Certificados de Abono Tributario), gave tax credits of 15% on goods having at least 35% value added. To further encourage exports, the United States introduced the Caribbean Basin Initiative (CBI) in 1984, offering duty-free status on export products coming into the United States with 35% value added.

Under the policy known as "Agriculture of Change," incentives for agricultural production for export included the removal of export taxes, exemption from income taxes on production for exports, preferential interest rates, and special access to foreign exchange for export producers (Hansen-Kuhn, 1993).

The main objective of such incentive policies was to encourage export diversification and increase production of new agricultural and industrial exports in order to generate foreign exchange (Horkan, 1996). Export-led growth strategies focused particularly on diversification into areas where Costa Rica supposedly had comparative advantage, replacing traditional commodities which some felt had become exhausted in terms of their market potential. As one example, USAID, the IMF, and the World Bank argued that existing basic food crop subsidies were "too costly and that it would be cheaper for Costa Rica to import basic grains rather than subsidize local producers" (Honey, 1994, p. 158). As a result, production and land area for certain basic grain crops decreased (Figures 23-2 and 23-3) and the country focused on those crops that Costa Rica was able to grow with high quality and greater efficiency.

A decline in world market prices of traditional agricultural export commodities, including coffee, cotton, sugar, bananas, and beef, led the Costa Rican government, with assistance from the international lending agencies, to support the development and expansion of nontraditional agricultural and manufactured commodities. These nontraditional products were defined as (1) not traditionally produced in Costa Rica, (2) traditionally produced for domestic consumption but now exported, or (3) traditional products now exported to a new market (Thrupp *et al.*, 1995). Between 1982 and 1988 USAID provided over $110 million in loans, designated for the encouragement

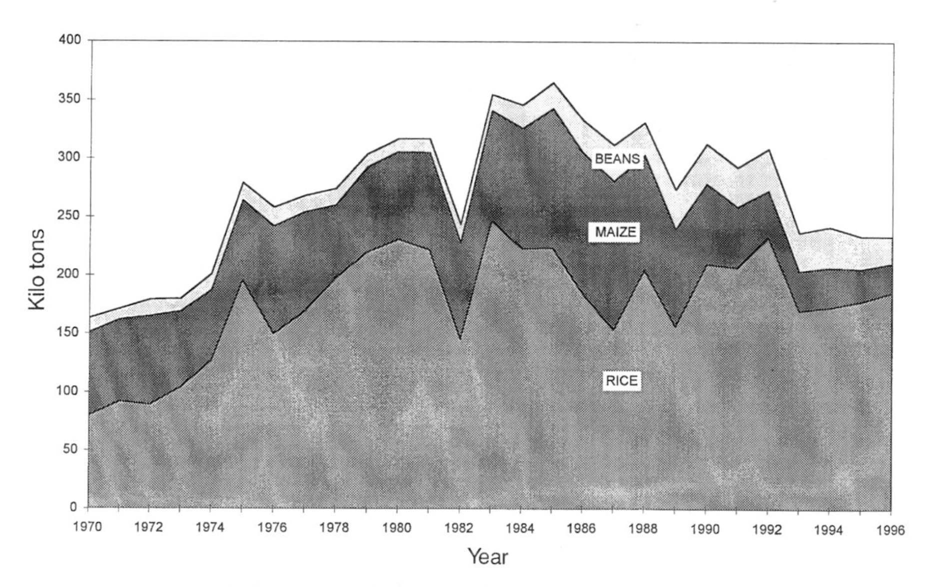


FIGURE 23-2 Volume production for basic grains.

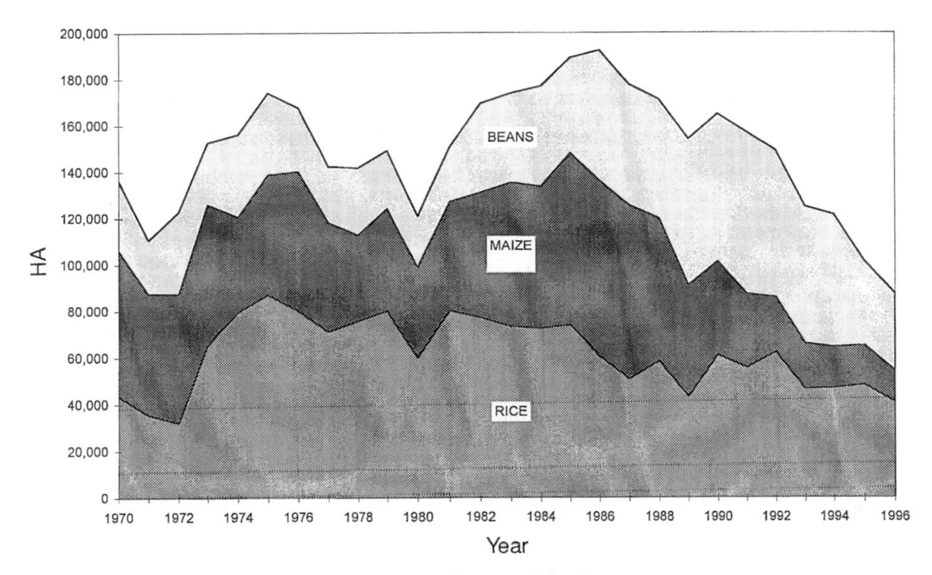


FIGURE 23-3 Area harvested for basic grains.

of nontraditional exports through technical assistance programs and by developing links to markets.

The strategy of nontraditional agricultural export promotion in particular revolved around diversification into areas that were thought to have the potential for high economic return, and would generate new jobs and foreign exchange earnings (USAID, 1995). The agricultural crops included African Palm, cut flowers and ferns, macadamia nuts, cassava, coconut and tropical fruits, cotton, and vegetables (EIU, 1995). Expansion of nontraditional commodities in the manufacturing sector included diversification into textiles, electrical equipment, and wood products.

Prompted by USAID policies, the Costa Rican government offered 100% tax allowances on the direct profits of nontraditional exporters (EIU, 1992). In addition, the Costa Rican Coalition and Development Initiatives (CINDE) were established through a grant from USAID with the goal of "stimulating growth in the production and export of nontraditional goods and services, thereby increasing levels of employment and foreign exchange earnings for Costa Rica" (Horkan, 1996, p. 9). The office was primarily "engaged in direct efforts to bring in export-oriented foreign investment" (Clark, 1995, p. 183), as well as provide technical assistance and training programs. With the help of investment attraction in areas such as agriculture, industry, tourism, electronics, metalworking, and the textile industry, CINDE generated approximately $150 million in non-traditional exports (CINDE, 1992).

CINDE's nontraditional agricultural export support project provided support both directly to producers and indirectly to country-based export federations (Hardesty, 1994). "Since nontraditional agricultural export crops are labor-intensive, it was argued that they could create employment opportunities for the rural poor. The region's tropical climate and the belief that these products would not likely compete with U.S. products were also important factors in the decision to promote nontraditional agricultural exports" (Horkan, 1996, p. 4). Together with textile exports, these nontraditional agricultural exports were perceived as having the potential to increase the country's export offerings and become the "economy's main engine for new growth" (Honey, 1994).

# II. RESULTS

## A. Results of Export-Led Growth Strategies

At first glance these policies seem to have been successful. Exports increased, with a substantial portion coming from nontraditional exports (Figure 23-4). From 1983 to 1990, the recorded average growth in GDP for Costa Rica was

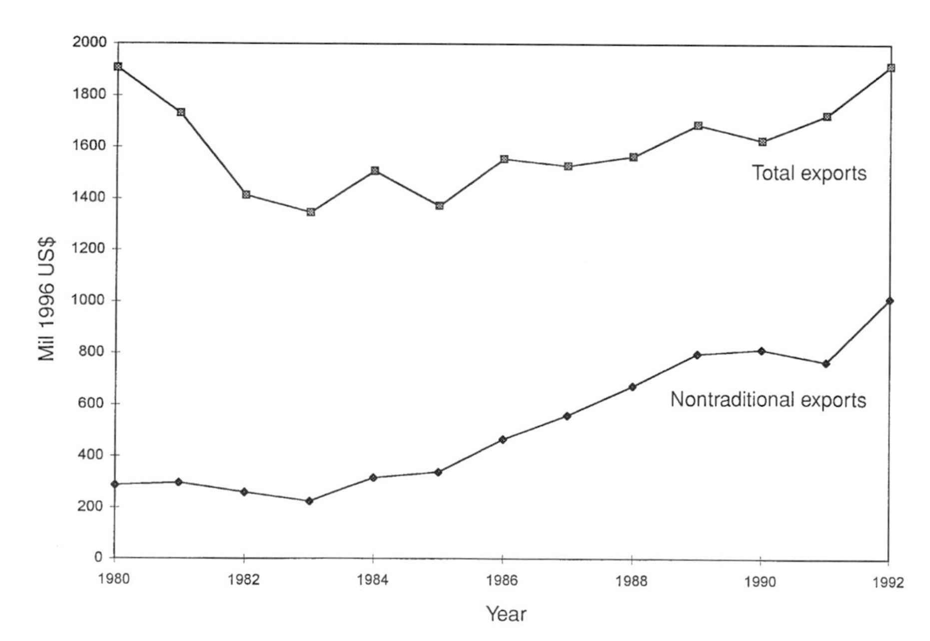


FIGURE 23-4 Nontraditional and total exports.

4.3%, reportedly one of the best performing economies in the region during that time (EIU, 1993b). USAID studies on the impact of the agency's policy reform efforts determined that the "increase in exports helped the country to recover from the economic crisis of the 1980–1982 period" (Horkan, 1996, p. 10).

Closer analysis confirms these findings. Exports of nontraditional agricultural products, which generated $142 million in 1983, rose to $678 million in 1990 (Figure 23-4). Even with a 38% fall in export shares of coffee and bananas in 1992, total export revenue rose to $1.829 billion. Nontraditional agricultural exports continued to account for a growing percentage of Costa Rica's foreign exchange from total exports and, according to a USAID evaluation, generated $98.6 million in foreign exchange from the beginning of the year through September 1994 (USAID, 1995).

The country's maquiladora sector generated revenues of over $100 million in export earnings in 1992 and attracted various multinational corporations to the country, including GTE, Motorola, Firestone, and Coca-Cola (EIU, 1993a).

The traditional agricultural sector, although growing at a slower rate than the nontraditional, provided the most production volume and foreign exchange. Bananas, the number one agricultural export since 1989, was one of Costa Rica's prime export revenue generators with volume production increasing from 48.9 million 18.14-kg boxes in 1980 to a record 101 million boxes in 1993 (USDA, 1994). Coffee also remained a primary export, with earnings second only to bananas (Figure 23-5).

## B. Assessment of Net Gain from Trade

While exports increased throughout the 1980s and early 1990s, imports also increased, leaving a trade deficit throughout this period (Figure 23-6). Among the most important nontraditional agricultural products were pineapples, melons, ornamental plants, foliage and flowers, roots and tubers, orange juice concentrate, and palm hearts (USDA, 1994). Many of these varieties required substantial investment in irrigation systems, fertilizer, and pesticides (Hansen-Kuhn, 1993). Nontraditional manufacturing, as well, required high amounts of imports of raw and intermediate materials for production.

Export-led growth strategies which called for intensified production also greatly increased the need for inputs. The annual cost of imported raw and capital inputs for agriculture rose from $290 million in 1982 to $566 million in 1992, with imports of fertilizer increasing from $16 million in 1982 to $34 million in 1992. Pesticide inputs also grew, with imports going from $35 million to $50 million during those same years (FAO *Trade Yearbook*, various years). Imports of raw and capital materials used for manufacturing

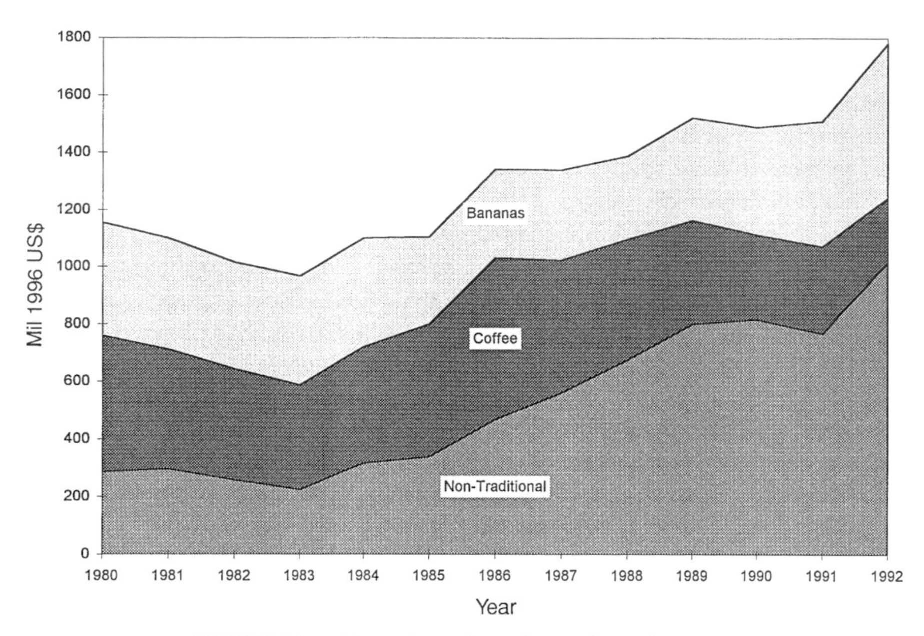


FIGURE 23-5 Nontraditional vs select traditional exports.

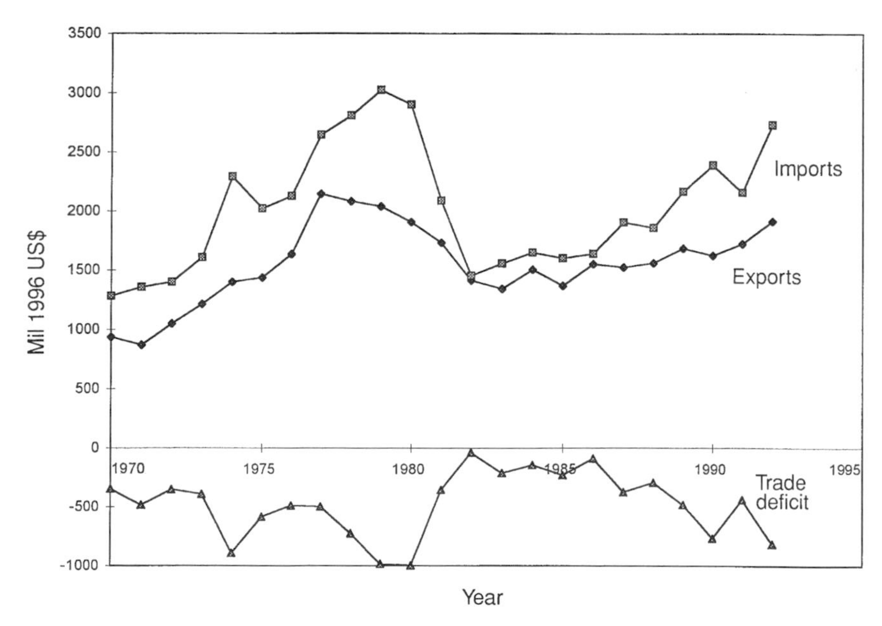


FIGURE 23-6 Trade balance.

also increased, rising from $706 million in 1982 to $1.417 billion in 1992 (Figure 23-7).

## C. Net Foreign Exchange

Although Costa Rica was able to pay $3.4 billion toward external debt service between 1982 and 1992, the external debt nevertheless grew from $675 million to $1.34 billion during this same period. Considering this growing debt burden, and the objective of increasing foreign exchange through outward-looking growth strategies, we calculated net foreign exchange to determine total revenue available for investment and consumption. The calculation involved subtracting payments for debt servicing and the imports used in export production from gross foreign exchange.

From 1982 to 1992, Costa Rica debt payments accounted for about 20% of total foreign exchange earnings. The value of imports used to produce exports was estimated from 25% to 30% (Figure 23-8). Based upon these values, the net foreign exchange generated from exports left only about 45–49% of foreign currency as net foreign exchange available to purchase the imports needed for the expanding internal economy, including fuel, food to feed the growing population, and transportation vehicles. All of these imports and

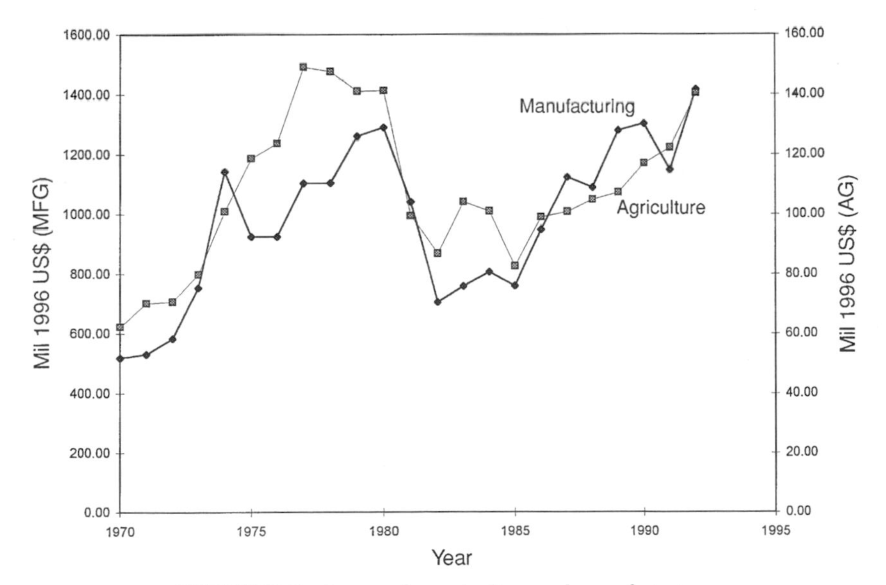


FIGURE 23-7 Imports for agriculture and manufacturing.

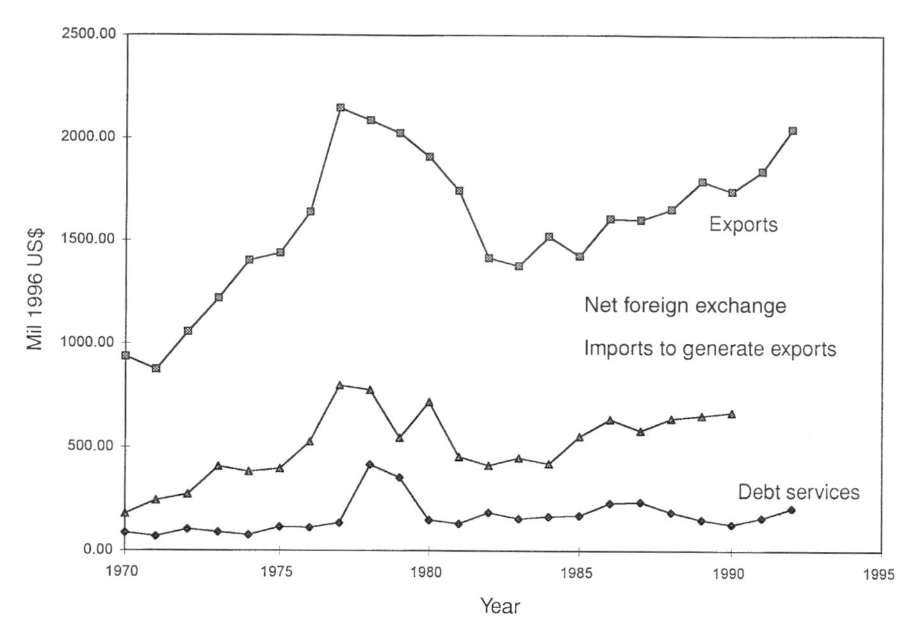


FIGURE 23-8 Net gains from trade.

much of the debt servicing can be considered as the costs of producing goods for export and both represent unavailable funds for purchases for consumption.

## D. Limits to Free Trade

"The Government of Costa Rica continues trade and economic policies in favor of open markets, international competition and freer trade supported through active IMF and World Bank programs" (U.S. State Dept., 1995). As a result of incentive programs for expanded trade in the early 1980s, Costa Rica expanded its sphere of trading partners. While the country is still involved in regional trade with Central America, the USA is now Costa Rica's most important trading partner, purchasing over half of its exports and providing over half of its imports (EIU, 1995). In 1994 North America (USA, Canada, and Mexico) imported Costa Rican goods worth $993 million while the European Union imported goods worth $665 million (EIU, 1995).

Although trade liberalization initiatives did open Costa Rica up to larger markets, there are external trade barriers that hamper the country's ability to find trading partners. The Caribbean Basin Initiative, introduced by the United States, offers duty-free status on value-added exports. But this initiative in fact has ended up excluding 50% of Costa Rican export products (Monge, 1994). Most of the CBI tariff reductions applied "mainly to those items already entering the United States duty free under the Generalized System of Preference" (Clark, 1995, p. 105). In addition the European Community has imposed restrictions on banana exports and the United States imposed its own restrictions on pineapple imports (U.S. State Dept., 1995; Sexton, 1997).

## E. Costs of Encouraging Export-Led Growth

Internal government trade incentive policies have served as an increasing economic drain on the country's already tightened financial resource base. Tax breaks deny the government revenues, while subsidies to export producers act as direct costs. Tax rebates or CATs were an increasing cost to the Costa Rican central government. When this incentive structure was fully implemented in 1983, the cost was $56 million. By 1992 it was $940 million per year (Figure 23-9). The tax-free zones that were meant to encourage new industries to locate in Costa Rica continue to deprive the government of customs duties and taxes on sales and profits (Clark, 1995).

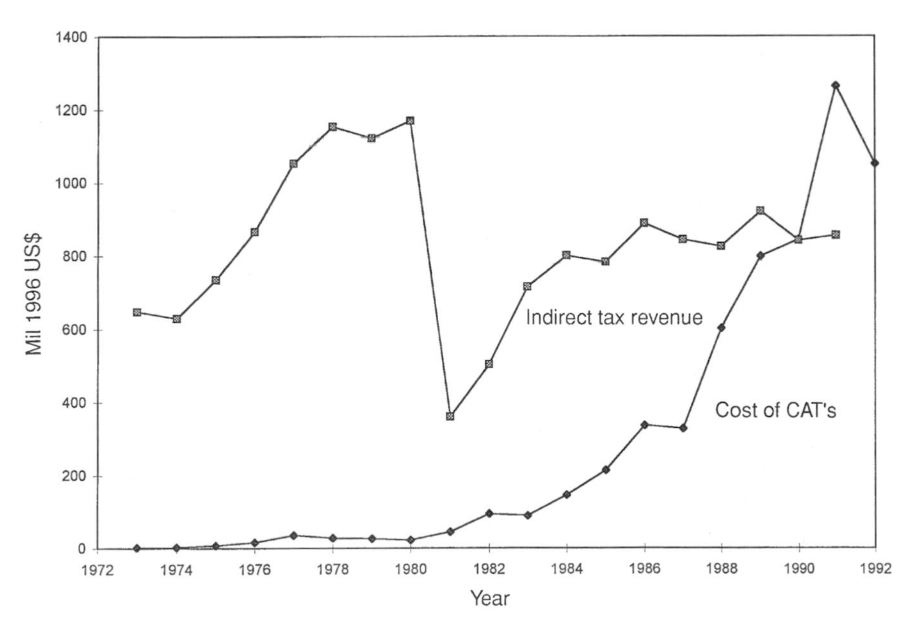


FIGURE 23-9 Cost of Certificados de Abono Tributario (CATs).

In addition, withdrawal of USAID support and the closing of CINDE in July of 1996 meant that Costa Rica would now have to sustain on its own the nontraditional export sector, which had previously received support, outreach, and educational services from the USAID country office.

## III. DISCUSSION

The choice of encouraging export-oriented growth required that Costa Rica obtain a steady stream of foreign currency in order to buy the inputs required to support these developing sectors. In the agricultural sector, fertilizer for nontraditional varieties costs two to three times as much per hectare (Figure 23-10) compared to that for traditional varieties. Pesticides are used more intensively for export crops than for domestic crops, especially for most nontraditional agricultural exports (Thrupp *et al.*, 1995). The large quantities of inputs required for agricultural production suggest the importance of biophysical limitations which must be compensated for by artificial inputs. These facts call into question the sustainability of agricultural production in the future, and the ability to repay the debt. Retiring the principle on the debt seems essentially impossible given these constraints.

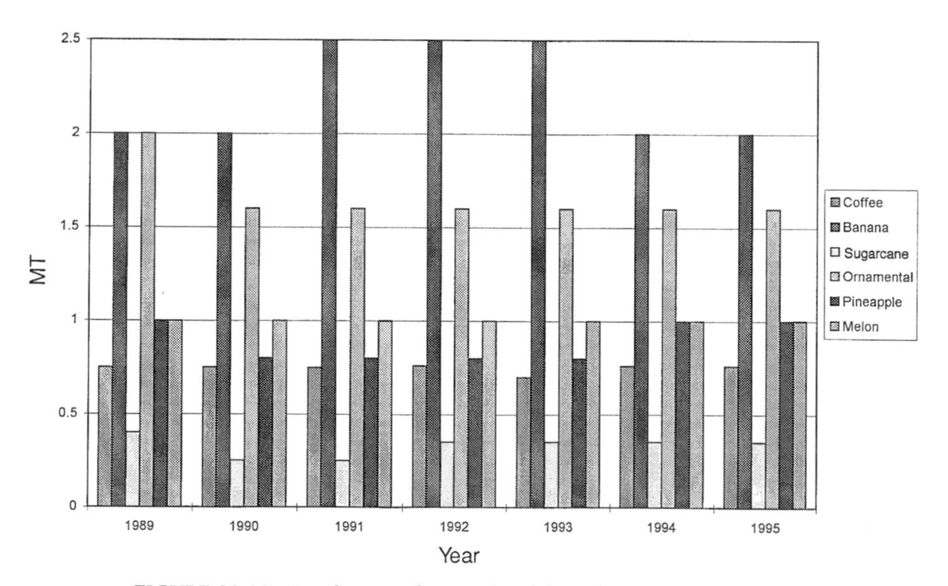


**FIGURE 23-10** Fertilizer per hectare (total formulation). Source: SEPSA.

## A. Debt Payments in Light of Net Foreign Exchange

The strategy of export-led growth also works against the repayment of loans because "while debt is principally in the public sector, the profits from non-traditional export agriculture and manufacturing accrue to the private sector" (Rosset, 1991, p. 31). The government is limited further fiscally by programs which include exemptions from export duties and tax allowances for nontraditional exporters. Consequently there is further pressure to increase total production in order to generate currency that will flow back into the Costa Rican government.

The increasing need for imports required for increasing both agricultural and manufacturing production undermines Costa Rica's ability to service its debt and reach economic stability. The calculation of net foreign exchange as exports minus inputs demonstrates that substantially less foreign exchange is available than gross revenue indicates. In essence, export-led growth strategies have led to less money for foreign debt payments, not more (Chapter 21).

## B. Relative Prices: The Achilles Tendon of Trade

We have shown that foreign exchange generated from increasing agriculture and manufacturing exports is much less when the costs of generating that foreign exchange are included in the equation. Net earnings of foreign exchange are compromised further by fluctuations in world market prices for goods such as fertilizers, pesticides, and fuel for farm machinery and transportation. The impact of relative prices is indicated in Costa Rica's trade balance. In 1990 the country's annual account deficit rose to a 10-year high of $584 million, largely as a result of high oil import prices and low prices and weaker global demand for coffee (EIU, 1992). Although increased exports were supposed to improve the negative trade balance that has plagued the country, it seems only to have exacerbated the situation because of their very high requirement for imported raw materials.

The country was able to achieve a trade surplus in 1982, 1984, and 1986, but that appears to be due to severely repressing imports (EIU, 1995). In subsequent years the annual trade deficit rose to $611 million in 1992 and $835 million in 1993 with resumed economic growth and an increase in imports. However, the slowdown in the economy in 1994 constrained the growth of imports and the trade deficit fell slightly (EIU, 1995). Although development objectives may have predicted otherwise, Costa Rica's stability ultimately rests on fluctuating world market prices and ironically its own level of economic growth.

## C. Increased Competition

While the lending agencies were encouraging Costa Ricans to shift agricultural production from staple products to "exotics" such as cut flowers and fruits for export, the same agencies were encouraging the production of similar commodities in other South and Central American countries. The result was an increase in the competition for foreign sales of these products. With so many vying for a share of world markets, prices fell, forcing governments to induce higher levels of exports in a desperate attempt to keep their hard currency revenues stable (George, 1992). The risk of such international competitiveness is market saturation, which can undermine the growth of the developing commodity sectors.

# IV. CONCLUSION

The international lending agencies that promoted export-led growth (USAID, World Bank, IMF) give a rather favorable report about the implications of this trade regime for growth and development. More comprehensive analyses, however, show that trade promotion programs of these agencies have failed to meet even their own objectives.

Specifically:

1. Costa Rica's debt continues to grow. Despite USAID's predictions for a reduction in the deficit, the debt growth rate increased nearly twofold between 1980 and 1985 and the trade deficit increased.
2. Increased exports have meant increased export earnings for Costa Rica. But when the direct costs of debt servicing (much of which was required to pay for loans incurred to develop the export industry) and the cost of the imports required to operate the export-generating industry are subtracted, the country is left with only about one-half of the foreign exchange earned from export sales.
3. While current export-led growth policies have diversified Costa Rica's export market, the result is a growing vulnerability to fluctuating world market prices for sale of their export goods and for the required inputs.

We have shown that trade reforms aimed at reducing Costa Rica's debt burden have instead worked to undermine the long-term and even the short-term economic health of the country. Continued programs that focus on blanket encouragement for the production of nontraditional goods, coupled with intensifying traditional agricultural and manufacturing production, are likely to further the financial burden already weighing heavily on this country.

What has become clear from our analysis is that the decision to diversify into nontraditional agriculture and increase yields in traditional agriculture did not take into account the biophysical constraints. The limitations of increasing crop production due to soil and climate constraints and the huge quantity of industrial inputs needed to compensate for that were not part of the decision-making equation. As a result what was required to overcome these limitations was a dramatic reliance on imported artificial inputs. The drawback, for Costa Rica in particular, is that it does not have the natural resource base from which to derive these industrial inputs and therefore must import them from industrialized nations. The result of such dependence is clear when one examines the growing trade deficit and increasing debt burden. If the objective for Costa Rica is, at the very least, economic stability, or at best some kind of sustainability, then an analysis must work within a biophysical framework to gain an accurate picture of the potential and limitations of increasing production.

## ACKNOWLEDGMENTS

We thank Timm Kroeger and Hong Qing Wang for review of this chapter.

## REFERENCES

CINDE. 1992. *International Guide of Costa Rica.* CINDE, San Jose, Costa Rica.

Clark, M. A. 1995. Nontraditional export promotion in Costa Rica: Sustaining export-led growth. *Journal of Interamerican Studies & World Affairs* 37:181–223.

Cottrell, F. 1955. *Energy and Society*. McGraw–Hill, New York. Reprint by Greenwood Press, Westport, CT.

George, S. 1992. *The Debt Boomerang: How Third World Debt Harms Us All.* Westview Press, Boulder, CO.

Gillis, M., D. H. Perkins, M. Roemer, and D. R. Snodgrass. 1996. *Economics of Development,* 4th ed. W. W. Norton, New York.

Hansen-Kuhn, K. 1993. Sapping the economy. Structural adjustment policies in Costa Rica. *Ecologist* 23:179–184.

Hardesty, S., and Taylor, T. G. 1994. An Analysis of the Economic Impacts of Nontraditional Agricultural Export Programs in Central America. USAID: Washington, DC.

Honey, M. 1994. *Hostile Acts: U.S. Policy in Costa Rica in the 1980s.* University of Florida Press, Gainesville, FL.

Horkan, K. M. 1996. *The Benefits and Costs of Nontraditional Agricutlural Export Promotion in Costa Rica.* Research & Reference Services Project. Washington, DC.

Inter-American Development Bank. 1996. *Economic and Social Progress in Latin America. 1996 Report. Special Edition: Making Social Services Work.* Inter-American Development Bank, Washington, DC.

Monge, G. and Gonzalez Vega, R. and C. 1994. *Political commercial, exportaciones y bienstar en Costa Rica.* Academia de Centroamerica y Centro Internacional para el desarrollo economico, San Jose, Costa Rica.

Norris, D. 1990. Free Trade: The Great Destroyer. *The Ecologist* **20**, 5.

Ritchie, M. 1992. Free Trade versus Sustainable Agriculture: The implications of NAFTA. *The Ecologist* **22**, 5.

Rosset, P. M. 1991. Sustainability, Economies of Scale, and Social Instability: Achilles Heel of Non-Traditional Export Agriculture? p.31 *Agriculture and Human Values.* Fall

Rottenberg, S. (Ed.) 1993. *The Political Economy of Poverty, Equity , and Growth: Costa Rica and Uruguay.* A World Bank Comparative Study. Oxford University Press, Oxford.

*Secretaria Ejecutiva de Planificacion Sectorial Agropecuaria* (SEPSA). Various years. *Boletin Estadistico* (statistical bulletin). Costa Rica.

Sexton, S. 1997. Going Bananas. *The Ecologist* **27**, 117–118.

Thrupp, L. A., G. Bergeron, and W. Waters, 1995. *Bittersweet Harvests for Global Supermarkets: Challenges in Latin America's Agricultural Export Boom.* World Resources Institute, Washington, DC.

USAID, 1995. *Nontraditional Agricultural Export Technical Support Project.* Evaluation Report. USAID, Washington, D.C.

U.S. Congress. House. Committee on Banking, Finance and Urban Affairs. 1994. *Oversight of the International Monetary Fund and the World Bank.* 103rd Cong., 2nd sess., 1994.

USDA. 1994. Agricultural Situation. *Agricultural Situation Annual-Costa Rica.* USDA/FAS, Washington, D.C.

US Department of State. 1995. Economic Policy and Trade Practices: Costa Rica. *Country Reports on Economic Policy and Trade Practices.* US Department of State, Washington, D.C.

CHAPTER 24

# An Assessment of the Effectiveness of Structural Adjustment Policies in Costa Rica

TIMM KROEGER AND DAWN MONTANYE



## I. THE COSTA RICAN STRATEGY OF EXPORT-LED GROWTH

This chapter is different from the rest of the book in that it is primarily about policy and not science. Although throughout the book the emphasis is on the biophysical aspects of Costa Rica, it is necessary to acknowledge the importance of policy as a frequent driver of biophysical events. The chapter is organized into two major parts. The first provides an overview of Costa Rica's experiences with the Structural Adjustment Policies (SAPs) demanded by the World Bank and the International Monetary Fund (IMF). Following this, the second part critiques the neoclassical economic theory underlying these policies, and gives an evaluation of the efficacy of these policies.

As developed in Chapter 2, the Costa Rican economy has never been truly autonomous, but has been influenced, and increasingly determined, by factors that lie beyond the control of decision-making processes inside the country. For reasons outlined in Chapter 22, Costa Rica accumulated massive international debt in the 1970s. The country temporarily terminated payments on its foreign debt in 1981. The effect of this action was compounded by Mexico defaulting on its debt payments the following year. Commercial banks stopped lending to Costa Rica. The country had to turn to International Financial Institutions (IFIs, i.e., the World Bank and the International Monetary Fund) for financial assistance, and had to accept the terms that accompanied it. The policies that these institutions imposed were based on promoting a strategy of *export-led growth* that fostered new export industries (Clark, 1995; Hamilton and Thompson, 1994), with the idea of generating sufficient foreign exchange so that the country's international debt could be repaid (Kahler, 1993).

These policies comprised what in economics is known as "stabilization" (the objective of the IMF) and "structural adjustment" (the objective of the World Bank), together known as "economic reform." Stabilization aims at "improving" the balance of payments and at controlling inflation via a reduction in public spending on consumption. Structural adjustment aims at (1) the reduction of a country's trade barriers in order to open it to foreign trade, (2) a change in the sectoral balance in a country's development strategy from production of heavy capital goods to consumer goods, such as textiles and agricultural products, (3) a shift from import-substitution industrialization to export promotion, and (4) a reduction in the size of the public sector. In short, stabilization and structural adjustment aim at trade liberalization, cutbacks in government expenditures, and privatization of government-owned industries. Together, these measures are supposed to improve a country's *allocative efficiency* and, ultimately, ensure the restoration of a positive balance of payments so that a country's debt service to northern governments, commercial banks, and multilateral institutions can be paid.

All of these adjustment measures are based on assumptions intrinsic to neoclassical economics, specifically that trade liberalization will encourage economic "efficiency" through the ensuing international division of the production of goods and services. According to neoclassical economic theory, the criterion by which this production is allocated between countries is cost efficiency, deriving from each country's comparative advantage. Neoclassical economic theory holds that through this process the aggregate welfare of all market participants will be enhanced (Chapter 3). Another assumption inherent in this argument is that this allocative efficiency is the key to the solution of all the other problems of a country.

The concept underlying these "adjustment" measures is commonly called "neoliberalism" (not to be confused with "liberalism" as often used in U.S. politics). Neoliberalism calls for privatization, deregulation, cuts in social spending, the replacement of institutional entitlements with a "residual" social policy, free capital flows, and free trade, all of which are argued for on grounds of neoclassical economic theory.

The IMF, the World Bank, and the U.S. Agency for International Development (USAID) all promote a strategy of export-led growth. Export-led growth is different from export promotion in that the former gives priority to exports over production for the domestic market, whereas the latter aims at eliminating biases against export but does not preclude strategies oriented to the development of the domestic market as well (Bhagwati, 1986; Barham *et al.*, 1992). Thus, the Costa Rican strategy was shifted, basically, from encouraging the domestic manufacturing of consumer goods to the encouragement of imports of such products (through tariff reductions), while promoting various exports to, in theory, pay for the imports. The overall strategy included the encouragement of *nontraditional export* products (Chapters 22 and 23).

By the early 1980s, the government of Costa Rica was encouraging the development of nontraditional exports by providing three types of incentives for export-oriented investment as outlined in the previous chapter:

1. Tax exemptions for maquila industries, which import their inputs and export their outputs (production in maquila industries is concentrated on textiles and apparel, with some recent diversification into jewelry, footwear, and electronics) (EIU, 1995)
2. The establishment of free trade zones to develop nontraditional exports that require greater domestic input
3. Export contracts, which provide tax exemptions, tariff reductions for imported inputs, and tax rebates (CATs) or credits of 15% for those exports having at least 35% of their value added in Costa Rica (Clark, 1995)

Such activities, along with a large influx of external finance, were supposed to encourage increased domestic production, which would allow for debt servicing and import expenditures. A main pillar in USAID's argument for increased nontraditional exports has been that export growth will help reduce Costa Rica's trade deficit. In a March 1988 report, for example, USAID projected that Costa Rica's trade deficit would fall from $264.4 million in 1987 (Figure 24-1) to $105.6 million in 1989 and a mere $84.4 million in 1990 (USAID, 1988).

## A. The Results of the Structural Adjustment Policies and the Export-Led Growth Strategy

Have the economic policies fulfilled the expectations of sufficient net economic growth to finance an industrializing Costa Rica while paying off the foreign debt required to generate that growth?

Although there were serious doubts expressed in material published by the IMF even as early as 1982 as to whether the Structural Adjustment Policies promoted would be able to fulfill their official goals (IMF, 1982), the develop-

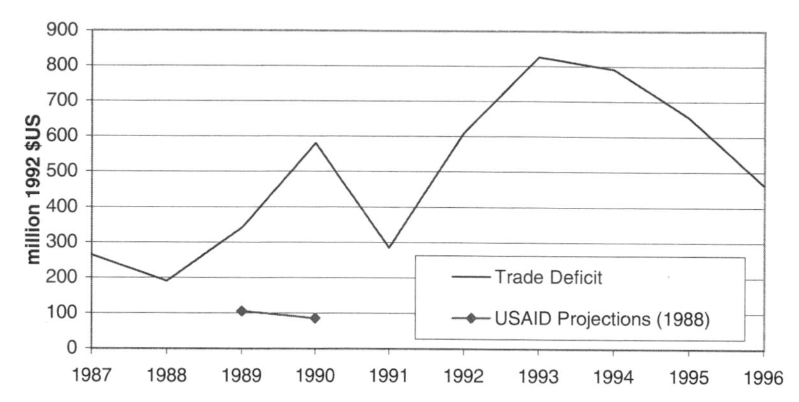


FIGURE 24-1 Costa Rica's trade deficit. (Source: IMF *International Financial Statistics*, 1998; USAID, 1988.)

ment institutions originally expected their interventions to be a finite process whereby appropriate policies, with the help of external aid flows, would permit countries to restore growth and tackle long-term development problems (UNCTAD, 1993).

Initially, the restructuring strategy in Costa Rica was considered an economic success by the criteria given by the IFIs. Production volume and, not surprisingly, the dollar value of the nontraditional exports increased, and the dollar value has been higher than that of the traditional export goods since 1992 (Figure 24-2; see also Chapter 22). Unemployment declined, and per capita income has increased gradually since 1984, following its precipitous decline by 44% between 1979 and 1983 (World Bank, 1981–1988). Moreover,

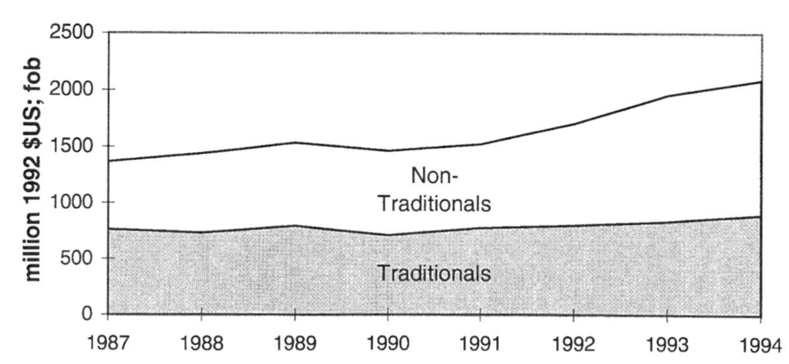


FIGURE 24-2 Dollar values of traditional and nontraditional exports. f.o.b., free on board. (Source: EIU, 1995.)

through the help of SAP-related loans from the IFIs, Costa Rica was able to come up with the debt service payments to its international creditor banks, thereby maintaining its international creditworthiness.

But not all of the original objectives were obtained. Otton Solis, chair of the Economic Subcommittee of the Costa Rican National Assembly and adviser to President Jose Figueres, describes the outcomes of the SAPs as follows: "We needed changes, we needed modernization, but structural adjustment policies were in a direction opposite from the changes we needed" (Solis, 1996, p. 20). To support this position, Solis points out that the economic objectives of the SAPs were not achieved: economic growth has remained very low, and where it has not, it appears unsustainable; international competitiveness has not been achieved; the fiscal deficit has grown sharply; and trade deficits became far greater during the structural adjustment period than before. For example, in 1990, the year for which the cited USAID report forecasted a trade deficit of $84.4 million, it actually approached $600 million (Figure 24-1). By 1993 it had risen to almost $760 million (IMF, 1996a). Specifically, Costa Rica's external debt rose from $2.7 billion in 1980 (five years before implementation of Structural Adjustment Policies) to $4.7 billion in 1987 (EIU, 1993b). As of December 1995, it was "down" to $3.9 billion (EIU, 1997), because the country had received a $1 billion debt forgiveness package in 1990. Although the country came up with $3.1 billion in debt service payments between 1982 and 1988 (and over $6.6 billion between 1982 and 1995), its debt burden grew during the same period by nearly $2 billion (Inter-American Development Bank, 1992, 1996) (all numbers in current dollars). The debt burden became so great that in 1986, four years after the first Costa Rican loan agreement was signed with the World Bank, the Costa Rican government suspended principal and interest payments on the country's external debt (EIU, 1992) due to an inability to pay, as it had done before in 1981. In 1986, service payments on this staggering debt were eating up 46% of all export earnings (Inter-American Development Bank, 1992).

Consequently, the country, like many others in Latin America, underwent a number of debt rescheduling and reduction programs. The total foreign debt stocks of Costa Rica were $3.856 billions as of December 1995 (EIU, 1997), a debt burden of nearly $1200 for every citizen of the country [3.3 million as of 1994 (World Bank, 1996a)], and equivalent to half a year's GNP (EIU, 1997). Additionally, the Central Government had an internal debt in excess of $1.4 billion, or about $420 per capita (Monge, 1995).

Solis is not alone in his critique of SAPs. Sinha (1995) has analyzed the outcomes of SAPs on a global scale. She concludes that "the hoped-for restoration of long-term growth remains an ever-receding goal for most countries undergoing the discipline [of structural adjustment]" (p. 557). Even the IMF admits that structural reform "remains incomplete" and external viability (i.e., ideally, a nonnegative balance of payments) "elusive" for most "adjusted" countries, at least for the near term (IMF, 1993). According to Stewart (1991) and Helleiner (1993), the overall balance of experience with SAPs in the 1980s was clearly negative. David Korten, formerly professor at the Harvard Business School, Harvard advisor to the Nicaragua-based Central American Management Institute (INCAE), and senior advisor on development management to the U.S. Agency for International Development, expands the experience with SAPs to World Bank and IMF policies in general:

> If measured by contributions to improving the lives of people or strengthening the institutions of democratic governance, the World Bank and the IMF have been disastrous failures—imposing an enormous burden on the world's poor and seriously impeding their development. In terms of fulfilling the mandates set for them by their original architects—advancing economic globalization under the domination of the economically powerful—they both have been a resounding success. (Korten, 1995, p. 171)

According to an insightful analysis by Taylor (1993), the World Bank itself, although not as strongly as Taylor, criticized its own free market, neoliberal approach toward development problems during the 1980s. It recognized it was not solving the problems of the poor [which the World Bank itself had identified as the most important goal of development (World Bank, 1992)] and that its lending policies were contributing to environmental problems that in turn made conditions even worse for the poor.

In Costa Rica the social effects of the SAPs have been very adverse: small entrepreneurs, small farmers, and the poor have been the losers, while at the same time wealth has been concentrated increasingly in the hands of fewer people (Figure 24-3) (see Gini Index, World Bank, 1996a; Hansen-Kuhn, 1993; Ghosh *et al.*, 1996). This includes especially transnational corporations, who were also the major beneficiaries of the export promotion incentives (Hamilton and Thompson, 1994; Clark, 1995). Tardanico and Lungo (1995) agree that "state policies of austerity and privatization involve a redistribution of the socio-economic burdens and benefits of fiscal policy . . . and have been a critical point in the unraveling of Costa Rica's social-democratic nexus of state, employment and social welfare" (pp. 225–226). The losers have been the poor. As Arnove *et al.* (1996) argue, the implementation of SAPs has provoked a number of crises throughout Latin America:

> In diminishing the role of the state in the provision of basic social services—part of the cost-cutting policies recommended by the World Bank and the IMF in order to reduce fiscal deficits and bring inflation under control, the social safety net provided the most marginalized populations has been effectively removed. The distance between the wealthy and the poor is increasing (p. 154).

Hamilton and Thompson (1994) point out that large, predominantly foreign corporations now have a major, if not dominant, role in the country's export sectors, because they were the major beneficiaries of export incentives, and

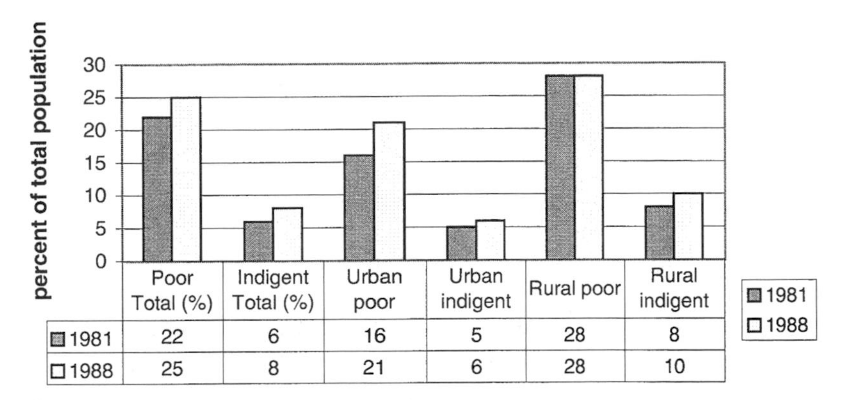


| | Poor Total (%) | Indigent Total (%) | Urban poor | Urban indigent | Rural poor | Rural indigent |
|---|---|---|---|---|---|---|
| 1981 | 22 | 6 | 16 | 5 | 28 | 8 |
| 1988 | 25 | 8 | 21 | 6 | 28 | 10 |

FIGURE 24-3 Poverty in Costa Rica. Note that in the UN source "poor" is defined as having incomes amounting to less than twice the cost of a basic basket of food. Includes indigent households. "Indigent" is defined as having incomes amounting to less than the cost of a basic basket of food. (Source: UN Economic Commission for Latin America and the Caribbean, *Statistical Yearbook, 1991*.)

because they were the only ones that possessed the financial resources necessary to build up large-scale infrastructure for nontraditional agroexports. The latter require large quantities of capital for initial investment in land, machinery, pesticides, fertilizer, irrigation and handling equipment, and technological knowledge, thus favoring large industrial agrocorporations (Thrupp *et al.*, 1995). According to Hansen-Kuhn (1993), a preliminary survey of crop budgets at the Costa Rican Banco National revealed that the initial investment required for 1 ha of any approved nontraditional crop varied from 170,000 to 85,000,000 colónes, while 97% of the agricultural population earned less than 180,000 annually. Hansen-Kuhn (1993) further points out that 40% of the macadamia nut production, 52% of the cut-flower production, and 46% of pineapple cultivation now are controlled by foreigners.

## B. The Continuing Import Dependence of Costa Rican Exports

As described in more detail in Chapter 22, another problem not resolved by, and perhaps exacerbated by, the policies of the IFIs was that Costa Rica's demand for manufactured imports and raw materials kept ahead of the value of its exports (World Bank, 1996a) (Figure 24-4). At the same time the per unit prices of these imports rose faster than the prices of the exports; i.e., the terms of trade deteriorated for Costa Rica (World Bank, 1996a; UN Economic

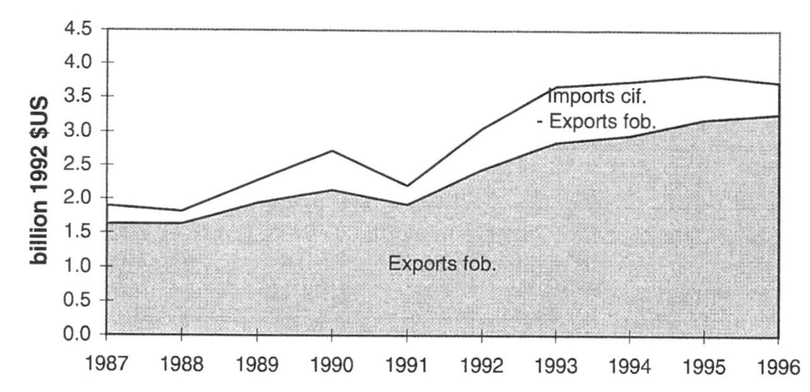


FIGURE 24-4 Dollar values of Costa Rican imports and exports. (Source: IMF *International Financial Statistics.*)

Commission for Latin America and the Caribbean, 1989, 1991) (Figure 24-5). The result of these two factors was that the import–export gap continued to increase (Figure 24-4). As Ghosh *et al.* (1996) point out, any stimulation of export-led growth cannot achieve a substantial reduction in international debt if it is heavily dependent on imported materials. This was the case for Costa Rica with respect to all the sectors selected for adjustment under SAPs: agroindustry, manufacturing in maquila industries and free trade zones, and consumer products and basic grains (Hamilton and Thompson, 1994). The continued and increasing need for imports to support production and industrialization undermines Costa Rica's ability to service its debt and reach economic stability.

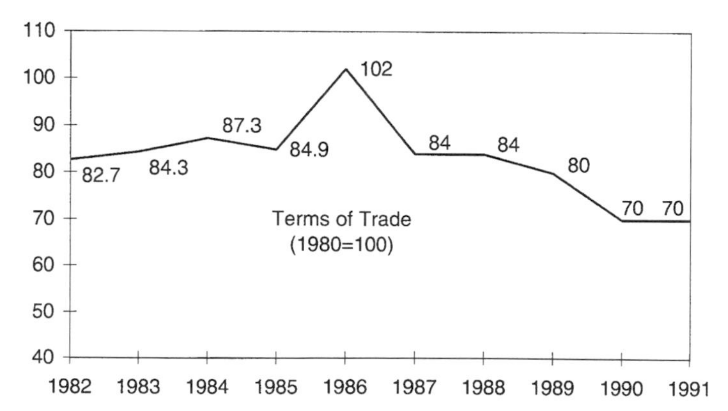


FIGURE 24-5 Costa Rica's terms of trade. CIF, cost, insurance, and freight. (Source: UN Economic Commission for Latin America and the Caribbean, 1989, 1991.)

According to Ghosh *et al.* (1996) what was needed instead was the development of an internal market for domestically produced goods, including the development of trade between agriculture and industry and between different industries within the country, all of which are preconditions for balanced domestic economic development. Instead, Costa Rica's continued industrialization was based on the importation of foreign capital goods, with the result that the country became even more dependent on imports (first, for acquiring these capital goods, and second, for the inputs necessary to maintain and operate them; see Montanye, 1998). The result was a long-term negative impact on the balance of trade induced by technology dependence (Hamilton and Thompson, 1994). As Annis (1990) puts it, "[t]he old dependence on finished foreign goods [was replaced] with a new dependence on foreign intermediate technology and raw materials" (p. 110).

Finally, a policy to constantly devalue the Costa Rican colón against the U.S. dollar in order to maintain a competitive exchange rate, using daily "crawling-peg" adjustments, i.e., minidevaluations (Diaz-Bonilla, 1990; EIU, 1997), was installed. This promoted Costa Rican exports by keeping them competitive on world markets. However, at the same time, it increased the price of imports that were needed as inputs to generate both Costa Rica's exports and its nonexport commodities, hurt the respective industries, and increased domestic inflation. The net effect of all these factors was the continued buildup of Costa Rica's trade deficit (Figure 24-1).

## C. Some Environmental Consequences of Increased Agricultural Production

Pressures to increase production of export crops also meant changes to the natural landscape of Costa Rica. Structural adjustment programs promoting the rapid growth of agroexport commodities have had many adverse environmental effects (Chapters 15, 18, and 20).

> The expansion of so-called non-traditional export crops...and the recent banana boom, notable in Costa Rica, have often resulted in negative environmental consequences, leading to increased deforestation and use of agrochemicals; the destruction of local peasant farming systems which, historically, had yielded important environmental, social and cultural benefits; and the displacement of peasant producers to more ecologically fragile areas. (Utting, 1994, p. 245)

In addition, intensified and expanded agricultural production has increased the stripping of soil nutrients and the loss of topsoil. According to a study by the World Resources Institute, soil losses averaged more than 300 tons per hectare from land in annual crop production and nearly 50 tons per hectare from pastures (Repetto, 1992; but see Chapter 15), necessitating increased fertilizer use. Forestlands have been cleared to provide additional open land for commodity production (Chapter 16).

Because of the lag time between initial setup and full production of plantations of nontraditional exports, additional efforts of the lending agencies went initially to increase the production of the existing export commodities, cattle, coffee, and bananas (Annis, 1990). These two arms of production, grazing and growing, have had different consequences for Costa Rica's natural environment.

## D. Environmental Impacts of Cattle, Coffee, and Bananas

Because the biomass yield of cattle per hectare of pasture is relatively low in comparison to that of grains, increasing cattle production meant clearing large areas of additional forested land, or converting agricultural land to pasture (Chapter 14). In 1990, cattle pastures occupied about 2.2 million ha, or 44% of the country's total land mass and about 54% of all agricultural land.

An international study by the World Resources Institute found a high correlation between the problems of debt and deforestation. Some of the conclusions from the research were that "Third World" countries that deforested the most or the fastest in the 1980s were also, on the whole, the largest debtors. In a number of smaller countries with less significant forest reserves, the fastest deforesters were also the most heavily indebted (George, 1992).

In particular, the countries with the highest annual rates of deforestation during the 1980s have on average a significantly higher ratio of total external debt to GNP (Repetto and Gillis, 1998; World Bank, 1986). Another study of developing nations determined that data indicate that both the logging and the conversion of forest to pasture or agricultural uses are largely the result of government policies. Many of these policies are driven by the severe economic pressures afflicting debt-burdened underdeveloped countries (Repetto, 1988). Costa Rica fits this profile as it had one of the highest rates of deforestation in relation to its land area (Repetto, 1988) (see Chapter 15) and also one of the highest per capita debts.

Prior to 1980, the land along Costa Rica's Atlantic coast produced over 70% of the country's maize. By 1990 this number had fallen to only 5%, much of the rest of the land being in pastures, forests, or pesticide-intensive banana plantations (some active, others abandoned), which now total 40,000 ha for the country as a whole (Hunter, 1994). Expanding banana plantations have two important negative effects other than degrading the ecosystem characteristics of the areas they occupy: first, they push crops previously planted in these areas onto the poorer soils of marginal lands, and second, by reducing the area available for crop production, they induce a switch from historically practiced swidden agriculture to permanent cultivation (Chapter 12). Both effects lead to increased application of synthetic fertilizers and pesticides to keep crop yields from declining steeply as a result of the reduced area available for the production of these crops (Figure 12-14).

It is feared that the pesticides, often applied by aircraft, will eliminate beneficial insects, thus adversely affecting the resident and migratory bird populations. Moreover, they pose acute health threats to the people living in the area (Chapter 20) (Lewis, 1992; Hunter, 1994; Thrupp *et al.*, 1995; Hernández and Witter, 1996). According to Lewis (1992) the use of pesticides in Costa Rica is seven times the world's per capita average. The increased use of pesticides and fertilizers, the resulting contaminative and, in some instances, toxic residues from extensive banana production, and the system of irrigation trenches required in banana plantations that annually flush millions of tons of soil and pesticide residues into the waterways that drain into the Caribbean Sea have been linked to destruction in both inland and coastal water environments (Hernández and Witter, 1996). For example, Costa Rica's formerly famous Caribbean coral reefs are nearly 90% dead, presumably in large part a result of pesticide runoff and sedimentation (Lewis, 1992; Bello, 1994).

In addition, the intensive use of chemicals in the production of bananas results in the degradation of the potential of the sites for agriculture, and creates a higher vulnerability to pests, often ultimately forcing the abandonment of the land altogether. Over 80,000 ha along the Caribbean coast have undergone this transformation since 1979 (Lewis, 1992).

## E. Environmental Impacts of Nontraditional Export Crops

Nontraditional agricultural products require many of the same inputs for production as do traditional crops, and in many cases at much higher levels (Thrupp *et al.*, 1995). Thus, the expansion of nontraditional crops not indigenous to Costa Rica requires large quantities of synthetic fertilizer and agrochemicals to compensate for suboptimal locations in gradient space and to ward off pests and diseases. Correspondingly, the value of imported pesticides and fertilizers increased from about $13 million in 1970 to almost $80 million in 1991 (Wilkie *et al.*, 1995), amounting to 28% of the country's 1991 trade deficit. Several peasant organizations have blamed improper and excessive pesticide use for causing various adverse health affects as well as contamination of groundwater, rivers, ocean shorelines, and soils (Honey, 1994). According to Repetto (1992), a conservative estimate of the depreciation of Costa Rica's natural resource capital, including forests, soils, and fisheries, amounts to 26%

of Costa Rica's gross capital formation in 1970; by 1989, this number had increased to 37%.

Repetto's analysis has profound implications for an accurate evaluation of SAPs. It casts a fundamentally different light on the increase in the country's GNP that is so often proclaimed as evidence for progress. If the depreciation of Costa Rica's natural resource capital is included in such an evaluation, the picture becomes rather gloomy. In the words of Repetto (1992, p. 98), "the conventional accounting framework overstated *actual* net capital formation (which subtracts depreciation of man-made *and* natural capital from gross capital formation) by more than 70 per cent in 1989" (emphasis added). Hence, the outcomes of SAPs must be considered even less positive for Costa Rica than they appear in the evaluation employed in this chapter, since this evaluation is primarily based on the conventional official data that do not consider natural resource depreciation.

The growing concern over observed damage to environmental health caused by the export-led growth strategy has prompted the IFIs to address the issue. Ironically, the lending agencies responded by advocating further participation in a market economy as the key for improved environmental health. As highlighted in 1994 U.S. Congressional hearings, both the IMF and the World Bank stated that as a country "develops" and gets wealthier, environmental standards tend to rise: "There can be no doubt that growth-inducing reforms, and trade liberalization in particular, bring long-run environmental as well as many economic benefits" (Subcommittee, 1994, p. 84). Judging from the empirical evidence accumulating around the world, specifically in (until recently) booming Southeast Asia (Worldwatch Institute, 1998), such a strongly positive evaluation of the environmental effects of economic growth appears somewhat unfounded. In any case, for Costa Rica it can only be hoped that these forecasted "long-run" environmental benefits start to materialize sometime in the not too distant future, before too much of the remaining natural wealth of the country is lost.

Summarizing the previous sections, we can describe the major problems caused by, or at least exacerbated by, SAPs and export-led growth as follows:

1. They have drained government revenues through excessively high export-promoting incentives which rob the government of much needed tax revenues. Examples are tax rebates on nontraditional export sales (CATs) that climbed to more than $80 million in 1993 (Clark, 1995), or a 100% tax exemption of all machinery and intermediate inputs, and up to 100% tax exemptions of profits.

2. They have increased trade deficits by encouraging imports through their emphasis on free trade and tariff reductions, the latter of which also have reduced government revenues.

3. They have increased vulnerability to external forces such as protectionism, recession, and/or saturation in target markets, particularly for the United States and the European Union.

4. They have not reduced Costa Rica's vulnerability to the volatility of international markets because the country became increasingly dependent on imports, whose relative prices have increased, and also because it has to compete with other "structurally adjusted" countries in the same limited export markets.

5. According to Hamilton and Thompson (1994) and Thrupp *et al.* (1995), they have concentrated wealth and economic power increasingly in the hands of large corporations because small producers could not afford the often prohibitively high costs of restructuring; much of this concentration of power was in the hands of the predominantly large foreign corporations.

6. They entailed negative environmental consequences for Costa Rica through increased and intensified agricultural production.

This is a pretty bleak assessment of SAPs, and we were unable to find any literature giving a much more positive perspective on a comprehensive evaluation of SAPs. What is not known, however, nor particularly discussed, is whether or to what degree alternative policies would have resulted in an equivalent requirement for externally derived inputs. The question of importance here is whether or for how long political decisions can override biophysical reality. Our view is that the policies imposed by the international financial institutions have led to a deterioration of Costa Rica's situation, and particularly that a careful biophysical assessment prior to the country's economic restructuring could have given a much better assessment of what policies would have a high probability of succeeding and those that would not. For example, an examination could have been carried out of the physical flows of resources into the manufacturing and agroindustry, using, for example, Leontief-type analyses for energy and material (e.g., Hannon, 1982; Hall *et al.*, 1986). This would have shown that building up final demand industries without building up domestic suppliers, to the degree that this was possible, would generate the foreign exchange problems that did indeed occur.

## II. DISCUSSION: THE POLITICAL IMPLICATIONS OF STRUCTURAL ADJUSTMENT POLICIES

The social and environmental consequences of the SAPs do not escape the notice of the Costa Ricans. Many of them fail to see the logic behind the changes that are occurring as a result of SAPs. Even after billions of dollars have been spent advocating "neoliberal" [read "ideological" as political scientist Richard Mansbach (1996) maintains] reforms, officials of international agencies promoting SAPs lament the unpopularity of their agenda. A case in point is one U.S. Department of Commerce official who in 1996 complained that

> in Costa Rica the direct participation of the state in certain economic activities is still widely accepted as legitimate and beneficial . . . [and] it remains difficult to enact serious reform of some of the principal state institutions because of the broad popular support they enjoy. Reform must be shown to be absolutely necessary to develop the needed consensus for a change. (U.S. Department of Commerce, 1996)

Apparently the author perceives little hypocrisy in his role as a state official whose job it is to undermine "broad popular and political support" in a sovereign foreign nation in order to manufacture "the needed consensus for a change" while supposedly promoting democracy, one of the stated goals of USAID and other development institutions. Neoliberalism's market freedom seemingly does not extend to the marketplace of ideas on governance and to accepting the way of self-governance a sovereign people collectively chooses.

By way of conclusion one might say that in the case of Costa Rica, as well as in other countries, external borrowing and Structural Adjustment Policies moved the nation from existing within its own particular national, social, and resource framework, with specific economic and social programs (Chapter 2), to being just another cog in the wheels of the global economy that is increasingly controlled by giant transnational corporations (EIU, 1993a) and global financial markets.

## A. Were the Negative Outcomes of the Structural Adjustment Policies Predictable?

This chapter outlines the arguments advanced by many prominent thinkers about development who have concluded that the policies of the International Financial Institutions (IFIs) were ill designed, for they did not solve the problems they were supposed to, while at the same time they generated many new ones. A further question is whether or not the actual outcomes could have been predicted on the basis of the neoclassical economic foundations underlying the policies of the IFIs. Our analysis of this question proceeds in three steps: first, we consider whether neoclassic economic theory could have predicted the adverse results of the SAPs now apparent. Second, we ask the more fundamental question as to whether neoclassic economic theory should have been applied to developing countries in the first place, and in so doing consider some highly relevant ideas in Makgetla and Seidman (1989). Finally, we ask whether the problems of Costa Rica were generated by processes not considered in traditional economic analysis and hence not amenable to solution within its domain.

The IFIs started out to help indebted "developing" countries. But even the use of the term "developing" implies a paternalistic and even arrogant attitude since, first, all countries are constantly developing, and second, there are grounds to question that the countries that happen to be fortunate enough to be wealthier are, therefore, somehow more "developed." For example, the "developed" countries also had their problems with debt servicing in the face of their balance-of-payments crises in the early 1980s, and prior to "development assistance" Costa Rica had higher literacy and lower crime rates, a more sophisticated and inclusive national health plan, higher health standards, and a more equitable distribution of wealth than it does today—and than almost any "highly developed" nation. Moreover, societies characterized by the term "developed" as it is used today appear to become less sustainable the more they become developed because, at a minimum, they require increasing amounts of nonrenewable energy. A fundamental redefinition of the terminology seems necessary.

The IFIs state that it is important to pay more attention to the costs of adjustment in terms of losses in production and employment. The authors agree that this is important, e.g., in terms of the effects of adjustment measures on the livelihood of those farmers who were affected by the promoted increase in agricultural export products and the reduced incentives for the production of basic food crops (Ghosh *et al.*, 1996). One might add that it is also necessary to consider:

1. The potential future social costs of the SAPs for Costa Rica in terms of the loss of domestic food self-sufficiency in the face of projected grain shortages on a global scale (L. R. Brown, 1997; Harris, 1996)
2. The reduced public health care and education expenditures and the resulting decline in public health and education (Hansen-Kuhn, 1993; Hamilton and Thompson, 1994; Arnove *et al.*, 1996)
3. The effect of the preceding on birth rates and, eventually, many consequent aspects of social well-being
4. The rising need for the importation of external goods and services that are no longer being provided by domestic suppliers
5. The losses in natural resource capital that accompany SAPs

The IFIs seem not, however, to question their very mission. As early as 1945, the year after their inception, U.S. Republican Senator Robert Taft did just that, by arguing that foreign aid based on large financial flows is a fundamentally flawed idea:

> I think we overestimate the value of American money and American aid to other nations. No people can make over another people. Every nation must solve its own

> problems, and whatever we do can only be of slight assistance to help it over its most severe problems. . . . A nation that comes to rely on gifts and loans from others is too likely to postpone the essential, tough measures necessary for its own salvation. (Senate Committee on Banking and Currency, 1945, in Korten, 1995, p. 167–168)

Taft expected the major beneficiaries of the international aid machine to be Wall Street Bankers: "It is almost a subsidy to the business of investment bankers, and will also undoubtedly increase the business to be done by the larger banks" (Korten, 1995, p.168).

Despite these early warnings, by now the World Bank and the IMF, in their roles as international debt collectors, "have become increasingly intrusive in dictating the public policies of indebted countries and undermining progress toward democratic governance and public accountability" (Korten, 1995, p. 165). Cahn (1993) argues in the *Harvard Human Rights Journal* that

> the World Bank must be regarded as a governance institution, exercising power through its financial leverage to legislate entire legal regimens and even to alter the constitutional structure of borrowing nations. Bank-approved consultants often rewrite a country's trade policy, fiscal policies, civil service requirements, labor laws, health care arrangements, environmental regulations, energy policy, resettlement requirements, procurement rules, and budgetary policy (p. 160).

Hence, the World Bank is making decisions for people to whom it is not accountable, decisions that in a democracy normally would be considered the responsibility of elected legislative bodies. In the face of such a undemocratic decision-making structure and considering the Bank's close ties to the local elite (Piddington, 1992) and transnational corporations (Finger and Kilcoyne, 1997; Feder, 1975), one cannot reasonably expect that World Bank projects are generally promoting the interest of the majority of the people in the developing countries. In particular, the poorest subsectors of these countries' populations lack any meaningful participatory influence in project design and implementation. One should expect them to incur negative "side effects" of the Bank's projects, and the figures seem to substantiate this expectation. By 1994, the Bank's dam projects alone had displaced more than 10 million people (Sklar, 1994).

## B. Does Free Trade Achieve Its Advertised Goals?

Another underlying assumption of the Structural Adjustment Policies is the supposed aggregate welfare-enhancing function of free trade. International free trade is defined as the trade that takes place without barriers, such as tariffs, quotas, "voluntary" self-restrictions, and exchange controls, being placed on the free movement of goods and services between countries. The aim of free trade is to secure the benefits of international specialization, i.e., the benefits that derive if each country concentrates its production according to its comparative advantages. However, one must not forget that two of the fundamental assumptions of the free trade theory are first, that the trade balance between two countries engaged in trade with each other is in equilibrium, i.e., neither country incurs a trade surplus or deficit, and second, that employment in each country remains constant (in the absence of the freedom for people to cross national borders and seek employment in other countries). Only then, according to the theory, do both countries gain an aggregate increase in welfare from free trade between them. Both assumptions are very restrictive, and usually never fulfilled in the real world.

The real world, in any case, has never followed that theory. From the historic evidence in Europe as well as the United States and Southeast Asia, it is evident that not many latecomers among the present-day industrialized countries (that is, basically every nation except England) experienced rapid economic growth within a free trade regime, nor could the governments of these nations correctly be characterized as noninterventionist. Almost all of the latecomers favored free trade, to varying degrees and always selectively, *only after* they reached technological maturity and international competitiveness in their industries (Sinha, 1995). England itself, often portrayed as the first nation to have embraced the principles of free trade, can be found to suddenly have done so only after it had destroyed the competitive superiority of the textile sector of its colony, India, and indeed, the whole subcontinent with it. Countries having undergone successful industrial development processes, such as South Korea, started with strict trade barriers. This is so because politicians in these nations were aware of the fact that trade liberalization can have negative consequences if not implemented wisely. Indeed, economists are well aware of this fact, too.

A standard part of theoretical economics focuses on this problem. The key term here is what economists denote as "first-mover advantage," and the argument can be sketched like this: domestic producers in so-called infant industries can hope to be competitive with long-standing "internationals" only after they have had time to gain experience and become efficient enough, if indeed ever. Powerful companies that have been in business longer and therefore have attained a high degree of knowledge and efficiency in their production usually produce at lower marginal costs than young competitors. If free trade were implemented abruptly, the more mature industries could theoretically take over the world markets for their products. This is the first-mover advantage.

The company that is able to take over a market (because it existed when the market was opened and because it was the only one that had the resources to take it over) thereafter, theoretically, can withstand any competition, even if its production function is characterized by a higher marginal cost curve than the production function of potential newcomers. This is so because any other company that subsequently wants to enter the market would have to produce its initially lower output (during the buildup of production capacity) at higher marginal and average costs than the dominating company produces its higher output. Hence, it cannot successfully enter into competition with the established company unless the newcomer has the necessary financial backup to survive the initial period up to the break-even point. At that point it would, if resources allowed, take over the market completely and become the new monopolist. In this scenario, one should add, the industry in question is characterized by diminishing marginal costs of production. This, however, is the case for virtually all of today's mass-produced consumer goods.

This phenomenon can be observed on national and global scales, with "free" markets becoming to an increasing degree highly monopolistic (EIU, 1993a). Thus it is understandable that the most emphatic calls for the opening of foreign countries' domestic markets to international competition come primarily from the highly developed and highly monopolistic, and frequently highly subsidized and protected, sectors in "developed" countries. That their demands for "free trade" are backed by widely accepted neoclassical economic arguments explains the aura of righteousness that usually surrounds their promoters. The fact that the assumptions on which the theory is built are not fulfilled in the real world does not seem to be of much concern to its advocates. John Kenneth Galbraith, as always, expresses it so eloquently: "The market has its own truth on which reality does not intrude" (Galbraith, 1992, p. 135). Hence, it is important to examine the ways in which free trade theory is abused in international trade and development politics.

## C. Free Markets for Whom?

There is a curious inequity in the application of free trade principles. While the industrialized nations and the development banks are pushing so-called free trade on countries like Costa Rica, they often continue to protect their own industries. To the present day many industries in the industrialized countries, particularly in the United States, Europe, and Japan, are protected from foreign competitors by either tariff or nontariff protectionist measures. Examples are abundant. To name a few, import restrictions, tariffs, and nontariff measures are in place in virtually all of the industrialized countries to protect domestic industries in sectors such as agriculture, coal and steel, and textiles. These barriers to free trade are maintained despite GATT (the General Agreement on Tariffs and Trade) and other trade agreements that ostensibly promote free trade. However, GATT and its regional sister treaties notwithstanding, managed trade is the rule. For the industrialized countries, that is. Globally, the rule is "free trade for the poor, managed trade for the wealthy."

In general, in each of the industrialized powers one can find a mixture of liberalization and protectionism, tailored to the needs of powerful domestic interests. The explanation for this is straightforward, and expressed so by Walter Russell Mead (1993), a senior analyst with the New York-based World Policy Institute: "[These treaties] lower some barriers to trade, keep others high, and contain vast and lucrative loopholes that benefit interest groups who were successful at lobbying powerful elected officials" (pp. 60–61). As a result of these policies, free trade is the exception rather than the rule, and only about 15% of all global trade is genuinely conducted in free market circumstances (Krause, 1992). The same pattern has been visible in the development of key industries in the newly industrializing Southeast Asian "little dragons" (Singapore, Hong Kong, South Korea, and Taiwan) as well as Japan, all of which facilitated the buildup of domestic (often import-substitution) industries, e.g., the national car industries, with the help of protective measures before they opened up (if indeed they did) their domestic markets for international competition. The Overseas Economic Cooperation Fund (OECF) explains and justifies this strategy. It states that it is too optimistic to expect that industries which would sustain the economy of the next generation will arise automatically through the activities of the private sector. Some measures for fostering their development are required. It further points out that "[t]he structural adjustment approach seems to lack the long-term viewpoint of how to develop such industries, perhaps because it assumes that activities of the private sector will attain this goal. This lack is very regrettable" (Overseas Economic Cooperation Fund, 1991).

Nevertheless, the IMF's *Annual Report* for 1996 mentions that "appropriate policies [are] needed to make these [developing] economies more responsive to the new trading opportunities arising from increased global integration" (IMF, 1996b, p. 26). Such increased responsiveness is obviously assumed to generate benefits for the respective countries. A closer look at the international trade flows in the example of Costa Rica, however, makes it apparent who benefits most from trade liberalization (Figure 24-3). Under the existing setup of the international trade system, it is the sectors in "developed" countries in which the imports to Costa Rica originate. A very instructive example is provided by the Costa Rican agricultural sector. As part of the SAPs, the World Bank aims at the removal of agricultural subsidies and tariffs (Piddington, 1992), if only in "developing" countries. This policy had adverse effects on the food self-sufficiency of Costa Rica and has destroyed the livelihood of many of its farmers (Hansen-Kuhn, 1993). Local grains are no longer competitive pricewise with highly subsidized grains imported from the industrial countries.

What is remarkable is that, for example, farmers in the United States received $56.6 billion (in constant 1987 dollars) in subsidies between 1990 and 1995 alone (U.S. Department of Commerce, 1995, No. 1109), and that the Agricultural Export Program of the Commodity Credit Corporation (covering, among others, the direct export credit sales program, the export guarantee program, the market promotion program, the export enhancement program, the dairy export incentive program, and technical assistance to emerging democracies) in 1993 reached net outlays of $2.2 billion (1995, No. 1113). Hence, it is hardly surprising that the main U.S. agricultural export commodities to Costa Rica include corn, soybeans, and wheat (1995, No. 1125), indirectly crowding out local competition by providing subsidized US products that change the dietary habits of Costa Ricans away from locally grown grains.

The subsidies to U.S. farmers just referred to amounted to an average of over $10 billion per year between 1990 and 1995, while the U.S. government, particularly through USAID, was at the same time aggressively promoting so-called "free trade" in Costa Rica and other "developing" countries. Carley and Christie (1992) call this the "double standards in trade policy." USAID- and World Bank-funded research is recurrently proclaiming the superiority of grain market liberalization. For example, Hazell and Stewart (1993) point out that government regulations in Costa Rican grain markets actually cause welfare losses to consumers and smaller farmers while benefiting mostly large-scale rice farmers. Based on this finding, they argue for further liberalization of Costa Rican grain markets, so that the country would be able to concentrate its agricultural production according to its comparative advantages.

In doing so, however, the authors miss a crucial step in their otherwise quite thorough analysis: Costa Rica is not isolated from the world market, and hence existing massive distortions in other markets (e.g., U.S. and EU grain markets) are bound to have spillover effects that preclude production according to comparative advantages. Therefore, Hazell and Stewart draw the wrong conclusion from their correct observation: domestic grain production is distorted between particular grains, with resulting welfare losses and socially adverse income transfers. The Costa Rican government would do well to address this fact. The liberalization of Costa Rican grain markets with respect to imports and exports, however, is a different issue, likely to result in a further loss of domestic production capacity, even higher international income transfers from Costa Rican to U.S. farmers, and, too often forgotten, an increased dependence of Costa Ricans on international grain supplies that are to a large extent subject to political decisions and changing biophysical conditions in other nations.

Tarp (1993) maintains that the economic models of the IMF and the World Bank are too simplistic to capture the complexities of the real world and to form the basis for policy prescriptions. As Sinha (1995) points out, the policies promoted by the IFIs suffered from inherent contradictions that worsened the problems of their implementation. One such example is the frequent currency devaluation of the Costa Rican colón that raised the costs of the increasing imports of food, petroleum, and machinery. It also increased the costs of the intermediate goods and hence the production costs of the export-producing industries, thereby aggravating the already pressing problem of inflation (World Bank, 1996a). As was already mentioned in the preceding section, an export-led growth strategy cannot succeed in improving the balance of trade if the exporting industries are heavily dependent on imported inputs.

Furthermore, the implementation of the IFIs' policies was dependent upon the domestic elite in administrative, technical, and political respects. Piddington (1992), a former high-ranking World Bank official, points out that the World Bank staff is "in dialogue at a very senior level with the governing elite." It could hardly have been expected that this elite would implement policies that would cut into their own privileges (Sinha, 1995). Hamilton and Thompson (1994) conclude with the OECF that export-led growth is too narrow a strategy for sustained development. When implemented unilaterally through nonselective trade liberalization policies in developing countries, it has resulted in increased trade imbalances and setbacks to these countries' domestic industries and agriculture due to cheaper imports. In addition, although this obviously seems to be of no excessive concern to the advocates of free trade, trade liberalization in developing countries has amassed a staggering record of social decay in almost all of Latin America, one of the prime subjects for its implementation. Owing to this fact, the region now stands out as an instructive example of the darker side of international monetary fundamentalism and its ideology, neoclassical liberalism, with Brazil being one of the worst cases in the region (Skidmore, 1988; Hecht and Cockburn, 1989; Leacock, 1990). The principle fate, however, is shared by most of the developing world subjected to structural adjustment. According to the World Bank, all poverty measures worsened in Sub-Saharan Africa, the Middle East and North Africa, and Latin America and the Caribbean (World Bank, 1992) during the 1980s. Not so in Southeast Asia, the only region whose development strategy did not follow free trade doctrine. That, naturally, can only be coincidence.

## D. Alternatives to the Neoclassical Development Model

In light of these arguments, it is difficult to understand how the World Bank can still claim that "[i]n the past twenty years, a consensus has emerged among economists on the best approach to economic development" (World Bank, 1993b, p. 85). The World Bank–IMF model of development is still principally based on a model advocated by Rostow (1960). According to Rostow, the process of "development" of so-called underdeveloped countries must be organized in five stages, the fifth stage being the completed conversion of the respective societies into consumer societies, with high mass consumption representing the desired end point of "development" activity. As Williamson (1993) argues, the World Bank-proclaimed consensus is clearly nonexistent. And no such consensus is in sight. One can distinguish broadly the economic critics of the neoclassical approach underlying the IFIs' development politics into four groups (based on Sinha, 1995):

1. The eclectics, who follow the Smith–Ricardo–Mill–Marshall–Pigou–Keynesian heritage, see the market as an institution facilitating the allocation and distribution processes, but recognize that it is fallible, and hence assign an important role to the state in correcting market failures
2. Those advocating "structural adjustment with a human face," as advocated by many public servants in the UN system, especially the UN Conference on Trade and Development (UNCTAD) and the UN Children Fund (UNICEF), and by many nongovernmental organizations (NGOs)
3. The dependency theorists, who are influenced by Marxian analysis and Latin American structuralism
4. The structuralists, who see underdevelopment in terms of the core–periphery theory with its uneven distribution of benefits from trade and economic progress

To this list we add the biophysical economists, such as the many represented in this book.

## E. Applicability of Neoclassic Economic Theory to "Developing" Countries

As discussed in Chapter 3, neoclassic economic theory is built on a number of assumptions. If these are not fulfilled, applying the theory and following its "advice" in an unconditional way does not necessarily lead to optimal economic (not to mention social and environmental) results in a real-world setting.

Sinha (1995) points out that the prescriptions given by the IFIs "amount to a social engineering on a vast scale to transform developing countries into 'market economies,' many elements of which even Adam Smith would have found unacceptable" (p. 557). She states, "the outcome of stabilization and

Structural Adjustment Policies, largely based on the neoclassical economic rationale, was justifiable neither in terms of the analytical nor the historical literature." She argues that the neoclassical perspective, mainly concerned with allocative efficiency, is selectively incomplete. If Adam Smith's principle of self-interest is not accompanied by a supporting social principle, then the former is incomplete as a social organizing device. As the eminent economist Schumpeter (1978) noted, Alfred Marshall (1920) was "the first theorist to prove theoretically that laissez-faire, even with perfect competition and independently of those evils of inequality, did not assure a maximum of welfare to the society as a whole."

As has been outlined, the influence of neoclassical economics on development policies occurs in the form of the policies of the International Monetary Fund, the World Bank, and the U.S. Agency for International Development. These policies are based on the neoclassical view that free market capitalism constitutes the means to realize the highest possible welfare for all, and on the assumption that Western-style progress is possible and desirable for all (Shiva, 1993). Therefore, in order for lower income countries to develop, they must adopt policies that, according to the economic theories developed in higher income countries, are expected to lead to the removal of the existing impediments to perfectly competitive markets (Makgetla and Seidman, 1989). This view, "increasingly popular among international economic agencies and Western governments" (Carley and Christie, 1992, p. 104), identifies non-free market approaches to economic policy as the reason for lower income countries' material poverty.

This sole emphasis on allocative efficiency completely ignores other normative standards of measurement, e.g., distributive and redistributive justice or equity (Ayres, 1996), and tends to overlook the frequent nonmarket roots of lower income countries' lesser welfare. Often these occur in the form of political and economic disempowerment (Makgetla and Seidman, 1989), a structure of social systems that are completely different from the ones in "developed" countries (Feder, 1975), or resource limitation, i.e., biophysical constraints. Moreover, the standard assumption of the existence of perfect competition, tenuous in higher income countries, is unjustified in many lower income countries, because the motives that underlay economic activity, as well as realities of market power, imperfect knowledge, barriers to market entry and exit, and factor immobility, are even further removed from the "ideal" as defined by neoclassic economic theory. Under such markedly different conditions, applications of neoclassical economic theory cannot be expected to lead to outcomes desirable to the majority of people in the recipient countries. The World Bank's claim that the primary task of development is to eliminate poverty (World Bank, 1992) cannot change this fact.

## F. Is the Problem the Neoclassical Economic Perspective?

Sinha (1995) assigns responsibility for the present situation to national leaders and particularly to private banks. If blame is due, however, perhaps "those neoclassical economists who provided the philosophical underpinnings for the IFIs' policies cannot escape blame either, for they advocated policies based on ideas which had limited intellectual or historical support" (p. 570) (see Chapter 3). "A major share of the blame rests also with the IFIs for maintaining their dogma at the cost of human suffering, thereby irrevocably damaging their credibility as intergovernmental organizations" (Sinha, 1995) that should, one might presume, serve the benefit of all people.

Underlying all of the SAPs has been a naively rigid and highly selective implementation of theoretical neoclassical economics. The central criterion by which economic decisions in neoclassical economic theory are made is the *efficiency* of the resource allocation process. The means offered to achieve this criterion is the market price mechanism, which is seen as the only instrument available to realize the optimal (defined as generating the maximum aggregate welfare) allocation of scarce resources. The optimal allocation of resources, however, is based on the prerequisite of perfect markets. Unfortunately, there exists abundant empirical evidence of inefficient markets, for reasons of what economists call market imperfections, e.g., massive concentration of power on the supply or/and demand side in most markets (*Business Week*, 1991; Krebs, 1992; EIU, 1993a), the "nonconforming" of people and societal structures to the assumptions of neoclassic economic theory (Makgetla and Seidman, 1989), imperfect information, externalities, and public goods.

Imperfect information, in particular, indicates the limits to optimal resource allocation by unfettered markets. Allocation, in order to be optimal, i.e., to achieve *social* efficiency, needs to take into account *all* externalities that occur in production and consumption processes, i.e., all costs and benefits of such processes that are not regulated by the market price mechanism. This would require the identification and valuation of such externalities in order to make their integration into the market mechanism possible. Valuation, however, will always be imperfect because of imperfect foresight regarding, e.g., cause–effect relationships, and of course for reasons of the intangibility of many of these effects that cannot be overcome by science. Hence, value judgments are impossible to avoid in the integration of externalities. Since values differ among people, such judgments require democratic procedures to overcome such differences. The unregulated market cannot solve these problems. Arrow (1974) showed that the nonexistence of, or imperfections in, a single market has spillover effects on other markets and thereby destroys the optimality of competitive equilibrium. Although this problem exists in industrialized countries as well, it is even more prevalent in the markets of industrializing countries with all their "imperfections" and rigidities (Makgetla and Seidman, 1989). Hence the validity of neoclassical economic theory for these countries is at best "questionable" (Sinha, 1995). Recognizing the existence of market imperfections, Stiglitz (1994) notes that

> The standard neoclassical model—the formal articulation of Adam Smith's invisible hand, the contention that market economies will ensure economic efficiency—provides little guidance for the choice of economic systems, since once information imperfections (and the fact that markets are incomplete) are brought into analysis, as surely they must be, there is no presumption that markets are efficient (p. 13).

These arguments question, to put it mildly, the operational value of neoclassical economic theory.

However, even if markets were efficient, there are fundamental philosophical arguments that question the validity of the market as an instrument for public policymaking. When defending neoclassical economic theory as a decision-making criterion it is often argued that this theory is value-neutral, and therefore in this regard superior to other, normative theories that, for instance, regard equity as an objective of policymaking. This, however, is a misconception. As Sagoff (1988) argues, efficiency as a goal is no less normative than equity (or any other, for that matter), because it has no intrinsic worth. He states this point as follows (1988, p. 45): "The [economic] analyst is neutral among our 'values'—having first assumed a view of what values are, that is, having assumed a particular theory of the good."

Summarizing these arguments, the most fundamental criticism of neoclassical economic theory can be expressed like this: in order to render the theory operational, economists are forced to try to adjust the behavior of people and institutions to conform to their theories, which, needless to say, violates the very logic of theory building as an attempt to develop an understanding of the processes that govern the real world.

A second major critique of neoclassic economic theory is that it fails to capture what may be the most fundamental part of the essence of the structure and functioning of economies. That is, it fails to address, let alone be based upon, the importance and the impact of the physical underpinnings of economic systems (Chapter 3). Time and again, this has led to policies that undermine the integrity of the physical world, and, consequentially, the integrity of those human economic systems dependent on it. Costa Rica shows this pattern: externally derived economic policies have pushed growth in agricultural production that undermines the country's soil resource base and results in higher dependency upon imported soil amendments. Both outcomes are unsustainable in that they move the country away from a renewable natural resource base.

Without falling victim to the arrogance of knowing better in hindsight, it should be obvious that a comprehensive biophysical analysis of the Costa Rican economy, including many of the analyses represented in the middle chapters of this book, would have yielded at least some of the information necessary to predict the general nature of the outcomes of the SAPs. This, however, does not imply that, had such an analysis been conducted and its findings been implemented in policy, Costa Rica would not have experienced any economic or social problems. Simply based on its biophysical constraints Costa Rica might not have been able to meet the demands exerted by a growing population, with or without the IFI-prescribed process of industrialization. A Costa Rican population that has grown by 2.8% per year in the 1980s and by 2.1% per year in the 1990s (World Bank, 1996a) requires a corresponding real growth in income just to maintain per capita standard of living. To achieve such growth represents a considerable challenge even for the "developed" countries. It is all the more difficult, and in many respects impossible, for "developing" nations. The programs enacted in Costa Rica under Structural Adjustment Policies were directed toward export-oriented industrialization. Whether or not these programs could have been theoretically successful in the pursuit of this type of industrialization (and for the provision of food for a rapidly increasing population), they were destined to fail because of biophysical limitations. The exclusive reliance on theoretical neoclassical economic models for policymaking has resulted at a minimum in a diagnosis that misspecified the country's problems. It focused selectively on narrowly defined theoretical (economic) problems, misled decision makers as to what the fundamental issues were, and, hence, caused the country to avoid addressing the real questions for over a decade.

A final point to be considered is that any measures now employed by the Costa Rican people, even if based on biophysical considerations, can be fully successful only if they take place in the context of an equitable international economic order. The fulfillment of this condition is, of course, beyond any Costa Rican's control, and therein lies the quintessential challenge any small country faces.

## ACKNOWLEDGMENTS

The authors express their thanks to Anke Wessels and Myrna Hall for critical reviews of earlier drafts of this chapter.

# REFERENCES


Annis, S. 1990. Debt and wrong-way resource flows in Costa Rica. *Ethics and International Affairs* 4:107–121.

Arnove, R. F., A. Torres, S. Franz, and K. Morse. 1996. A political sociology of education and development in Latin America. The conditioned state, neoliberalism, and educational policy. *International Journal of Comparative Sociology* 37(1,2):140–158.

Arrow, K. J. 1974. Limited knowledge and economic analysis. *American Economic Review* 64:1–10.

Ayres, R. U. 1996. Limits to the growth paradigm. *Ecological Economics* 19:117–134.

Barham, B., M. Clark, E. Katz, and R. Schurman. 1992. National agricultural exports in Latin America. *Latin American Research Review* 27:43–82.

Bello, W. 1994. *Dark Victory: The US, Structural Adjustment, and Global Poverty*. Pluto Press, London, UK.

Bhagwati, J. N. 1986. Rethinking trade strategy. In J. P. Lewis and V. Kallab (Eds.), *Development Strategies Reconsidered*. Transaction Books, New Brunswick, NJ.

Brown, L. R. 1997. Can we raise grain yields fast enough? *Worldwatch* 10:8–17.

*Business Week*. 1991. The age of consolidation. Oct. 14, 86–94.

Cahn, J. 1993. Challenging the new imperial authority: The World Bank and the democratization of development. *Harvard Human Rights Journal* 6:160.

Carley, M., and I. Christie, 1992. *Managing Sustainable Development*. Earthscan Publications, London.

Clark, M. A. 1995. Nontraditional export promotion in Costa Rica: Sustaining export-led growth. *Journal of Interamerican Studies & World Affairs* 37:181–223.

Diaz-Bonilla, E. 1990. *Structural Adjustment Programs and Economic Stabilization in Central America*. Economic Development Institute of the World Bank, Washington, DC.

Economist Intelligence Unit. 1993. *A Survey of Multinationals: Everybody's Favourite Monsters*, special supplement (March 27, 1993). EIU, London, UK.

Feder, E. 1975. *The New Penetration of the Agricultures of the Underdeveloped Countries by the Industrial Nations and Their Multinational Concerns*. Institute of Latin American Studies, Occasional Paper 19, University of Glasgow, Scotland.

Finger, M., and J. Kilcoyne. 1997. Why transnational corporations are organizing to "save the global environment." *Ecologist* 27(4):138–142.

Galbraith, J. K. 1992. *The Culture of Contentment*. Houghton Mifflin, Boston/New York.

George, S. 1992. *The Debt Boomerang: How Third World Debt Harms Us All*. Westview Press, Boulder, CO.

Ghosh, S. M., J. M. Tanski, and C. E. Enomoto, 1996. Macroeconomic adjustments for debt-laden economies: The Central American experience. *Social Science Journal* 33:121–136.

Hall, C. A. S., C. J. Cleveland, and R. K. Kaufmann. 1986. *Energy and Resource Quality: The Ecology of the Economic Process*. Wiley–Interscience, New York. Reprinted 1992, University Press of Colorado, Boulder, CO.

Hamilton, N., and C. Thompson. 1994. Export promotion in a regional context: America and Southern Africa. *World Development* 22(9):1379–1392.

Hannon, B. 1982. Analysis of the energy costs of economic activities: 1963–2000. *Energy Systems Policy Journal* 6:249–278.

Hansen-Kuhn, K. 1993. Sapping the economy. Structural adjustment policies in Costa Rica. *Ecologist* 23:179–184.

Hazell, P., and R. Stewart, 1993. Should Costa Rica's grain markets be liberalized? *Food Policy* 18:471–481.

Hecht, S., and A. Cockburn. 1989. *The Fate of the Forest: Developers, Destroyers, and Defenders of the Amazon.* Verso, London/New York.
Helleiner, G. K. 1993. Trade strategy in medium-term adjustment. *World Development* 18:879–897.
Hernández, C. E., and S. G. Witter. 1996. Evaluating and managing the environmental impact of banana production in Costa Rica: A systems approach. *Ambio* 25(3):171–178.
Hunter, J. R. 1994. Is Costa Rica truly conservation-minded? *Conservation Biology* 8:592–595.
Inter-American Development Bank. 1992. *Economic and Social Progress in Latin America. 1992 Report.* Inter-American Development Bank, Washington, DC.
Inter-American Development Bank. 1996. *Economic and Social Progress in Latin America. 1996 Report. Special Edition: Making Social Services Work.* Inter-American Development Bank, Washington, DC.
IMF (International Monetary Fund). 1982. *Adjustment and Financing in the Developing World. The Role of the International Monetary Fund.* Edited by T. Killick. IMF, Washington, DC.
IMF. 1993. *Economic Adjustment in Low-Income Countries: Experience under the Enhanced Structural Adjustment Facility.* IMF, Washington, DC.
IMF. 1996c. *Direction of Trade Statistics. September 1996.* IMF, Washington, DC.
IMF. 1996e. *Annual Report 1996.* IMF, Washington, DC.
Kahler, M. 1993. Bargaining with the IMF: Two-level strategies and developing countries. In P. Evans, H. Jacobson, and R. Putnam (Eds.), *Double-Edged Diplomacy: International Bargaining and Domestic Policies.* UC Press, Berkeley/Los Angeles.
Krause, L. B. 1992. Managed trade: The regime of today and tomorrow. *Journal of Asian Economics* 3:301–313.
Krebs, A. V. 1992. *The Corporate Reapers: The Book of Agribusiness.* Essential Books, Washington, DC.
Leacock, R. 1990. *Requiem for the Revolution: The United States and Brazil, 1961–1969.* Kent State University, Kent, OH.
Lewis, S. A. 1992. Banana bonanza. Multinational fruit companies in Costa Rica. *Ecologist* 22:289–290.
Makgetla, N. S., and R. B. Seidman. 1989. The applicability of law and economics to policy making in the third world. *Journal of Economic Issues* 23:35–78.
Mansbach, R. W. 1996. Neo-this and neo-that: Or, "play it Sam" (again and again). In *Mershon International Studies Review,* supplement to the *International Studies Quarterly,* Vol. 40, pp. 90–95. Blackwell, Cambridge, MA.
Marshall, A. 1920. *Principles of Economics.* Macmillan, London.
Mead, W. R. 1993. Why the deficit is a godsend, and five other economic heresies. *Harper's Magazine,* May, 56–63.
Monge, C. 1995. Costa Rica se Ahoga en Deudas. *La Pensa Libre,* July 11, p. 3.
Montanye, D. R. 1998. Examining sustainability: An evaluation of US AID policies for agricultural export-led growth in Costa Rica. Unpublished Master's thesis. SUNY-ESF, Syracuse, NY.
Ortiz, F. 1998. Costa Rica has Intel inside, but no place for toxic waste. *Eco. Americas* 1, 1–8.
Overseas Economic Cooperation Fund (OECF). 1991. *Issues Related to the World Bank's Approach to Structural Adjustment: Proposal from a Major Partner.* OECF Occasional Paper no. 1, Oct. 1991, OECF, Tokyo.
Piddington, K. 1992. The role of the World Bank. In A. Hurrell and B. Kingsbury (Eds.), *The International Politics of the Environment. Actors, Interests, and Institutions,* pp. 212–227. Clarendon Press, Oxford.
Repetto, R. 1988. *The Forest for the Trees? Government Policies and the Misuse of Forest Resource.* World Resources Institute, Washington, DC.
Repetto, R. 1992. Accounting for environmental assets. *Scientific American,* June, 94–100.
Rostow, W. W. 1960. *The Stages of Economic Growth.* Pergamon, New York.

Sagoff, M. 1988. *The Economy of the Earth. Philosophy, Law, and the Environment.* Cambridge University Press, Cambridge, UK.

Schumpeter, J. A. 1978. *History of Economic Analysis,* 10th ed. Oxford University Press, New York.

Senate Committee on Banking and Currency. 1945. *Participation of the United States in the International Monetary Fund and the International Bank for Reconstruction and Development.* 79th Cong., 1st sess., 1945, S. Rpt. 452, pt. 2, "Minority Views," p. 9.

Shiva, V. 1993. The impoverishment of the environment: Women and children last. In M. Mies and V. Shiva (Eds.), *Ecofeminism,* pp. 70–90. Zed Books, London, UK.

Sinha, R. 1995. Economic reform in developing countries: Some conceptual issues. *World Development* 23:557–575.

Skidmore, T. 1988. *The Politics of Military Rule in Brazil, 1964–85.* Oxford University Press, New York.

Sklar, L. 1994. *Damming the Rivers: The World Bank's Lending for Large Dams, 1994.* International Rivers Network, Berkeley, CA.

Solis, O. 1996. Defending the state, empowering the people. Interview. *Multinational Monitor,* Sept., 20–23.

Stewart, F. 1991. The many faces of adjustment. *World Development* 19:1847–1864.

Stiglitz, J. E. 1994. *Wither Socialism?* MIT Press, Cambridge, MA.

Subcommittee on International Development, Finance, Trade, and Monetary Policy, Committee on Banking, Finance, and Urban Affairs. 1994. *Oversight of the International Monetary Fund and the World Bank.* House. 103rd Cong., 2nd sess., Nov. 21, 1994.

Tardanico, R., and M. Lungo. 1995. Local dimensions of global restructuring: Changing labour-market contours in urban Costa Rica. *International Journal of Urban and Regional Research* 19:223–249.

Tarp, F. 1993. *Stabilization and Structural Adjustment: Macroeconomic Frameworks for Analyzing the Crisis in Sub-Saharan Africa.* Routledge, London.

Taylor, L. 1993. The World Bank and the environment: The World Development Report 1992. *World Development* 21(5):869–881.

Thrupp, L. A., G. Bergeron, and W. Waters, 1995. *Bittersweet Harvests for Global Supermarkets: Challenges in Latin America's Agricultural Export Boom.* World Resources Institute, Washington, DC.

UN Conference on Trade and Development (UNCTAD). 1993. *Trade and Development Report 1993.* UN, Geneva.

UN Economic Commission for Latin America and the Caribbean. 1989. *Economic Survey of Latin America and the Caribbean, 1988. Santiago 1989.* UN, New York.

U.S. Agency for International Development. 1988. *Strategy Update.* USAID Mission, San Jose, Costa Rica.

U.S. Department of Commerce, Economics and Statistics Administration, Bureau of the Census. 1995. *Statistical Abstract of the United States. The National Data Book,* 115th ed. Department of Commerce, Washington, DC.

U.S. Department of Commerce. 1996. *National Trade Data Bank. The Export Connection.* On CD-ROM. U.S. Department of Commerce, Economics and Statistics Administration, Washington, DC.

Utting, P. 1994. Social and political dimensions of environmental protection in Central America. In *Development,* pp. 231–259. Blackwell, Cambridge, MA.

Wilkie, J. W., C. A. Contreras, and C. Komisaruk (Eds.). 1995. *Statistical Abstract of Latin America,* Vol. 31. University of California, Los Angeles.

Williamson, J. 1993. Democracy and the Washington consensus. *World Development* 21(8): 1329–1336.

World Bank. 1981. *World Development Report 1981.* Oxford University Press, New York.

World Bank. 1982. *World Development Report 1982*. Oxford University Press, New York.
World Bank. 1983. *World Development Report 1983*. Oxford University Press, New York.
World Bank. 1984. *World Development Report 1984*. Oxford University Press, New York.
World Bank. 1985. *World Development Report 1985*. Oxford University Press, New York.
World Bank. 1986. *World Development Report 1986*. Oxford University Press, New York.
World Bank. 1987. *World Development Report 1987*. Oxford University Press, New York.
World Bank. 1988. *World Development Report 1988*. Oxford University Press, New York
World Bank. 1992. *World Development Report 1992*. Oxford University Press, New York.
World Bank. 1996. *World Development Report 1996. From Plan to Market*. Oxford University Press, New York.
Worldwatch Institute. 1998. State of the World 1998. W. W. Norton, London/New York.

CHAPTER 25

# Comparative Estimates of Sustainability

## Economic, Resource Base, Ecological Footprint, and Emergy

MARK BROWN, CHARLES A. S. HALL, AND MATHIS WACKERNAGEL



## I. INTRODUCTION

Questions concerning sustainability, carrying capacity, and the welfare of developing nations tend to center on the role of natural resources, or what economists term "natural capital," in economies. In recent years, a number of economists have taken a hard look at resource depletion and its effect on GNP and net capital formation, suggesting that when taken into account, many otherwise growing economies may in fact be declining since declining levels of natural capital threaten long-term economic sustainability (e.g., Solorzano *et al.*, 1991). Carrying capacity and the welfare of growing populations in developing nations may be tied, ultimately, to natural resources much more tightly than industrial capital formation. This chapter summarizes several of the assessments of the resource base of Costa Rica given in other chapters and compares these to several relatively novel approaches to estimating the resource base of the economy of Costa Rica, the relation of that base to sustainability and carrying capacity, and the long-term implications of policies that favor exports of resources. We use several biophysical economic analyses, a land area-based "footprint" analysis, and an energy-based "emergy" analysis. The question is, do the results differ when very different approaches are used to estimate sustainability?

## II. SUMMARY FROM OTHER CHAPTERS OF THIS BOOK

Traditionally sustainability has been estimated using economic criteria. So we start our summary there. Chapters 2 and 4 show that the economy is growing, but little more than the population, so that per capita income is remaining about constant although it is being distributed increasingly to the wealthier component of society. These chapters, plus Chapter 23, also show that debt has become a very large and difficult issue, and has become nearly impossible to retire for a variety of reasons, including the need for imports to pay for the exports (e.g., Figure 23-7). Domestic food production is increasingly insufficient to feed Costa Ricans (Figures 5-5 and 23-2), and increasing yields have reached serious problems with diminishing agronomic returns to inputs of fertilizers or land area (Figure 12-4). Erosion continues to eat away at the productive potential of the nation's agriculture (Chapter 15). Since per hectare yields are stable or declining for most crops, it appears likely that the combined effects of declining returns on inputs and the cumulative losses due to erosion are compensating or more than compensating for any increases in agricultural technology. The number of people that can be fed without external resources appears to be no more than roughly 500,000, and with inputs of fertilizer and other technologies from the industrialized nations, perhaps about 4 million, if the inputs are paid for (Chapter 4, Figure 12-8). Natural forests continue to be destroyed (Chapter 16), although at a slower rate than in the recent past, lumber and paper are being imported (Chapter 18), and carbon emissions from deforestation are about as large as the quantity released from fossil fuel burning, and enormously greater than the carbon sequestered through reforestation (Chapter 17). Other tradeoffs are presented in Bouman *et al.* (1998) and Haines and Peterson (1998). None of these patterns represents sustainability, as is developed further in the final chapter.

There are some positive signs related to sustainability as well: tourism and ecotourism are increasingly important, population growth rates may (or may not) be declining, and the hydroelectric potential is large. The potential of the forests to supply energy also is large, but not in the forms needed. High-tech industries are beginning to locate in Costa Rica.

Generally both the positive and the negative trends are consuming increasing amounts of fossil fuel, which does not contribute to sustainability. It is very difficult to pay for the imports of fuels or the machines in which they are used (Figure 23-7). We conclude that Costa Rica left behind any possibility of sustainability at a modern standard of living using its own resources when the population passed about 1 to 2 million. Expressed differently, at the present standard of living Costa Rica is using 2 to 4 times the resources, many from external sources and much of that paid for with debt, that could be sustainably supported from the resource base that exists, which is mostly land, climate, and soils. Other Costa Rican scientists have reached more or less the same conclusions, although often expressed in very different ways (i.e., Hartshorn *et al.*, 1982; Monge, 1995). But there may be flaws in our (or their) analyses, and it is useful to compare our results with other techniques that attempt to measure sustainability from quite different perspectives. Thus we believe it useful to compare our results with those of others. There are two very clever techniques "out there" that attempt to measure sustainability in quite different ways. These analyses were made quite independently from the analyses presented in this book. What do they find?

## III. CALCULATING THE ECOLOGICAL FOOTPRINT OF COSTA RICA: DOES COSTA RICA FIT INSIDE COSTA RICA?

### A. Why Measure the Ecological Footprint of Costa Rica?

Sustainability means securing people's quality of life within the carrying capacity of nature. The preservation of nature is not only an ethical call in favor of other species' survival, but it is also the precondition for humane living conditions. People exist in, and are a part of, ecosystems. They depend on ecosystems for the steady supply of the basic requirements for life—food, water, energy, fibers, waste sinks, and life-support services. The "ecological footprint" concept enables us to quantify the human use of nature and compare it to nature's carrying capacity. It thereby summarizes society's "ecological impacts" and provides an indicator of ecological sustainability (Wackernagel *et al.*, 1999; Wackernagel and Rees, 1996).

### B. The Calculation

People require land to grow their food, produce energy they use, grow the timber they consume, capture the water they drink and otherwise use and assimilate their wastes and those from their industrial activity. The ecological footprint keeps track of these uses based on two simple facts: first, we can monitor most of the resources we consume and many of the wastes we generate; second, most of these resource and waste flows can be converted to a corresponding biologically productive area. Thus, the *ecological footprint* of any defined population (from a single individual to a whole city or country) is the total area of ecologically productive land and water occupied exclusively to produce all the resources consumed and to assimilate all the wastes generated by that population, using prevailing technology. As people use resources from all over the world and affect far away places with their wastes, footprints sum up these ecological areas wherever that land and water may be located on the planet.

We prepared a comprehensive ecological footprint calculation for Costa Rica on a spreadsheet of 135 lines and 15 columns (Table 25-1). The calculation consists of a consumption analysis of 15 main resources (lines 8–46), a section where all uses of commercial energy are listed (lines 49–57), and an energy balance of traded goods (lines 61–114). All figures are taken from published UN statistics (Table 25-2). Consumption is translated into land areas using world average yield figures. To make these figures comparable with local ecological productivity, local areas are adjusted with yield factors. These factors indicate how much more (and less) productive Costa Rican areas are compared to world average areas. By standardizing the calculation with world average figures, ecological load and local ecological capacity can be compared among nations. Equivalence factors adjust the ecological categories for their productivity to make them mutually comparable.

The footprint and the available ecological capacity are composed of six types of ecologically productive areas: arable land, pasture, forest, sea space, built-up land, and energy land, all of which represent competing or potentially competing uses of nature. Fossil energy land is the land that we should be putting aside for $CO_2$ absorption. However, this is something that is not done in today's world.

The box on lines 117–130 summarizes the results using 1992 data. They show that Costa Rica uses 2.5 ha (25,000 $m^2$ or five football fields) of nature per person. However, there was only a little less than 2.5 ha of biotically productive space per person available within its perimeter in 1992, including sea space and assuming 12% is set aside for other species. On the land alone, Costa Ricans are using 0.1 ha more of biotically productive space than what is available. With increasing national consumption (because of larger population numbers and higher per capita consumption), the ecological deficit of Costa Rica will continue to grow. Assuming a growth rate of 3% for national consumption (of which 2.4% is used for demographic growth and 0.6% for per capita increase in consumption), it will take only 10 years to develop a 35% ecological deficit. In other words, the footprint would be one-third larger than the ecological capacity of Costa Rica.

TABLE 25-1 Calculation of the Costa Rican's Average Ecological Footprint (1992 Data; Population of Costa Rica, 3,270,000)

| | CATEGORIES | Yield [kg/ha][a] (global average) | Ref.-yield[b] | Production [t/yr] | Ref.-prod. | Import [1000 $] | Import [t/yr] | Ref.-imp. | Export [1000 $] | Export [t/yr] | Ref.-export | Apparent consumption [t/yr] | Footprint component [ha/cap] |
|---|---|---|---|---|---|---|---|---|---|---|---|---|---|
| 3 | LAND AND SEA AREA ACCOUNTING | | | | | | | | | | | | |
| 7 | FOODS | | | | | 172,596 | | 1:214#0 | 1,093,972 | | 1:217#0 | | |
| 8 | meat. (yield for animal products from pasture; expressed in average units) | 74 | 3:212#98, 3:228#105, 3:215#99, 3:3#1 | 152,000 | 3:209#97 | | 430 | 4:27#12 | | 23,550 | 4:27#12 | 128,880 | |
| 9 | bovine, goat, mutton, and buffalo meat | 33 | | 82,000 | 3:197#92 | 200 | 100 | 4:30–34#13 | 56,000 | 23,048 | 4:30–34#13 | 59,382 | 0.5532 pasture |
| 10 | nonbovine, nongoat, nonmutton, nonbuffalo | 457 | | 70,000 | calc. (equiv. in cereals) | | 330 | calc. | | 502 | calc. | 69,828 | −0.0001 arable land |
| 11 | dairy (milk equiv.) | | | 470,000 | | | 6,300 | | | 6,150 | | 470,150 | 0.2863 pasture |
| 12 | milk | 502 | | 470,000 | 3:215#99 | 5,520 | | 4:60#27 | 5,350 | 5,350 | 4:60#27, 28 (est.) | | |
| 13 | cheese | 50 | | | | 2,300 | 600 | 4:76#33 | 150 | 30 | 4:76#33 | | |
| 14 | butter | 50 | | | | 60 | 30 | 4:73#32 | 130 | 50 | 4:73#32 | | |
| 15 | marine fish | 29 | | | | | | | | | | 9[d] | 0.3140 sea |
| 16 | cereals | 2,744 | 3:65#15 | 204,000 | 3:65#15 | 83,361 | 476,815 | 1:214#04 (est. of subcategories) | 11,326 | 8,325 | 1:217#04 (est. 048) | 672,490 | 0.0749 arable land |
| 17 | wheat | | | | | 23,371 | 130,462 | 1:214#041 | | | | | |
| 18 | rice | | | | | 13,924 | 48,804 | 1:214#042 | | | | | |
| 19 | maize | | | | | 32,079 | 217,545 | 1:214#044 | | | | | |
| 20 | veg. & fruit | 18,000 | | 2,573,000 | 3:123#49 | 18,967 | 39,126 | 1:214#05 (est.) | 645,104 | 2,185,326 | 1:217#05 (est.) | 426,800 | 0.0073 arable land |
| 21 | veg., etc. | | | | | 2,934 | 5,010 | 1:214#054 | 33,941 | 87,868 | 1:217#054 | | |
| 22 | fresh fruit | | | | | 4,100 | 9,500 | 4:123–150#58, #59 | 568,743 | 1,953,758 | 1:217#057 | | |

*(continues)*

**TABLE 25-1** *(Continued)*

LAND AND SEA AREA ACCOUNTING

| | CATEGORIES | Yield [kg/ha][a] (global average) | Ref.-yield[b] | Production [t/yr] | Ref.-prod. | Import [1000 $] | Import [t/yr] | Ref.-imp. | Export [1000 $] | Export [t/yr] | Ref.-export | Apparent consumption [t/yr] | Footprint component [ha/cap] |
|---|---|---|---|---|---|---|---|---|---|---|---|---|---|
| 23 | roots and tubers | 12,607 | 3:86#25 | 142,000 | 3:86#25 | 100 | 200 | 4:112#49 | 30 | 150 | 4:112#49 | 142,050 | 0.0034 arable land |
| 24 | pulses | 852 | 3:97#31 | 33,000 | 3:97#31 | 1,500 | 2,500 | 4:115#50 | 480 | 1,200 | 4:115#50 | 34,300 | 0.0123 arable land |
| 25 | coffee & tea | 566 | 3:171 | 74,000 | 3:171#78 | | | | 215,304 | 157,980 | 1:217#07 (est. 071) | −83,980 | −0.0454 arable land |
| 26 | cocoa | 454 | 3:173#79 | | | | | | | | | 0[d] | 0.0000 arable land |
| 27 | sugar | 4,893 | 3:153–156#67, 68, 69 | 293,000 | 3:156#89 | | | | 29,000 | 95,000 | 4:151#68 | 198,000 | 0.0124 arable land |
| 28 | | | | | | | | | | | | | |
| 29 | oil seed (incl. soya) | 1,856 | 3:106, 111, 112, 114 | | 3:106–121#38, #39, #40, #41, #42, #43, #44, #45, #46, #47 | 39,252 | 142,640 | 4:213–227#99, #100, #107 | 8 | 4 | 4:213–227#99 | 142,636 | 0.0235 arable land |
| 30 | TIMBER [in roundwood equivalent, m³] | 1.99 | FAO-calc. | 2,794,474 | | | 259,976 | | | 14,250 | | 3,040,200 | 0.4672 forest |
| 31 | roundwood [m³/ha,m³] | waste factors | | 4,315,000 | 5:03 | | 5,000 | 5:05 | | 0 | 5:10 | 4,320,000 | |
| 32 | firewood | 0.53 | | 3,210,000 | 5:20 | | | | | | | 3,210,000 | 48% of consumption firewood |
| 33 | direct roundwood consumption [m³] | 1 for RWE (roundwood equivalent in m³) | | | | | | | | | 5:103 | 227,000 | 7.04% of consumption mines |
| 34 | sawed wood [m³] | 1.50 | for RWE | 798,000 | 5:109 | | 2,000 | 5:111#248 | | 2,000 | 5:117 | 798,000 | 34% of consumption sawed wood |

| | | | | | | | | | | | | | |
|---|---|---|---|---|---|---|---|---|---|---|---|---|---|
| 35 | wood-based panels [m³] | 2.25 | for RWE | 74,000 | 5:146 | | 5,000 | 5:148#634 | | 5,000 | 5:154#634 | 74,000 | 5% of consumption panels |
| 36 | wood pulp [t] | 1.98 | for RWE | 3,000 | 5:223 | | 6,000 | 5:225 | | | 5:229#251 | 9,000 | |
| 37 | paper and paper board [t] | 0.94 | for RWE | 19,000 | 5:285 | | 244,000 | 5:287#641 | | | 5:293#641 | 263,000 | 7% of consumption paper |
| 38 | | | | | | | | | | | | | |
| 39 | | | | | | | | | | | | | |
| 40 | OTHER CROPS | | | | | | | | | | | | |
| 41 | tobacco | 1,548 | 3:176#82 | 0 | 3:176#82 | 600 | 1,015 | 4:213#98 | 8 | 4 | 4:213#98 | 1,001 | 0.0002 arable land |
| 42 | cotton | 1,000 | IIED, p. 64 | | | | ref. of apparent cotton consumption; 2:247, 255 | | | | | 2,998 | 0.0009 arable land |
| 43 | jute | 1,500 | gov. of Vietnam | | | | ref. of apparent jute consumption, 2:263, 267 | | | | | 0 | 0.0000 arable land |
| 44 | rubber | 1,000 | gov. of Vietnam | | | | ref. of apparent rubber consumption, 2:231, 234 | | | | | 0 | 0.0000 arable land |
| 45 | wool | 15 | Wackernagel *et al.* (1993, 67) | | | | ref. of apparent wool consumption, 2:280, 287 | | | | | 0 | 0.0000 pasture |
| 46 | hide | 74 | like bovine meat | | | | ref. of apparent hide consumption, 2:227, 228 | | | | | 12,648 | 0.0523 pasture |
| 47 | | | | | | | | | | | | | |

| 48 | | | | | | | [GJ/yr/cap] | | | |
|---|---|---|---|---|---|---|---|---|---|---|
| 49 | ENERGY BALANCE: | Glob. ave. | Specific energy footprint[d] | Energy type | [GJ/yr/cap] | | for 1992 | Ref. | Footprint component in [ha/cap] | |
| 50 | | 55 | [GJ/ha/yr] coal | coal consumption | 0 | calc. | 0 | 6#5:116 | 0.0000 | fossil energy land for coal |
| 51 | | 71 | [GJ/ha/yr] liquid fossil fuel | liquid fossil fuel consumption | 15 | calc. | 5 | 6#14:172 | 0.2111 | fossil energy land for liquid fuel |
| 52 | | 93 | [GJ/ha/yr] fossil gas | fossil gas consumption | 0 | calc. | 0 | 6#28:332 | 0.0000 | fossil energy land for fossil gas |
| 53 | | | | total fossil fuel consumption | 15 | calc. & WRI (1996, 287) | 5 | calc. | | |
| 54 | | 71 | [GJ/ha/yr] nuclear energy (thermal) | nuclear energy consumption (thermal) | 0 | WRI (1996, 285) | | | 0.0000 | fossil energy land for nuclear energy |
| 55 | | 71 | [GJ/ha/yr] assumed to be fossil energy | energy embodied in net imported goods | 12 | calc. | | | 0.1657 | fossil energy land for embodied energy in net imp. goods |
| 56 | | 1,000 | [GJ/ha/yr] hydroelectric energy | hydroelectricity consumption | 4 | WRI (1996, 285) | | | 0.0043 | built-up area for hydropower |
| 57 | | | | | | | | | | |
| 58 | | | | | | | | | | |
| 59 | | | | | | | | | | |
| 60 | | | | | | | | | | |

*(continues)*

TABLE 25-1 (*Continued*)

| 61<br>62 | CATEGORIES<br>units if not specified | Energy intensity<br>([Gj/t] of<br>embodied energy) | Import<br>[1000 $] | Import<br>[t/yr] | Ref-imp. | Export<br>[1000 $] | Export<br>[t/yr] | Ref.-export | Embodied energy in net import<br>[PJ/yr] | |
|---|---|---|---|---|---|---|---|---|---|---|
| 63 | Beverages | 10 | 10,881 | 6,102 | 1:214#1 | | | | 0.06 | |
| 64 | alcoholic beverages | | 7,996 | 4,484 | 1:214#112 | | | | | |
| 65 | crude materials | | 71,158 | | 1:241#2 | 91,091 | | 1:217#2 | | |
| 66 | hides, skin | 5 | | | | 8,530 | 1,051 | 1:217#21 (est. 211) | −0.01 | |
| 67 | textile fibers | 5 | 8,355 | 7,996 | 1:214#26 (est. of Mexico) | | | | 0.04 | |
| 68 | minerals | 1.5 | 9,254 | 190,923 | 1:214#27 (est. of Brazil) | | | | 0.29 | |
| 69 | Biological fat | 39 | | | | 9,391 | | 1:217#4 | 0.00 | |
| 70 | Chemicals | | 454,057 | | 1:214#5 | 102,834 | | 1:217#5 | | |
| 71 | chem. organics | 40 | 54,246 | 87,212 | 1:214#51 (est.)[g] 512, 514, 515 | 11,472 | 30,042 | 1:217#51 (est. 512) | 2.29 | |
| 72 | alcohol, phenols | | 14,005 | 48,487 | 1:214#512 | 11,038 | 28,905 | 1:217#512 | | |
| 73 | nitrogen | | 7,330 | 2,345 | 1:214#514 | | | | | |
| 74 | org.-inorg. compounds | | 11,235 | 1,531 | 1:214#515 | | | | | |
| 75 | chem. inorganics | 20 | 34,428 | 120,707 | 1:214#52 (est. 522, 523) | | | | 2.41 | |
| 76 | oxides | | 19,763 | 78,465 | 1:214#522 | | | | | |
| 77 | other inorganics | | 14,338 | 41,096 | 1:214#523 | | | | | |
| 78 | dyes, tanning, color products | | 22,759 | 7,203 | 1:214#53 (est. 533) | | | | 0.00 | |
| 79 | medicinal, pharm. products | | 80,579 | 3,100 | 1:214#541 | 39,577 | 2,521 | 1:217#541 | 0.00 | |
| 80 | plastic materials | 50 | 99,345 | 95,625 | 1:214#58 (est. 582, 583) | 14,510 | 9,701 | 1:217#58 (est. 583) | 4.30 | |
| 81 | Basic manufactures | | 567,976 | | 1:214#6 | 174,597 | | 1:217#6 | | |
| 82 | rubber manufactures | 35 | 28,632 | 8,563 | 1:214#62 (est. 625) | 39,540 | 13,303 | 1:217#62 (est. 625) | −0.17 | |
| 83 | paper, paperboard | 35 | 159,071 | 234,976 | 1:214#64 (est. 641, 642) | 28,232 | 18,739 | 1:217#64 (est. 642) | 7.57 | (included in line 42) |
| 84 | textile | 20 | 102,133 | 23,470 | 1:215#65 (est. 651, 652) | 25,244 | 7,494 | 1:218#65 (est. 651, 653) | 0.32 | 0.0024 arable land) |
| 85 | iron and steel | 30 | 117,146 | 276,024 | 1:215#67 (est. 672, 673, 674) | 13,873 | 15,613 | 1:218#67 (est. 674) | 7.81 | |
| 86 | metal manufactures | 60 | 77,803 | 29,187 | 1:215#69 (est. 691, 692, 699) | 15,455 | 4,685 | 1:218#69 (est. 692) | 1.47 | |
| 87 | base metal manufactures | | 21,404 | 4,933 | 1:215#699 | | | | | |

| | | | | | | | | | | |
|---|---|---|---|---|---|---|---|---|---|---|
| 88 | Indust. products | | 710,123 | 1:215#7 | | 60,325 | | 1:218#7 | | |
| 89 | power generating | | 46,097 | 7,421 | 1:215#71 (est. 713, 716) | | | | | |
| 90 | internal combustion | 140 | 13,043 | 1,712 | 1:215#713 | | | | 0.24 | |
| 91 | rotating electric plant | 100 | 20,328 | 3,660 | 1:215#716 | | | | 0.37 | |
| 92 | machines for special industries | 100 | 124,521 | 24,979 | 1:215#72 (est. with subcategories) | | | | 2.50 | |
| 93 | tractors, nonroad | | 27,149 | 6,734 | 1:215#722 | | | | | |
| 94 | civil engineering equip. | | 20,821 | 5,374 | 1:215#723 | | | | | |
| 95 | textile, leather | | 20,235 | 3,787 | 1:215#724 | | | | | |
| 96 | other machinery for special industries | | 20,909 | 1,981 | 1:215#728 | | | | | |
| 97 | metalworking machinery | 100 | 7,220 | 919 | 1:215#73 (est. 736) | | | | 0.09 | |
| 98 | general industrial | 100 | 111,386 | 11,236 | 1:215#74 (est. of subcategories) | 9,106 | 2,363 | 1:218#74 (est. of subcategories) | 0.89 | |
| 99 | heating/cooling | | 21,133 | 3,361 | 1:215#741 | 5,600 | 1,453 | 1:218#741 | | |
| 100 | pumps for liquids | | 9,130 | 816 | 1:215#742 | | | | | |
| 101 | pumps centrifuges | | 14,672 | 1,963 | 1:215#743 | | | | | |
| 102 | nonelec. machinery | | 21,980 | 1,273 | 1:215#745 | | | | | |
| 103 | nonelec. machinery parts | | 31,724 | 2,537 | 1:215#749 | | | | | |
| 104 | office machines | 140 | 48,298 | 1,044 | 1:215#75 (est. 752, 759) | | | | 0.15 | |
| 105 | automatic data proc. | | 32,253 | 698 | 1:215#752 | | | | | |
| 106 | office accessories | | 12,260 | 264 | 1:215#759 | | | | | |
| 107 | telecom., sound | 140 | 55,992 | 2,997 | 1:215#76 (est. 761, 762, 764) | | | | 0.42 | |
| 108 | electric machinery | 100 | 85,902 | 11,825 | 1:215#77 (est. 772, 773, 775) | 42,486 | 7,781 | 1:218#77 (est. 772, 773, 778) | 0.40 | |
| 109 | road vehicles | 140 | 225,475 | 52,088 | 1:216#78 (est. 781, 784) | | | | 7.29 | |
| 110 | Misc. manufactured goods | | 241,717 | | 1:216#8 | 137,582 | | 1:218#8 | | |
| 111 | clothing and accessories | 20 | 23,875 | | 1:216#84 | 69,360 | 5,604 | 1:218#84 (est. 842, 843, 844) | −0.11 | −0.0009 arable land |
| 112 | misc. manufactured | 100 | 146,922 | 14,459 | 1:216#89 (est. 892, 893, 894) | 38,136 | 15,922 | 1:218#89 (est. 892, 893) | −0.15 | |
| 113 | Goods not classed by kind | 50 | 313,883 | | 1:216#9 | 148,912 | | 1:218#9 | 0.00 | |
| 114 | special transactions | | 309,725 | 58,690 | 1:216#931 | 138,930 | 44,777 | 1:218#931 | 0.00 | |
| 115 | | | | | Energy embodied in net import per capita | | | | 11.76 [GJ/cap/yr] | |
| 116 | | | | | | | | | | |

*(continues)*

**TABLE 25-1** *(Continued)*

| | | | | | | | | | | |
|---|---|---|---|---|---|---|---|---|---|---|
| 117 | SUMMARY | | | | | | | | | |
| 118 | DEMAND | | | | SUPPLY | | | | | |
| 119 | FOOTPRINT (per capita) | | | | EXISTING BIOCAPACITY WITHIN COUNTRY (per capita) | | | | OTHER INDICATORS | |
| 120 | | | Equivalence | Equivalent | | | National | Yield adjusted | (average land with world average productivity in [ha/capital]) | |
| 121 | | Total | factor | total | | Yield | area | equiv. area | 2.4 | footprint on the land |
| 122 | Category | [ha/cap] | [−] | [ha/cap] | Category | factor | [ha/cap] | [ha/cap] | 2.3 | existing land-based capacity within Costa Rica |
| 123 | fossil energy | 0.4 | 1.1 | 0.4 | $CO_2$ absorption land | | 0.00 | 0.00 | | |
| 124 | built-up area | 0.0 | 2.8 | 0.0 | built-up area | 1.13 | 0.01 | 0.05 | 0.0 | Costa Rica's national ecological deficit |
| 125 | arable land | 0.1 | 2.8 | 0.3 | arable land | 1.13 | 0.16 | 0.52 | 100% | Costa Rica's capacity as percentage of its footprint |
| 126 | pasture | 0.9 | 0.5 | 0.5 | pasture | 1.89 | 0.72 | 0.73 | | |
| 127 | forest (incl. deforestation) | 1.1 | 1.1 | 1.2 | forest | 1.83 | 0.48 | 1.00 | −0.4 | Costa Rica's global deficit (for 1993) |
| 128 | sea | 0.3 | 0.2 | 0.1 | sea | 1.00 | 2.60 | 0.57 | 121% | Costa Rica's per capita footprint compared to the global per capita biocapacity (for 1993) |
| 129 | | | | | TOTAL existing | | 4.0 | 2.9 | | |
| 130 | TOTAL used | 2.8 | | 2.5 | TOTAL available (minus 12% for biodiversity) | | | 2.5 | | |
| 131 | | | | | | | | | | |
| 132 | | | | | | | | | | |
| 133 | | | | | | | | | | |
| 134 | | | | | | | | | | |
| 135 | | | | | | | | | | |

[a] Units if not specified.

[b] See Table 25-2 for references 1 to 6. For example, "3 : 212 #98" means book 3, page 212, category 98.

[c] Apparent consumption is calculated by adding imports to production and subtracting exports. All the production, import, and export data stem from United Nations' documents.

[d] Expressed in harvested fish in Kg/capita. From WRI (1996, 311).

[e] Reference of apparent cocoa consumption, 2:162, 169.

[f] The energy footprint for fossil fuel is expressed as area necessary for $CO_2$ absorption. Replacing the fossil fuel with biotically productive substitutes would occupy even larger areas.

[g] Est., derived from following sources.

TABLE 25-2 Source of Data Used to Calculate Ecological Footprint

(1) United Nations. 1995. *1993 International Trade Statistics Yearbook*, Vol. 1. Department for Economic and Social Information and Policy Analysis, Statistical Division, New York.

(2) United Nations Conference on Trade and Development (UNCTAD). 1994. *UNCTAD Commodity Yearbook 1994*. United Nations, New York/Geneva.

(3) Food and Agriculture Organization of the United Nations (FAO). 1995. *FAO Yearbook: Production 1994*, Vol. 48. FAO, Rome.

(4) Food and Agriculture Organization of the United Nations (FAO). 1994. *FAO Yearbook: Trade 1993*, Vol. 47, FAO, Rome.

(5) Food and Agriculture Organization of the United Nations (FAO). 1995. *FAO Yearbook: Forest Production 1993*. FAO, Rome.

(6) World Resources Institute. 1996. *World Resources 1996–1997*. World Resources Institute, UNEP, UNDP, The World Bank, Washington, DC.

## C. Interpreting the Results

The numbers in Table 25-1 illustrate which consumption type occupies how much ecological space. In the end, the footprint assessment comes to the conclusion that Costa Rica's population is already consuming more than what the country can regenerate. This means that it has to import natural capital or deplete its own stocks, both of which it is doing. The current trend of a growing population and declining ecological capacity signals more severe conditions in the future. Because of its persistent international debt, Costa Rica may find it increasingly difficult to pay for its imports. In addition, if current worldwide tendencies of increasing resource consumption continue, the ecological capacity available "elsewhere" that is exported to countries with ecological deficits (such as Costa Rica) will diminish.

A similar, but less comprehensive, analysis of the Costa Rican footprint was completed by Ko *et al.* (1998). They found, as well, that the footprint was larger than the national area and increasing for the nation as a whole.

Costa Rica is not alone with its ecological deficit; similar calculations for Switzerland show that its average citizen occupies approximately 5 ha of ecologically productive space to provide for his or her current level of consumption. In comparison, the average American lives on a footprint twice the size of the Swiss one, or over 4 times that of the average Costa Rican. The United States is endowed with more biological capacity per capita than Costa Rica because of its relatively low population density and high biotic productivity. But, because of high levels of consumption, the U.S. national ecological deficit is 3.6 ha per capita, which is 45% larger than the average Costa Rican footprint.

As these calculations include basic ecological functions only, the results may fall short of what the actual ecological deficit is. Footprints underestimate the actual use of nature's ecological capacity because they assume that current yields are sustainable and do not include freshwater use and some human wastes. However, these assessments document the magnitude of humanity's impact on the earth. To improve these assessments, more resources and waste products will be included in later studies.

In conclusion, the ecological footprint is a simple resource accounting tool for measuring the ecological aspects of sustainability. As such, it indicates that the nation of Costa Rica, together with many others on the globe, is not sustainable by this criterion and is becoming less so.

Throughout history nations have enhanced their own ability to support their human population or its material affluence by bringing in resources from outside their borders through trade or conquest. How important is this today? To what degree can a nation be supported by its own resources?

## IV. EMERGY, FOREIGN TRADE, AND CARRYING CAPACITY IN COSTA RICA

### A. EMERGY, WEALTH, AND ECONOMIC VITALITY

A new unit of evaluation, called "emergy" (which is the energy required to make something), offers the potential to evaluate *all* resources in a national economy in the same units so that comparisons and judgments concerning sustainability can be made.

Emergy is a quantitative measure of all of the energy resources required to develop a product (whether those resources are mineral resources that result from biogeologic processes, renewable resources such as wood, a fuel source used directly, or an economic product that results from industrial processes). All these resources are expressed in units of one type of energy (usually solar energy). We suggest that evaluations using emergy may help to clarify policy options, because the use of emergy as a measure of value overcomes four important limitations of other methods for evaluating alternative fuels and technologies.

These limitations are as follows: (1) A truly comparative analysis cannot be undertaken by mixing units of measure such as weight, volume, heat capacity, or economic market price. (2) Evaluations that use only the heat value of (energy) resources (such as kcal or joules) for quantification assume that the only value of a resource is the heat derived from its combustion. In this way, for example, human services are evaluated as the calories expended doing work and, when compared to other inputs to a given process, are generally several orders of magnitude smaller than, for example, fossil fuel use. Thus human labor inputs are often considered irrelevant when in fact they are critical because of their quality. (3) Nonmonetary resources and processes (i.e., those outside the monetized economy) are often considered externalities and not quantified. Most processes, and all economies, are driven by a combination of renewable and nonrenewable energy. (4) An assumption that price determines value is often made. The price of a product or service reflects human preferences, often called "willingness-to-pay." It can also reflect the amount of human services "embodied" in a product. A valuing system based on human preference alone assigns either relatively arbitrary values, or no value at all, to necessary resources or environmental services. Emergy as a unit of evaluation avoids these problems.

Emergy is a quantitative measure of the ability to cause work that is independent of price (Odum, 1984; Odum, 1996). New energy sources often are evaluated based on dollar costs per unit of energy produced. Economic theory is based on the assumption that price is proportional to value, and suggests that if prices rise, a new source may become economical and thus competitive. However, price merely suggests what humans are willing to pay for something. The true value of something to society is determined by the ability of a resource to cause work, that is, the effect it has in stimulating an economy. For example, a gallon of gasoline will power a car the same distance no matter what its price. Its value to the driver is the number of miles (work) that can be driven, regardless of price. Its price reflects both the scarcity of gasoline and how important it is to do the work. Price is often inverse to a resource's contribution to an economy. When a resource is plentiful, its price is low, yet it contributes much to the economy. When a resource is scarce, its total contribution to the economy is small, yet its price is high. For example, 200 years ago salmon was very abundant in New England and contributed enormously to the welfare of society even though its price was very low. Today the converse is true.

Emergy may be a measure of equivalence when one resource is substituted for another. Sunlight and fossil fuels are very different energies; yet when their heat values are used the difference is not elucidated. A joule of sunlight is not equivalent to a joule of fossil fuel in any system other than a laboratory heat engine. In the realm of the combined system of humanity and nature, sunlight and fuels are not equally substitutable joule for joule. However, when a given amount of fuel energy is expressed as the amount of solar energy required to make it (solar emergy), its equivalence to sunlight energy is defined. Since emergy is a measure of the work that goes into a product expressed in units of one type of energy (sunlight), it is also a measure of what the product should contribute in useful work in relation to sunlight.

## B. Emergy Analysis

We used emergy analysis to evaluate the economy of Costa Rica, including its sustainability. It is a method of energy analysis that accounts for the direct and indirect use of energy in producing a commodity, resource fuel, or service, in energy of one type.

Consider a fine wooden chair. In Marxist terms the embodied labor in the chair is the hours of labor that went into the production of that chair. A comprehensive analysis should also include the hours of labor that went into cutting the tree and making the tools that the logger and the artisan used. Likewise there was energy used to make the chair. But that was not the only work processes required to make the chair. Solar energy grew the tree, evaporated water from the ocean to generate rain, generated the winds that blew the rain from the ocean to the land, and so on. Centuries of solar energy went into making the soil that the tree grew in, and thousands or millions of years to make the oil that powered the saws that cut the tree and produced the boards the artisan used. The sum of all of these energies represents the emergy input required to make the chair, and they all were necessary.

Thus the solar emergy in a resource, product, or service is the sum of the solar energies required to make it. Emergy includes both fossil fuel energies and environmental energies (like sunlight, rain, tides, etc.) that are necessary inputs of most processes of energy transformation. Cumulative emergy inputs of the past are found embodied in the soils, forests, petroleum, educated people, and other resources that constitute the resource base of a nation. Of particular importance here, the emergy resources of a region or a country largely define and limit the amount of economic work that can be done by that region or country, and that resource base divided by the number of people is the per capita resource base. The emergy base is related to the ecological footprint previously discussed in that the footprint is largely related to the energy-capturing potential of land, and is corrected at a national level for differing productivities. The emergy analysis is more comprehensive but more controversial (e.g., Brown and Herendeen, 1996).

Emergy can be conceptualized as energy memory (Scienceman, 1987, 1989), since it is a measure of all of the energy previously required to produce a given product or process. The term "emergy" differs somewhat from embodied energy as defined by other schools of thought. For example, environmental inputs and labor are omitted in the analyses of IFIAS (1974) and Slesser (1978), energies are added without using transformities (except for electricity) by Hall *et al.* (1986), and energies are assigned by input–output data (usually based on money flows) with different results by Hannon *et al.* (1976), Herendeen *et al.* (1975), and Costanza (1978).

We now provide some of our definitions:

- *Energy*. Sometimes referred to as the ability to do work, although this definition does not include the different qualities of different energies. Energy is a property of all things that can be turned into heat, and is measured in heat units (BTUs, calories, or joules)

- *Emergy*. An expression of all the energy used in the work processes that generate a product or service expressed in units of one type of energy. The solar emergy of a product is the emergy of the product expressed in the equivalent solar energy required to generate it. Sometimes it is convenient to think of emergy as energy memory.
- *Emjoule*. The unit of measure of emergy, or emergy joule. It is expressed in the units of energy previously used to generate the product; for instance, the solar emergy of wood is expressed as joules of solar energy that were required to produce the wood. Solar emjoule is abbreviated "sej."
- *Macroeconomic Dollar* (Emdollar or EM$). A measure of the money that circulates in an economy as the result of some energy-driven process. In practice, to obtain the macroeconomic dollar value of an emergy flow or storage, the emergy is divided by the ratio of total emergy to gross domestic product for the national economy.
- *Nonrenewable Energy*. Energy and material storages such as fossil fuels, mineral ores, and soils that are consumed at rates that far exceed the rates at which they are produced by geologic processes.
- *Renewable Energy*. Energy flows of the biosphere that are more or less constant and reoccurring, which ultimately drive the biological and chemical processes of the earth and contribute to geologic processes.
- *Resident Energy*. Renewable energies that are characteristic of a region.
- *Transformity*. The ratio obtained by dividing the total emergy that was used in a process by the energy yielded in the process. Transformities have the dimensions of emergy/energy (sej/J). A transformity for a product is calculated by summing all the emergy inflows to the process and dividing by the energy of the product. Transformities are used to convert energies of different types to emergy of the same type.

A complete description of the methods employed to evaluate the economy of Costa Rica is beyond the scope of this short section but can be found in Odum (1978, 1984, 1995, and 1996). Data sources overlap with those in Table 25-2. The methodology evaluates the main flows of resources, energy, and human services within the economy for a given year and converts them into common units of emergy using transformities. Mark Brown calculated transformities for the most important resources, fuels, and human resources within the Costa Rican economy directly, but some others were calculated based on data for other areas, all based on the procedures in Odum (1996).

Our analysis was done for 1994, the latest year for which we had sufficient data. The flows of resources, energy, and services throughout, into, and out of the economy, evaluated in emergy terms, were compared and several ratios calculated to assess sustainability and carrying capacity of the economy as a whole. A more complete picture would emerge if data for several years were evaluated and compared, since long-term trends would be revealed.

## V. RESULTS AND DISCUSSION OF EMERGY ANALYSIS

Per capita emergy use in Costa Rica was almost one-quarter that of the United States (Table 25-3; the U.S. per capita emergy consumption in 1987 was $29.2 \times 10^{15}$ sej/yr). (An emergy analysis of Costa Rica is given in Table III-1 of Appendix III and summarized in Tables III-2 and III-3.) Emergy use per unit area (sej/$m^2$) was about one-half that of the United States (U.S. emergy use per unit area was $7.0 \times 10^{11}$ sej/$m^2$/yr), reflecting the greater population density. Almost 90% of total emergy use is derived from within Costa Rica, compared to about 78% for the United States, showing that Costa Rica was still highly dependent on its own resources in 1987, although this appears to be changing rapidly. Costa Rica has a negative emergy balance of payments, exporting about twice as much emergy as it imports.

From these data we derived an index, called the Environmental Loading Ratio (ELR), defined as the ratio of nonrenewable emergy flow to renewable emergy flow in the economy. This ratio relates the use of nonrenewable sources of energy and materials to the ability of the environment to absorb the wastes and disorder that result from their use. In essence, the ELR reflects the intensity of use of nonrenewable energy as compared to that which is renewable. In Costa Rica the ELR was 0.51/1, meaning that nearly twice as much of the total emergy driving the economy comes from renewable sources as from nonrenewable ones. This also means that even though the economy of Costa Rica is industrializing rapidly, still some two-thirds of the emergy running it comes from solar-powered sources.

By way of comparison, in the United States the ELR ratio was about 7.2 : 1. One consequence of ratios this high is that the environment can no longer

**TABLE 25-3** Summary of Emergy Indices of Costa Rica

| Index | Value |
|---|---|
| Per capita emergy consumption | $1.21 \times 10^6$ sej/capita/yr |
| Emergy use per unit area | $7.82 \times 10^{11}$ sej/$m^2$ |
| Emergy use from within Costa Rica | 59% |
| Emergy use from renewable sources | 53% |
| Environmental loading ratio | 0.9 : 1 |
| Ratio of exports to imports | 1.39 : 1 |

assimilate wastes, and more and more nonrenewable resources must be expended to generate wealth and to process waste by-products, and a greater "load" is placed on the environment. Thus by this means of calculation, in the United States the load on the environment is nearly 14 times that in Costa Rica. When ELRs are relatively low, such as in Costa Rica, the environment assimilates wastes, processing and recycling them without enormous negative consequences. So, for example, many coffee wastes can be transported back to the fields to decompose there, and the potential for using "ecologically engineered" natural systems to process human sewage is great (although hardly developed).

The picture that emerges of Costa Rica's economy in 1987 is one that:

- Derives 59% of its resources and materials from within the country (about 41% of Costa Rica's total emergy use comes from imports)
- Could support only about one-quarter of its 1994 population at U.S. levels of consumption
- Could support about 78% of its 1987 population on its renewable resource base
- Was exporting about one and a half times the emergy it was importing, suggesting that continued negative emergy balance of payments may threaten long-term sustainability

This analysis is a relatively incomplete picture of the overall state of emergy balance of the economy of Costa Rica. Not accounted for in this overview are the many interior balances of emergy that could have serious impacts on long-term sustainability, such as agricultural sectors and their soil losses, deforestation, or overexploitation of coastal fisheries. For example, we have seen in other developing economies that the emergy lost in eroded soils from agricultural lands neutralizes gains that may be had from imports of resources obtained from the export of agricultural commodities grown on the soil (Doherty *et al.*, 1993). Thus even our present incomplete analysis suggests that the Costa Rican population is approaching or has arrived at the ability of the biophysical resources of Costa Rica to support them at a moderate standard of living, and that even this analysis neglects long-term erosion of national capital. More can be squeezed out of the economy by importing more oil, but the intrinsic resources must be squeezed even tighter to pay for that oil.

Policies that enhance external trade increase population and economic carrying capacity but work to decrease long-term sustainability by destroying soils, forests, and other basic emergy resources. These analyses provide some insights into the hard choices that face developing nations, probably the most important of which is the short-term vs the long-term perspective. To develop

TABLE 25-4 Summary of Different Analyses of the Number of People That Could Be Supported Sustainably in Costa Rica

| Criteria | Method | No. of people | Source |
|---|---|---|---|
| BIOPHYSICAL: FOOD ONLY (does not account for other resources used) | | | |
| Agronomic | Indigenous resources only | 400,000 | Chapters 2, 12 |
| Agronomic | Inputs bought with coffee | 2,000,000 | Chapters 2, 12 |
| Agronomic | Land limitations, with inputs | 4,000,000 | Chapters 2, 12 |
| BIOPHYSICAL: (present standard of living) | | | |
| Footprint | All resources used | 2,400,000 | This chapter (1987 population times 0.8) |
| Emergy | All resources used | 1,750,000 | This chapter (1987 population times 0.53) |

their economies, developing nations tend to increase external trade for short-term gains in assets and carrying capacity, but by doing so they may be compromising long-term sustainability.

It is rather remarkable that each of these analyses, biophysical, economic, ecological footprint, and emergy, reaches approximately the same conclusion: that the resource base of Costa Rica has become in the past decade or two inadequate to support sustainably (that is, without using up environmental capital) the present population density at the present standard of living (Table 25-4). None of these analyses give very much hope for any future scheme that might generate a truly sustainable economy, although of course there is always the chance that some unforeseen technology or global market shift could make things much better (or much worse). Again these conclusions, done here in increasing detail, are hardly news to thoughtful Costa Ricans who have been warning of the dangers of continued population growth and high levels of resource exploitation for decades (i.e., Hartshorne *et al.*, 1982; Alvarado and Monge, 1997). And there is nothing special about Costa Rica. For example, the United States, while it has many resources and a low population density, is extremely and increasingly vulnerable to probable future oil shortages that may begin within a decade (Campbell and Laherrére, 1998; Kerr, 1998).

## VI. SUMMARY AND CONCLUSIONS

Resource availability should be at the heart of questions concerning carrying capacity and sustainability. Developing nations are particularly susceptible to changes in resource availability since there is little luxury consumption and less "fat to trim." In addition population densities tend to be high and increase rapidly. Sustainability should be defined quantitatively to include depletions of resources, and it should include considerations of resource balance of payments at all scales within an economy. The use of monetary measures of resource use and trade does not value resources and energy adequately, and may give the false impression that economic activities are sustainable in the long run, when in fact they may be seriously degrading local environments and carrying capacity, and the ability to sustain local populations. In addition money can give false signals through inflation and speculation, which real resources do not. These lessons are hard to learn, as we saw with many Asian economies in 1998.

The seemingly hard choice for developing nations is to temper consumption, resource use, and exports now, and instead use domestic resources at home to foster long-term sustainability. This is a conclusion contrary to the increasing worldwide emphasis on free trade, the net effect of which is to strip less developed countries of their basic resources, such as forests and soils, and to encourage the unsustainable consumption of industrially-derived products through market penetration and advertising. The net result is debt and a destruction of the productive capacity of environments and hence the ability to ever pay back that debt, not to mention providing increasing populations with basic goods and services.

## REFERENCES

Bouman, B. A. M., R. A. Schipper, A. Nieuwenhuyse, H. Hengsdijk, and H. G. P. Janson. 1998. Quantifying, economic and biophysical sustainability trade-offs in land-use exploitation: a case study for the Northern Atlantic Zone of Costa Rica. *Ecological Modelling* 114:95–109.

Brown, M. T. and R. A. Herendeen. 1996. Embodied Energy Analysis and Emergy Analysis: A comparative view. *Ecological Economics* 19:219–235.

Campbell, C. 1997. Depletion patterns show change due for production of conventional oil. Oil and Gas Journal (Special Publication. December 29:33–37.

Campbell, C. and J. H. Laherrére. 1998. The end of cheap oil. *Scientific American:* 78–83.

Costanza, R. 1978. Energy Costs of Goods and Services in 1967 Including Solar Energy Inputs and Labor and Government Service Feedbacks. Urbana, IL: Center for Advanced Computation, University of Illinois at Urbana-Champaign. 46 pp.

Doherty, S. J., M. T. Brown, R. C. Murphy, H. T. Odum, and G. A. Smith. 1993. *Emergy Synthesis Perspectives, Sustainable Development, and Public Policy Options for Papua New Guinea. Final Report to the Cousteau Society.* Center for Wetlands and Water Resources, University of Florida, Gainesville, FL. 182 pp. [CFWWR-93-06]

Haines, B. and C. Peterson. 1998. El desarrollo sustentable en montañas des de la perspectiva de un ecologo: el caso del "proyecto charral" en Costa Rica. Geografia aplicado 'y desarrollo. III Simposio Internacional, Quito, Ecuador.

Hall, C. A. S., C. J. Cleveland, and R. K. Kaufmann. 1986. *Energy and Resource Quality: The Ecology of the Economic Process.* Wiley–Interscience, New York. Reprinted 1992, University Press of Colorado, Boulder, CO.

Hannon, B. M., C. Harrington, R. W. Howell, and K. Kirkpatrick. 1976. *The Dollar, Energy, and Employment Costs of Protein Consumption.* Center for Advanced Computation, University of Illinois at Urbana-Champaign, Urbana, IL. 81 pp.

Hartshorn, G. S., L. Hartshorn, A. Atmella, L. D. Gomez, A. Mata, R. Morales, R. Ocampo, D. Pool, C. Quesada, C. Solera, R. Solarzano, G. Stiles, J. Tosi, A. Umaña, C. Villalobos, and R. Wells. 1982. *Costa Rica: Country Environmental Profile: A Field Study.* Tropical Science Center, San Jose, Costa Rica.

Herendeen, R. A., B. Z. Segal, and D. L. Amado. 1975. *Energy and Labor Impact of Final Demand Expenditures, 1963 and 1967.* Center for Advanced Computation, University of Illinois at Urbana-Champaign, Urbana, IL. 17 pp.

IFIAS. 1974. Energy Analysis. International Federation of Institutes of Advanced Study, Report 6. Ulriksdal Slott, Solna, Sweden.

Ko, J-Y, C. A. S. Hall, and L. G. López Lemus. 1998. Resource use rates and efficiencies as indicators of regional sustainability: An examination of five countries. *Environmental Monitoring and Assessment* 51:571–593.

Monge, C. 1995. Costa Rica se ahoga en deudas. *La Prensa Libre,* Martes, 11 de julio.

Odum, H. T. 1978. Energy analysis, energy quality and environment. In M. W. Gilliland (Ed.), *Energy Analysis: A New Public Policy Tool,* pp. 55–87. Selected Symposia of American Association for Advancement of Science. Westview Press, Boulder, CO.

Odum, H. T. 1984. Embodied energy, foreign trade and welfare of nations. In A-M. Jansson (Ed.), *Integration of Economy and Ecology—An Outlook for the Eighties,* pp. 185–199. Proceedings of the Wallenberg Symposium. Asko Laboratory, Stockholm, Sweden.

Odum, H. T. 1986. Environmental Accounting: Emergy and Environmental Decision Making. John Wiley and Sons. New York. 370 p.

Odum, H. T. 1995. Self organization and maximum empower. In C. Hall (Ed.), *Maximum Power. The Ideas and Applications of H. T. Odum,* pp. 311–329. University Press of Colorado, Niwot, CO.

Scienceman, D. 1987. Energy and emergy. In G. Pillet and T. Murota (Eds.), *Environmental Economics—The Analysis of a Major Interface,* pp. 257–276. Leimgruber, Geneva, Switzerland.

Scienceman, D. M. 1989. The emergence of emonomics. In *Proceedings of the International Society for Social Systems Science, 33rd Meeting, 1989, Edinburgh, Scotland,* Vol. 3, pp. 62–68.

Slesser, M. 1978. *Energy in the Economy.* Macmillan, London.

Wackernagel, M., and W. Rees. 1996. Our ecological footprint. New Society Publishers, Gabriolie Island, B.C.

Wackernagel, M., L. Onisto, P. Bello, A. C. Linares, I. S. L. Falfán, J. M. García, A. S. Guerrero, and M. G. S. Guerrero, "National Natural Capital Accounting with the Ecological Footprint Concept." *Ecological Economics* (June 1999 Vol. 29:375–390).

CHAPTER 26

# The Myth of Sustainable Development

CHARLES A. S. HALL



## I. THE IMPOSSIBILITY OF SUSTAINABLE DEVELOPMENT

We asked in Chapter 2 whether it is possible to design a "sustainable" economy in the tropics, what the characteristics of such an economy might be, and the degree to which sustainability can or cannot work. From the analysis presented in the intervening chapters we have concluded that it is not possible, or probably even desirable, to construct a sustainable economy for Costa Rica, at least for anything like the present population level and standard of living (see also Willers, 1994). The basic reasons for this are now reviewed.

### A. ECONOMIC AND SOCIAL DIFFICULTIES

1. The existing public debt burden makes it extremely difficult for the government of Costa Rica to generate a viable economic plan for addressing the major resource and social problems of the nation. For recent decades about 15% of imports has been based on new borrowing, and an equivalent part of

exports has gone to pay interest on external debt (Figure 23-8). In the simplest terms debt is not sustainable, for it makes any future economy even less sustainable as real resources have to be diverted to service that debt. Approximately one year's total economic production would need to be diverted to pay off this debt.

2. The Costa Rican economy is in many ways paralyzed by the need to generate foreign exchange. The energy price shocks of the 1970s live on as foreign aid and debt have been diverted increasingly from production to consumption, to meet governmental payrolls, and to maintain government social programs, such as health care and education, in turn necessary to keep birth rates down. In other words the concept of debt should be to invest in development of resources that would in turn generate an increased future revenue stream. But increasingly the aid has been needed for "maintenance metabolism," that is, for maintaining consumption rather than enhancing production. Although the concept of investing in infrastructure and productive capital equipment seems to make sense, this strategy, as we have seen in Costa Rica, has not always produced the desired results (e.g., Chapters 20 and 22). The problem is that *modern investments require fuel and other industrial inputs to make them work*. For example, it might seem very straightforward and logical to borrow money to clear land and grow bananas. And, indeed, this does work. But the revenue to Costa Rica is far less than commonly believed because every year about half of the revenue must leave the country to pay for the production process. Similar analyses apply to industry, tourism, ecotourism, textiles, and so on. The previous chapter concluded that the net gain from exports was only about half the gross once the interest on loans and the cost of the imports necessary to make the exports were subtracted.

3. Probably the majority of Costa Ricans are no more interested in having a truly sustainable economy—based on solar energy, agriculture, animals for transportation, and hydropower—than are North Americans. They, too, like the comforts, conveniences, high yields, and high labor productivity that come with a fossil-fuel-intensive, "modern" industrial society. It seems absurd to ask Costa Ricans to give up television sets, automobiles, trucks, and computers in order to have a "sustainable" society that is unlikely even to feed the present population.

## B. Biophysical Problems

1. Costa Rica essentially can no longer produce the food it needs to feed its own growing population. Even though it is possible to envision scenarios where domestic food production barely could feed the present or even future populations, the dependence upon imported fuels and chemicals for sufficient

agriculture production to do this severely limits these possibilities (Chapters 5, 12, and 23).

2. The topography, climate, and soils of the nation are of only moderate quality for agricultural production (Chapters 8, 9, and 10). The majority of the land is simply too steep for any truly sustainable agriculture. Only about 15% of the nation's soils have real sustained potential for agriculture (Chapter 10). In addition some of the soil resource base has been seriously and significantly degraded by erosion (Chapter 15) and past toxins (Thrupp *et al.*, 1995), and it requires inputs from the industrial world to maintain present fertility levels, let alone to regenerate soil quality. Perhaps most important, the length of night in the tropics appears to preclude high agricultural yields (Chapter 12).

3. About 30% of food consumed is imported (Chapter 5). This food is produced in highly industrialized and often soil-depleting agriculture in other parts of the world, requires fossil fuel for transport outside and inside Costa Rica, and requires intensive import-requiring commercial production of export crops to pay for it.

4. The present economy is based more on fossil fuels, by definition nonrenewable, than on renewable energy (Chapter 4). There has been and continues to be a very tight relation between energy use and economic activity, and there is no indication that greater efficiency is in the offing (Figures 4-5e and 12-7).

5. The oil price shocks of the 1970s, far from being an event of the distant past, are still being felt in Costa Rica every day. The debt accumulated during that period, as Costa Rica attempted to maintain private and government consumption while the price of imports rose dramatically relative to the price of exports, still dominates much of the economic life of Costa Rica (Chapters 2 and 4). It also has allowed foreign agencies to dictate social and political policies even when those policies clearly were unable to resolve the problems they were supposed to (Chapter 24).

6. Deforestation and land degradation continue despite strong laws and a national commitment to the contrary (Chapters 15 and 16).

7. While it is possible to continue agricultural production on eroded soils, this can be done only through the continued intensive use of industrial inputs such as fertilizers. The alternative of moving on to even steeper slopes would provide some production for some time, but would presumably be limited eventually by very great erosion.

8. Industries touted as "sustainable" and non-resource-degrading (such as tourism and ecotourism) are in fact extremely energy and material requiring, and often consume large quantities of scarce water and prime seashore property (Chapter 22). They are sustainable only while world economies are strong and jet fuel is cheap.

9. With the exception of its stable government, high literacy, and excellent climate, Costa Rica has few competitive advantages compared to many other

countries (Chapter 23). Costa Rica cannot readily outcompete (a) other large and oil-rich heavily industrialized nations, such as Mexico, for industrial production, (b) highly sophisticated high-tech countries such as the "Asian tigers" and, increasingly, China simply because these other countries "got there first" and have a much more sophisticated educational system, or (c) other tropical agroexport nations, because Costa Rican wages are too high compared to many other poorer tropical nations with similar climate and export structures. On the other hand, Costa Rica can be a minor player in all of those markets, and its political stability and climate allow Costa Rica to be very competitive in the tourist industry, although many other nations, many with cheaper fares, are vying for the same tourist dollars.

10. Population growth has been essentially identical to the growth of the economy, so that there has been little or no increase in the per capita economic level.

11. All of the previous issues make Costa Rica extremely vulnerable to the future prices of fossil fuels and industrial products.

In summary, sustainability implies living off interest and not capital (Ehrenfeld, 1996). While issues of interest and capital are normally considered in terms of money, they ultimately must be based on physical resources. For a green tropical country without fossil fuels, interest implies living off of solar energy without degrading photosynthetic capacity. But clearly in Costa Rica photosynthetic capacity is only moderate and, furthermore, is not being maintained sustainably: there is still net destruction of forests (Chapter 16 and 17), continued loss of soils (Chapters 10 and 15), and a decrease in yield per unit input of crops (Chapter 12). Some of these factors can be compensated for with increased imports of, e.g., fertilizers, but this strategy is not sustainable from the perspective of the ability of Costa Rica to pay for them or to replace them, since they are derived from nonrenewable resources. Fossil fuels can and must be used to enhance the use of Costa Rica's solar resources, but it is a very expensive proposition in monetary and biophysical terms. Indeed, the use of fossil fuels for personal luxury consumption is extremely expensive to the nation in many ways.

## C. Possibilities for Energy Resources

Our analysis has emphasized the importance of fossil fuels, the tremendous impact of their price increase in the 1970s, and their unsustainability. Another increase in the price of fossil fuels would devastate again (presumably) the economy of Costa Rica. How long will relatively cheap fossil fuels last? What are the energy alternatives?

It is hard to answer the first question exactly, but if global use continues to grow exponentially, oil is projected to be seriously depleted in one or two decades, and natural gas soon thereafter (Hubbert, 1968; Hall *et al.*, 1986; Cleveland and Kaufmann, 1991; Kerr, 1998). A particularly relevant and sobering analysis is that of Campbell (1997) in the authoritative and conservative *Oil and Gas Journal.* Campbell examined each producer nation carefully and found that about half had reached peak production prior to 1997 and most of the rest would soon follow. He predicted global oil production would begin to decline by 2008, in agreement with Hubbert's prediction in 1968.

On the other hand, new discoveries and increased recovery rates may stretch oil supplies significantly at an unknown price. High-quality coal might last from three to six generations, depending on the degree to which rates of consumption do or do not change as the other fossil fuels are exhausted (Campbell, 1997, Campbell and Laherrere, 1998). Reserves of low-grade, low-energy-return-on-investment, "dirty" fossil fuels, including subbituminous coal, oil shale, and tar sands, are large. The actual availability of oil, while it lasts, probably has more to do with unpredictable future political events, since most of the remaining oil is in the potentially volatile Middle East. Kaufmann (1992) predicts that oil prices will increase slightly through the years 2000 to 2010, after which the price is anyone's guess. A big unknown is the degree to which China, with few oil resources of its own and an economy sometimes growing at more than 10% per year, will enter global markets. Campbell found China paying large sums for Venezuelan oil fields, apparently indicating that domestic exploration was not going very well.

## 1. Indigenous Energy

Although Costa Rica apparently has little or no fossil fuel of its own, biomass and hydroelectric power currently constitute about half of its energy use (Figure 4-5). To what extent can these indigenous and nominally renewable sources substitute for imported energy and contribute to a sustainable energy base? (A background for each of these possibilities is found in Hall *et al.*, 1986).

Costa Rica has tremendous hydroelectric potential because of its sharp relief and very high rainfall. It has been suggested that Costa Rica use trains or buses powered by hydroderived electricity to replace imported transportation fuels. But electric transportation is not always easy in Costa Rica because electric motors do not have the torque required to negotiate the frequently steep slopes. Only a small part of Costa Rica's hydroelectric potential has been tapped. The main problems are the cost of creating the infrastructure and the environmental effects of reservoirs and alterations of rivers. Since the Costa

Rican government is nearly broke in many ways, there is little large-scale national investment capital available. A law passed in 1995 allowed for the first time private generation of electricity. This means that many private small plants may be built with, presumably, fewer environmental problems than the large proposed national-level projects. This approach also encourages private investments and the promotion of small businesses, both politically popular at this time. The public utility, ICE, has chosen this alternative as the number one solution to the need for expansion. The major investors are from foreign nations, as they have more capital. Thus, unfortunately, such financial schemes will act as a continual siphon of foreign exchange away from Costa Ricans as investors reap their economic return on their investments.

There are many environmental issues associated with the construction of dams. One critical issue is the degree to which dams contributed to the large-scale loss of coral reefs that has occurred along the east coast of Costa Rica. A related issue is whether dams have impacted fish migration at the coast or in the interior. Additionally there has been developed recently a "rafting" tourist (and ecotourist) industry dependent upon free-flowing rivers. This industry is very much opposed to further dam construction. Costa Rica is prone to earthquakes, which could cause dam failure and devastating loss. The engineers who built the existing Arenal Dam took into consideration seismic problems. It has an impermeable clay core and rock exterior that provides mass. This design is very flexible and adjusts to movements of the earth's crust. But if dams are to be the energy salvation of the Costa Rican economy, one can expect that over decades to centuries there is the real possibility of a giant earth tremor too large for any dam's absorptive capacity.

Finally, dams provide renewable energy only in the short to medium term. Over time all dams fill with sediments brought by the rivers emptying into them. Costa Rica has very high rainfall, much higher than what is experienced in most of the temperate world. Consequently the potential for infilling is very large. The Cachi Project, already partly filled in, requires very expensive sediment removal every few years. This process causes considerable sedimentation in the river below the dam. The Arenal Project, the country's largest, is not expected to fill in for one to several hundred years, but this could be much sooner if agriculture develops on the slopes above the project. On the other hand these watersheds are ideal areas for parks and nature conservation areas. Overall dams, usually considered a renewable energy source, have a finite life not unlike fossil fuels.

There are some prospects for *geothermal* and *wind* development. The highest sustained winds are in Guanacaste Province, where electrical demand is not large. Biomass produces a larger proportion of energy for Costa Rica (about 20%) than for most other nations. But much of the "easy" biomass fuel is developed already; for example, bagasse (sugarcane waste) is used to fuel many sugar mills. It is possible that the biomass digesters being developed at Earth College could fuel rural domestic cooking in a cost-effective way. Chapter 18 explores the considerable potential but large difficulties of forest-derived fuels. For example, Costa Rica presently is a wood-products importer (Chapter 18).

## 2. Efficiency

It is commonly argued that through technology people can use less energy over time to create the same amount of output (wealth). This is a very popular approach in the sustainable development literature (Reddy and Goldemberg, 1989). We examined several aspects of efficiency for Costa Rica and found that even though the efficiency of land and labor have been enhanced, it has been at the expense of using more energy, and with a decrease in the efficiency of turning energy into economic product [Figures 4-5e and 4-6; see also Ko *et al.* (1998)].

## 3. Discussion

Overall there are many possibilities for using indigenous fuels, but none seem to be catching on, except for hydropower. This is partly because oil and electricity are already in place, are reasonably cheap, do things that people want done, and use existing and available infrastructure. It is also because alternative energy sources may not be economic on a significant scale, which probably means that they simply do not return enough energy profit to make them economically viable. Until these questions are researched much more thoroughly it is simply not possible to determine what the potential is for indigenous energy.

This result is similar to many other findings: that despite a great deal of talk about economic growth based on improving energy *efficiency*, almost all economic growth is based on increasing energy *use*, often at an approximately one for one rate ratio (Nilsson, 1993; Ko *et al.*, 1998). There is some evidence that all nations in the world are converging on roughly the same energy use per unit of economic production—about $3000 1980 U.S. dollars of production generated per ton of oil (Nilsson, 1993). Jose Goldemberg, once a supreme advocate of energy efficiency in developing countries, has more recently called for increases in energy use to solve these country's problems (Reddy and Goldemberg, 1989; Goldemberg, 1995). Although there are certainly many gains to be realized here and there in the efficiency of using energy, overall the concept of a path to development through efficiency does not seem to be happening in actual practice. It certainly is no panacea.

One conclusion is that for Costa Rica there does not seem to be a ready substitute for imported petroleum except for hydroelectricity, which increasingly is generating the same problem that oil does—it drains foreign exchange out of the country and it has important environmental problems. Further research and development might lead to another resource that would make use of indigenous energy resources, but there are precious few resources available for that research and development. It seems most likely that the use of petroleum will continue to increase, at least for as long as oil is cheap.

## D. The Undesirability of Using the Term "Sustainable Development"

### 1. Calling a Spade a Spade

In summary, I believe that just using the term "sustainable development" is likely to cause more damage than not using it. The reasons are:

1. It simply is not physically possible to sustain today's levels of population and affluence without using depletable resources.
2. It is time to call a spade a spade. All organisms on the face of the earth support themselves, their progeny, and their cultures through the exploitation of resources. Humans are no different. What economics is about, at least from a biophysical perspective, is the exploitation of resources by humans. To say otherwise is simply to deny reality.
3. It makes no sense for humans not to live off depletable resources such as fossil fuels while they are available, as long as it is acknowledged that the situation is temporary and the costs large, although probably not as large as not using them. For example, if somehow fossil fuels were to be withdrawn from the Costa Rican economy in the name of sustainability there would be enormous loss of economic activity, including deprivation and loss of positive government activity. We also believe—and of greater concern—that it is not possible to develop such a situation in Costa Rica without creating substantial soil erosion. Ironically, most efforts toward sustainability would result in the replacement of tree crops with subsistence food crops which have enormously greater impact on soil erosion (Chapters 5 and 15).
4. It serves as a smoke screen that hides people from the reality of the impacts of their actions, environmental and otherwise, such as living affluent lifestyles or having many children (e.g., Hall *et al.*, 1992). We will never address the real problems that growing populations and affluence generate if we keep believing that some sort of technology will make it all to the better, however we may define that. (See also Cleveland *et al.*, in press).

Thus the idea of sustainable development is probably ill-conceived, moreover misleading, for it implies that we have power over what we do not, that we should deliberately make humans poorer by not using up nonrenewable resources, and that we not be held accountable for our actions. There is no need or rationale for developing a truly sustainable society, although there are critically important reasons for doing what we do better.

None of this argues against the continued maintenance of environmental resources to the largest degree possible, or for making sustainable that which can be (for example, soils and hydroelectric capacity). Nor do we wish to undermine the thoughtful, sometimes idealistic and sometimes very useful national program for sustainable development (i.e., Quesada-Mateo and Solis-Rivera, 1990). This plan includes many excellent components that are implementable and that can lead to greater protection of forests, biodiversity, soils, and human quality of life. But other components of this plan seem not to be occurring at all, or if they could be implemented, they could be done only with some harm to another dimension of sustainability. Their recommendations include intensification of agriculture, restricting agriculture's incursion into marginal land, reducing deforestation, slowing the rate of immigration into cities, increasing the efficiency of energy use, reducing pollution, and many other programs. Perhaps it can be argued that it is too early for these plans to have taken hold, and without a plan there would have been even less progress. But wishing and planning for sustainability, and listing its virtues, is not the same as causing it to happen. From the perspective of its total economy, Costa Rica is clearly becoming ever more dependent upon fossil fuels and other industrial resources, and ever less dependent upon its own indigenous renewable resources. I think it critical to examine which components of sustainability are possible and which are not, and to examine each strategy from the perspective of trade-offs rather than from some absolute good or virtue.

### 2. The Faustian Bargain

Our principal conclusion is that population growth within Costa Rica has generated an almost impossible situation for the economy and the natural environment. While it is possible that the present population of Costa Rica might (barely) be supported (i.e., fed) on the present land mass, this is not possible without dedicating at least a nearly equal area of land to generating foreign exchange. The foreign exchange is necessary for the inputs to subsistence agriculture plus all of the other components of the economy.

## II. WHY CONVENTIONAL DEVELOPMENT DOES NOT WORK

The first issue to be confronted is whether there is indeed a problem with development. If there is no problem, then there is no need for a solution. A visitor to Costa Rica sees a vibrant, growing society, in many ways similar to that experienced in the developed world. The problem, of course, is that this seeming affluence is based in large part on debt, especially external debt, and also on environmental debt. But that too may be little different from, e.g., the United States. Foreign exchange is needed to pay for the interest on this debt for both countries. For Costa Rica especially this international governmental debt is exacerbated by private international debt of individuals with credit cards.

The difference is that the United States has much greater per capita productive capacity to pay off that debt, or its interest. Much of this capacity is due simply to relatively low U.S. population levels relative to the agricultural potential. The United States has roughly 150 million ha of prime farmland in production, and another 125 million ha that could be put into production. The former is about 0.6 ha per person. Costa Rica has about 250,000 ha in crop production, and about 780,000 ha of soils with good crop potential (the difference is the good land being used for pastures or urban regions). This works out to be 0.07 and 0.2 ha per capita, respectively. Average yields of grains in the United States are about 5 tons per hectare, and for Costa Rica about 1.5 (Chapter 12). Thus the agricultural potential per capita for the United States is about 30 times that of Costa Rica. In addition, of course, the United States has tremendous industrial and technical infrastructure, produces half of the oil it burns, and has tremendous coal reserves. Curiously the federal debt load relative to GDP of both countries is similar. On the one hand it is possible to say that the United States has a good potential to pay off its debt relative to Costa Rica. On the other hand Costa Rica has been just as responsible (or perhaps irresponsible) in paying off its debt as the United States, despite having far fewer resources to do so. Thus a relevant question seems to be why the United States has failed to better balance its books than Costa Rica!

## A. The Failure of Neoclassical Economics

Most of Latin America is paralyzed by debt, much generated by "free trade." It has become impossible for governments to pay off this debt, and the payment of interest and such capital as is repaid is undermining governments in many ways. The big problem in this issue is that there is not an intellectual argument to counter that of the neoclassical economists who have, with the downfall of communism, been "proven correct" and against whom there is no coherent, synthetic, broadly accepted argument. It is important to have new arguments to counter the perspective of the bank technocrats and others who force the supposed virtues of neoclassical economics and unquestioned free markets down the throats of small countries, and it is important to state that no matter how important it may be to repay debts and maintain international credit, there may be even more important issues that need to be faced. These include:

1. The destruction of the function of government. Debt service has been destroying the beneficial functions of governments as the debtors increasingly insist upon reduced government expenditures to meet debt obligation, and as they impose a "free market world view" on the debtor nations. In Costa Rica this has had the following results:

a. Destruction of the health provision service. Health spending was greatly reduced in the early 1980s, and as a consequence there was a resurgence in some diseases, including cholera and malaria, and an increase in teenage pregnancies. Subsequently, much health service was reinstalled.
b. Massive reduction in the internal capacity to produce staple grains and pulses
c. Enhanced deforestation to generate foreign exchange to pay debt

2. Almost complete destruction of national autonomy. The increased requirement for fossil fuels and all the accouterments of industrial society that cannot be supplied from within Costa Rica means that Costa Rica increasingly will be beholden to outside forces. But perhaps this is the fate of most of the world.

## B. The Options

What then, is a country to do? There are several possibilities:

1. Encourage further industrialization. Some other nations, faced with similar problems of large populations, relatively little good land, and no fossil fuel (e.g., Taiwan, Korea, and Singapore), have been very successful at building industries based on imported oil. This is a genuine alternative for Costa Rica. There are three general requirements for a nation to be a candidate for enhanced industrialization aimed at generating products for the general international market in these times: a well-educated working class, a stable government, and low wages. These characteristics have been used by successful developing countries such as those already listed as well as Puerto Rico and Japan. Costa Rica certainly has a stable government, a reasonably well-educated workforce, and moderately low wages. The problem is that Costa Rica is a Johnny-come-lately in this arena. It defies the imagination to think of Costa Rica competing with these other nations for major markets of industrial products.

2. Renege on outstanding debt. This has two strong positive advantages: (a) the debt burden, a real problem, would be eliminated, and (b) no one would loan Costa Rica money again, which would force Costa Rica to live on its real resource base. It has one particularly negative aspect: the lending institutions,

and the nations wherein they reside, would likely do everything they could to undermine the Costa Rican economy, so that it would be unlikely that a local economy could be made to work in any reasonable way even if the burden of debt was removed. It might be possible, for example, for all Latin American nations to do this and to generate a "Latin American common market" or "developing nation common market" in order to share economic possibilities. This was tried in the 1960s and 1970s by Costa Rica, with marginal results.

3. Forgive debt. An associated idea is that the international lending banks need to rethink their role in development. Rather than improving the economic situation of Costa Rica, their intervention is destroying the soils and other resource bases that could provide Costa Rica with any hope for a viable economy in the future. My own view is that the loans need to be forgiven as humanitarian and environmental aid to Costa Rica (after all, the loans have been repaid many times over through extended interest payments). As part of this bargain I also believe that Costa Rica should agree to never borrow from international banks again, but must instead learn to live within the country's biophysical limits. This concept was suggested as a "millennium" issue by Pope John Paul in late 1999.

4. Continue to muddle along. This is probably what happens most frequently. When there are years with good prices for coffee or bananas, things look hopeful, more hectares are planted, money is borrowed, and hope abounds. Then come a few years of bad prices, overproduction in Brazil, or whatever, and the profits evaporate. At least this has been the historical pattern. The major threat to this approach is an energy price increase, such as happened in 1979. This could eliminate plantation profits, making it impossible to generate net foreign exchange. Will this happen again? Almost certainly—when is anyone's guess. My best guess is that cheap petroleum will no longer be available in roughly 20 years for physical reasons, and on any given day for political reasons. Although there is a lot of oil left in the world, most remains in the Middle East, and this is a region with many animosities both between and within nations. Anything could happen, or nothing at all. If the latter, the muddle through approach might just work, although the increasing need for food imports as the population grows is a concern.

There are a number of characteristics of Costa Rica that could assist in generating some sustainability or easing the problems if and when fuels become too expensive to use freely. These include a tremendous potential for enhanced food production by reducing pastureland and replacing it with annual or possibly perennial crops. The advantage of this is that annual crops yield far more food per hectare per year than do pastures (Chapter 4, Table 26-1). While this concept is intriguing, its actual implementation is difficult because the particular advantage of pastures and beef production is that they do not require industrial inputs, of necessity derived from outside the nation. A second

TABLE 26-1 Potentials of Land Use: What Could Be Obtained if All Land Class I and II Were Used for Different Activities

| Land use | Yield (t/ha) | Production (106 tons) | People fed (million) | Required inputs ($10^6$\$) | Internat. value ($10^6$\$) | Erosion* (tons/ha/yr) |
|---|---|---|---|---|---|---|
| Maize | 0.5 | 0.75 | 2.5 | 0 | little | 21 |
| | 1.5 | 2.25 | 7.5 | 115 | little | 21 |
| | 3.0 | 4.50 | 15.0 | 750 | little | 21 |
| Beans | 0.5 | 0.75 | 2.5 | 20 | little | 21 |
| Pasture | 0.2 | 0.30 | 1.0 | 0 | 390 | 4 |
| Coffee | 1.5 | 2.25 | — | 1125 | 7875 | 11 |

*From Table 15-2. Values are minimal, and would be much larger on steep land which may occur with expanded production/urbanization.
Assumption: 1.5 million hectares of land that could be used for cattle, annual or perennial crops. People need about 300 kg grain apiece per year. Values are approximate. Input costs are calculated as fertilizer costs times a factor of 5 to represent pesticides, transportation and storage. These are a low estimate. Also assumed is that there would be a market for the coffee.

problem is that row crops degrade land much more rapidly than cows. Thus, in a peculiar sense pastures are the most "sustainable" of food-producing land uses. Such an assessment depends, however, upon whether your assessment is based on a per hectare or per unit production base: in per ton of food produced, row crops actually generate less erosion (Table 26-1).

## III. CONCLUSION: THERE IS HOPE

The most important conclusion is that we cannot foresee any "silver bullets," that is, magic schemes that by themselves resolve the major problems facing Costa Rica, at least without creating new ones. The past of Costa Rica is littered with such schemes (pp. 51–52). None of them have resolved the problems they set out to, and in many cases they have made the problems worse (Chapter 24). To seek such a resolution through some new investment scheme, especially one that comes with large strings attached, is to court continued disappointment, and perhaps even disaster.

On the positive side, Costa Rica has found a way to generate an extremely high standard of human well-being on a modest resource base. Perhaps more than any country in the world Costa Rica and its leaders have generated a reasonable standard of living, still relatively equitably distributed, extraordinary levels of health and education by any nation's criteria, peace, and a high degree of environmental protection on a modest resource budget. Even though the possibilities of making the economy "truly sustainable" seem slim, there is a great deal that the rest of the world, including the industrial nations, can learn from Costa Ricans about how to generate quality human life relatively efficiently. Lack of armed conflict or oppressive police, a well-operated, democratic, and quite honest government, and stable internal financial institutions have all contributed to this. Within this context, whether we wish to call it sustainable or not within the context of fossil fuels and the need for inputs, conservation can flourish thanks to the hard efforts of many individuals and a well-educated public. But we should not confuse conservation of, e.g., biodiversity with sustainability, for they are only partly related. The problem with this somewhat rosy perspective is the possibility or inevitability, depending on one's view, of future dramatic increases in the price of needed products from the industrial world—from fertilizers to computers to ecotourists. If Campbell (1997) is correct then "the end of cheap oil" may be with us within the decade. If or when this occurs it is a whole new ball game for sustainability.

This leaves us with the only realistic possibility if sustainability is the goal: to make expectations consistent with the capability of the national biophysical resources to deliver them. This includes the absolute necessity of stabilizing or reducing population levels if there is to be any hope of increasing mean per capita wealth, and a reduction in the aspirations and advertisement for affluent lifestyles. This is a bitter pill for a reasonably successful nation with high ambitions. Many will not swallow it, but for those who do and who want to convince others of the need to plan a "realistic" future, they must undertake biophysical analyses at least as detailed and thorough as those presented here. Slogans, whether "economic theory," "sustainable development," or "efficiency," are not powerful enough to make the impossible happen.

But even should any of these approaches prove to be extremely effective, which we think has a vanishingly small probability, it is unlikely to be a long-term solution if population continues to grow. According to the analyses summarized in this chapter and the model described in Chapter 5, in the long run the best and perhaps only way for Costa Rica as a nation to hedge its bets is by decreasing the number of future Costa Ricans. We reject completely the perspective of those who believe that "free markets" or "technological advances" (or "human capital") will compensate for increased crowding and reduced resource availability, because those arguments fail to consider declining resource quality with increased exploitation rates or the present role of cheap petroleum in mitigating the effects of increasing population densities. In the long run, even as now, policies based on such "belief" systems will only exacerbate enormously unnecessary human suffering.

On the other hand it is not clear which policies, other than stabilizing population, offer the greatest probability of avoiding future problems. Whatever those policies should or should not be is for Costa Ricans themselves to decide and implement. Unfortunately, although Costa Ricans are basically very supportive of protecting their environment, only half of those polled recently thought there was a link between population and the environment, and, of those, only a third could articulate what that link might be, e.g., that a reduction in resources (such as good land) per capita would effect them personally (Holl *et al.*, 1993). It would seem that the first step in implementation of any sustainability policy will have to be educational.

It is not my business or intent to tell any nation or any individual that they should practice population control, for that is their business. I find coercive approaches to population control as distasteful (and frequently counterproductive) as anyone. But developing nations cannot escape this issue, and must bring it to the forefront of their populace and their economic development planning discussions. Certainly all of the socially progressive ways to inform public opinion need to be encouraged to the hilt, including a much greater emphasis on the link between resources and economics. Likewise reducing aspirations for affluence is a difficult issue within democratic nations with traditions of free speech and extensive advertising. That does not reduce its importance.

I conclude with the only explicit recommendation that I have for this book and all of the issues it addresses. I find the impact of population growth to be pervasive and the most critical for all the many reasons explicit and implicit in the previous 25 chapters. If a nation is to have any possibility for maintaining an economy not tied to foreign debt, with some hope for raising living standards while protecting the environment and maintaining its own economic and political autonomy, it has to become less dependent upon international banks, foreign loans, imported industrial inputs, and, in general, the developed world. But as we have demonstrated clearly in this book, a developing country cannot feed an ever-increasing human population without increasingly relying on massive and expensive industrialization of agriculture—if indeed it can at all. And this leads in turn to all of the other problems addressed in this volume.

## A. A Final Thought

This book has focused on the biophysical aspects of economies, and it has calculated many biophysical dimensions and ratios of efficiency and sustainability. The results are not always encouraging for the future of the Costa Rican economy, or indeed for many other tropical economies. But there is another dimension to sustainability, one that may be more important than much of what we have calculated even though it is difficult to put explicit numbers on it. The people of Costa Rica have constructed an extremely free, healthy, and just society that provides nearly all of its citizens with at least the essentials of life on a relatively modest resource base. They have done this through hard work, relatively honest government, a functional democracy that allows for the free expression of divergent opinions, and a political system that traditionally did not allow for an enormous division of wealth between rich and poor. Poll after poll has shown that Costa Ricans believe in their government and that they have been happy to pay taxes to provide the basics of a decent life: education, health care, and basic infrastructure. In return their government, through thousands of skilled and dedicated professionals, has provided these services. The results are that Costa Rica has higher standards of health and literacy than nearly any other country in the world, including the United States. As a nation they are not rich materially, but they are rich in many ways.

It is painful for me to read that bank officials and bureaucrats in Washington and other financial capitals insist that Costa Rica "put its fiscal house in order" by reducing governmental expenditures and eliminating subsidies that once allowed their own farmers to feed their own people. It is even more painful to realize that they have the power to do this, the power to disrupt a wonderfully functioning government, and equally painful to know that my own government is likewise blindly pushing free markets around the world. What are the objectives of the bank policymakers? Are they beholden only to their stockholders? Or do they believe so strongly and blindly in the principles of neoclassical economics and free markets that they will implement government-reducing policies simply on the basis of ideology? Where is their long view, both economically and environmentally? As government health and education services are reduced, and as traditional small farming families are displaced by massive new farms or, worse, food production in the temperate world, one result is that more girls become pregnant, many of them unmarried. Is this the fiscally responsible solution?

I end by beginning the process of restructuring the "sustainable development" dialogue into a more realistic and useful form. The question is not whether Costa Rica can have sustainable development. It cannot. But it can have a society that can function for a long time within whatever resources are available, renewable and nonrenewable alike. These resources can be used to maintain what is presently a very viable and democratic society. A critical issue, which is beyond the scope of this book, is "what is Costa Rica doing right?" How has it managed to have a reasonably honest and well-functioning government and society, along with a decent material standard of living, on such a limited per capita resource base? How can other countries implement such a government? Can good government survive in Costa Rica when it is not physically possible to do what most politicians promise, that is, provide more affluence? How should whatever fossil fuels as will be available in the future be invested to generate a workable, if not affluent, economy over the long term? What crops can provide the best economic return with the least loss of soil capital?

These are not easy questions to answer, but they cannot be answered at all by simply putting blind faith into markets or any system of economic ideology. This book has begun the process of showing how economies in developing countries need to be examined biophysically in order to assess realistically what is and is not possible. It has emphasized that even such simple biophysical properties as the ratio of arable land to the number of people put severe constraints on what is and is not possible. Hopefully some readers will take our analyses a step further to design a future for other developing countries that can emulate the best social characteristics of Costa Rica within those countries' biophysical limits.

## ACKNOWLEDGMENT

I thank Carlos Hernandez for contributions to this chapter, and Troy Belden for literature help throughout.

## REFERENCES

Campbell, C. J. 1997. Depletion Patterns Show Change Due For Production of Conventional Oil. *Oil and Gas Journal Special.* Dec. 29:33–37.

Campbell, C. J., and J. H. Laherrere. 1998. The end of cheap oil. *Scientific American.* 278:78–83.

Cleveland, C., Costanza, R., Hall, C., and Kaufmann, R. 1984. Energy and the United States economy: a biophysical perspective. *Science* 225:890–897.

Cleveland, C., and R. Kaufmann. 1991. Forecasting ultimate oil recovery and its rate of production: Incorporating forces into the models of M. King Hubbert. *The Energy Journal:* 12:17–46.

Cleveland, C. J., C. A. S. Hall, and R. Kaufmann. (in press). Climate change: Human and biophysical driving forces and probable impacts. In P. Vartia (ed). *Climate Change—Socioeconomic Dimensions and Consequences of Mitigation Measures.* VTT, Helsinki.

Goldemberg, J. 1995. Energy needs in developing countries and sustainability. *Science* 269:1058–1059.

Hall, C. A. S., C. J. Cleveland, and R. K. Kaufmann. 1986. *Energy and Resource Quality: The Ecology of the Economic Process.* Wiley-Interscience, New York. (Reprinted 1992. University Press of Colorado, Boulder, CO).

Hall, C. A. S. 1992. Economic development or developing economics: What are our priorities? pp. 101–126. In M. Wali (Ed.), *Ecosystem Rehabilitation,* Vol I. Policy issues. SPB Publishing. The Hague, The Netherlands.

Hall, C. A. S., R. G. Pontius, J. Y. Ko, and L. Coleman. 1994. The environmental impact of having a baby in the United States. *Population and Environment* 15:505–524.

Hubbert, M. K. 1968 Energy Resources. pp 157–242. In *Resources and Man.* National Acad. of Science. W. H. Freeman, San Francisco.

Kaufmann, R. 1992. A biophysical analysis of the energy/real GDP ratio: Implications for substitution and technical change. *Ecological Economics* 6:35–36.

Kerr, R. 1998. The next oil crisis looms large—and perhaps close. *Science* 281:1128–1131.

Ko, J-Y., C. A. S. Hall, and L. G. Lopez Lemus. 1998. Resource use rates and efficiency as indicators of regional sustainability: An examination of five countries. *Environmental Monitoring and Assessment* 51:571–593.

Nilsson, L. J. 1993. Energy Intensity Trends in 31 Industrial and Developing Countries 1950–1988. *Energy*, 18:309–322.

Reddy, A., and Goldemberg, J. 1989. Energy for the developing world. *Scientific American*. 263:1–10.

Thrupp, L. A., G. Bergeron, and W. Waters. 1995. Bittersweet Harvests for Global Supermarkets: Challenges in Latin America's Agricultural Export Boom. Washington, DC: World Resources Institute.

APPENDIX **I**

# Low-Cost Mapmaking

ERIC DUDLEY

## THE NEED

With all the sophisticated technology and analytical techniques now available it is easy to forget that a simple map is also a geographic model. It is a representation of a particular view of reality. For hundreds—indeed thousands—of years it has been recognized that maps can be extraordinarily powerful and versatile tools for planning and understanding. In the 1970s and 1980s the computer revolution and the advent of GIS held out a new promise for an even greater and more powerful use for maps. However, it has become apparent that, at least in the field of development projects, this promise has not been met.

Around the world, in medium to large development projects, both governmental and nongovernmental, one can find air conditioned rooms in which lurk GIS units. These rooms are full of expensive computers and plotters (many not working for lack of a spare part) and shelves bending under the weight of GIS manuals. More importantly, they are populated by perplexed, well-meaning computer boffins who have been on long GIS courses but remain uncertain what they are actually meant to be doing in practice. Elsewhere in the institution development professionals—aware that hundreds of thousands, if not millions, of dollars have been spent on a mapping system—are equally perplexed as to why their requests for an apparently simple map take weeks to be done. They are often also frustrated that the results are not really what they wanted anyway. In many instances, the development professionals now ignore the presence of the GIS unit, whose only remaining function seems to be a couple of maps for the glossy annual report and maybe maps for a student's thesis.

In the 1960s and 1970s many development professionals would commonly reach for a roll of tracing paper and sketch out maps of project areas, vegetation, and the like as a matter of course. Now, because of the expectation created by GIS, such simple techniques are largely rejected as crude, yet practical alternatives have not replaced them.

There is a new generation of GIS packages which are based on the Windows interface. Windows offers the potential for more user-friendly manipulation of graphics. Programs such as Map-Info, ArcView, and Map Viewer pioneered this approach and now some of the more advanced systems like AcrInfo, IDRISI, and Atlas GIS have brought out Windows versions. These programs are a major advance over their predecessors, yet they still belong to the world of GIS, not that of development.

In response to the needs of development projects and the potential for user-friendly personal computer map processing offered by Windows, a program called *Map Maker* has been developed by the author of this appendix. The program has been largely developed and tested in the field through projects in Asia, Latin America, and Africa. The program has been placed in the public domain and users are encouraged to copy it and pass it on to other potential users. It is available over the World Wide Web at "http://www.mapmaker.com. The belief is that the power of maps is so great that the wider use of maps in day to day project management and monitoring will in itself lead to better considered development. Good maps can help to avoid mistakes, save public funds, and facilitate participation in decision making.

This appendix describes how the Map Maker software can be used to facilitate low-cost and useful mapmaking.

## FROM MUDDY BANKS TO THE WORLD BANK

The mapping process is a long one that stretches from a junior field worker standing in the muddy field wondering how many hectares of seedlings have been planted to the pristine report on the desk of the banker who wants to know how the money has been spent. Specifically, the mapping process involves:

1. Field surveys
   - Spatial objects: boundaries, roads, rivers, houses, etc.
   - Data associated with spatial objects
   - Deciding on the appropriate level of precision
   - Reference to a larger system of spatial coordinates (e.g., latitude–longitude, national grid)

2. Importing data
   - Aerial or satellite photographs
   - Tracing on a digitizing tablet
   - Using purchased vector data
   - Exporting data from another GIS system
   - Processing numerical survey data
3. Drawing the map
   - Making the map up of thematic layers
   - Drawing and editing spatial objects on the screen
   - Associating a database with spatial objects
   - Map furniture: scale bar, title, north point, legend, etc.
4. Analysis (for some users)
   - Determining areas
   - Observing the interaction of variables across space
   - Determining densities of distributed points
5. Printing the map
   - Printing to scale on conventional printers
   - Printing poster size wall maps
   - Exporting images for inclusion in documents

For many projects, none of these activities requires a high level of accuracy. It often does not matter if your field areas are only accurate to within 10%. One is looking at an order of magnitude. Each project needs to decide what level of accuracy it really needs so that it can do its job. Few projects are involved in establishing national cartographic institutes. Rather, they require an adequate level of accuracy for the task at hand. Often this means little more than a sketch map so long as the person making the sketch knows what the sketch is for.

As long as mapmaking is delegated to GIS units, maps will inevitably be overdetailed and overaccurate because their makers do not know what can be omitted. The key to useful and low-cost mapping is that the person using the map should also make it. While word processors have made this personalized approach to document production commonplace, a similar acceptance of map processing has yet to emerge.

Once one has accepted a more relaxed approach to mapmaking that does not always demand high levels of accuracy, then the technological options become more open. Traditional sources of precision data can still be used, but also maps can be entered using increasingly common desktop scanners or data collected directly from the field.

## SIMPLE FIELD SURVEYS

The mainstay of many GIS projects is satellite images. Often these are wildly out of date. Similarly paper maps from which digitized images are copied are also often unreliable due to their age. No serious GIS project can reasonably avoid the need for doing firsthand surveys in the field if only to update specific project sites, but remarkably few seem to actually do it. Cartography seems to have retreated into the office. Modern surveying makes use of lasers and precision theodolites, and more recently geographic positioning systems. But for centuries remarkably fine and accurate maps have been made with embarrassingly simple tools. All of surveying is essentially about two things: measuring the length of straight lines, and measuring the angles of triangles. But there are two obstacles to do-it-yourself surveying: making reasonably accurate measuring instruments, and processing the data.

Medieval mapmakers needed skilled craftsmen to make the measuring scales of their instruments for them even though the instruments were conceptually simple. And processing the data meant either careful draftsmanship or else complex trigonometry—often both. However, today the computer can readily do both these tasks. Any normal desktop printer is a precision instrument which, given the right instructions, can print out accurate scales for measuring angles, and also distances. Processing the data is a trivial exercise for computers, so field measurements can be fed straight into the computer, which then produces the electronic drawing file just as if it had been digitized. By using the computer in this way, medieval ideas for surveying instruments can take on a new lease of life and make surveying accessible to nonexperts and to those without ready access to expensive surveying equipment.

The surveying instruments which are produced by the Map Maker program are of six types (see website).

1. *Plane table.* The plane table is the simplest of surveying instruments. It is a flat surface fixed at a point, usually on a tripod. At its simplest a piece of paper is stuck to the table and then lines are drawn through a fixed point on the paper in the directions of the objects in the landscape which one wishes to survey. The table is then moved to a second known reference point. This is represented on the paper with a second point and the exercise is repeated. The points on the paper where the lines of sight intersect describe the objects in the landscape. A variant of this system requires the user to measure the angles of the objects rather than drawing lines. Using Map Maker, these angles can be processed and converted into a drawing. The Map Maker program simply prints out a large protractor which can be used as the base for a plane table so that angles can be read directly using a wooden or metal straight edge revolving around a central nail or other pivot.

2. *Solar compass.* In order to use the plane table it is necessary to orient it with respect either to a landmark or else to the north. Large-diameter compasses used for surveying are expensive and any magnetic compass is subject to both

magnetic variation and local disturbances such as power lines and buried metal objects or rock formations. The Map Maker program will print out a page which is a solar compass corrected for any latitude and longitude and time of year. One places a vertical object at the center, which might be as simple as a pack of cigarettes, checks the time on one's watch, then turns the page until the shadow from the sun falls on the mark corresponding with that time. The page is then facing toward true north.

3. *Angle measurer*. The sextant is an ancient instrument designed to measure the angle between the horizon and a heavenly body. The user lines the horizon up with a mirror which is then rotated on an arm until the reflection of the heavenly body corresponds with the horizon. The angle can then be read by reading the position of the arm against an angular scale. The angle read off the scale is double the actual angle of the arm since the mirror halves the angle. Map Maker can print out an angle measurer for use with a mirrored arm which is essentially a horizontal sextant for measuring angles between two points. This can be useful where one does not wish to set up a fixed plane table but simply to read an angle from a handheld device.

4. *Clinometer*. A simple device for measuring vertical angles can be made by sticking a protractor to a piece of wood and hanging a weighted string from its center. The Map Maker clinometer is little more than this but it has an additional two scales which show the tangent and the sine of the angle. Thus, if the distance to the object whose height is being measured is known, the height can be derived simply by multiplying the distance by the tangent, or else the distance along the slope by the sine.

5. *Distance estimator*. As anyone knows, any object appears smaller the further away it is. Less well understood, is that the apparent size is easily predictable. If one holds a piece of wood vertically in one's hand with the arm fully stretched, the distance from the wood to one's eye can be measured by an assistant. This distance is then entered into the computer and out of the printer comes a scale which can be stuck to the piece of wood. If one looks at an object 1 m high and sees how big it appears against the scale, the numbers on the scale give the distance in meters. If the object is actually 3 m high the distance can simply be multiplied by three. Alternately, a scale can be printed out by Map Maker for an object of a given height. Thus, if your assistant is 1.65 m high a scale can be produced for an object 1.65 m high. Your assistant can then walk about the site and by sighting the scale against the assistant the distances can be read directly. The distances recorded using this technique are inevitably of poor accuracy but in many circumstances may be perfectly adequate. In trials accuracies of within 4% were consistently achieved for distances up to 150 m.

A variation on the distance estimator, now being used in projects in Burma and India, uses two pieces of wood, the second of which is attached at a right angle to the first. The end of the second piece of wood is placed on the cheek just below the eye so that the length of the wood, rather than one's arm, determines the distance of the scale from the eye so that the scale is standard for all people and anyone can use it. These can then be produced in quantity.

6. *Photograph measurer*. A camera is little more than a means of recording relative angles between objects in the landscape onto a piece of film. If one can calibrate the photograph there is no reason why one cannot read those angles from the photograph just as if one were in the field. Many people now use pocket cameras with a fixed lens and get their pictures processed at a laboratory which processes identically sized prints. For such users, they need to only calibrate one photograph and the rest will be the same. Once the picture is calibrated Map Maker will print out an angle-measuring scale which measures the horizontal or vertical angles from the center of the photograph. The precision of the technique will depend on the size of the photograph and the quality of the lens.

## PROCESSING SURVEY DATA

All of the preceding techniques produce numerical survey data—angles and distances. These need to be translated into maps. Map Maker offers two routes for doing this. First there is a spreadsheet-like graphic interface which allows users to enter the data following a defined sequence. The spreadsheet can subsequently be edited and added to. Alternately, and more quickly for experienced users, data can be typed directly into an ASCII text file of a prescribed format [$(x,y)$ format]. The $(x,y)$ file has a variety of simple subformats, allowing a large range of surveying techniques to be used:

- Simple $(x,y)$ coordinates (default)
- Distances from two reference points
- Compass bearings from two reference points
- Angles measured from two ends of a reference line
- Polar coordinates
- Relative polar coordinates
- Compass traverse in which each point is defined by a compass bearing from the previous point
- Angle traverse where the angle of each line segment is measured from the previous line segment
- Offsets from a line between two points

A simple example from an $(x,y)$ file could be

COMPASS TRAVERSE
Forest path, 0, 0
23.45, 123.56
67.78, 234.22
176.4, 12.3
65.20, 450.56

The first line tells us what kind of survey has been done, which determines the meaning of the subsequent numbers—in this case a compass traverse. The next line starts with the name of the object, in this case a line called "Forest path," and the coordinates of the start point. The subsequent pairs of numbers each define a compass bearing followed by a distance. Angles can either be expressed in degrees and minutes or as decimal degrees. Distances and coordinates within an $(x,y)$ file can be described in various units: meters (default), yards, feet, or chains.

Using a simple text editor, a complex survey may be entered which can be processed by Map Maker. In Burma, agricultural census surveyors have to prepare cadastral maps of the whole country. The data collected in the field are recorded on data sheets and then processed in the office so that a drawing may be made from which an area can be measured. A typical data sheet takes about a day to process. Using Map Maker the same operation, including data entry, was completed in three minutes.

While Map Maker includes all the tools to process the survey data, these tools do not actually need the graphic interface of Windows. For this reason a companion program to Map Maker, called *Survey Maker*, has been written to operate under DOS, which can function perfectly well on a computer with an 8086 processor. In many developing countries a 386 or Pentium computer with Windows and a mouse is a rarity while basic 8086 or 286 computers are relatively common. By introducing the Survey Maker program, out stations with less sophisticated equipment can start to benefit from computerized mapping while at the same time producing data which are compatible with Map Maker.

The surveying tools already described may be used with the survey processing techniques but they do not have to be. If an accurate theodolite survey is required it may be done while still processing the data in the ways previously described. Similarly users with no access to computers can still use the simple surveying tools and produce conventional paper drawings from the results.

In Burma, where modern surveying equipment is scarce, trainee surveyors are now being taught how to make and use Map Maker's survey tools, and while only the offices in the capital currently have the necessary hardware to run Map Maker, some of the out stations are starting to use Survey Maker.

Through the intervention of UNDP in a "Human Development Initiative," more modern computers will soon be available in remoter areas and then the survey tools will be used with Map Maker to chart things like water sources and human settlements.

## CONCEPTUAL MAPS

The surveying tools described here are not going to give immensely accurate results, but for many purposes they still may be more detailed than is actually necessary. Planners, policymakers, and researchers are often dealing with little more than bubble diagrams—conceptual maps which describe general dispositions and relationships rather than the stuff of engineering drawings. With modern Windows-based, mouse-driven graphics packages, on-screen sketch drawing is easy. Programs such as Freelance Plus or Corel Draw provide excellent tools for freehand drawing. Map Maker also is designed for simple drawing and the sketch maps may gradually be improved to become more comprehensive and accurate. In Map Maker the user may cut existing polygons into parts or join adjacent polygons together. In this way a crude polygon representing, say, a forest, may be subdivided by rivers, roads, and political divisions depending on the level of detail required or known. By working on a grid, if desired, the sketch maps may be drawn approximately to scale. In Map Maker you can both show a grid on the screen and print out a grid on tracing paper to a chosen scale, allowing the user to transfer data from a paper map to the computer with neither a digitizer or a scanner.

Conceptual maps are not only the province of planners and researchers. Villagers and other ordinary people have their own geographical models of how their surroundings are organized. The making of community maps is a well-known technique of participatory rural appraisal (PRA) in which people as individuals or as groups draw a map of their village or valley. Often these maps are very distorted in "objective" physical terms but they describe the villagers' own reality.

These sketch maps can then be transferred to the computer and if need be brought back as printouts to the next meeting with the community so that details may be discussed and new projects located on the map which they themselves have drawn. When printing a map, Map Maker has a facility called "poster print" which will print the map out on several sheets—from 4 to 100. The pages can then be stuck together, after trimming the margins, to form one large wall map. In this way a field office with a simple office printer can rapidly make wall maps for use in community meetings without having to wait for a central office with a large plotter to do it for them.

In Pakistan, in a water and sanitation project, anthropologists are using simple models of houses and pipes to help villagers describe their village and where they think the water supply should go. Once these layouts are complete they are photographed, and recently the anthropologists have been experimenting with transferring the resulting plans to the computer so that they can then be passed on to the engineers who need to implement the work.

In Zimbabwe, a field worker with the World Conservation Union (IUCN) got villagers to make sketch maps of their villages showing things like grazing lands and communal woodlots. Back in the office with Map Maker he was able to relate these sketch maps to a scanned map of the region. Even though the dimensions of the villagers' maps were distorted, the topology was more or less correct—rivers could be identified and road junctions found. By comparing the conceptual maps of different villages with the scanned map he was able to identify things like clashes of interest where two villages were claiming grazing rights to the same land. Armed with this information he could go back to the villages with an idea about the priority issues for discussion.

## LINKING DATA TO MAPS

While most makers of simple maps will not require all the power of a full-blown GIS there are still many people who want to display data on a map. Sometimes it is sufficient to plot data directly onto a map, placing symbols or ascribing particular colors to individual polygons and lines. But more generally it is useful to be able to attach a database to a map so that entries in the database are linked to individual graphic objects (polygons, lines, or points) on the map. In principle, all that is required is that one column in the database, generally the first, contain a unique identifying name or number which is also used to identify an individual object in the graphics file. In Map Maker, DBF files (the industry standard format originally used in dBase) or simple comma-separated text files may be attached to graphic files in this way. The advantage of using comma-separated data files is that they may be edited from any simple text editor. The values in any column in the database may be used to determine which color, texture, and typeface are used for a particular object. Alternately a data value can be used to determine the appearance of a "data point" such as a circle or triangle, the size of which is governed by the value. Also, values in the same or a different column may be used to label the objects. For example, areas of a forest may be colored according to tree age and labeled according to the number of species to be found in the area. By the use of histograms and pie charts, values from several columns may be shown at once.

Apart from attaching data to an existing graphic file, many forms of data can best be shown as an overlay showing how one phenomenon interacts with other features. Such overlays can readily be drawn straight onto the screen and, if need be, several overlays may be superimposed. For example, if a certain socioeconomic group is known to live in a certain region, that region can be drawn as a crude indicative "bubble" over a more detailed base map.

## SIMPLE BUT POWERFUL MODELS FROM SURFACES

There are many kinds of variables which vary continuously over the horizontal plane of a map, the obvious example being altitude, but there are many others such as temperature and rainfall. In practice, data rarely come to us as a complete continuous surface. Generally it is as point data so we may know the number of endemic species in a forest very precisely at a number of dispersed points, but not at the places in between.

Map Maker has the facility to rapidly generate models of three-dimensional surfaces from randomly scattered point values. The more points there are, the more highly articulated the surface is likely to be. A three-dimensional surface representing rainfall, for example, is a type of continuous database in which a single variable is located by its $x$ and $y$ coordinates. By overlaying several surfaces representing different variables, the relationships between different variables may be observed even when we do not have precise values for the geographical location in question. In Map Maker several surfaces can be loaded, each as a layer. By clicking on the screen you can find the value for all the surfaces at that point. Further, you can generate a single composite surface from two or more surfaces, combining the surfaces using a variety of different combining rules (adding, averaging, always taking the lowest, etc.), and each surface may be given a weighting as in a simple cost–benefit analysis. In this way quite sophisticated geographical models may be created rapidly.

This kind of simple technique for geographic modeling can help to remove some of the mystique of modeling so that practitioners who would not consider themselves academic researchers can start to create their own models and apply the theoretical models of others to their own situation.

## NEW LIFE FOR WHITE ELEPHANTS

New, simple programs such as Map Maker are not designed to replace the major GIS systems such as Arc-Info, IDRISI, and Atlas-GIS currently used in development projects, but rather to complement them. At present these rather complex programs are failing to meet certain real needs; on the other hand their potential is not being utilized. There are many projects around the world in which several man-years worth of digitized data, and millions of dollars of raster imagery, are sitting unused simply because the data are not readily accessible to the people who could benefit from them. Through a marriage with simple map processors these systems could become important central resources of data for decentralized mapmaking. Investment which, until recently, had seemed doomed to be lost can be given a new breath of life.

For example, in Afghanistan, where Map Maker is used by the United Nations, extensive basic geographic data covering the whole country from a large GIS system have been converted and saved to three diskettes in a form which can be used directly by Map Maker. In this way, mapmaking which before was done in a centralized office is now done on the laptops of individual professionals who need maps for reports, proposals, and evaluations today rather than in six weeks.

By liberating the base map data, a wide range of institutions, both within and without the UN system, have been able to prepare their own maps to show their own data, whether it be on water supply, vaccination, forestry, or land mines. Because all the institutions started from the same base data their maps are compatible, which makes reporting across the system easier.

The UN Centre for Human Settlements (UNCHS-Habitat), which has played an important role in encouraging the development of Map Maker, is using the software as a tool for diagnosis, monitoring, and reporting in its neighborhood upgrading program in Kabul. Using relatively crude and largely out-of-date digitized maps of Kabul as a starting point, workers produced skeleton maps of each urban district. Using the poster print facility these were printed out as wall maps and taken to public meetings in Kabul. At these meetings they were marked up with felt tip pens showing errors and new construction as well as destruction. Water sources were indicated as were the boundaries between the different neighborhood groups. They were also marked up to show the communities' priorities for action.

The marked up maps were taken back to the office where they were used as a basis for correcting the digital maps and for indicating the intended areas of action. Subsequently, as work progressed the maps were used for monitoring activities on the ground and for reporting progress to donors and sister agencies. This process was only possible by having a mapping system which was capable of quick and easy updating and printing in the field.

## STRAIGHTFORWARD MAPS WITH A MESSAGE

In the same way that a word processor cannot write a good document for you, having a map processor will not guarantee a clear, concise map. But it helps. Many map novices make the mistake of trying to put too much information on a map so that it becomes cluttered and the key information gets lost. Designing a clear map involves skills of editing, emphasis, and omission—the same as any document. These skills are not particularly difficult to acquire so long as one has practice. Map processing software allows individual professionals to gain these skills and so develop a vocabulary for straightforward maps which tell a clear message and save a thousand words.

## USING APPROPRIATE TECHNOLOGY

When talking of development projects today, there is a whole range of development institutions in which "appropriate technology" really means laptop computers, photocopiers, A4 or letter-sized paper, conventional desktop printers, and bits of string and sticky-tape. This is the stuff of the normal office. Any technique which requires equipment outside of this universal family is going to require special attention, special funding, and a special effort. If surveying requires theodolites, map printing requires A0 plotters, and report production requires color photocopies, then mapmaking is not going to enter the range of everyday activities in most development projects. By using conventional office infrastructure, programs such as Map Maker can make mapmaking commonplace, low-cost, and useful.

APPENDIX II

# A Free Micro GIS Program

CHARLES A. S. HALL

We found a need for displaying data easily in both geographical and gradient space. This was the impetus for developing the computer program ASSGNDI2.FOR. The program, available on the CD, will do the following: (1) read tabular data on, e.g., agricultural output (yield per hectare for a particular crop) and plot it on a map template provided by the user that has corresponding map coordinates for the map corners, and (2) plot the same data as a function of environmental parameters (gradients). To use this program for a new area the user has to prepare files similar to those given in Table 6-3. Basically this requires a list for each political unit considered that includes the geographic (longitude–latitude) coordinates of the center of that unit and the mean values for, e.g., temperature and rainfall, or any other physical or climatic variables for the region of interest. Once this is prepared go to the CD with an editing program and click on PROGRAMS and then GRADIENTS. Once there, click on "new" and follow the directions. You should be able to get output figures similar to Figures 12-10 to 12-14, which are readable and printable with PEDIT and any printer.

# APPENDIX III

**TABLE A-III.1 Emergy Evaluation of Resource Basis for Costa Rica (c. 1994)**

| Note | Item | Raw units | Transformity (sej/unit) | Solar emergy (E20 sej) | EmDollars (E9 1994 US$) |
|---|---|---|---|---|---|
| Renewable resources: | | | | | |
| 1 | Sunlight | 2.95E+20 J | 1 | 2.95 | 0.06 |
| 2 | Rain, chemical | 6.30E+17 J | 18199 | 114.60 | 2.37 |
| 3 | Rain, geopotential | 3.49E+17 J | 27874 | 97.36 | 2.02 |
| 4 | Wind, kinetic energy | 4.69E+17 J | 1496 | 7.02 | 0.15 |
| 5 | Waves | 1.99E+17 J | 30550 | 60.79 | 1.26 |
| 6 | Tide | 5.87E+16 J | 16842 | 9.88 | 0.20 |
| 7 | Earth cycle | 5.11E+16 J | 34377 | 17.57 | 0.36 |
| Indigenous renewable energy: | | | | | |
| 8 | Hydroelectricity | 1.51E+16 J | 1.65E+05 | 24.95 | 0.52 |
| 9 | Agriculture production | 7.23E+16 J | 2.00E+05 | 144.67 | 3.00 |
| 10 | Livestock production | 1.09E+15 J | 2.00E+06 | 21.77 | 0.45 |
| 11 | Fisheries production | 7.41E+13 J | 2.00E+06 | 1.48 | 0.03 |
| 12 | Fuelwood production | 1.93E+16 J | 18700 | 3.61 | 0.07 |
| 13 | Forest extraction | 1.15E+16 J | 18700 | 2.14 | 0.04 |
| Nonrenewable sources from within system: | | | | | |
| 14 | Natural gas | 0.00E+00 J | 4.80E+04 | 0.00 | 0.00 |
| 15 | Oil | 0.00E+00 J | 5.40E+04 | 0.00 | 0.00 |
| 16 | Coal | 0.00E+00 J | 4.00E+04 | 0.00 | 0.00 |
| 17 | Calcium carbonate | 1.49E+12 g | 1.00E+09 | 14.90 | 0.31 |
| 18 | Metals | 0.00E+00 g | 1.00E+09 | 0.00 | 0.00 |
| 19 | Top soil | 2.70E+15 J | 7.40E+04 | 2.00 | 0.04 |
| Imports and outside sources: | | | | | |
| 20 | Oil derived products | 6.41E+16 J | 6.60E+04 | 42.27 | 0.88 |
| 21 | Metals | 4.61E+11 g | 1.80E+09 | 8.30 | 0.17 |
| 22 | Minerals | 1.92E+11 g | 1.00E+09 | 1.92 | 0.04 |
| 23 | Food & ag. products | 1.06E+16 J | 2.00E+05 | 21.13 | 0.44 |
| 24 | Livestock, meat, fish | 2.99E+14 J | 2.00E+06 | 5.99 | 0.12 |

(continues)

TABLE A-III.1 (*continued*)

| Note | Item | Raw units | Transformity (sej/unit) | Solar emergy (E20 sej) | EmDollars (E9 1994 US$) |
|---|---|---|---|---|---|
| 25 | Plastics & rubber | 3.93E+16 J | 6.60E+04 | 25.94 | 0.54 |
| 26 | Chemicals | 9.13E+11 g | 3.80E+08 | 3.47 | 0.07 |
| 27 | Wood, paper, textiles | 4.40E+15 J | 3.49E+04 | 1.54 | 0.03 |
| 28 | Mech. & trans. equip. | 7.26E+10 g | 6.70E+09 | 4.86 | 0.10 |
| 29 | Service in imports | 3.88E+09 $ | 1.24E+12 | 48.11 | 1.00 |
| 30 | Tourism | 9.92E+08 $ | 1.24E+12 | 12.30 | 0.25 |
| Exports | | | | | |
| 31 | Food & ag. products | 3.49E+16 J | 2.00E+05 | 69.89 | 1.45 |
| 32 | Livestock, meat, fish | 3.29E+14 J | 2.00E+06 | 6.59 | 0.14 |
| 33 | Wood, paper, textiles | 3.93E+14 J | 3.49E+04 | 0.14 | 0.00 |
| 34 | Oil derived products | 6.53E+15 J | 5.30E+04 | 3.46 | 0.07 |
| 35 | Metals | 2.01E+10 g | 1.00E+09 | 0.20 | 0.00 |
| 36 | Minerals | 3.75E+10 g | 1.00E+09 | 0.38 | 0.01 |
| 37 | Chemicals | 2.09E+11 g | 3.80E+08 | 0.79 | 0.02 |
| 38 | Mech. & trans. equip. | 1.59E+10 g | 6.70E+09 | 1.07 | 0.02 |
| 39 | Plastics & rubber | 4.59E+14 J | 6.60E+04 | 0.30 | 0.01 |
| 40 | Service in exports | 2.88E+09 $ | 4.83E+12 | 139.00 | 2.88 |

RENEWABLE RESOURCES:

1 SOLAR ENERGY:

| | | | |
|---|---|---|---|
| Cont Shelf Area | = | 1.60E+10 $m^2$ at 200 m depth | Espenshade, 1990 |
| Land Area | = | 5.11E+10 $m^2$ | Espenshade, 1990 |
| Insolation | = | 1.50E+02 $Kcal/cm^2/yr$ | Rodenwaaldt & Jusatiz, 1965 |
| Albedo | = | 0.3 (% given as decimal) | * estimate |
| Energy(J) | = | (area incl shelf)*(avg. insolation)*(1-albedo) | |
| | = | (__$m^2$)*(__$Cal/cm^2/y$)*(E+04 $cm^2/m^2$)* | |
| | = | (1-0.30)*(4186 J/kcal) | |
| | = | 2.95E+20 J/yr | |

2 RAIN, CHEMICAL POTENTIAL ENERGY:

| | | | |
|---|---|---|---|
| Land Area | = | 5.11E+10 $m^2$ | |
| Cont Shelf Area | = | 1.60E+10 $m^2$ AT 200 m d. | |
| Rain (land) | = | 2.80 m/yr | Rodenwaaldt & Jusatiz, 1965 |
| Rain (shelf) | = | 1.26 m/yr (est. as 45% of tot. rain) | Estimate (45% rain) |
| Evapotrans rate | = | 2.10 m/yr | Rodenwaaldt & Jusatiz, 1965 |
| Engery (land) (J) | = | (area)(Evapotrans)(rainfall)(Gibbs no.) | |
| | = | (__$m^2$)*(__m)*(1000 $kg/m^3$)*(4.94E+03 J/kg) | |
| | = | 5.30E+17 J/yr | |
| Energy (shelf) (J) | = | (area of shelf)(Rainfall)(Gibbs no.) | |
| | = | 9.96E+16 J/yr | |
| Total energy (J) | = | 6.30E+17 J/yr | |

3 RAIN, GEOPOTENTIAL ENERGY:

| | | | |
|---|---|---|---|
| Area | = | 5.11E+10 $m^2$ | |
| Rainfall | = | 2.80 m | |
| Avg. Elev. | = | 996.40 m | Espenshade, 1990 |
| Runoff rate | = | 0.25% (percent, given as a decimal) | |
| Energy (J) | = | (area)(% runoff)(rainfall)(avg. elevation)(gravity) | |
| | = | (__$m^2$)(__m)*(1000 $kg/m^3$)*(__m)*(9.8 $m/s^2$) | |
| | = | 3.49E+17 J/yr | |

4 WIND ENERGY:

Energy (J) = 4.69E+17 J/yr — Rodenwaldt & Jusatiz, 1965

5 WAVE ENERGY:

Energy (J) = 1.99E+17 J/yr — Rodenwaldt & Jusatiz, 1965

6 TIDAL ENERGY:

Cont Shelf Area = 1.60E+10 $m^2$
Avg Tide Range = 1.00 m — Espenshade, 1990
Density = 1.03E+03 $kg/m^3$
Tides/year = 7.30E+02 (est. of 2 tides/day in 365 days)
Energy (J) = (shelf)(0.5)(tides/y)(mean tidal range)$^2$ (density of seawater)(gravity)
= (__$m^2$)*(0.5)*(__/yr)*(__m)$^2$*(__ $kg/m^3$) *(9.8 $m/s^2$)
= 5.87E+16 J/yr

7 EARTH CYCLE:

Land Area = 5.11E+10 $m^2$
Heat flow = 1.00E+06 $J/m^2$
Energy (J) = (5.11E+10)(1.00E+06)
= 5.11E+16

**INDIGENOUS RENEWABLE ENERGY**

8 HYDROELECTRICITY:

Kilowatt Hrs/yr = 4.20E+09 KwH/yr (assume 80% load) — World Bank (1999)
Energy (J) = (4.29E+09 KwH/yr)*(3.6E+06 J/KwH)
= 1.51E+16 J/yr

9 AGRICULTURAL PRODUCTION:

Production = 5.40E+06 MT — (Europa, 1997)
Energy (J) = (5.4E+06 MT)*(1E06 g/MT)*(80%)*(4.0 Cal/g)* (4186 J/Cal)
= 7.23E+16 J/yr

10 LIVESTOCK PRODUCTION:

L'stock Production: = 2.60E+05 MT — (Europa, 1997)
Energy (J) = (2.6E+05 MT)*(1E+06 g/MT)*(20%)*(5 Cal/g)* (4186 J/Cal)
= 1.09E+15 J/yr

11 FISHERIES PRODUCTION:

Fish Catch = 1.77E+04 MT — (Europa, 1997)
Energy (J) = (1.77E+07 MT)*(1E+06 g/MT)*(5 Cal/g)*(20%)* (4186 J/Cal)
= 7.41E+13 J/yr

12 FUELWOOD PRODUCTION:

Fuelwood Prod = 3.20E+06 $m^3$ — (Europa, 1997)
Energy (J) = (3.2E6 $m^3$)(0.5E6 $g/m^3$)(3.6 Cal/g)(80%)(4186 J/Cal)
= 19.3E+16 J/yr

13 FOREST EXTRACTION:

Harvest = 1.90E+06 $m^3$ — (Europa, 1997)
Energy (J) = (1.9E+06 $m^3$)(0.5E+06 $g/m^3$)(80%)(3.6 Cal/g) (4186 J/Cal)
= 1.15E+16 J/yr

**NONRENEWABLE RESOURCE USE FROM WITHIN COSTA RICA:**

14 NATURAL GAS: 0.00E+00

Consumption = 0.00E+00 $ft^3/yr$
Energy (J) = (0.0 $ft^3/yr$)*(1.055E+6 Joules/$ft^3$)
= 0.00E+00 J/yr

15 OIL:

Consumption = 0.00E+00 barrels
Energy (J) = (5.7E06 b)*(6.1E9 Joules/barrel)
= 0.00E+00 J/yr

16 COAL:

Consumption = 0.00E+00 MT/yr
Energy (J) = (0.0 Mt/yr)*(2.9E+10 J/Mt)
= 0.00E+00 J/yr

17 CALCIUM CARBONATE:

Consumption = 1.49E+06 MT/yr (Europa, 1997)
Mass (g) = (1.49E6 MT/yr)*(1E6 g/MT)
= 1.49E+12 g/yr

18 METALS:

Production = 0.00E+00 MT/yr
Consumption = 0.00E+00
Mass (g) = (5.84E5 MT)*(1E6 g/MT)
= 0.00E+00 g/yr

19 TOPSOIL:

Soil loss = 7.78E+01 $g/m^2/yr$ (Oldman, 1994)
Energy (J) = (77.8 $g/m^2/yr$)*(5.11E10 $m^2$)*(0.03 organic)
*(5.4 Kcal/g)(4186 J/Kcal)
= 2.70E+15 J/yr

**IMPORTS OF OUTSIDE ENERGY SOURCES:**

20 OIL DERIVED PRODUCTS:

Imports = 1.05E+07 bb (UN, 1997)
Energy (J) = (10.15E6 b/d)*(6.1E9 Joules/Barrel)
= 6.41E+16J/yr

21 METALS:

Imports = 4.61E+05 MT/yr (UN, 1997)
Mass (g) = (4.61E5 MT/yr)*(1E6 g/MT)
= 4.61E+11 g/yr

22 MINERALS:

Imports = 1.92E+05 MT/yr (UN, 1997)
Mass (g) = (1.92E5 MT/yr)*(1E+6 g/MT)
= 1.92E+11 g/yr

23 FOOD AND AGRICULTURAL PRODUCTS:

Imports = 9.02E+05 MT/yr (UN, 1997)
Energy (J) = (9.02E5 MT/yr)*(1E6 g/MT)*(3.5 Kcal/g)
*(4186 J/Kcal)*(80%)
= 1.06E+16 J/yr

24 LIVESTOCK, MEAT, FISH:

Imports = 6.50 E+04 MT/yr (UN, 1997)
Energy (J) = (6.5E4 MT/yr)*(1E6 g/MT)*(5 Kcal/g)
*(4186 J/Kcal)*(.22 protein)
= 2.99E+14 J/yr)

25 PLASTICS AND RUBBER:

Imports = 1.31E+06 MT/yr (UN, 1997)
Energy (J) = (1.31E6 MT/yr)*(1000 Kg/MT)*(30.0E6 J/kg)
= 3.93E+16

26 CHEMICALS:

Imports = 9.13E+05 MT/yr (UN, 1997)
Mass (g) = (9.13E5 MT/yr)*(1E6 g/MT)
= 9.13E+11 g/yr)

27 WOOD, PAPER, TEXTILES, LEATHER:

Imports = 2.93E+05 MT/yr (UN, 1997)

Energy (J) = (2.93E5 MT/yr)*(1E6 g/MT)*(15E3 J/g)

= 4.40E+15 J/yr

28 MACHINERY, TRANSPORTATION, EQUIPMENT:

Imports = 7.26E+04 MT/yr (UN, 1997)

Mass (g) = (7.26E4 MT/yr)*(1E6 g/MT)

= 7.26E+10 g/yr

29 IMPORTED SERVICES:

Dollar Value = 3.88E+09 $US (UN, 1997)

30 TOURISM:

Dollar Value = 9.92E+08 $US (Europa, 1997)

**EXPORTS OF ENERGY, MATERIALS, AND SERVICES:**

31 FOOD AND AGRICULTURAL PRODUCTS:

Exports = 2.98E+06 MT/yr (UN, 1997)

Energy (J) = (2.98E6 MT)*(1E+06 g/MT)*(80%)*(3.5 Cal/g)
*(4186 J/Cal)

= 3.49E+16 J/yr

32 LIVESTOCK, MEAT, FISH:

Exports = 7.15E+04 MT/yr) (UN, 1997)

Energy (J) = (7.15E4 MT)(1E+06 g/MT)(5 Cal/g)(4187 J/Cal)
(.22% prot)

= 3.29E+14 J/yr

33 WOOD, PAPER, TEXTILES, LEATHER:

Exports = 3.26E+04 MT/yr) (UN, 1997)

Energy (J) = (3.26E4 MT)(1.0E+06 g/MT)(80%)(3.6 Cal/g)
(4186 J/Cal)

= 3.93E+14 J/yr

34 OIL DERIVED PRODUCTS:

Exports = 1.07E+06 Barrels/yr (UN, 1997)

Energy (J) = (1.07E6 B/yr)*(6.1E9 J/Barrel)

= 6.53E+15 J/yr

35 METALS:

Exports = 2.01E+04 MT/yr (UN, 1997)

Mass (g) = (2.01E4 MT)*(1E6 g/MT)

= 2.01E+10 g/yr

36 MINERALS:

Exports = 3.75E+04 MT/yr (UN, 1997)

Mass (g) = (3.75E+4 MT)(1.0E+06 g/MT)

= 3.75E+10 g/yr

37 CHEMICALS:

Exports = 2.09E+05 MT/yr) (UN, 1997)

Mass (g) = (2.09E5 MT)*(1E6 g/MT)

= 2.09E+11 g/yr

38 MACHINERY, TRANSPORTATION, EQUIPMENT:

Exports = 1.59E+04 MT/yr (UN, 1997)

Mass (g) = (1.59E4 MT/yr)*(1E6 g/MT)

= 1.59E+10 g/yr

39 PLASTICS AND RUBBER:

Exports = 1.53E+04 MT/yr (UN, 1997)

Energy (J) = (1.53E4 MT/yr)*(1000 Kg/MT)*(30.0E6 J/kg)

= 4.59E+14

40 SERVICES IN EXPORTS:

Dollar Value = 2.88E+09 $US (UN, 1997)

**References**

Espenshade, E. B., 1990. Goode's World Atlas. Rand McNally. Chicago. 367 p.

Europa Publications, 1997. The Europa World Year Book 1996. Vol 1. Europa Publications Ltd. London.

Oldeman, L. R. 1994. The Global Extent of Soil Degradation. pp. 99–118. In D. J. Greenland and I. Szabolcs (eds). Soil Resilience and Sustainable Land Use. CAB International, Wallington, UK. 561 p.

Rodenwaaldt, E. & H. J. Jusatiz (eds), 1965. World Maps of Climatology, 2nd ed. Spronger, New York.

United Nations, 1997. 1996 International Trade Statistics yearbook. Vol. 1 Trade by country. United Nations, New York.

World Bank, 1999. World Development Indicators. The World Bank, New York.


TABLE A.III.2. Summary of Flows in Costa Rica, (c. 1994)

| Variable | Item | Solar emergy (E20 sej/y) | Dollars |
|---|---|---|---|
| R | Renewable sources (rain, tide, earth cycle) | 211.96 | |
| N | Nonrenewable resources from within Costa Rica | 24.13 | |
| N0 | Dispersed Rural Source | 9.23 | |
| N1 | Concentrated Use | 14.90 | |
| N2 | Exported without Use | 4.83 | |
| F | Imported Fuels and Minerals | 52.49 | |
| G | Imported Goods | 62.93 | |
| I | Dollars Paid for Imports | | 3.88E+09 |
| P21 | Emergy of Services in Imported Goods and Fuels | 48.11 | |
| E | Dollars Received for Exports | | 2.88E+09 |
| P1E | Emergy Value of Goods and Service Exports | 221.81 | |
| x | Gross National Product | | 8.28E+09 |
| P2 | World emergy/$ ratio, used in imports | 1.24E+12 | |
| P1 | Costa Rica Emergy/$ ratio | 4.83E+12 | |

# INDEX

# D



# E

# F

## G



## H



## I

## Y



## Z